T0235117

Classical and Quantum Dissipative Systems

Second Edition

Classical and Quantum Dissipative Systems

Second Edition

MOHSEN RAZAVY

University of Alberta, Canada

 World Scientific

NEW JERSEY · LONDON · SINGAPORE · BEIJING · SHANGHAI · HONG KONG · TAIPEI · CHENNAI · TOKYO

Published by

World Scientific Publishing Co. Pte. Ltd.

5 Toh Tuck Link, Singapore 596224

USA office: 27 Warren Street, Suite 401-402, Hackensack, NJ 07601

UK office: 57 Shelton Street, Covent Garden, London WC2H 9HE

Library of Congress Cataloging-in-Publication Data

Names: Razavy, Mohsen, author.

Title: Classical and quantum dissipative systems / Mohsen Razavy, University of Alberta, Canada.

Description: Second edition. | Singapore ; Hackensack, NJ : World Scientific, [2017] |
    Includes bibliographical references and index.

Identifiers: LCCN 2016058647| ISBN 9789813207905 (hardcover ; alk. paper) |
    ISBN 9813207906 (hardcover ; alk. paper) | ISBN 9789813207912 (pbk. ; alk. paper) |
    ISBN 9813207914 (pbk. ; alk. paper)

Subjects: LCSH: Energy dissipation. | Quantum theory. | Mechanics.

Classification: LCC QC173.458.E53 R39 2017 | DDC 530.12--dc23

LC record available at https://lccn.loc.gov/2016058647

British Library Cataloguing-in-Publication Data

A catalogue record for this book is available from the British Library.

Printed in Singapore

To the memory of my father

M.T. Modarres Razavy

Preface to the Second Edition

This updated revision of the original edition carries substantial additions. Among the changes made are the addition of the following important topics:

(1) On the classical formulation of the dynamics of dissipative motions, the fractional derivative methods for the construction of the Lagrangian and Hamiltonian have been introduced in Sections 2.6 and 3.18 respectively. In this way, one can enlarge the group of problems which can be studied using the variational technique.

(2) An elegant method, "the dynamical matrix theory" proposed by Enz is applied to the classically damped motion and to its corresponding quantum system, Sections 3.8 and 11.10.

(3) The classical formulation of a tuned circuit coupled to a transmission line advanced by Stevens and Josephson is treated in Section 7.6, and then it is applied to the quantum theory of damped spin system. This problem is discussed in Section 18.4.

(4) Quantum dissipative forces arising from vacuum fluctuations, Section 19.1, and the retarding force acting on a spherical mirror moving in vacuum, Section 19.4, are among the many examples of purely quantum mechanical dissipative forces considered in this edition.

(5) In nuclear scattering theory one describes dissipative forces in terms of the optical potentials, but for heavy ion collisions one uses a semi-classical approach with velocity-dependent damping forces. In the last chapter of this book the connection between the two approaches have been investigated in detail.

I would like to express my indebtedness to a number of my colleagues for their messages of encouragement, and for their critical and constructive comments concerning the first edition of this book. To my daughter, Maryam, for her continued help in the writing of my books. As always, my wife has been a source of inspiration and support and without her help this work would not have come to fruition.

Mohsen Razavy
Edmonton, Canada, December 2016

Preface to the First Edition

The aim of this book is to elucidate the origin and nature of dissipative forces and to present a detailed account of various attempts to study the phenomena of dissipation in classical mechanics and in quantum theory. From the early days of the old quantum theory it was realized that the quantization of the velocity-dependent dissipative forces are problematic, partly because of the inconsistencies in the mathematical formulation and partly due to the problem of interpretation. Recent works on the quantum theory of measurement, the theory of collision of heavy ions and the macroscopic quantum tunneling has necessitated a more critical examination of the nature of the frictional forces and their derivations from conservative many-body systems. In this book, we discuss the basic concepts of these forces without discussing any particular application. There are monographs available dealing with these applications. For instance *"Quantum Theory of Dissipative Systems"* Second Edition, by U. Weiss (World Scientific, 2012) gives an excellent account of the role of dissipative forces in condensed matter physics.

The scope of this book is also limited to the discussion of regular motions since the very important subject of chaotic dissipative motion deserves a completely different approach. Furthermore, the emphasis of the present work is on the solvable models where the ideas of symmetries and the conservation laws, the way that the time asymmetry arises in the equations of motion, and the classical-quantum correspondence can be discussed without the ambiguity which is often associated with approximate solutions.

This text is divided into three parts. In the first part, we present a detailed coverage of the classical dissipative systems using the canonical formalism. This includes a description of the inverse problem of analytical dynamics, i.e. determination of Lagrangians and Hamiltonians from the equations of motion specifically for the cases involving dissipative forces. Important theoretical concepts such as the Noether theorem and the minimal coupling rule in the presence of frictional forces are presented and applied to simple examples. In addition to the phenomenological frictional forces, a large part is devoted to the problem of derivation of frictional forces from solvable many-body problems. Chapters 1 through 10 cover the classical description of the subject.

In the second part of the book, Chapters 11-17 lay the groundwork for the problem of quantization of classical systems with phenomenological velocity-dependent forces. Various attempts to find a consistent theory satisfying the basic postulates of quantum theory, and their successes and failures are examined. This is followed by a more basic approach in which we try to derive a wave equation for the motion of the damped system, again from a closed conservative many-body system.

Finally in the third part of the book, Chapters 18-20, we discuss a number of problems where the starting point is quantum mechanical, but only in some cases we can associate a classically damped system to these quantal problems.

In selecting references I have tried to cite a number of interesting albeit forgotten works. Unfortunately the task of achieving a complete list of important contributions is an impossible one. I apologize to the authors of the papers that I have overlooked or have failed to give the proper credit to. In writing this monograph I have benefited greatly from the help of my colleagues and my former students. I should thank my wife for sympathetic understanding of the ups and downs of writing a book.

Edmonton, Alberta, Canada, September 2004.

Contents

Introduction

In classical dynamics the second law of motion is used both as a definition of the force and also as the equation for predicting the position and the momentum of the particle as a function of time. This dual role of the equation of motion has been criticized by a number of eminent scientists [1]. In order to investigate the nature of the force law in many cases we are helped by the so called "theories of forces" [2], where theories like gravitational and electrodynamics specify the force function. Based on these and other well-defined theories we can derive the interaction between complicated system of particles and fields from the inter-particle forces. For instance we can determine the dipole-dipole interaction from the Coulomb force, or the radiation reaction force from the coupling of an electron to the electromagnetic field.

The idea that the force function is dependent on the position of the particle and not on time nor on its velocity has been suggested by a group of philosophers of science [3]. Let us quote Nigel's observation in this regard [4]:

"In point of fact, the force-function employed in many of the familiar applications of the equations of motion is specified in a manner analogous to the Newtonian hypothesis, in so far as it does not contain the time-variable explicitly. Indeed, though there are numerous cases for which the time-variable enters explicitly into the force function (as in the case of damped vibrations), it is commonly assumed that the explicit presence of the time-variable can in principle be eliminated if the initial system of interacting bodies is suitably enlarged by including other bodies into it. For reasons that will be presently apparent, what is called the "principle of causality" (as distinguished from special causal laws) in fact usually construed in classical physics as the maxim that should be force-function for a given physical system contain the time-variable explicitly, the system is to be enlarged in such a manner as to allow a specification of the force-function in which the time-variable does not appear. And it is a matter of historical fact that in the main the search for such enlarged systems that do not coincide with the entire cosmos has been successful."

Thus certain forces such as the damping force exerted on a particle while it is moving in a viscous medium can be postulated as a velocity-dependent force proportional to a given power of the velocity of the particle. The dependence of the force law on velocity in this case is determined from the observed state of the motion. On the other hand one may consider the damping to be due to

1

the collision of the particle in question with the smaller particles forming the environment, e.g. gas molecules. In the latter case the force of friction can be derived and its dependence on the momentum of the particle can be determined.

This idea of dividing a large conservative system into two parts, one part being the heat bath and the other forming the dissipative system is a useful one, particularly in quantum theory. In this way we can avoid certain difficulties in the formulation of the problem, but in turn, nearly in all cases we can only solve the problem approximately.

For a given closed quantum many-particle system we can decompose the system into two parts: an open system, e.g. a central particle, and a separated quantum system, which we call the heat bath. This separation must be such that the time evolution of the closed system can be described in terms of the free evolution of the heat bath, a free time evolution of the open system, and an interaction part of the evolution. If such a decomposion is possible, then what should be the structure of its components. This basic question has been studied by Kümmerer and we refer the reader to his work [5].

Usually we assume that the interaction between the two parts causes the transfer of energy from the dissipative system to the heat bath. However we can generalize the models of dissipation to include those cases where the coupling between the two parts requires the transfer of mass or the number of particles from one part to the other.

In quantum theory we borrow the idea of force (or preferably the potential) from classical dynamics. When the force is conservative and derivable from a potential function, the formulation of the quantum analogue of the classical motion is straightforward. But if we try to quantize a non-conservative classical system we find inconsistencies between the equations of motion and the canonical commutation relations. In addition the result depends on our choice of the Hamiltonian or Lagrangian, since for non-conservative systems the Hamiltonian, in general, does not represent the energy of the system.

We define dissipative forces in classical dynamics as any and all types of interaction where the energy is lost when the motion takes place (usually in the form of heat to a heat bath) [6]. Frequently the magnitude of the force, f, on a particle or a body may be closely represented, over a limited range of velocity by a power law, $f_d = av^\nu$, where v is the velocity of the particle or body and a and ν are constants.

Depending on the value of ν we have the following types of dissipative forces [6]:

(1) Frictional force. This work is usually required to slide one surface over another, and once the motion is started the magnitude of the force is independent of the speed. Thus in this case $\nu = 0$.

(2) Viscous force. When the force is proportional to the speed of the particle, i.e. $\nu = 1$ we call the force to be viscous force (see also Sec. 2.1).

(3) Newtonian dissipative force. For high speed motion of an object in the air, the force is proportional to the square of the velocity, i.e. $\nu = 2$.

In this book we will follow the present usage of these terms and occasionally use frictional force and viscous force to indicate dissipative forces of the general type av^ν with ν, a positive but not necessarily an integer, taking different values.

References

[1] B. Russell, *The Principles of Mathematics*, (W.W. Norton, London, 1996) pp. 482-488.

[2] E.J. Konopinski, *Classical Description of Motion*, (W.H. Freeman, San Francisco, 1969) p. 38.

[3] M. Jammer, *Concepts of Force*, (Harvard University Press, Cambridge, Massachusetts, 1957) Chapter 11.

[4] E. Nigel in *The World of Physics*, Vol. 3. edited by J.H. Weaver (Simon and Schuster, ew York, 1987) p. 741.

[5] B. Kümmerer, in *On Three Levels: Micro-, Meso-, and Macro- Approaches in Physics*, Edited by M. Fannes, C. Maes and A. Verbeure, (Plenum Press, New York, N.Y. 1994).

[6] D.A. Wells, *Theory and Problems of Lagrangian Dynamics*, (McGraw-Hill, New York, N.Y. 1967) Chapter 6.

Chapter 1

Phenomenological Equations of Motion for Dissipative Systems

The simplest way of describing a damped motion in classical dynamics is by adding a resisting force, generally velocity-dependent, to the Newtonian equation of motion. In principle we can derive the damped motion by coupling the system to a heat bath or to the motion of other particles (or to a field) and then eliminate the degrees of freedom of these other particles (or the field) to obtain the equation of the system that we want to study.

We will begin our discussion with an examination of various resistive force laws of common occurrence in physics, some of them are found phenomenologically and others are derived from the fundamental laws governing mechanics, electrodynamics, and hydrodynamics.

In this chapter two topics will be discussed at some length. First the old problem of a radiating charged particle like an electron. This subject is not only of great historical interest, but it also illustrates certain difficulties associated with the interpretation of the solutions of the classical equations of motion.

Another set of dynamical problems of considerable interest in classical and quantum theories of chaotic motions are those expressible as linear and nonlinear difference equations [1]. These problems have received a great deal of attention in recent years. But here we will limit our discussion to very simple classical systems without touching the extensive and important advances in mathematical understanding of these motions [1]–[4].

1.1 Frictional Forces Depending on Velocity

One of the earliest reference to the frictional or drag forces proportional to a power of velocity is in Newton's Principia [5],[6]. In Proposition 23 of Principia,

Newton calculated the resistance on bodies moving through a fluid and showed that such a resistances were "in ratio compounded of the squared ratio of their velocities and the squared ratio of their diameters, and the simple ratio of the density of the parts of the system". That is the drag force D is proportional to

$$D \propto \rho S v^2, \tag{1.1}$$

where ρ is the fluid density, S is the body reference area and v is the velocity (see Sec. 1.4) [7].

Motion of a Bullet in Air — Let us consider a bullet of mass m moving in the air where the coefficient of friction is λ. The equation for the force acting on this particle is given by [8]

$$\mathbf{F} = m\mathbf{g} - m\lambda \mathbf{v}|\mathbf{v}|^{n-1}, \tag{1.2}$$

where $\mathbf{v}$ denotes the velocity of the particle. For speeds less than $24 \frac{m}{s}$, $n \approx 1$ (Stokes' law of resistance) and for speeds between $30 \frac{m}{s}$ and $330 \frac{m}{s}$, n is approximately equal to 2 (Newton's law of resistance).

When n is not an integer, for the one-dimensional motion of a particle subject to a force of friction $\lambda \operatorname{sgn}(\dot{x}) |\dot{x}|^\alpha$, where

$$\operatorname{sgn}(\dot{x}) = \begin{cases} 1 & \text{for } \dot{x} > 0 \\ -1 & \text{for } \dot{x} < 0 \end{cases}, \tag{1.3}$$

we write the equation of motion as

$$m\ddot{x} = -\lambda \operatorname{sgn}(\dot{x}) |\dot{x}|^\alpha. \tag{1.4}$$

We note that in this case λ has the dimension $\left[\text{mass} \, (\text{time})^{\alpha-2} \, (\text{length})^{1-\alpha}\right]$. The nonlinear differential equation (1.4) for any value of α can be integrated and $x(t)$ can be found analytically. Assuming that for $t > 0$, $\dot{x} > 0$, then the solution of (1.4) subject to the initial condition

$$\dot{x}(t=0) = v_0, \quad \text{and} \quad x(t=0) = x_0, \tag{1.5}$$

is given by

$$x(t) = x_0 + \frac{m v_0^{2-\alpha}}{\lambda(2-\alpha)} - \frac{m}{\lambda(2-\alpha)} \left\{ (\alpha-1) \left(-\frac{v_0^{1-\alpha}}{1-\alpha} + \frac{\lambda t}{m} \right) \right\}^{\frac{2-\alpha}{1-\alpha}}. \tag{1.6}$$

If we expand (1.6) and retain terms linear and quadratic in t we obtain

$$x(t) \approx x_0 + v_0 t - \frac{\lambda}{2m} v_0^\alpha t^2. \tag{1.7}$$

The third term on the right-hand side of (1.7) is negative and shows a deceleration of the particle which is proportional to v_0^α.

A Simple Derivation of the Stokes Law — In the case of the Stokes law of resistance we can use the following simple picture to show that the drag force is linear in velocity. Later in the next section we will give a full derivation of the motion of a spherical object in a viscous fluid.

Consider an object e.g. a plate moving with a velocity V in a direction normal to the surface of the plate in a gas which is at a very low pressure. We assume that the average velocity of gas molecules, v, is much larger than V. The rate that the gas molecules hit the plate is proportional to the relative velocity of the incoming molecules and the plate. On the two sides of the plate the relative velocities are $v + V$ and $v - V$ respectively. The pressure is proportional to the product of average momentum transfer per molecule and the rate at which these molecules hit the plate. Since the momentum transfer is also proportional to the relative velocity, therefore the pressures on the two sides of the plates are

$$P_1 \propto (v + V)^2, \quad P_2 \propto (v - V)^2. \tag{1.8}$$

The net pressure P is the difference between P_1 and P_2

$$P = P_1 - P_2 \propto 4vV, \tag{1.9}$$

therefore the force of friction on the plate is proportional to its speed V and opposes the motion.

Drag Force on a Wall Moving through a Gas of Fermions — A more elaborate derivation of this result which takes the quantum nature of the particles into account is the piston model proposed by Gross [9],[10]. This simple model has been used to explain the reason as to how the damping force arises in heavy-ion collisions.

Again let us examine the force exerted on a rigid wall when the wall is moving with a velocity V through a gas of fermions. The flux $dj_L(\mathbf{p})$ of the particles with relative momenta $\mathbf{p} = m\mathbf{u}$ hitting the wall from the left is given by

$$dj_L(\mathbf{p}) = \frac{1}{(2\pi\hbar)^3} \frac{p_x}{m} g(\mathbf{p} + m\mathbf{V}) d^3 p, \tag{1.10}$$

where $g(\mathbf{p}) = g(-\mathbf{p})$ is the isotropic single particle probability. Since the wall is assumed to be rigid, the reflection coefficient is equal to one and the flux undergoing reflection at the wall is

$$dj'_L(\mathbf{p}) = dj_L(\mathbf{p}) \left[1 - g\left(\mathbf{p}' + m\mathbf{V}\right)\right]. \tag{1.11}$$

The term in the square bracket in (1.11) includes the Pauli blocking factor, i.e. the final state for a relative component $\mathbf{p}'$, differing from $\mathbf{p}$ by the reversal of the x component must be unoccupied. The expression for the flux reflected from the rigid wall at right can be found by changing $\mathbf{p}$ to $\mathbf{p}'$. Since the momentum transfer to the wall for a single collision is given by $\Delta p_x = 2p_x$, therefore the force exerted per unit area of the wall is

$$f = \int 2p_x \left[dj'_L(\mathbf{p}) - dj_R(\mathbf{p}) \right]$$

$$= \frac{2}{m(2\pi\hbar)^3} \int p_x^2 \left[g\left(\mathbf{p} + m\mathbf{V}\right) - g\left(\mathbf{p} - m\mathbf{V}\right) \right] d^3 p. \tag{1.12}$$

By expanding $g\left(\mathbf{p}' \pm m\mathbf{V}\right)$ in powers of V in (1.12) we observe that the leading term in the force f is proportional to V.

1.2 Drag Force on a Sphere Moving through Viscous Fluid

The pioneering work of Stokes in hydrodynamics is an important example of the derivation of dissipative force laws. Stokes showed that to the first order in Reynolds number the force of friction is linear in velocity [11]–[15]. The system that we will be considering consists of two parts: an object which is a sphere of radius R, and a field, which in this case is an incompressible fluid. We assume that the fluid is viscous of uniform density ρ and of dynamical viscosity η. The equation of motion of the fluid is the Navier-Stokes equation [13],

$$(\mathbf{v} \cdot \nabla)\mathbf{v} = -\frac{1}{\rho}\nabla p + \left(\frac{\eta}{\rho}\right)\nabla^2 \mathbf{v}. \tag{1.13}$$

where p and $\mathbf{v}$ are the pressure and fluid velocity at any given point $\mathbf{r}$ in the fluid respectively. Here we will outline the approach of Stokes and derive the damping force exerted by the fluid on the sphere, as this sphere moves uniformly in the fluid with constant velocity $\mathbf{u}$. This problem is equivalent to that of flow past a fixed sphere with the fluid having a velocity $-\mathbf{u}$ at infinity. In this picture of the system, the moving fluid passes around the fixed sphere and assumes a uniform motion with velocity $\mathbf{v} = \mathbf{u}$ at infinity. We introduce a velocity $\mathbf{v}'$ by $\mathbf{v} = \mathbf{v}' + \mathbf{u}$. Since the fluid is incompressible we have

$$\nabla \cdot \mathbf{v} = \nabla \cdot \mathbf{v}' = 0. \tag{1.14}$$

Now from $\nabla \cdot \mathbf{v} = 0$, it follows that $\mathbf{v}$ can be written as the **curl** of a vector;

$$\mathbf{v}' = \nabla \wedge \mathbf{A} + \mathbf{u}, \tag{1.15}$$

where $\wedge$ denotes the vector product.

Both $\mathbf{v}'$ and $\mathbf{u}$ are polar vectors, therefore $\mathbf{A}$ has to be an axial vector. Taking the origin at the center of sphere, we observe that as the symmetry of the problem implies, from the two polar vectors $\mathbf{r}$ and $\mathbf{u}$, we can only construct one axial vector, $\mathbf{r} \wedge \mathbf{u}$. This means that the most general form of $\mathbf{A}$ is

$$\mathbf{A} = f'(r)\mathbf{n} \wedge \mathbf{u}, \tag{1.16}$$

where $f'(r)$ is a scalar function of r, and $\mathbf{n}$ is the unit vector in the direction of radius vector. Thus we can write $f'(r)\mathbf{n} = \nabla f(r)$ and

$$\mathbf{A} = \nabla f(r) \wedge \mathbf{u}. \tag{1.17}$$

Nothing that $\mathbf{u}$ is a constant vector, we have $\nabla f(r) \wedge \mathbf{u} = \nabla \wedge (f(r)\mathbf{u})$ and therefore we get the following vector relation

$$\mathbf{v} = \nabla \wedge \nabla \wedge (f(r)\mathbf{u}) + \mathbf{u}. \tag{1.18}$$

In order to find the function $f(r)$, we try to solve the Navier-Stokes equation for the moving fluid subject to the boundary condition that at $r = R$ the velocity must be zero. To this end we start with (1.13) and note that in this equation for small Reynolds number $(\mathbf{v} \cdot \nabla)\mathbf{v} \ll \left(\frac{\eta}{\rho}\right)\nabla^2\mathbf{v}$ and may be neglected. Then (1.13) reduces to a linear equation for $\mathbf{u}$

$$\eta\nabla^2\mathbf{v} = \nabla p. \tag{1.19}$$

This equation together with the equation of continuity $\nabla \cdot \mathbf{v} = 0$ completely determines the motion. By taking the **curl** of the two sides of (1.18) we get

$$\nabla^2(\nabla \wedge \mathbf{v}) = 0. \tag{1.20}$$

Thus from (1.17) and (1.19) we obtain

$$\nabla \wedge \mathbf{v} = \nabla \wedge \nabla \wedge \nabla \wedge [f(r)\mathbf{u}] = \left(\nabla\nabla \cdot -\nabla^2\right)[\nabla \wedge (f(r)\mathbf{u})]$$
$$= -\nabla^2[\nabla \wedge (f(r)\mathbf{u})], \tag{1.21}$$

where we have used the vector identity

$$\nabla \wedge \nabla \wedge \mathbf{W} = \nabla\nabla \cdot \mathbf{W} - \nabla^2\mathbf{W}. \tag{1.22}$$

Now from Eqs. (1.19) and (1.21) we find

$$\nabla^4(\nabla(f(r) \wedge \mathbf{u})) = 0. \tag{1.23}$$

and since $\mathbf{u}$ is a constant vector it follows that

$$\nabla^4(\nabla(f(r) \wedge \mathbf{u})) = \nabla^4(\nabla(f(r))) \wedge \mathbf{u} = 0. \tag{1.24}$$

This last relation shows that

$$\nabla^4(\nabla(f(r))) = \text{a constant.} \tag{1.25}$$

The constant on the right-hand side of (1.25) must be zero, since both $\mathbf{v}$ and its derivatives must vanish at infinity. By integrating (1.25) we obtain

$$\nabla^4 f(r) = \frac{1}{r^2}\frac{d}{dr}\left(r^2\frac{d}{dr}\right)\nabla^2 f(r) = 0. \tag{1.26}$$

Solving this equation we find

$$\nabla^2 f(r) = \frac{2a}{r} + B, \tag{1.27}$$

where again to satisfy the boundary condition $\mathbf{v} \to 0$ as $r \to \infty$, we must choose $B = 0$. From the equation $\nabla^2 f(r) = \frac{2a}{r}$, we find $f(r)$ to be

$$f(r) = ar + \frac{b}{r}. \tag{1.28}$$

Again we have set the constant of integration equal to zero, because $\mathbf{v}$ is given by the derivatives of $f(r)$. By substituting (1.28) in (1.27) we find

$$\mathbf{v} = \mathbf{u} - \frac{a}{r}\left[\mathbf{u} + \mathbf{n}(\mathbf{u} \cdot \mathbf{n})\right] + \frac{b}{r^3}\left[3\mathbf{n}(\mathbf{u} \cdot \mathbf{n}) - \mathbf{u}\right]. \tag{1.29}$$

Now we impose the boundary condition $\mathbf{v}(r = R) = 0$, and we note that when we set $r = R$ in (1.29), the coefficients of $\mathbf{u}$ and $\mathbf{n}(\mathbf{u} \cdot \mathbf{n})$ must each vanish so that Eq. (1.29) is satisfied for all $\mathbf{n}$. In this way we find the values of a and b to be

$$a = \frac{3}{4}R, \qquad b = \frac{1}{4}R^3. \tag{1.30}$$

using these values in (1.29) and noting that $\mathbf{u} \cdot \mathbf{n} = u \cos\theta$, we can determine the components of velocity in spherical coordinates

$$v_r = u \cos\theta \left[1 - \frac{3R}{2r} + \frac{R^3}{2r^3}\right], \tag{1.31}$$

$$v_\theta = -u \sin\theta \left[1 - \frac{3R}{4r} - \frac{R^3}{4r^3}\right]. \tag{1.32}$$

Next we calculate the pressure p from Eq. (1.19);

$$\nabla p = \eta \nabla^2 \mathbf{v} = \eta \nabla^2 \left[\nabla \wedge \nabla \wedge (f(r)\mathbf{u})\right]$$
$$= \eta \nabla^2 \left[\nabla \nabla \cdot (f(r)\mathbf{u}) - \mathbf{u}\nabla^2 f(r)\right]. \tag{1.33}$$

Since $\nabla^4 f(r) = 0$, we have

$$\nabla p = \nabla \left[\eta \nabla^2 \nabla \cdot (f(r)\mathbf{u})\right] = \nabla \left[\eta (\mathbf{u} \cdot \nabla)\nabla^2 f(r)\right]. \tag{1.34}$$

Therefore

$$p = \eta (\mathbf{u} \cdot \nabla)\nabla^2 f(r) + p_0, \tag{1.35}$$

where p_0 is the fluid pressure at infinity. By substituting for $\nabla^2 f(r)$ from (1.27) and (1.28) in (1.35) we find p to be

$$p = p_0 - \frac{3}{2}\frac{\eta R}{r^2}(\mathbf{u} \cdot \mathbf{n}). \tag{1.36}$$

Now the force per unit surface area of the sphere, P is given by its components P_i:

$$P_i = pn_i - \sum_k \sigma_{ik} n_k, \tag{1.37}$$

where the first term is the ordinary pressure of the fluid and other terms are due to the viscosity acting on the surface of the body immersed in the fluid, σ_{ik} being the viscosity tensor [13].

By choosing the spherical coordinates so that the polar axis is parallel to **u** then all quantities will be functions of r and of the polar angle θ, and the components of σ are:

$$\sigma_{rr} = 2\eta \frac{\partial v_r}{\partial r}, \tag{1.38}$$

$$\sigma_{r\theta} = \eta \left(\frac{1}{r} \frac{\partial v_r}{\partial \theta} + \frac{\partial v_\theta}{\partial r} - \frac{v_\theta}{r} \right). \tag{1.39}$$

Substituting for v_r and v_θ and their derivatives from Eqs. (1.31) and (1.32) we find the components of σ at the surface of the sphere to be

$$\sigma_{rr}(r = R) = 0, \tag{1.40}$$

$$\sigma_\theta = -\frac{3}{2R} \eta u \sin \theta. \tag{1.41}$$

The infinitesimal force on an element of the surface of the sphere, dS, is given by

$$dF = (-p \cos \theta + \sigma_{rr} \cos \theta - \sigma_{r\theta} \sin \theta) \, dS, \tag{1.42}$$

therefore by substituting σ_{rr} and $\sigma_{r\theta}$ from (1.38) and (1.39) in (1.42) we obtain

$$dF = \frac{3\eta u}{2R} dS. \tag{1.43}$$

Thus

$$F = 6\pi R \eta u, \tag{1.44}$$

and this is the well-known Stokes' formula [11],[13]. The Stokes law agrees with experiments only for very small Reynolds numbers. Thus if we express the Reynolds number, $\mathcal{R}$, in terms of the diameter of the sphere and velocity then we have

$$\mathcal{R} = \frac{2vR\rho}{\eta}. \tag{1.45}$$

The Reynolds number must be less than one for the validity of the Stokes law [16]. Such a small Reynolds numbers can occur in highly viscous fluids or for very small spheres such as droplets of mist in atmosphere.

Correction to the Drag Force of Stokes — We can calculate this drag force to the second order in the Reynolds number (or the velocity of the sphere) and find that [13],[17]

$$F = 6\pi \eta R v \left(1 + \frac{3vR}{8\nu} \right). \tag{1.46}$$

Here ν is the kinematic viscosity $\nu = (\eta/\rho)$ and ρ is the density of the fluid. This expression first derived by Oseen [17] extends the range of validity of the drag force law to Reynolds number $\mathcal{R} \leq 5$, a result that has been confirmed by experiment [16].

The drag force can also be calculated for a plane circular disk of radius R moving perpendicular to its plane. This force turns out to be [13]

$$F = 16\eta R v. \tag{1.47}$$

Another interesting problem is that of the motion of a spherical drop of fluid of viscosity η' moving under gravity in a fluid of viscosity η [13]. In this case the drag force is given by

$$F = 2\pi v \eta R \left(\frac{2\eta + 3\eta'}{\eta + \eta'} \right). \tag{1.48}$$

This expression reduces to (1.44) as $\eta' \to \infty$ (i.e. a solid sphere) and to

$$F = 4\pi \eta R v, \tag{1.49}$$

when $\eta' \to 0$ (i.e. the case of a gas bubble). For a motion subject to the drag force F, Eq. (1.48), the terminal velocity u_T is given by

$$u_T = \frac{2R^2 g \left(\rho - \rho' \right) \left(\eta + \eta' \right)}{3\eta \left(2\eta + 3\eta' \right)}. \tag{1.50}$$

Motion of a Simple Pendulum in a Viscous Medium — The small oscillations of a pendulum consisting of a solid sphere of radius R and mass M attached to a fine wire in air or vacuum is well known. Now if such a pendulum is immersed in a viscous medium of density ρ and of viscosity η then the equation of motion takes the following form

$$M \frac{d^2 s}{dt^2} = -\frac{g}{\ell} \left(M - M' \right) s - M' \left(\frac{1}{2} + \frac{9}{4\beta R} \right) \frac{d^2 s}{dt^2}$$
$$- \frac{9}{4} M' \left(\frac{2\pi}{T_0} \right) \left(\frac{1}{\beta R} + \frac{1}{\beta^2 R^2} \right) \frac{ds}{dt}, \tag{1.51}$$

where s is the displacement of the center of the solid sphere in time, M' is the mass of displaced fluid, ℓ is the length of the pendulum, T_0 is the period of oscillation in the absence of damping and β is given by

$$\beta = \sqrt{\frac{\pi \rho}{\eta T_0}}. \tag{1.52}$$

Here we have assumed that that the motion is limited to small oscillations about the equilibrium. The period of the damped oscillation determined from the solution of (1.51) is given by

$$T = 2\pi \left[\left(\frac{M - kM'}{M - M'} \right) \frac{\ell}{g} \right]^{\frac{1}{2}}, \tag{1.53}$$

where

$$k = \left(\frac{1}{2} + \frac{9}{4\beta R}\right).$$ (1.54)

This result for T reduces to the well known result of $T_0 = 2\pi\sqrt{\frac{\ell}{g}}$ when $M' = 0$.

By measuring T, we can determine the coefficient of viscosity η [14]

$$\eta = \frac{1}{9\pi\rho T R^4}\left\{(M - M')\frac{T^2}{T_0^2} - \left(M + \frac{1}{2}M'\right)\right\}^2.$$ (1.55)

1.3 Raleigh's Oscillator

Let us consider a tuning fork vibrating in vacuum with a given frequency. We want to know how the vibration of this tuning fork will be changed if we immerse it in a viscous fluid, e.g. in air or in a gas. Raleigh observed that the fork and the air surrounding it constitute a single system whose parts cannot be treated separately [18]. However he noted that we can simplify the problem when the effect of the medium on the fork, during few periods is very small, and only becomes important by accumulation. Thus he was led to consider the effect as a perturbation, where the disturbing force is periodic (roughly with the periodicity of the fork) and may be divided into two parts. One part is proportional to the acceleration $(m' - m)\ddot{x}$, where $(m' - m)$ is a small mass, and the other part is proportional to the velocity $\dot{x}$ and opposes the motion. The first part can be absorbed in $m\ddot{x}$, with the result that the actual mass m is replaced by the effective mass m'. Thus Rayleigh found the equation of motion to be

$$m'\ddot{x} + m'\lambda\dot{x} + m'\omega_0^2 x = 0,$$ (1.56)

where λ and ω_0^2 are defined in such a way that $m'\lambda\dot{x}$ is the frictional and $m'\omega_0^2 x$ is the harmonic for force acting on the fork.

1.4 Frictional Forces Quadratic in Velocity

The drag force on an object which moves with high speed through a gas (e.g. the motion of a meteoroid through atmosphere [19]) can be calculated in a simple way if we assume that the mean free path of the molecule is large compared to the linear dimensions of the object. Let us denote the mass of the object by M, its effective cross-sectional area by S and the density of the gas by ρ. For simplicity we assume that the collisions are inelastic and the molecules stick to the object. Then the law of conservation of momentum implies that

$$Mv = (M + \rho Svdt)(v + dv),$$ (1.57)

from which it follows that the force of friction is proportional to the square of velocity

$$\frac{dv}{dt} = -\frac{\rho S}{M}v^2.$$

(1.58)

For the general case the last equation can be written as

$$\frac{dv}{dt} = -\Gamma\frac{\rho S}{M}v^2,$$

(1.59)

where Γ is a dimensionless constant.

The atmospheric drag force on an artificial satellite which enters the Earth atmosphere also depends quadratically on the velocity of the satellite, i.e.

$$F = -\frac{1}{2}C\rho Sv^2,$$

(1.60)

where ρ is the air density which varies exponentially with height above Earth's surface, S is the effective cross section of the satellite and C is the drag coefficient based on the shape of satellite. The energy loss due to the dissipative force affects the orbital motion by a minor contraction of the orbit.

Air drag tends to make the elliptic orbit closer to the circular orbit by constantly reducing the apogee distance while having a minor effect on the perigee distance [20]. The motion of a projectile in the atmosphere with a quadratic drag force is an interesting example of the modification of the orbit caused by the dissipative forces [21].

A Simple Microscopic Model for Quadratic Damping — From a microscopic model we can derive a resistive force proportional to the square of the velocity. Let us consider the elastic collision between a body of mass M and a group of identical molecules, each of mass m ($M > m$). These molecules form a heat bath which absorbs the energy of the body. Since we want to study a one-dimensional problem, we can take these molecules as well as the massive body to be a collection of " billiard rods" [22]. We assume all collisions to be elastic and conserve kinetic energy and momentum. Now in elastic collision between a body of mass M moving with velocity V_0 and a molecule of mass m moving with velocity v_0, from the conservation laws we find the velocities after the collision to be

$$V_1 = V_0 - \left(\frac{2m}{M+m}\right)(V_0 - v_0),$$

(1.61)

$$v_1 = V_0 + \left(\frac{M-m}{M+m}\right)(V_0 - v_0).$$

(1.62)

Let us assume that all of the molecules are to the right of the body of mass M, and they are all at rest, whereas the body moves to the right with the initial velocity V_0. We label these molecules from the left to the right by *1, 2, 3,* etc.

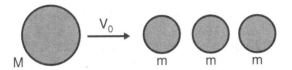

Figure 1.1: A one-dimensional model which shows how quadratic damping can be experienced by an object through successive collisions. This model consists of a heavy body of mass M colliding elastically with a large number of lighter molecules each of mass m. For this system from the laws of conservation of kinetic energy and momentum, we find that the energy of the massive body reduces like $-\gamma \dot{x}^2$. The initial energy of the body is distributed among a large number of molecules [22].

After the first collision between the body and *1*, the velocities of the two become

$$V_1 = \left(\frac{M-m}{M+m}\right) V_0, \tag{1.63}$$

$$v_1 = \left(\frac{2M}{M+m}\right) V_0, \tag{1.64}$$

and these show that $v_1 > V_1$.

After the first collision, molecule *1* moves away from the body and collides with *2*. Since the two molecules have the same mass, *1* comes to rest whereas *2* moves in the direction of *3* with velocity v_1. This collision between the two molecules repeats itself, and thus the momentum lost by the body after its first collision with *1* is transferred to the right. Now the body, after its first collision, according to (1.63) moves to the right and for a second time collides with *1* which is at rest. This time the body emerges from the collision with a speed

$$V_2 = \left(\frac{M-m}{M+m}\right)^2 V_0, \qquad V_2 < V_1, \tag{1.65}$$

while *1* has acquired a speed v_2

$$v_2 = \frac{2M(M-m)}{(M+m)^2} V_0, \qquad v_2 > V_2. \tag{1.66}$$

The momentum mv_2 of the molecule *1* will be completely transferred to *2*, and then to the end of the system. Thus the speed of the body after N collisions, all with molecule *1* becomes

$$V_N = \left(\frac{M-m}{M+m}\right)^N V_0 = \left(\frac{1-r}{1+r}\right)^N V_0, \tag{1.67}$$

with $r = \frac{m}{M}$.

Assuming that the molecules are spread uniformly along the x-axis and have an average density ρ, then the number of collisions after traversing a distance x will be ρx. Therefore we can write (1.67) as [22]

$$V(x) = \left(\frac{1-r}{1+r}\right)^{\rho x} V_0. \tag{1.68}$$

To calculate the drag force acting on the body we use Newton's second law for the motion of the body

$$F(x) = M\frac{dV(x)}{dt} = M\frac{dV(x)}{dx}V(x). \tag{1.69}$$

Finally by substituting from (1.68) in (1.69) and defining a constant γ by

$$\gamma = \frac{m\rho}{r}\ln\left(\frac{1+r}{1-r}\right), \tag{1.70}$$

we find

$$F(x) = -\gamma V^2. \tag{1.71}$$

In Sec. 1.1 we discussed the motion of an object in the air where we assumed that the retarding force is proportional to $-\dot{x}|\dot{x}|^{n-1}$, with $1 \leq n \leq 2$. Therefore this model gives us the maximum resisting force, and for the motion of an object in the air it is not realistic. The reason being the one-dimensionality of the model. Since all collisions are "head-on" collisions, the body compresses the air in front of it as it moves. In two or three dimensions because of the glancing collisions between the body and molecules, there is a reduction of the forward momentum loss rate compared to the case that we solved, thus reducing the quadratic dependence of the resistive force [22]. In Chapter 7 we will consider other models where a particle M interacts with a group of particles and the resulting resistive force has a complicated dependence on the velocity of M.

1.5 Non-Newtonian and Nonlocal Dissipative Forces

Dissipative forces which explicitly depend on acceleration, i.e. $\mathbf{F} = \mathbf{F}(\mathbf{r}, \dot{\mathbf{r}}, \ddot{\mathbf{r}}, t)$ are called non-Newtonian [23]. These forces violate some of the basic principles of the Newtonian dynamics. For instance, in the presence of these forces, the total acceleration is not given by the vector sum of accelerations produced by each individual force. As we will see later, Eq. (1.101) describing the relativistic motion of an electron provides an example of a non-Newtonian force law.

When a particle interacts with a system of particles, or when an extended object moves in a resistive medium, then the effective force on the particle or on the extended object is nonlocal. This nonlocality can be spatial or temporal.

Examples of the latter type of nonlocality are given in Chapter 7, where the effective force felt by the particle at the time t is given by

$$F(x,t) = \int_0^t K\left(t-t'\right) x\left(t'\right) dt'. \tag{1.72}$$

Here the motion is assumed to be one-dimensional.

Nonlocal forces which occur in the description of the motion of an extended object depend on the shape and structure of the object and have the general form

$$\mathbf{F}\left(\mathbf{r}, \dot{\mathbf{r}}, t\right) = \int \mathbf{Q}\left(\mathbf{r}, \dot{\mathbf{r}}; \mathbf{r}', \dot{\mathbf{r}}', t\right) d^3 r', \tag{1.73}$$

where the integral is taken over the volume of the object. Under certain conditions we can expand the integrand in (1.73) in powers of velocity with the result that

$$\mathbf{F}\left(\mathbf{r}, \dot{\mathbf{r}}, t\right) \approx \mathbf{f}(\mathbf{r}) - \lambda_1(\mathbf{r})\dot{\mathbf{r}} - \lambda_2(\mathbf{r})|\dot{\mathbf{r}}|\dot{\mathbf{r}} - \cdots \tag{1.74}$$

The analogue of this type of nonlocal force (or potential) is important in the quantum theory of nuclear structure where it is obtained from the coupling of the object to a large number of different channels (see Chapters 18 and 20).

1.6 One-Dimensional Dissipative Motion and the Problem of Harmonically Bound Electron

For a one-dimensional motion (say along the x-axis) and for frictional forces linear in velocity Newton's equation of motion is

$$m\frac{d^2 x}{dt^2} + m\lambda\frac{dx}{dt} = f(x). \tag{1.75}$$

In this equation $f(x)$ represents all of the other conservative forces which act on the particle and are derivable from potentials. For instance for a simple pendulum which is oscillating in a liquid Eq. (1.75) takes the following form:

$$m\frac{d^2 x}{dt^2} + m\lambda\frac{dx}{dt} = -m\omega_0^2 x. \tag{1.76}$$

Denoting the time derivative of x by $\dot{x}$, the initial conditions of the motion are $x(0) = x_0$ and $\dot{x}(0) = \dot{x}_0$. With these initial conditions Eq. (1.76) has a simple solution, and if λ is sufficiently small while the particle is oscillating, the amplitude of x decreases exponentially and asymptotically we have $\dot{x}(\infty) = x(\infty) = 0$, i.e. after a very long time we know both the position and the momentum of the particle precisely.

As we will see in the following subsection the classical equation of motion for an harmonically bound electron coupled to the electromagnetic field

(Abraham-Lorentz equation [24]–[30]) which contains the third order derivative of x, can be transformed to an equation similar to (1.76).

Harmonically Bound Radiating Electron — In the pioneering works carried out by Planck, by Abraham and by Lorentz [25]–[27] the starting point is the analysis of the energy balance. From the Lienard-Wiechart potentials it follows that the energy radiated during the time T is given by

$$\frac{2}{3}\frac{e^2}{c^3} \int_t^{t+T} \left(\frac{d^2x}{dt^2}\right) dt. \tag{1.77}$$

Since the energy of the oscillator is

$$E = \frac{1}{2}m\left(\frac{dx}{dt}\right)^2 + \frac{1}{2}kx^2, \tag{1.78}$$

and that the energy lost by radiation has to be compensated by a decrease in E, therefore by balancing the energy flow, we obtain

$$\int_t^{t+T} \left\{ \frac{dE}{dt} + \frac{2e^2}{3c^3}\left(\frac{d^2x}{dt^2}\right)^2 \right\} dt = 0. \tag{1.79}$$

From this equation alone we cannot specify the dynamics of the system completely. However we can obtain a differential equation by setting the integrand in (1.79) equal to zero. Thus from (1.78) and (1.79) we find

$$m\frac{d^2x}{dt^2} = -kx - \frac{2e^2}{3c^3}\left(\frac{d^2x}{dt^2}\right)^2\left(\frac{dx}{dt}\right)^{-1}. \tag{1.80}$$

This nonlinear equation can be linearized if we make the following approximation: Let us assume that we can write (1.79) as

$$\int_t^{t+T} \left\{ \frac{d}{dt}\left[E + \frac{2e^2}{3c^3}\left(\frac{dx}{dt}\right)\left(\frac{d^2x}{dt^2}\right)\right] - \frac{2e^2}{3c^3}\left(\frac{dx}{dt}\right)\left(\frac{d^3x}{dt^3}\right) \right\} dt = 0, \tag{1.81}$$

then noting that

$$\sqrt{\frac{k}{m}} \ll \frac{mc^2}{e^2} \sim 10^{23}\ (\text{sec})^{-1}, \tag{1.82}$$

we get

$$E \gg \frac{2e^2}{3c^3}\left(\frac{dx}{dt}\right)\left(\frac{d^2x}{dt^2}\right). \tag{1.83}$$

Now by neglecting the last term in the integrand in (1.81), we find the equation of motion for x;

$$m\frac{d^2x}{dt^2} = -kx + m\tau\frac{d^3x}{dt^3}, \tag{1.84}$$

where

$$\tau = \left(\frac{2e^2}{3mc^3}\right) = 6.26 \times 10^{-24}\ \text{s}. \tag{1.85}$$

A different way of getting a linear equation for $x(t)$ without making the foregoing approximation is to assume the existence of a reaction force F_{react}. This force when added to $F = m\frac{d^2x}{dt^2}$, would produce the reaction losses. Now we replace the equation for the energy flow balance (1.79) by the equation for force balance.

$$\frac{2e^2}{3c^3} \int_t^{t+T} \left(\frac{d^2x}{dt^2}\right)^2 dt = -\int_t^{t+T} F_{\text{react}} \left(\frac{dx}{dt}\right) dt. \tag{1.86}$$

Integrating the left-hand side of (1.86) by parts and choosing T such that the acceleration $\frac{d^2x}{dt^2}$ vanishes at each end of the interval, we find the following result [28]

$$\int_t^{t+T} \left\{ \frac{2e^2}{3c^3} \left(\frac{d^3x}{dt^3}\right) - F_{\text{react}} \right\} \frac{dx}{dt} dt = 0. \tag{1.87}$$

Again the trivial solution of (1.87) is obtained by setting the expression inside the curly bracket equal to zero, i.e.

$$F_{\text{react}} = \frac{2e^2}{3c^3} \left(\frac{d^3x}{dt^3}\right). \tag{1.88}$$

We observe that this result does not depend on the nature of the oscillator, so we may argue that (1.88) is in fact generally valid. Thus we can express the entire radiation reaction theory in terms of a single equation

$$m\frac{d^2x}{dt^2} = F_{\text{ext}} + \frac{2e^2}{3c^3}\frac{d^3x}{dt^3}. \tag{1.89}$$

The Bopp Transformation — For a nonrelativistic harmonically bound radiating electron we can transform the third order differential equation (1.84) to a second order equation. Thus setting $k = m\omega_0^2$, we introduce the following transformation [31]

$$X = \left[\frac{dx}{dt} - \left(\frac{1}{a}\right)\left(\frac{d^2x}{dt^2}\right)\right], \tag{1.90}$$

and

$$Y = \left[X - \left(\frac{1}{2a}\right)\frac{dX}{dt} - \left(\frac{1}{2a^2}\right)\left(\frac{d^2X}{dt^2}\right)\right], \tag{1.91}$$

where

$$a = \frac{\nu}{\sqrt{(\lambda\tau)}} = \frac{1}{\tau} + \lambda, \tag{1.92}$$

and λ is the real positive root of the equation

$$\lambda\left(1 + \lambda\tau\right)^2 - \nu^2\tau = 0. \tag{1.93}$$

By differentiating (1.91) we find the coupled set of equations [32]

$$\dot{Y} + \frac{1}{2}\lambda Y - \frac{\left(1 + \frac{3}{4}\lambda\tau\right)}{1 + \lambda\tau}X = 0, \tag{1.94}$$

and

$$\dot{X} + \frac{1}{2}\lambda X + \frac{\lambda}{\tau}(1 + \lambda\tau)Y = 0. \tag{1.95}$$

If we eliminate X or Y between these equations we obtain the equation of motion for X which is the same as Eq. (1.76) with $\omega_0^2 = \frac{\nu^2}{a\tau}$.

1.7 Abraham-Lorentz-Dirac Equation for Radiating Electron

According to classical electrodynamics, an electron of mass m_0 moving in an external electromagnetic field experiences the Lorentz force [4]

$$m_0\frac{d\mathbf{v}}{dt} = e\left(\mathbf{E} + \frac{1}{c}\mathbf{v}\wedge\mathbf{B}\right). \tag{1.96}$$

If we assume that the electron is uniformly charged sphere of radius r_0, each part of this charged sphere repels every other part with a Coulomb force, and this repulsion is responsible for the self force $\mathbf{F}_S$

$$\mathbf{F}_S = \int\left(\mathbf{E} + \frac{1}{c}\mathbf{v}\wedge\mathbf{B}\right)\rho dV. \tag{1.97}$$

In this expression $\mathbf{E}$ and $\mathbf{B}$ are the fields produced by the electron itself and ρ is the charge density. A lengthy calculation yields the following result, first derived by Lorentz, [27],[34]:

$$\mathbf{F}_S = -\frac{2}{3c^2}\sum_{n=0}^{\infty}\frac{(-1)^n}{n!c^n}\frac{d^n\mathbf{a}}{dt^n}\int\frac{\rho(\mathbf{r})\rho(\mathbf{r}')}{|\mathbf{r}-\mathbf{r}'|}|\mathbf{r}-\mathbf{r}'|^n d^3rd^3r'$$

$$= -\frac{4}{3c^2}\left(\frac{1}{2}\int\frac{\rho(\mathbf{r})\rho(\mathbf{r}')}{|\mathbf{r}-\mathbf{r}'|}d^3rd^3r'\right)\times\frac{d\mathbf{v}}{dt} + \frac{2e^2}{3c^3}\frac{d\mathbf{a}}{dt}$$

$$- \frac{2e^2}{3c^2}\sum_{n=2}^{\infty}\frac{(-1)^n}{n!c^n}\frac{d^n\mathbf{a}}{dt^n}\mathcal{O}\left(r_0^{n-1}\right), \tag{1.98}$$

where $\mathbf{a}$ is the acceleration of the electron. The first term on the right side which is proportional to acceleration represents the electromagnetic mass of the electron which we denote by m'. Thus if we ignore all of the terms proportional to r_0 and its higher powers, then we have the equation of motion

$$m_0\frac{d\mathbf{v}}{dt} = -m'\frac{d\mathbf{v}}{dt} + \frac{2e^2}{3c^3}\frac{d\mathbf{a}}{dt} + \mathbf{F}_{\text{ext}}, \tag{1.99}$$

where $\mathbf{F}_{\text{ext}}$ is the external force. Even in this first order of approximation we find that the equations of motion are non-Newtonian and there are problems associated with the interpretation of their solutions [35]. We note that of these terms we have included only the first term in (1.98). We can inquire whether the first order correction to $\mathbf{F}_S$ is of order r_0 or some higher power of r_0. This problem has been investigated by Gally, Leibovich and Rothstein using effective field theory approach [34]. These authors show that as a consequence of Poincaré and gauge symmetries, the leading term due to the finite size r_0 of a spherically symmetric charge is proportional to r_0^2 rather than r_0.

The relativistic generalization of this equation first obtained by von Laue in 1909, in covariant notation is [36],[37]

$$(m_0 + m') \frac{dv^\mu}{ds} = \frac{2e^2}{3c^3} \left[\frac{da^\mu}{ds} - \frac{1}{c^2} \sum_\nu a_\nu a^\nu v^\mu \right] + F^\mu, \qquad (1.100)$$

where again in (1.100) higher order derivatives of v^μ have been ignored. In this relation s denotes the proper time, and $a^\nu = dv^\nu/ds$.

Finally Dirac in 1938 found the exact relativistic form of this equation without ignoring higher order terms [38],[39]. In this seminal work, he attempted to construct a classical relativistic radiation reaction theory from Maxwell's equation based on the following principles:

(a) By choosing an appropriate linear combination of retarded and advanced potential so that the divergences of the point electron theory be eliminated.

(b) By deriving the relativistic equations of motion from conservation laws by means of Gauss's theorem.

This classical Abraham-Lorentz-Dirac equation in the presence of the external electromagnetic field is given by

$$ma^\mu = \frac{e}{c} F^{\mu\nu} v_\nu + \left(\frac{2e^2}{3c^3} \right) \left[\frac{da^\mu}{ds} - \frac{1}{c^2} \sum_\nu a^\nu a_\nu v^\mu \right]. \qquad (1.101)$$

In this relation m is the physical mass (or renormalized mass) of the electron, a^μ and v^μ are the μ components of the acceleration and velocity respectively and s is the proper time. The force $F^{\mu\nu}$ is related to the electromagnetic 4-potential by

$$F^{\mu\nu} = \frac{\partial A^\nu}{\partial x_\mu} - \frac{\partial A^\mu}{\partial x_\nu}. \qquad (1.102)$$

1.8 The Abraham-Lorentz-Dirac Equations of Motion for a Charged Particle

An elementary derivation of the relativistic classical Abraham-Lorentz-Dirac equation is given by Rohrlich [40]. This derivation is based on the fact that when the loss due to radiation is taken into account, for the consistency of the equations of motion, an additional term "Schott force" must be added to these equations [41]. Let us consider the Dirac equation which we write as

$$m_0 \frac{dv^\mu}{ds} = F^\mu_{\text{ext}} + F^\mu_{\text{self}}, \tag{1.103}$$

where F^μ_{ext} which is the external force can be the Lorentz force or any other force law, and where m_0 is the bare mass of the electron. The other term in (1.103), F^μ_{self}, is given by

$$
\begin{aligned}
F^\mu_{\text{self}} &= F^\mu_{\text{Sch}} + F^\mu_{\text{rad}} \\
&= F^\mu_{\text{Sch}} - \left(\frac{2e^2}{3c^3}\right) v^\mu \sum_\alpha \left(\frac{dv^\alpha}{ds} \frac{dv_\alpha}{ds}\right).
\end{aligned} \tag{1.104}
$$

This self force arises from the action of the charge's own field on itself. The first term on the right-hand side of Eq. (1.104) is the Schott force and the second term is the rate at which momentum and energy are lost in the form of radiation (Larmor's relativistic formula [42]). Now the presence of F_{Sch} is necessary for the consistency of the equations of motion. Otherwise by multiplying

$$m_0 \frac{dv^\mu}{ds} = F^\mu_{\text{rad}}, \tag{1.105}$$

by v^μ and summing over μ we find that contrary to the laws of classical electrodynamics F^μ_{rad} must vanish.

To find this additional force, F^μ_{Sch}, we observe that if $P^{\mu\nu}$ denotes the projection operator into hyper-plane orthogonal to the 4-velocity v^μ, i.e.

$$\sum_\nu P^{\mu\nu} v_\nu = 0,, \tag{1.106}$$

then $P^{\mu\nu}$ has the general form

$$P^{\mu\nu} = \left(\eta^{\mu\nu} + \frac{1}{c^2} v^\mu v^\nu\right), \tag{1.107}$$

where $\eta^{\mu\nu}$ is the diagonal matrix

$$\eta^{\mu\nu} = (1,\ 1,\ 1,\ -1). \tag{1.108}$$

Therefore the added term, F^μ_{Sch}, to the Eq. (1.104) must be of the form

$$\sum_\nu P^{\mu\nu} X_\nu = \sum_\nu \left[\left(\eta^{\mu\nu} + \frac{1}{c^2} v^\mu v^\nu\right) X_\nu\right]. \tag{1.109}$$

Now we assume that X^μ is a linear function of the velocity v^ν and in addition it does not contain terms higher than the second order time derivative of v^μ, i.e.

$$X^\mu = A v^\mu + B \frac{dv^\mu}{ds} + C \frac{d^2 v^\mu}{ds^2}. \tag{1.110}$$

By substituting X^μ in Eq. (1.109) we observe that the first term Av^μ does not contribute to the sum, and the second term is just like the inertial term in (1.103), and thus can be absorbed in it if we define the physical mass by $m = m_0 + m' = m_0 - B$. Thus Eq. (1.109) will take the form

$$m\frac{dv^\mu}{ds} = F^\mu_{\text{ext}} + C\sum_\nu P^{\mu\nu}\frac{d^2 v_\nu}{ds^2}, \tag{1.111}$$

Substituting for $P^{\mu\nu}$ from (1.109) in (1.111) and differentiating by parts will give us the result [40]

$$m\frac{dv^\mu}{ds} = F^\mu_{\text{ext}} + F^\mu_{\text{self}}, \tag{1.112}$$

where

$$F^\mu_{\text{self}} = \frac{2e^2}{3c^3}\left[\frac{d^2 v^\mu}{ds^2} - \frac{v^\mu}{c^2}\sum_\alpha\left(\frac{dv^\alpha}{ds}\frac{dv_\alpha}{ds}\right)\right], \tag{1.113}$$

provided that we choose

$$C = \frac{2e^2}{3c^3}. \tag{1.114}$$

Relativistic Form of the Reaction Force — The same method of obtaining the nonrelativistic form of F_{react} from the radiated energy, Eq. (1.77), can be used to obtain the relativistic form of this force. For the relativistic regime we use the well known relativistic Larmor formula [42]

$$P = \frac{2e^2}{3c^3}\gamma^6\left[\left(\frac{d\mathbf{v}}{dt}\right)^2 - \frac{1}{c^2}\left(\mathbf{v}\times\frac{d\mathbf{v}}{dt}\right)^2\right], \tag{1.115}$$

where $\mathbf{v}$ is the velocity of the particle and

$$\gamma = \frac{1}{\sqrt{1 - \frac{v^2}{c^2}}}. \tag{1.116}$$

Thus for the relativistic formulation we replace (1.77) by

$$\frac{2e^2}{3c^3}\gamma^4\int_t^{t+T}\left\{\left(\frac{d^2 x}{dt^2}\right)^2 + \left[\frac{\gamma}{c}\left(\frac{dx}{dt}\right)\left(\frac{d^2 x}{dt^2}\right)\right]^2\right\}dt, \tag{1.117}$$

for one-dimensional case and by

$$\frac{2e^2}{c^3}\gamma^4\int_t^{t+T}\left\{\dot{v}^2 + (\mathbf{v}\cdot\dot{\mathbf{v}})^2\right\} \tag{1.118}$$

for the three-dimensional motion.

By retracing the steps leading to (1.87) and (1.88) but now with (1.115) rather than (1.77) we get [26]

$$\mathbf{F}_{\text{react}} = \frac{2e^2}{3c^3}\gamma^2\left\{\frac{d^2\mathbf{v}}{dt^2} + \left(\frac{\gamma}{c}\right)^2\left(\mathbf{v}\cdot\frac{d^2\mathbf{v}}{dt^2}\right)\mathbf{v}\right\}$$

$$+ \frac{2e^2}{3c^3}\gamma^2\left\{3\left(\frac{\gamma}{c}\right)^2\left(\mathbf{v}\cdot\frac{d\mathbf{v}}{dt}\right)\frac{d\mathbf{v}}{dt} + 3\left(\frac{\gamma}{c}\right)^4\left(\mathbf{v}\cdot\frac{d\mathbf{v}}{dt}\right)^2\mathbf{v}\right\}.$$

$$(1.119)$$

With the help of (1.119) we can write a generalization of (1.89) in the form of [28]

$$F^\mu + \frac{2e^2}{3c^3}\left\{\frac{d^2v^\mu}{ds^2} - \frac{v^\mu}{c^2}\sum_\mu\left(\frac{dv_\mu}{ds}\right)\left(\frac{dv^\mu}{ds}\right)\right\} = m\frac{dv^\mu}{ds},\qquad(1.120)$$

where F^μ and v^μ are the Minkowski force and the four velocity respectively.

As we can see from the expansion of $\mathbf{F}_S$, Eq. (1.98), the coefficient of r_0^2 contains the derivatives $x^{(5)}$, $x^{(4)}\cdots$ of $x(t)$. An examination of the coefficients also shows that there is a divergent contribution to $\mathbf{F}_S$ and there is a finite part. The latter corrections can be added to the Abraham-Lorentz-Dirac equation.

Derivation of the Dirac Equation Using Coordinate Transformation — An alternative way of getting the Dirac equation is by observing that the field of an arbitrarily moving charged particle in an inertial frame can be related to the field of uniformly accelerated charged particle in its rest frame [43]. The latter field is static, therefore we can derive the field of an arbitrarily moving charged particle by a coordinate transformation. By applying this technique we can also calculate the self-force on a charged particle, and thus we derive Dirac equation for radiating electron in a simple way [43].

Problems Related to the Abraham-Lorentz-Dirac Equation — According to Rohrlich there are the following major defects of the Abraham-Lorentz-Dirac equation [44]:

The presence of the self-acceleration term

$$\frac{d^2v^\mu}{ds^2} = \frac{d^3x^\mu}{ds^3},\qquad(1.121)$$

makes the Abraham-Lorentz-Dirac equation a third order equation in x^μ, therefore giving the initial position and velocity of the particle is not enough to determine the solution uniquely.

The second problem is that in the absence of external force, there are solutions with exponentially increasing velocity, known as "runaway solutions", in addition to the trivial solution of $v^\mu = $ a constant. These solutions do not obey the law of inertia [44].

The third problem is that according to the classical electrodynamics, a moving point charge radiates if and only if it is accelerated, i.e. only when an external force acts on it. But when we set $F^\mu_{\text{ext}} = 0$, still the radiation reaction term F^μ_{self} survives [44].

We observe that the self-force which includes the radiation reaction force is always a small term in the equation of motion

$$m\frac{dv^\mu}{ds} = F^\mu_{\text{ext}}, \tag{1.122}$$

therefore we can use an approximate expression for the self-force. To this end we write (1.112) and (1.113) as the following equation

$$m\frac{dv^\mu}{ds} = F^\mu_{\text{ext}} + \frac{2e^2}{3mc^3}\sum_\mu\left[\left(\eta^{\mu\nu} + \frac{1}{c^2}v^\mu v^\nu\right)\frac{d}{ds}\left(\frac{dv_\mu}{ds}\right)\right]. \tag{1.123}$$

Treating the last term on the right-hand side of (1.123) as a small perturbation, we replace $\frac{dv^\mu}{ds}$ by $\frac{1}{m}F_{\text{ext}}$, to get

$$m\frac{dv^\mu}{ds} = F^\mu_{\text{ext}} + \frac{2e^2}{3mc^3}\sum_\mu\left[\left(\eta^{\mu\nu} + v^\mu v^\nu\right)\frac{d}{ds}F^{\text{ext}}_\nu\right]. \tag{1.124}$$

Rohrlich claims that for a point particle subject to the Lorentz force

$$F^\mu_{\text{ext}} = \sum_\nu F^{\mu\nu}v_\nu, \tag{1.125}$$

equation (1.124) is exact. Using the expression for the time derivative of $\left(\sum_\alpha F^{\mu\alpha}v_\alpha\right)$

$$\frac{d}{ds}\left(e\sum_\alpha F^{\mu\alpha}v_\alpha\right) = e\sum_{\alpha,\beta}\left\{v^\beta\frac{\partial}{\partial x^\beta}F^{\mu\alpha}v_\alpha + \frac{e}{m}F^{\mu\alpha}F_{\alpha\beta}\,v^\beta\right\}, \tag{1.126}$$

we can write (1.124) as

$$\begin{aligned}
m\frac{dv^\mu}{ds} &= e\sum_\alpha F^{\mu\alpha}v_\alpha + \frac{2e^3}{3mc^3}\sum_{\alpha,\beta}\left[v^\beta\frac{\partial}{\partial x^\beta}F^{\beta\alpha}v_\alpha\right.\\
&\quad + \left.\frac{e}{m}F^{\mu\alpha}F_{\alpha\beta}v^\beta + \frac{e}{m}v^\mu v_\alpha F^{\alpha\beta}\sum_\gamma F_{\beta\gamma}v^\gamma\right].
\end{aligned} \tag{1.127}$$

This claim has been refuted by O'Connell who points out that Eq. (1.124) proposed by Rohrlich to describe the motion of a point charge is in fact for the motion of a charge with structure [45]–[47].

The Nonrelativistic Limit of the Abraham-Lorentz-Dirac Equation
Let us now consider the nonrelativistic limit of Eq. (1.111). In this limit we find the component of the forces to be

$$\mathbf{F}_{\text{Sch}} \to \frac{2e^2}{3c^3}\frac{d^2\mathbf{v}}{dt^2}, \tag{1.128}$$

$$F^0_{\text{Sch}} \rightarrow \frac{2e^2}{3c^3} \left[\mathbf{v} \cdot \frac{d^2\mathbf{v}}{dt^2} - \left(\frac{d\mathbf{v}}{dt} \right)^2 \right],$$ (1.129)

$$\mathbf{F}_{\text{rad}} \rightarrow 0,$$ (1.130)

and

$$F^0_{\text{rad}} \rightarrow -\frac{2e^2}{3c^3} \left(\frac{d\mathbf{v}}{dt} \right)^2.$$ (1.131)

The equation of motion in this limit becomes

$$m\frac{d\mathbf{v}}{dt} = \mathbf{F}_{\text{ext}} + \mathbf{F}_{\text{self}} = \mathbf{F}_{\text{ext}} + \mathbf{F}_{\text{Sch}}.$$ (1.132)

In the following section we will study the solution of (1.132) for the one-dimensional motion.

For the general solution of the Abraham-Lorentz-Dirac equation we are faced with the following difficulties:

The lack of the stability of solutions with respect to small variations of the initial conditions. This means that the Abraham-Lorentz-Dirac does not represent a well posed problem, i.e. the solution is highly sensitive to the changes in the initial data.

Let us assume that we know exactly the initial conditions, which in the case of the Abraham-Lorentz-Dirac equation means knowing not only the initial position and velocity, but also the initial acceleration. If numerically integrated, the runaway contribution to the solution grows exponentially due to numerical error and makes the solution unacceptable.

1.9   A Method for Solving Abraham-Lorentz-Dirac Equation

A technique for finding a stable solution of the Abraham-Lorentz-Dirac equation called "order reduction" is as follows [48]:

The basic assumption in this method is that the true equation of motion should be a second order equation of the form

$$\ddot{x}^\mu = \xi^\mu \left(s, x, \dot{x}, \tau \right),$$ (1.133)

where dots denote derivatives with respect to s. The 4-vector ξ^μ defined by (1.133) satisfies the Abraham-Lorentz-Dirac equation

$$\xi^\mu = f^\mu + \tau \left[\frac{\partial \xi^\mu}{\partial s} + \sum_\nu \left(\frac{\partial \xi^\mu}{\partial x^\nu} \dot{x}^\nu + \frac{\partial \xi^\mu}{\partial \dot{x}^\nu} \xi^\nu - \frac{1}{c^2} \xi^\nu \xi_\nu \dot{x}^\mu \right) \right].$$ (1.134)

In this equation f^μ denotes the force per unit mass. Using Eq. (1.134) the physical solution can be found for a set of given initial position and velocity.

Noting that the runaway solutions are singular in the limit of $\tau \to 0$ (or $e \to 0$), we impose the condition

$$\lim_{\tau \to 0} \xi^\mu = f^\mu. \tag{1.135}$$

For an iterative solution of the Abraham-Lorentz-Dirac equation consider the following scheme which has been suggested by Aguirregabiria for solving the differential equation [48]

$$\ddot{x}^\mu = \xi_n^\mu \qquad n = 0, \ 1, \ 2 \cdots . \tag{1.136}$$

In the zeroth and the first order we have

$$\ddot{x}^\mu = \xi_0^\mu \equiv f^\mu, \tag{1.137}$$

$$\ddot{x}^\mu = \xi_1^\mu \equiv f^\mu + \tau \left[\frac{\partial f^\mu}{\partial s} + \sum_\nu \left(\frac{\partial f^\mu}{\partial x^\nu} \dot{x}^\nu + \frac{\partial f^\mu}{\partial \dot{x}^\nu} f^\nu - \frac{1}{c^2} f^\nu f_\nu \dot{x}^\mu \right) \right]. \tag{1.138}$$

In the $(n+1)$th order of iteration we obtain

$$\ddot{x}^\mu = \xi_{n+1}^\mu \equiv f^\mu + \tau \left[\frac{\partial \xi_n^\mu}{\partial s} + \sum_\nu \left(\frac{\partial \xi_n^\mu}{\partial x^\nu} \dot{x}^\nu + \frac{\partial \xi_n^\mu}{\partial \dot{x}^\nu} \xi_n^\nu - \frac{1}{c^2} \xi_n^\nu \xi_{\nu n} \dot{x}^\mu \right) \right]. \tag{1.139}$$

The limit of these successive solutions, if it exists, will satisfy Eqs. (1.134)-(1.135) and thus will be a Newtonian equation of motion with no runaway solutions. Of course we can generate a power series expansion for the solution of (1.134), but the second, third and higher order terms become very complicated. For this reason the iterative solutions given by (1.139) are more appropriate for the numerical solution than the power series. Later we will investigate the convergence of the iterative solution for a one-dimensional Abraham-Lorentz-Dirac equation [49].

Solution of the One-Dimensional Nonrelativistic Abraham-Lorentz Equation — For the one-dimensional motion Eq. (1.132) reduces to

$$\ddot{x} = f(x, t) + \tau \frac{d}{dt} \ddot{x}, \tag{1.140}$$

where a dot denotes derivative with respect to time t, and where $f(x, t)$ is the force per unit mass. This equation which is called Abraham-Lorentz equation can be solved analytically for some special cases. We will consider two of these solvable cases to illustrate the method of order reduction discussed earlier in this chapter.

(1) If the force is only a function of time, then the $n+1$th iteration (see Eq. (1.139)) is given by

$$\ddot{x} = \xi_{n+1}(t) = \sum_{k=0}^{n+1} \tau^k f^{(k)}(t), \tag{1.141}$$

where $f^{(k)}(t)$ denotes the kth time derivative of $f(t)$. Therefore the exact solution of the problem using the order reduction method is

$$\ddot{x} = \sum_{k=0}^{\infty} \tau^k f^{(k)}(t), \tag{1.142}$$

However the direct integration of Eq. (1.140) yields the following expression for $\ddot{x}$:

$$\ddot{x} = \int_0^{\infty} e^{-u} f(t + \tau u) du + C e^{\frac{t}{\tau}}, \tag{1.143}$$

where C is the integration constant. This exact solution shows that $\ddot{x}(t) \to \infty$ as $t \to \infty$ and is also singular in the limit of $\tau \to 0$. The acceptable solution of (1.140) is obtained if we set $C = 0$ in (1.143). With this choice of C, we observe that (1.143) reduces to (1.142).

(2) When $f(x) = -\omega_0^2 x$, we can transform the equation of motion to a linear second order equation using the Bopp transformation Eqs. (1.90),(1.91) [49].

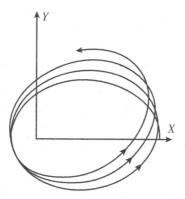

Figure 1.2: The orbit of a relativistic point charged particle in the absence of the radiation reaction force [48].

Classical Capture of a Relativistic Point Charge — Aguirregabiria has used the classical Eqs. (1.137)-(1.139) to obtain the solution to the equations of motion of a relativistic point charge e by static Coulomb center of charge $(-e)$. For simplicity one can choose a system of units where the unit of length is the classical electron radius

$$r_0 = \frac{e^2}{m_0 c^2}, \tag{1.144}$$

and also set $c = 1$. The initial position and velocity of the particle are assumed to be $x_0 = (50, 0)$ and $\dot{x}_0 = (0, 0.085)$. Now if the radiation reaction force is completely ignored, then the resulting trajectory will be the same as that of a relativistic Kepler problem, i.e. a precessing ellipse (Fig. 1.2) [50]–[53]. With the radiation reaction force, with the same initial conditions the trajectory is a spiral

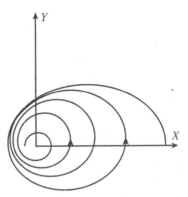

Figure 1.3: The trajectory of a point charge captured by a fixed charge of opposite sign. The orbit is computed by integrating forward the Abraham-Lorentz-Dirac equation [48].

shown in Fig. 1.3. In this case the result of the numerical integration becomes unreliable because of the lack of convergence of the iteration procedure when the radial distance is less than $3r_0$. For details of the numerical computation and the associated problem of convergence see the paper of Aguirregabiria [48].

1.10 The Classical Theory of Line Width

The spectral lines observed in a spectrograph have an observable width and this is related to the width of the entrance slit. By reducing the width of the slit further we reach a limiting value which is determined by diffraction. The line width can be reduced further by increasing the resolving power of the instrument. Under the best set of conditions we find that each line consists of a distribution of the frequencies and is not monochromatic, and this is because of the finite lifetime of the excited state [54].

Let us study the classical theory of the line width emitted by an isolated stationary atom [55]. Thus if we remove the emitting atom to an infinite distance from all other atoms and reducing its effective temperature so that the atom is at rest, then under these conditions a spectral line emitted by this atom displays its "natural " shape [56]. We start with Eq. (1.84), i.e. we assume that the radiating electron is harmonically bound $\omega_0^2 = k/m$ [57]. We write (1.84) as

$$\frac{d^2 X}{dt^2} - \tau \frac{d^3 X}{dt^3} + \omega_0^2 X = 0. \tag{1.145}$$

This equation can be solved by taking X to be of the form $\exp(-\alpha t)$ and substituting this in Eq. (1.145) to find a cubic equation for α;

$$\tau \alpha^3 + \alpha^2 + \omega_0^2 = 0. \tag{1.146}$$

The cubic equation (1.146) has a real negative root, (the runaway solution [55]) and two other complex roots, one being the complex conjugate of the other.

Since τ is very small, an accurate solution of (1.146) can be found by assuming that $\tau\omega_0 \ll 1$, and then we have

$$\alpha = \frac{\Gamma}{2} \pm i\left(\omega_0 + \Delta\omega\right), \tag{1.147}$$

where

$$\Gamma = \omega_0^2 \tau, \tag{1.148}$$

is the decay constant and

$$\Delta\omega = -\frac{5}{8}\omega_0^3 \tau^2, \tag{1.149}$$

is the level shift. The energy of the oscillator which is proportional to X^2 decays as $e^{-\Gamma t}$ because of the radiation damping. Thus the emitted radiation is like a wave train of an effective length c/Γ. But the wave train of a finite length cannot be monochromatic. We can find the shape of the frequency spectrum through the Fourier transform of $\ddot{X}$ which is proportional to the electric field $E(\omega)$;

$$E(\omega) \propto \int_0^\infty e^{-\alpha t} e^{i\omega t} dt = \frac{1}{\alpha - i\omega}. \tag{1.150}$$

The distribution of the intensity, $I(\omega)$, that would be obtained from a spectrograph of infinite resolving power is proportional to $|E(\omega)|^2$ and is given by the Lorentzian distribution [58]

$$I(\omega) = I_0 \left(\frac{\Gamma}{2\pi}\right) \frac{1}{(\omega - \omega_0 - \Delta\omega)^2 + \frac{1}{4}\Gamma^2}, \tag{1.151}$$

where I_0 is the total energy radiated. If we express the classical line width $\Gamma/2$ for an oscillator in terms of the wavelength we get a universal constant

$$\Delta\lambda = \frac{2\pi c}{\omega_0^2}\Gamma = 2\pi c\tau = 1.2 \times 10^{-4} \quad \text{Angstrom}. \tag{1.152}$$

In Chapter 18 we will study the quantum theory of the natural line width.

1.11 Dynamical Systems Expressible as Linear Difference Equations

For certain dynamical systems the momentum of the system (or the particle) can change discontinuously as a function of time. This discontinuous change can be described simply in terms of difference equations. Among the well-known examples of this type of motion one can cite the kicked rotator and the Fermi accelerator [59],[60]. Both of these systems have been studied to determine, among other characteristics, conditions for their regular and chaotic behaviors.

Before we start on the Hamiltonian description of motions expressible in terms of difference equations we must emphasize that in these systems the forces acting on the particle are impulsive forces or forces of constraint. These are idealization of stiff elastic forces and therefore we have to be careful when we take the limit of the stiffness going to infinity. In classical mechanics taking this limit is nontrivial, whereas in quantum theory such a limit in general does not exist [61].

Classically there are at least two distinct forms of problems where the momentum changes discontinuously:

(1) Those cases where the discrete times at which the momentum changes are predetermined and are independent of the state of motion, e.g. a rotator which is kicked periodically.

(2) For a number of problems the time interval between the events that causes the change in momentum may depend on the state of the motion of the particle. The bouncing motion of a ball between two rigid plane, when these planes are moving relative to each other is an example of the latter case. Here we want to examine the question related to the formulation of such a motion and later study the problems that we encounter when we try to quantize this type of motion.

The general form of the equation of motion for the momentum p is given by the linear difference equation [62]

$$p(t_{n+1}) = -\rho p(t_n) + 2mV(t_n). \tag{1.153}$$

In this equation ρ is the coefficient of restitution, m is the mass of the particle and $V(t)$ is a given function of time.

The Fermi-Ulam accelerator [63],[64] can be described in the following way: A rigid wall is fixed at $x = 0$ and a second rigid wall is moving and its position is given by $x = L(t)$. Between these walls the particle moves freely, but at the boundaries we have the conditions

$$p \to -p \quad \text{at} \quad x = 0,. \tag{1.154}$$

and

$$p \to -p + 2m\dot{L}(t) \quad \text{at} \quad x = L(t). \tag{1.155}$$

The behavior of the system is completely specified by giving the values of p, x and t at the nth bounce from the moving wall. The time between two successive bounces from the moving wall is

$$t_{n+1} - t_n = \frac{m}{|p_n|} \left[L(t_n) + L(t_{n+1}) \right], \tag{1.156}$$

where $p_n = p(t_n)$ is the momentum of the ball after the nth bounce from the moving wall and this momentum is always negative;

$$p_n = p_{n-1} + 2m\dot{L}(t_n). \tag{1.157}$$

This equation is a special case of (1.153) for $\rho = 1$ and $V(t_n) = \dot{L}(t_n)$.

Equations (1.156) and (1.157) form a set of difference equations. Except for a few special forms of $L(t)$, these equations can only be solved numerically. The most general form of $L(t)$ for which we can solve (1.156) and (1.157) recursively is when

$$L(t) = \left(At^2 + Bt + C\right)^{\frac{1}{2}}, \tag{1.158}$$

where A, B and C are arbitrary constants subject to the condition that $L(t)$ has to be real. For this case we have (see also Sec. 19.2)

$$t_{n+1} = \frac{B + \left[A + \left(\frac{p_n}{m}\right)^2\right]t_n + 2\left(\frac{|p_n|}{m}\right)\left(At_n^2 + Bt_n + C\right)^{\frac{1}{2}}}{\left(\frac{p_n}{m}\right)^2 - A}, \tag{1.159}$$

and

$$p_n = p_{n-1} + 2m\frac{At_n + \frac{1}{2}B}{\left(At_n^2 + Bt_n + C\right)^{\frac{1}{2}}}. \tag{1.160}$$

1.12   The Fermi Accelerator

In 1949 Fermi advanced a hypothesis on the origin of cosmic rays in which he assumed that these rays are originated and accelerated in the interstellar space of the galaxy by collisions against moving magnetic field [63],[64]. More recent experiments on atoms bouncing off a modulated atomic mirror demonstrates an atomic optics analogue of the Fermi accelerator [65]. In such an experiment one can show that there exists, within a range of parameters, situations where many features of the Fermi accelerator are exhibited [66]. In what follows we will study two discrete models of a particle colliding with time-dependent boundaries. The first one which has a simple geometry and can be solved numerically without any approximation is used to determine the accuracy of a simplified model. We then use the same simplifying technique for the motion of the particle in a more complicated geometry. For a confined two-dimensional case one of the simpler models can be described by a time-dependent annular billiard where a point particle moves freely along a straight line until it hits a boundary. If the boundary is static the particle's energy remains constant, however when the boundary moves periodically in time, the energy may decrease or increase. One can inquire whether the particle can gain unlimited energy, the unlimited gain was the reason for Fermi's proposal.

A Particle Bouncing Between A Fixed Wall and an Oscillating Wall (The Zaslavskii-Chirikov Model) — Let us first consider the dynamics of a particle bouncing elastically between a fixed and a periodically oscillating wall. The moving wall oscillates in such a way that during each half period the velocity varies linearly with time, i.e. the velocity is a sawtooth function of time. We denote the maximum and the minimum distances between the two plates by $\ell + a$ and $\ell - a$ respectively, where $2a$ is the peak amplitude of oscillations of the moving wall. Also let $\frac{1}{4}V$ be the amplitude of the velocity of the wall.

The integer n measures the number of collisions with the oscillating wall and ϕ_n is the phase of oscillating wall at the moment of collision. This phase changes from 0 to $\frac{1}{2}$ as the oscillating wall moves from position ℓ to $\ell + \frac{1}{2}$ and from $\frac{1}{2}$ to 1 during the reverse motion.

Now we can write the exact equations describing the motion of the particle in terms of the following equations [1],[67]:

$$v_{n+1} = \pm v_n + V\left(\phi_n - \frac{1}{2}\right), \tag{1.161}$$

$$\phi_{n+1} = \frac{1}{2} - \frac{2v_{n+1}}{V} + \sqrt{\left(\frac{1}{2} - \frac{2v_{n+1}}{V}\right)^2 + \frac{4v_{n+1}}{V}\varphi_n}, \quad \text{if} \quad v_{n+1} > \frac{V\phi_n}{4}, \tag{1.162}$$

$$\phi_{n+1} = 1 - \phi_n + \frac{4v_{n+1}}{V}, \quad \text{if} \quad v_{n+1} < \frac{V\phi_n}{4}, \tag{1.163}$$

where φ_n in Eq. (1.162) is defined by

$$\varphi_n = \left\{\phi_n + \frac{\phi_n(1 - \phi_n) + \frac{\ell}{4a}}{\frac{4v_{n+1}}{V}}\right\}. \tag{1.164}$$

In the last relation the curly brackets denote the fractional part of the argument.

We can simplify the set of four equations by assuming that the oscillating wall imparts a momentum to the particle according to its velocity without changing its position in space. This approximation works very well as has been shown by Lieberman and Lichtenberg. These authors compared the exact numerical solution of Eqs. (1.161)-(1.164) with the simplified problem [68] and found good agreement. The simplification amounts to replacing the above equations with a set of two coupled differential equations:

$$v_{n+1} = \left|v_n + V\left(\phi_n - \frac{1}{2}\right)\right|, \tag{1.165}$$

$$\phi_{n+1} = \left\{\phi_n + \frac{\ell V}{16av_{n+1}}\right\}. \tag{1.166}$$

The absolute sign in Eq. (1.165) corresponds to the velocity reversal, when $v_{n+1} < \frac{1}{4}V\varphi_n$ and this appears in (1.161) and (1.163). But for the interesting case of $v_{n+1} > V$, this assumption is trivially satisfied. We can also reduce (1.161)-(1.164) to (1.165),(1.166) if the following conditions are satisfied:

$$\frac{\ell}{a} \gg 1, \quad \frac{v_n}{V} \gg 1. \tag{1.167}$$

If in addition the inequality

$$\frac{\ell V}{16av_{n+1}} \ll 1, \tag{1.168}$$

is true, then (1.165)-(1.166) can be approximated by the coupled differential equations

$$\frac{dv_n}{dn} = V\left(\phi(n) - \frac{1}{2}\right), \qquad \frac{d\phi(n)}{dn} = \frac{\ell V}{16av(n)}. \tag{1.169}$$

Now let us introduce two functions of n, which we denote by $\Phi(n)$ and $\Omega(n)$

$$\Phi(n) = \phi_n - \frac{1}{2}, \qquad \Omega^2(n) = \frac{\ell V^2}{16av^2(n)}, \tag{1.170}$$

where from (1.167) and (1.168) it follows that $\Omega^2(n) \ll 1$. Rather than (1.169), we can study the phase equation given by [67]

$$\frac{d^2\Phi(n)}{dn^2} + \Omega^2(n)\Phi(n) = 0, \tag{1.171}$$

Solutions to equations of this type will be discussed in detail in Sec. 12.14. Here we consider the first order approximation where Ω^2 is assumed to be a constant, i.e. set $v_n = v_0$. Then the solution of (1.171) shows that the velocity of the particle is oscillatory with an amplitude proportional to $v_0\sqrt{\frac{a}{\ell}}$ around its initial value with the frequency Ω. Under these conditions, the particle cannot reach a state with very large energy. By considering the correlation coefficient for phases ϕ_n, Zaslavskii and Chirikov have concluded that there are three regimes for the possible motion of the particle:

(1) When the Fermi mechanism works, the mean velocity is $v \le \frac{1}{4}V\sqrt{\frac{\ell}{a}}$ (i.e. for $\Omega^2 \gg 1$) and the motion is stochastic.

(2) The intermediate case where the average velocity satisfies the inequality

$$\frac{1}{4}V\sqrt{\frac{\ell}{a}} < v < \frac{V\ell}{16a}.$$

(3) The average velocity is $v \ge \frac{V\ell}{16a}$, and this is the case that we studied earlier.

For simplicity let us introduce the dimensionless variable u defined by $v = uV$ for the region (1), the velocity distribution function can be described by stochastic process obeying the Fokker-Planck equation [68],[69]

$$\frac{\partial f}{\partial n} = -\frac{\partial}{\partial u}(B(u)f) + \frac{1}{2}\frac{\partial^2}{\partial u^2}(D(u)f), \tag{1.172}$$

where the frictional coefficient is given by

$$B(u) = -\frac{1}{n}\int \Delta u P(v - \Delta u, \, n|u)d(\Delta u). \tag{1.173}$$

In this relation P is the conditional probability of a particle being at u if it were at $u - \Delta u$, n collisions earlier [68]. The diffusion coefficient $D(u)$ is also related to P;

$$D(u) = \frac{1}{n}\int (\Delta u)^2 P(u - \Delta u, \, n|u)d(\Delta u). \tag{1.174}$$

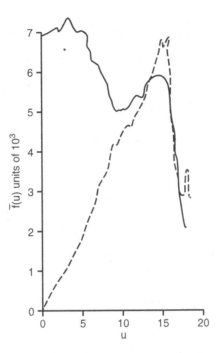

Figure 1.4: The distribution function integrated over collisions, $\bar{f}(u)$ in units of 10^3, obtained from the exact solution of Eqs. (1.161)–(1.164) is shown by the dashed line in this graph. The approximate solution given by (1.165),(1.166) is shown by the solid line. For these calculations the dimensionless parameter $\frac{\ell}{16a}$ is chosen to be 1000 [68].

The condition for the validity of the Fokker-Planck equation for this type of stochastic motion is discussed in detail by Lieberman and Lichtenberg [1],[68]. The solution of the Fokker-Planck equation (1.172) is obtained using the following boundary conditions: We assume perfectly reflecting barriers at $v = 0$ and $v = Vu_s$, such that

$$-Bf = \frac{1}{2}\frac{\partial}{\partial u}(Df) \tag{1.175}$$

at $u = 0$ and $u = u_s$. The steady state solution $\frac{\partial f}{\partial n} \equiv 0$ of (1.172) with these boundary conditions and with $\bar{f}(u_0)$ is given by

$$f(u, n \to \infty) = \bar{f}(u) = \bar{f}(u_0)D(u_0)D^{-1}(u)$$
$$\times \int_{u_0}^{u} 2B\left(u'\right)D^{-1}\left(u'\right)du'. \tag{1.176}$$

Assuming that $D(u)$ can be approximated by a constant value, $D(u) = \frac{1}{12}$, we find $B(u)$;

$$B(u) = \frac{1}{24}\frac{d}{du}\ln\left(\bar{f}\right). \tag{1.177}$$

As we can see from Fig. 1.4, $(\bar{f}) \propto u$ and thus $B = \frac{1}{24u}$, a result which is in agreement with the analytical calculation for $n = 2$ [68].

We can compare the result of the exact solution with the approximate one (see Fig. 1.3). In this figure the solution of the Fokker-Planck equation for the exact problem is shown by the dashed line and the corresponding distribution using the approximate solution is shown by solid line. In Fig. 1.3 the abscissa is the velocity of the particle in units of the maximum velocity of the wall V, $u = \frac{v}{V}$ and the ordinate is the distribution function integrated over collisions, $\bar{f}(u)$ in units of 10^3.

$$B(u) = \frac{1}{24} \frac{d}{du} \ln \left(\bar{f} \right). \qquad (1.178)$$

As it can be seen from Fig. 1.4, $\bar{f}(u) \propto u$ and thus $B = \frac{1}{24u}$, a result which is in agreement with the analytical calculation for $n = 2$ [68].

By studying the solution of Zaslavskii-Chirikov model, we conclude that for the stochastic regime the simplified equations (1.165),(1.166) provide us with a reasonable approximation for the calculation of $\bar{f}(u)$. Based on this conclusion, we can apply the simplified equation when the motion of one of the plates is more complicated. For instance for sinusoidal motion, we can replace (1.165),(1.166) by

$$v_{n+1} = |v_n + V \sin \phi_n|, \qquad (1.179)$$

$$\phi_{n+1} = \phi_n + \frac{2\pi \ell V}{16 a v_{n+1}}. \qquad (1.180)$$

The maximum speed reached by the particle compared to the velocity of the wall is bounded as long as the motion is one-dimensional. For instance for the sawtooth model it is about 100, and this speed is reached after 10^5 collisions [67].

When we consider motions in two or three dimension, we observe that they are, in an essential way, different from the one-dimensional motion. This difference arises from the fact that in the case of several degrees of freedom, the stochasticity of the motion of the particle can be the result of a new distribution of the energy over the degrees of freedom. Thus for large angle scattering $V \approx \Delta V \approx v$ and consequently the criterion for stochastisity (region 2) is always fulfilled and a stationary distribution does not exist [67]. Therefore we are led to examine the simplest two-dimensional motion of a point particle, the so-called annular billiard, to see whether or not the Fermi acceleration is a possible outcome of such a motion.

Motion of a Particle Bouncing Between Two Oscillating Circular Walls — Let us consider the specific case where the two boundaries are two eccentric circles of radii R and r, $r < R$, and the centers are separated by a distance d. The special case of motion between two concentric circles will be obtained if we set $d = 0$. Now depending on the combination of both velocity of the particle and the phase of moving boundary, it is possible for the particle, after its first collision with the boundary, to suffer additional collisions before leaving the collision area. For instance if the larger circle is oscillating, one can encounter a situation where the particle hits the smaller circle after a number of collisions with the inner boundary of the larger circle. We first study the dynamics of collisions with the inner boundary of the larger circle.

We first consider the specific case where the two boundaries are two eccentric circles of radii R and r, $r < R$, and the centers are separated by a distance d. Now depending on the combination of both velocity of the particle and the phase of moving boundary, it is possible for the particle, after its first collision with the boundary, suffers additional collisions before leaving the collision area. For instance if the larger circle is oscillating, one can encounter a situation where the particle hits the smaller circle after a number of collisions, and we want to discuss the dynamics of collisions with the inner boundary of the larger circle.

To simplify the problem further, one can consider a model where, in two dimensions, both internal and external boundaries are fixed, but when the particle collides with the walls, it suffers a change in its velocity as if the boundaries were moving [70],[71]. Here we assume that the particle is reflected in such a way that the angle of incident is equal to the angle of reflection.

Let us choose the radius R to be of unit length, $R = 1$, and let n denote the nth collision with the external boundary. The dynamical and geometrical variables are the velocity v_n and the angular variables θ_n, α_n, φ_n and ϕ_n, all shown in Fig. 1.5. We want to find the mapping M such that $M(v_n, \theta_n \cdots, \phi_n) = (v_{n+1}, \theta_{n+1} \cdots, \phi_{n+1})$. The angle of incident α_n is measured with respect to the normal at the point of impact. We use the convention where α_n is positive if measured counterclockwise and negative if clockwise and has the range $-\frac{\pi}{2} \le \alpha_n \le \frac{\pi}{2}$. Similarly for θ_n which denotes the position of the particle on the boundary we use the same convention, except that θ_n has the range $-\pi \le \theta_n \le \pi$. The two angles φ_n and ϕ_n represent the phases generated in the exchange of momentum with external and internal boundaries respectively.

Now after hitting the external boundary, the components of the position of the particle in rectangular coordinates are given by

$$x_n = \cos\theta_n, \qquad y_n = \sin\theta_n, \tag{1.181}$$

and the components of the velocity written in polar coordinates will be

$$v_{n\eta} = -v_n \cos\alpha_n, \qquad v_{n\tau} = v_n \sin\alpha_n. \tag{1.182}$$

In the last equation η denotes the normal component of the velocity (the unit vector is defined as pointing outward) and τ is the tangential component of the velocity. This velocity when expressed in rectangular coordinates can be written as

$$v_{nx} = v_{n\eta} \cos\theta_n - v_{n\tau} \sin\theta_n, \tag{1.183}$$

$$v_{ny} = v_{n\eta} \sin\theta_n + v_{n\tau} \cos\theta_n. \tag{1.184}$$

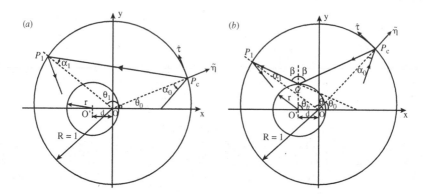

Figure 1.5: A simplified model of a time-dependent annular billiard: (a) type (A) collision and (b) type (B) collision as described in the text [70].

We observe that there are two distinct types of motion depending on the condition whether

$$I_n = |\sin \alpha_n - d \sin(\theta_n - \alpha_n)|, \qquad (1.185)$$

is greater or less than r.

The Case where the Particle Does Not Hit the Inner Boundary — If $I_n > r$, then the motion is of type (A) shown in Fig. 3(a), i.e. no collision with the sphere of radius r takes place. The coordinates of the particle are given by

$$\begin{cases} x_{n+1} = \cos \theta_{n+1} \\ y_{n+1} = \sin \theta_{n+1}, \end{cases} \qquad (1.186)$$

where

$$\theta_{n+1} = \theta_n + \pi - 2\alpha_n. \qquad (1.187)$$

Since between two successive collisions the velocity remains constant, the corresponding travel time is given by

$$t_f = \frac{1}{v_n} \sqrt{(x_{n+1} - x_n)^2 + (y_{n+1} - y_n)^2}. \qquad (1.188)$$

The components of the velocity of the particle just before collision with the outer boundary are;

$$\begin{cases} v_{n\eta} = v_{nx} \cos \theta_{n+1} + v_{ny} \sin \theta_{n+1} \\ v_{n\tau} = -v_{nx} \sin \theta_{n+1} + v_{ny} \cos \theta_{n+1}. \end{cases} \qquad (1.189)$$

After the collision, the components of the radial velocity, for the simplified model that we are discussing, are given by

$$v_{(n+1)\eta} = -|-v_{n\eta} - 2\epsilon_R \sin(t_f + \varphi_n)|, \quad v_{(n+1)\tau} = v_{n\tau}, \qquad (1.190)$$

where, in the complete model ϵ_R is the amplitude of oscillations of the moving outer boundary. For the simplified form that we are considering, with the fixed boundaries, $2\epsilon_R$ beomes the change of momentum as if the boundary were moving. The dynamical and geometrical variables after the collision are related to those before collision by the following relations:

$$v_{n+1} = \sqrt{v_{(n+1)\eta}^2 + v_{(n+1)\tau}^2}, \tag{1.191}$$

$$\varphi_{n+1} = \varphi_n + t_f, \quad \text{mod } 2\pi, \tag{1.192}$$

$$\phi_{n+1} = \phi_n + \omega t_f, \quad \text{mod } 2\pi, \tag{1.193}$$

and

$$\alpha_{n+1} = \tan^{-1}\left[-\frac{v_{(n+1)\tau}}{v_{(n+1)\eta}}\right]. \tag{1.194}$$

Equations (1.190)–(1.194) show us how these variables just before and after the $n+1$th collision are related to each other by the nonlinear mapping.
Considering the Motion where the Particle Collides with the Inner Boundary After Being Bounced from the Outer Boundary — Here we introduce n as the number of hits with the outer circle, and therefore we label the variables with the integer n. But now the iterations of the map are computed only when the particle reaches the outer boundary. The coordinates of the particle when it hits the inner boundary are given by (see Fig. 1.5(b))

$$x_r = r \cos\theta_b - d, \quad y_r = r \sin\theta_b, \tag{1.195}$$

with

$$\theta_b = \beta + \theta_n - \alpha_n, \tag{1.196}$$

where β is found from

$$\sin\beta = \frac{1}{r}[\sin\alpha_n - d\sin(\theta_n - \alpha_n)]. \tag{1.197}$$

The travel time, up to the internal boundary, can be obtained from

$$t_f = \frac{1}{v_n}\sqrt{(x_r - x_n)^2 + (y_r - y_n)^2}, \tag{1.198}$$

where v_n is the velocity of the particle before the collision. This velocity can be resolved into radial and tangential components:

$$v_{n\eta}^r = v_{nx}\cos\theta_b + v_{ny}\sin\theta_b, \quad v_{n\tau}^r = -v_{nx}\sin\theta_b + v_{ny}\cos\theta_b. \tag{1.199}$$

Choosing the frame associated with the internal circle, only the radial component of the velocity changes. With this choice, immediately after the collision with the internal boundary, the components of the velocity will be given by

$$v_{(n+1)\eta}^r = |-v_{n\eta}^r - 2\epsilon_r\omega\sin(\omega t_f + \phi_n)|, \quad v_{(n+1)\tau}^r = v_{n\tau}^r. \tag{1.200}$$

Here the superscript r shows that the position of the particle, as it is measured on the internal circle, and ϵ_r is the change of momentum caused by collision with this circle. If we were solving the problem exactly, ϵ_r would have been the amplitude of oscillation of the internal time varying boundary. Then in our simplified version, after the impact, the expression of the particle's velocity is the same as the one in the complete model. The module function here is used to avoid the particle from moving into the forbidden regions. In Eq. (1.200) ω represents the ratio of the internal to external oscillating frequencies $\omega = \frac{\omega_r}{\omega_R}$. The path of the particle is a line extending to the boundary at $R = 1$, and when the particle reaches this point at time t, then its position is

$$R_p(t) = 1. \tag{1.201}$$

Here $R_p(t)$ is the norm of the radius vector of the particle and is given by

$$R_p(t) = \left[\left(x_r^2 + y_r^2\right) + 2\left(x_r v_x + y_r v_y\right)t + \left(v_x^2 + v_y^2\right)t^2\right]^{\frac{1}{2}}. \tag{1.202}$$

The velocities v_x and v_y are given by

$$v_x = v_{(n+1)\eta}^r \cos\theta_b - v_{(n+1)\tau}^r \sin\theta_b, \tag{1.203}$$

$$v_y = v_{(n+1)\eta}^r \sin\theta_b + v_{(n+1)\tau}^r \cos\theta_b. \tag{1.204}$$

From Eqs. (1.201) and (1.202) we find the following quadratic equation for t:

$$\left(v_x^2 + v_y^2\right)t^2 + 2\left(x_r v_x + y_r v_y\right)t + \left(x_r^2 + y_r^2\right) - 1 = 0. \tag{1.205}$$

Denoting the larger root of (1.205) by t_c, where t_c is the time that the particle spends after a collision with the internal circle untill it reaches the external boundary.

Now for the $(n+1)$th collision the coordinates of the particle are:

$$x_{n+1} = x_r + v_x t_c, \qquad y_{n+1} = y_r + v_y t_c, \tag{1.206}$$

and the new angle θ_{n+1} is obtained from

$$\theta_{n+1} = \tan^{-1}\left(\frac{y_{n+1}}{x_{n+1}}\right). \tag{1.207}$$

The corresponding components of the particle's velocity immediately before the impact with the outer circle are:

$$v_{n\eta} = v_x \cos\theta_{n+1} + v_y \sin\theta_{n+1}, \tag{1.208}$$

$$v_{n\tau} = -v_x \sin\theta_{n+1} + v_y \cos\theta_{n+1}. \tag{1.209}$$

Just after the collision the components of velocity become

$$v_{(n+1)\eta} = -|-v_{n\eta} - 2\epsilon_R \sin(t_c + t_f + \varphi_n)|, \quad v_{(n+1)\tau} = v_{n\tau}, \tag{1.210}$$

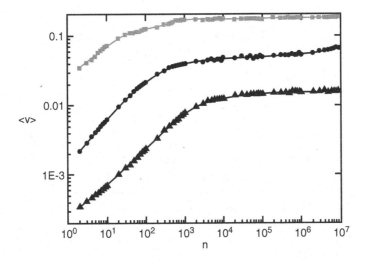

Figure 1.6: The average velocity of the particle $\langle v \rangle$ bouncing between two concentric circles plotted as a function of the number of iterations n. For this graph the following parameters have been used: $r = 0.5$, $\omega = 2$, $\varphi_0 = \phi_0$ and, $\epsilon_R = 0.01$, $0,001$, 0.0001. These are shown by squares, circles and triangles respectively. For details see [70].

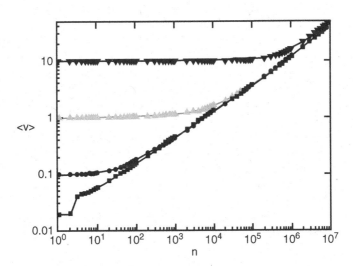

Figure 1.7: The log log plot of the average velocity of the particle $\langle v \rangle$ shown as a function of the number of iterations n for two eccentric circles and for different velocities. For this calculation the following set of parameters have been used: $r = 0.05$, $d = 0.65$, $\epsilon_R = 0.01$. The initial velocities are chosen to be $v_0 = (2\epsilon_R, 10\epsilon_R, 100\epsilon_R, 1000\epsilon_R)$, with the corresponding results starting with the lowest curve and going up. Details can be found in ref. [70].

where as we mentioned earlier only the radial component of the velocity is changed. the magnitude of the new velocity is $v_{n+1} = \sqrt{v_{(n+1)\eta}^2 + v_{(n+1)\tau}^2}$, and the corresponding phases, φ_n, and reflection angles, ϕ_n are:

$$\varphi_{n+1} = \varphi_n + t_c + t_f, \quad \mathrm{mod}\ 2\pi, \qquad (1.211)$$

$$\phi_{n+1} = \phi_n + \omega(t_c + t_f), \quad \mathrm{mod}\ 2\pi. \qquad (1.212)$$

Note that These relations complete the equations needed for calculating the velocities and the phases for the motion of the particle.

Numerical Results for the Simplified Annular Billiard — We first consider the case of two concentric circles. For this case we set $d = 0$ ia all of the equations derived for this problem. Then we calculate the set of variables $(\theta_n,\ v_n,\ \phi_n,\ \phi_n)$ by successive iterations. However, in this case, as a consequence of angular momentum conservation, the initial condition α_n can be found from

$$\sin \alpha_n = \frac{l}{(1 + \epsilon_R \cos \varphi_n)v_n}, \qquad (1.213)$$

where we have set $R = 1$ and where we allow l to vary between 0 and ϵ_R. In Fig. 1.6 the average velocity of the particle as a function of the number of iterations n is plotted for three different values of ϵ_R, $\epsilon_R = (0.01, 0.001, 0.0001)$. For a given value of ϵ_R, $\langle v \rangle$ increases for small n, and then reaches a plateau. Thus in this model there is no Fermi acceleration. The next figure, Fig. 1.7 is a log log plot of $\langle v \rangle$ versus n, using the parameters $r = 0.05$, $d = 0.65$, $\epsilon_r = 0.01$, $\epsilon_r = \epsilon_R \frac{\sqrt{5}+1}{2}$ and $\omega = \frac{1}{2}(\sqrt{5} + 1)$. The four curves shown in this figure are for four starting velocities and from the bottom to the top these values of v_0 are $(2\epsilon_R, 10\epsilon_R, 100\epsilon_R, 1000\epsilon_R)$ respectively. The result of this calculation shows that $\langle v \rangle$ increases without bound, and this result supports the concept of the Fermi accelerator.

Another example of the Fermi accelerator which has been studied in detail is that of the motion of particles bouncing off from a two-dimensional smoothly oscillating polygon. The interesting aspect of this model is that the energy growth is exponential in time [72].

The collection of different dissipative systems considered in this chapter can be formulated directly in terms of equations of motion without any reference to the variational principle and the canonical formalism.

References

[1] A.J. Lichtenberg and M.A. Lieberman, *Regular and Stochastic Motions*, Second Edition, (Springer, New York, N.Y. 1991).

[2] R.L. Devaney, *An Introduction to Chaotic Dynamical Systems*, (Westview Press, Boulder, 2003).

[3] E. Ott, *Chaos in Dynamical Systems*, Second Edition, (Cambridge University Press, Cambridge, 2002).

[4] A.M.O. de Almeida, Hamiltonian Systems: Chaos and Quantization, (Cambridge University Press, Cambridge, 1988).

[5] For a recent and indepth account of the history of the problem of motion in a resistive medium see the article of W.W. Hackborn in Am. J. Phys. 84, 127 (2016).

[6] I. Newton, *The Principia*, translated by I. Bernard Cohen and A. Whitman, (University of California Press, Berkley, 1999) p. 177.

[7] J.D. Anderson, *A History of Aerodynamics*, (Cambridge University Press, Cambridge, 1997).

[8] See for instance J.B. Marion, *Classical Dynamics of Particles and Systems*, (Academic Press, New York, N.Y. 1965) p. 65.

[9] D.H.E. Gross, Nucl. Phys. A 240, 472 (1975).

[10] W.E. Schröder and J.R. Huizenga in *Treatise on Heavy-Ion Scattering, Fusion and Qusi-Fusion Phenomena*, Edited by D.A. Bromley, Vol. 2, (Plenum Press, New York, N.Y. 1984) p. 192.

[11] G. Stokes, Tran. Cambridge Phil. Soc. 9, 8 (1850) part II.

[12] A detailed and well-documented account of the history of the Stokes' drag force can be found in a paper by E.R. Lindgren in Phys. Scripta, 60, 97 (1999).

[13] L.D. Landau and E.M. Lifshitz, *Fluid Mechanics*, (Pergamon Press, London, 1959) pp. 67-68.

[14] F.E. Hoare, Phil. Mag. 8, 899 (1929).

[15] H. Lamb, *Hydrodynamics*, (Dover Publications, New York, N.Y. 1945) p. 642.

[16] W. Kaufmann, *Fluid Mechanics*, (McGraw-Hill, New York, N.Y. 1963) p. 236.

[17] C.W. Oseen, Arkiv Mat. Astron. Fys. Vol. 6. no. 29 (1910).

[18] J.W.S. Rayleigh, *The Theory of Sound*, Vol. I (Dover Publications, New York, N.Y. 1945) p. 45.

[19] C. Kittel, W.D. Knight and M.A. Rutherman, *Berkeley Physics Course* Vol, 1 (McGraw-Hill, New York, N.Y. 1962) p. 188.

[20] See for instance R. Deutsch, *Orbital Dynamics of Space Vehicles*, (Prentice-Hall, Englewood Cliffs, N.J. 1963) p. 207.

[21] D. Hesteness, *New Foundations for Classical Mechanics*, (D. Reidel Publishing Company, 1986, Dordrecht) p. 140.

[22] M.I. Molina, Am. J. Phys. 66, 973 (1998).

[23] R.M. Santilli, *Foundations of Theoretical Mechanics II*, (Springer-Verlag, New York, N.Y. 1983) p. 2.

[24] M. Abraham, Ann. Physik, 10, 105 (1903).

[25] M. Planck, *Vorlesungen über die Theorie der Wärmstrahlung*, (J. Barth, Leipzig, 1906). Sec. III.

[26] M. Abraham, *Electromagnetische Theorie der Strahlung*, (B.G. Teubner, Berlin, 1908).

[27] H.A. Lorentz, *The Theory of Electrons and Its Applications to the Phenomena of Light and Radiant Heat*, Second Edition (Dover Publications, New York, N.Y. 1952).

[28] T. Erber. A comprehensive review of Abraham-Lorentz-Dirac equation from 1909 to 1960 can be found in the article by T. Erber, Fortsch. Physik, 9, 343 (1961).

[29] B. Leaf, Phys. Rev. 127, 1369 (1962).

[30] B. Leaf, Phys. Rev. 132, 1321 (1963).

[31] F. Bopp, Zeit. Angw. Phys. 14, 699 (1962).

[32] M. Razavy, Can J. Phys. 56, 1372 (1978).

[33] See for instance A.O. Barut, *Electrodynamics and Classical Theory of Fields and Particles*, (The McMillan Company, New York, N.Y. 1964).

[34] C.R. Galley, A.K. Leibovich, and I.Z. Rothstein, Phys. Rev. Lett. 105, 094802 (2010).

[35] For an interesting argument due to Kirchhoff regarding equations of motion with higher derivatives see Max Jammer's *Concepts of Force*, (Harvard University Press, Cambridge, Massachusetts, 1957) p. 223.

[36] M. von Laue, Annalen der Physik, 28, 436 (1909).

[37] For the history of the theories leading to the derivation of the Abraham-Lorentz-Dirac equation see F. Rohrlich, *Classical Charged Particles*, (Addison-Wesley, Reading, 1965).

[38] P.A.M. Dirac, Proc. Roy. Soc. A 167, 148 (1938).

[39] P.A.M. Dirac, Ann. Inst. Poincarè, 9, 13 (1938).

[40] F. Rohrlich, Am. J. Phys. 68, 1109 (2000).

[41] G.A. Schott, *Electromagnetic Radiation*, (Cambridge University Press, Cambridge, 1912).

[42] J.D. Jackson, *Classical Electrodynamics*, Third Edition (John-Wiley & Sons, New York, N.Y. 1999) Chapter 14.

[43] A. Gupta and T. Padmanabham, Phys. Rev. D 57, 7241 (1998).

[44] F. Rohrlich, Phys. Lett. A 283, 276 (2001).

[45] R.F. O'Connell, Phys. Lett. 313, 491 (2003).

[46] A.M. Steane, Am. J. Phys. 83, 256 (2015).

[47] A.M. Steane, Am. J. Phys. 83, 703 (2015).

[48] J.M. Aguirregabiria, J. Phys. A 30, 2391 (1997).

[49] D. Vogt and P.S. Letelier, Gen. Rel. Grav. 35, 2261 (2003).

[50] Yoshida, Phys. Rev. A 39, 19 (1988).

[51] T.H. Boyer, Am. J. Phys. 72, 992 (2004).

[52] L. Jia, Int. J. Astronomy and Astrophys. 3, 29 (2013).

[53] S.C. Bell, Comp. Phys. 9, 281 (1995).

[54] W. Heitler, *The Quantum Theory of Radiation*, Third Edition (Oxford University Press, London, 1954) p. 33.

[55] J.D. Jackson, *Classical Electrodynamics*, Third Edition (John-Wiley & Sons, New York, N.Y. 1999).

[56] R.G. Breene Jr. Rev. Mod. Phys. 29, 94 (1957).

[57] R.H. Good Jr. and T.J. Nelson, *Classical Theory of Electric and Magnetic Fields*, (Academic Press, New York, N.Y. 1971) p. 509.

[58] A. Corney, *Atomic and Laser Spectroscopy*, (Oxford University Press, London, 1977) Chapter 8.

[59] G.M. Zaslavsky, *Chaos in Dynamics Systems*, (Harwood Academic, New York, N.Y. 1985) Chapter 3.

[60] G. Casati, B.L. Chirikov, J. Ford and F. Israilev, in *Stochastic Behavior in Classical and Quantum Hamiltonian Systems*, Edited by G. Casati and J. Ford (Springer-Verlag, New York, 1979) p. 334.

[61] N.G. van Kampen and J.J. Lodder, Am. J. Phys. 52, 419 (1984).

[62] M. Razavy, Hadronic J. 17, 515 (1994).

[63] E. Fermi, Phys. Rev. 75, 1169 (1949).

[64] S.M. Ulam, *Proceedings of the Fourth Berkeley Symposium Mathematics, Statistics and Probability*, Vol. 3, (University of California Press, Berkeley, 1961).

[65] F. Saif, I. Bialynicki-Birula, M. Fortunato and W.P. Schleich, Phys. Rev. A 58, 4779 (1998).

[66] F. Saif, Phys. Rep. 419, 207 (2005).

[67] G.M. Zaslavskii and B.V. Chirikov, Sov. Phys. Doklady 9, 989 (1965).

[68] M.A. Lieberman and A.J. Lichtenberg, Phys. Rev. A 5, 1852 (1972).

[69] H. Risken, *The Fokker-Planck Equation: Methods of Solutions and Applications*, Second Edition, (Springer-Verlag, Berlin, 1996).

[70] R. E. de Carvalho, F. C. de Souza and E.D. Leonel, J. Phys. A. 39, 3561 (2006).

[71] R. E. de Carvalho, F. C. de Souza and E.D. Leonel, Phys. Rev. E 73, 066229 (2006).

[72] K. Shah, D. Turaev and V. Rom, Phys. Rev. E 81, 056205 (2010).

Chapter 2

Lagrangian Formulation

In the first chapter we studied a variety of problems related to the dissipative forces formulated using the second law of motion. Concerning the variational formulation of damped motions we observe that there are certain advantages in Lagrangian and Hamiltonian formulations of dissipative systems, among them the following are the important ones:

(a) Rather than writing $3N$ equations of motion for a system of $3N$ degrees of freedom, we can we can state and express the problem in terms of a single Lagrangian and or a Hamiltonian function.

(b) It is easier to find invariances and conservation laws of the motion from the variational formulation.

(c) Minimizing the action to find the exact or in most cases, the approximate solution, is a powerful tool for finding positions and momenta of the particles and their trajectories.

(d) To find the quantum mechanical system corresponding to a given classical dissipative system, we quantize the motion using either the Lagrangian or the Hamiltonian, but rarely the equation of motion (see Sec. 11.3 for the direct quantization starting from the second law).

In this chapter we want to review the canonical formulation of the simplest forms of dissipative systems. The methods of construction of the Lagrangian and of the Hamiltonian for the dissipative systems have been discussed in a number of papers and books [1]-[16]. Once a Lagrangian or a Hamiltonian for the system is found we can obtain conserved quantities associated with the motion as well as the solution – for the latter either by using the conventional technique of integrating the equations of motion or by applying methods based on the calculus of variation. If in the formulation of a problem in addition to the Lagrangian (or Hamiltonian), a dissipative function is also introduced, then there will be two independent functions required to specify the motion.

But the main purpose of the Hamiltonian (or the Lagrangian) formulation studied in this book is to find the quantum description of the dissipative motion.

2.1 Dissipative Functions of Rayleigh, Lur'e and Sedov

The simplest way of incorporating dissipative forces in the Lagrangian formalism is by introducing Raleigh's dissipation function $\mathcal{F}$, where this $\mathcal{F}$ is defined in such a way that $-\frac{\partial \mathcal{F}}{\partial \dot{x}_j}$ is the jth component of the dissipative force [17],[18]. The equations of motion in this case are given by

$$\frac{d}{dt}\left(\frac{\partial L}{\partial \dot{x}_j}\right) - \frac{\partial L}{\partial x_j} + \frac{\partial \mathcal{F}}{\partial \dot{x}_j} = 0, \quad j = 1, 2, \cdots \tag{2.1}$$

For instance if the force of friction $\mathbf{Q}$ is linear in velocity

$$\mathbf{Q} = -m\lambda \dot{\mathbf{r}} = (\mathcal{Q}_1 \mathbf{i} + \mathcal{Q}_2 \mathbf{j} + \mathcal{Q}_3 \mathbf{k}), \tag{2.2}$$

with $\mathbf{i}$, $\mathbf{j}$ and $\mathbf{k}$ denote unit vectors in the directions of x, y and z respectively, then $\mathcal{F}$ is given by

$$\mathcal{F} = \frac{m}{2}\lambda \dot{\mathbf{r}}^2. \tag{2.3}$$

In this approach $2\mathcal{F}$ represents the rate of energy dissipation due to friction since

$$dW_Q = -\mathbf{Q} \cdot d\mathbf{r} = -\mathbf{Q} \cdot \mathbf{v}dt = m\lambda \dot{\mathbf{r}}^2 dt. \tag{2.4}$$

Lur'e's Dissipation Function — One can generalize Rayleigh's dissipation function to include nonlinear damping forces [18]. Lur'e has formulated this problem by considering the damping force with the components [19]

$$f_j = -k_j(x_1, x_2, \cdots x_N)g_j(\dot{x}_i), \tag{2.5}$$

where k_j s are positive functions of their arguments and

$$\dot{x}_j g_j(\dot{x}_i) \geq 0. \tag{2.6}$$

Here x_j s are the Cartesian components of the N-dimensional configuration space. We can transform the coordinates $x_j(t)$ by the transformation

$$x_j = x_j(q_1, q_2 \cdots q_N), \tag{2.7}$$

to the generalized coordinates $q_j(t)$. Then the generalized force becomes

$$Q_s^L = -\sum_{j=1}^{N} k_j(x_1, x_2, \cdots x_N)g_j(\dot{x}_i)\frac{\partial x_j}{\partial q_s}. \tag{2.8}$$

From Eq. (2.7) it follows that

$$\dot{x}_j = \sum_{s=1}^{N} \frac{\partial x_j}{\partial q_s}\dot{q}_s, \tag{2.9}$$

and therefore

$$\frac{\partial \dot{x}_j}{\partial \dot{q}_s} = \frac{\partial x_j}{\partial q_s}. \tag{2.10}$$

Substituting (2.10) in (2.8) we find Q_s^L to be

$$Q_s^L = -\sum_{j=1}^{N} k_j(x_1, x_2, \cdots x_N) g_j(\dot{x}_i) \frac{\partial \dot{x}_j}{\partial \dot{q}_s}. \tag{2.11}$$

Now Lur'e defines the dissipation function $\mathcal{F}^L$ by

$$\mathcal{F}^L = \sum_{j=1}^{N} k_j(x_1, x_2, \cdots x_N) \int_0^{\dot{x}_j} g_j(y) dy. \tag{2.12}$$

From this definition and Eq. (2.11) we have

$$\frac{\partial \mathcal{F}^L}{\partial \dot{q}_s} = -Q_s^L, \quad s = 1, 2, \cdots N. \tag{2.13}$$

This result should be compared with Eq. (2.3) for the Rayleigh function. The latter case is a special case of (2.12) if we choose

$$g_j(\dot{x}_j) = \dot{x}_j, \quad k_j = \text{constant}. \tag{2.14}$$

Let us consider a one-dimensional motion where the force of friction is of the form

$$f = -k|\dot{x}|^n. \tag{2.15}$$

In this relation n is an even integer and k is a positive constant. Then from Eqs (2.11) and (2.12) it follows that

$$\mathcal{F}^L = \frac{k}{n+1}|\dot{x}|^{n+1}, \tag{2.16}$$

and

$$Q^L = -k\dot{x}^n \text{sgn}(\dot{x}), \tag{2.17}$$

where the function $\text{sgn}(\dot{x})$ is defined by

$$\text{sgn}(\dot{x}) = \begin{cases} 1 & \text{for } \dot{x} > 0 \\ -1 & \text{for } \dot{x} < 0 \end{cases}. \tag{2.18}$$

Sedov's Approach to the Variational Formulation of Damped Systems
— Let us define the principle of least action by

$$\delta S(x_i) \equiv \delta \int L(x_i, v_i, t) \, dt = 0, \tag{2.19}$$

where $v_i = \frac{dx_i}{dt}$ is the ith component of the velocity of the particle and L is the Lagrangian for the conservative part of the system, $L = T - V$, i.e. $S(x_i)$

is a holonomic functional. Holonomic systems are characterized by the fact that the number of degrees of freedom is equal to the number of independent coordinates required to specify the configuration of the system [20]. For the definition of nonholonomic functional see [21], The nonholonomic contribution to the action coming from dissipative forces will be denoted by $W(x_i)$. With these definitions of S and W, according to Sedov, the principle of the least action becomes [22]–[24]

$$\delta S(x_i) + \delta W(x_i) = 0. \tag{2.20}$$

Here we will consider the special case where $\delta W(x_i)$ which is the variation of the nonholonomic functionals is linear in variations δx_i and δv_i:

$$\delta W = \delta \int w(x,v)dt = \sum_i \int \left[w_i^{(1)}(x_i, v_i)\delta x_i + w_i^{(2)}(x_i, v_i)\delta v_i \right]. \tag{2.21}$$

Here the equations of motion for the dissipative system is obtained from the variation of $S(x_i) + W(x_i)$, i.e. from

$$\delta \int \left[S(x_i) + W(x_i) \right] dt = 0, \tag{2.22}$$

and are given by

$$\frac{d}{dt} \left\{ \frac{\partial L(x,v)}{\partial v_i} + w_i^{(2)}(x,v) \right\} = \frac{\partial L(x,v)}{\partial x_i} + w_i^{(1)}(x,v). \tag{2.23}$$

For example for a linearly damped, one-dimensional motion subject to the potential $V(x)$ we choose $w_i^{(1)} = -m\lambda v$ (λ having the dimension of $time^{-1}$) and $w_i^{(2)} = 0$, then we have

$$m\frac{d^2x}{dt^2} + \frac{\partial V(x)}{\partial x} + m\lambda v = 0. \tag{2.24}$$

In this formalism, the canonical momentum which is defined as the partial derivative of L with respect to v_i is given by

$$p_i = \frac{\partial L(x,v)}{\partial v_i} + w_i^{(2)}(x,v). \tag{2.25}$$

From p_i and L we can find the Hamiltonian of the system and the canonical equations of motion.

Generalized Euler-Lagrange Equations — The dissipative functions of Rayleigh, of Lur'e and of Sedov, will account for the damping forces proportional to a given power of velocity, but not to the radiation reaction force which is proportional to the time derivative of the acceleration.

There are physical systems such as Drude model for the motion of a valence electron where both of these forces are present [25],[26]. For these systems we replace $\mathcal{F}$ by $\mathcal{L} = \mathcal{F} + \mathcal{G}$ where

$$\mathcal{F} = \frac{1}{2} \sum_{i,j=1}^{3} \dot{x}_i R_{ij} \dot{x}_j, \tag{2.26}$$

and

$$\mathcal{G} = \frac{1}{2} \sum_{i,j=1}^{3} \ddot{x}_i P_{ij} \ddot{x}_j. \tag{2.27}$$

The equations of motion in this case are given by the generalized Euler-Lagrange equations:

$$\left\{ \frac{\partial \mathcal{L}}{\partial \dot{x}_i} - \frac{d}{dt}\left(\frac{\partial \mathcal{L}}{\partial \ddot{x}_i} \right) \right\} - \left\{ \frac{\partial L}{\partial x_i} - \frac{d}{dt}\left(\frac{\partial L}{\partial \dot{x}_i} \right) \right\} = f_i, \tag{2.28}$$

where

$$L = T - V(x_i), \tag{2.29}$$

and f_i is the ith component of the external time-dependent force. Note that L has the dimension of energy, but $\mathcal{L}$ has the dimension of power. For small displacements we can take V to be a quadratic function of the coordinates,

$$\frac{1}{2} \sum_{i,j=1}^{3} x_i V_{ij} x_j, \tag{2.30}$$

and T as the sum of the quadratic terms in velocities,

$$\frac{1}{2} \sum_{i,j=1}^{3} \dot{x}_i M_{ij} \dot{x}_j, \tag{2.31}$$

and in addition we can have a "gyroscopic" term, which is dependent on both $\dot{x}_i$ and x_j;

$$\frac{1}{2} \sum_{i,j=1}^{3} \dot{x}_i G_{ij} x_j. \tag{2.32}$$

Here the matrix elements M_{ij} and V_{ij} are real and symmetrical, while G_{ij} s are asymmetrical, but are all constants. Using these quadratic forms we find the linear equations of motion for x_i s to be

$$\sum_{j=1}^{3} \left\{ -P_{ij}\frac{d^3 x_j}{dt^3} + M_{ij}\frac{d^2 x_j}{dt^2} + B_{ij}\frac{dx_j}{dt} + V_{ij}x_j \right\} = f_i, \quad i = 1, 2, 3, \tag{2.33}$$

where

$$B_{ij} = R_{ij} + G_{ij}. \tag{2.34}$$

The Role of Frictional Forces in Heavy Ion Scattering — The theory of damped heavy ion scattering offers an interesting example of the application of the above-mentioned Lagrangian formulation. While we expect that in the domain of nuclear scattering the problem has to be treated quantum mechanically, but under the conditions where semi-classical approximations such as WKB are valid, we can justify a classical description for the motion of the nuclei.

As a specific case let us consider the classical collision of two rigid spherical nuclei of radii R_P and R_T, where P and T refer to the projectile and target

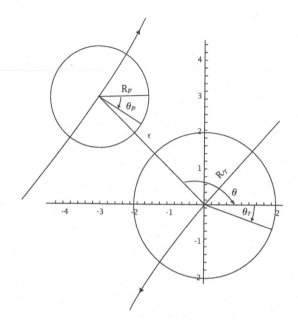

Figure 2.1: Scattering of the projectile P from the target T showing the degrees of freedom of the motion. Both P and T are assumed to be rigid spheres.

nuclei respectively. We formulate this scattering using the generalized coordinates q_i s instead of x_i s in Eq. (2.1), and write the Raleigh dissipation function as [27]

$$\mathcal{F} = \frac{1}{2} \sum_{i,j} k_{ij} \dot{q}_i \dot{q}_j, \tag{2.35}$$

where k_{ij} is the friction tensor. We note that there are four degrees of freedom for the collision. These are: the distance r between the centers of the projectile and the target, their relative orientation angles θ_P and θ_T, and the angle of orientation of the total system θ. We assume that the interaction potentials which consists of the nuclear and Coulomb are only functions of r. Thus the Lagrangian depends on r, $\dot{r}$, $\dot{\theta}$, $\dot{\theta}_P$ and $\dot{\theta}_T$

$$L = \frac{\mu}{2}\left(\dot{r}^2 + r^2\dot{\theta}^2\right) + \frac{1}{2}I_P\dot{\theta}_P^2 + \frac{1}{2}I_T\dot{\theta}_T^2 - V_{Coul}(r) - V_N(r). \tag{2.36}$$

Here I_P and I_T denote the moments of inertia of the projectile and the target about their centers, i.e.

$$I_P = \frac{2}{5}M_P R_P^2, \quad I_T = \frac{2}{5}M_T R_T^2, \tag{2.37}$$

μ is the reduced mass

$$\mu = \frac{M_P M_T}{(M_P + M_T)}, \qquad (2.38)$$

and V_{Coul} and V_N are Coulomb and nuclear potentials respectively.

Associated with these four degrees of freedom we can introduce a symmetric 4×4 matrix for the friction tensor k_{ij}. However to fit the observed data it seems that we need a coefficient of friction k_r for the radial motion, and a coefficient k_t for the relative sliding motion of the two nuclear surfaces [27]–[29]. Thus $\mathcal{F}$ can be written as

$$\mathcal{F} = \frac{1}{2}\left\{ k_r \dot{r}^2 + k_t \left(\frac{r}{R_P + R_T} \right)^2 \left[R_P \left(\dot{\theta}_P - \dot{\theta} \right) + R_T \left(\dot{\theta}_T - \dot{\theta} \right) \right]^2 \right\}. \qquad (2.39)$$

The equations of motion derived from (2.36) and (2.39) are:

$$\mu \ddot{r} - \mu r \dot{\theta}^2 + k_r \dot{r} + \frac{\partial V_{Coul}}{\partial r} + \frac{\partial V_N}{\partial r} = 0, \qquad (2.40)$$

$$\mu r^2 \ddot{\theta} + I_P \ddot{\theta}_P + I_T \ddot{\theta}_T = 0, \qquad (2.41)$$

$$I_P \ddot{\theta}_P = -k_t \left(\frac{r}{R_P + R_T} \right)^2 R_P \left(\dot{\theta}_P - \dot{\theta} \right), \qquad (2.42)$$

and

$$I_T \ddot{\theta}_T = -k_t \left(\frac{r}{R_P + R_T} \right)^2 R_T \left(\dot{\theta}_T - \dot{\theta} \right). \qquad (2.43)$$

These equations can only be solved numerically.

A similar but simpler model for friction which is of the form

$$\frac{d}{dt}(\mu \dot{r}) - \dot{\theta}^2 + k_r \dot{r} + \frac{\partial V_{Coul}}{\partial r} + \frac{\partial V_N}{\partial r} = 0, \qquad (2.44)$$

and

$$\frac{d}{dt}\left(\mu r^2 \dot{\theta} \right) + k_\theta r^2 \dot{\theta} = 0, \qquad (2.45)$$

has been used to study the classical trajectory of the scattered ion [30]–[33].

2.2   The Inverse Problem of Analytical Dynamics

While the simple formulation discussed in Sec. 2.1 is useful for solving certain classical problems, for the purpose of quantizing dissipative systems and for questions relating to the compatibility of the commutation relations with the equations of motion we need to find a Lagrangian describing the damped motion and at the same time defining the canonical coordinates. In what follows we will see that for a given set of equations of motion we can construct infinitely many Lagrangians or in some cases we cannot find a single Lagrangian which will give us the exact equations of motion.

Existence of the Variational Principle for Dissipative Systems — The question of the existence of variational principles for dissipative systems has been carefully analysed by Synge [34]. His work was motivated by by the apparent contradictory results found by Bauer and by Bateman [25]–[38]. According to Bauer , for a linear dissipative system where the damping force is proportional to velocity, the Lagrangian cannot be derived from 'a variational formulation. Bauer's argument is as follows: Let us consider the class of dynamical system which can be expressed as linear differential equations of the second order, i.e. they satisfy the equations of motion

$$\sum_{j=1}^{n} \{M_{ij}\ddot{x}_j + B_{ij}\dot{x}_j + V_{ij}x_j\} = Q_i(t), \quad i = 1, 2, \cdots n. \tag{2.46}$$

The coefficients M_{ij}, B_{ij} and V_{ij} in (2.46) are, in general, functions of time. In the first part we restrict our discussion to the cases where the coefficients satisfy the symmetry properties

$$M_{ij} = M_{ji}, \quad B_{ij} = B_{ji}, \quad V_{ij} = V_{ji}, \quad \text{and} \quad |M_{ij}| \neq 0. \tag{2.47}$$

Physically M_{ij} s are related to masses, B_{ij} s dissipation constants and V_{ij} s elastic coefficients. Defining the kinetic and potential energies of the system by

$$T = \sum_{i,j=1}^{n} \int_0^t M_{ij}\ddot{x}_i\dot{x}_j dt, \tag{2.48}$$

$$V = \sum_{i,j=1}^{n} \int_0^t V_{ij}x_i\dot{x}_j dt. \tag{2.49}$$

and also defining the Raleigh dissipation function by

$$\mathcal{F} = \sum_{i,j=1}^{n} B_{ij}\dot{x}_i\dot{x}_j, \tag{2.50}$$

then (2.46) can be derived from the Euler-Lagrange equation

$$\frac{d}{dt}\left[\frac{\partial(T+V)}{\partial\dot{x}_i}\right] + \frac{\partial\mathcal{F}}{\partial\dot{x}_i} = 0. \tag{2.51}$$

Bauer Conditions for the Existence of Variational Principle — If the system of differential equations

$$J_i(x_i, x_j, \dot{x}_i, \dot{x}_j, \ddot{x}_i, \ddot{x}_j, t) = 0, \quad i, j = 1, 2 \cdots n, \tag{2.52}$$

are the results of minimizing the integral

$$I = \int_{t_1}^{t_2} L(x_i, x_j, \dot{x}_i, \dot{x}_j, t)\, dt, \tag{2.53}$$

then their equations of variation must be a self-adjoint system along every curve $x_i = x_i(t)$. The necessary and sufficient conditions for the self-adjointness for a system given by (2.46) are [36],[37]:

$$M_{ij} = M_{ji}, \tag{2.54}$$

$$B_{ij} + B_{ji} = 2\frac{dM_{ij}}{dt} \tag{2.55}$$

and

$$\frac{d^2 M_{ij}}{dt^2} - \frac{dB_{ij}}{dt} = V_{ji} - V_{ij}. \tag{2.56}$$

In the absence of the damping force, $B_{ij} = 0$ and (2.54)-(2.56) simplify to

$$M_{ij} = M_{ji}, \tag{2.57}$$

$$2\frac{dM_{ij}}{dt} = 0, \tag{2.58}$$

and

$$V_{ij} = V_{ji}. \tag{2.59}$$

Thus for such a linear system, variational principle exists if and only if $M_{ij} = $ a constant.

Next suppose that $B_{ij} \neq 0$, then from Eqs. (2.54)-(2.56) it follows that

$$M_{ij} = M_{ji}. \tag{2.60}$$

$$B_{ij} = \frac{dM_{ji}}{dt}. \tag{2.61}$$

From these results Bauer arrived at the following surprising conclusions:

(a) "The equations of motion of a conservative dynamical systems are given by a variational if and only if the masses of the system are constants", and

(b) "The equations of motion of a dissipative linear dynamical system are given by a variational principle if and only if the dissipative coefficients are identically equal to the rate of change of corresponding masses".

Thus Bauer concludes that the equations of motion of a dissipative linear dynamical system with constant coefficients cannot be given by a variational principle.

As we will see later in this section, we can construct a Lagrangian for a linearly damped damped harmonic oscillator, Eq. (2.175), and this contradicts the result found by Bauer [35]. Now Synge who has analyzed the argument of Bauer and resolved this contradiction, distinguishes two types of variational principles which he calls type (A) and type (B) [34]. In type (A), the Euler-Lagrange equation

$$\frac{d}{dt}\frac{\partial L}{\partial \dot{x}_i} - \frac{\partial L}{\partial x_i} = 0, \tag{2.62}$$

is formally identical with Eq. (2.52). On the other hand, in type (B) the totality of trajectories

$$x_i = g_i\left(t; \beta_1, \beta_2, \cdots \beta_{2n}\right), \tag{2.63}$$

with β_i s constant, i.e. the extremal of $\int L dt$, is the same as the totality of the trajectories

$$x_i = f_i\left(t; \alpha_1, \alpha_2, \cdots \alpha_{2n}\right), \tag{2.64}$$

obtained from Eq. (2.52). We observe that while (A) implies (B), the converse is not true. For a general linear equations of motion describing dissipative systems, Synge has found the necessary and sufficient conditions for the existence of a Lagrangian.

Synge's Conditions for the Existence of a Lagrangian Function for Linear Damping — Let us consider a linear system, Eq. (2.46) subject to the restriction that $|M_{ij}| \neq 0$. Here we do not need to assume that the coefficients M_{ij}, B_{ij} and V_{ij} are symmetric with the respect to the indices i and j.

We first reduce (2.46) to a normal form by multiplication by M^{ik}, the cofactor of M_{ik} in the determinant $|M_{ij}|$, and then dividing by the determinant. Therefore we have

$$\sum_i M^{ik} M_{ij} = \delta_j^k, \quad \delta_j^k = \begin{cases} 1 & \text{if } k = j \\ 0 & \text{if } k \neq j \end{cases}, \tag{2.65}$$

and we find the normal form of (2.46) to be

$$\ddot{x}_k + \sum_j \mathcal{B}_{kj} \dot{x}_j + \sum_j \mathcal{C}_{kj} x_j = \mathcal{G}_k, \tag{2.66}$$

where

$$\mathcal{B}_{kj} = \sum_i M^{ik} B_{ij}, \quad \mathcal{C}_{kj} = \sum_i M^{ik} V_{ij}, \quad \text{and} \quad \mathcal{G}_k = \sum_i M^{ik} Q_i. \tag{2.67}$$

Now we want to find conditions where Eq. (2.46) admits a variational principle of the form

$$\delta \int L\left(\dot{x}, x, t\right) dt = 0, \tag{2.68}$$

in the sense of (B). Here we are not aiming to find the most general Lagrangian for Eq. (2.46), rather we want to find the simplest L, and for this we assume that L is a quadratic function of the form

$$L = \frac{1}{2} \sum_{i,j} \alpha_{ij} \dot{x}_i \dot{x}_j + \sum_{i,j} \beta_{ij} \dot{x}_i x_j + \frac{1}{2} \sum_{i,j} \gamma_{ij} x_i x_j + \sum_i \left(\delta_i \dot{x}_i + \epsilon_i x_i\right) + \eta, \tag{2.69}$$

where all coefficients in (2.69) are functions of time. By substituting (2.69) in the Euler-Lagrange equation we find

$$\frac{d}{dt}\left[\sum_j \left(\alpha_{ij} \dot{x}_j + \beta_{ij} x_j\right) + \delta_i\right] - \left[\sum_j \left(\beta_{ji} \dot{x}_j + \gamma_{ij} x_j\right) + \epsilon_i\right] = 0. \tag{2.70}$$

In order that the trajectories defined by these equations be identical with those defined by (2.66), there are two conditions that must be satisfied:

(a) It is necessary that (2.70) can be solved for the second derivatives, therefore

$$|\alpha_{ij}| \neq 0 \qquad (2.71)$$

(b) It is necessary that when the values of the second derivatives are substituted from (2.66) into (2.70), the resulting equations must be true for arbitrary values of $\dot{x}_1, \dot{x}_2, \cdots \dot{x}_n$; $x_1, x_2 \cdots x_n$, t. This condition yields the following set of equations:

$$\dot{\alpha}_{ij} - \sum_k \alpha_{ik}\mathcal{B}_{kj} + \beta_{ij} - \beta_{ji} = 0, \qquad (2.72)$$

$$-\sum_k \alpha_{ik}\mathcal{C}_{kj} + \dot{\beta}_{ij} - \gamma_{ij} = 0, \qquad (2.73)$$

and

$$\sum_k \alpha_{ik}\mathcal{G}_k + \dot{\delta}_i - \epsilon_i = 0. \qquad (2.74)$$

After simplifying we find two essential conditions:

$$2\dot{\alpha}_{ij} = \sum_k (\alpha_{ik}\mathcal{B}_{kj} + \alpha_{jk}\mathcal{B}_{ki}), \qquad (2.75)$$

and

$$\sum_k \left[\frac{d}{dt}(\alpha_{ik}\mathcal{B}_{kj} - \alpha_{jk}\mathcal{B}_{ki}) \right] = 2\sum_k (\alpha_{ik}\mathcal{C}_{kj} - \alpha_{jk}\mathcal{C}_{kj}). \qquad (2.76)$$

A System with One Degree of Freedom — For this very simple system where the equation of motion is

$$\ddot{x} + \mathcal{B}\dot{x} + \mathcal{C}x = \mathcal{G}, \qquad (2.77)$$

we take the Lagrangian to be

$$L = \frac{1}{2}\alpha\dot{x}^2 + \beta\dot{x}x + \frac{1}{2}\gamma x^2 + \delta\dot{x} + \epsilon x + \eta. \qquad (2.78)$$

Solving Eqs. (2.72)–(2.74) for α, γ and ϵ and choosing $\beta = \delta = 0$, we obtain

$$\begin{cases} \alpha = \exp\left(\int \mathcal{B}dt\right) \\ \gamma = -\mathcal{C}\exp\left(\int \mathcal{B}dt\right) \\ \epsilon = \mathcal{G}\exp\left(\int \mathcal{B}dt\right). \end{cases} \qquad (2.79)$$

Thus the resulting Lagrangian becomes

$$L = e^{\int \mathcal{B}dt}\left(\frac{1}{2}\dot{x}^2 - \frac{1}{2}\mathcal{C}x^2 + \mathcal{G}x\right). \qquad (2.80)$$

Systems with Two Degrees of Freedom— For a linear damped system with two degrees of freedom where the equations of motion are

$$\ddot{x} + \lambda \dot{x} + \omega^2 y = 0. \tag{2.81}$$

$$\ddot{y} + \lambda \dot{y} + \omega^2 x = 0, \tag{2.82}$$

we have the constant coefficients

$$\mathcal{B}_{11} = \mathcal{B}_{22} = \lambda, \quad \mathcal{B}_{12} = \mathcal{B}_{21} = 0, \tag{2.83}$$

$$\mathcal{C}_{11} = \mathcal{C}_{22} = 0, \quad \mathcal{C}_{12} = \mathcal{C}_{21} = \omega^2, \tag{2.84}$$

and

$$\mathcal{G}_1 = \mathcal{G}_2 = 0. \tag{2.85}$$

By substituting these coefficients in Eqs. (2.72)–(2.74) we get

$$\alpha_{11} = \alpha_{22}, \tag{2.86}$$

$$\dot{\alpha}_{11} = \lambda \alpha_{11}, \quad \dot{\alpha}_{12} = 0, \quad \dot{\alpha}_{22} = \lambda \alpha_{22}, \tag{2.87}$$

The solution of these equations are given by

$$\alpha_{11} = e^{\lambda t} = \alpha_{22}, \quad \alpha_{12} = 0. \tag{2.88}$$

Therefore the condition for the existence of the Lagrangian is satisfied. Again setting $\epsilon_i = \delta_i = 0$, we obtain the Lagrangian [25],[34];

$$L = e^{\lambda t} \left[\frac{1}{2} \left(\dot{x}^2 + \dot{y}^2 \right) - \omega^2 xy \right]. \tag{2.89}$$

Whittaker's Problem— An interesting problem with two degrees of freedom is the Whittaker's system of two coupled equations: [38], [39]

$$\ddot{x} - x = 0, \tag{2.90}$$

$$\ddot{y} - \dot{x} = 0. \tag{2.91}$$

Whittaker believed that this system might not be derivable from a Lagrangian. Bateman, Bauer and Synge have studied this or similar systems and have reached the the same conclusion as Whittaker's. Since these equations are linear, we can use the criteria for the existence of Lagrangian that we have studied [34]. For these equations the coefficients are

$$\mathcal{C}_{11} = -1, \quad \mathcal{B}_{21} = -1, \tag{2.92}$$

and therefore from (2.72)–(2.74) we find

$$\dot{\alpha}_{22} = 2\alpha_{12}, \tag{2.93}$$

and

$$\dot{\alpha}_{11} = -\alpha_{12}, \quad 2\dot{\alpha}_{12} = -\alpha_{12} - \alpha_{22}, \quad \dot{\alpha}_{22} = 0. \tag{2.94}$$

Since here $|\alpha_{ij}| = 0$, therefore we conclude that there exists no quadratic Lagrangian for Whittaker's equations. This system will be considered again when we discuss an extended Lagrangian formulation with fractional derivatives.

Questions Regarding the Existence and Nonuniqueness of the Lagrangian Function — The non-uniqueness of the Lagrangian and Hamiltonian formalism for conservative as well as dissipative systems have been discussed extensively. Excellent surveys of the essential works in construction of q-equivalent Lagrangians, i.e. Lagrangians generating the same equation of motion in coordinate space are given by Dodonov *et al.* [40] and by Santilli [8]. Here we will briefly mention conditions under which Lagrangians for a given set of equations of motion do exist and a method for the construction of these Lagrangians.

In classical dynamics for a conservative system the simple expression of

$$L\left(\dot{x}_i, t\right) = T\left(x_i, \dot{x}_i\right) - V(x_i, t), \tag{2.95}$$

or in terms of generalized coordinates $q_j(t)$ s and generalized velocities $\dot{q}_j(t)$ s [7]

$$L\left(q_i, \dot{q}_i, t\right) = T\left(q_i, \dot{q}_i\right) - V(q_i, t), \tag{2.96}$$

is generally regarded as the Lagrangian of the system [41]. Here $T\left(x_i, \dot{x}_i\right)$ is the kinetic and $V(x_i, t)$ is the potential energy of the system and x_i s and $\dot{x}_i$ s are the $2N$ coordinates and velocities of the particles in the system and if we define $L\left(x, \dot{x}, t\right)$ by (2.95), then it is unique. However we may define $L\left(x, \dot{x}, t\right)$ as the generator of the equation of motion, i.e. a twice differentiable function of the coordinates x_i, and of velocities $\dot{x}_i$ in such a way that when this L is substituted in the Euler-Lagrange equation

$$\frac{\delta L}{\delta x_i} \equiv \frac{\partial L}{\partial x_i} - \frac{d}{dt}\left(\frac{\partial L}{\partial \dot{x}_i}\right) = 0, \quad i = 1, 2, \cdots N, \tag{2.97}$$

the equations of motion are obtained. That the first definition (2.95) is a special case of L satisfying Eq. (2.97) for conservative systems is well-known [41]. When the system is not conservative then we may or may not have Lagrangians and when we have, then the Lagrangian is not unique.

For the equations of motion of second order the question of existence of the Lagrangian and its dependence on the number of degrees of freedom has been studied by a number of authors [3],[42]–[46]. If the motion is one-dimensional, we can always find a Lagrangian whose variational derivative yields the equation of motion. When the number of degrees of freedom is equal or greater than two, then we may or may not be able to have a Lagrangian formulation [43].

As an example of the systems with no Lagrangian let us consider the coupled system with the equations of motion

$$\begin{cases} m_1 \ddot{x}_1 = A\left(x_1 - x_2\right)^{\alpha} \\ m_2 \ddot{x}_2 = B\left(x_2 - x_1\right)^{\beta} \end{cases}. \tag{2.98}$$

This system, for arbitrary values of A, B, α and β has no Lagrangian. As a second example of a motion which cannot be derived from a Lagrangian we can cite the oscillations of two coupled oscillators with different frequencies and different damping constants:

$$\begin{cases} \ddot{x}_1 + \lambda_1 \dot{x}_1 + \omega_1^2 x_1 - c_2 x_2 = 0 \\ \ddot{x}_2 + \lambda_2 \dot{x}_2 + \omega_2^2 x_2 - c_1 x_1 = 0 \end{cases}, \qquad (2.99)$$

where c_1 and c_2 are coupling constants.

Helmholtz's Conditions for Existence of a Lagrangian — The non-uniqueness of the Lagrangians function for dynamical systems will be studied in detail later. But first let us examine the conditions for the existence of a Lagrangian function and how it can be constructed [3],[42],[47].

The laws governing the motion of a dynamical system can be formulated according to Hamilton's principle and this principle states that there is a function (or functions) $L(x_1, \cdots x_N; \dot{x}_1 \cdots \dot{x}_N, t)$ and this L satisfies the following condition [48]:

If the system at the instant t_1 has the coordinates $x_i^{(1)}$ and at t_2 has the coordinates $x_i^{(2)}$ then this condition implies that the functional differential

$$\delta S = \delta \int_{t_1}^{t_2} L(x_i, \dot{x}_i, t)\, dt = \int_{t_1}^{t_2} \sum_{j=1}^{N} \frac{\delta L(x_i, \dot{x}_i, t)}{\delta x_j} \delta x_j(t) dt$$

$$= \int_{t_1}^{t_2} \sum_{j,k} \mu_{jk}(x_i, \dot{x}_i, t) \left[\ddot{x}_k - \frac{1}{m} f_k(x_i, \dot{x}_i, t) \right] \delta x_j(t) dt,$$

$$j = 1, 2, \cdots N, \qquad (2.100)$$

must vanish identically for any continuous variation of the path consistent with the requirement

$$\delta x_j(t_1) = \delta x_j(t_2) = 0. \qquad (2.101)$$

In Eq. (2.100) $\mu_{jk}(x_i, \dot{x}_i, t)$ is a positive definite matrix which may be viewed as an integrating factor [3],[47],[49]. Thus the requirement of the extremum of action S implies the vanishing of the integrand between any two arbitrary instances t_1 and t_2. Since we have imposed the condition of positive definiteness on the integrating factor μ_{jk} and since the variation $\delta x_j(t)$ is arbitrary (apart from the continuity condition and the variational conditions of (2.101)), therefore we conclude that

$$\ddot{x}_k(t) = \frac{1}{m} f_k(x_i, \dot{x}_i, t), \qquad k = 1, 2, \cdots N, \qquad (2.102)$$

which is Newton's second law of motion.

The Lagrangian function determined from the requirement of $\delta S \equiv 0$ is not unique. We can add the total time derivative of any function $\Phi(x, \dot{x}, t)$ to the Lagrangian without affecting the resulting equations of motion (2.102). For the special case of $\mu_{jk} = \delta_{jk}$ we find the Lagrangian to be the same as that given by (2.95).

The second variation of S with respect to $\delta x_j(t)$ and $\delta x_k(t')$ must satisfy the identity:

$$\frac{\delta^2 S}{\delta x_j(t)\,\delta x_k(t')} = \frac{\delta^2 S}{\delta x_k(t')\,\delta x_j(t)}. \tag{2.103}$$

By imposing this condition on (2.100) we find the following equations which μ_{jk} has to satisfy:

$$\mu_{jk} = \mu_{kj}, \tag{2.104}$$

$$\frac{\partial \mu_{kj}}{\partial \dot{x}_l} = \frac{\partial \mu_{jl}}{\partial \dot{x}_k} = \frac{\partial \mu_{lk}}{\partial \dot{x}_j}, \tag{2.105}$$

$$\hat{\mathcal{D}}\mu_{jk} + \frac{1}{2m}\sum_l \left(\mu_{jl}\frac{\partial f_l}{\partial \dot{x}_k} + \mu_{kl}\frac{\partial f_l}{\partial \dot{x}_j} \right) = 0, \tag{2.106}$$

and

$$\hat{\mathcal{D}}\sum_k \left(\mu_{ik}\frac{\partial f_k}{\partial \dot{x}_j} - \mu_{jk}\frac{\partial f_k}{\partial \dot{x}_i} \right) - 2\sum_k \left(\mu_{ik}\frac{\partial f_k}{\partial \dot{x}_j} - \mu_{jk}\frac{\partial f_k}{\partial \dot{x}_i} \right) = 0, \tag{2.107}$$

where in Eqs. (2.106) and (2.107) $\hat{\mathcal{D}}$ denotes the following differential operator

$$\hat{\mathcal{D}} = \frac{\partial}{\partial t} + \sum_k \left[\dot{x}_k\frac{\partial}{\partial x_k} + \frac{1}{m}f_k(x_i,\dot{x}_i,t)\frac{\partial}{\partial \dot{x}_k} \right]. \tag{2.108}$$

These are the Helmholtz conditions [3]–[6],[42].

When the equations of motion are given by

$$g_n(\ddot{x}_j,\dot{x}_j,x_j,t) = 0, \tag{2.109}$$

then Helmholtz conditions become

$$\frac{\partial g_n}{\partial \ddot{x}_j} = \frac{\partial g_j}{\partial \ddot{x}_n}, \tag{2.110}$$

$$\frac{\partial g_n}{\partial \dot{x}_j} + \frac{\partial g_j}{\partial \dot{x}_n} = \frac{d}{dt}\left(\frac{\partial g_n}{\partial \ddot{x}_j} + \frac{\partial g_j}{\partial \ddot{x}_n} \right), \tag{2.111}$$

$$\frac{\partial g_n}{\partial x_j} - \frac{\partial g_j}{\partial x_n} = \frac{1}{2}\frac{d}{dt}\left(\frac{\partial g_n}{\partial \dot{x}_j} - \frac{\partial g_j}{\partial \dot{x}_n} \right). \tag{2.112}$$

If we apply these conditions to the motion of a damped oscillator where g is given by

$$m\ddot{x}_k = f_k(x_i,\dot{x}_i,t), \quad k = 1,\,2,\cdots N, \tag{2.113}$$

$$g(x,\dot{x},\ddot{x}) = m\ddot{x} + m\lambda\dot{x} + m\omega_0^2 x, \tag{2.114}$$

then they fail. However if we apply it to $e^{-\lambda t}g_n(\ddot{x}_j,\dot{x}_j,x_j,t)$, they are satisfied.

We can construct the Lagrangian from the equations of motion by solving a linear partial differential equation. To this end we first write the following identity:

$$\frac{\delta L}{\delta x_k} = \frac{\partial L}{\partial x_k} - \frac{d}{dt}\left(\frac{\partial L}{\partial \dot{x}_k}\right) \equiv \frac{\partial L}{\partial x_k} - \frac{\partial^2 L}{\partial t \partial \dot{x}_k}$$

$$- \sum_j \dot{x}_j \frac{\partial^2 L}{\partial x_j \partial \dot{x}_k} - \sum_j \ddot{x}_j \frac{\partial^2 L}{\partial \dot{x}_j \partial \dot{x}_k} = 0, \qquad (2.115)$$

and then we substitute for $\ddot{x}$ from the equation of motion to obtain

$$\frac{\partial L}{\partial x_k} - \frac{\partial^2 L}{\partial t \partial \dot{x}_k} - \sum_j \dot{x}_j \frac{\partial^2 L}{\partial x_j \partial \dot{x}_k} - \frac{1}{m}\sum_j \frac{\partial^2 L}{\partial \dot{x}_j \partial \dot{x}_k} f_j\left(x_i, \dot{x}_i, t\right) = 0. \qquad (2.116)$$

This equation together with the set of equations (2.104)–(2.107) determines both $L(x, \dot{x}, t)$ and $\mu(x, \dot{x}, t)$.

For another method of constructing Lagrangians and Hamiltonians for nonlinear systems the reader is referred to a paper by Musielak *et al* [10] and for the Lagrangian formulation of the motion of radiating electron see [11].

For a one-dimensional motion Eq. (2.116) simplifies and later we will use it to find the Lagrangian for simple damping systems. But there are other ways of constructing the Lagrangian for motions confined to one dimension [12],[13]. Let us start with the equation of motion

$$m\ddot{x} = f\left(x, \dot{x}\right), \qquad (2.117)$$

and write the Lagrangian for this motion as

$$L\left(x, \dot{x}\right) = \dot{x}\int^{\dot{x}} \frac{1}{v^2}G(v, x)dv, \qquad (2.118)$$

where $G(v, x)$ is a function to be determined later. By substituting $L\left(x, \dot{x}\right)$ in the Euler-Lagrange equation we find

$$\frac{d}{dt}\left(\frac{\partial L}{\partial \dot{x}}\right) - \frac{\partial L}{\partial x} = \frac{\ddot{x}}{\dot{x}}\left(\frac{\partial G\left(\dot{x}, x\right)}{\partial \dot{x}}\right) + \frac{\partial G\left(\dot{x}, x\right)}{\partial x} \equiv m\ddot{x} - f\left(\dot{x}, x\right). \qquad (2.119)$$

From Eq. (2.119) it follows that $G\left(\dot{x}, x\right)$ is the solution of the following differential equation

$$\frac{1}{m\dot{x}}f\left(x, \dot{x}\right)\frac{\partial}{\partial \dot{x}}G\left(\dot{x}, x\right) = -\frac{\partial}{\partial x}G\left(\dot{x}, x\right), \qquad (2.120)$$

provided that

$$\frac{1}{\dot{x}}\frac{\partial}{\partial \dot{x}}G\left(\dot{x}, x\right) \neq 0. \qquad (2.121)$$

Now let us consider two examples using this method of construction of the Lagrangian. When the equation of motion for a damped system is of the form

$$\ddot{x} + \frac{\dot{x}}{\left(\frac{dg(\dot{x})}{d\dot{x}}\right)}\frac{dV\left(x\right)}{dx} = 0, \qquad (2.122)$$

then the Lagrangian which is found by solving Eq. (2.120) is;

$$L\left(x, \dot{x}\right) = \dot{x} \int^{\dot{x}} \frac{1}{v^2} F\left[z(x, v)\right] dv, \qquad (2.123)$$

where F is an arbitrary function of its argument and

$$z(v, x) = g(v) + V(x). \qquad (2.124)$$

However the condition

$$\frac{1}{v} \frac{dg(v)}{dv} \frac{dF(z)}{dz} \neq 0, \qquad (2.125)$$

must be satisfied. Thus for the very simple case of $\ddot{x} + \lambda \dot{x} = 0$ and for $F(z) = z$ we find

$$L\left(x, \dot{x}\right) = \dot{x} \ln \dot{x} - \lambda x. \qquad (2.126)$$

As a second example let us consider the damped harmonic oscillator

$$\ddot{x} + \lambda \dot{x} + \omega_0^2 x = 0. \qquad (2.127)$$

By calculating $G\left(\dot{x}, x\right)$ we find that $G\left(\dot{x}, x\right) = F(z)$, where now

$$z = \frac{1}{2} \frac{\left[x + \left(\frac{\lambda}{2} - i\omega\right) \dot{x}\right]^{\nu^*}}{\left[x + \left(\frac{\lambda}{2} + i\omega\right) \dot{x}\right]^{\nu}} = z^*, \qquad (2.128)$$

and

$$\nu^2 = \frac{\frac{\lambda}{2} + i\omega}{\frac{\lambda}{2} - i\omega}, \qquad (2.129)$$

with $\omega^2 = \omega_0^2 - \frac{\lambda^2}{4}$. In the following section we study some special cases where L can be obtained analytically.

2.3 Some Examples of the Lagrangians for Dissipative Systems

In this and in the next section we illustrate method of constructing the Lagrangian function for simple one-dimensional systems. We will also address the question of non-uniqueness of the Lagrangian by providing specific examples. Our starting point will be Eq. (2.116) for a single degree of freedom.

(1) The simplest choice of μ_{jk} is $\mu_{jk} = \delta_{jk}$. Then Eqs. (2.106) and (2.107) imply that $\frac{\partial f_k}{\partial \dot{x}_j} = 0$ for all k s and j s. Hence f_k can be a function of the coordinate and time but not of velocity.

(2) For the one-dimensional motion Eqs. (2.107) and (2.115) reduce to the following two equations:

$$\frac{\partial \mu}{\partial t} + \dot{x} \frac{\partial \mu}{\partial x} + \frac{1}{m} \left[\frac{\partial \mu}{\partial \dot{x}} f(x, \dot{x}, t) + \mu \frac{\partial f(x, \dot{x}, t)}{\partial \dot{x}} \right] = 0, \qquad (2.130)$$

and

$$\frac{\partial L}{\partial x} - \frac{\partial^2 L}{\partial t \partial \dot{x}} - \dot{x}\frac{\partial^2 L}{\partial x \partial \dot{x}} - \frac{1}{m}\frac{\partial^2 L}{\partial \dot{x}^2}f(x,\dot{x},t) = 0. \qquad (2.131)$$

We note that μ is a dimensionless function and L has the dimension of energy. If L in Eq. (2.131) is known, then μ can be obtained from L. Thus if we substitute for μ from

$$\mu = \frac{1}{m}\left(\frac{\partial^2 L}{\partial \dot{x}^2}\right), \qquad (2.132)$$

in Eq. (2.130) and use Eq. (2.131) we find that (2.132) is a solution of (2.130).

For analytically solvable cases of Eq. (2.131) when the motion is dissipative we have the following cases:

(a) If μ is only a function of time then from (2.130) we have

$$\frac{d\mu(t)}{dt} + \frac{\mu(t)}{m}\left(\frac{\partial f}{\partial \dot{x}}\right) = 0, \qquad (2.133)$$

or f has to have the general form of

$$f = -m\lambda g(t)\dot{x} - \frac{\partial V(x)}{\partial x}, \qquad (2.134)$$

where

$$\mu(t) = m\lambda \int^t g(t')\,dt'. \qquad (2.135)$$

In Eq. (2.135) λ is a constant and $g(t)$ is an arbitrary function of time.

From this result we conclude that a Lagrangian which reduces to $L = T-V$ in the limit of $\lambda \to 0$ is of the form

$$L = \left[\frac{1}{2}m\dot{x}^2 - V(x,t)\right]\exp\left(\lambda \int^t g(t')\,dt'\right). \qquad (2.136)$$

The particular case of $g(t) = 1$ corresponds to the damping force linear in velocity where the force is defined by (1.75).

(b) When μ depends on x only, i.e. $\mu = \mu(x)$, then Eq. (2.130) reduces to

$$\dot{x}\frac{d\mu(x)}{dx} + \left(\frac{\mu(x)}{m}\right)\frac{\partial f}{\partial \dot{x}} = 0. \qquad (2.137)$$

Thus the force $f(x,\dot{x})$ is of the form

$$f(x,\dot{x}) = -m\gamma\dot{x}^2 - \frac{\partial V(x)}{\partial x}, \qquad (2.138)$$

and from Eq. (2.137) we find $\mu(x)$ to be

$$\mu(x) = \exp(2\gamma x). \qquad (2.139)$$

In Eq. (2.138) we have written the x-dependent part of the force f in terms of the potential $V(x)$. From Eq. (2.132) and (2.139) we obtain the following expression for the Lagrangian

$$L = \frac{1}{2}m\dot{x}^2 \exp(2\gamma x) - W(x). \tag{2.140}$$

Now if we substitute (2.138) and (2.140) in Eq. (2.131) for L, we find $W(x)$ to be

$$W(x) = \int^x e^{-2\gamma y} \frac{\partial V(y)}{\partial y} dy, \tag{2.141}$$

and this agrees with $L = T - V$ in the limit of $\gamma \to 0$.

(c) As another example let us obtain the Lagrangian for the motion of a particle subject to the resistive force $m\beta \dot{x}^\nu$ where ν is a positive number $1 < \nu < 2$. We can formulate this problem as that of the motion of a particle with a velocity-dependent mass moving in a constant potential. Let us write the equation of motion as

$$m\mu''(\dot{x}) \frac{d\dot{x}}{dt} + m\beta = 0, \tag{2.142}$$

where primes denote derivatives with respect to $\dot{x}$, and

$$\mu''(\dot{x}) = \dot{x}^{(-\nu)}. \tag{2.143}$$

The Lagrangian for this motion is given by [50]

$$L(\dot{x}, x) = m\mu(\dot{x}) - m\beta x. \tag{2.144}$$

The Euler-Lagrange equation for $L(\dot{x}, x)$, Eq. (2.144) yields the equation of motion (2.142).

(d) The motion of a rocket is another example of the motion of a system with variable mass moving in a viscous medium. Consider a rocket with the initial mass $m_v + m_f$, where m_v is the mass of the rocket and m_f is the mass of fuel. Assuming a constant exhaust rate and denoting the exhaust velocity by u, we have the equation of motion [51];

$$\frac{dv}{dt} + \lambda v = -u \frac{d}{dt} \ln[m(t)], \tag{2.145}$$

where

$$m(t) = \begin{cases} m_v + m_f \left(1 - \frac{t}{T}\right) & \text{for } 0 \le t \le T \\ m_v & \text{for } t > T. \end{cases} \tag{2.146}$$

Using (2.146) we can write the right hand side of (2.145) as

$$-u \frac{d}{dt} \ln[m(t)] = \frac{\alpha u}{1 - \alpha t} = F(t), \tag{2.147}$$

where

$$\alpha = \frac{1}{T}\left(\frac{m_f}{m_v + m_f}\right). \tag{2.148}$$

Equation (2.145) can be integrated to yield $v(t)$ [51]

$$v(t) = u\exp\left[\frac{\lambda}{\alpha}(1-\alpha t)\right]\left[Ei\left(-\frac{\lambda}{\alpha}\right) - Ei\left(-\frac{\lambda}{\alpha}(1-\alpha t)\right)\right], \tag{2.149}$$

where

$$Ei(y) = -\int_{-y}^{\infty}\frac{e^{-z}}{z}dz, \tag{2.150}$$

is the exponential integral function [52]. The position of the rocket as a function of time can be found by integrating $v(t)$ with respect to t.

We can construct the Lagrangian for this system using equation (2.132) with $\mu(t) = e^{\lambda t}$. Thus we obtain the following expression for the Lagrangian

$$L = \left[\frac{1}{2}\dot{x}^2 + xF(t)\right]e^{\lambda t}. \tag{2.151}$$

2.4 Non-Uniqueness of the Lagrangian

If $L(x_i, \dot{x}_i, t)$ is the Lagrangian describing a dissipative system then the functional derivative of L, i.e. $\frac{\delta L}{\delta x_i}$, Eq. (2.115), must vanish. Now if we multiply L by a constant number, α, and add the total time derivative of a function of x_i and t to it we find a new Lagrangian $\bar{L}$,

$$\bar{L} = \alpha L + \frac{dg(x_i, t)}{dt}. \tag{2.152}$$

This Lagrangian is equivalent to L, i.e. $\frac{\delta \bar{L}}{\delta x_i} = 0$ gives us the same equation of motion as Eq. (2.115). However the Lagrangian $\bar{L}$ is not the most general Lagrangian for the motion. We can show this in a simple way for one-dimensional motion of a particle. However we should note that this method, in general, cannot be extended to multi-dimensional motions. We start with Eq (2.115) for a single generalized coordinate x, and solve it for $\ddot{x}$, assuming that the coefficient of $\ddot{x}$ is not zero,

$$\ddot{x} = \left(\frac{\partial^2 L}{\partial \dot{x}^2}\right)^{-1}\left[\frac{\partial L}{\partial x} - \frac{\partial^2 L}{\partial \dot{x}\partial t} - \frac{\partial^2 L}{\partial \dot{x}\partial x}\dot{x}\right]. \tag{2.153}$$

Now we assume that $\bar{L}$ is equivalent to L, i.e. $\frac{\delta \bar{L}}{\delta x} = 0$ yields the same equation of motion as $\frac{\delta L}{\delta x} = 0$

$$\ddot{x} = \left(\frac{\partial^2 \bar{L}}{\partial \dot{x}^2}\right)^{-1}\left[\frac{\partial \bar{L}}{\partial x} - \frac{\partial^2 \bar{L}}{\partial \dot{x}\partial t} - \frac{\partial^2 \bar{L}}{\partial \dot{x}\partial x}\dot{x}\right]. \tag{2.154}$$

Next we define Λ called "fouling function" by [53]

$$\Lambda = \left(\frac{\partial^2 \bar{L}}{\partial \dot{x}^2}\right)\left(\frac{\partial^2 L}{\partial \dot{x}^2}\right)^{-1}, \qquad (2.155)$$

so that

$$\frac{\partial \bar{L}}{\partial x} - \frac{\partial^2 \bar{L}}{\partial \dot{x}\partial t} - \frac{\partial^2 \bar{L}}{\partial \dot{x}\partial x}\dot{x} = \Lambda\left(\frac{\partial L}{\partial x} - \frac{\partial^2 L}{\partial \dot{x}\partial t} - \frac{\partial^2 L}{\partial \dot{x}\partial x}\dot{x}\right). \qquad (2.156)$$

For any $x(t)$ which is the solution of the equation of motion, Eq. (2.156) is satisfied.

Now we want to show that Λ is a constant of motion. First we note that

$$\frac{d\Lambda}{dt} = \frac{\partial \Lambda}{\partial \dot{x}}\ddot{x} + \frac{\partial \Lambda}{\partial x}\dot{x} + \frac{\partial \Lambda}{\partial t}, \qquad (2.157)$$

and for the moment we do not assume that $x(t)$ is a solution of $\frac{\delta L}{\delta x} = 0$. By taking the partial derivatives of Λ, Eq. (2.155), with respect to x and t, and substituting the results in (2.157) and also eliminating $\ddot{x}$ between (2.157) and (2.153) we find

$$\frac{d\Lambda}{dt} = \frac{\partial \Lambda}{\partial \dot{x}}\left(\frac{\partial^2 L}{\partial \dot{x}^2}\right)^{-1}\left[\frac{\delta L}{\delta x} + \frac{\partial L}{\partial x} - \frac{\partial^2 L}{\partial \dot{x}\partial t} - \frac{\partial^2 L}{\partial \dot{x}\partial x}\dot{x}\right] - $$
$$- \left(\frac{\partial^2 L}{\partial \dot{x}^2}\right)^{-2}\left(\frac{\partial^2 \bar{L}}{\partial \dot{x}^2}\right)\left(\frac{\partial^3 L}{\partial \dot{x}^2 \partial x}\dot{x} + \frac{\partial^3 L}{\partial \dot{x}^2 \partial t}\right)$$
$$+ \left(\frac{\partial^2 L}{\partial \dot{x}^2}\right)^{-1}\left(\frac{\partial^3 \bar{L}}{\partial \dot{x}^2 \partial x}\dot{x} + \frac{\partial^3 \bar{L}}{\partial \dot{x}^2 \partial t}\right). \qquad (2.158)$$

We also take the partial derivative of Eq. (2.156) with respect to $\dot{x}$

$$\left(\frac{\partial^3 \bar{L}}{\partial \dot{x}^2 \partial x}\dot{x} + \frac{\partial^3 \bar{L}}{\partial \dot{x}^2 \partial t}\right) = \frac{\partial \Lambda}{\partial \dot{x}}\left[\frac{\partial^2 L}{\partial \dot{x}\partial t} + \frac{\partial^2 L}{\partial \dot{x}\partial x}\dot{x} - \frac{\partial L}{\partial x}\right]$$
$$+ \Lambda\left(\frac{\partial^3 L}{\partial \dot{x}^2 \partial x}\dot{x} + \frac{\partial^3 L}{\partial \dot{x}^2 \partial t}\right). \qquad (2.159)$$

From Eqs. (2.158) and (2.159) it follows that

$$\frac{d\Lambda}{dt} = \frac{\partial \Lambda}{\partial \dot{x}}\left(\frac{\delta L}{\delta x}\right), \qquad (2.160)$$

and this is true whether $\delta L/\delta x$ is zero or not.

When $x(t)$ is a solution of Euler-Lagrange equation, $\frac{\delta L}{\delta x} = 0$, then $\frac{d\Lambda}{dt} = 0$ and Λ is a constant of motion. Conversely for any constant of motion Λ, there exists a Lagrangian $\bar{L}$ such that [53]

$$\frac{\delta \bar{L}}{\delta x} = \Lambda\frac{\delta L}{\delta x}. \qquad (2.161)$$

The multi-dimensional case is very different from the one-dimensional problem which we have studied in this section. For details the reader is referred to the book of Matzner and Shepley [53].

Examples of Generalized Lagrangians — As a specific example let us consider the simple case of a particle moving in a viscous medium with linear damping, where the Lagrangian is

$$L = \frac{1}{2} m \dot{x}^2 e^{\lambda t}. \tag{2.162}$$

The generalized Lagrangian $\bar{L}$ can be found from Eq. (2.162),

$$\frac{\partial^2 \bar{L}}{\partial \dot{x}^2} = \Lambda \frac{\partial^2 L}{\partial \dot{x}^2} = m \Lambda e^{\lambda t}. \tag{2.163}$$

Since Λ is a constant of motion, we can find it from L. By noting that

$$\frac{d}{dt} \left(\frac{\partial L}{\partial \dot{x}} \right) = 0, \quad \text{therefore} \quad m \dot{x} e^{\lambda t} = \text{constant}, \tag{2.164}$$

we have a possible choice of Λ

$$\Lambda = \frac{m \dot{x} e^{\lambda t}}{p_0}, \tag{2.165}$$

with p_0 being a constant ($p_0 \neq 0$). From Eqs. (2.163) and (2.165) we find

$$\bar{L} = \frac{1}{6 p_0} m^2 \dot{x}^3 e^{2\lambda t} + g_1(x,t) \dot{x} + g_2(x,t), \tag{2.166}$$

where $g_1(x,t)$ and $g_2(x,t)$ are functions of their arguments. By substituting $\bar{L}$ in the Euler-Lagrange equation we get the result

$$\frac{\partial g_1(x,t)}{\partial t} = \frac{\partial g_2(x,t)}{\partial x}, \tag{2.167}$$

that is $g_1(x,t) \dot{x} + g_2(x,t)$ is the total time derivative of a function $g(x,t)$.

In a similar way we find the Lagrangian

$$\bar{L} = \frac{1}{24} m^2 \dot{x}^4 e^{4\gamma x} + \frac{dg(x,t)}{dt}, \tag{2.168}$$

for the damping force $\left(-\gamma \dot{x}^2 \right)$ acting on the particle. Here we have used

$$\Lambda = \frac{1}{2} m \dot{x}^2 e^{2\gamma x}, \tag{2.169}$$

as the constant of motion.

Examples of Construction of Lagrangians Using Fouling Functions — The fouling method can be used to construct the Lagrangian for a given frictional force law provided that a first integral of motion is known [12]. Let $\Lambda(x, \dot{x})$ be a first integral for a one-dimensional motion in a viscous medium and let $F(\Lambda)$ be an arbitrary but differentiable function of its argument, $\Lambda(x, \dot{x})$, then the Lagrangian for this motion is expressible as (see also Eq. (2.118))

$$L = \dot{x} \int^{\dot{x}} \frac{G(\Lambda(x, v))}{v^2} dv. \tag{2.170}$$

To show that L is a solution of the Euler-Lagrange equation we calculate $\frac{\partial L}{\partial x}$ and $\frac{d}{dt}\left(\frac{\partial L}{\partial \dot{x}}\right)$, and we simplify the results using the fact that $\frac{dF(\Lambda)}{dt} = 0$. By substituting these in (2.97) we observe that L is indeed a Lagrangian.

If we apply this method to the problem of the damped harmonic oscillator, $\ddot{x} + \lambda \dot{x} + \omega_0^2 x = 0$, we find that $\Lambda(x, \dot{x})$ is a solution of the partial differential equation

$$\frac{d\Lambda}{dt} = \dot{x}\frac{\partial \Lambda}{\partial x} - \left(\lambda \dot{x} + \omega_0^2 x\right)\frac{\partial \Lambda}{\partial \dot{x}} = 0. \tag{2.171}$$

This equation can be solved by the method of characteristics, i.e. by solving the set of ordinary differential equations [54]

$$\frac{dx}{\dot{x}} = -\frac{d\dot{x}}{\lambda \dot{x} + \omega_0^2 x} = d\Lambda, \tag{2.172}$$

and these yield the result

$$\Lambda(x, \dot{x}) = \exp\left\{\frac{1}{2}\ln\left(\dot{x}^2 + \lambda x\dot{x} + \omega_0^2 x\right) - \frac{\lambda}{2\omega}\tan^{-1}\left[\frac{1}{2\omega}\left(\frac{2\dot{x}}{x} + \lambda\right)\right]\right\}. \tag{2.173}$$

Thus the general form of the time-independent Lagrangian for the damped oscillator is expressed by the integral Eq. (2.170) where $\Lambda(x, \dot{x})$ is given by (2.173).

2.5 Acceptable Lagrangians for Dissipative Systems

By solving the set of differential equations (2.116) we find a finite or infinite set of Lagrangians L_i. However not all of these L_i s are acceptable, some because of their singular or defective nature, and others because they violate other physical requirements. We have already obtained an explicitly time-dependent Lagrangian for the damped harmonic oscillator,

$$\ddot{x} + \lambda \dot{x} + \omega_o^2 x = 0, \tag{2.174}$$

for which from Eqs. (2.134) and (2.136) we have

$$L = \frac{m}{2}\left(\dot{x}^2 - \omega_0^2 x^2\right)e^{\lambda t}. \tag{2.175}$$

Among the explicitly time-independent Lagrangians we can obtain the following Lagrangian if we choose $F(\Lambda) = -\Lambda$ in (2.170) (see also Chapter 6 Eq. (6.56))

$$
L_2(x, \dot{x}) = -\frac{1}{2} \ln \left[\dot{x}^2 + \lambda x \dot{x} + \left(\omega^2 + \frac{\lambda^2}{4} \right) x^2 \right]
$$
$$
+ \left(\frac{1}{2\omega} \right) \left(\frac{2\dot{x}}{x} + \lambda \right) \arctan \left[\left(\frac{1}{2\omega} \right) \left(\frac{2\dot{x}}{x} + \lambda \right) \right]. \tag{2.176}
$$

This Lagrangian is defective in the sense that it does not reduce to the undamped Lagrangian when $\lambda \to 0$. A time-independent Lagrangian which has the correct form as $\lambda \to 0$ is obtained if we choose $F(\Lambda) = \frac{1}{2} \exp(2\Lambda)$ in (2.170) but then L is given as an integral [12]

$$
L(x, \dot{x}) = \dot{x} \int^{\dot{x}} \left(\frac{1}{2} \right) \left[\left(1 + \frac{\lambda x}{y} + \frac{\omega^2 x^2}{y^2} \right) \right.
$$
$$
\times \exp \left[- \left(\frac{\lambda}{\omega} \right) \arctan \left\{ \left(\frac{1}{2\omega} \right) \left(\frac{2y}{x} + \lambda \right) \right\} \right] \right] dy. \tag{2.177}
$$

Some Other Forms of Lagrangian for the Damped Harmonic Oscillator — A different way of constructing Lagrangians for the damped harmonic oscillator is to start with the Lagrange's equation (2.1) with the damping force (2.2),(2.3), and write it as

$$
\frac{d}{dt} \left(\frac{\partial L'}{\partial \dot{x}'} \right) - \frac{\partial L'}{\partial x'} = -m\lambda \dot{x}', \tag{2.178}
$$

and try to find an L' which satisfies this equation. Here we have denoted this Lagrangian by L' and the coordinate by x' to distinguish it from the explicitly time-dependent Lagrangian (2.136) or the $L(x, \dot{x})$ given by (2.177). The total time derivative of L' is given by

$$
\frac{dL'}{dt} = \frac{\partial L'}{\partial x'} \frac{dx'}{dt} + \frac{\partial L'}{\partial \dot{x}'} \frac{d\dot{x}'}{dt} + \frac{\partial L'}{\partial t}. \tag{2.179}
$$

Now if we solve (2.179) $\frac{\partial L'}{\partial x'}$ and substitute it in (2.178) we find

$$
\frac{d}{dt} \left(\dot{x}' \frac{\partial L'}{\partial \dot{x}'} - L' \right) = -m\lambda \dot{x}'^2 - \frac{\partial L'}{\partial t}, \tag{2.180}
$$

Here we will not try to consider the general form of the Lagrangian which satisfies (2.180), but we will discuss the particular case where L' is given by

$$
L' = L(x', \dot{x}') + k(x')^a (\dot{x}')^b. \tag{2.181}
$$

In this relation L is the Lagrangian for the oscillator in the absence of friction

$$
L(x', \dot{x}') = \frac{m}{2} \left(\dot{x}'^2 - \omega^2 x'^2 \right), \tag{2.182}
$$

and k , a and b are all constants. If we substitute (2.181) in (2.180) we get

$$k(b-1)\left(x'\right)^{a-1}\left(\dot{x}'\right)^{b-1}\left(bx'\ddot{x}'+a\dot{x}'^2\right)=0. \tag{2.183}$$

Clearly to satisfy (2.183) for all x', $\dot{x}'$ and $\ddot{x}'$ we choose $b=1$. For L' to have the dimension of energy, the simplest choice is to take $a=1$ and k to be proportional to λ, e.g. $k=-\frac{1}{2}m\lambda$. This Lagrangian reduces to L when λ goes to zero, and is given by

$$L'=\frac{m}{2}\left(\dot{x}'^2-\omega^2x'^2+kx'\dot{x}'\right). \tag{2.184}$$

But unlike Caldirola-Kanai form of the Lagrangian, this L' does not depend explicitly on time [55]. The Hamiltonian corresponding to this Lagrangian can be found by the Legendre transformation and is given by Eq. (15.73).

2.6 Complex or Leaky Spring Constant

The complex classical Lagrangian that we want to consider is modeled after the complex optical potentials used in nuclear physics [56]. Let us study a two-dimensional harmonic oscillator which is subject to a damping force linear in velocity, i.e.

$$\frac{d^2\mathbf{r}}{dt^2}+2\alpha\omega\frac{d\mathbf{r}}{dt}+\omega^2\left(1+\alpha^2\right)\mathbf{r}=0. \tag{2.185}$$

The x and y components of this motion are

$$x(t)=r_0\cos(\omega t)e^{-\alpha\omega t},\quad y(t)=r_0\sin(\omega t)e^{-\alpha\omega t}, \tag{2.186}$$

where we have used the initial conditions $x(0)=r_0$ and $y(0)=0$. Now let us consider a complex time variable τ defined by the relation

$$\tau=(1-i\alpha)t, \tag{2.187}$$

and then write the equation of motion as

$$\frac{d^2z}{d\tau^2}=-\omega^2z, \tag{2.188}$$

or changing τ to t we have

$$\frac{d^2z}{dt^2}=-\Omega^2z=-\omega^2(1-i\alpha)^2z, \tag{2.189}$$

where

$$\mathrm{Im}\,\Omega^2=-2\alpha\omega^2<0. \tag{2.190}$$

Of the two solutions of (2.188) the one which is of the form $e^{-i\omega\tau}$ is acceptable, i.e. z remains finite as $t\to\infty$. If we write

$$z(t)=x(t)+iy(t)=r_0\exp(-i\omega t-\alpha t), \tag{2.191}$$

we observe that $x(t)$ and $y(t)$ are the same as those given by (2.186). Equation (2.188) has a simple Lagrangian

$$L = \left(\frac{m}{2}\right)\left[\left(\frac{dz}{d\tau}\right)^2 - \omega^2 z^2\right], \tag{2.192}$$

from which we can determine the Hamiltonian and also the classical action.

2.7 A Fractional Derivative Approach to the Lagrangian for the Damped Motion

A different approach for including the damping force in the Lagrangian is by allowing the Lagrangian to depend on the fractional time derivatives. In Sec. 2.1 we observed that one way of introducing a linear damping force in the equations of motion is by adding a term $\frac{\partial \mathcal{F}}{\partial x_i}$ to the Euler-Lagrange equation. Thus the frictional force is not included in the Lagrangian and therefore the equations of motion cannot be described by a single Lagrangian. From the Euler-Lagrange equation it follows that if L contains a term proportional to $\left(\frac{d^n x}{dt^n}\right)^2$, then the equation of motion contains a term proportional to $\left(\frac{d^{2n} x}{dx^{2n}}\right)$. Therefore the force $\lambda\left(\frac{dx}{dt}\right)$ can be obtained directly from a Lagrangian containing a term proportional to the fractional derivative $\left(\frac{d^{\frac{1}{2}} x}{dt^{\frac{1}{2}}}\right)$ [39],[57].

In order to introduce the fractional derivative, we first consider the fractional integral of order ν:

$$\frac{d^{-\nu}}{d(t-c)^{-\nu}} f(t) = \frac{1}{\Gamma(\nu)} \int_c^t (t-t')^{\nu-1} f(t')\, dt', \quad \mathrm{Re}\,\nu > 0. \tag{2.193}$$

Let k be the smallest integer greater than $\mathrm{Re}(\nu)$ and $\nu = k - u$, then we define the fractional derivative of order u of a function $f(t)$ by

$$_cD_t^u f(t) \equiv \frac{d^u}{d(t-c)^u} f(t) = \frac{d^k}{dt^k}\left(\frac{d^{-\nu} f(t)}{d(t-c)^{-\nu}}\right). \tag{2.194}$$

We note that if u is an integer, then (2.194) will reduce to the standard definition of derivative. We also note that the fractional operator $_aD_t^\alpha$ can be written as

$$_aD_t^u f(t) = \frac{d^k}{dt^k}\frac{1}{\Gamma(k-u)} \int_a^t (t-\tau)^{k-u-1} f(\tau) d\tau, \tag{2.195}$$

or alternatively as

$$_tD_b^u f(t) = \frac{d^k}{dt^k}\frac{1}{\Gamma(k-u)} \int_t^b (t-\tau)^{k-u-1} f(\tau) d\tau. \tag{2.196}$$

If the Lagrangian is a function of the coordinate x_j, the time and the derivative of x_j with respect to t, then we use $s(n)$ to denote the order of nth derivative appearing in the Lagrangian.

Euler-Lagrange Equation with Fractional Derivatives — In fractional calculus of variation we encounter two types of derivatives. For instance if the lowest order derivative is

$$\frac{d^{\frac{1}{2}} x}{d(t-b)^{\frac{1}{2}}},\qquad(2.197)$$

then $s(1) = \frac{1}{2}$. Thus for each order of derivative in the Lagrangian we define the generalized coordinates $q_{j,s'(n),a}$ and $q_{j,s(n),b}$ by :

$$q_{j,s(n)} = q_{j,s(n),b} = \frac{d^{s(n)} x_j}{d(t-b)^{s(n)}},\qquad(2.198)$$

and

$$q_{j,s'(n)} = q_{j,s'(n),a} = \frac{d^{s'(n)} x_j}{d(t-a)^{s'(n)}},\qquad(2.199)$$

where $s(n)$ and $s'(n)$ are any non-negative real numbers. If the variational principle is applied over the interval $t = a$ to $t = b$, both $q_{j,s'(n),a}$ and $q_{j,s(n),b}$ will appear in the Lagrangian. Here we will derive the Euler-Lagrange for those cases where L does not contain the derivative of the second kind, and for the sake of simplicity we will omit the dependence of $q_{j,s(n)}$ on b. We can use this notation and write $\alpha = s(n)$ and $\beta = s'(n)$ in Eqs. (2.195) and (2.196).

If the Lagrangian depends on N different derivatives of x_j, then n takes the values $n = 1, 2 \cdots N$, and the Lagrangian will be a function of t as well as all $(q_{j,s(n),b})$ s and $(q_{j,s'(n),a})$ s:

$$L = L\left(\{q_{j,s'(n),a},\ q_{j,s(n),b}\},t\right).\qquad(2.200)$$

Therefore when the Lagrangian depends on x, $\frac{d^{\frac{1}{2}} x}{dx^{\frac{1}{2}}}$, and $\frac{dx}{dt}$ then we have $N = 2$, $s(0) = 0$, $s(1) = \frac{1}{2}$ and $s(2) = 1$. By varying the action integral

$$J(\alpha) = \int_a^b L\left(\{q_{j,s(n),b}\},t\right)dt,\qquad(2.201)$$

which depends on the parameter α and then setting $\left.\left|\frac{\partial J(\alpha)}{\partial \alpha}\right.\right|_{\alpha=0} = 0$ we obtain

$$\delta J = \frac{\partial J(\alpha)}{\partial \alpha}d\alpha = \int_a^b \sum_{n=0}^{N} \frac{\partial L}{\partial q_{r,s(n)}}\frac{\partial q_{r,s(n)}}{\partial \alpha}\,d\alpha dt$$

$$= \int_a^b \sum_{n=0}^{N} \frac{\partial L}{\partial q_{r,s(0)}}\frac{\partial^{s(n)}}{\partial (t-b)^{s_n}}\frac{\partial q_{r,s(n)}}{\partial \alpha}\,d\alpha\,dt$$

$$= \int_a^b \sum_{n=0}^{N}\left[(-1)^n \frac{d^{s(n)}}{d(t-b)^{s(n)}}\frac{\partial L}{\partial q_{r,s(0)}}\right]\delta q_{r,s(0)}\,dt,\qquad(2.202)$$

where

$$\delta q_{r,s(0)} = \frac{\partial q_{r,s(0)}}{\partial \alpha}. \tag{2.203}$$

Since the variations are independent, $\delta J(\alpha)$ can be zero if all $q_{r,s(0)}$ are zero. In this way we find the Euler-Lagrange equation

$$\frac{\partial L}{\partial q_{j,b}} + \sum_{n=1}^{N} (-1)^{s(n)} \frac{d^{s(n)}}{d(x-a)^{s(n)}} \frac{\partial L}{\partial q_{j,s(n),b}} +$$

$$\sum_{n=1}^{N'} (-1)^{-s'(n)} \frac{d^{s'(n)}}{d(x-b)^{s'(n)}} \frac{\partial L}{\partial q_{j,s'(n),a}} = 0. \tag{2.204}$$

As we noted earlier, in this approach not only we have a single L for generating the equations of motion, but in addition here the damping term is simply added to the difference between kinetic and potential energies. For instance in the case of linear damping the Lagrangian is given by

$$L = \frac{1}{2} m \left(\frac{dx}{dt} \right)^2 - V(x) + \frac{i}{2} \lambda \left(\frac{d^{\frac{1}{2}} x}{d(t-b)^{\frac{1}{2}}} \right)^2. \tag{2.205}$$

Writing this Lagrangian in terms of q variables, Eqs. (2.198) and (2.199) we find

$$L = \frac{1}{2} m q_1^2 - V(q_0) + \frac{i}{2} q_{\frac{1}{2}}^2. \tag{2.206}$$

Again we note the appearance of i directly in the classical Lagrangian and consequently in the Hamiltonian formulation (see Sec. 3.17 and Eq. (3.337). This is an essential feature of a large class of damped motions.

Substituting this L in the Euler-Lagrange equation, Eq. (2.204), i.e.

$$\frac{\partial L}{\partial q_0} + i \frac{d^{\frac{1}{2}} L}{d(t-b)^{\frac{1}{2}}} \frac{\partial L}{\partial q_{\frac{1}{2}}} - \frac{d}{dt} \frac{\partial L}{\partial q_1} = 0, \tag{2.207}$$

and replacing q s by x, $\dot{x}$ and $\ddot{x}$ we obtain

$$m\ddot{x} + \lambda\dot{x} + \frac{\partial V(x)}{\partial x} = 0. \tag{2.208}$$

We will discuss the result of quantization of the Lagrangian (2.200) in Chapter 11.

A Lagrangian Formulation of the Whittaker Problem — Earlier in this chapter we proved the nonexistence of a quadratic Lagrangian for the Whittaker problem. Now as Riewe has shown, by introducing the concept of Lagrangian with functional derivatives, we can enlarge the set of dissipative motions where variational formulation is applicable. For instance, an extension of Lagrangian formulation which allows the inclusion of fractional derivatives, enables us to write the following Lagrangian for the dissipative system given by Whittaker (2.90)-(2.91)[57],

$$L = \left(\frac{dx}{dt}\right)^2 + = \left(\frac{dy}{dt}\right)^2 + x^2 + y\left(\frac{dx}{dt}\right) - ix_{\frac{1}{2}}y_{\frac{1}{2}}. \qquad (2.209)$$

Let us emphasize that while this extension of the Lagrangian formulation (and its corresponding Hamiltonian) is a novel concept, its usefulness for finding an acceptable wave equation is questionable. In particular the presence of frictional time derivatives would result in having negative probability densities. For additional information regarding equations of motion with fractional derivatives see the following books and papers [57]–[61].

References

[1] P. Caldirola, Nuovo Cimento 18, 393 (1941).

[2] E. Kanai, Prog. Theor. Phys. 3, 440 (1948).

[3] P. Havas, Nuovo Cimento Supp. 5, 363 (1957).

[4] W. Sarlet, J. Phys. A 15, 1503 (1982).

[5] O. Krupkova, *The Geometry of Ordinary Variational Equations*, (Springer Berlin, 1997) p. 5.

[6] B.D. Vujanovic and S.E. Jones, *Variational Methods in Nonconservative Phenomena*, (Academic Press, New York, 1989) Sec. 2.5.

[7] H. Goldstein, *Classical Mechanics*, Second Edition (Addison-Wesley Reading, 1980) Chapter 8.

[8] R.M. Santilli, *Foundations of Theoretical Mechanics*, Vol. 1 (Springer-Verlag, New York, N.Y. 1978).

[9] M. Razavy, Z. Phys. B 26, 201 (1977).

[10] Z.E. Musielak, D. Roy and L.D. Swift, Chaos, Solitons & Fractals, 38, 894 (2008).

[11] V.G. Kupriyanov, Intl. J. Theor. Phys. 45, 1129 (2006).

[12] J.A. Kobussen, Acta Phys. Austriaca, 59, 293 (1979).

[13] S. Okubo Phys. Rev. A 23, 2776 (1981).

[14] J.L. Cieśliński and T. Nikiciuk, J. Phys. A 43, 175205 (2010).

[15] N.A. Lemov, Phys. Rev. D 24, 1036 (1981).

[16] P. Caldirola, Rend. Ist. Lomb. Sc., A 93, 439 (1959).

[17] L. Meirovitch, *Methods of Analytical Dynamics*, (McGraw-Hill, New York, N.Y. 1970) p. 88.

[18] R.M. Rosenberg, *Analytical Dynamics of Discrete Systems*, (Plenum Press, 1977) p. 229.

[19] A.I. Lur'e, *Analytical Mechanics*, (Springer-Verlag, Berlin, 2002) Sec. 5.11.

[20] E.T. Whittaker, *A Treatise on the Analytical Dynamics of Particles and Rigid Bodies*, Fourth Edition (Cambridge University Press, London, 1965)

[21] C. Monforte, *Geometric, Control and Numerical Aspects of Nonholonomic Systems*, (Springer-Verlag, Berlin, 2004) p. 53.

[22] See for instance, V.A. Zhelnorvich and L.I. Sedov, J. App. Math. Mech. 42, 771 (1978).

[23] L.I. Sedov, "Applied mechanics" in : *Proc. of the 11th Inter. Congr. Appl. Mech. Munich, 1964*, (Springer-Verlag 1966), p. 9.

[24] V.E. Tasarov, Theor. Math. Phys. 100, 1100, (1994).

[25] B.R. Gossick, *Hamilton's Principle and Physical Systems*, (Academic Press, New York, N.Y. 1967) p. 102.

[26] B.R. Gossick, IEEE Trans. on Education, v E-10, 37 (1967).

[27] W.E. Schröder and J.R. Huizenga in *Treatise on Heavy-Ion Scattering, Vol. 2, Fusion and Quasi-Fusion Phenomena*, Edited by D.A. Bromley, (Plenum Press, New York, N.Y. 1984) p. 138.

[28] W.U. Schöder and J.R. Huizenga, Annu. Rev. Nucl. Sci. 27, 465 (1977).

[29] J.R. Birkelund, J.R. Huizenga, J.N. De and D. Sperber, Phys. Rev. Lett. 40, 1123 (1978).

[30] D.H.E. Gross and H. Kalinowski, Phys. Rep. 45, 175 (1978).

[31] D. Sperber, Phys. Scr. Supp. 10, 115 (1974).

[32] J.R. Birkelund, Phys. Rep. 56, 107 (1979).

[33] R.W. Hasse, Rep. Prog. Phys. 41, 1027 (1978).

[34] J.L. Synge, Trans. Roy. Soc. Canada, 26, 49 (1932).

[35] P.S. Bauer, Proc. Nat. Acad. Sci. USA, 17, 311 (1931).

[36] M. Morse, Math. Annalen, 52, 103 (1930).

[37] D.R. Davis, Trans. Am. Math. Soc. 30, 711 (1928).

[38] H. Bateman, Phys. Rev. 38, 815 (1931).

[39] F. Riewe, Phys. Rev. E 56, 3581 (1997).

[40] V.V. Dodonov, V.I. Man'ko and V.D. Skarzhinskiy in *Quantization, Gravitation and Group Methods in Physics*, Edited by A.A. Komar (Nova Science, Commack, 1988) p. 57.

[41] L.A. Pars, *A Treatise on Analytical Dynamics*, (John Wiley & Sons, New York, N.Y. 1965).

[42] H. Helmholtz, J. Reine Angew. Math. 100, 137 (1887).

[43] J. Douglas, Trans. Am. Math. Soc. 50, 71 (1940).

[44] W. Sarlet, J. Phys. A 15, 1503 (1982).

[45] R. de Ritis, G. Marmo, G. Platania and P. Scudellaro, Int. J. Theo. Phys. 22, 931 (1983).

[46] M. Henneaux, Ann. Phys. (NY), 140, 45 (1982).

[47] V. Dodonov, V.I. Man'ko and V.D. Skarzhinsky, Lebedev Physical Institute Preprint No (216), Moscow (1978).

[48] L.D. Landau and E.M. Lifshitz, *Mechanics*, (Pergamon Press, London, 1960) p. 2.

[49] V. Dodonov, V.I. Man'ko and V.D. Skarzhinsky, Hadronic J. 4, 1734 (1981).

[50] J. Geicke, unpublished (2000).

[51] I. Campos, J.L. Jimenez and G. del Valle, Eur. J. Phys. 24, 469 (2003).

[52] I.S. Gradshetyn and I.M. Ryzhik, *Tables of Integrals, Series and Products*, (Academic Press, New York, N.Y. 1965) p. 925.

[53] R.A. Matzner and L.C. Shepley, *Classical Mechanics*, (Prentice Hall, Englewood Cliffs, New Jersey, 1991) Chapter 5.

[54] C.R. Chester, *Techniques in Partial Differential Equations*, (McGraw-Hill, New York, N.Y. 1971) Chapter 8.

[55] S. Srivastava, Vishwamittar and I.S. Minhar, J. Math. Phys. 32, 1510 (1991).

[56] M. Razavy, Hadronic J. 10, 7 (1987).

[57] F. Riewe, Phys. Rev. E 53, 1890 (1996).

[58] O.P. Agrawal, J. Math. Anal. Appl. 272, 368 (2002).

[59] E.M. Rabei, J. Math. Anal. Appl. 344, 799 (2008).

[60] K.S. Miller and B. Ross, *An Interoduction to the Functional Calculus and Fractional Differential Equations*, (John Wiley & Sons, New York, N.Y. 1993).

[61] R.H. Herrmann, *Fractional Calculus: An Introduction for Physicists*, (World Scientific, Singapore, 2011) Chapter 4.

Chapter 3

Hamiltonian Formulation

For conservative as well as certain dissipative motions the Lagrange method provides an elegant and concise formulation of dynamics. However as we have seen in the last chapter, not all forms of dissipative systems can be described by means of a Lagrangian. Whereas in the Lagrangian description, for a system of N degrees of freedom, there are N differential equations of second order, in the Hamiltonian formulation of the same system we have $2N$ first order differential equations.

As in the case of conservative systems, for the dissipative systems the Hamiltonian formulation has the following advantages:

(1) The equal footing of the canonical variables provides for a larger choice of transformations.

(2) The Hamiltonian is a function of the variables alone and not of their derivatives.

(3) For quantizing the system the Hamiltonian form is more convenient. As we will see in the following sections, we can introduce the Hamiltonian in a number of ways, and for a dissipative motion, just as in the Lagrangian formulation, there will be no unique Hamiltonian form.

3.1   Inverse Problem for the Hamiltonian

In the last chapter we observed that we can solve the inverse problem of classical dynamics by finding the solution of a linear partial differential equation for the Lagrangian L if the forces acting on the system are known. We now turn to an investigation of the Hamiltonian formulation, where for a system of N degrees of freedom, we find $2N$ first order differential equations. But unlike the previous inverse problem, i.e. determination of the Lagrangian, the partial differential equation for obtaining the Hamiltonian is nonlinear and is of second order. As in the case of Lagrangian, the Hamiltonian for a given force law is not unique. However for conservative systems we can choose the Hamiltonian to be the generator of motion as well as a specific constant of motion, viz, the energy of the system. But when dissipative forces are present, the second condition cannot be met, if our aim is to describe the complete motion by a single Hamiltonian. Having already defined the canonical momentum by

$$p_j = \frac{\partial L}{\partial x_j}, \quad j = 1, 2, \cdots N, \tag{3.1}$$

we define the Hamiltonian $H\,(x_j, p_j, t)$ by the Legendre transformation

$$H\,(x_j, p_j, t) = \left(\sum_k \dot{x}_k p_k - L \right)_{\dot{x}_k = \dot{x}_k(p_j)}, \tag{3.2}$$

where the subscript $\dot{x}_k = \dot{x}_k(p_j)$ means that we replace all $\dot{x}_j$ s on the right hand side of (3.2) by p_k s using Eq. (3.1).

We can also write the action S in terms of the Hamiltonian

$$S\,(x_j, p_j) = \int_{t_1}^{t_2} \left[\sum_j \{ \dot{x}_j\,(p_k)\,p_j \} - H\,(x_j, p_j, t) \right] dt. \tag{3.3}$$

Here x_j s and p_j s are assumed to be independent variables [1],[2]. By setting the functional derivatives of $S(x_j, p_j)$ with respect to x_j and p_j equal to zero we find the Hamilton canonical equations. For a system of particles interacting via the potential $V(x_1 \cdots x_N, t)$ with the Lagrangian (2.95), the Hamiltonian (3.2) reduces to

$$H = \sum_{j=1}^{N} \frac{p_j^2}{2m_j} + V(x_1, x_2 \cdots x_N, t) = \sum_{j=1}^{N} \frac{p_j^2}{2m_j} + V(x_1, x_2 \cdots x_N, t). \tag{3.4}$$

But we can also define H as a function of p_j s, x_j s and t in such a way that by eliminating p_j s between Hamilton's canonical equations

$$\dot{p}_j(t) = -\frac{\partial H(x_k, p_k, t)}{\partial x_j}, \tag{3.5}$$

$$\dot{x}_j(t) = \frac{\partial H(x_k, p_k, t)}{\partial p_j}, \tag{3.6}$$

we get the equations of motion in the coordinate space

$$m\ddot{x}_j(t) = f_j\,(x_1, x_2 \cdots x_N; \dot{x}_1, \dot{x}_2 \cdots \dot{x}_N, t)\,, \quad j = 1, 2 \cdots N. \tag{3.7}$$

This definition of H contains (3.4) as a special case. Using the latter definition we can extend the Hamiltonian formulation to dissipative systems, where unlike Eq. (3.4), H does not represent the energy of the system, and p_j, in general is not the same as the mechanical momentum $m_j \dot{x}_j$. Here as in the case of Lagrangian formulation there are infinitely many Hamiltonians generating the same equation of motion in coordinate space. We call these Hamiltonians q-equivalent, implying that when H is substituted in the canonical equations (3.5) and (3.6) and p_j s are eliminated we obtain Eq. (3.7) [3]–[6]. We note that in this formulation p_j s play the role of dummy variables. In classical dynamics these q-equivalent Hamiltonians are all acceptable for the variational formulation and the solution of the equations of motion. However we can impose other conditions on the Hamiltonian function. For instance we may require that $H(x_j, p_j, t)$ should generate the equations of motion in phase space in which case only a subset of q-equivalent Hamiltonians will be acceptable.

Examples of q-equivalent Hamiltonian for an underdamped harmonic oscillator are given later in this chapter, but first let us consider the partial differential equation for the most general Hamiltonian when the force $f(x, \dot{x})$ does not depend explicitly on time. We note from (3.5) and (3.6) that for one-dimensional motion the acceleration can be written in terms of the partial derivatives of H:

$$\ddot{x}(t) = \left[\frac{\partial^2 H}{\partial t \partial p} + \left(\frac{\partial^2 H}{\partial x \partial p} \right) \left(\frac{\partial H}{\partial p} \right) - \left(\frac{\partial^2 H}{\partial p^2} \right) \left(\frac{\partial H}{\partial x} \right) \right]. \tag{3.8}$$

Substituting this expression and (3.6) in the equation of motion $m\ddot{x} = f(x, \dot{x})$ we find

$$m \left[\frac{\partial^2 H}{\partial t \partial p} + \left(\frac{\partial^2 H}{\partial x \partial p} \right) \left(\frac{\partial H}{\partial p} \right) - \left(\frac{\partial^2 H}{\partial p^2} \right) \left(\frac{\partial H}{\partial x} \right) \right] = f \left(x, \frac{\partial H}{\partial p} \right). \tag{3.9}$$

For instance when $f(x, \dot{x})$ is given by (2.138) we obtain the equation for H to be

$$m \left[\left(\frac{\partial^2 H}{\partial x \partial p} \right) \left(\frac{\partial H}{\partial p} \right) - \left(\frac{\partial^2 H}{\partial p^2} \right) \left(\frac{\partial H}{\partial x} \right) \right] + \frac{\partial V(x)}{\partial x} + m\gamma \left(\frac{\partial H}{\partial p} \right)^2 = 0, \tag{3.10}$$

and in this case one of the solutions of (3.10) is given by

$$H = \frac{1}{2m} p^2 e^{-2\gamma x} + \int^x e^{2\gamma y} \frac{\partial V(y)}{\partial y} dy. \tag{3.11}$$

We can generalize (3.9) when there are a number of degrees of freedom. In this general case (3.9) is replaced by

$$\frac{\partial^2 H}{\partial t \partial p_k} + \sum_j \left[\left(\frac{\partial^2 H}{\partial x_j \partial p_k} \right) \left(\frac{\partial H}{\partial p_j} \right) - \left(\frac{\partial^2 H}{\partial p_j \partial p_k} \right) \left(\frac{\partial H}{\partial x_j} \right) \right] = \frac{1}{m} f_k \left(x_i, \frac{\partial H}{\partial p_i} \right). \tag{3.12}$$

Since Eqs. (3.9) and (3.12) are nonlinear partial differential equations their most general solutions are not known. However for very special cases we can find solutions for (3.9).

3.2 Nonuniqueness of the Hamiltonian for Dissipative Motions in the Coordinate Space and in the Phase Space

Now we want to illustrate the nonuniqueness of the Hamiltonian by constructing a number of Hamiltonians for simple systems. Let us start with the case where the frictional force present is linear in velocity and there are no conservative forces, i.e. $f = -m\lambda\dot{x}$. In this case we can find a number of q-equivalent Hamiltonians, where q-equivalent Hamiltonians were defined in the previous section.

(1) If we choose H to be independent of x, then for a one-dimensional motion (3.9) reduces to [7],[8]

$$\frac{\partial}{\partial p}\left(\frac{\partial H}{\partial t} + \lambda H\right) = 0, \tag{3.13}$$

and the most general solution of this equation is

$$H(p,t) = h(p)\exp(-\lambda t), \tag{3.14}$$

where $h(p)$ is a differentiable but otherwise an arbitrary function of p.

(2) If we assume that H can be written as the sum of two terms

$$H(p,x) = h_1(p) + h_2(x), \tag{3.15}$$

then the solution of (3.9) is

$$H(p,x) = \mathcal{E}\exp\left(\frac{p}{p_0}\right) + \lambda p_0 x, \tag{3.16}$$

where $\mathcal{E}$ and p_0 are constants.

(3) Finally Dodonov and collaborators have considered a Hamiltonian which is of the form $H(t,xp)$. For this Hamiltonian Eq. (3.9) reduces to

$$\frac{\partial^2 H}{\partial t \partial \xi} + \left(\frac{\partial H}{\partial \xi}\right)^2 + \lambda\left(\frac{\partial H}{\partial \xi}\right) = 0, \tag{3.17}$$

where $\xi = xp$. By changing $\frac{\partial H}{\partial \xi}$ to u we find the differential equation satisfied by u

$$\frac{du}{dt} + \lambda u + u^2 = 0. \tag{3.18}$$

This equation can be integrated and the result is

$$u = \frac{\partial H}{\partial \xi} = \frac{\lambda}{\exp[\lambda(t-\alpha)] - 1}, \tag{3.19}$$

or

$$H = \lambda \int^{xp} \frac{\lambda d\xi}{\exp[\lambda(t - \alpha(\xi))] - 1},\tag{3.20}$$

where $\alpha(\xi)$ is an arbitrary function of its argument.

Since the phase space formulation plays an essential role in quantum-classical correspondence, we need to study the non-uniqueness of the Hamiltonian formalism for the damped motion in phase space. For this group of problems it is more difficult to construct qp-equivalent Hamiltonians (for the definition of these see below).

For the simple systems with Stokes type resistive force, i.e. a force proportional to $\dot{x}(t)$, Eq. (1.75), we can find explicitly time-dependent Lagrangian and Hamiltonian or time-independent Hamiltonian. Let us consider the Lagrangian function given by Cardirola [9] and by Havas [10] (see Eq. (2.136))

$$L = \frac{m}{2} \left[\dot{x}^2 + \int^x f(x') \, dx' \right] \exp(\lambda t),\tag{3.21}$$

where $\dot{x}(t)$ is the velocity of the particle and $f(x)$ is the conservative force acting on it. Substituting this Lagrangian in Euler-Lagrange differential equation, i.e. [1]

$$\frac{\partial L}{\partial x} - \frac{d}{dt}\left(\frac{\partial L}{\partial \dot{x}}\right) = 0,\tag{3.22}$$

we find the equation of motion (1.75). We note that in this formulation the canonical momentum p is explicitly time-dependent and is given by

$$p = \left(\frac{\partial L}{\partial \dot{x}}\right) = m\dot{x}e^{\lambda t}.\tag{3.23}$$

Hamiltonian for One-Dimensional Linearly Damped Motion — We find the Hamiltonian from (3.21) and (3.23);

$$H(x,p,t) = (p\dot{x} - L)_{\dot{x}=\dot{x}(p,t)} = \frac{p^2}{2m}e^{-\lambda t} - e^{\lambda t}\int^x f(x')\, dx'.\tag{3.24}$$

In particular, for a harmonically bound particle $V(x) = \frac{1}{2}m\omega_0^2 x^2$ we have

$$H_1(x,p,t) = \frac{p^2}{2m}e^{-\lambda t} + \left(\frac{m}{2}\right)e^{\lambda t}\omega_0^2 x^2.\tag{3.25}$$

For these systems the Hamiltonian function is not unique. For instance another Hamiltonian which does not explicitly depend on time and generates the same motion in coordinate space as (3.25) is:

$$H_2(x,p) = -\left(\frac{\lambda}{2}\right)xp - \ln\cos(\omega px) + \ln x,\tag{3.26}$$

where $\omega^2 = \omega_0^2 - \frac{\lambda^2}{4}$. This Hamiltonian corresponds to the Lagrangian given by Eq. (2.176).

Let us discuss some of the similarities and differences of the Hamiltonians $H_1(x, p, t)$ and $H_2(x, p)$. Both of these Hamiltonians generate the same equations of motion in coordinate space, viz,

$$m\ddot{x} + m\lambda\dot{x} + m\left(\omega^2 + \frac{\lambda^2}{4}\right)x = 0, \tag{3.27}$$

but if we eliminate x and $\dot{x}$ between the Hamilton canonical equations

$$\dot{p} = -\frac{\partial H}{\partial x}, \text{ and } \dot{x} = \frac{\partial H}{\partial p}, \tag{3.28}$$

then we find an equation for p which is different from the standard one obtained by differentiating (3.27) with respect to t.

From the Hamiltonian $H_1(x, p, t)$ and the Hamilton canonical equations we find the equation of motion for the canonical momentum to be;

$$m\ddot{p} - m\lambda\dot{p} + m\left(\omega^2 + \frac{\lambda^2}{4}\right)p = 0. \tag{3.29}$$

It is important to note that the equations for x and p are not the same, and therefore p, the canonical momentum, is not the same as the mechanical momentum $m\dot{x}$, nor it satisfies the equation for p derived from $H_2(x, p)$.

Thus $H_1(x, p, t)$ and $H_2(x, p)$ are q-equivalent Hamiltonians, but they give rise to different motions in phase space. Both $H_1(x, p, t)$ and $H_2(x, p)$ are not invariant under time-reversal transformation $t \to -t$, and this is expected since the equation of motion (3.27) is not invariant under this transformation. However $H_2(x, p)$ is invariant under time translation transformation, i.e. $t \to t + t_0$ whereas $H_1(x, p, t)$ is not. In fact for $H_2(x, p)$ we have

$$\frac{dH_2}{dt} = \frac{\partial H_2}{\partial t} + \{H_2, H_2\} = 0. \tag{3.30}$$

This result implies that H_2 is a constant of motion and therefore cannot be the energy of the system.

We can also regard $H_1(x, p, t)$ as the Hamiltonian describing an oscillating system whose mass is increasing with time

$$m(t) = m \exp(\lambda t). \tag{3.31}$$

For example consider an empty bucket attached with a rope to a fixed point and is oscillating while it is collecting raindrops. If the mass of this system increases exponentially like (3.31) then its motion can be described by $H_1(x, p, t)$.

q-Equivalent Hamiltonians — Different Hamiltonians that generate the same equations of motion in coordinate space, called q-equivalent Hamiltonians are related by canonical transformations. For instance consider the two Hamiltonians given by (Eqs. (3.16) and (3.24))

$$H_1 = \frac{p^2}{2m}e^{-\lambda t}, \tag{3.32}$$

and

$$H_2 = \mathcal{E}_2 \exp\left(\frac{P}{p_0}\right) + \lambda p_0 X. \tag{3.33}$$

Both of these give us the equation of motion $m\ddot{x} + m\lambda\dot{x} = 0$, and therefore are q-equivalent. The generator of the canonical transformation relating these two is given by [1],[7]

$$F_3(p, X, t) = X(\lambda p_0 t - p) + \frac{1}{\lambda}\left(1 - e^{-\lambda t}\right)\left[\mathcal{E}\exp\left(\frac{p}{p_0}\right) - \frac{p^2}{2m}\right]. \tag{3.34}$$

Using this generating function $F_3(p, X, t)$ we can connect the old coordinate and the old momentum, x and p to the new ones X and P;

$$P = -\frac{\partial F_3}{\partial X} = p - \lambda p_0 t, \tag{3.35}$$

$$x = -\frac{\partial F_3}{\partial p} = X - \frac{1}{\lambda}\left(1 - e^{-\lambda t}\right)\left[\frac{\mathcal{E}}{p_0}\exp\left(\frac{p}{p_0}\right) - \frac{p}{m}\right], \tag{3.36}$$

and

$$H_2 = H_1 + \frac{\partial F_3}{\partial t}. \tag{3.37}$$

Since

$$\frac{\partial F_3}{\partial t} = \lambda X p_0 + e^{-\lambda t}\left[\mathcal{E}\exp\left(\frac{p}{p_0}\right) - \frac{p^2}{2m}\right], \tag{3.38}$$

we have

$$H_1(x, p, t) + \frac{\partial F_3}{\partial t} = \lambda p_0 X + \mathcal{E}e^{-\lambda t}\exp\left(\frac{P + \lambda p_0 t}{p_0}\right) = H_2(X, P). \tag{3.39}$$

We note that the new coordinate X is a combination of the old coordinate x and the old momentum p, as well as time.

Examples of qp-Equivalent Hamiltonians — Other q-equivalent Hamiltonians can be found from the equivalent Lagrangians, for example from (2.162) and (2.168) we obtain;

$$H = \frac{1}{3mp_0}(2p_0 p)^{\frac{3}{2}} e^{-\lambda t}, \tag{3.40}$$

and this Hamiltonian is qp-equivalent to H_1, Eq. (3.32), for linear damping. For quadratic damping we have

$$H = \frac{m^2}{8}\left(\frac{6p}{m^2}\right)^{\frac{4}{3}} e^{\frac{-4\gamma x}{3}}. \tag{3.41}$$

which is qp-equivalent to

$$H = \frac{1}{2m}p^2 e^{-2\gamma x}. \tag{3.42}$$

3.3   Ostrogradsky's Method

We have already seen that the equations of motion for a radiating electron has a damping term proportional to $\left(\frac{d^3 x}{dt^3}\right)$ and possibly higher derivatives. For dissipative dynamical systems where forces do depend on higher derivatives of acceleration, we can construct the Hamiltonian function using a method due to Ostrogradsky [11]-[13].

The starting point in this construction is to obtain the extremum of the action integral

$$S = \int L\left(x, x^{(1)}, x^{(2)}, \cdots x^{(j)}, t\right) dt. \tag{3.43}$$

where $x^{(j)}$ denotes the jth time derivative of x. Here for the sake of simplicity we consider a one-dimensional motion, but the method can be generalized to more dimensions in a straightforward way.

By requiring that the functional S in (3.43) be stationary, i.e. $\delta S \equiv 0$, we find that the integrand L must satisfy the differential equation

$$\frac{\partial L}{\partial x} - \frac{d}{dt}\left(\frac{\partial L}{\partial x^{(1)}}\right) + \cdots + (-1)^j \frac{d^j}{dt^j}\left(\frac{\partial L}{\partial x^{(j)}}\right) = 0. \tag{3.44}$$

To construct the Hamiltonian we first introduce the canonical momenta $p_1 \cdots p_j$ by the following relations:

$$p_1 = \frac{\partial L}{\partial x^{(1)}} - \frac{d}{dt}\left(\frac{\partial L}{\partial x^{(2)}}\right) + \cdots + (-1)^{j-1}\frac{d^{j-1}}{dt^{j-1}}\left(\frac{\partial L}{\partial x^{(j)}}\right), \tag{3.45}$$

$$p_2 = \frac{\partial L}{\partial x^{(2)}} - \frac{d}{dt}\left(\frac{\partial L}{\partial x^{(3)}}\right) + \cdots + (-1)^{j-2}\frac{d^{j-2}}{dt^{j-2}}\left(\frac{\partial L}{\partial x^{(j)}}\right), \tag{3.46}$$

$$\cdots\cdots\cdots\cdots\cdots\cdots$$

$$p_j = \frac{\partial L}{\partial x^{(j)}}. \tag{3.47}$$

We also define a set of canonical coordinates $q_1 \cdots q_j$ by

$$q_1 = x, \quad q_2 = x^{(1)}, \quad \cdots \quad q_j = x^{(j-1)}. \tag{3.48}$$

In terms of these p_i s and q_i s we can write the Hamiltonian function $H\left(q_1, \cdots q_j; p_1 \cdots p_j, t\right)$ as

$$H = -L + p_1 q_2 + p_2 q_3 + \cdots + p_{j-1}q_j + p_j x^{(j)}, \tag{3.49}$$

where in the last term of (3.49), $x^{(j)}$ should be replaced in terms of other p_i s and q_i s by solving

$$p_j = \frac{\partial L}{\partial x^{(j)}}, \tag{3.50}$$

for $x^{(j)}$.

Let us now consider the small variations $\delta q_1, \cdots \delta q_j, \delta p_1 \cdots \delta p_j$ of q_i s and p_i s and the corresponding variation in H,

$$\delta H = -\sum_{i=0}^{j-1} \frac{\partial L}{\partial x^{(j)}} \delta q_{j+1} - \frac{\partial L}{\partial x^{(j)}} \delta x^{(j)} + \sum_{i=1}^{j-1} p_j \delta q_{i+1} + p_j \delta x^{(j)}$$

$$+ \sum_{i=1}^{j-1} q_{i+1} \delta p_i + x^{(j)} \delta p_j. \tag{3.51}$$

From Eqs. (3.44)-(3.47) we have

$$\frac{\partial L}{\partial x} = \dot{p}_1, \quad \frac{\partial L}{\partial x^{(1)}} = \dot{p}_2 + p_1, \quad \frac{\partial L}{\partial x^{(2)}} = \dot{p}_3 + p_2, \quad \frac{\partial L}{\partial x^{(j)}} = p_j. \tag{3.52}$$

Substituting these in (3.51) we find

$$\delta H = -\sum_{i=1}^{j} \dot{p}_i \delta q_i + \sum_{i=1}^{j} \dot{q}_i \delta p_i. \tag{3.53}$$

Now we compare (3.53) with the variation of H, i.e. with

$$\delta H \left(q_1 \cdots q_j; p_1, \cdots p_j, t \right) = \sum_{i=1}^{j} \left(\frac{\partial H}{\partial q_i} \delta q_i + \frac{\partial H}{\partial p_i} \delta p_i \right), \tag{3.54}$$

and we obtain the Hamilton canonical equations

$$\dot{q}_i = \frac{\partial H}{\partial p_i}, \quad \dot{p}_i = -\frac{\partial H}{\partial q_i}. \tag{3.55}$$

Ostrogradsky's Method Applied to a Simple Problem — As an example of the application of this method let us consider a slightly different model for radiating electron than the one given by Eq. (1.89). Here we want to find the Hamiltonian for the three-dimensional motion of an electron subject to the potential $V(q,t)$, where the equations of motion are given by [14],[15]

$$m \left[\left(\frac{d^2 x_i}{dt^2} \right) - \tau \left(\frac{d^3 x_i}{dt^3} \right) + \left(\frac{\tau}{2} \right)^2 \left(\frac{d^4 x_i}{dt^4} \right) \right] = -\frac{\partial V}{\partial x_i}, \quad i = 1, 2, 3. \tag{3.56}$$

These equations can be derived from the Lagrangian

$$L = -\left[\frac{\tau^2}{8} \sum_i m \left(\frac{d^2 x_i}{dt^2} \right)^2 + V \right] \exp\left[-\frac{2t}{\tau} \right]. \tag{3.57}$$

Since the equation of motion is of the fourth order we can construct the generalized canonical variables by Ostrogradsky method [11],[12]

$$q_i = x_i, \quad p_i = \frac{\partial L}{\partial \dot{q}_i} - \frac{d}{dt} \left(\frac{\partial L}{\partial \ddot{q}_i} \right),$$

$$Q_i = \dot{x}_i, \quad P_i = \frac{\partial L}{\partial \ddot{q}_i} - \frac{d}{dt} \left(\frac{\partial L}{\partial \frac{d^3 q_i}{dt^3}} \right). \tag{3.58}$$

With the help of these canonical variables we can write the Hamiltonian as

$$H = \sum_i p_i Q_i - \frac{2}{m\tau^2} \exp\left(\frac{2t}{\tau}\right) \sum_i' P_i^2 + \exp\left(-\frac{2t}{\tau}\right) V(q,t). \tag{3.59}$$

For other attempts to formulate the Lagrangian and the Hamiltonian for a radiative electron see Infeld [16] and also Englert [17]. In the latter work, using Ostrogradsky's method the Hamiltonian for the third order equation of motion is constructed and then quantized.

3.4   Dekker's Complex Coordinate Formulation

Another method using complex coordinates was suggested by Dekker [18],[19], which unlike the model studied in Sec. 2.6 has no direct connection to the optical potential model.

Consider the underdamped harmonic oscillator of unit mass with the equation of motion

$$\ddot{x} + \lambda\dot{x} + \left(\omega^2 + \frac{\lambda^2}{4}\right) x = 0, \tag{3.60}$$

and let us introduce the complex coordinate $q(t)$ by

$$q(t) = \frac{1}{\sqrt{\omega}}\left[p(t) + \left(\frac{\lambda}{2} - i\omega\right) x(t)\right], \tag{3.61}$$

where

$$p(t) = \dot{x}. \tag{3.62}$$

By differentiating $q(t)$ and using Eq. (3.62) we find that $q(t)$ is the solution of the first order differential equation

$$\dot{q}(t) + i\omega q(t) + \frac{\lambda}{2}q(t) = 0. \tag{3.63}$$

Assuming that (3.63) rather than (3.60) is the equation of motion, then the equation for $q(t)$ can be derived from the complex Lagrangian

$$\mathsf{L} = \frac{i}{2}[q^*(t)\dot{q}(t) - q(t)\dot{q}^*(t)] - \left(\omega - i\frac{\lambda}{2}\right)q^*(t)q(t). \tag{3.64}$$

Next we find the conjugate to the complex coordinate $q(t)$ which we denote by $\pi(t)$

$$\pi(t) = \frac{\partial \mathsf{L}}{\partial \dot{q}} = \frac{i}{2}q^*(t). \tag{3.65}$$

From the expression for the Lagrangian L and the canonical momentum $\pi(t)$ we find the Hamiltonian H in terms of $\pi(t)$ and $q(t)$;

$$\mathsf{H} = H_1 + i\Gamma = -\left(i\omega + \frac{\lambda}{2}\right)\pi(t)q(t). \tag{3.66}$$

We can express this Hamiltonian in terms of the original variables $x(t)$ and $p(t)$. To obtain such an expression for H, we note that $q(t)$ is given by Eq. (3.61) and $\pi(t)$ according to (3.65) is expressible as

$$\pi(t) = \frac{i}{2\sqrt{\omega}} \left[p(t) + \left(i\omega + \frac{\lambda}{2} \right) x(t) \right], \qquad (3.67)$$

according to (3.65). Now by substituting for $\pi(t)$ and $q(t)$ in (3.66) we find that

$$\mathsf{H} = \frac{1}{2} \left\{ p^2 + \left(\omega^2 + \frac{\lambda^2}{4} \right) x^2 + \left(\frac{\lambda}{2} xp \right) + i\Gamma \right\}, \qquad (3.68)$$

is the complex Hamiltonian of the system. From this Hamiltonian we obtain the equation of motion (3.60) for $x(t)$ and a similar equation for $p(t)$;

$$\ddot{p} + \lambda\dot{p} + \left(\omega^2 + \frac{\lambda^2}{4} \right) p = 0. \qquad (3.69)$$

We note that the Hamiltonian (3.68), apart from a constant term, can be found by other methods, e.g. see Eq. (15.73).

3.5 Hamiltonian Formulation of the Motion of a Particle with Variable Mass

We have seen that the Kanai-Caldiroa Hamiltonian generates the classical equation of motion for the coordinate of the particle when the motion takes place in a viscous medium with a drag force proportional to the velocity of the particle. However a different interpretation can be given to this Hamiltonian, viz, that it generates the equation for motion of a particle with a variable (time-dependent) mass [20]-[23].

We write the Hamiltonian for the variable mass system as

$$H = H(x, p; \ \theta, m; t), \qquad (3.70)$$

where in this generalized form we consider the mass m (measured in units of energy, i.e. mc^2, with $c = 1$), and the "proper time" θ as its conjugate variables. That is p and m should be treated as momenta and x and θ as their conjugates. Thus we have the following canonical equations for such a Hamiltonian:

$$\dot{x} = \frac{\partial H}{\partial p}, \quad \dot{\theta} = \frac{\partial H}{\partial m}, \qquad (3.71)$$

and

$$\dot{p} = -\frac{\partial H}{\partial x}, \quad \dot{m} = -\frac{\partial H}{\partial \theta}. \qquad (3.72)$$

According to this formulation a potential which depends on x provides a force, thus causing a change in momentum, and similarly a potential which depends on θ causes a change in the mass of the particle.

Let us consider the specific example of a harmonically bound particle with a variable mass which increases exponentially with time

$$m = m_0 e^{\lambda t}. \tag{3.73}$$

The Hamiltonian in this case is

$$H = \frac{p^2}{2m} + \frac{1}{2} m \omega_0^2 x^2 - \lambda m \theta, \tag{3.74}$$

and from this Hamiltonian we find the canonical equations of motion to be

$$\dot{x} = \frac{\partial H}{\partial p} = \frac{p}{m}, \tag{3.75}$$

$$\dot{\theta} = \frac{\partial H}{\partial m} = -\frac{p^2}{2m^2} + \frac{1}{2} \omega_0^2 x^2 - \lambda \theta, \tag{3.76}$$

$$\dot{m} = -\frac{\partial H}{\partial \theta} = \lambda m, \tag{3.77}$$

and

$$\dot{p} = -\frac{\partial H}{\partial x} = -m \omega_0^2 x. \tag{3.78}$$

Equation (3.77) can be integrated and the result is given by (3.73). By differentiating (3.75) with respect to t and substituting for $\dot{p}$ and $\dot{m}$ from (3.77) and (3.78) we find that the equation of motion for x is;

$$\ddot{x} + \lambda \dot{x} + \omega_0^2 x = 0. \tag{3.79}$$

Similarly we obtain the equation of motion for p to be

$$\ddot{p} + \lambda \dot{p} + \omega_0^2 p = 0. \tag{3.80}$$

Thus H given by (3.74) is qp-equivalent to Dekker's complex Hamiltonian and gives us the correct equations for the motion in phase space.

3.6   Variable Mass Oscillator

A system with variable mass for which the equation of motion is solvable and can be tested in laboratory is that of an oscillator where the mass is decreasing uniformly and is subject to a frictional force proportional to velocity.

Let us consider a container filled with sand and attached to a fixed point by means of a spring with a spring constant K. The equation of motion of this system is [24]

$$m(t) \frac{d^2 x}{dt^2} + b \frac{dx}{dt} + K x = 0, \tag{3.81}$$

where

$$m(t) = m_0 - \beta t, \tag{3.82}$$

and b is the coefficient of friction. Thus $\beta = -dm/dt$ is the rate with which the system is losing its mass. For the equation of motion (3.81) with $m(t)$ defined by (3.82) we find a Hamiltonian similar to H defined by (3.74), i.e.

$$H = \frac{1}{2m(t)}p^2 + \frac{1}{2}Kx^2 - b\theta. \tag{3.83}$$

Using the canonical equations (3.75)-(3.78) and eliminating p and $\dot{p}$ between the resulting equations we obtain (3.81). Also by multiplying Eq. (3.81) by dx/dt we can find the rate of change of mechanical energy of the system;

$$\frac{d}{dt}(E_k + E_p) = \left(\frac{1}{2}\frac{dm(t)}{dt} - b\right)\left(\frac{dx}{dt}\right)^2 = 0, \tag{3.84}$$

where $E_k = (m(t)/2)\dot{x}^2$ and $E_p = (K/2)x^2$ are the kinetic and potential energies respectively. Equations (3.81),(3.82) can be solved analytically. We write the solution as the product of an amplitude and a time-dependent phase [24]

$$x(t) = A_0 f(t) \sin[h(t) + \phi], \tag{3.85}$$

where A_0 and ϕ are constants, and we substitute $x(t)$, Eq. (3.85) in (3.81) and equate the coefficients of $\sin[h(t)]$ and $\cos[h(t)]$ separately equal to zero to find

$$m(t)\left[\ddot{f} - \left(\dot{h}\right)^2 f\right] + b\dot{f} + Kf = 0, \tag{3.86}$$

and

$$m(t)\left[2\dot{f}\dot{h} + f\ddot{h}\right] + bf\dot{h} = 0. \tag{3.87}$$

The solutions of these differential equations are given by:

$$f(t) = \left(1 - \frac{\beta t}{m_0}\right)^\alpha, \tag{3.88}$$

and

$$h(t) = \frac{2\mu}{\beta}\left[\tan^{-1}\sqrt{\frac{1 - a - \frac{\beta t}{m_0}}{a}} - \sqrt{\frac{1 - a - \frac{\beta t}{m_0}}{a}}\right], \tag{3.89}$$

where

$$\alpha = \frac{b}{2\beta} + \frac{1}{4}, \tag{3.90}$$

$$\mu = \frac{1}{2}\sqrt{\left(b + \frac{\beta}{2}\right)\left(b + \frac{3}{2}\beta\right)}, \tag{3.91}$$

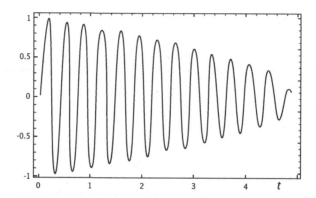

Figure 3.1: Damped oscillations of a system with variable mass described by Eqs. (3.81) and (3.82).

and

$$a = \frac{\mu^2}{m_0 K}.$$ (3.92)

In Fig. 3.1 the damped oscillations of a container with variable mass connected to a spring is shown. For this calculation the constants, $b = 0.1$, $\beta = 0.2$, $K = 2.5$ and $m_0 = 1$ have been used.

3.7 Bateman's Damped-Amplified Harmonic Oscillators

A different two-dimensional model which as a whole is conservative is the Bateman's system which is described by the Lagrangian [25]-[28]

$$L = m\dot{x}\dot{y} + \frac{m}{2}\lambda(x\dot{y} - \dot{x}y) - m\left(\omega^2 + \frac{\lambda^2}{4}\right)xy.$$ (3.93)

From this Lagrangian we find the classical equations of motion to be

$$m\ddot{x} + m\lambda\dot{x} + m\left(\omega^2 + \frac{\lambda^2}{4}\right)x = 0,$$ (3.94)

and

$$m\ddot{y} - m\lambda\dot{y} + m\left(\omega^2 + \frac{\lambda^2}{4}\right)y = 0.$$ (3.95)

We note that the equation for y is the time-reversed of the x motion. Thus while the x oscillator is losing energy, the y oscillator is gaining, so that the total energy of the system is conserved. This also follows from the fact that the Hamiltonian

$$H = (p_x\dot{x} + p_y\dot{y} - L)_{\dot{x}(p_x,p_y),\dot{y}(p_x,p_y)},$$ (3.96)

is a constant of motion. We can write H in terms of the canonical momenta p_x and p_y as

$$H = \frac{p_x p_y}{2m} - \frac{\lambda}{2}(x p_x - y p_y) + m\omega^2 xy, \qquad (3.97)$$

and since H does not depend on time explicitly, therefore $dH/dt = 0$.

An Alternative Formulation of Bateman's Hamiltonian — We can also write the Bateman Lagrangian (3.93) in a compact form

$$L = \frac{m}{2} \sum_{i,j=1}^{2} \left(g_{ij} \dot{x}_i \dot{x}_j - \lambda \epsilon_{ij} x_i \dot{x}_j - g_{ij} \omega_0^2 x_i x_j \right). \qquad (3.98)$$

In this equation g_{ij} is pseudo-Euclidean metric;

$$g_{ij} = \begin{bmatrix} 1 & 0 \\ 0 & -1 \end{bmatrix}, \qquad (3.99)$$

and ϵ_{ij} is the Levi-Civita symbol

$$\epsilon_{ij} = \begin{bmatrix} 0 & 1 \\ -1 & 0 \end{bmatrix}, \qquad (3.100)$$

From this Lagrangian we find the following Hamiltonian [29]

$$H = \frac{1}{2m}\left(p_1 - \frac{1}{2}m\lambda x_2 \right)^2 + \frac{1}{2}m\omega_0^2 x_1^2$$
$$- \frac{1}{2m}\left(p_2 + \frac{1}{2}m\lambda x_1 \right)^2 - \frac{1}{2}m\omega_0^2 x_2^2, \qquad (3.101)$$

where

$$p_1 = m\left(\dot{x}_1 + \frac{1}{2}\lambda x_2 \right), \qquad (3.102)$$

and

$$p_2 = -m\left(\dot{x}_2 + \frac{1}{2}\lambda x_1 \right), \qquad (3.103)$$

are the canonical momenta conjugate to x_1 and x_2 respectively. By introducing the canonical transformation from $(x_1, x_2; p_1, p_2)$ to $(x_+, x_-; p_+, p_-)$ where

$$p_{\pm} = \left(\frac{\omega_{\pm}}{2m\omega} \right)^{\frac{1}{2}} p_1 \pm i \left(\frac{m\omega_{\pm}\omega}{2} \right)^{\frac{1}{2}} x_2, \qquad (3.104)$$

and

$$x_{\pm} = \left(\frac{m\omega}{2\omega_{\pm}} \right)^{\frac{1}{2}} x_1 \pm i \left(\frac{1}{2m\omega\omega_{\pm}} \right)^{\frac{1}{2}} p_2, \qquad (3.105)$$

with ω and $\omega_{\pm}$ given by

$$\omega = \left(\omega_0^2 - \frac{1}{4}\lambda^2 \right)^{\frac{1}{2}}, \qquad (3.106)$$

and

$$\omega_\pm = \omega \pm i\frac{\lambda}{2}, \tag{3.107}$$

we can transform the Hamiltonian and write it as a sum of two terms:

$$H = H_+ + H_-, \tag{3.108}$$

where

$$H_\pm = \frac{1}{2}p_\pm^2 + \frac{1}{2}\omega_\pm^2 x_\pm^2. \tag{3.109}$$

We will see later how this Hamiltonian can be quantized and what are the eigenvalues of this system.

Bateman's Lagrangian for a Radiating Electron — The Bateman approach for constructing the Lagrangian for a dissipative system has been extended to include the classical motion of a radiating electron by Englert [17]. Again let us consider the motion of a harmonically bound radiating electron Eq. (1.84) which we write as

$$m\frac{d^2x}{dt^2} = -m\omega_0^2 x + m\tau\left(\frac{d^3x}{dt^3}\right). \tag{3.110}$$

We can write the Lagrangian either as

$$L = -\frac{1}{2}m\tau\left(\ddot{x}\dot{y} - \dot{x}\ddot{y}\right) + m\dot{x}\dot{y} - m\omega_0^2 xy, \tag{3.111}$$

or alternatively the equivalent Lagrangian which is asymmetric in x and y

$$L_1 = -m\tau\dot{x}\ddot{y} + m\dot{x}\dot{y} - m\omega_0^2 xy. \tag{3.112}$$

Both of these Lagrangians generate the same equation of motion for x and y, viz, Eq. (3.110) and its time-reversed motion.

$$m\frac{d^2y}{dt^2} = -m\omega_0^2 y - m\tau\left(\frac{d^3y}{dt^3}\right). \tag{3.113}$$

A Hamiltonian corresponding to the Lagrangian (3.112) which has three degrees of freedom or a six-dimensional phase space (x, p_1), (y, p_2) and (z, p_3) can be constructed with the result that

$$H = \frac{p_1 p_3 + p_2 z}{\sqrt{m\tau}} - \frac{p_3 z}{\tau} + m\omega_0^2 xy. \tag{3.114}$$

A Hamiltonian for the Damped Harmonic Oscillator with One Degree of Freedom — Following the works of Chandrasekar *et al.* and of Bender *et al.* we can replace the two Hamiltonians H_{pm}, Eq. (3.109) by H_1 and H_2 [30]-[32]. These are defined by

$$H_1 = -i\omega_1 xp + \frac{\omega_1}{\omega_1 - \omega_2}p^{1-\frac{\omega_2}{\omega_1}}, \tag{3.115}$$

and

$$H_2 = -i\omega_2 xp + \frac{\omega_2}{\omega_2 - \omega_1} p^{1 - \frac{\omega_1}{\omega_2}}, \tag{3.116}$$

where we have set the mass of the oscillator equal to unity and where $\omega_{1,2}$ are given by the complex roots of the frequency equation

$$\nu^2 + i\lambda\nu - \omega_0^2 = 0, \tag{3.117}$$

which for an underdamped oscillator are

$$\omega_{1,2} = -i\frac{\lambda}{2} \pm \bar{\omega} = -i\frac{\lambda}{2} \pm \sqrt{\omega_0^2 - \frac{\lambda^2}{4}}. \tag{3.118}$$

The canonical equations of motion for H_1 yield the following relations:

$$\dot{x} = \frac{\partial H_1}{\partial p} = -i\omega_1 x + p^{-\frac{\omega_2}{\omega_1}}, \tag{3.119}$$

and

$$\dot{p} = -\frac{\partial H_1}{\partial x} = i\omega_1 p. \tag{3.120}$$

Now by eliminating p between (3.119) and (3.120) we obtain

$$\ddot{x} + i\omega_1 \dot{x} = -i\omega_2 \left(\dot{x} + i\omega_1 x \right), \tag{3.121}$$

or

$$\ddot{x} + i(\omega_1 + \omega_2)\dot{x} - \omega_1 \omega_2 x = \ddot{x} + \lambda\dot{x} + \left(\omega^2 + \frac{\lambda^2}{4} \right) = 0. \tag{3.122}$$

We note that in this formulation H_1 and H_2 are constants of motion. By solving (3.119) for p and substituting the result in H_1 we find that

$$C_1 = \frac{(\dot{x} + i\omega_2 x)^{\omega_2}}{(\dot{x} + i\omega_1 x)^{\omega_1}}, \tag{3.123}$$

is a constant of motion. Had we used H_2 rather than H_1, then we would have obtained $C_2 = (C_1)^{-1}$, i.e. there is just one constant of motion. In the absence of damping $\lambda = 0$, we have $\omega_0 = \omega_1 = -\omega_2$, so that

$$H_1 = -i\omega_0 xp + \frac{1}{2}p^2, \tag{3.124}$$

and then the conserved quantities C_1 and C_2 become equal to $(\dot{x}^2 + \omega_0^2 x^2)^{\pm\omega_0}$, that is both are are functions of the total energy of the oscillator.

A Different Formulation of the Problem of Linearly Bound Radiating Electron — We can apply this technique to the problem of a harmonically bound radiating electron Eq.(1.84) and construct a Hamiltonian for it. The equation of motion here is of third order and we can factorize it and write it as

$$\left(\frac{d}{dt} + i\omega_1 \right) \left(\frac{d}{dt} + i\omega_2 \right) \left(\frac{d}{dt} + i\omega_3 \right) x = 0, \tag{3.125}$$

where ω_i s are the frequencies related to the the third-order differential equation

$$m\dddot{x} + m\omega_0^2 x - m\tau \ddot{x} = 0, \tag{3.126}$$

The general solution of this equation is

$$x = a_1 e^{-i\omega_1 t} + a_2 e^{-i\omega_2 t} + a_3 e^{-i\omega_3 t}, \tag{3.127}$$

where a_k s are arbitrary constants. Now from the product of the second and third factors in (3.125) i.e. $\ddot{x} + i(\omega_2 + \omega_3)\dot{x} - \omega_2\omega_3 x$ we obtain

$$a_1 e^{-i\omega_1 t} = -\frac{\left(\frac{d}{dt} + i\omega_2\right)\left(\frac{d}{dt} + i\omega_3\right)}{(\omega_1 - \omega_2)(\omega_1 - \omega_3)} x, \tag{3.128}$$

in which a_2 and a_3 do not appear. By considering the two other factors in (3.125) we find a_2 and a_3:

$$a_2 e^{-i\omega_2 t} = -\frac{\left(\frac{d}{dt} + i\omega_3\right)\left(\frac{d}{dt} + i\omega_1\right)}{(\omega_2 - \omega_3)(\omega_2 - \omega_1)} x, \tag{3.129}$$

$$a_3 e^{-i\omega_3 t} = -\frac{\left(\frac{d}{dt} + i\omega_1\right)\left(\frac{d}{dt} + i\omega_2\right)}{(\omega_3 - \omega_1)(\omega_3 - \omega_1)} x. \tag{3.130}$$

Therefore if ω_k s are all different, here we obtain two distinct constants of motion [32]

$$C_2 = \frac{[\ddot{x} + i(\omega_2 + \omega_3)\dot{x} - \omega_2\omega_3 x]}{[\ddot{x} + i(\omega_1 + \omega_2)\dot{x} - \omega_1\omega_2 x]^{\frac{\omega_1}{\omega_3}}}, \tag{3.131}$$

$$C_3 = \frac{[\ddot{x} + i(\omega_2 + \omega_3)\dot{x} - \omega_2\omega_3 x]}{[\ddot{x} + i(\omega_1 + \omega_3)\dot{x} - \omega_1\omega_3 x]^{\frac{\omega_1}{\omega_2}}}, \tag{3.132}$$

This is a remarkable result since from a single Hamiltonian depending on x and p one can derive the equations of motion plus the two constants of motion C_2 and C_3. The Hamiltonian in this case is

$$H = -i\omega_1 x p + \frac{b_2 \omega_1}{\omega_1 - \omega_2} p^{1 - \frac{\omega_2}{\omega_1}} + \frac{b_3 \omega_1}{\omega_1 - \omega_3} p^{1 - \frac{\omega_3}{\omega_1}}, \tag{3.133}$$

where b_2 and b_3 are arbitrary constants.

From the canonical equation $\dot{p} = -\partial H/\partial x = i\omega_1 p$ we obtain $p \propto e^{i\omega_1 t}$, thus $1/p$ is directly related to the products shown in (3.128). Also from $\dot{x} = \frac{\partial H}{\partial p}$ it follows that

$$\dot{x} = -i\omega_1 x + b_2 p^{-\frac{\omega_2}{\omega_1}} + b_3 p^{-\frac{\omega_3}{\omega_1}}, \tag{3.134}$$

$$\ddot{x} = -i\omega_1 \dot{x} - i\omega_2 b_2 p^{-\frac{\omega_2}{\omega_1}} - i\omega_3 b_3 p^{-\frac{\omega_3}{\omega_1}}, \tag{3.135}$$

$$\dddot{x} = -i\omega_1 \ddot{x} - \omega_2^2 b_2 p^{-\frac{\omega_2}{\omega_1}} - \omega_3^2 b_3 p^{-\frac{\omega_3}{\omega_1}}, \tag{3.136}$$

Now by eliminating b_2 and b_3 between these equations we obtain a third order equation for $x(t)$ which is independent of b_2 and b_3:

$$\dddot{x} + i(\omega_1 + \omega_2 + \omega_3)\ddot{x} - (\omega_1\omega_2 + \omega_2\omega_3 + \omega_1\omega_3)\dot{x} - i\omega_1\omega_2\omega_3 x = 0, \qquad (3.137)$$

On the other hand if we take just two equations (3.134) and (3.135), we can solve for b_2 or b_3 in terms of $\ddot{x}$, $\dot{x}$ and x and find

$$b_2 p^{-\frac{\omega_2}{\omega_1}} = \frac{i}{\omega_2 - \omega_3}\left[\ddot{x} + i(\omega_1 + \omega_3)\dot{x} - \omega_1\omega_3 x\right], \qquad (3.138)$$

$$b_3 p^{-\frac{\omega_3}{\omega_1}} = \frac{i}{\omega_3 - \omega_2}\left[\ddot{x} + i(\omega_1 + \omega_2)\dot{x} - \omega_1\omega_2 x\right]. \qquad (3.139)$$

Observing the two expression in the brackets in Eqs. (3.138) and (3.139) and their dependence on the pairs (ω_1, ω_3) and on (ω_1, ω_2), we can inquire about the third relation where (ω_2, ω_3) appears. Calculating the same bracket with (ω_2, ω_3), we get

$$\ddot{x} + i(\omega_2 + \omega_3)\dot{x} - \omega_2\omega_3 x = -(\omega_1 - \omega_2)(\omega_1 - \omega_3)x$$
$$+ ib_2(\omega_3 - \omega_1)p^{-\frac{\omega_2}{\omega_1}} + ib_3(\omega_2 - \omega_1)p^{-\frac{\omega_3}{\omega_1}}$$
$$= \frac{i}{\omega_1 p}(\omega_1 - \omega_2)(\omega_3 - \omega_1)H \qquad (3.140)$$

From this relation we find two constants C_2 and C_3

$$C_2 = [i(\omega_1 - \omega_2)(\omega_3 - \omega_1)H/(\omega_1 p)]$$
$$\times p\,[i(\omega_2 - \omega_3)b_3]^{-\frac{\omega_1}{\omega_3}} \propto Hb_3^{-\frac{\omega_1}{\omega_3}}, \qquad (3.141)$$

and

$$C_3 \propto Hb_2^{-\frac{\omega_1}{\omega_2}}. \qquad (3.142)$$

The Hamiltonian which is conserved can be expressed as a function x, $\dot{x}$, $\ddot{x}$ and p.

$$H = i\omega_1 p\frac{\ddot{x} + i(\omega_2 + \omega_3)\dot{x} - \omega_2\omega_3 x}{(\omega_1 - \omega_2)(\omega_1 - \omega_3)}. \qquad (3.143)$$

The Case of Degenerate Frequencies — For a damped harmonic oscillator, in the case of critical damping $\omega_2 \to \omega_3$, we can make the following replacement in the Hamiltonian H_1, Eq. (3.115)

$$\frac{\omega_1}{\omega_1 - \omega_2}p^{1-\frac{\omega_2}{\omega_1}} \to \ln p, \qquad (3.144)$$

where by making this replacement we have shifted the Hamiltonian by an infinite constant [32]. In this limit H_1 is reduced to

$$H_1 = -i\omega_1 xp + \ln p, \qquad (3.145)$$

and this Hamiltonian generates the equation of motion

$$\ddot{x} + 2i\omega_1 \dot{x} - \omega_1^2 x = 0. \tag{3.146}$$

Similarly for the third-order differential equation (3.137), if we set $\omega_2 = \omega_1$, then the Hamiltonian (3.133) becomes

$$H = -i\omega_1 x p + b_2 \ln p + \frac{b_3 \omega_1}{\omega_1 - \omega_3} p^{1 - \frac{\omega_3}{\omega_1}}. \tag{3.147}$$

Finally if $\omega_1 = \omega_2 = \omega_3 = \omega$, then the Hamiltonian (3.133) reduces to

$$H = -i\omega x p + b \ln p + \frac{c}{2}(\ln p)^2, \tag{3.148}$$

where the two constants b and c should be determined from the initial conditions. From H defined by (3.148) we obtain the equation of motion

$$\dddot{x} + 3i\omega \ddot{x} - 3\omega^2 \dot{x} - i\omega^3 x = 0. \tag{3.149}$$

This method can be generalized and applied to a linear differential equation of the nth order and for such a case the Hamiltonian will be

$$H = -i\omega_1 x p + \sum_{j \neq 1}^{n} \frac{b_j \omega_1}{\omega_1 - \omega_j} p^{1 - \frac{\omega_1}{\omega_j}}. \tag{3.150}$$

3.8   Group-Theoretical Approach to the Solution of the Damped Harmonic Oscillator

A very different physical interpretation of the solution of Caldirola-Kanai eigenvalue problem has been proposed by Cerveró and Villarroel [33],[34]. In the following section we will study the classical basis of this group-theoretical approach and later in Chapter 13 we will see the consequences of this formulation. Let us introduce the dimensionless quantities $\tau = \omega_0 t$ and $\gamma_0 = \frac{\lambda}{\omega_0}$, and write the equation of motion as

$$\frac{d^2 x}{d\tau^2} + \gamma_0 \frac{dx}{d\tau} + x = 0. \tag{3.151}$$

The Lagrangian for this motion is given by (2.175), provided that we write it in terms of γ_0 and τ. In the following discussion we denote the derivative with respect to τ with a dot. It is convenient to replace γ_0 by γ,

$$\gamma = \sqrt{\gamma_0^2 - 4}, \tag{3.152}$$

where γ can be real or pure imaginary. The critical value of damping is obtained by setting $\gamma = 0$. For non-critical values, γ is real for overdamped case and is pure imaginary for underdamped oscillators. Now according to the Lie theory the group of transformations which leaves the differential equation (3.151) invariant is the eight-parameter group $SL(3, R)$ [33]–[35]. In this case we have the following set of eight generators:

$$G_1 = \frac{2}{\gamma} \left[\sinh(\gamma\tau) \frac{\partial}{\partial\tau} + \left(\frac{\gamma}{2} \cosh(\gamma\tau) - \frac{\gamma_0}{2} \sinh(\gamma\tau) \right) x \frac{\partial}{\partial x} \right], \qquad (3.153)$$

$$G_2 = \frac{2i}{\gamma} \left[\cosh(\gamma\tau) \frac{\partial}{\partial\tau} + \left(\frac{\gamma}{2} \sinh(\gamma\tau) - \frac{\gamma_0}{2} \cosh(\gamma\tau) \right) x \frac{\partial}{\partial x} \right], \qquad (3.154)$$

$$G_3 = \frac{2i}{\gamma} \left[\exp\left(-\frac{\gamma_0\tau}{2} \right) \cosh\left(\frac{1}{2}\gamma\tau \right) \frac{\partial}{\partial x} \right], \qquad (3.155)$$

$$G_4 = \frac{2}{\gamma} \left[\exp\left(-\frac{\gamma_0\tau}{2} \right) \sinh\left(\frac{1}{2}\gamma\tau \right) \frac{\partial}{\partial x} \right], \qquad (3.156)$$

$$G_5 = \frac{2i}{\gamma} \left(\frac{\partial}{\partial\tau} - \frac{1}{2}\gamma_0 x \frac{\partial}{\partial x} \right), \qquad (3.157)$$

$$G_6 = x \frac{\partial}{\partial x}, \qquad (3.158)$$

$$G_7 = -i \exp\left(\frac{\gamma_0\tau}{2} \right) \left\{ x \sinh\left(\frac{\gamma\tau}{2} \right) \frac{\partial}{\partial\tau} \right.$$
$$\left. + \left[\frac{\gamma}{2} \cosh\left(\frac{\gamma\tau}{2} \right) - \frac{\gamma_0}{2} \sinh\left(\frac{\gamma\tau}{2} \right) \right] x^2 \frac{\partial}{\partial x} \right\}, \qquad (3.159)$$

and

$$G_8 = \exp\left(\frac{\gamma_0\tau}{2} \right) \left\{ x \cosh\left(\frac{\gamma\tau}{2} \right) \frac{\partial}{\partial\tau} \right.$$
$$\left. + \left[\frac{\gamma}{2} \sinh\left(\frac{\gamma\tau}{2} \right) - \frac{\gamma_0}{2} \cosh\left(\frac{\gamma\tau}{2} \right) \right] x^2 \frac{\partial}{\partial x} \right\}, \qquad (3.160)$$

These group elements simplify to the corresponding group elements for the simple harmonic oscillator found by Lutzky [36]. Thus by setting $\gamma_0 = 0$ and $\gamma = 2i$ we get:

$$G_1 = \sin(2\omega_0 t) \frac{\partial}{\partial t} + x \cos(2\omega_0 t) \frac{\partial}{\partial x}, \qquad (3.161)$$

$$G_2 = \cos(2\omega_0 t) \frac{\partial}{\partial t} - x \sin(2\omega_0 t) \frac{\partial}{\partial x}, \qquad (3.162)$$

$$G_3 = \cos(2\omega_0 t) \frac{\partial}{\partial x}, \qquad (3.163)$$

$$G_4 = \sin(2\omega_0 t) \frac{\partial}{\partial x}, \qquad (3.164)$$

and

$$G_5 = \frac{\partial}{\partial t}. \tag{3.165}$$

Returning to the problem of damped harmonic oscillator, we note that for the Lagrangian (2.175), the symmetry generators are $\{G_1, G_2, G_3, G_4 \text{ and } G_5\}$, and from Noether's theorem we find the five invariants associated with these generators:

$$I_1 = e^{\gamma_0\tau}\left\{\left(\dot{x}^2 + x^2\right)\sinh(\gamma\tau) - x\dot{x}\left[\gamma\cosh(\gamma\tau) - \gamma_0\sinh(\gamma\tau)\right]\right\}$$
$$+ \frac{1}{2}\left\{e^{\gamma_0\tau}x^2\left[\gamma^2\sinh(\gamma\tau) - \gamma\gamma_0\cosh(\gamma\tau)\right]\right\}, \tag{3.166}$$

$$I_2 = e^{\gamma_0\tau}\left\{\left(\dot{x}^2 + x^2\right)\cosh(\gamma\tau) - x\dot{x}\left[\gamma\sinh(\gamma\tau) - \gamma_0\cosh(\gamma\tau)\right]\right\}$$
$$+ \frac{1}{2}\left\{e^{\gamma_0\tau}x^2\left[\gamma^2\cosh(\gamma\tau) - \gamma\gamma_0\sinh(\gamma\tau)\right]\right\}, \tag{3.167}$$

$$I_3 = \exp\left(\frac{\gamma_0 + \gamma}{2}\tau\right)\left(\dot{x} + \frac{\gamma_0 - \gamma}{2}x\right), \tag{3.168}$$

$$I_4 = \exp\left(\frac{\gamma_0 - \gamma}{2}\tau\right)\left(\dot{x} + \frac{\gamma_0 + \gamma}{2}x\right), \tag{3.169}$$

and

$$I_5 = \exp\left(\gamma_0\tau\right)\left(\dot{x}^2 + x^2 + \gamma_0 x\dot{x}\right). \tag{3.170}$$

Let us note that the conserved quantities $\{I_1, \cdots I_5\}$ are not all independent of each other for this one-dimensional motion. Thus we find the following relations among them

$$I_1 = \frac{1}{2}\left(I_3^2 - I_4^2\right), \quad I_2 = \frac{1}{2}\left(I_3^2 + I_4^2\right), \quad \text{and} \quad I_5 = I_3 I_4. \tag{3.171}$$

Among the eight infinitesimal generators, Eq. (3.153)–(3.160), G_5 plays an important role, viz, the action remains invariant under G_5, and this generator also represents a non-conventional Hamiltonian. We call it non-conventional since it is not obtained via a Legendre transformation, as we find standard Hamiltonians, and in addition it is a conserved quantity.

Noting that the action is invariant under G_5, and considering G_5 as a vector field associated with the Lagrangian (2.175), we have the differential system

$$d\tau = -\frac{2}{\gamma_0}\frac{dx}{x} = -\frac{2}{\gamma_0}\frac{d\dot{x}}{\dot{x}} = \frac{2}{\gamma_0}\frac{dp}{p}, \tag{3.172}$$

where we have used $p = m\omega_0\exp(\gamma_0\tau)\dot{x}$.

The first integrals of the system (3.172) are

$$Q = \exp\left(\frac{\gamma_0\tau}{2}\right)x, \quad \text{and} \quad P = \exp\left(-\frac{\gamma_0\tau}{2}\right)p, \tag{3.173}$$

with the Poisson bracket $\{Q, P\} = 1$. We denote the resulting Hamiltonian, after making this canonical transformation by H^*

$$H^* = \frac{1}{2m}P^2 + \frac{1}{2}m\omega_0^2 Q^2 + \frac{1}{2}\omega_0\gamma_0 QP. \tag{3.174}$$

We note that H^* is invariant under the action of I_5, therefore

$$H^* = \left(\frac{1}{2}m\omega_0^2\right)I_5 = \left(\frac{1}{2}m\omega_0^2\right)\left[e^{\gamma_0\tau}\left(\dot{x}^2 + x^2 + \gamma_0 x\dot{x}\right)\right]. \tag{3.175}$$

The Hamiltonian H^* defined by (3.174) which is a first integral of motion for the damped oscillator, unlike Caldirola-Kanai Hamiltonian does not depend explicitly on time. For a different derivation of H^* from a Lagrangian see Eq. (15.73). The quantal solution for the damped motion of a harmonic oscillator will be considered in Chapter 11.

The Caldirola-Kanai Hamiltonian and the Oscillator with Variable Frequency— A large number of interesting physical problems, classical as well as quantal, can be reduced to a system of harmonic oscillators with time-dependent frequencies [37]–[39]. A special case which is of interest to our present discussion is that of Caldirola-Kanai Hamiltonian Eq. (7.39) which can be written as

$$H(t) = \frac{1}{2m(t)}p^2 + \frac{1}{2}m(t)\omega_0^2 x^2, \quad m(t) = m\exp(\gamma_0\omega_0 t). \tag{3.176}$$

To write this Hamiltonian as a Hamiltonian with time-dependent frequency we introduce a time transformation of the form [40]

$$\bar{t} = \int_0^t \frac{dt}{m(t)} = -\frac{1}{m\gamma_0\omega_0}\exp(-\gamma_0\omega_0 t), \tag{3.177}$$

and express the dynamical variables and t in terms of $\bar{t}$. Thus we obtain

$$H\left(\bar{t}\right) = \frac{1}{2}p^2 + \frac{1}{2}\Omega^2\left(\bar{t}\right)x^2, \tag{3.178}$$

where

$$\Omega\left(\bar{t}\right) = \frac{1}{\gamma_0\bar{t}} = -m\omega_0 e^{\gamma_0\omega_0 t}, \tag{3.179}$$

and

$$\Omega^2\left(\bar{t}\right) = \omega_0^2\left(me^{\gamma_0\omega_0 t}\right)^2. \tag{3.180}$$

This result is the same as (3.73).

3.9 Dissipative Forces Quadratic in Velocity

In the last section we studied the one-dimensional motion of a particle when it is affected by a force linear in velocity. Now let us consider the motion of a particle of mass m in a medium where in addition to the conservative force a Newtonian resistive force $m\gamma\dot{x}^2$ is acting on it. Here the equation of motion is given by [41]

$$m\left(\frac{d^2x}{dt^2}\right) + m\gamma\left(\frac{dx}{dt}\right)^2 = f(x), \tag{3.181}$$

where the constant γ has the dimension of length^{-1}. Again the Hamiltonian for this system is not unique, but a simple time-independent Hamiltonian for this motion is given by

$$H(x,p) = \left(\frac{p^2}{2m}\right)e^{-2\gamma x} - \int^x f(y)e^{2\gamma y}dy. \tag{3.182}$$

Note that (3.181) remains invariant under time-reversal transformation. If the motion is such that $\dot{x}$ changes its sign, then for a dissipative motion we must replace $(dx/dt)^2$ by $\left(\frac{dx}{dt}\right)\left|\frac{dx}{dt}\right|$ (see Sec. 1.1).

3.10 Resistive Forces Proportional to Arbitrary Powers of Velocity

As we have already mentioned in Sec. (1.1) in the case of a bullet in the air the resistance can be proportional to $\dot{x}^\nu$, $(\nu \approx 2)$. In the following discussion we assume ν to be a number greater than one, but not equal to two

The equation of motion of the particle in the absence of a conservative force is

$$m\left(\frac{d^2x}{dt^2}\right) + m\beta\left(\frac{dx}{dt}\right)^\nu = 0, \tag{3.183}$$

or if x changes sign we have

$$m\left(\frac{d^2x}{dt^2}\right) + m\beta\left(\frac{dx}{dt}\right)\left|\frac{dx}{dt}\right|^{\nu-1} = 0. \tag{3.184}$$

For Eq. (3.183) the Hamiltonian is given by the implicit relation

$$p(H,x) = \int^H [C(H) - \beta(2-\nu)x]^{\frac{1}{(\nu-2)}}\,dH, \quad \nu \neq 2, \tag{3.185}$$

where C is an arbitrary function of H. The simplest choice is that of C being a constant independent of H. Then from (3.185) we find

$$H = p\,[C - \beta(2-\nu)x]^{\frac{1}{(2-\nu)}} = pf(x). \tag{3.186}$$

Noting that H does not explicitly depend on time, and therefore it is a constant of motion, the Hamilton canonical equations yield

$$\dot{x} = \frac{H}{p} \quad \text{and} \quad \dot{p} = \beta p\dot{x}^{(\nu-1)}. \tag{3.187}$$

By eliminating p between these two relations we obtain Eq. (3.183).

While Eq. (3.185) gives us the most general although an implicit Hamiltonian, we can use the Lagrangian formulation Eq. (2.144) to construct the Hamiltonian. The canonical momentum p found from the Lagrangian is [42]

$$p = \frac{\partial L}{\partial \dot{x}} = m\mu'(\dot{x}) = (1 - \nu)\dot{x}^{1-\nu}, \quad 1 < \nu < 2, \qquad (3.188)$$

where prime denotes derivative with respect to the argument and $\mu''(\dot{x}) = \dot{x}^{(-\nu)}$, (see Eq. (2.143)). The Hamiltonian of the system is found from the definition of H;

$$H = p\dot{x} - L = m\left[\dot{x}\mu'(\dot{x}) - \mu(\dot{x})\right] + m\beta x. \qquad (3.189)$$

By eliminating $\dot{x}$ between (3.188) and (3.189) we have

$$H = \left(\frac{m}{2 - \nu}\right)\left(\frac{1 - \nu}{m}p\right)^{\frac{\nu-2}{\nu-1}} + m\beta x, \quad 1 < \nu < 2. \qquad (3.190)$$

The above Hamiltonian can also be obtained from (3.185) if we choose $C(H)$ to be

$$C(H) = \frac{2 - \nu}{m}H, \qquad (3.191)$$

and after carrying out the integration over H, solve for H in terms of p.

3.11 Constrained Lagrangian and Hamiltonian Formulation of Damped Systems

In Chapter 2 we studied the Helmholtz condition for the existence of a Lagrangian for the set of equations of motion

$$m\ddot{x}_k - f_k(x_i, \dot{x}_i, t) \equiv G(x_i, \dot{x}_i, \ddot{x}_i, t) = 0, \quad k = 1, 2 \cdots N, \qquad (3.192)$$

and we noticed that in general, for dissipative systems, we have to introduce a set of integrating factor $\mu_{jk}(x_i, \dot{x}_i, t)$ in order to satisfy the Helmholtz conditions. We can inquire about the possibility of constructing Lagrangian (or Hamiltonian) functions for open systems where, like simple conservative systems, there are no integrating factors. This possibility has been studied by Cawley who has called such a Lagrangian a "universal Lagrangian" [43]. We introduce the Lagrangian $L_u(x_i, z_i, \dot{x}_i, \dot{z}_i, y_i, t)$ with $2N$ additional coordinates z_i and y_i and additional velocities $\dot{z}_i$ by

$$L_u(x_i, z_i, \dot{x}_i, \dot{z}_i, y_i, t) = \sum_{k=1}^{N}\left[f_k(x_i, \dot{x}_i, t)z_k + m\dot{x}_k\dot{z}_k + \frac{1}{2}y_k z_k^2\right]. \qquad (3.193)$$

The Euler-Lagrange equation for L yields the N equations of motion (3.192) together with a set of N subsidiary conditions

$$z_k = 0, \quad k = 1, \cdots N. \qquad (3.194)$$

If we try to construct a Hamiltonian for Eq. (3.192) we obtain N first class constraints [44],[45]:

$$\frac{\partial L_u}{\partial \dot{y}_k} = p_{y_k} \approx 0, \quad k = 1, \cdots N, \tag{3.195}$$

where the wavy equality sign "$\approx$" denotes the weak equality. By weak equality we mean that these constraints should not be used before working out the Poisson brackets [44]. For the momenta p_{z_k} from (3.192) we find p_{z_k};

$$p_{z_k} = m\dot{x}_k, \quad k = 1, \cdots N. \tag{3.196}$$

Using these momenta, we obtain the Hamiltonian H_u corresponding to the Lagrangian L_u;

$$H_u = \sum_{k=1}^{N} \left[\frac{p_{z_k} p_{x_k}}{m} - f_k \left(x_i, \frac{1}{m} p_{z_i}, t \right) z_k - \frac{1}{2} y_k z_k^2 + \phi_k(t) p_{y_k} \right], \tag{3.197}$$

where ϕ_k s are arbitrary functions of time. Expressing the time derivative in terms of the Poisson bracket

$$\dot{g}_k = \{g_k, H_u\}, \tag{3.198}$$

we get the following relations:

$$\dot{p}_{y_k} \approx \{p_{y_k}, H_u\} \approx \frac{1}{2} z_k^2 \approx 0, \quad k = 1, \cdots N, \tag{3.199}$$

$$z_k \approx 0, \quad k = 1, \cdots N, \tag{3.200}$$

and

$$z_k^2 \approx 0, \quad k = 1, \cdots N. \tag{3.201}$$

We observe that Eqs. (3.200) are secondary constraints and the consistency conditions are [44],[46]:

$$\dot{z}_k \approx \{z_k, H_u\} = \frac{1}{m} p_{x_k} \approx 0, \quad k = 1, \cdots N. \tag{3.202}$$

One-Dimensional Motion Subject to Linear Friction — Let us apply this method to the problem of motion of a particle subject to an applied force $F(x, t)$ in addition to the damping force $-m\lambda\dot{x}$. According to the present formulation the Lagrangian of the system is

$$L_u(x, \dot{x}, z, \dot{z}, y, t) = \left[m\dot{x}\dot{z} + (F(x, t) - m\lambda\dot{x}) z + \frac{1}{2} y z^2 \right]. \tag{3.203}$$

From this Lagrangian we find

$$p_z = m\dot{x}, \tag{3.204}$$

and

$$p_x = m\dot{z} - m\lambda z. \qquad (3.205)$$

Knowing the momenta p_x and p_z and L we obtain the Hamiltonian H_u

$$H_u = \frac{1}{m}p_z p_x + (\lambda p_z - F(x,t)) z - \frac{1}{2}yz^2. \qquad (3.206)$$

This Hamiltonian can be related to the Birkhoff's Hamiltonian defined in the following way [47]: Let us write the classical equations of motion as first order coupled equations

$$\frac{dq_k}{dt} = X_k(q_i, t), \quad k = 1, \cdots 2N. \qquad (3.207)$$

The Birkhoff Hamiltonian has the form

$$H_B = \sum_{k=1}^{2N} X_k(q_i, t) p_k, \qquad (3.208)$$

where $p_1, \cdots, p_{2N}$ are the $2N$ momenta conjugate to the coordinates $q_1, \cdots, q_{2N}$. Equations (3.207) are one-half of the Hamilton canonical equations. The rest are given by

$$\frac{dp_k}{dt} == -\frac{\partial H_B}{\partial q_k} = -\sum_{j=1}^{2N} \frac{\partial X_j(q_i, t)}{\partial q_k} p_j, \quad k = 1, \cdots 2N. \qquad (3.209)$$

From the Hamiltonian H_B we can find the Lagrangian L_B to be

$$L_B = \sum_{k=1}^{2N} \frac{dq_k}{dt} p_k - H_B \equiv 0. \qquad (3.210)$$

Next we identify the canonical variables x_j, z_j, p_{x_j} and p_{x_j} by the following relations:

$$q_1, \cdots q_N \rightarrow x_1, \cdots x_N, \qquad (3.211)$$

$$q_{N+1}, \cdots q_{2N} \rightarrow p_{z_1}, \cdots p_{z_N}, \qquad (3.212)$$

$$p_1, \cdots p_N \rightarrow p_{x_1}, \cdots p_{x_N}, \qquad (3.213)$$

$$p_{N+1}, \cdots p_{2N} \rightarrow -z_1, \cdots - z_N, \qquad (3.214)$$

$$X_1, \cdots X_N \rightarrow \dot{x}_1(p_{z_i}), \cdots \dot{x}_N(p_{z_i}), \qquad (3.215)$$

and

$$X_{N+1}, \cdots X_{2N} \rightarrow f_1\left(x_i, \frac{1}{m}p_{z_i}, t\right) \cdots f_N\left(x_i, \frac{1}{m}p_{z_i}, t\right). \qquad (3.216)$$

Substituting for q_k s and p_k s in H_B, Eq. (3.208) we find

$$H_B = \sum_{j=1}^{N} \dot{x}_j(p_{z_i}) p_{x_j} - \sum_{j=1}^{N} f_j\left(x_i, \frac{1}{m}p_{z_i}, t\right) z_j. \qquad (3.217)$$

This H_B is the same as H_u, except for the two terms,

$$\left(-\frac{1}{2}y_k z_k^2 + \phi(t)p_{y_k}\right),\tag{3.218}$$

which do not affect the equations of motion.

For the one-dimensional damped motion we have $N = 1$ and

$$X_1 = \dot{q}_1 = \frac{p_z}{m} = \dot{x}, \quad p_1 = p_x,\tag{3.219}$$

and

$$X_2 = -\lambda q_2 + F(q_1, t), \quad p_2 = -z.\tag{3.220}$$

By substituting these in H_B, we find the same expression as H_u, Eq. (3.206) without the last term.

3.12 Hamiltonian Formulation in Phase Space of N-Dimensions

The Abraham-Lorentz equation for the non-relativistic motion of a radiating electron is of third order, Eq. (1.84). The Hamiltonian formulation of equations of motion containing higher derivatives of the coordinate(s) of a particle has been studied in connection with nonlocal field theories [12]. The time-reversal invariance imposed on these fields allows only for even derivatives of the coordinate(s) to appear in the equation of motion. However there are other generalizations of the Hamiltonian dynamics where odd derivatives can occur in the equations of motion and that the phase space can have odd dimensions [5],[13],[48].

The equations of motion for a system of particles can be written as a set of first order differential equations

$$\dot{\eta}_i = X_i(\eta_1, \eta_2 \cdots \eta_N), \quad i = 1, 2, \cdots N,\tag{3.221}$$

where $\dot{\eta}_i = \frac{d\eta_i}{dt}$ and η_i s are the dynamical variables of the system. Next let us suppose that the system is completely integrable [2] and that $H_1, H_2 \cdots H_{N-1}$ are the $N - 1$ constants of motion of this system, then

$$\frac{dH_i}{dt} = \sum_{k=1}^{N}\left(\frac{\partial H_i}{\partial \eta_k}\right)\dot{\eta}_k = 0, \quad i = 1, 2, \cdots N - 1.\tag{3.222}$$

This set of $N - 1$ equations can be solved for the variables $\dot{\eta}_1, \dot{\eta}_2 \cdots \dot{\eta}_{N-1}$ in terms of $\dot{\eta}_N$;

$$\dot{\eta}_j = (-1)^N \frac{d\eta_N}{dt}\left(\frac{\Delta_j}{\Delta_N}\right),\tag{3.223}$$

where

$$\Delta_j = \sum_{[j]} \varepsilon_{i_1, i_2, \cdots i_{N-1}} \frac{\partial H_1}{\partial \eta_{i_1}} \cdots \frac{\partial H_{N-1}}{\partial \eta_{i_{N-1}}}. \tag{3.224}$$

In Eq. (3.224) $\varepsilon_{i_1, i_2, \cdots i_{N-1}}$ is the Levi-Civita tensor, and the summation is over all integers $1, 2 \cdots N$ except j. Thus Δ_j is the determinant obtained from $N \times (N-1)$ matrix elements $\left(\frac{\partial H_j}{\partial \eta_k} \right)$ by deleting the jth column. Now if we compare (3.221) and (3.223) we find

$$\frac{X_1}{\Delta_1} = \frac{X_2}{-\Delta_2} = \cdots = \frac{X_N}{(-1)^{N-1}\Delta_N} = \frac{1}{M(\eta_1 \cdots \eta_N)}, \tag{3.225}$$

where M is, in general, a function of η_i and is called the multiplier of the system [13]. From Eq. (3.224) it follows that

$$\frac{\partial \Delta_1}{\partial \eta_1} - \frac{\partial \Delta_2}{\partial \eta_2} + \cdots + (-1)^{N-1}\frac{\partial \Delta_N}{\partial \eta_N} = 0, \tag{3.226}$$

or in terms of X_i s we have

$$\sum_{i=1}^{N} \frac{\partial}{\partial \eta_i} (MX_i) = 0. \tag{3.227}$$

This relation can be used to determine $M(\eta_1 \cdots \eta_N)$.

If we consider the phase space of three dimensions when the forces acting on the system satisfy the relation

$$\sum_{i=1}^{3} \frac{\partial X_i}{\partial \eta_i} = \nabla \cdot \mathbf{X} = 0, \tag{3.228}$$

then the présent formulation reduces to Nambu's dynamics [5],[48] . Thus from (3.223) and (3.224) we find $\dot{\eta}_i$ to be

$$\dot{\eta}_i = \sum_{j,k} \varepsilon_{ijk} \frac{\partial H_1}{\partial \eta_j} \frac{\partial H_2}{\partial \eta_k}, \tag{3.229}$$

and Eq. (3.226) simplifies to

$$\nabla \cdot (\nabla H_1 \wedge \nabla H_2) = 0, \tag{3.230}$$

where ∇ acts on the coordinates η_1, η_2 and η_3 of the phase space.

In an N-dimensional phase space let us take H_{N-1} to be one of the generators of motion for a set of dynamical variables $\eta_1 \cdots \eta_N$. Then we define the Poisson bracket for any two functions of the dynamical variables ϕ and ψ by the relation

$$\{\phi, \psi\} = \frac{1}{M} \sum_{j,k} \mathcal{J}_{jk} \frac{\partial \phi}{\partial \eta_j} \frac{\partial \psi}{\partial \eta_k}, \tag{3.231}$$

where M is defined by (3.225) and (3.227) and $\mathcal{J}_{jk}$ is the antisymmetric matrix

$$\mathcal{J}_{jk} = \sum_{i_3 \ldots i_N} (-1)^N \varepsilon_{jk\, i_3 \cdots i_N} \frac{\partial H_1}{\partial \eta_{i_3}} \cdots \frac{\partial H_{N-2}}{\partial \eta_{i_N}}. \tag{3.232}$$

From the definition of $\mathcal{J}_{jk}$ it follows that

$$\mathcal{J}_{jk} + \mathcal{J}_{kj} = 0. \tag{3.233}$$

However the Jacobi identity [1]

$$\sum_s^N \left[\mathcal{J}_{ks} \frac{\partial}{\partial \eta_s} \left(\frac{\mathcal{J}_{lm}}{M} \right) + \mathcal{J}_{ls} \frac{\partial}{\partial \eta_s} \left(\frac{\mathcal{J}_{mk}}{M} \right) + \mathcal{J}_{ms} \frac{\partial}{\partial \eta_s} \left(\frac{\mathcal{J}_{kl}}{M} \right) \right], \tag{3.234}$$

does not in general vanish.

By replacing ψ in (3.231) by H_{N-1} we find

$$\{\phi, H_{N-1}\} = \frac{1}{M} \left[\frac{\partial \phi}{\partial \eta_1} \Delta_1 - \frac{\partial \phi}{\partial \eta_2} \Delta_2 + \cdots + (-1)^{N-1} \frac{\partial \phi}{\partial \eta_N} \Delta_N \right] = 0. \tag{3.235}$$

In particular we have

$$\{\eta_i, H_{N-1}\} = (-1)^{i-1} \left(\frac{\Delta_i}{M} \right) = X_i = \dot{\eta}_i, \quad i = 1, \cdots N. \tag{3.236}$$

The first integrals of motion, $H_1, H_2 \cdots H_{N-1}$ are found by writing the equations of motion (3.221) as

$$\frac{d\eta_1}{X_1} = \frac{d\eta_2}{X_2} = \cdots = \frac{d\eta_N}{X_N}, \tag{3.237}$$

and integrating them. Here we assume that X_i s do not depend explicitly on time. This set of equations are the Lagrange's subsidiary equations for the partial differential equations

$$\sum_{i=1}^N X_i \frac{\partial H_k}{\partial \eta_i} = 0, \quad k = 1, 2 \cdots N - 1, \tag{3.238}$$

and H_k s are the first integrals of motion. We also observe that the Poisson brackets of the canonical variables $\eta_1 \cdots \eta_N$ are given by

$$\{\eta_j, \eta_k\} = \frac{1}{M} \mathcal{J}_{jk}, \quad j, k = 1, 2 \cdots N. \tag{3.239}$$

A Hamiltonian for the Motion of Self-Accelerating Electron — Returning to the problem of a radiating electron of mass m moving in a constant external field, $(-ma)$, we first write the equation of motion as

$$m\tau \frac{d^3 \eta_1}{dt^3} - m \frac{d^2 \eta_1}{dt^2} = ma, \tag{3.240}$$

where the constant external field is written as ma. We can also write (3.240) as three coupled first order equations:

$$\frac{d\eta_1}{dt} = \eta_2, \quad \frac{d\eta_2}{dt} = \eta_3, \quad \text{and} \quad \frac{d\eta_3}{dt} = \frac{1}{\tau}(\eta_3 + a). \tag{3.241}$$

By integrating these equations, we can construct two constants of motion:

$$H_1 = \eta_2 - \tau\eta_3 + a\tau \ln(\eta_3 + a), \tag{3.242}$$

and

$$H_2 = \frac{1}{\tau}\eta_1 - \frac{a\tau}{2}[\ln(\eta_3 + a)]^2 - \tau\eta_3 + [\tau(\eta_3 + a) - \eta_2]\ln(\eta_3 + a). \tag{3.243}$$

These two constants H_1 and H_2 are the generators of the equation of motion (3.241) with the multiplier M given by

$$M = -\frac{1}{(\eta_3 + a)}. \tag{3.244}$$

From Eq. (3.239) we calculate the Poisson bracket for the dynamical variables η_1, η_2 and η_3:

$$\{\eta_1, \eta_2\} = -\tau\eta_3, \quad \{\eta_2, \eta_3\} = 0, \quad \text{and} \quad \{\eta_3, \eta_1\} = \eta_3 + a. \tag{3.245}$$

These relations together with the Hamilton's equations

$$\dot{\eta}_i = \{\eta_i, H_2\}, \tag{3.246}$$

determine the equations of motion (3.241).

3.13 Classical Brackets for Damped Systems

The Dirac rule of association establishes a relation between the classical Poisson brackets of dynamical variables and the quantum mechanical commutators of the corresponding operators [49]–[51]. According to the Poisson theorem the bracket of any two functions of dynamical variables q_i and p_i remains constant in time, a result which can be proved for a conservative system described by a Hamiltonian (e.g. when second or higher order perturbation are used in the calculation). For a nonconservative system we will show that in general the Poisson theorem and also the Jacobi identity are not satisfied. In what follows we will discuss two rather similar approaches to the canonical formulation of dissipative systems. In the first one we study the pseudo-Hamiltonian formulation of Duffin [52],[53]. Then we consider Sedov's approach to the variational principle for these systems as has been described by Tarasov [54],[55].

Properties of the Poisson Bracket for a Conservative System — For a conservative Hamiltonian the Poisson brackets satisfy the following properties:

(1) Linearity

$$\{\alpha f + \beta g, \, h\} = \alpha \{f, \, h\} + \beta \{g, \, h\}, \tag{3.247}$$

(2) Antisymmetry

$$\{f, \, g,\} = - \{g, \, f\} \tag{3.248}$$

(3) Jacobi identity

$$\{ \{f, \, g\}, \, h\} + \{ \{g, \, h\}, \, f\} + \{ \{h, \, f\}, \, g\} = 0, \tag{3.249}$$

where α and β are constants. Thus the Poisson bracket is also a Lie bracket. The bracket $\{f, \, g\}$ also satisfies the Leibnitz rule

$$\frac{d}{dt} \{f, \, g\} = \left\{ \frac{df}{dt}, \, g \right\} + \left\{ f, \, \frac{dg}{dt} \right\}. \tag{3.250}$$

This property implies the vanishing of the Jacobi identity for f, g, h. That is from the definition of time derivative $\frac{df}{dt} = \{f, \, H\}$ and (3.250) we have

$$\{ \{f, \, g\}, \, H\} = \{ \{f, \, H\}, \, g\} + \{ \{f, \, g\}, \, H\}$$
$$= - \{ \{H, \, f\}, \, g\} - \{ \{g, \, H\}, \, f\}, \tag{3.251}$$

where we have used the antisymmetry property (3.248). Equation (3.250) is equivalent to the constancy of the Poisson bracket in time for conservative systems and (3.251) is the Jacobi identity for f, g and H. Consider $\{f, \, g\}(t)$ as a function of time and let us expand it in the neighbourhood of $t = 0$ [53];

$$\{f, \, g\}(t) = \{f, \, g\}(0) + t \left(\frac{d}{dt} \{f, \, g\} \right)_{t=0}$$
$$= \left\{ f + t \left(\frac{df}{dt} \right)_{t=0}, \, g + t \left(\frac{dg}{dt} \right)_{t=0} \right\}$$
$$+ \{f, \, g\}_{t=0} + t \left[\left\{ f, \, \frac{dg}{dt} \right\} + \left\{ \frac{df}{dt}, \, g \right\} \right]_{t=0}, \tag{3.252}$$

and thus leading to (3.250).

Pseudo-Hamiltonian and Duffin's Bracket — For certain types of dissipative systems Duffin has suggested a pseudo-Hamiltonian mechanics where the canonical equations are given by

$$\dot{q}_i = \frac{\partial H}{\partial p_i}, \qquad \epsilon \dot{p}_i = \frac{\partial H}{\partial q_i}, \tag{3.253}$$

where ϵ is a constant $\epsilon \neq -1$. In this way rather than using two functions L and $\mathcal{F}$ to describe the motion as we did in the Lagrangian approach, Eqs.

(2.1), (2.3), we have a single function H. Now if we compare (3.253) with the Hamilton equations derived from the Lagrangian (2.1), viz,

$$\dot{q}_i = \frac{\partial H}{\partial p_i}, \quad \dot{p}_i = -\frac{\partial H}{\partial q_i} + \frac{\partial \mathcal{F}}{\partial \dot{q}_i} = -\frac{\partial H}{\partial q_i} + Q_i, \tag{3.254}$$

then we can say that the generalized external forces are given by

$$Q_i = -\left(1 + \frac{1}{\epsilon}\right) \frac{\partial H}{\partial q_i}. \tag{3.255}$$

For a dissipative system the time derivative of a dynamical variable $g(q_i, p_i)$ is defined by

$$\frac{dg}{dt} = \sum_i \left(\frac{\partial g}{\partial q_i} \frac{\partial H}{\partial p_i} + \epsilon \frac{\partial g}{\partial p_i} \frac{\partial H}{\partial q_i} \right) = \{g, H\}_{\mathcal{D}}. \tag{3.256}$$

Duffin has shown that this bracket $\{\,,\,\}_{\mathcal{D}}$ remains invariant under specific representations of canonical transformations [52]. For the type of frictional force derived from $\mathcal{F}$, we can express Duffin's bracket in terms of Poisson bracket;

$$\{g, H\}_{\mathcal{D}} = \sum_i \left(\frac{\partial g}{\partial q_i} \frac{\partial H}{\partial p_i} - \frac{\partial g}{\partial p_i} \frac{\partial H}{\partial q_i} - \frac{\partial g}{\partial p_i} f_i \right)$$

$$= \{g, H\} - \sum_i Q_i \frac{\partial g}{\partial p_i}, \tag{3.257}$$

where Q_i s are defined by (2.2).

The Duffin bracket (3.257) represents a nonconservative system and as such the properties (3.248)-(3.252) do not hold for it. In particular $\{f, g\}_{\mathcal{D}} =$ as a function of time is not a constant as can be seen from the inequality

$$\frac{d}{dt} \{f, g\}_{\mathcal{D}} \neq \left\{ \frac{df}{dt}, g \right\}_{\mathcal{D}} + \left\{ f, \frac{dg}{dt} \right\}_{\mathcal{D}}, \tag{3.258}$$

and consequently

$$\{f, g\}_{\mathcal{D}}(t) \neq \{f(t), g(t)\}_{\mathcal{D}}, \tag{3.259}$$

therefore the Poisson theorem is violated.

Now let us consider two other forms of brackets for expressing time derivatives. The first one which we denote by the subscript L, i.e. the bracket $\{\,,\,\}_{\mathcal{D}}$ is defined by

$$\{f, g\}_L = \{f, g\}_{\mathcal{D}} - \{g, f\}_{\mathcal{D}}, \tag{3.260}$$

and is a Lie bracket, since it is constant [56]–[58]

$$\{f, g\}_L = (1 - \epsilon) \{f, g\}. \tag{3.261}$$

For this case we call $\{f, g\}_{\mathcal{D}}$ a Lie admissible bracket and we have the Jacobi identity

$$\{\{f, g\}_L, h\}_L + \{\{g, h\}_L, f\}_L + \{\{h, f\}_L, g\}_L = 0. \tag{3.262}$$

Now if the generalized force Q_i is not equal to $- \left(1 + \frac{1}{\epsilon}\right)\left(\frac{\partial H}{\partial q_i}\right)$, then $\{g,\, H\}$ given by (3.257) is not Lie-admissible, i.e. the Jacobi identity does not hold for $\{f,\, g\}_{\mathcal{D}} - \{g,\, f\}_{\mathcal{D}}$. For such a case we introduce a second form of bracket $\{\,,\,\}_S$ for the symmetrized equations of motion

$$\left(\frac{df}{dt}\right)_S = \{f,\, H\}_S = \{f,\, H\}_{\mathcal{D}} + \{H,\, f\}_{\mathcal{D}}. \tag{3.263}$$

This bracket is trivially Lie-admissible for $\{f,\, H\}_S - \{H,\, f\}_S = 0$.

In the case of a linearly damped oscillator this symmetrization property of the Hamiltonian changes the Lie-admissible algebra into a Lie algebra. Thus for Bateman's damped-amplified osillator Sec. 3.7 for each of x and y motion, Eqs. (3.94) and (3.95), taken separately, the corresponding algebra is Lie-admissible. However if we consider the symmetrized Hamiltonian (3.96), H is a constant of motion and one can formulate the problem in terms of Lie brackets [53].

Canonical Formulation According to Sedov's Variational Principle — We have seen the equations of motion derived from Sedov's variational principle are Eqs. (2.23) and the canonical momentum is given by (2.25). In these equations L is the Lagrangian for the conservative forces acting on the system. The Hamiltonian can be determined by the standard method used for conservative systems [59]

$$H(x, p) = \sum_i p_i v_i(x, p) - L(x, v(x, p)). \tag{3.264}$$

From the variation of the Hamiltonian (3.264) we obtain the canonical equations of motion

$$\frac{dq_i}{dt} = \frac{\delta(H - w)}{\delta p_i}, \qquad \frac{dp_i}{dt} = -\frac{\delta(H - w)}{\delta q_i}, \tag{3.265}$$

where

$$\delta w(x, p) = \delta w(x, v(x, p)) = \sum_i \left[(w_q)_i \delta q_i + (w_p)_i \delta p_i \right]. \tag{3.266}$$

Let us consider the $(2n + 2)$-dimensional phase space composed of the following coordinates and momenta $(z_1, \cdots z_{2n}, w, t)$, where $z_i = x_i$ and $z_{n+1} = p_i$, $i = 1,\, 2, \cdots n$, and let us assume that δw and δz_k are related to each other by

$$\delta w - \sum_k a_k(z, t) \delta z_k = 0. \tag{3.267}$$

In this relation $a_k(z, t)$ is a vector function in this phase space. If the vector function satisfies the following relations

$$\frac{\partial a_k(z)}{\partial z_l} = \frac{\partial a_l(z)}{\partial z_k}, \qquad k, l = 1,\, 2 \cdots 2n, \tag{3.268}$$

then $w(z)$ describes a holonomic system, otherwise it is nonholonomic (for the definitions of these systems see Sec. 2.1).

Poisson Brackets for Sedov's Variational Principle — We can replace the Poisson bracket for dissipative system, as formulated by Tarasov, by Duffine bracket [54],[55]

$$\{f,\, g\}_{\mathcal{D}} = \sum_i \left(\frac{\delta f}{\delta x_i} \frac{\delta g}{\delta p_i} - \frac{\delta f}{\delta p_i} \frac{\delta g}{\delta x_i} \right). \tag{3.269}$$

Thus the Leibnitz rule has to be modified in the following way:

$$\frac{d}{dt}\{f,\, g\}_{\mathcal{D}} = \left\{ \frac{df}{dt},\, g \right\}_{\mathcal{D}} + \left\{ f,\, \frac{dg}{dt} \right\}_{\mathcal{D}} + J\{f,\, w,\, g\}, \tag{3.270}$$

where $J\{f,\, w,\, g\}$ is the Jacobian;

$$J\{f,\, w,\, g\} = \{f,\, \{w,\, g\}\} + \{w,\, \{g,\, f\}\} + \{g,\, \{f,\, w\}\}. \tag{3.271}$$

The nonvanishing of the Jacobian $J(f, w, g)$ indicates the dissipative nature of the motion.

This definition of bracket, for holonomic systems is identical with the standard definition of Poisson brackets and satisfies the same properties [59].

As we have seen Sedov's variational principle allows us to express the equations of motion as (3.265), or write them in terms of the Poisson brackets:

$$\frac{dq_i}{dt} = \{q_i,\, H\} - G_i(t, q, p), \tag{3.272}$$

$$\frac{dp_i}{dt} = \{p_i,\, H\} + F_i(t, q, p), \tag{3.273}$$

where the Poisson brackets $\{q_i,\, H\}$ and $\{p_i,\, H\}$ are the standard forms of the brackets [59]. The two functions F_i and G_i describe the contributions of dissipative part, w, to the conservative system given by H. For instance these functions can express the effects of the coupling of the system to a heat bath. If we define $\Omega_{ij}^{(s)}$, $s = 1,\, 2,\, 3$ by the relations;

$$\Omega_{ij}^{(1)}(t, q, p) = \frac{\partial G_j}{\partial p_i} - \frac{\partial G_i}{\partial p_j}, \tag{3.274}$$

$$\Omega_{ij}^{(2)}(t, q, p) = \frac{\partial G_i}{\partial q_j} - \frac{\partial F_j}{\partial p_i}, \tag{3.275}$$

and

$$\Omega_{ij}^{(3)}(t, q, p) = \frac{\partial F_j}{\partial q_i} - \frac{\partial F_i}{\partial q_j}, \tag{3.276}$$

then we can write these $\Omega_{ij}^{(s)}(t, q, p)$ s as Poisson brackets:

$$\Omega_{ij}^{(1)}(t, q, p) = \{q_i,\, G_j\} - \{q_j,\, G_i\}, \tag{3.277}$$

$$\Omega_{ij}^{(2)}(t, q, p) = \{G_i,\, p_j\} - \{q_i,\, F_j\}, \tag{3.278}$$

$$\Omega_{ij}^{(3)}(t, q, p) = \{p_j, F_i\} - \{p_i, F_j\}. \tag{3.279}$$

In the case of a conservative closed system all $\Omega_{ij}^{(s)}(t, q, p)$ s will be zero.

Now with the help of the functions F_i and G_i, we can write the total time derivative of any dynamical variable A as

$$\frac{dA}{dt} = \frac{\partial A}{\partial t} + \{A, H\} + D_0(A), \tag{3.280}$$

where $D_0(A)$ is defined by

$$D_0(A) = \sum_i \left(F_i \frac{\partial A}{\partial p_i} - \frac{\partial A}{\partial q_i} G_i \right). \tag{3.281}$$

We note that the operator $D_0(A)$ cannot be represented as $\{A, H'\}$, when we restrict H' to be a function of q_i s and p_i s.

For the one-dimensional motion of a particle in a potential $V(x)$ and subject to a damping force $-\lambda p$, we have $G = 0$ and $F = -\lambda p$, and from Eq. (3.280) and (3.281) we find the equation of motion to be

$$m\ddot{x} + \lambda m\dot{x} + \frac{\partial V(x)}{\partial x} = 0. \tag{3.282}$$

There are others ways of formulating the Hamiltonian of a dynamical system with the generalized canonical equations of motion similar to what we have discussed here. Such formulations can also be used in the case of dissipative classical motions (for instance see reference [60]).

3.14   Dynamical Matrix Formulation of the Classical Hamiltonian and the Generalized Poisson Bracket

One of the most interesting generalizations of the Hamiltonian formulation which is applicable to dissipative systems has been proposed by Enz [61]. Let us consider the canonical equations of motion in any number of dimensions n:

$$\dot{p}_i = -\frac{\partial H}{\partial q_i}, \quad \dot{q}_i = \frac{\partial H}{\partial p_i}, \quad i = 1, 2 \cdots n. \tag{3.283}$$

We write these equations as a single matrix equation (see also Sec. 3.13)

$$\dot{x}_i = D_0 \frac{\partial H}{\partial x_i}, \quad D_0 = \begin{bmatrix} 0 & -1 \\ 1 & 0 \end{bmatrix}, \quad i = 1, 2 \cdots n, \tag{3.284}$$

where $x_i = (p_i, q_i)$. Now we replace the symplectic matrix D_0 by a general, real dynamical matrix D [61]

$$D = D^a - D^s, \tag{3.285}$$

where D^a and D^s are antisymmetric and semi-definite positive symmetric matrices respectively. Then H which is defined by

$$\dot{x}_i = D\frac{\partial H}{\partial x_i}, \quad i = 1, 2 \cdots n, \tag{3.286}$$

represents the Hamiltonian for a dissipative system.

Now supposing that $\frac{\partial H}{\partial t} = 0$, for the total time derivative of H we find

$$\dot{H} = \sum_{i=1}^{2n} \frac{\partial H}{\partial x_i}\dot{x}_i = \sum_{i=1}^{2n} \frac{\partial H}{\partial x_i}D_{ij}\frac{\partial H}{\partial x_j}$$

$$= -\sum_{i,j=1}^{2n} D_{ij}^s \frac{\partial H}{\partial x_i}\frac{\partial H}{\partial x_j} \le 0. \tag{3.287}$$

The inequality in (3.287) follows from the fact that D^s is semi-definite positive and symmetric matrix. Therefore $\dot{H} = 0$ implies that $D^s = 0$.

Generalized Poisson Bracket — We define the generalized Poisson bracket with the help of the matrix D as

$$\{A, B\}_D = \sum_{i,j=1}^{2n} \frac{\partial A}{\partial x_i}D_{ij}\frac{\partial B}{\partial x_j}. \tag{3.288}$$

This expression is similar to Eq. (3.231) defined earlier. Using this definition we observe that for any function of x, say $f(x)$ we have

$$\dot{f}(x) = \sum_{j=1}^{2n} \frac{\partial f(x)}{\partial x_j}\dot{x}_j = \{f, H\}_D. \tag{3.289}$$

Let us investigate the question of time-reversal invariance in this formulation. Assuming that both D and H are invariants under $t \to -t$, and in particular $\frac{\partial H}{\partial t} = 0$, we want to find conditions on D so that the resulting equations of motion are also invariant under time-reversal. For this we write Eq. (3.286) as

$$\begin{bmatrix} \dot{p} \\ \dot{q} \end{bmatrix} = \begin{bmatrix} D_{pp} & D_{pq} \\ D_{qp} & D_{qq} \end{bmatrix} \begin{bmatrix} \frac{\partial H}{\partial p} \\ \frac{\partial H}{\partial q} \end{bmatrix}, \tag{3.290}$$

where we have suppressed the subscripts of $\dot{q}, \dot{p} \cdots$. Then the requirement of the time-reversal invariance obtained from (3.290) is the vanishing of the matrix elements

$$D_{pp} = 0 \quad \text{and} \quad D_{qq} = 0 \tag{3.291}$$

Now comparing Eqs. (3.290) with (3.285) and (3.286) we find that

$$\begin{bmatrix} D_{pp} & 0 \\ 0 & D_{qq} \end{bmatrix} = -D^s, \quad \begin{bmatrix} 0 & D_{pq} \\ D_{qp} & 0 \end{bmatrix} = D^a. \tag{3.292}$$

Thus $D^s = 0$ implies that (a) - $\dot{H} = 0$, that is the Hamiltonian is conserved and (b) - the equations of motion are invariant under the transformation $t \to -t$.

Two Examples — As a first example we consider the one-dimensional motion of a particle of unit mass in a potential $V(q)$ subject to a damping force $-\lambda\dot{q}$. Here we choose D to be

$$D = \begin{bmatrix} -\lambda & -1 \\ 1 & 0 \end{bmatrix}, \tag{3.293}$$

then the resulting equation of motion is

$$\ddot{q} + \lambda\dot{q} + \frac{\partial V(q)}{\partial q} = 0. \tag{3.294}$$

As a second example we examine the Hamiltonian for the Lotka-Volterra model with the equations of motion [62]

$$\dot{x}_1 = a_1 x_1 (1 - r_1 x_2), \quad x_1 > 0, \quad a_1 > 0, \quad r_1 > 0, \tag{3.295}$$

$$\dot{x}_2 = -a_2 x_2 (1 - r_2 x_1), \quad x_2 > 0, \quad a_2 > 0, \quad r_2 > 0. \tag{3.296}$$

These equations express the rate of change of concentrations of substances X_1 and X_2. If we choose $r_1 = r_2 = 1$ then we can write the Hamiltonian as

$$H(x_1, x_2) = \frac{1}{a_1} h(x_1) + \frac{1}{a_2} h(x_2), \tag{3.297}$$

where

$$h(x) = x - 1 - \ln x \ge 0. \tag{3.298}$$

This Hamiltonian enables us to write the equations of motion (3.295) and (3.296) in terms of the dynamical matrix Eq. (3.286) as

$$D = D^a = \begin{bmatrix} 0 & -a_1 a_2 x_1 x_2 \\ a_1 a_2 x_1 x_2 & 0 \end{bmatrix} \tag{3.299}$$

Since D is antisymmetric then from Eq. (3.287) it follows that $\dot{H} = 0$. The simplest way of adding dissipation to the Lotka-Volterra model is to add the following symmetric matrix D^s to Eq. (3.299) where

$$D^s = \begin{bmatrix} a_1 x_1 & 0 \\ 0 & a_2 x_2 \end{bmatrix}. \tag{3.300}$$

With this addition the resulting equations of motion become

$$\dot{x}_1 = a_1 x_1 (1 - x_2) + 1 - x_1, \tag{3.301}$$

$$\dot{x}_2 = -a_2 x_2 (1 - x_1) + 1 - x_2. \tag{3.302}$$

In this case the rate of change of the Hamiltonian with respect to time is negative definite [61]

$$\dot{H} = -\sum_{i=1,2} \frac{(1 - x_i)^2}{a_i x_i} \le 0, \quad x_i > 0. \tag{3.303}$$

3.15   Symmetric Phase Space Formulation of the Damped Harmonic Oscillator

A different but symmetric extension of the phase space, again suitable for studying dissipative motions, can be found by treating the phase space coordinates on the same footing by introducing canonical momenta conjugate to these coordinates in the Lagrangian formalism [63]. Here we start with the extended Lagrangian $L(q, p, \dot{q}, \dot{p})$ defined by

$$L(q, p, \dot{q}, \dot{p}) = -\dot{q}p - \dot{p}q + L_q(q, \dot{q}) + L_p(p, \dot{p}), \qquad (3.304)$$

where L_q and L_p are the q and p space Lagrangians, i.e. they give correct equations of motion in coordinate and momentum space respectively. From the Lagrangins we find the canonical conjugate variables corresponding to q and p:

$$\pi_q = \frac{\partial L}{\partial \dot{q}} = \frac{\partial L_q}{\partial \dot{q}} - p, \qquad (3.305)$$

and

$$\pi_p = \frac{\partial L}{\partial \dot{p}} = \frac{\partial L_p}{\partial \dot{p}} - q. \qquad (3.306)$$

We note that π_p has the dimension of the coordinate q whereas π_q has the dimension of the canonical momentum p.

From these canonical momenta we obtain the extended phase space Hamiltonian

$$H(q, p, \pi_q, \pi_p) = \dot{q}\pi_q + \dot{p}\pi_p - L = H(p + \pi_q, q) - H(p, q + \pi_p). \qquad (3.307)$$

This formulation can be applied to describe the motion of a damped harmonic oscillator using successive canonical transformations [64]. To this end we start with the problem of the undamped harmonic oscillator for which the extended Hamiltonian is

$$H = \frac{1}{2m}\pi_q^2 + \frac{1}{m}p\pi_q - \frac{m\omega^2}{2}\pi_p^2 - m\omega^2 q\pi_p. \qquad (3.308)$$

Now we make the canonical transformation

$$q \to q, \quad p \to p, \quad \pi_q \to -\pi_q - p, \quad \pi_p \to -\pi_p - q, \qquad (3.309)$$

with the result that

$$H_1 = \left(\frac{\pi_q^2}{2m} + \frac{1}{2}m\omega^2 q^2\right) - \left(\frac{p^2}{2m} + \frac{1}{2}m\omega^2 \pi_p^2\right). \qquad (3.310)$$

A second transformation

$$q \to q, \quad p \to p, \quad \pi_q \to \pi_q + \frac{m\lambda q}{2}, \quad \pi_p \to \pi_p - \frac{\lambda p}{2m\omega^2}, \qquad (3.311)$$

changes H_1 to H_2, where

$$H_2 = \left[\frac{\pi_q^2}{2m} + \frac{1}{2}m\left(\omega^2 + \frac{\lambda^2}{4}\right)q^2\right] - \left[\frac{p^2}{2m}\left(1 + \frac{\lambda^2}{4\omega^2}\right) + \frac{1}{2}m\omega^2\pi_p^2\right]$$
$$+ \frac{\lambda}{2}\left(q\pi_q + p\pi_p\right). \tag{3.312}$$

Finally we use a third transformation with the generator

$$F_2\left(q, \mathcal{P}_q, p, \mathcal{P}_p, t\right) = q\mathcal{P}_q e^{\frac{-\lambda t}{2}} + p\mathcal{P}_p e^{\frac{\lambda t}{2}}. \tag{3.313}$$

From this last transformation, (3.313), we obtain the following relations between the old and the new canonical variables:

$$Q = \frac{\partial F_2}{\partial \mathcal{P}_q} = qe^{\frac{-\lambda t}{2}}, \quad P = \frac{\partial F_2}{\partial \mathcal{P}_p} = pe^{\frac{\lambda t}{2}}, \tag{3.314}$$

and

$$\pi_q = \frac{\partial F_2}{\partial q} = \mathcal{P}_q e^{\frac{-\lambda t}{2}}, \quad \pi_p = \frac{\partial F_2}{\partial p} = \mathcal{P}_p e^{\frac{\lambda t}{2}}. \tag{3.315}$$

Substituting these in the new Hamiltonian H_3 we find

$$H_3 = H_2\left(Q, \mathcal{P}_q, P, \mathcal{P}_p, t\right) + \frac{\partial F_2}{\partial t}, \tag{3.316}$$

or

$$H_3 = \left[\frac{\mathcal{P}_q^2}{2m}e^{-\lambda t} + \frac{1}{2}m\left(\omega^2 + \frac{\lambda^2}{4}\right)Q^2 e^{\lambda t}\right]$$
$$- \left[\frac{P^2}{2m}\left(1 + \frac{\lambda^2}{4\omega^2}\right)e^{-\lambda t} + \frac{1}{2}m\omega^2\mathcal{P}_p^2 e^{\lambda t}\right]. \tag{3.317}$$

The expression in the first bracket of (3.317) is the Caldirola-Kanai Hamiltonian and the terms in the second bracket give us the classical evolution for the image oscillator [64]. The equation of motion for the image oscillator found from H_3 is

$$\ddot{P} - \lambda P + \left(\omega^2 + \frac{\lambda^2}{4}\right)P = 0, \tag{3.318}$$

and this is the same as that of the equation for y of the Bateman model we obtained earlier (see Sec. 3.7).

3.16 Hamiltonian Formulation of Dynamical Systems Expressible as Difference Equations

We can find a Hamiltonian description for the equation of motion (1.153) when t_n s are defined by an expression similar to (1.156) or preferably given as a set of numbers independent of p_n s. To this end we write Eq. (1.153) as a differential equation

$$\frac{dp}{dt} = -f(t)p + g(t), \tag{3.319}$$

where $f(t)$ and $g(t)$ are defined by

$$f(t) = (i\pi - \ln \rho) \sum_n \delta(t - t_n), \tag{3.320}$$

and

$$g(t) = -2m \sum_n V(t)\delta(t - t_n). \tag{3.321}$$

Equation (3.319) shows that the impulsive forces are time- and velocity-dependent, and the system, in general, is not conservative.

From the equation of motion (3.319) which does not explicitly depend on the x-coordinate, we can construct the Hamiltonian. Noting that the mechanical momentum $p = m\frac{dx}{dt}$ is not the same as the canonical momentum, P, we write the Hamiltonian as a function of x, P and t;

$$H = \frac{P^2}{2m} \exp\left[-\int_0^t f(t') \, dt'\right] - xg(t) \exp\left[\int_0^t f(t') \, dt'\right]. \tag{3.322}$$

The canonical equations of motion found for H are:

$$p = m\frac{dx}{dt} = m\frac{\partial H}{\partial P} = P \exp\left[-\int_0^t f(t') \, dt'\right], \tag{3.323}$$

and

$$\frac{dP}{dt} = -\frac{\partial H}{\partial x} = g(t) \exp\left[\int_0^t f(t') \, dt'\right]. \tag{3.324}$$

By eliminating P between these two equations we find the equation of motion (3.319). In order to relate the Hamiltonian H to the energy of the particle, $E(t)$, we substitute for P in terms of p in H, Eq. (3.322) and thus we find;

$$H = \left[\frac{p^2}{2m} - xg(t)\right] \exp\left[2\int_0^t f(t') \, dt'\right] = E \exp\left[2\int_0^t f(t') \, dt'\right], \tag{3.325}$$

where E is the sum of kinetic and potential energies of the particle. When the coefficient of restitution ρ is one, then $f(t) = i\pi \sum_n \delta(t - t_n)$, and $H = E$.

3.17 Fractional Derivatives in the Hamiltonian Formulation of Damped Motion

We have already observed that the equation for the damped motion of a dynamical system can be derived from a single Lagrangian containing the fractional time derivative, i.e. $L\left(\left\{q_{j,s'(n),a}, \ q_{j,s(n),b}\right\}, t\right)$ (Chapter 2). From this

Lagrangian we can find the generalized momenta as we found earlier, Eq. (3.1). Thus we define the momentum conjugate to the coordinate $q_{j,s(n)}$ by [65],[66]

$$
p_{j,s(n)} = p_{j,s(n),b} = \sum_{k=0}^{N-n-1} (-1)^{[s(k+n+1)-s(n+1)]}
$$

$$
\times \frac{d^{s(k+n+1)-s(n+1)}}{d(t-a)^{s(k+n+1)-s(n+1)}} \times \left(\frac{\partial L}{\partial q_{j,s(k+n+1),b}} \right), \tag{3.326}
$$

If the Lagrangian L depends on $q_{j,s'(n),a}$ as well as $q_{j,s(n),b}$, then we have a set of additional momenta $p_{r,s'(n),a}$, where

$$
p_{j,s'(n),a} = \sum_{k=0}^{N'-n-1} (-1)^{[s'(k+n+1)-s'(n+1)]}
$$

$$
\times \frac{d^{s'(k+n+1)-s'(n+1)}}{d(t-a)^{s'(k+n+1)-s'(n+1)}} \times \left(\frac{\partial L}{\partial q_{j,s'(k+n+1),a}} \right), \tag{3.327}
$$

and $n =$ in this relation takes $0, 1, \cdots N - 1$.

The Hamiltonian for this system is given by the Legendre transformation

$$
H = \sum_{j} \sum_{n=1}^{N} q_{j,s(n),b}\, p_{j,s(n-1),b} + \sum_{j} \sum_{n=1}^{N'} q_{j,s'(n),a}\, p_{j,s'(n-1),a} - L. \tag{3.328}
$$

By applying the variational principle (2.201) to the Hamiltonian H (3.328) we get

$$
\delta I(\alpha) = \delta \int_{a}^{b} \left(\sum_{j} \sum_{n=1}^{N} q_{j,s(n),b}\, p_{j,s(n-1),b} + \sum_{j} \sum_{n=1}^{N'} q_{j,s'(n),a}\, p_{j,s'(n-1),b} - H \right) dt
$$

$$
= 0, \tag{3.329}
$$

and thus derive Hamilton's canonical equations of motion

$$
\frac{\partial H}{\partial q_{j,s(n),b}} = \left((-1)^{[s(n+1)-s(n)]} \right) \frac{d^{s(n+1)-s(n)}}{d(t-a)^{s(n+1)-s(n)}} p_{j,s(n),b}, \tag{3.330}
$$

$$
\frac{\partial H}{\partial p_{j,s(n),b}} = q_{j,s(n+1),b}, \tag{3.331}
$$

$$
\frac{\partial H}{\partial t} = -\frac{\partial L}{\partial t}, \tag{3.332}
$$

and

$$
\frac{\partial H}{\partial q_{j,s'(n),a}} = \left[(-1)^{[s'(n+1)-s'(n)]} \right] \frac{d^{s'(n+1)-s'(n)}}{d(t-a)^{s'(n+1)-s'(n)}} p_{j,s'(n),a} \tag{3.333}
$$

$$\frac{\partial H}{\partial p_{j,s'(n),a}} = q_{j,s'(n+1),a}. \tag{3.334}$$

These are the canonical equations of motion for systems expressible in terms of fractional derivatives.

Fractional Hamiltonian for a Frictional Force Proportional to Velocity — Let us again consider the one-dimensional motion of a particle of mass m subject to a linear frictional force with the Lagrangian given by Eq. (2.206). From the latter Lagrangian and (3.326) we find the canonical momenta

$$p_0 = \frac{\partial L}{\partial q_{\frac{1}{2}}} + i\frac{d^{\frac{1}{2}}}{d(t-b)^{\frac{1}{2}}}\left(\frac{\partial L}{\partial q_1}\right)$$

$$= i\lambda x_{\frac{1}{2},b} + imx_{\frac{3}{2},b}, \tag{3.335}$$

and

$$\frac{\partial L}{\partial q_1} = m\dot{x}. \tag{3.336}$$

By substituting p_0 and $p_{\frac{1}{2}}$ in (3.328) we obtain the Hamiltonian

$$H = q_{\frac{1}{2}}p_0 + q_1 p_{\frac{1}{2}} - L$$

$$= \frac{p_{\frac{1}{2}}^2}{2m} + q_{\frac{1}{2}}p_0 + V(q_0) - \frac{i\lambda}{2}q_{\frac{1}{2}}^2. \tag{3.337}$$

The canonical equations of motion obtained from (3.337) are

$$\frac{\partial H}{\partial q_0} = i\frac{d^{\frac{1}{2}}p_0}{d(t-b)^{\frac{1}{2}}} = \frac{\partial V(q_0)}{\partial q_0}, \tag{3.338}$$

$$\frac{\partial H}{\partial p_0} = q_{\frac{1}{2}}, \tag{3.339}$$

$$\frac{\partial H}{\partial q_{\frac{1}{2}}} = i\frac{d^{\frac{1}{2}}}{d(t-b)^{\frac{1}{2}}}p_{\frac{1}{2}} = p_0 - i\lambda q_{\frac{1}{2}}, \tag{3.340}$$

and

$$\frac{\partial H}{\partial p_{\frac{1}{2}}} = q_1. \tag{3.341}$$

From Eqs. (3.338) and (3.340) we find that q_0 satisfies (2.208). However if we change our notations and write (3.337) as

$$H = \frac{1}{2m}P^2 + Qp + V(q) - \frac{i}{2}\lambda Q^2, \tag{3.342}$$

then from the Hamilton canonical equations and (3.342) we obtain the equation of motion for q which is of the fourth order

$$m\frac{d^4q}{dt^4} - \frac{\partial V(q)}{\partial q} - i\lambda\frac{d^2q}{dt^2} = 0 \tag{3.343}$$

This result shows that we cannot use the classical Hamiltonian (3.342) for the purpose of quantization.

Let us also note that while the motion described by (2.208) is one-dimensional and has a single degree of freedom, in the Hamiltonian formulation we have two degrees of freedom represented by q_0 and $q_{\frac{1}{2}}$. This aspect of the damped motion will be studied further in Chapter 20.

References

[1] H. Goldstein, *Classical Mechanics*, Second Edition (Addison-Wesley Reading, 1980) Chapter 8.

[2] See for instance, J.L. McCauley, *Classicl Mechanics*, (Cambridge University Press, Cambridge, 1997) p. 75.

[3] F.J. Kennedy and E.H. Kerner, Am. J. Phys. 33, 463 (1965).

[4] N.A. Lemos, Phys. Rev. D 24, 1036 (1981).

[5] M. Razavy and F.J. Kennedy, Can. J. Phys. 52, 1532 (1974).

[6] V.V. Dodonov, V.I. Man'ko and V.D. Skarzhinskiy in *Quantization, Gravitation and Group Methods in Physics*, Edited by A.A. Komar (Nova Science, Commack, 1988) p. 73.

[7] V.V. Dodonov, V.I. Man'ko and V.D. Skarzhinsky, Hadronic J. 4, 1734 (1981).

[8] V.V. Dodonov, V.I. Man'ko and V.D. Skarzhinsky, Nuovo Cimento B 69, 185 (1982).

[9] P. Cardirola, Nuovo Cimento 18, 393 (1941).

[10] P. Havas, Nuovo Cimento Supp. 5, 363 (1957).

[11] M. Ostrogradsky, Mem. Ac. St. Petersbourg, VI, 385 (1850).

[12] A. Pais and G. Uhlenbeck, Phys. Rev. 79, 145 (1950).

[13] E.T. Whittaker, *A Treatise on the Analytical Dynamics of Particles and Rigid Bodies*, Fourth Edition (Cambridge University Press, London, 1965) Chapter X.

[14] P. Caldirola, Rend. Ist. Lomb. Sc., A 93, 439 (1959).

[15] G. Valentini, Nuovo Cimento, XIX, 1280 (1961).

[16] L. Infeld, Acta Physica Hungaricae XVII, 7 (1964).

[17] B-G. Englert, Ann. Phys. (NY), 129, 1 (1980).

[18] H. Dekker, Z. Phys. B 26, 273 (1977).

[19] H. Dekker, Phys. Rev. A 16, 2126 (1977).

[20] D.M. Greenberger J. Math. Phys. 20, 762 (1979).

[21] D.M. Greenberger J. Math. Phys. 20, 771 (1979).

[22] J.R. Ray, Am. J. Phys. 47, 626 (1979).

[23] V.G. Ibarra-Sierra, A. Anzaldo-Meneses, J.L. Cardosso, H. Hernandez-Saldana, A. Kunold and J.A.E. Roa-Neri, Ann. Phys (NY). 335, 86 (2013).

[24] J. Flores, G. Solovey and S. Gil, Am. J. Phys. 71, 721 (2003).

[25] H. Bateman, Phys. Rev. 38, 815 (1931).

[26] P.M. Morse and H. Feshbach, *Methods of Theoretical Physics*, Part I (McGraw-Hill, New York, N.Y. 1953) p. 298.

[27] H. Dekker, Phys. Rep. 80, 1 (1981).

[28] H. Feshbach and Y. Tikochinsky, Tran. N.Y. Acad. Sci. Ser. II, 38, 44 (1977).

[29] R. Banerjee and P. Mukherjee, J. Phys. A 35, 5591 (2002).

[30] V.K.Chandrasekar, M. Senthilvelan and M. Lakshmanan, Proc. Roy. Soc. A461, 2451 (2005).

[31] V.K.Chandrasekar, M. Senthilvelan and M. Lakshmanan, J. Math. Phys. 48, 032701 (2007).

[32] C.M. Bender, M. Gianfreda, N. Hassanpour and H.F. Jones, J. Math. Phys. 57, 084101 (2016).

[33] J.M. Cerveró and J. Villarroel, J. Phys. 17, 1777 (1984).

[34] J.M. Cerveró and J. Villarroel, J. Phys. 17, 2963 (1984).

[35] S. Lie, *Vorlesungen über Differentialgleichungen mit bekannten infinitesimalen Transformationen*, ed. G. Sheffers (Tubner, Leipzig, 1891).

[36] M. Lutzky, J. Phys. 11, 249 (1978).

[37] H.R. Lewis, Phys. Rev. Lett. 18, 635 (1967).

[38] H.R. Lewis and P.G.L. Leach, J. Math. Phys. 23, 2371 (1982).

[39] H.R. Lewis and W.B. Riesenfeld, J. Math. Phys. 10, 1458 (1969).

[40] P.G.L. Leach, J. Phys. A 16, 3261 (1983).

[41] M. Razavy, Phys. Rev. A 36, 482 (1987).

[42] J. Geicke, Rev. Brasil. Fisica, 21, 429 (1991).

[43] R. Cawley, Phys. Rev. A 20, 2370 (1979).

[44] P.A.M. Dirac, *Lectures on Quantum Mechanics*, (Yeshiva University, New York, 1965) Lecture No. 1.

[45] H.J. Rothe and K.D. Rothe, *Classical and Quantum Dynamics of Constrained Hamiltonians*, (World Scientific, Singapore, 2010).

[46] R. Cawley, Phys. Rev. Lett. 42, 413 (1979).

[47] G.D. Birkhoff, *Dynamical Systems*, (American Mathematical Society, New York, N.Y. 1927) p. 57.

[48] Y. Nambu, Phys. Rev. D 7, 2405 (1973).

[49] P.A.M. Dirac, *The Principles of Quantum Mechanics*, 4th Edition, (Oxford University Press, London, 1981).

[50] H.J. Groenewold, Physica, 12, 405 (1946).

[51] J.R. Shewell, Am. J. Phys. 27, 16 (1959).

[52] R.J. Duffin, Arch. Rational Mech. Anal., 9, 309 (1962).

[53] M.D. Srinivas and T.S. Shankara, Lettere Al Nuovo Cimento, 6, 321 (1973).

[54] V.E. Tasarov, Theor. Math. Phys. 100, 1100, (1994).

[55] V.E. Tarasov, Theor. Math. Phys. 110, 57, (1997).

[56] F. Iachello, *Lie Algebras and Applications*, (Springer, Berlin, 2006).

[57] A.A. Albert, Trans. Am. Math. Soc. 64, 552 (1948).

[58] R.M. Santilli, Supp. al Nuovo Cimento, 6, 1225 (1968).

[59] See for instance L.D. Landau and E.M. Lifshitz, *Mechanics*, (Addison-Wesley, Reading, 1960) p. 134.

[60] S.Q.H. Nguyen and L.A. Turski, J. Phy. A 34, 8281 (2001).

[61] C.P. Enz, Found. Phys. 24, 1281 (1994).

[62] V.I. Arnol'd, *Ordinary Differential Equations*, (Springer-Verlag, 1992) p. 44.

[63] Y. Sobouti and S. Nasiri, Int. J. Mod. Phys. B 7, 3255 (1993).

[64] S. Nasiri and H. Safari, Proc. Inst. Math. NAS, Ukraine, 43, 654 (2002).

[65] F. Riewe, Phys. Rev. E 53, 1890 (1996).

[66] F. Riewe, Phys. Rev. E 56, 3581 (1997).

[67] M. Razavy, *Heisenberg's Quantum Mechanics*, (World Scientific, Singapore, 2011).

[68] J. Hartley and J.R. Ray, Phys. Rev. D 25, 385 (1982).

[69] See, for example, M.A. Morrison *Understanding Quantum Mechanics: A User's Manual*, (Prentice-Hall, Englewood Cliffs, 1990) Chapter 9.

[70] M. Razavy, Can. J. Phys. 50, 2037 (1972).

[71] F. Calogero and A. Degasperis, Am. J. Phys. 72, 1202 (2004).

[72] C.M. Binder and S. Boettcher, Phys. Rev. Lett. 80, 5243 (1998).

[73] A. Mostafazadeh, J. Math. Phys. 43, 205 (2002).

[74] N.Gisin, J. Phys. A 14, 2259 (1981).

Chapter 4

Hamilton-Jacobi Formulation

Our next task is to study the simplest forms of the Hamilton-Jacobi equation for damped systems. The idea here is to construct a generating function $F_2\,(q_1\cdots q_N; P_1\cdots P_N)$ which transforms an arbitrary set of canonical variables $(q_1\cdots q_N;\ P_1\cdots P_N)$ to a new set of canonical variables $Q_i\ =\ \beta_i\ =$ constant and $P_i = \alpha_i =$ constant [1],[2]. Here we assume that the system is integrable, and that the generating function exists globally [3]. The condition of integrability excludes most of the dissipative systems, yet for few simple but interesting cases that we want to consider the Hamilton-Jacobi (H-J) is separable and thus integrable.

The Hamilton-Jacobi equation which is a nonlinear equation of first order admits a great variety of complete integrals, i.e. solutions expressed in terms of N coordinates, q_j s and of time t and N parameters $\alpha_1\cdots\alpha_N$. Now if $S(q_1...q_N;\ \alpha_1\cdots\alpha_N)$ is a complete integral of the Hamilton-Jacobi equation then the integrals of the Hamilton's equations of motion are given by [2]

$$\frac{\partial S}{\partial \alpha_j} = -\beta_j, \quad \text{and} \quad \frac{\partial S}{\partial q_j} = p_j, \quad j = 1, 2\cdots N. \tag{4.1}$$

The close connection between the Schrödinger equation and the Hamilton-Jacobi (H-J) equation of classical dynamics enables us to study the semi-classical solution of the wave equation and also use H-J formulation to quantize the classical dissipative motion [4]–[7].

127

4.1   The Hamilton-Jacobi Equation for Linear Damping

Again let us consider the Hamiltonian for the linearly damped motion in three dimensions [6];

$$H\left(\mathbf{p},\mathbf{r},t\right) = \left(\frac{\mathbf{p}^2}{2m}\right)\exp(-\lambda t) + V(\mathbf{r})\exp(\lambda t), \tag{4.2}$$

where $V(\mathbf{r})$ is the potential which acts on the particle of mass m.

The standard Hamilton-Jacobi equation for this Hamiltonian is given by [1]

$$\left(\frac{1}{2m}\right)(\nabla\zeta)^2 e^{-\lambda t} + V(\mathbf{r})e^{\lambda t} + \frac{\partial\zeta}{\partial t} = 0. \tag{4.3}$$

A different form of the H-J equation can be found by utilizing the transformation

$$\zeta(\mathbf{r},t) = S(\mathbf{r},t)e^{\lambda t}, \tag{4.4}$$

and substituting for ζ in (4.3) to get

$$\left(\frac{1}{2m}\right)(\nabla S)^2 + V(\mathbf{r}) + \lambda S + \frac{\partial S}{\partial t} = 0. \tag{4.5}$$

Both of these forms can be used to obtain solutions for the motion of a particle in the presence of a linear dissipative force. For instance let us consider the application of the Hamilton-Jacobi equation to the problem of the damped harmonic oscillator. If we write $V(x) = \frac{1}{2}m\omega_0^2 x^2$, then Eq. (4.3) for one-dimensional motion reduces to

$$\left(\frac{1}{2m}\right)\left(\frac{\partial\zeta}{\partial x}\right)^2 e^{-\lambda t} + \frac{1}{2}m\omega_0^2 x^2 e^{\lambda t} + \frac{\partial\zeta}{\partial t} = 0. \tag{4.6}$$

Changing the variable x to y where

$$y = x\exp\left(\frac{\lambda t}{2}\right), \tag{4.7}$$

we transform (4.6) to an equation with variables y and t

$$\left(\frac{1}{2m}\right)\left(\frac{\partial\zeta}{\partial y}\right)^2 + \frac{1}{2}m\omega_0 y^2 = -\left(\frac{\partial\zeta}{\partial t} + \frac{1}{2}\lambda y\frac{\partial\zeta}{\partial y}\right). \tag{4.8}$$

Now we can expand the variables in the usual way of separation used in the Hamilton-Jacobi equation by assuming that ζ is the sum of two terms [1]

$$\zeta(y,t) = -\alpha t + W(y). \tag{4.9}$$

By substituting (4.9) in (4.8) we obtain the following differential equation for $W(y)$

$$\frac{1}{2m}\left[\left(\frac{dW}{dy}\right) - \frac{1}{2}\lambda my\right]^2 + \frac{1}{2}m\omega^2 y^2 = \alpha, \tag{4.10}$$

where $\omega^2 = \omega_0^2 - \frac{\lambda^2}{4}$. Now if we integrate (4.10) and substitute for $W(y)$ in (4.9) we find $\zeta(y,t)$

$$\zeta(y,t) = -\alpha t + \frac{1}{4}\lambda my^2 + \int \left\{2m\left[\alpha - \frac{m}{2}\omega^2 y^2\right]\right\}^{\frac{1}{2}} dy. \tag{4.11}$$

From this equation we can obtain y as a function of time t by noting that [1]

$$\beta = \frac{\partial \zeta(y,t)}{\partial \alpha} = -t + \sqrt{\frac{m}{2\alpha}} \int \frac{dy}{\sqrt{1 - \frac{m\omega^2 y^2}{2\alpha}}}, \tag{4.12}$$

where β is a constant. Carrying out the integration in (4.12) and then substituting from (4.7) we finally determine x as a function of t

$$x(t) = \sqrt{\frac{2\alpha}{m\omega^2}} e^{\frac{-\lambda t}{2}} \sin\left[\omega(t+\beta)\right]. \tag{4.13}$$

We can also solve Eq. (4.5) for the damped harmonic oscillator. For the Jacobi function S we can separate the time dependent part by writing

$$S(x,t) = W_1(x) + \frac{q^2}{2m\lambda}\left(e^{-\lambda t} - 1\right), \tag{4.14}$$

and substitute this in the equation

$$\left(\frac{1}{2m}\right)\left(\frac{\partial S}{\partial x}\right)^2 + V(x) + \lambda S + \frac{\partial S}{\partial t} = 0, \tag{4.15}$$

to find an ordinary differential equation for W_1

$$\left(\frac{1}{2m}\right)\left(\frac{dW_1}{dx}\right)^2 + V(x) + \lambda W_1 = \frac{q^2}{2m}. \tag{4.16}$$

In Eqs. (4.14) and (4.16) q^2 is the separation constant.

For the harmonic oscillator $V(x) = \frac{1}{2}m\omega_0^2 x^2$ and by using this method, we can obtain a parametric solution for W_1. Thus

$$W_1(z,q) = \frac{q^2}{2m\lambda}\left\{1 - \frac{1+\omega_0^2 z^2}{1+\lambda z + \omega_0^2 z^2} \exp\left[-\frac{\lambda}{\omega}\tan^{-1}\left(\frac{\lambda + 2\omega_0^2 z}{\omega}\right)\right]\right\},$$

where

$$x(z,q) = \frac{qz}{m\sqrt{(1+\lambda z + \omega_0^2 z^2)}} \exp\left[-\frac{\lambda}{2\omega}\tan^{-1}\left(\frac{\lambda + 2\omega_0^2 z}{\omega}\right)\right]. \tag{4.17}$$

Equation (4.14) is not the only way that we can write S. Another interesting form of S which is complex is given by [8]

$$S_J(x,t) = \frac{m}{2}\left(-\frac{\lambda}{2} \pm i\omega\right)(x - \xi(t))^2 + mx\dot{\xi}(t) + C(t), \qquad (4.18)$$

where $\xi(t)$ and $C(t)$ are solutions of the differential equations:

$$\ddot{\xi}(t) + \lambda\dot{\xi}(t) + \omega_0^2\xi(t) = 0, \qquad (4.19)$$

and

$$\dot{C}(t) + \lambda C(t) + \frac{1}{2}m\left(\dot{\xi}^2(t) - \omega_0^2\xi^2(t)\right) = 0. \qquad (4.20)$$

As we will see later such a complex action is obtained for the classical limit of a particular type of the solution of the Schrödinger-Langevin equation (Sec. 12.7).

4.2   Classical Action for an Oscillator with Leaky Spring Constant

In Sec. 2.6 we found the Lagrangian for an oscillator with complex spring constant, Eq. (2.192). The classical action S for this motion can be obtained directly from the definition of S;

$$S = \int_{\tau_1}^{\tau_2} L\left(z, \frac{dz}{d\tau}, \tau\right)d\tau. \qquad (4.21)$$

By substituting for L and carrying out the integration we find

$$S = \frac{m\omega}{2\sin[\omega(\tau_2 - \tau_1)]}\left[(z_2^2 + z_1^2)\cos\omega(\tau_2 - \tau_1) - 2z_1z_2\right], \qquad (4.22)$$

where z_1 and z_2 refer to the coordinates of the particle at τ_1 and τ_2 respectively.

We can determine the position of the particle directly from $S(\tau, 0)$ by noting that

$$p = m\frac{dz}{d\tau} = \frac{\partial S(\tau, 0)}{\partial z} = m\omega\left[z\cot(\omega\tau) - \frac{z_0}{\sin(\omega\tau)}\right]. \qquad (4.23)$$

There is another way that we can formulate the action for the damped harmonic oscillator with leaky spring constant. Consider the classical equation of motion

$$m\left(\frac{d^2x}{dt^2}\right) = m\left[-\omega^2\left(1 + \alpha^2\right)x + 2\alpha\omega\left(\frac{d\xi(t)}{dt}\right)\right], \qquad (4.24)$$

where $\xi(t)$ is given as the solution of the differential equation

$$\ddot{\xi} + 2\alpha\omega\dot{\xi} + \omega^2\left(1 + \alpha^2\right)\xi = 0. \qquad (4.25)$$

In this case x denotes the coordinate of a damped harmonic oscillator, $\xi(t)$, superimposed on the background of a harmonic oscillator of frequency $\omega\sqrt{1+\alpha^2}$. The Hamiltonian and the Hamilton-Jacobi equation for the motion (4.24) are given by:

$$H = \frac{p^2}{2m} + \frac{1}{2}m\omega^2\left(1+\alpha^2\right)x^2 + 2\alpha\omega m\dot{\xi}(x-\xi), \qquad (4.26)$$

and

$$\frac{\partial S}{\partial t} + \frac{1}{2m}\left(\frac{\partial S}{\partial x}\right)^2 + \frac{m}{2}\left(1+\alpha^2\right)\omega^2 x^2 + 2\alpha\omega m\dot{\xi}(x-\xi) = 0, \qquad (4.27)$$

respectively.

We can solve (4.27) with the result that

$$S(x,t) = \frac{1}{2}m\omega\sqrt{1+\alpha^2}\left[x-\xi(t)\right]^2 \cot\left(\sqrt{1+\alpha^2}\omega t\right) + mx\dot{\xi} + g(t), \qquad (4.28)$$

where $g(t)$ is a function of time and is defined by

$$g(t) = m\int^t\left(2\alpha\omega\xi\dot{\xi} + \frac{1}{2}m\omega\sqrt{1+\alpha^2}\xi^2 - \frac{1}{2}m\dot{\xi}^2\right)dt. \qquad (4.29)$$

4.3 More About the Hamilton-Jacobi Equation for the Damped Motion

Now we want to point out three interesting facts about the Hamilton-Jacobi equation for damped systems:

(a) The non-uniqueness of the Hamilton-Jacobi equation and thus the principal function.

(b) The separability of the principal function for a system of two interacting particles moving in a viscous medium.

The fact that λS appears linearly in (4.5) is important since the motion of a particle in the viscous medium should not affect the motion of another particle moving in the same medium as long as there is no interaction between the two particles.

(c) The effective mass in classical systems and its connection to the quadratic damping force.

Let us discuss these three points in detail:

(a) For q-equivalent Hamiltonians, e.g. H_1 and H_2 , Eq. (3.32),(3.33) we have different Hamilton-Jacobi equations. For H_1 the Hamilton-Jacobi equation is given by (4.2) with $V = 0$, but for H_2 the corresponding equation for the principal function S is

$$\mathcal{E}_2 \exp\left[\frac{1}{p_0}\left(\frac{\partial S}{\partial x}\right)\right] + \lambda p_0 x + \frac{\partial S}{\partial t} = 0. \qquad (4.30)$$

This equation is separable, and by writing it as

$$S = -\alpha t + W(\alpha, x), \tag{4.31}$$

we find the differential equation for W and by integrating it we get

$$W(\alpha, x) = p_0 \int^x \ln\left(\frac{\alpha - \lambda p_0 x'}{\mathcal{E}}\right) dx'. \tag{4.32}$$

Since $(\partial S/\partial \alpha) = \beta$ is a constant, from (4.31) and (4.32) we obtain

$$-\lambda(\beta + t) = \int^x \frac{dx'}{x' - \left(\frac{\alpha}{\lambda p_0}\right)}, \tag{4.33}$$

which is the solution of the equation of motion.

(b) Let $S(\mathbf{r}_1, \mathbf{r}_2, t)$ be the solution of the modified Hamilton-Jacobi for the two particles, then

$$\frac{1}{2m_1}[\nabla_1 S(\mathbf{r}_1, \mathbf{r}_2, t)]^2 + \frac{1}{2m_2}[\nabla_2 S(\mathbf{r}_1, \mathbf{r}_2, t)]^2 + V_1(\mathbf{r}_1) + V_2(\mathbf{r}_2) + \lambda S + \frac{\partial S}{\partial t} = 0. \tag{4.34}$$

This equation can be decomposed into two equations:

$$\frac{1}{2m_i}[\nabla_i S_i(\mathbf{r}_i, t)]^2 + V_i(\mathbf{r}_i) + \lambda S_i(\mathbf{r}_i, t) + \frac{\partial}{\partial t} S_i(\mathbf{r}_i, t) = 0, \quad i = 1, 2, \tag{4.35}$$

where

$$S(\mathbf{r}_1, \mathbf{r}_2, t) = S_1(\mathbf{r}_1, t) + S_2(\mathbf{r}_2, t). \tag{4.36}$$

Thus each particle moves independently of the other in the medium. This additivity property of the S function in the corresponding quantum mechanical problem corresponds to the separability of the wave function.

(c) Finally we observed that the Hamiltonian for linear damping can also be interpreted as the Hamiltonian for a system where the mass of the system increases with time. We can inquire whether the same is true about the quadratic damping or not. Let us consider the following Hamiltonian

$$H = \left[\frac{p_r^2}{2m^*(r)} + \frac{p_\theta^2}{2m^*(r)r^2}\right] - \frac{1}{m}\int^r m^*(r')F(r')dr', \tag{4.37}$$

where $m^*(r)$ or the effective mass is given by

$$m^*(r) = m\left(1 - Ae^{-2\gamma r}\right), \quad A < 1. \tag{4.38}$$

This Hamiltonian gives us the equations of motion

$$m^*\ddot{r} = -\frac{1}{2}\frac{dm^*}{dr}\dot{r}^2 + \frac{rp_\theta^2}{m^{*2}(r)r^4}\left(m^* + \frac{r}{2}\frac{dm^*}{dr}\right) + \frac{m^*}{m}F(r), \tag{4.39}$$

and

$$\dot{\theta} = \frac{p_\theta}{m^*(r)r^2}, \quad p_\theta = \text{constant}. \tag{4.40}$$

Equation (4.39) shows that the radial motion of the particle is subject to a dissipative force proportional to $e^{-2\gamma r}\dot{r}^2$, i.e. this force becomes important at short distances. The Hamilton-Jacobi equation for the Hamiltonian (4.37) is found from H;

$$\frac{1}{2m^*(r)}(\nabla S)^2 + V(r) + \frac{\partial S}{\partial t} = 0, \tag{4.41}$$

where

$$V(r) = -\int^r \frac{m^*(r')}{m} F(r')\,dr'. \tag{4.42}$$

Later we will see that Eq. (4.41) is the classical limit of the Schrödinger equation for certain types of velocity-dependent potentials (see Sec. 11.9).

We can inquire whether for quadratic damping it is possible to give an interpretation in terms of variable mass also. As we will see it is possible to assume that the mass of the particle increases as it moves away from the center of force. Such a motion is mathematically equivalent to a dissipative force proportional to the square of the radial velocity.

4.4 The Hamilton-Jacobi Equation with Fractional Derivative

In Chapter 3 we found a Hamiltonian with fractional derivative for the motion of a particle in a medium where the dissipative force is proportional to velocity, Eq. (3.337). As a result of two successive canonical transformations similar to ones used for conservative systems, we find the canonical momenta as derivatives of the Hamilton's principal function S [9]–[11]

$$p_{\frac{1}{2m}} \to \frac{d^{\frac{1}{2}}}{d(t-a)^{\frac{1}{2}}} \frac{\partial S}{\partial q_{\frac{1}{2}}}, \tag{4.43}$$

and

$$p_0 \to (-1)^{\frac{1}{2}} \frac{d^{\frac{1}{2}}}{d(t-a)^{\frac{1}{2}}} \frac{\partial S}{\partial q_0}. \tag{4.44}$$

Substituting these in the Hamiltonian we obtain the Hamilton-Jacobi equation

$$-\frac{1}{2}\left[\frac{d^{\frac{1}{2}}}{d(t-a)^{\frac{1}{2}}}\frac{\partial S}{\partial q_{\frac{1}{2}}}\right]^2 + (-1)^{\frac{1}{2}}q_{\frac{1}{2}}\left(\frac{d^{\frac{1}{2}}}{d(t-a)^{\frac{1}{2}}}\frac{\partial S}{\partial q_0}\right)$$
$$+ V(q_0) - \left(\frac{i\lambda}{2}\right)q_{\frac{1}{2}}^2 = -\frac{\partial S}{\partial t}. \tag{4.45}$$

The presence of fractional time derivative in (4.45) makes this equation unsuitable for finding Hamilton's principal function. A simpler expression for S can be found by introducing the canonical coordinates $\bar{q}_0$ and $\bar{q}_{\frac{1}{2}}$, and expressing S in terms of these coordinates. Let us define these coordinates by the following relations [9]:

$$\bar{q}_0 = q_{-\frac{1}{2}} = \frac{1}{\Gamma\left(\frac{1}{2}\right)} \int_c^t (t-t')^{-\frac{1}{2}} q(t') \, dt', \tag{4.46}$$

and

$$\bar{q}_{\frac{1}{2}} = q_0, \tag{4.47}$$

then the momenta conjugate to these coordinates will be

$$p_{\frac{1}{2}} \rightarrow \frac{\partial S}{\partial q_0}, \tag{4.48}$$

and

$$p_0 \rightarrow \frac{\partial S}{\partial q_{-\frac{1}{2}}}. \tag{4.49}$$

We note that $p_{\frac{1}{2}}$ and p_0 are not canonically conjugate momenta to the coordinates $q_{\frac{1}{2}}$ and q_0.

If we introduce these momenta in H, Eq. (3.337), we obtain a Hamilton-Jacobi equation closer to the standard form, i.e. without mixed derivatives (see e.g. [1],[9]);

$$\frac{1}{2m}\left(\frac{\partial S}{\partial q_0}\right)^2 + q_{\frac{1}{2}}\left(\frac{\partial S}{\partial q_{-\frac{1}{2}}}\right) + V(q_0) - \left(\frac{i\lambda}{2}\right)q_{\frac{1}{2}}^2 + \frac{\partial S}{\partial t} = 0. \tag{4.50}$$

References

[1] H. Goldstein, *Classical Mechanics*, Second Edition (Addison-Wesley, Reading, Mass. 1980) Chapter 10.

[2] L.A. Pars, *A Treatise on Analytical Dynamics*, (John Wiley & Sons, New York, N.Y. 1965).

[3] J.L. McCauley, *Classical Mechanics*, (Cambrdge University Press, 1997) Chapter 16.

[4] M. Razavy, Z. Phys. B 26, 201 (1977).

[5] M. Razavy, Can. J. Phys. 56, 311 (1978).

[6] H.H. Denman and L.H. Buch, J. Math. Phys. 14, 326 (1973).

[7] L. Herrera, L. Nuñez, A Patiño, and H. Rago, Am. J. Phys. 54, 273 (1986).

[8] M. Razavy, Can. J. Phys. 56, 1372 (1978).

[9] F. Riewe, Phys. Rev. E 56, 3581 (1997).

[10] E.M. Rabei, I. Almayteh, S.I. Muslih and D. Beleamu, Physica Scr. 77, 1402 (2008).

[11] E.M. Rabei and B.S. Ababneh, J. Math. Anal. Appl. 344, 799 (2008).

Chapter 5

Motion of a Charged Damped Particle in an External Electromagnetic Field

When a particle of charge e moves in an external electromagnetic field and at the same time is subject to a conservative force $(-\nabla V(\mathbf{r}))$, we can either add the Lorentz force $e\left(\mathbf{E} + \frac{1}{c}\mathbf{v} \wedge \mathbf{B}\right)$ to the conservative force in the equation of motion or alternatively use the principle of minimal coupling and replace the canonical momentum $\mathbf{p}$ by $\left(\mathbf{p} - \frac{e}{c}\mathbf{A}\right)$ in the Hamiltonian $H(\mathbf{r}, \mathbf{p})$ and add a term $e\phi$ to H. Here $\mathbf{A}(\mathbf{r}, t)$ and $\phi(\mathbf{r}, t)$ are the vector and the scalar electromagnetic potentials respectively [1].

The physical meaning of the concept of minimal coupling which plays an essential role in the quantum theory of fields has been carefully analyzed and explained by Wentzel [2] (see also references [3],[4]. When the Hamiltonian H represents the energy of the particle, the two formulations lead to the same result [1]. However for the systems where H is not the energy of the system, and also for the dissipative systems in general, the minimum coupling rule does not give us the correct equation of motion. Let us illustrate this point by the following examples:

Minimal Coupling for Motion of a Charged Particle in a Viscous Medium — In the case of a charged particle moving in an external electromagnetic field $(\mathbf{E}(\mathbf{r}, t), \mathbf{B}(\mathbf{r}, t))$, and at the same time experiencing a linear damping force $(-\lambda\dot{\mathbf{r}})$ the equation of motion is

$$m\ddot{\mathbf{r}} + m\lambda\dot{\mathbf{r}} = \left(e\mathbf{E} + \frac{e}{c}\dot{\mathbf{r}} \wedge \mathbf{B}\right). \tag{5.1}$$

We have already seen that the Hamiltonian

$$H = \frac{\mathbf{p}^2}{2m} \exp(-\lambda t), \tag{5.2}$$

generates the equation of motion

$$m\ddot{\mathbf{r}} + m\lambda\dot{\mathbf{r}} = 0. \tag{5.3}$$

Now by applying the minimal coupling rule to H, Eq. (5.2), we find

$$H = \frac{1}{2m} \left(\mathbf{p} - \frac{e}{c}\mathbf{A}\right)^2 \exp(-\lambda t) + e\phi. \tag{5.4}$$

But this Hamiltonian will not give us the equation of motion (5.1). The Hamiltonian which is the generator of the equation of motion (5.1) is [5]

$$H = \frac{1}{2m} \left[\mathbf{p} - \frac{e}{c}\mathbf{A}_\lambda(\mathbf{r}, t)\right]^2 \exp(-\lambda t) + e\phi_\lambda(\mathbf{r}, t), \tag{5.5}$$

where

$$\mathbf{A}_\lambda(\mathbf{r}, t) = \mathbf{A}(\mathbf{r}, t) e^{\lambda t} - \lambda \int^t \mathbf{A}(\mathbf{r}, t') e^{\lambda t'} dt', \tag{5.6}$$

and

$$\phi_\lambda(\mathbf{r}, t) = \phi(\mathbf{r}, t) e^{\lambda t}. \tag{5.7}$$

This Hamiltonian with the generalized form of minimal coupling will lead to the equation of motion (5.1). Associated with this definition of the electromagnetic potentials we have the corresponding gauge transformation

$$\phi'_\lambda(\mathbf{r}, t) \to \phi_\lambda(\mathbf{r}, t) - \frac{1}{c}\frac{\partial\chi_\lambda(\mathbf{r}, t)}{\partial t} e^{\lambda t}, \tag{5.8}$$

for the electric field and

$$\mathbf{A}'_\lambda(\mathbf{r}, t) \to \mathbf{A}_\lambda(\mathbf{r}, t) + \nabla\chi_\lambda(\mathbf{r}, t), \tag{5.9}$$

for magnetic field where

$$\chi_\lambda(\mathbf{r}, t) = \chi(\mathbf{r}, t) e^{\lambda t} - \lambda \int^t \chi(\mathbf{r}, t') e^{\lambda t'} dt'. \tag{5.10}$$

Motion in a Constant Magnetic Field with Linear Damping Force —
As a second example let us study the time-independent Hamiltonian formulation for the motion of a particle of charge e which is moving in a viscous medium and is under the influence of a constant magnetic field $\mathbf{B}$ which is along the z axis. Here the vector potential is given by

$$\mathbf{A} = \frac{1}{2}B(\mathbf{j}x - \mathbf{i}y), \tag{5.11}$$

where $\mathbf{i}$, $\mathbf{j}$ and $\mathbf{k}$ are the unit vectors along the x, y and z axes respectively. The x
and y components of the position of the particle, which we assume is constrained
to move in the $z = $ constant plane, are the solutions of the equations of motion

$$\ddot{x} + \lambda \dot{x} - \Omega \dot{y} = 0, \tag{5.12}$$

and

$$\ddot{y} + \lambda \dot{y} + \Omega \dot{x} = 0, \tag{5.13}$$

where ω is the cyclotron frequency and is given by

$$\Omega = \frac{eB}{mc}. \tag{5.14}$$

We want to construct a time-independent Hamiltonian. This Hamiltonian
will be a constant of motion, a result which follows from Hamilton's canonical
equations.

Writing $v_x = \dot{x}$ and $v_y = \dot{y}$, from Eqs. (5.12) and (5.13) we have

$$\frac{dv_x}{dv_y} = \frac{\lambda v_x - \Omega v_y}{\Omega v_x + \lambda v_y}. \tag{5.15}$$

We can integrate this equation to yield

$$(v_x^2 + v_y^2) \exp \left[\frac{2\lambda}{\Omega} \tan^{-1} \left(\frac{v_x}{v_y} \right) \right] = \text{constant}. \tag{5.16}$$

Next we replace v_x and v_y in (5.16) by

$$v_x = \frac{\partial H}{\partial p_x}, \quad v_y = \frac{\partial H}{\partial p_y}, \tag{5.17}$$

to get

$$\left[\left(\frac{\partial H}{\partial p_x} \right)^2 + \left(\frac{\partial H}{\partial p_y} \right)^2 \right] \exp \left[\frac{2\lambda}{\Omega} \tan^{-1} \left(\frac{\frac{\partial H}{\partial p_x}}{\frac{\partial H}{\partial p_y}} \right) \right] = C(H), \tag{5.18}$$

where C is an arbitrary function of H. When $\lambda = 0$ and $C(H) = \frac{2}{m} H$, Eq.
(5.18) can be solved and we find the general form of H to be

$$H = \frac{1}{2m} [p_x - f(x, y)]^2 + \frac{1}{2m} [p_y - g(x, y)]^2, \tag{5.19}$$

where $f(x, y)$ and $g(x, y)$ are arbitrary functions of x and y. For the special case
of

$$f(x, y) = -m\Omega By, \quad \text{and} \quad g(x, y) = m\Omega Bx, \tag{5.20}$$

we have a Hamiltonian which can be found from the free particle Hamiltonian
by minimal coupling. But when $\lambda \neq 0$ the solution of (5.18) for $C(H) = \frac{2}{m} H$ is
not known.

The classical equations (5.12) and (5.13) must be modified if we want that in the corresponding quantum description the particle relaxes to a well-defined equilibrium state [6]. Rather than using Eqs. (5.12) and (5.13) one can consider the following, classical equations:

$$v_x = \dot{x} + \frac{\lambda}{2}x, \tag{5.21}$$

$$v_y = \dot{y} + \frac{\lambda}{2}y, \tag{5.22}$$

$$\ddot{x} = \Omega\dot{y} - \lambda\dot{x} - \omega_0^2 x + \frac{1}{2}\lambda\Omega y, \tag{5.23}$$

and

$$\ddot{y} = -\Omega\dot{x} - \lambda\dot{y} - \omega_0^2 y - \frac{1}{2}\lambda\Omega x. \tag{5.24}$$

These equations can be derived from the time-dependent Hamiltonian

$$H = \frac{1}{2m}\left(\mathbf{p} - \frac{e}{2c}\mathbf{B}\wedge\mathbf{r}\,e^{\lambda t}\right)^2 e^{-\lambda t} + \frac{1}{2}m\omega_0^2\mathbf{r}^2 e^{\lambda t}, \tag{5.25}$$

in which the minimal coupling rule is violated [7]. For a constant $\mathbf{B}$ the minimal coupling rule implies that $\mathbf{p}$ should be replaced by

$$\mathbf{p} \to \mathbf{p} - \frac{e}{2c}\mathbf{B}\wedge\mathbf{r}, \tag{5.26}$$

rather than

$$\mathbf{p} - \frac{e}{2c}\mathbf{B}\wedge\mathbf{r}e^{\lambda t}, \tag{5.27}$$

as was assumed in writing down the Hamiltonian (5.25).

Complex Coordinate Formulation — The second method which can be used to construct a Hamiltonian for the equations of motion (5.12) and (5.13) is to utilize the complex coordinate z. Thus if we define z and α by

$$z = x + iy, \quad \alpha = \lambda + i\Omega, \tag{5.28}$$

then we can combine (5.12) and (5.13) into a single equation

$$\ddot{z} + \alpha\dot{z} = 0. \tag{5.29}$$

By integrating (5.29) with respect to time t we find

$$v_z + \alpha z = \text{constant} = C(H). \tag{5.30}$$

For different choices of $C(H)$ we get different Hamiltonians. However if we want a Hamiltonian to have the dimension of energy and be a quadratic function of momentum, we choose

$$C(H) = \sqrt{\frac{2}{m}H}, \tag{5.31}$$

and substitute this and $v_z = \frac{\partial H}{\partial p_z}$ in (5.30) to find

$$\sqrt{\frac{2}{m}H} = \frac{\partial H}{\partial p_z} + \alpha z. \tag{5.32}$$

This equation can be integrated to yield an implicit equation for H

$$p_z = \pm\sqrt{2mH} + m\alpha z \ln \left[\frac{\pm\sqrt{2mH}}{m\alpha z} - 1 \right]. \tag{5.33}$$

Another simple Hamiltonian for (5.29) is found by taking $C(H) = \beta H$ where β is a real constant. In this case the result of integration of (5.30) is

$$H = A \exp \left(\beta p_z \right) + \left(\frac{\alpha}{\beta} \right) z. \tag{5.34}$$

If we write $p_z = p_x + i p_y$ and take the real and imaginary parts of H, we find the new Hamiltonian to be [8]

$$H_R = A \exp \left(\beta p_x \right) \cos \left(\beta p_y \right) + \frac{1}{\beta} \left(\lambda x - \Omega y \right). \tag{5.35}$$

The Lagrangian corresponding to (5.35) is given by

$$L_R = \frac{1}{2\beta} \dot{x} \ln \left(\dot{x}^2 + \dot{y}^2 \right) - \frac{1}{\beta} \dot{y} \tan^{-1} \left(\frac{\dot{y}}{\dot{x}} \right) - \frac{\dot{x}}{\beta} + \frac{1}{\beta} \left(\Omega y - \lambda x \right), \tag{5.36}$$

and this Lagrangian yields the following equations of motion:

$$\left(\frac{1}{\dot{x}^2 + \dot{y}^2} \right) \left[\dot{y} \left(\ddot{x} + \lambda \dot{x} - \Omega \dot{y} \right) - \dot{x} \left(\ddot{y} + \lambda \dot{y} + \Omega \dot{x} \right) \right] = 0, \tag{5.37}$$

$$\left(\frac{1}{\dot{x}^2 + \dot{y}^2} \right) \left[\dot{x} \left(\ddot{x} + \lambda \dot{x} - \Omega \dot{y} \right) + \dot{y} \left(\ddot{y} + \lambda \dot{y} + \Omega \dot{x} \right) \right] = 0.. \tag{5.38}$$

These equations of motion are equivalent to (5.12) and (5.13) provided that $\left(\frac{1}{\dot{x}^2 + \dot{y}^2} \right) \neq 0$. The imaginary part of H gives us Eqs. (5.12) and (5.13), but with $\lambda \to \Omega$ and $\Omega \to -\lambda$. A microscopic model of the interaction between a charged particle moving in a uniform magnetic field and at the same time is coupled to a heat bath will be discussed later in Chapter 12 [9].

References

[1] H. Goldstein, *Classical Mechanics*, Second Edition, (Addison-Wesley, Reading, 1980).

[2] G. Wentzel in *Preludes in Theoretical Physics*, edited by A. De-Shalit, H. Feshbach and L. van Hove,(North-Holland, Amsterdam, 1966) p. 199.

[3] J-M. Levy-Lablond, Ann. Phys. 57, 481 (1970).

[4] A.O. Barut and S. Malin, Ann. Phys. 69, 463 (1972).

[5] A. Pimpale and M. Razavy, Phys. Rev. A 36, 2739 (1987).

[6] V.V. Dodonov and O.V. Man'ko, Theor. Math. Phys. Vol. 65, 1033 (1986).

[7] M.D. Tokman, Phys. Rev. A 79, 053415 (2009).

[8] V. Dodonov, V.I. Man'ko and V.D. Skarzhinsky, Nuovo Cimento, 69 B, 185 (1982).

[9] X.L. Li, G.W. Ford and R.F. O'Connell, Phys. Rev. A 41, 5287 (1990).

Chapter 6

Noether and Non-Noether Symmetries and Conservation Laws

The Lagrangian formulation of dissipative systems enable us to determine the constants of motion, if they exist, with the help of an important theorem due to Noether [1]–[4]. When the dynamics of a dissipative system cannot be formulated in terms of the variational principle, there are other methods which can be used to obtain the constants of motion directly from the equations of motion [5],[6]. To begin with, we will follow a very simple approach based on the transformation properties of the generalized coordinates and time in the Lagrangian formulation to obtain the conserved quantities, and then examine other possibilities [7],[8].

Consider the following set of transformations for a system of N degrees of freedom:

$$X_i = X_i\left(x_j, \dot{x}_j, t; \epsilon\right), \quad i, j = 1, 2, \cdots N, \tag{6.1}$$

and

$$T = T\left(x_j, \dot{x}_j, t; \epsilon\right), \quad j = 1, 2, \cdots N. \tag{6.2}$$

We assume that these transformations have the property that

$$X_i \to x_i, \quad \text{as} \quad \epsilon \to 0, \tag{6.3}$$

and

$$T \to t, \quad \text{as} \quad \epsilon \to 0. \tag{6.4}$$

The Lagrangian for the system is given by $L\left(x_i, \dot{x}_i, t\right)$ from which the equation of motion for the damped system can be derived via the Euler-Lagrange equation.

If we replace $x_i, \dot{x}_i$ and t in the Lagrangian by $X_i, \dot{X}_i$ and T, then we find the new Lagrangian $L\left(X_i, \frac{dX_i}{dT}, T; \epsilon\right)$, where

$$\frac{dX_i}{dT} = \left(\frac{dX_i}{dt}\right)\left(\frac{dT}{dt}\right)^{-1} = \frac{\dot{X}_i\left(x_i, \dot{x}_i, \ddot{x}_i, t; \epsilon\right)}{\dot{T}\left(x_i, \dot{x}_i, \ddot{x}_i, t; \epsilon\right)}. \tag{6.5}$$

Now let us calculate

$$\left[\frac{\partial}{\partial \epsilon}\left\{L\left(X_i, \frac{dX_i}{dT}, T; \epsilon\right)\dot{T}\left(x_i, \dot{x}_i, \ddot{x}_i, t; \epsilon\right)\right\}\right]_{\epsilon=0}. \tag{6.6}$$

If we write this expression which is a function of $x_i, \dot{x}_i$ and t as $dF(x_i, \dot{x}_i, t)/dt$ then the Noether theorem states that the quantity

$$L\Lambda + \sum_{j=1}^{N} \frac{\partial L\left(x_i, \dot{x}_i, t\right)}{\partial x_j}\left(\Delta_j - \dot{x}_j \Lambda\right) - F, \tag{6.7}$$

is a constant of motion. In this relation

$$\Lambda = \left[\frac{\partial T\left(x_i, \dot{x}_i, t; \epsilon\right)}{\partial \epsilon}\right]_{\epsilon=0}, \tag{6.8}$$

and

$$\Delta_j = \left[\frac{\partial X_j\left(x_i, \dot{x}_i, t; \epsilon\right)}{\partial \epsilon}\right]_{\epsilon=0}. \tag{6.9}$$

In order to prove that (6.7) is indeed a constant, we write $\dot{F}$ which is defined by (6.6) as the limit of

$$\left(\frac{\partial L}{\partial T}\right)\left(\frac{\partial T\left(x_j, \dot{x}_j, t; \epsilon\right)}{\partial \epsilon}\right)\left(\frac{dT\left(x_j, \dot{x}_j, t; \epsilon\right)}{dt}\right)$$

$$+ \sum_k \left(\frac{\partial L\left(X_j, \frac{dX_j}{dT}, T\right)}{\partial X_k}\right)\left(\frac{\partial X_k\left(x_j, \dot{x}_j, t; \epsilon\right)}{\partial \epsilon}\right)\left(\frac{dT\left(x_j, \dot{x}_j, t; \epsilon\right)}{dt}\right)$$

$$+ \sum_k \left(\frac{\partial L\left(X_j, \frac{dX_j}{dT}, T\right)}{\partial\left(\frac{dX_k}{dT}\right)}\right)\left[\frac{\partial}{\partial \epsilon}\left(\frac{dX_k\left(x_j, \dot{x}_j, \ddot{x}_j, t; \epsilon\right)}{dT}\right)\right]\left(\frac{dT\left(x_j, \dot{x}_j, t; \epsilon\right)}{dt}\right)$$

$$+ L\left(X_j, \dot{X}_j, T\right)\frac{\partial}{\partial \epsilon}\left(\frac{dT\left(x_j, \dot{x}_j, t; \epsilon\right)}{dt}\right), \tag{6.10}$$

as ϵ tends to zero.

For calculating the derivatives in (6.10) we first expand T and X_j in powers of ϵ;

$$T = t + \Lambda\epsilon + \cdots, \tag{6.11}$$

and

$$X_j = x_j + \Delta_j\epsilon + \cdots. \tag{6.12}$$

Using these expansions we obtain the following relations:

$$\dot{T} = 1 + \dot{\Lambda}\epsilon + \cdots , \tag{6.13}$$

$$\dot{X}_j = \dot{x}_j + \dot{\Delta}_j\epsilon + \cdots , \tag{6.14}$$

$$\left(\frac{dX_j}{dT}\right)_{\epsilon=0} = \left(\frac{dX_j}{dt}\right)\left(\frac{dT}{dt}\right)^{-1} = \dot{x}_j, \tag{6.15}$$

$$\left[\frac{\partial T\left(x_j, \dot{x}_j, t; \epsilon\right)}{\partial \epsilon}\right]_{\epsilon=0} = \Lambda, \tag{6.16}$$

$$\left[\frac{\partial X_j\left(x_j, \dot{x}_j, t; \epsilon\right)}{\partial \epsilon}\right]_{\epsilon=0} = \Delta_j, \tag{6.17}$$

$$\left[\frac{\partial}{\partial \epsilon}\left(\frac{dT\left(x_j, \dot{x}_j, t; \epsilon\right)}{dt}\right)\right]_{\epsilon=0} = \dot{\Lambda}, \tag{6.18}$$

$$\left[\frac{\partial}{\partial \epsilon}\left(\frac{dX_j\left(x_j, \dot{x}_j, t; \epsilon\right)}{dt}\right)\right]_{\epsilon=0} = \dot{\Delta}_j, \tag{6.19}$$

and finally

$$\left[\frac{\partial}{\partial \epsilon}\left(\frac{dX_j\left(x_j, \dot{x}_j, t; \epsilon\right)}{dT}\right)\right]_{\epsilon=0} = \dot{\Delta}_j - \dot{x}_j\dot{\Lambda}. \tag{6.20}$$

We note also that the partial derivatives of $L\left(X_j, \frac{dX_j}{dT}, T\right)$ with respect to T, X_j and $\frac{dX_j}{dT}$ in the limit of $\epsilon \to 0$ tend to the corresponding derivatives of $L\left(x_j, \dot{x}_j, t\right)$ with respect to t, x_j and $\dot{x}_j$. Also in the limit of $\epsilon \to 0$, $L\left(X_j, \frac{dX_j}{dT}, T\right)$ goes over to $L\left(x_j, \dot{x}_j, t\right)$. From Eqs. (6.13)-(6.20) and these limits we find the following relation;

$$\frac{\partial L\left(x_i, \dot{x}_i, t\right)}{\partial t}\Lambda + \sum_j \frac{\partial L\left(x_i, \dot{x}_i, t\right)}{\partial x_j}\Delta_j$$

$$+ \sum_j \frac{\partial L\left(x_i, \dot{x}_i, t\right)}{\partial \dot{x}_j}\left(\dot{\Delta}_j - \dot{x}_j\dot{\Lambda}_j\right) + L\left(x_i, \dot{x}_i, t\right)\dot{\Lambda} = \dot{F}. \tag{6.21}$$

Since now we can write all the arguments in terms of x_i, $\dot{x}_i$ and t, we will suppress the arguments in L appearing in the differential relations.

Next we find the total derivatives of $\frac{dL}{dt}$, $\frac{d}{dt}\left(\frac{\partial L}{\partial \dot{x}_j}\Delta_j\right)$ and $\frac{d}{dt}\left(\frac{\partial L}{\partial \dot{x}_j}\dot{x}_j\Lambda\right)$, and from these we calculate $\frac{\partial L}{\partial t}$, $\frac{\partial L}{\partial \dot{x}_j}\dot{\Delta}_j$ and $\frac{\partial L}{\partial \dot{x}_j}x_j\dot{\Lambda}$ respectively. We substitute for these partial derivatives in the above equation for $\dot{F}$ and after simplifying the result we obtain

$$\sum_j \left\{ \frac{\partial L}{\partial x_j} - \frac{d}{dt}\left(\frac{\partial L}{\partial \dot{x}_j}\right) \right\} (\Delta_j - \dot{x}_j \Lambda)$$

$$+ \frac{d}{dt}\left[L\Lambda + \sum_j \frac{\partial L}{\partial x_j}(\Delta_j - \dot{x}_j \Lambda) - F \right] = 0. \tag{6.22}$$

The curly bracket in (6.22) is zero since L satisfies the Euler-Lagrange equation, therefore the total time derivative in (6.22) is also zero. This shows that the quantity in the square bracket in (6.22) is constant in agreement with (6.7).

Conservation Laws Found from Noether's Theorem — Now let us study some specific examples of the application of the Noether theorem.

(1) In cases where the Lagrangian L is not explicitly dependent on time, i.e. $\frac{\partial L}{\partial t} = 0$, we have a constant of motion. This is the case of invariance of the Lagrangian under time translation. We choose X_j and T to be

$$X_j = x_j, \quad j = 1, 2, \cdots N, \quad \text{and} \quad T = t + \epsilon, \tag{6.23}$$

then $\frac{dX_j}{dT} = \dot{x}_j, \Lambda = 1$, and $\Delta_j = 0$. Also from (6.6) it follows that $\frac{dF}{dt} = 0$ or $F =$ constant. Substituting these values of Δ_j and Λ in (6.7) we find that

$$L - \sum_j \dot{x}_j \frac{\partial L}{\partial \dot{x}_j} = \text{constant} = -H, \tag{6.24}$$

where H is the Hamiltonian of the system.

For dissipative motions such as the one given by Eqs. (2.140),(2.141) the Hamiltonian H is not the energy of the system.

(2) If the Lagrangian does not depend on a particular coordinate x_k, then the momentum conjugate to this coordinate p_k is conserved . In this case we choose

$$T = t, \quad \text{and} \quad X_k = x_k + \epsilon, \quad X_j = x_j, \quad j \neq k. \tag{6.25}$$

From Eqs. (6.8) and (6.9) we obtain Δ_k and Λ;

$$\Lambda = 0, \quad \Delta_k = 1, \quad \Delta_j = 0, \quad j \neq k. \tag{6.26}$$

Now from (6.6) it follows that $\dot{F} = 0$, and therefore F is a constant. We can set this constant equal to zero, and hence Noether's theorem, Eq. (6.7), implies that

$$\frac{\partial L}{\partial \dot{x}_k} = p_k = \text{constant}. \tag{6.27}$$

For the one-dimensional motion of a particle in a viscous medium in the absence of any other force except linear damping, as we have seen before, the Lagrangian is

$$L = \frac{1}{2}m\dot{x}^2 e^{\lambda t}. \tag{6.28}$$

The transformation (6.25) with $x_k = x$ shows us that according to Noether's theorem

$$p(t) = \frac{\partial L}{\partial \dot{x}} = m\dot{x}e^{\lambda t} = p_0, \tag{6.29}$$

where p_0 is a constant.

A Case for Transformation of Time — The Lagrangian L in (6.28) is not invariant under time translation, Eq. (6.23). However for this motion we can still find a Lagrangian which is not explicitly dependent on time. If we want to transform the time t to another time t^* then the action S, not the Lagrangian must remain invariant.

Let us examine the infinitesimal action Ldt:

$$Ldt = \frac{1}{2}m\left(\frac{dx}{dt}\right)^2 e^{\lambda t}dt, \tag{6.30}$$

and let $t^* = t^*(t)$ be the new time variable, then we want

$$dS = L(\dot{x},t)\,dt = L\left(\frac{dt}{dt^*}\right)dt^*, \tag{6.31}$$

to be invariant or

$$\frac{1}{2}m\left(\frac{dx}{dt^*}\right)^2\left(\frac{dt^*}{dt}\right)^2 e^{\lambda t}dt = \frac{1}{2}m\left(\frac{dx}{dt^*}\right)^2 dt^*. \tag{6.32}$$

From this relation we find that

$$\frac{dt^*}{dt} = e^{-\lambda t} \quad \text{or} \quad t^* = \frac{1-e^{-\lambda t}}{\lambda}, \tag{6.33}$$

where the integration constant is chosen in such a way that when $\lambda \to 0$ then $t^* \to t$. Now $L(\dot{x}^*)\,dt^*$ is invariant under the new time translation, i.e. $t^* \to t^* + \epsilon$, and this means that the original Lagrangian with the variable t is invariant under the time translation given by [9]

$$t \to t - \frac{1}{\lambda}\ln\left(1 - \lambda\epsilon e^{\lambda t}\right). \tag{6.34}$$

General Linear Transformations for One-Dimensional Systems — For a number of problems involving dissipative forces, the direct approach of investigating the invariance under a general linear transformation is simpler [10]-[15]. Thus let us consider the case of a one-dimensional motion under the action of a force $f(x,\dot{x},t)$ when the coordinate x and the time t are transformed according to

$$X = \alpha x + \beta t + x_0, \tag{6.35}$$

and

$$T = \delta x + \nu t + t_0, \tag{6.36}$$

where $\alpha, \beta, \delta, \nu, x_0$ and t_0 are all constants and

$$\alpha\nu - \beta\delta \neq 0, \tag{6.37}$$

i.e. the transformation is not singular and x and t can be obtained in terms of X and T. Let us denote the time derivatives with respect to T by primes, e.g.

$$X' = \frac{dX}{dT} = \frac{\alpha\dot{x} + \beta}{\delta\dot{x} + \nu}, \tag{6.38}$$

and write the equation of motion using the new variables X and T. By substituting for x, $\dot{x}, \ddot{x}$ and t in terms of X, X', X'' and T in the equation of motion $m\ddot{x} = f(x, \dot{x}, t)$ and requiring that the transformed equation of motion be the same, i.e.

$$mX'' = f(X, X', T), \tag{6.39}$$

we find that [44]

$$f(x, \dot{x}, t) = \frac{(\delta\dot{x} + \nu)^3}{(\alpha\nu - \beta\delta)} f(X, X', T). \tag{6.40}$$

Now let us suppose that the Lagrangian $L(x, \dot{x}, t)$ remains invariant under this transformation, i.e.

$$L(x, \dot{x}, t) = L(X, X', T), \tag{6.41}$$

then Eq. (2.116) yields

$$\left[\frac{\partial^2 L}{\partial \dot{x}^2} \left(\frac{\partial L}{\partial t} + \dot{x} \frac{\partial L}{\partial x} \right) - 2\frac{\partial L}{\partial \dot{x}} \left(\dot{x} \frac{\partial^2 L}{\partial x \partial \dot{x}} + \frac{\partial^2 L}{\partial \dot{x} \partial t} - \frac{\partial L}{\partial x} \right) \right] \delta = 0. \tag{6.42}$$

This equation can be satisfied apart from some uninteresting Lagrangians only when $\delta = 0$. When this is the case, then (6.40) reduces to

$$f(x, \dot{x}, t) = \frac{\nu^2}{\alpha} f(X, X', T). \tag{6.43}$$

If $g(x, \dot{x}, t)$ is a first integral of motion, i.e. $(dg/dt) = 0$, then g is a solution of the partial differential equation

$$\frac{\partial g}{\partial t} + \dot{x}\frac{\partial g}{\partial x} + \frac{1}{m} f(x, \dot{x}, t) \frac{\partial g}{\partial \dot{x}} = 0. \tag{6.44}$$

Now let us investigate the effect of the transformations (6.35) and (6.36) with $\delta = 0$ on the function $g(x, \dot{x}, t)$. To this end we write $g(x, \dot{x}, t) = G(X, X', T)$ and show that $G(X, X', T)$ is also a solution of (6.44):

$$\begin{aligned}
\frac{\partial g}{\partial t} &+ \dot{x}\frac{\partial g}{\partial x} + \frac{1}{m} f(x, \dot{x}, t) \frac{\partial g}{\partial \dot{x}} \\
&= \alpha\frac{\partial G}{\partial X}\dot{x} + \frac{\alpha}{m\nu} f(x, \dot{x}, t) \frac{\partial G}{\partial X'} + \beta\frac{\partial G}{\partial X} + \nu\frac{\partial G}{\partial T} \\
&= \nu\left[\left(\frac{\alpha\dot{x} + \beta}{\nu} \right) \frac{\partial G}{\partial X} + \frac{\alpha}{m\nu^2} f\frac{\partial G}{\partial X'} + \frac{\partial G}{\partial T} \right] \\
&= \nu\left[\frac{\partial G}{\partial X}Q' + \frac{1}{m} f(X, X', T) \frac{\partial G}{\partial X'} + \frac{\partial G}{\partial T} \right] = 0. \tag{6.45}
\end{aligned}$$

Before applying these ideas to the problems involving dissipative systems, we observe that there are the following invariances that can be studied using the transformations (6.35) and (6.36):
(1) Time translation invariance:

$$\alpha = \nu = 1, \quad \beta = x_0 = 0, \quad t_0 \text{ arbitrary.} \tag{6.46}$$

(2) Coordinate translation:

$$\alpha = \nu = 1, \quad \beta = t_0 = 0, \quad x_0 \text{ arbitrary.} \tag{6.47}$$

(3) Time-scale transformation:

$$\alpha = 1, \quad \beta = x_0 = t_0 = 0, \quad \nu \text{ arbitrary.} \tag{6.48}$$

(4) Coordinate-scale transformation:

$$\nu = 1, \quad \beta = x_0 = t_0 = 0, \quad \alpha \text{ arbitrary,} \tag{6.49}$$

and finally
(5) Galilean transformation:

$$\alpha = \nu = 1, \quad x_0 = t_0 = 0, \quad \beta \text{ arbitrary.} \tag{6.50}$$

These are not the only possible transformations which lead to the corresponding conservation laws, but they are the important ones in our study of the dissipative systems [16],[44].

As an example of the application of these invariances as applied to the damped motion, let us consider the damped harmonic oscillator

$$f(x, \dot{x}) = -m\lambda\dot{x} - m\omega_0^2 x, \tag{6.51}$$

for which $f(x, \dot{x})$ is a homogeneous function of first degree in x and $\dot{x}$. Here we set $\nu = 1$ but we allow α to be a free coordinate scale factor. If we are looking for a Lagrangian which remains invariant under this particular transformation, thus independent of α, then we search for a Lagrangian which is a function of $z = \dot{x}/x$ and t [44]. In terms of z and t Eqs. (2.116) and (6.44) can be written as

$$\frac{\partial^2 L}{\partial z \partial t} - z^2 \frac{\partial^2 L}{\partial z^2} + \frac{1}{m} f(1, z, t) \frac{\partial^2 L}{\partial z^2} = 0, \tag{6.52}$$

and

$$\frac{\partial g}{\partial t} - \left[z^2 - \frac{1}{m} f(1, z, t) \right] \frac{\partial g}{\partial z} = 0. \tag{6.53}$$

For the velocity dependent force $f(x, \dot{x})$ given by (6.51) these equations reduce to:

$$\frac{\partial^2 L}{\partial z \partial t} - z^2 \frac{\partial^2 L}{\partial z^2} - (\lambda z + \omega_0^2) \frac{\partial^2 L}{\partial z^2} = 0, \tag{6.54}$$

and

$$\frac{\partial g}{\partial t} - \left(z^2 + \lambda z + \omega_0^2\right) \frac{\partial g}{\partial z} = 0. \tag{6.55}$$

The particular solutions of (6.54) and (6.55) are given by [44]:

$$L = \frac{\lambda + 2z}{2\omega} \left[\tan^{-1}\left(\frac{2z + \lambda}{2\omega}\right) + \omega t \right] - \frac{1}{2} \ln \left[1 + \left(\frac{2z + \lambda}{2\omega}\right)^2 \right], \tag{6.56}$$

and

$$g = \tan^{-1}\left(\frac{2z + \lambda}{2\omega}\right) + \omega t, \tag{6.57}$$

respectively, where $\omega = \left(\omega_0^2 - \frac{\lambda^2}{4}\right)^{\frac{1}{2}}$. The Lagrangian L, Eq. (6.56), and the Lagrangian L_2 given by (2.176) differ from each other by a total time derivative.

As an application of Noether's theorem, symmetries and first integrals for Caldirola-Kanai Hamiltonian have been obtained for linearly damped harmonic oscillator. Such an approach leads to the determination of all the time-dependent potentials which allow for such symmetries, both in one-dimensional and also three-dimensional in the latter case with potentials depending only on the radial distance [17].

6.1 Non-Noether Symmetries and Conserved Quantities

The foregoing discussion of the symmetries was based on the existence of Lagrangians for conservative as well as non-conservative dynamical systems. Now we want to study the symmetries which can be derived directly from the equations of motion which we assume to be of the second order [5],[18].

We write these equations as

$$\frac{d^2 x_j}{dt^2} = f_j\left(x_i, \dot{x}_i, t\right), \quad j = 1, 2, \cdots n, \tag{6.58}$$

where we have set the mass of the particle(s) equal to unity. In terms of the generalized dissipative forces discussed in Chapter 2, we can derive Eq. (6.58) from the Euler-Lagrange equation

$$\frac{d}{dt}\left(\frac{\partial L}{\partial \dot{x}_j}\right) - \frac{\partial L}{\partial x_j} = Q_j. \tag{6.59}$$

As before we introduce the infinitesimal time and coordinate transformations (6.11) and (6.12) with Λ and Δ_j being functions of the coordinates x_j and time t, and the small parameter ϵ. If the equations of motion (6.58) remain invariant under the transformation (6.11) and (6.12) then by direct substitution for X_i

and T in $\frac{d^2 X_j}{dT^2} = f_j\left(X_i, \frac{dX_i}{dT}, T\right)$ and subsequent expansion in powers of ϵ, we find that Λ and Δ_j must satisfy the differential equation

$$\Pi_j \equiv \frac{d^2 \Delta_j}{dt^2} - \dot{x}_j \frac{d^2 \Lambda}{dt^2} - 2f_j\left(x_i, \dot{x}_i, t\right)\frac{d\Lambda}{dt} - \hat{\mathcal{D}}(f_j) = 0. \qquad (6.60)$$

Here $\hat{\mathcal{D}}$ is the partial differential operator

$$\hat{\mathcal{D}} = \Lambda \frac{\partial}{\partial t} + \sum_j \left[\Delta_j \frac{\partial}{\partial x_j} + \left(\dot{\Delta}_j - \dot{x}_j \dot{\Lambda}\right)\frac{\partial}{\partial \dot{x}_j}\right], \qquad (6.61)$$

which operates on $f_j(x_i, \dot{x}_i, t)$. In Eq. (6.61) a dot indicates the total time derivative along the trajectory, i.e.

$$\frac{d}{dt} = \frac{\partial}{\partial t} + \sum_j \left(\dot{x}_j \frac{\partial}{\partial x_j} + f_j \frac{\partial}{\partial \dot{x}_j}\right). \qquad (6.62)$$

Let us define D by

$$D = \det\left[\frac{\partial^2 L}{\partial \dot{x}_j \partial \dot{x}_k}\right], \qquad (6.63)$$

and let M_{jk} be the cofactor of $\frac{\partial^2 L}{\partial \dot{x}_j \partial \dot{x}_k}$ in the matrix formed by the second order derivatives. Writing Eq. (6.59) in the expanded form, viz,

$$\sum_k \frac{\partial^2 L}{\partial \dot{x}_j \partial \dot{x}_k}\ddot{x}_k = \frac{\partial L}{\partial x_j} - \frac{\partial^2 L}{\partial \dot{x}_j \partial t} - \sum_k \dot{x}_k \frac{\partial^2 L}{\partial \dot{x}_j \partial x_k} + Q_j, \qquad (6.64)$$

we can show that

$$\sum_k \left[\frac{\partial f_k}{\partial \dot{x}_k} - \sum_j \frac{\partial}{\partial \dot{x}_k}\left(\frac{M_{kj}}{D}Q_j\right)\right] + \frac{d}{dt}(\ln D) = 0. \qquad (6.65)$$

Now we state the conservation law in the following way: Let $g = g(x_j, \dot{x}_j, t)$ satisfy the equation

$$\frac{dg}{dt} = \sum_{k,j} \frac{\partial}{\partial \dot{x}_j}\left(\frac{M_{jk}}{D}Q_k\right), \qquad (6.66)$$

and Λ and Δ satisfy (6.60), then equations (6.58) admit a conserved quantity which we denote by Φ;

$$\Phi = 2\sum_j \left(\frac{\partial \Delta_j}{\partial x_j} - \dot{x}_j \frac{\partial \Lambda}{\partial x_j}\right) - N\dot{\Lambda} + \hat{\mathcal{D}}(\ln D) - \hat{\mathcal{D}}(g). \qquad (6.67)$$

In order to show that Φ is a constant of motion, we try to relate it to $\sum_j \frac{\partial \Pi_j}{\partial \dot{x}_j}$. To this end we differentiate (6.60) with respect to $\dot{x}_j$ and then using (6.66) we substitute for

$$\sum_{k,j}\left[\frac{\partial}{\partial \dot{x}_j}\left(\frac{M_{jk}}{D}Q_k\right)\right] \qquad (6.68)$$

in the resulting expression to find

$$\sum_j \frac{\partial \Pi_j}{\partial \dot{x}_j} = \frac{d}{dt}\left[2\sum_j\left(\frac{\partial \Delta_j}{\partial x_j} - \dot{x}_j\frac{\partial \Lambda}{\partial x_j}\right) - N\dot{\Lambda} + \hat{\mathcal{D}}(\ln D) - \hat{\mathcal{D}}(g)\right] = \frac{d\Phi}{dt} = 0,$$

(6.69)

where for one-dimensional motion $N = 1$. Thus Φ is a conserved quantity.

Conserved Quantities for Motion Subject to a Newtonian Dissipative Force — As an example let us consider the simple case of a dissipative force which is quadratic in velocity, $\ddot{x} + \gamma\dot{x}^2 = 0$. For this damping force Eq. (6.60) becomes

$$\begin{aligned}
\Delta_{tt} &- 2\gamma\dot{x}\Delta_{tx} + \gamma^2\dot{x}^2\Delta_{xx} + \gamma^2\dot{x}^2\Delta_x + \gamma\dot{x}\Lambda_{tt} \\
&- 2\gamma^2\dot{x}^2\Lambda_{tx} + \gamma^3\dot{x}^3\Lambda_{xx} + \gamma^3\dot{x}^3\Lambda_x - 2\gamma^2\dot{x}^2\Lambda_t + 2\gamma^3\dot{x}^3\Lambda_x \\
&= -2\gamma\dot{x}\Delta_t + 2\gamma^2\dot{x}^2\Delta_x - 2\gamma^2\dot{x}^2\Lambda_t + 2\gamma^3\dot{x}^3\Lambda_x.
\end{aligned}$$

(6.70)

Here subscripts denote partial derivatives. Since Δ and Λ and their partial derivatives do not depend on $\dot{x}$, therefore the coefficients of different powers of $\dot{x}$ on the two sides of (6.70) should match. Thus we find four partial differential equations for Δ and Λ. The most general solution of these equations are:

$$\Lambda = a_0 + a_1 t + a_2 e^{\gamma x} + a_3 t e^{\gamma x},$$

(6.71)

and

$$\Delta = b_0 + b_1 e^{\gamma x} + b_2 t e^{-\gamma x}.$$

(6.72)

The nontrivial Φ s obtained from these results are given by [18]

$$\Lambda = e^{\gamma x}, \quad \Delta = 0, \quad \Phi = -2\gamma\dot{x}e^{\gamma x},$$

(6.73)

and

$$\Lambda = -te^{\gamma x}, \quad \Delta = e^{-\gamma x}, \quad \Phi = -3\left(1 - \gamma t\dot{x}\right)e^{\gamma x}.$$

(6.74)

6.2   Noether's Theorem for a Scalar Field

We have already seen that by applying the discrete version of the Noether's theorem to the Lagrangian L, we can find certain conserved quantities for a dissipative system. In this section we want to study the field theoretical description of this theorem which can be used to study the invariances and the conservation laws of the wave equation [19]-[21].

Let $\mathcal{L}\left[\psi, \frac{\partial\psi}{\partial x_j}, \frac{\partial\psi}{\partial t}, x_j, t\right]$ be the Lagrangian density which is the generator of the wave equation for a dissipative system. Then we make an infinitesimal change of coordinates x_j s and the time t by writing

$$x_j \to X_j(x_j), \quad \text{and} \quad t \to T(t),$$

(6.75)

and at the same time change the wave function

$$\psi(x_j, t) \rightarrow \Psi[\psi(x_j, t), x_j, t] = \psi(x_j, t) + \delta\psi[\psi(x_j, t), x_j, t], \tag{6.76}$$

where

$$\delta\psi(x_j, t) = \Psi[\psi(x_j, t), x_j, t] - \psi(x_j, t), \tag{6.77}$$

measures the effect of both the changes in x_j s and t and also in $\psi(x_j, t)$. This transformation changes the Lagrangian density to $\mathcal{L}'$ where

$$\mathcal{L}\left[\psi, \frac{\partial\psi}{\partial x_j}, \frac{\partial\psi}{\partial t}, x_j, t\right] \rightarrow \mathcal{L}'\left[\Psi(X_j, T), \frac{\partial\Psi}{\partial X_j}, \frac{\partial\Psi}{\partial T}, X_j, T\right]. \tag{6.78}$$

The fact that the action $\mathcal{S}$ should remain invariant, i.e.

$$\mathcal{S}'[\Psi] = \mathcal{S}[\psi], \tag{6.79}$$

implies that the new Lagrangian density $\mathcal{L}'$, as a function of the new variables X_j and T, must be given by

$$\mathcal{L}'[\Psi, X_j, T] = \left(\frac{\partial t}{\partial T}\right) J \mathcal{L}[\psi, x_j, t], \tag{6.80}$$

where

$$J = \left(\frac{\partial(x_1, x_2, x_3)}{\partial(X_1, X_2, X_3)}\right), \tag{6.81}$$

is the Jacobian of the transformation. As in the case of the infinitesimal transformation of the Lagrangian for particles, Eq. (6.6), here we have the infinitesimal change of the Lagrangian which is given by

$$\mathcal{L}'\left[\Psi, X_j, \frac{\partial\Psi}{\partial T}, \frac{\partial\Psi}{\partial X_j}, T\right] - \mathcal{L}\left[\psi, \frac{\partial\psi}{\partial t}, \frac{\partial\psi}{\partial x_j}, x_j, t\right] = \frac{d(\Delta\Lambda^0)}{dt} + \nabla \cdot (\Delta\boldsymbol{\Lambda}). \tag{6.82}$$

That is for the case of a field the total time derivative is replaced by a 4-divergence of a 4-vector $(\Delta\Lambda^0, \Delta\boldsymbol{\Lambda})$.

Assuming that the transformation is infinitesimal

$$X_j(x_j) = x_j + \delta x_j, \tag{6.83}$$

and

$$T = t + \delta t, \tag{6.84}$$

and

$$\Psi(x_j, t) = \psi(x_j, t) + \delta\psi(x_j, t), \tag{6.85}$$

then the variations of dt and d^3r are expressible as

$$\delta(dt) = dt\frac{d\delta t}{dt}, \quad \text{and} \quad \delta(d^3r) = d^3r\nabla \cdot (\delta\mathbf{r}). \tag{6.86}$$

Thus we obtain the following results:

$$\delta\psi = \delta_0\psi + \delta t\dot{\psi} + \delta\mathbf{r}\cdot\nabla\psi, \tag{6.87}$$

$$\delta\dot{\psi} = \delta_0\dot{\psi} + \delta t\ddot{\psi} + \delta\mathbf{r}\cdot\nabla\dot{\psi}, \tag{6.88}$$

and

$$\delta\left(\nabla\psi\right) = \delta_0\nabla\psi + \delta\nabla\dot{\psi} + \left(\delta\mathbf{r}\cdot\nabla\right)\nabla\psi, \tag{6.89}$$

where dots denote partial derivatives with respect to t.
The variation of action can be written down as

$$\delta S = \int \mathcal{L}'\left(\Psi, \frac{\partial\Psi}{\partial T}, \frac{\partial\Psi}{\partial X_j}, X_j, T\right) d^3X\,dT$$

$$- \int \mathcal{L}\left(\psi, \frac{\partial\psi}{\partial t}, \frac{\partial\psi}{\partial x_j}, x_j, t\right) d^3x\,dt. \tag{6.90}$$

Now by substituting from Eqs. (6.83)-(6.85) in (6.90) we obtain

$$\delta S = \int \left[\{dt d^3r + (\delta dt)\,d^3r + \left(\delta d^3r\right)dt\}\{(\mathcal{L}+\delta\mathcal{L})-\mathcal{L}\}\right]$$

$$= \int \left[\left\{\frac{d(\delta t)}{dt} + \nabla\cdot\delta\mathbf{r}\right\}\mathcal{L} + \frac{d\mathcal{L}}{dt}\delta t + \delta\mathbf{r}\cdot\nabla\mathcal{L} + \delta_0\mathcal{L}\right]dt d^3r$$

$$= \int \left[\frac{d\rho}{dt} + \nabla\cdot\mathbf{j} + \left(\frac{\partial\mathcal{L}}{\partial\psi} - \frac{d}{dt}\frac{\partial\mathcal{L}}{\partial\dot{\psi}} - \nabla\cdot\frac{\partial\mathcal{L}}{\partial(\nabla\psi)}\right)\right.$$

$$\left. \times\left(\delta\psi - \delta t\dot{\psi} - \delta\mathbf{r}\cdot\nabla\psi\right)\right]d^3r\,dt, \tag{6.91}$$

where the quantities $\delta\rho$ and $\delta\mathbf{j}$ are defined by

$$\delta\rho = \mathcal{L}\delta t + \frac{\partial\mathcal{L}}{\partial\dot{\psi}}\delta_0\psi, \tag{6.92}$$

and

$$\delta\mathbf{j} = \mathcal{L}\delta\mathbf{r} + \frac{\partial\mathcal{L}}{\partial\left(\nabla\psi\right)}\delta_0\psi. \tag{6.93}$$

Thus from (6.91) it follows that

$$\frac{d\rho}{dt} + \nabla\cdot\mathbf{j} + \left[\frac{\partial\mathcal{L}}{\partial\psi} - \frac{d}{dt}\left(\frac{\partial\mathcal{L}}{\partial\dot{\psi}}\right) - \nabla\cdot\frac{\partial\mathcal{L}}{\partial(\nabla\psi)}\right]\left(\delta\psi - \delta t\dot{\psi} - \delta\mathbf{r}\cdot\nabla\psi\right) = 0. \tag{6.94}$$

For a stationary path the sum of the terms in the square bracket vanishes (Euler-Lagrange equation) and Eq. (6.94) becomes the equation of continuity

$$\frac{d\rho}{dt} + \nabla\cdot\mathbf{j} = 0. \tag{6.95}$$

Next we define the momentum density $\pi(\mathbf{r}, t)$ by

$$\pi(\mathbf{r}, t) = \frac{\delta\mathcal{L}}{\delta\dot{\psi}}, \tag{6.96}$$

where $\frac{\delta \mathcal{L}}{\delta \psi}$ is the functional derivative of $\mathcal{L}$. Then the Hamiltonian density for this ψ field is given by

$$\mathcal{H} = \pi(\mathbf{r}, t)\dot{\psi}(\mathbf{r}, t) - \mathcal{L}, \tag{6.97}$$

and the action can be written as

$$S = \int dt \int \left[\pi(\mathbf{r}, t)\dot{\psi}(\mathbf{r}, t) - \mathcal{L}\right]_{\dot{\psi} = \dot{\psi}(\pi)} d^3 r. \tag{6.98}$$

Now the functional derivative of S with respect to variation $\delta_0 \psi(\mathbf{r}, t)$ with no variation of $\mathbf{r}$ and t must vanish. Since $\pi(\mathbf{r}, t)$ and $\psi(\mathbf{r}, t)$ are independent functions, the functional derivatives of S yield the canonical equations:

$$\dot{\psi}(\mathbf{r}, t) = \frac{\delta \mathcal{H}}{\delta \pi}, \quad \text{and} \quad \dot{\pi}(\mathbf{r}, t) = -\frac{\delta \mathcal{H}}{\delta \psi}. \tag{6.99}$$

In Sec. 12.11 we will discuss the application of the Noether theorem to obtain conservation laws for linear and nonlinear wave equations for dissipative systems.

Noether theorem can be extended and applied to those problems of calculus of variation which involve fractional derivatives (see for instance references [22], [23]).

References

[1] E. Noether, Nachr. Ges. Wiss. Göttingen, 235 (1918).

[2] E.T. Whittaker, *A Treatise on the Analytical Dynamics of Particles and Rigid Bodies*, Fourth Edition (Cambridge University Press, London 1965) pp. 54-69.

[3] L.D. Landau and E.M. Lifshitz, *Mechanics*, (Addison-Wesley, Reading, 1960) Chapter 2.

[4] H. Goldstein, *Classical Mechanics*, Second Edition (Addison-Wesley Reading, 1980) Chapter 8.

[5] S. Hojman, J. Phys. A 25, L 291 (1992).

[6] M. Lutzky, J. Phys. A 28, L 637 (1995).

[7] E.A. Desloge and R.I. Karsh, Am. J. Phys. 45, 336 (1977).

[8] E.A. Desloge, *Classical Mechanics*, Vol. 2 (John Wiley & Sons, New York, N.Y. 1982) p. 581.

[9] H-J Wagner, Z. Physik, B 95, 261 (1994).

[10] H.H. Denman, J. Math. Phys. 6, 1611 (1965).

[11] H.H. Denman, J. Math. Phys. 7, 1910 (1966).

[12] H.H. Denman, Am. J. Phys. 36, 516 (1968).

[13] W. Sarlet, J. Math. Phys. 19, 1049 (1978).

[14] W. Sarlet, SIAM Reviews, 23, 467 (1981).

[15] W. Sarlet, J. Phys. A 17, 1999 (1984).

[16] P. Havas, Acta Phys. Austriaca, 38, 145 (1973).

[17] R. Leone and T. Gourieux, preprint, hal-01097517.

[18] J-L. Fu and L-Q. Phys. Lett. A 317, 255 (2003).

[19] See for instance, N.A. Doughty, *Lagrangian Interaction*, (Addison-Wesley, Reading, 1990) p. 214.

[20] D.E. Soper, *Classical Field Theory*, (John Wiley & Sons, New York, N.Y. 1976) pp. 101-108.

[21] J. Rosen, Ann. Phys. 69, 349 (1972).

[22] M. Klimak, J. Phys. A 35, 6675 (2002).

[23] G.S.F. Frederico and D.F.M. Torres, J. Math. Anal. App. 334, 834 (2007).

Chapter 7

Dissipative Forces Derived from Many-Body Problems

Consider a large system S of N interacting particles which is isolated from the rest of the universe and therefore has a constant total energy. Now let us divide this system into two parts S_1 and S_2 and study the development of S_1 in time. Because of the coupling between S_1 and S_2 the two parts will exchange energy with each other. For instance initially S_1 will be losing, and S_2 will be gaining energy respectively. When both S_1 and S_2 have finite degrees of freedom then this energy flow from S_1 to S_2 follows by a flow in the opposite direction, i.e. from S_2 to S_1, and therefore there is no fixed direction for the flow of energy for all times. Furthermore, unlike the case of some of the phenomenological frictional forces, the motion of S_1 as well as S_2 will be invariant under the time reversal transformation $t \to -t$, $\mathbf{r}_i \to \mathbf{r}_i$ and $\dot{\mathbf{r}}_i = -\dot{\mathbf{r}}_i$, where i refers to the ith particle of the system S $(i = 1, 2, \cdots N)$. However if S_1 contains a few and S_2 has an infinite number of degrees of freedom, and if the initial conditions are such that the direction of the energy flow is from S_1 to S_2, then S_1 will be losing energy at all times and thus it becomes a dissipative system. In this chapter we study a number of exactly solvable classical systems, where a particle called the "central particle" loses energy to a large number of oscillators usually referred to as a "heat bath".

In the last part of this chapter we study some interesting aspects of damping in tuned circuits.

7.1 The Schrödinger Chain

This many-body problem was originally formulated and solved by Schrödinger before his discovery of wave mechanics [1],[2]. The remarkable feature of this chain is that the decay law of the central particle is non-exponential and is proportional to $t^{-\frac{1}{2}}$.

The classical as well as the quantal motion of a finite long chain has also been investigated in order to examine the recurrence time and the quantum spectrum of such a system [3],[4] but in this chapter we study an infinite chain.

Here the large system S consists of a linear chain of infinite mass points each coupled to its nearest neighbor by elastic springs. Let us denote the displacement of the jth particle in the chain from its equilibrium position by ξ_j, then we can write the equation of motion of the jth particle as

$$m \left(\frac{d^2 \xi_j}{dt^2} \right) = \frac{1}{4} m \nu^2 \left(\xi_{j+1} + \xi_{i-1} - 2\xi_j \right), \quad j = 0, \pm 1, \pm 2 \cdots, \tag{7.1}$$

where $\frac{1}{4} m \nu^2$ is the spring constant.

For the initial conditions we assume that the central particle, $j = 0$, which we choose as the decaying part S_1 is displaced by a distance A and then released with zero initial velocity while other particles are at rest in their equilibrium positions

$$\xi_0(t = 0) = A, \quad \xi_j(t = 0) = 0, \quad j \neq 0, \tag{7.2}$$

$$\dot{\xi}_j(t = 0) = 0, \quad j = 0, \pm 1, \pm 2 \cdots. \tag{7.3}$$

We can solve the system of equations (7.1) by the following method [5],[6]. Let $G(z,t)$ be the generating function defined by

$$G(z,t) = \sum_{j=-\infty}^{+\infty} \xi_j(t) z^{2j}. \tag{7.4}$$

We multiply (7.1) by z^{2j} and sum over all j s using Eq. (7.4). This yields the differential equation for the generating function

$$\frac{d^2}{dt^2} G(z,t) - \frac{1}{4} \nu^2 \left(\frac{1}{z} - z \right)^2 G(z,t) = 0. \tag{7.5}$$

The function $G(z,t)$ is subject to the following initial conditions which can be derived from Eqs. (7.2)-(7.4):

$$G(z, t = 0) = A, \tag{7.6}$$

and

$$\left(\frac{dG(z,t)}{dt} \right)_{t=0} = 0. \tag{7.7}$$

The solution of (7.5) which satisfies (7.6) and (7.7) is given by

$$G(z, t = 0) = A \cosh \left[\frac{1}{2} \nu t \left(\frac{1}{z} - z \right) \right]. \tag{7.8}$$

Equation (7.8) shows that $G(z,t)$ is the generating function for the Bessel function of even order [7]

$$\cosh \left[\frac{1}{2} \nu t \left(\frac{1}{z} - z \right) \right] = \sum_{j=-\infty}^{+\infty} J_{2j}(\nu) z^{2j}, \tag{7.9}$$

and therefore

$$\xi_j(t) = AJ_{2j}(\nu) \tag{7.10}$$

is the solution of (7.1) with the initial conditions (7.2) and (7.3).

The motion of the central particle, $k = 0$, can also be obtained from the Hamiltonian

$$H_0 = \frac{1}{2m}p_0^2 + \frac{1}{4}m\nu^2 x_0^2 + x_0 \left\{ \frac{1}{4}m\nu^2 A \left(J_{-1}(\nu t) + J_1(\nu t) \right) \right\}. \tag{7.11}$$

Similar single particle Hamiltonian can be written down for other particles in the system. However, classically we can obtain other q-equivalent Hamiltonians for the motion of the central particle which is damped. For instance we can write

$$H = \frac{p_0^2(t)}{2m\nu t} + \frac{1}{2}m\nu^3 x_0^2(t)t. \tag{7.12}$$

This Hamiltonian gives us the equation of motion for $x_0(t)$ which is the same as the equation for the Bessel function, i.e. $x_0(t) = AJ_0(\nu t)$, but note that the relation between H and the energy of the system E is given by

$$E = H\nu t. \tag{7.13}$$

Thus the Hamiltonian does not represent the energy, and we cannot proceed with the quantization of this motion using (7.12).

7.2 A Particle Coupled to a Chain

A model similar to the Schrödinger chain is that of a massive particle, M, coupled to a semi-infinite chain of oscillators (Rubin's model) [8],[9]. In this model the total Hamiltonian is given by

$$H = \frac{p^2}{2M} + V(x) + \sum_{n=1}^{\infty} \left[\frac{p_n^2}{2m} + \frac{1}{2}m\omega_0^2 \left(x_{n+1} - x_n \right)^2 \right] + \frac{1}{2}m\omega_0^2 \left(x - x_1 \right)^2. \tag{7.14}$$

We can diagonalize this Hamiltonian using the Fourier transform method. Let us denote the highest frequency mode of the oscillators by ω_R, and introduce $X(k)$ and $P(k)$ by

$$x_n = \sqrt{\frac{2}{\pi}} \int_0^\pi \sin(kn) X(k) dk, \tag{7.15}$$

and

$$p_n = \sqrt{\frac{2}{\pi}} \int_0^\pi \sin(kn) P(k) dk. \tag{7.16}$$

By substituting (7.15) and (7.16) in (7.14) and simplifying the result we obtain

$$H = \frac{p^2}{2M} + V(x) + \frac{1}{2}m\omega_0^2 x^2 + \int_0^\pi \left(\frac{P^2(k)}{2m} + \frac{m}{2}\omega^2(k)X^2(k) - c(k)X(k) \right) dk, \tag{7.17}$$

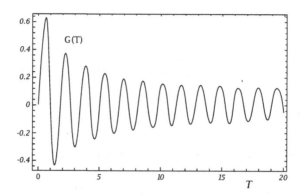

Figure 7.1: The kernel $G(T)$, Eq. (7.22), is shown as a function of T.

where the eigenfrequencies $w(k)$ and the coupling $c(k)$ are related to w_R by

$$w(k) = w_R \sin\left(\frac{k}{2}\right) \tag{7.18}$$

and

$$c(k) = \sqrt{\frac{2}{\pi}} \left(\frac{m w_R^2}{4}\right) \sin(k). \tag{7.19}$$

Next we write the equations of motion for the particles forming the chain, and also for the massive particle, M. Using the initial conditions

$$x_n(0) = 0, \quad \text{and} \quad p_n(0) = 0, \quad n = 1, 2, 3 \cdots, \tag{7.20}$$

and then eliminating the degrees of freedom of the particles in the chain, i.e. $X(k)$ and $P(k)$ (see Sec. 7.4 for the details), we obtain the equation of motion for the particle M,

$$M\frac{d^2 x(t)}{dt^2} + \frac{\partial V}{\partial x} + \frac{1}{2} m w^2 x^2(t) - \int_0^t G\left(t - t'\right) x\left(t'\right) dt' = 0, \tag{7.21}$$

where the kernel $G(T)$ is given by

$$G(T) = \begin{cases} \frac{m w_R}{2\pi} \int_0^\infty \frac{\sin k}{\sin\left(\frac{k}{2}\right)} \sin\left[w_R \sin\left(\frac{k}{2}\right) T\right] dk & \text{for } T > 0 \\ 0 & \text{for } T < 0 \end{cases}. \tag{7.22}$$

In Fig. (7.1) the kernel $G(T)$ as a function of T is shown. This kernel is an oscillating function of time with decreasing amplitude similar to the Bessel function $J_1(w_R t)$.

7.3 Dynamics of a Non-Uniform Chain

As a classical model exhibiting the exponential decay law for $t \to \infty$ we want to discuss the special case of a non-uniform chain which is again exactly solvable [10]. Denoting the mass of the jth particle in the chain by m_j and the spring constant connecting the particles j and $j+1$ by K_j, the equation of motion for the jth particle can be written as

$$m_j \left(\frac{d^2 \xi_j(t)}{dt^2} \right) = K_j \left[\xi_{j+1}(t) - \xi_j(t) \right] + K_{j-1} \left[\xi_{j-1}(t) - \xi_j(t) \right], \quad j = 1, 2, \cdots N. \tag{7.23}$$

Next we introduce the variables η_j by

$$\eta_j(t) = \sqrt{m_j} \xi_j(t), \tag{7.24}$$

and the constants $\omega_1, \omega_2, \cdots \omega_{2N-2}$ by the following relations [11]

$$\omega_{2j-1}^2 = \frac{K_j}{m_j}, \quad \omega_{2j}^2 = \frac{K_j}{m_{j+1}}, \tag{7.25}$$

then the equations of motion become

$$\left(\frac{d^2 \eta_j(t)}{dt^2} \right) = (\omega_{2j-1}\omega_{2j}) \, \eta_{j+1}(t) + (\omega_{2j-3}\omega_{2j-2}) \, \eta_{j-1}(t)$$
$$- \left(\omega_{2j-1}^2 + \omega_{2j-2}^2 \right) \eta_j(t). \tag{7.26}$$

Now we change η_j s to a new set of variables $z_1(t), z_2(t), \cdots z_{N-1}(t)$ where

$$\left(\frac{dz_j(t)}{dt} \right) = \omega_{2j}\eta_{j+1}(t) - \omega_{2j-1}\eta_j(t), \tag{7.27}$$

and then write (7.26) as

$$\left(\frac{d\eta_j(t)}{dt} \right) = \omega_{2j-1}z_j(t) - \omega_{2j-2}z_{j-1}(t). \tag{7.28}$$

We can combine (7.27) and (7.28) to form a single set of equations:

$$\left(\frac{dX_j(t)}{dt} \right) = \omega_j X_{j+1}(t) - \omega_{j-1}X_{j-1}(t), \tag{7.29}$$

where $X_1(t), X_2(t) \cdots X_{2N-1}(t)$ are defined by

$$X_{2j-1}(t) = \eta_j(t), \quad X_{2j}(t) = z_j(t). \tag{7.30}$$

Equation (7.29) is a linear differential-difference equation that we want to solve for a semi-infinite chain, i.e. in the limit of $N \to \infty$.

The set of equations (7.29) are quite general and are valid for any set of K_j s and m_j s. For an exactly solvable case we choose

$$\omega_j = \alpha j. \tag{7.31}$$

where α is a positive constant with the dimension of time^{-1}. Then from (7.25) we obtain

$$\frac{m_{j+1}}{m_j} = \left(\frac{\omega_{2j-1}}{\omega_{2j}}\right)^2 = \left(\frac{2j-1}{2j}\right)^2, \quad j = 1, 2, \cdots . \tag{7.32}$$

For this special case we have

$$\omega_{2j-1}^2 = \frac{K_j}{m_j} = \alpha^2(2j-1)^2, \quad \omega_{2j}^2 = \frac{K_{j-1}}{m_j} = 4\alpha^2(j-1)^2, \tag{7.33}$$

and therefore

$$\frac{d^2\xi_j}{dt^2} = \alpha^2(2j-1)^2(\xi_{j+1} - \xi_j) + \alpha^2(2j-2)^2(\xi_{j-1} - \xi_j). \tag{7.34}$$

For the initial conditions we assume that the particle $j = 1$ is displaced by one unit length and zero initial speed, while all of the other particles are at rest with no initial displacement, i.e.

$$\xi_1(0) = 1, \quad \left(\frac{d\xi_j}{dt}\right)_{t=0} = 0, \quad j = 2, 3, \cdots , \tag{7.35}$$

$$\xi_j(0) = 0, \quad j \neq 1. \tag{7.36}$$

Thus we can reduce (7.34) to a first order differential equation for $X_j(t)$;

$$\frac{dX_j(t)}{dt} = \alpha\left[jX_{j+1}(t) - (j-1)X_{j-1}(t)\right], \tag{7.37}$$

which is subject to the initial conditions

$$X_1(0) = \sqrt{m_1}, \quad X_j(0) = 0, \quad j \neq 1. \tag{7.38}$$

Displacement of the jth Particle in the Chain — Now we use the method of generating function to find the solution of (7.37) with the initial conditions (7.38). To this end we define $G(t, z)$ by the infinite series:

$$G(t, z) = \sum_{j=1}^{\infty} X_j(t)z^{j-1}. \tag{7.39}$$

By differentiating (7.39) we find that

$$z\frac{\partial}{\partial z}[zG(t, z)] = \sum_{j=1}^{\infty}(j-1)X_{j-1}(t)z^{j-1}, \tag{7.40}$$

and

$$\frac{\partial}{\partial z}G(t,z) = \sum_{j=1}^{\infty} jX_{j+1}(t)z^{j-1}. \tag{7.41}$$

Next we multiply (7.37) by z^{j-1} and sum over all j s and thus we obtain the following first order linear partial differential equation for $G(t,z)$

$$\frac{\partial}{\partial t}G(t,z) = \alpha\left[\frac{\partial G(t,z)}{\partial z} - z\frac{\partial}{\partial z}(zG(t,z))\right]. \tag{7.42}$$

The value of $G(t,z)$ at $t=0$ is independent of z, this follows from the initial conditions (7.38) and Eq. (7.39)

$$G(t=0,z) = \sqrt{m_1}. \tag{7.43}$$

We can obtain the exact solution of (7.42) with the initial value (7.43) and the result is

$$G(t,z) = \frac{\sqrt{m_1}}{\cosh(\alpha t)}[1 + z\tanh(\alpha t)]^{-1}. \tag{7.44}$$

Now if we expand (7.44) in powers of z and compare it to (7.39) we find $X_j(t)$ to be;

$$X_j(t) = \frac{(-1)^{j-1}\sqrt{m_1}\,[\tanh(\alpha t)]^{j-1}}{\cosh(\alpha t)}. \tag{7.45}$$

Thus the displacement of the jth particle in this model is given by

$$\xi_j(t) = \sqrt{\frac{m_1}{m_j}}\frac{[\tanh(\alpha t)]^{2j-2}}{\cosh(\alpha t)}, \tag{7.46}$$

where

$$\sqrt{\frac{m_1}{m_j}} = \frac{2\times 4\times 6...\times(2j-2)}{1\times 3\times 5...\times(2j-3)}. \tag{7.47}$$

As Eq. (7.46) shows the motion of all of the particles will be damped, and the jth particle has its maximum displacement at the time t_j where

$$t_j = \left(\frac{1}{\alpha}\right)\cosh^{-1}(\sqrt{2j-1}), \tag{7.48}$$

and its maximum displacement is

$$\max \xi_j = \sqrt{\frac{m_1}{m_j}}\left(\frac{1}{\sqrt{2j-1}}\right)\left(\frac{2j-2}{2j-1}\right)^{j-1}. \tag{7.49}$$

We can find an approximate uncoupled equation for the solution of the jth particle by replacing for $\xi_{j+1}(t)$ and $\xi_{j-1}(t)$ in terms of the derivatives of $\xi_j(t)$ in Eq. (7.34) to get

$$\ddot{\xi}_j(t) + \alpha\left[2j(2j-1) + 2(j-1)(2j-3)\right]\dot{\xi}_j(t)$$
$$+ \alpha^2\left[(2j-1)^2 + 4(j-1)^2\right]\xi_j(t) = 0, \tag{7.50}$$

or in the simpler form of

$$\ddot{\xi}_j(t) + \alpha\gamma(j)\dot{\xi}_j(t) - \frac{1}{m_j}f_j(\xi) = 0, \qquad (7.51)$$

with $\gamma(j)$ and $f_j(\xi)$ defined by comparing the two equations (7.50) and (7.51). This last equation shows that the motion of each particle is damped and that the damping constant is proportional to α. We note that while the original equation (7.34) is invariant under time-reversal transformation, the approximate form (7.50) is not. This approximate differential equation (7.50) for $\xi_j(t)$ is valid for $t > \frac{1}{\alpha}$.

7.4   Mechanical System Coupled to a Heat Bath

A heat bath is a collection of harmonic oscillators all of them coupled to the central system and this collection must have the following properties:

(1) The spectrum of the oscillator frequencies must be dense.

(2) In almost all cases studied up to now a linear coupling is assumed between the central particle (or system) and the heat bath.

(3) The coupling constant must be a smooth function of the frequencies of the oscillators.

In cases where a particle is coupled to a field, such as the interaction of an electron with the electromagnetic field (e.g. van Kampen's model [12]) all of these conditions are satisfied. Furthermore we assume that the coupling of a small system to a field is a local coupling, that is the coupling depends on the value of the field at a single point, but occasionally the coupling can be nonlocal in space or in time (e.g. the version of the Wigner-Weisskopf model discussed in Sec. 18.5).

For the same problem a Lagrangian formulation in terms of the real space coordinates has been advanced by Dekker, in which he uses the diploe approximation to obtain and solve the equations of motion of the electron . The quantized version of this model has a logarithmic singularity in the electron's momentum fluctuation [13]–[15].

A similar work on the radiation damping of a harmonic oscillator has been formulated and solved by Castrigiano and Kokiantonis. In their work, they use two renormalizations: a mass renormalization and a redefinition of the zero point of the potential energy. Thus they find a charged oscillator which is in thermal equilibrium with the surrounding radiation field and moves as a mechanical damped harmonic oscillator [16].

Another model of a damped system which has been studied extensively consists of a central particle with the coordinate Q and momentum P, (system S_1), coupled to a large number of harmonic oscillators and these oscillators form the heat bath S_2 (Ullersma's model) [17]–[19].

Ullersma's Model — Here the coupling between the two parts is linear and the Hamiltonian for the entire system, S, is given by

$$H = \frac{P^2}{2M} + V(Q) + \sum_{n=1}^{N} \frac{1}{2}\left(\frac{p_n^2}{m_n} + m_n\omega_n^2 q_n^2\right) + \sum_{n=1}^{N} \epsilon_n q_n Q. \tag{7.52}$$

This Hamiltonian has two defects:

(a) For the special case of $V(x) = 0$ there is no lower bound on the energy [19].

(b) The Hamiltonian is not invariant under spatial transformation. We can remedy these defects by replacing the interaction term $\sum_n \epsilon_n q_n Q$ by $\frac{1}{2}\sum_n (q_n - Q)^2$. This modification will not affect the main conclusion of the work regarding the dissipative nature of the motion of the central particle. In the following discussion, for the sake of simplicity, we set all masses appearing in (7.52) equal to one, i.e.

$$M = m_1 = m_2 = \cdots = m_N = 1. \tag{7.53}$$

As for the initial conditions, we assume that at $t = 0$ the oscillators are all at rest at their equilibrium positions. These conditions guarantee that the flow of energy is from the central body or S_1 to the heat bath.

$$q_n(0) = p_n(0) = 0, \quad n = 1, 2, \cdots N. \tag{7.54}$$

From (7.52) we find the equations of motion for q_n and p_n to be

$$\dot{q}_n = p_n. \tag{7.55}$$

and

$$\dot{p}_n = -\left(\omega_n^2 q_n + \epsilon_n Q\right) = \ddot{q}_n. \tag{7.56}$$

The formal solution of (7.56) is given by

$$q_n(t) = q_n(0)\cos(\omega_n t) + \frac{1}{\omega_n}\dot{q}_n(0)\sin(\omega_n t)$$

$$- \frac{\epsilon_n}{\omega_n}\int_0^t \sin[\omega_n(t - t')]\, Q(t')\, dt'. \tag{7.57}$$

By imposing the initial conditions (7.54) we can simplify (7.57)

$$q_n(t) = -\frac{\epsilon_n}{\omega_n}\int_0^t \sin[\omega_n(t - t')]\, Q(t')\, dt'. \tag{7.58}$$

Similarly for the motion of the central oscillator we have

$$\dot{Q}(t) = P(t), \quad \text{and} \quad \dot{P}(t) = -\frac{\partial V}{\partial Q} - \sum_{n=1}^{N} \epsilon_n q_n. \tag{7.59}$$

Thus the equation of motion for the central particle can be obtained from Eqs. (7.58) and (7.59) by eliminating q_n:

$$\frac{d^2Q(t)}{dt^2} + \frac{\partial V}{\partial Q} - \sum_{n=1}^{N} \left(\frac{\epsilon_n^2}{\omega_n}\right) \int_0^t \sin\left[\omega_n\left(t - t'\right)\right] Q\left(t'\right) dt' = 0. \tag{7.60}$$

This is an integro-differential equation for Q which can be written as

$$\frac{dP(t)}{dt} + \frac{\partial V}{\partial Q} - \int_0^t K\left(t - t'\right) Q\left(t'\right) dt' = 0, \tag{7.61}$$

and

$$\frac{dQ(t)}{dt} = P(t), \tag{7.62}$$

where the kernel $K\left(t - t'\right)$ is given by

$$K\left(t - t'\right) = \sum_{n=1}^{N} \left(\frac{\epsilon_n^2}{\omega_n}\right) \sin\left[\omega_n\left(t - t'\right)\right] dt'. \tag{7.63}$$

We also note that in the time-reversed motion $t \to -t$, $t' \to -t'$ we have an equation similar to the integro-differential equation for $Q(t)$, i.e.

$$\frac{d^2Q(-t)}{dt^2} + \frac{\partial V}{\partial Q} - \int_0^{-t} K\left(t - t'\right) Q\left(-t'\right) dt' = 0. \tag{7.64}$$

An important relation in this case is the time derivative of the equal-time Poisson bracket of $P(t)$ and $Q(t)$;

$$\frac{d}{dt}\{P(t), Q(t)\} = \left\{\frac{dP(t)}{dt}, Q(t)\right\} + \left\{P(t), \frac{dQ(t)}{dt}\right\}$$

$$= \int_0^t K\left(t - t'\right) \{Q(t), Q\left(t'\right)\} dt'. \tag{7.65}$$

Since the Poisson bracket $\{Q(t), Q\left(t'\right)\}$ is not zero for all values of t and t', therefore $\frac{d}{dt}\{P, Q\}\}$ is not zero. In other words the equal-time Poisson bracket of $P(t)$ and $Q(t)$ is not equal to (-1) but is given by

$$\{P(t), Q(t)\} = -1 + \int_0^t dt' \int_0^{t'} K\left(t' - t''\right) \{Q\left(t'\right), Q\left(t''\right)\} dt''. \tag{7.66}$$

In general the coupling between the bath of oscillators and the central particle is nonlinear, i.e. instead of the interaction $\sum_n \epsilon_n q_n(t)Q$, we have

$$H_I = \sum_n \varepsilon_n q_n(t)\Phi(Q), \tag{7.67}$$

where $\Phi(Q)$ is a given function of Q. In this case the equation of motion for the central particle takes the form

$$\frac{dP}{dt} + \frac{\partial V}{\partial x} + \sum_n \varepsilon_n q_n \frac{\partial \Phi(Q)}{\partial Q} = 0, \qquad (7.68)$$

or by eliminating $q_n(t)$ as before we get

$$\frac{dP}{dt} + \frac{\partial V}{\partial x} - \left(\frac{\partial \Phi(Q)}{\partial Q}\right) \int_0^t K(t-t') \Phi[Q(t')] dt' = 0. \qquad (7.69)$$

For the nonlinear coupling we find the time derivative of the Poisson bracket to be

$$\frac{d}{dt}\{P(t), Q(t)\} = \int_0^t K(t-t') \{\Phi[Q(t')], Q(t)\} dt' \qquad (7.70)$$

Thus in general the equal-time Poisson bracket of $P(t)$ and $Q(t)$ will depend on time as well as on the variables $P(t)$ and $Q(t)$. This result has important consequences in the problem of quantization of the motion of the central particle.

The expression for the Poisson bracket Eq. (7.66) can be used to calculate the effect of dissipation by the heat bath as a perturbation on the motion. Thus to the zeroth order we ignore the effect of the coupling of the central particle to the system of oscillators, and we calculate the bracket $\{Q(t'), Q(t'')\}$ from the equations of motion

$$\frac{dP^{(0)}}{dt} + \frac{\partial V(Q)}{\partial Q} = 0, \quad P^{(0)} = \dot{Q}^{(0)}. \qquad (7.71)$$

Then the bracket $\{P^{(1)}(t), Q^{(1)}(t)\}$ to the first order is obtained by substituting $\{Q^{(0)}(t'), Q^{(0)}(t'')\}$ on the right hand side of Eq. (7.66) and carrying out the integrals. Even in this first order we can see whether the bracket $\{P(t), Q(t)\}$ depends on time or depends on the initial values P_0 and Q_0 as well as t.

Time-Dependence of the Commutator $\{P(t), Q(t)\}$ — We can determine the commutator $\{P(t), Q(t)\}$ analytically only for few cases. For instance let us consider the special case where $K(t-t')$ is of the form

$$K(t-t') = 2\lambda \frac{d}{dt'} \delta(t'-t) \qquad (7.72)$$

(see below) and when $V(Q) = 0$. Then we have

$$\frac{dP}{dt} + \lambda \dot{Q} = 0, \quad P = \dot{Q}, \qquad (7.73)$$

and therefore

$$Q(t) = \frac{P_0}{\lambda}\left(1 - e^{-\lambda t}\right) + Q_0, \qquad (7.74)$$

where Q_0 and P_0 are the initial values of $Q(t)$ and $P(t)$. For this case we can calculate the Poisson bracket

$$\{P(t), Q(t)\}_{P_0, Q_0} = -e^{-\lambda t}, \tag{7.75}$$

or use the perturbation form mentioned above to first calculate

$$Q^{(0)} = P_0 t + Q_0. \tag{7.76}$$

Once $Q^{(0)}$ is found we can use it to determine $\{Q^{(0)}(t'), Q^{(0)}(t'')\}$. Substituting this bracket in Eq. (7.66) we obtain

$$\{P^{(1)}(t), Q^{(1)}(t)\}_{P_0, Q_0} = -1 + \lambda t. \tag{7.77}$$

In this case as well as the problem of damped harmonic oscillator, i.e. $V(Q) = \frac{1}{2} m \omega_0^2 x^2$ the bracket depends only on time. However for the nonlinear forces $\{P(t), Q(t)\}$ will also depend on Q_0 and P_0. For instance if $V(Q) = \frac{A}{Q^2}$, then in the absence of damping

$$\left[Q^{(0)}(t) \right]^2 = Q_0^2 + 2 P_0 Q_0 t + \left(P_0^2 + \frac{2A}{Q_0^2} \right) t^2, \tag{7.78}$$

and $\{Q^{(0)}(t'), Q^{(0)}(t'')\}_{Q_0, P_0}$ and consequently $\{P^{(1)}(t), Q^{(1)}(t)\}$ will depend on Q_0, P_0 and t. In the latter case we know that in the quantum formulation of the problem the commutator will not be a c-number.

A Special Case where $K(t - t')$ Can Be Found in Closed Form — For the special case of

$$\omega_n = \frac{n \pi c}{L}, \quad n = 1, 2, \cdots \infty, \tag{7.79}$$

and for one of the following choices of ϵ_n

$$\epsilon_n(\pm) = \frac{\lambda \alpha \omega_n \Omega}{(\omega_n^2 \pm \alpha^2)^{\frac{1}{2}}}, \tag{7.80}$$

we can find $K(t - t')$ analytically. The ω_n s are the characteristic frequencies of a wave confined in an enclosure of length L, and α and Ω are constants having the dimension of frequency and λ is a dimensionless constant. Using the summation formulae [20]

$$\sum_{n=1}^{\infty} \frac{n \sin(nx)}{(n^2 + A^2)} = \frac{\pi}{2} \frac{\sinh[A(\pi - x)]}{\sinh(A\pi)}, \quad 0 < x < 2\pi, \tag{7.81}$$

and

$$\sum_{n=1}^{\infty} \frac{n \sin(nx)}{(n^2 - A^2)} = \left(\frac{\pi}{2} \right) \frac{\sin\{A[(2m+1)\pi - x]\}}{\sin(A\pi)},$$

$$0 < x < 2\pi, \quad A \text{ not an integer}, \tag{7.82}$$

we find that for $\epsilon_n(+)$ we have

$$K\left(t - t'\right) = 0, \quad t - t' = 0,$$

$$K\left(t - t'\right) = \left(\frac{\lambda^2 \alpha^2 \Omega^2 L}{2c}\right) \frac{\sinh\left[\frac{\alpha L}{c} - \alpha\left(t - t'\right)\right]}{\sinh\left(\frac{L\alpha}{c}\right)},$$

$$0 < c\left(t - t'\right) < 2L.$$

$$(7.83)$$

Similarly for $\epsilon_n(-)$ we obtain

$$K\left(t - t'\right) = 0, \quad t - t' = 0,$$

$$K\left(t - t'\right) = \left(\frac{\lambda^2 \alpha^2 \Omega^2 L}{2c}\right) \frac{\sin\left[\frac{\alpha L}{c} - \alpha\left(t - t'\right)\right]}{\sin\left(\frac{L\alpha}{c}\right)},$$

$$0 < c\left(t - t'\right) < 2L.$$

$$(7.84)$$

There are two special cases that we can simplify the problem of the motion of the central particle:

Harmonically Bound Particle Coupled to a Heat Bath — For the special case when the central particle is harmonically bound, the integro-differential equation will be linear in $Q(t)$ and there are two independent solutions for the equation (7.61) for a given $K\left(t - t'\right)$. The first one, $Q_1(t)$, is obtained by imposing the initial conditions

$$Q(0) = 1, \quad \dot{Q}(0) = 0,$$

$$(7.85)$$

and the second one which we denote by $Q_2(t)$ satisfies the conditions

$$Q(0) = 0, \quad \dot{Q}(0) = 1.$$

$$(7.86)$$

Writing the potential in the form of $V(Q) = \frac{1}{2}m\Omega^2 Q^2$, we can convert the integro-differential equation (7.60) to a differential equation of fourth order. We first note that $K\left(t - t'\right)$ is discontinuous at $t - t'$, and we take K at $\left(t - t'' = 0\right)$ to be $\frac{1}{2}C$, where $C = \frac{\lambda^2 \alpha^2 \Omega^2 L}{2c}$. The integro-differential equation for Q in this reduces to

$$\ddot{Q}(t) + \Omega^2 Q(t) = \int_0^t K\left(t - t'\right) Q\left(t'\right) dt'.$$

$$(7.87)$$

By differentiating (7.87) twice and noting that $K(0) = \frac{1}{2}C$, we get

$$\frac{d^4 Q(t)}{dt^4} + \left(\Omega^2 + \alpha^2\right)\frac{d^2 Q(t)}{dt^2} - \frac{1}{2}C\frac{dQ(t)}{dt} + \left[\alpha C \cot\left(\frac{\alpha L}{c}\right) + \alpha^2 \Omega^2\right] Q(t) = 0.$$

$$(7.88)$$

If we impose the initial conditions of

$$Q(0) = 1, \quad \text{and} \quad \left(\frac{dQ(t)}{dt}\right)_{t=0} = 0,$$

$$(7.89)$$

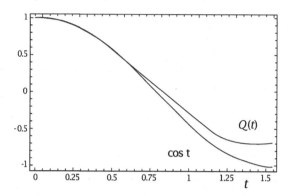

Figure 7.2: The damped oscillation of the central particle $Q(t)$ when it is coupled to the heat bath, and the coupling is small $C = 0.2$. The undamped motion $\cos t$ is also shown.

on Eq. (7.87), then (7.88) must satisfy the initial conditions

$$Q(0) = 1, \quad \dot{Q}(0) = 0, \quad \ddot{Q}(0) = -\Omega^2, \quad \text{and} \quad \left(\frac{d^3Q(t)}{dt^3}\right)_{t=0} = \frac{1}{2}C. \qquad (7.90)$$

The condition for the nonexistence of self-accelerating solutions is that (see Sec. 9.1) [12]

$$\Omega^2 - \sum_{n=1}^{\infty} \frac{\epsilon_n^2}{\omega_n} \geq 0, \qquad (7.91)$$

and we assume that this condition holds. Next we observe that the characteristic roots of the differential equation (7.88) are given by the quartic equation

$$r^4 + \left(\Omega^2 + \alpha^2\right) r^2 - \frac{1}{2C} r + \left[\alpha C \cot\left(\frac{\alpha L}{c}\right) + \alpha^2\Omega^2\right] = 0. \qquad (7.92)$$

This equation has four complex roots and we write them as

$$r'_{\pm} = -\lambda \pm i\beta, \quad \text{and} \quad r''_{\pm} = -\lambda \pm i\nu. \qquad (7.93)$$

Thus we can express Q as

$$Q(t) = e^{-\lambda t}\left[D_1 \sin(\beta t) + D_2 \cos(\beta t)\right] + e^{\lambda t}\left[D_3 \sin(\nu t) + D_4 \cos(\nu t)\right], \quad t \leq \frac{L}{c}, \qquad (7.94)$$

where $D_1 \cdots D_4$ are four arbitrary constants which can be determined from the initial conditions (7.90). In Fig. (7.2) the result of the integration of differential equation (7.88) with the boundary conditions (7.90) is shown for small damping. For comparison the undamped solution, $\cos t$, is also displayed.

A Special Case Leading to Dissipation which Is Linear in Velocity — For the general form of the potential $V(Q)$ we can choose ϵ_n and ω_n in such a way that [21]

$$\sum_{n=0}^{\infty} \left(\frac{\epsilon_n^2}{\omega_n^2} \right) \cos \left[\omega_n \left(t - t' \right) \right] = 2\lambda \delta \left(t - t' \right), \tag{7.95}$$

therefore

$$\sum_{n=0}^{\infty} \left(\frac{\epsilon_n^2}{\omega_n} \right) \sin \left[\omega_n \left(t - t' \right) \right] = 2\lambda \frac{d}{dt'} \delta \left(t - t' \right). \tag{7.96}$$

Then the equation of motion is reduced to

$$\frac{d^2 Q(t)}{dt^2} + \lambda \frac{dQ(t)}{dt} + \frac{\partial V(Q)}{\partial Q} = 0, \tag{7.97}$$

in which the damping force is proportional to the velocity (see also Sec. 17.5).

7.5   Euclidean Lagrangian

When the central particle moves in a potential $V(Q)$, and this potential forms a barrier to the motion of the particle, then only by the mechanism of tunnelling the particle can pass through the barrier. In the region where the classical momentum becomes imaginary, we can use an imaginary time formulation for the motion of the particle [22],[23].

By introducing the imaginary time τ, where $\tau = it$, we write the Lagrangian for the interaction between the central particle and the heat bath as

$$L = - \left[\frac{1}{2} \left(\frac{dQ}{d\tau} \right)^2 + V(Q) + \sum_{0}^{\infty} \frac{1}{2} \left\{ \left(\frac{dq_n}{d\tau} \right)^2 + \omega_n^2 q_n^2 \right\} + \sum_{n=0}^{\infty} \epsilon_n q_n \right]. \tag{7.98}$$

The equations of motion derived from this Lagrangian are:

$$\frac{d^2 Q}{d\tau^2} - \frac{\partial V(Q)}{\partial Q} - \sum_{n=0}^{\infty} \epsilon_n q_n = 0, \tag{7.99}$$

and

$$\frac{d^2 q_n}{d\tau^2} - \omega_n^2 q_n - \epsilon_n Q = 0. \tag{7.100}$$

Again we solve (7.100) for $q_n(\tau)$, but now we assume that

$$q_n(-\infty) = q_n(\infty) = 0, \tag{7.101}$$

and substitute the result in (7.99) to get

$$\frac{d^2 Q}{d\tau^2} - \frac{\partial V(Q)}{\partial Q} + \int_{-\infty}^{\infty} \tilde{K} \left(\tau - \tau' \right) Q \left(\tau' \right) d\tau', \tag{7.102}$$

where

$$\tilde{K}\left(\tau - \tau'\right) = -\sum_{n=0}^{\infty} \frac{\epsilon_n^2}{2\omega_n} \exp\left(-\omega_n \left|\tau - \tau'\right|\right), \qquad (7.103)$$

and where the dots refer to derivatives with respect to τ.

When $V(Q)$ is a double-well potential, then Eq. (7.102) can be solved numerically.

7.6 Hamiltonian Formulation of a Tuned Circuit Coupled to a Transmission Line

We have seen different mechanical models where, as a whole, the system is conservative but the subsystem exhibits damped motion. The close analogy between mechanical and electrical systems suggests that instead of point particles connected with springs, we may consider electrical circuits where charges and currents show damping effects [24]–[27]. For instance let us consider a tuned circuit (shown by $\underline{C}$ and $\underline{L}$ in Fig. 7.3) which is coupled to an artificial (or transmission) line of the form of a series of identical L and C parts. We will show that the initial charge in $\underline{C}$ has a damped sinusoidal dependence on t. Since the system as a whole is conservative we can derive the equations for the time dependence of the charges in successive capacitors from a Lagrangian or a Hamiltonian. Here we find it more convenient to work with a Hamiltonian. Thus we identify the dynamical variables as charges and then allow the conjugate momenta to be obtained from the canonical equations.

Let H be defined by

$$H = \frac{1}{2\underline{C}}q^2 + \frac{(p - p_1)^2}{2\underline{L}} + \frac{1}{2C}\sum_n q_n^2 + \frac{1}{2L}\sum_n (p_n - p_{n-1})^2, \qquad (7.104)$$

where $\underline{L}$ and $\underline{C}$ refer to the tuned circuit and L and C denote elements of the artificial line Fig. 7.3.

The conservation of the total charge in the system follows from the canonical equation of motion

$$\frac{dq}{dt} + \sum_n \frac{dq_n}{dt} = \frac{\partial H}{\partial p} + \sum_n \frac{\partial H}{\partial p_n} = 0, \qquad (7.105)$$

Since

$$\frac{dq}{dt} = \frac{(p - p_1)}{\underline{L}}, \qquad (7.106)$$

and as the current in $\underline{L}$ is $\frac{dq}{dt}$, therefore $\frac{(p-p_1)^2}{2\underline{L}}$ is the energy in the capacitor, and H represents the total energy in the circuit. Now in the Hamiltonian formulation we can use the canonical transformation with the generating function $F = qQ + \sum_n q_n Q_n$ to get

$$p = \frac{\partial F}{\partial q} = Q, \quad p_n = \frac{\partial F}{\partial q_n} = Q_n, \qquad (7.107)$$

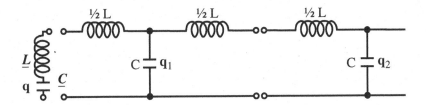

Figure 7.3: An L, C circuit coupled to an artificial transmission line.

and

$$P = -\frac{\partial F}{\partial Q} = -q, \quad P_n = \frac{\partial F}{\partial Q_n} = -q_n. \tag{7.108}$$

With this transformation the Hamiltonian becomes

$$H = \frac{1}{2\underline{C}}P^2 + \frac{(Q - Q_1)^2}{2\underline{L}} + \frac{1}{2C}\sum_n P_n^2 + \frac{1}{2L}\sum_n (Q_n - Q_{n+1})^2. \tag{7.109}$$

The physical meaning of these new variables can be understood if we consider the Hamilton canonical equations;

$$\frac{\partial H}{\partial P} = \frac{dQ}{dt} = \frac{P}{\underline{C}} = \frac{q}{\underline{C}}, \quad \text{and} \quad \frac{\partial H}{\partial P_n} = \frac{dQ_n}{dt} = \frac{P_n}{C} = \frac{q_n}{C}. \tag{7.110}$$

From these relations we can see that P, $P_1, \cdots P_n, \cdots$ are charges Q, $Q_1, \cdots, Q_n, \cdots$ are integrals with respect to time of the charges. Now for any value of n, Hamilton's other canonical equations are:

$$\frac{dP_n}{dt} = -\frac{\partial H}{\partial Q_n} = -\frac{1}{L}(2Q_n - Q_{n+1} - Q_{n-1}), \tag{7.111}$$

and by eliminating P_n between (7.110) and (7.111) we find

$$\frac{d^2 Q_n}{dt^2} = -\omega_0^2(2Q_n - Q_{n+1} - Q_{n-1}), \tag{7.112}$$

where

$$\omega_0^2 = \frac{1}{LC}. \tag{7.113}$$

Equation (7.112) is exactly like the equation of motion for the Schrödinger chain Eq. (7.1) but now the initial conditions are different.

Let us write the general solution of (7.112) as

$$Q_n = \sum_\lambda \left\{ A_\lambda e^{i(\omega t - n\lambda)} + \bar{A}_\lambda e^{-i(\omega t - n\lambda)} + A_{-\lambda} e^{i(\omega t + n\lambda)} + \bar{A}_{-\lambda} e^{-i(\omega t + n\lambda)} \right\}, \tag{7.114}$$

where

$$\omega^2 = 4\omega_0^2 \sin^2\left(\frac{\lambda}{2}\right), \tag{7.115}$$

and where ω and λ are assumed to be positive numbers. From Q_n we can find the time dependent form of P_n;

$$P_n = C\frac{dQ_n}{dt} = C\sum_\lambda \left\{ i\omega A_\lambda e^{i(\omega t - n\lambda)} + \cdots\cdots \right\}, \quad n \geq 2, \qquad (7.116)$$

where in (7.116) we have written just the first term, and the other terms found by taking the time derivative of (7.114) are denoted by dots. In particular we are interested in the flow of charges in the tuned circuit. For this we calculate Q and P from

$$\frac{dQ}{dt} = \frac{1}{C}P, \quad \frac{dP}{dt} = -\frac{Q - Q_1}{L} \qquad (7.117)$$

$$\frac{dQ_1}{dt} = \frac{1}{C}P_1, \quad \frac{dP_1}{dt} = -\frac{Q - Q_1}{L} + \frac{Q_2 - Q_1}{L}. \qquad (7.118)$$

Again by eliminating P s we obtain Q_1

$$Q_1 = 2Q_2 - Q_3 + \frac{1}{\omega_0^2}\frac{d^2 Q_2}{dt^2}$$

$$= \sum_\lambda \left[A_\lambda e^{i(\omega t - 2\lambda)} \left\{ 2 - e^{-i\lambda} - \left(\frac{\omega}{\omega_0}\right)^2 \right\} + \cdots\cdots \right]. \qquad (7.119)$$

Similarly for P_1 we find

$$P_1 = C\sum_\lambda \left[i\omega A_\lambda e^{i(\omega t - 2\lambda)} \left\{ 2 - e^{-i\lambda} - \frac{\omega^2}{\omega_0^2} \right\} \cdots\cdots \right]. \qquad (7.120)$$

Now we want to isolate the equation for the time dependence of the charge in the tuned circuit. From Eq. (7.117) we find that

$$\frac{d}{dt}(P + P_1) = \frac{Q_2 - Q_1}{L}, \qquad (7.121)$$

so that

$$\frac{dP}{dt} = \sum_\lambda \left[A_\lambda e^{i(\omega t - 2\lambda)} \left\{ \frac{1 - e^{i\lambda}}{L} + \omega^2 C \left(2 - e^{-i\lambda} - \frac{\omega^2}{\omega_0^2} \right) \right\} + \cdots\cdots \right]. \qquad (7.122)$$

Next from (7.117) we calculate $\frac{d^2 P}{dt^2}$

$$L\frac{d^2 P}{dt^2} = -\left(\frac{dQ}{dt} - \frac{dQ_1}{dt} \right) = -\frac{P}{C} + \frac{dQ_1}{dt}. \qquad (7.123)$$

and thus we obtain

$$L\frac{d^2 P}{dt^2} + \left(\frac{L}{C}\right)^{\frac{1}{2}}\frac{dP}{dt} + \frac{P}{C} = \mathcal{J}(t), \qquad (7.124)$$

where

$$\mathcal{J}(t) = \sum_\lambda \left[A_\lambda e^{i(\omega t - 2\lambda)} i\omega \left(2 - e^{-i\lambda} - \frac{\omega^2}{\omega_0^2} \right) \right]$$

$$+ \sum_\lambda \left\{ A_\lambda e^{i(\omega t - 2\lambda)} \left(\frac{L}{C} \right)^{\frac{1}{2}} \left[\frac{1 - e^{i\lambda}}{L} + \omega^2 C \left(2 - e^{-i\lambda} \frac{\omega^2}{\omega_0^2} \right) \right] + \cdots \cdots \right\}$$

$$= \left[A_\lambda e^{i(\omega t - 2\lambda)} \left\{ \omega_0 \left(1 - e^{i\lambda} \right) + \left(2 - e^{-i\lambda} - \frac{\omega^2}{\omega_0^2} \right) \left(i\omega + \frac{\omega^2}{\omega_0^2} \right) \right\} + \cdots \cdots \right].$$

$$(7.125)$$

Equations (7.124) and (7.125) can also be written in the following form, viz,

$$\left\{ L \frac{d^2}{dt^2} + R \frac{d}{dt} + \frac{1}{C} \right\} (P + X) = 0, \tag{7.126}$$

where

$$R = \sqrt{\frac{L}{C}}, \tag{7.127}$$

and

$$X = -\sum_\lambda \left[e^{-2i\lambda} A_\lambda \left\{ \omega_0 \left(1 - e^{i\lambda} \right) + \left(2 - e^{-i\lambda} - \frac{\omega^2}{\omega_0^2} \right) \left(i\omega + \frac{\omega^2}{\omega_0} \right) \right\} \right.$$

$$\times \left\{ \frac{e^{i\omega t} - \beta e^{-\gamma t + i\Omega t} - (1 - \beta) e^{-\gamma t - i\Omega t}}{-L\omega^2 + iR\omega + C^{-1}} \right\} \cdots \cdots \right]. \tag{7.128}$$

In this relation

$$\gamma = \frac{R}{2L}, \quad \text{and} \quad \Omega = \sqrt{\left(\frac{1}{LC} - \frac{R^2}{4L^2} \right)}, \tag{7.129}$$

and β in (7.128) is a constant to be determined later. The time dependent function X contains linear combinations of the solutions $\exp(-\gamma t \pm i\Omega t)$ of the differential equation

$$\left(L \frac{d^2}{dt^2} + R \frac{d}{dt} + \frac{1}{C} \right) Z = 0, \tag{7.130}$$

and it vanishes at $t = 0$.

We choose the initial conditions such that

$$P(t = 0) = P_0 \quad \text{and} \quad Q(t = 0) = Q_0. \tag{7.131}$$

By imposing these conditions we find

$$P + X = e^{-\gamma t} \left[P_0 \cos(\Omega t) - \frac{\sin(\Omega t)}{\omega} \left(\frac{R P_0}{2L} + \frac{Q_0}{L} \right) \right]. \tag{7.132}$$

Also we require that

$$\frac{dQ}{dt} = \frac{P}{\underline{C}},$$

(7.133)

i.e. Eq. (7.110) be satisfied. This can be achieved if we integrate (7.126) and choose β in such a way that

$$\frac{1}{i\omega} - \frac{\beta}{i\Omega - \gamma} + \frac{1 - \beta}{\gamma + i\Omega} = 0.$$

(7.134)

With this choice of β we find

$$Q + Y = e^{-\gamma t}\left[Q_0 \cos(\Omega t) + \frac{\sin(\Omega t)}{\omega}\left(\frac{P_0}{C} + \frac{R}{2L}Q_0\right)\right],$$

(7.135)

where

$$Y = -\frac{1}{\underline{C}}\sum_\lambda\left[e^{-2i\lambda}A_\lambda\left\{\omega_0\left(1 - e^{i\lambda}\right) + \left(2 - e^{-i\lambda} - \frac{\omega^2}{\omega_0^2}\right)\left(i\omega + \frac{\omega^2}{\omega_0}\right)\right\}\right.$$
$$\left.\times\left\{\left[\frac{e^{i\omega t}}{i\omega} - \frac{\beta e^{-\gamma t + i\Omega t}}{i\Omega - \gamma} + \frac{(1 - \beta)e^{-\gamma t - i\Omega t}}{\gamma + i\Omega}\right]\left[-\underline{L}\omega^2 + i\omega R + \frac{1}{\underline{C}}\right]^{-1}\right\}\cdots\cdots\right].$$

(7.136)

Next let us introduce two new variables $\tilde{Q}$ and $\tilde{P}$ by the relations

$$\tilde{Q} = Q + Y = e^{-\gamma t}\left[Q_0 \cos(\Omega t) + \frac{\sin(\Omega t)}{\Omega}\left(\frac{P_0}{C} + \frac{RQ_0}{2\underline{L}}\right)\right]$$

(7.137)

and

$$\tilde{P} = e^{2\gamma t}(P + X) = e^{\gamma t}\left[P_0 \cos(\Omega t) - \frac{\sin(\Omega t)}{\Omega}\left(\frac{RP_0}{2\underline{L}} + \frac{Q_0}{L}\right)\right].$$

(7.138)

The important property of these conjugate quantities is the fact that their Poisson bracket remains constant in time

$$\left\{\tilde{Q}, \tilde{P}\right\} = 1.$$

(7.139)

According to Stevens, the variable $Q = \underline{C}^{-1}\int q\,dt$, where q is the true charge, includes the noise fluctuations. But the new variable $\tilde{Q}$ which varies smoothly with time may be interpreted as $\underline{C}^{-1}\int\tilde{q}\,dt$ where $\tilde{q}$ is the "smoothed" charge [24]. We will see later that this model for the damped system is useful in formulating the coupling between a particle with spin and the electromagnetic field in a cavity with lossy walls.

References

[1] E. Schrödinger, Ann. Phys. (Leipzig) 44, 916 (1914).

[2] H. Levine, *Unidirectional Wave Motions*, (North-Holland, Amsterdam, 1978) p. 301.

[3] H.J. Kreuzer, *Nonequilibrium Thermodynamics and its Statistical Foundations*, (Oxford University Press, London, 1981) p. 341.

[4] D.H. Zanette, Am. J. Phys. 62, 404 (1994).

[5] R.E. Bellman and K.L. Cooke, *Differential-Difference Equations*, (Academic Press, New York, N.Y. 1963).

[6] M. Razavy, Can. J. Phys. 57, 1731 (1979).

[7] I.S. Gradshteyn and I.M. Ryzhik, *Tables of Integrals, Series, and Products*, (Academic Press, New York, N.Y. 1965) p. 973.

[8] R.J. Rubin, Phys. Rev. 131, 964 (1963).

[9] U. Weiss, *Quantum Dissipative Systems*, Fourth Edition (World Scientific, Singapore, 2012) p. 27.

[10] M. Razavy, Can J. Phys. 58, 1019 (1980).

[11] F.J. Dyson, Phys. Rev. 92, 1331 (1953).

[12] N.G. van Kampen, Dans. mat.-fys. Medd. 26, Nr. 15 (1951).

[13] H. Dekkar, Physica 133 A, 1 (1985).

[14] H. Dekker, Phys. Lett. 107 A, 255 (1985).

[15] H. Dekker, Phys. Rep. 80, 1 (1981).

[16] D.P.L. Castrigiano and N. Kokiantonis, J. Phys. A 20 4237 (1987).

[17] P. Ullersma, Physica 32, 27 (1966).

[18] M. Razavy, Phys. Rev. 41, 1211 (1990).

[19] G.W. Ford, J.T. Lewis and R.F. O'Connell, Phys. Rev. A 37, 4419 (1988).

[20] E.R. Hanson, *A Table of Series and Products*, (Prentice-Hall, Inc. Englewood Cliffs, 1975) p. 222.

[21] A. Pimpale and M. Razavy, Phys. Rev. A 36, 2739 (1987).

[22] A.O. Caldeira and A. Leggett, Ann. Phys. (NY) 149, 374 (1983).

[23] M. Razavy, *Quantum Theory of Tunneling*, Second Edition, (World Scientific, Singapore, 2014) Chapter 13.

[24] K.W.H. Stevens, Proc. Phys. Soc. 77, 515 (1961).

[25] The analogy between mechanical systems and electrical circuits are discussed in many books, e.g. D. Findeisen's *System Dynamics and Mechanical Vibrations: An Introduction*, (Springer-Verlag, Berlin, 2000). Chapter 4.

[26] M.H. Devoret in *Les Houches Session LXIII*, Edited by S. Reynaud, E. Giacobino and J. Zinn-Justin (Elsevier, Amsterdam, 1997).

[27] M.J. Morgan, Am. J. Phys. 56, 643 (1988).

Chapter 8

The Equation of Motion for an Oscillator Coupled to a Field

In this chapter we will study a number of examples of mechanical systems coupled to vector or scalar fields. When a system is coupled to a quantum field, the vacuum fluctuations are always present and should be a part of the equations of motion. The effect of the vacuum noise is to ensure that in the Langevin equation there is a time-dependent driving force in addition to the damping force.

8.1 Harmonically Bound Radiating Electron

We have already seen that a classical radiating electron is an interesting example of a damped system. Now we want to derive a non-relativistic form of this dissipative system from a large and conservative system S consisting of an electron, (system S_1) interacting with an external electromagnetic field (system S_2) [1]–[4]. Here the field is the transverse electromagnetic field described by the vector potential $\mathbf{A}$ with the constraint $\nabla \cdot \mathbf{A} = 0$ (van Kampen's model). The Hamiltonian for the entire system is

$$H = \frac{1}{2m_0} \left(\mathbf{P} - e\tilde{\mathbf{A}} \right)^2 + V(\mathbf{R}) + \frac{1}{8\pi} \int \left[\mathbf{E}^2 + (\mathrm{curl}\mathbf{A})^2 \right] d^3r. \qquad (8.1)$$

In this expression we have set the velocity of light $c = 1$, and m_0 is the bare mass of the electron and e is its charge, and the coupling term, $e\tilde{\mathbf{A}}$, is defined by

$$e\tilde{\mathbf{A}}(\mathbf{R}, t) = \int \mathbf{A}(\mathbf{r}, t)\rho(|\mathbf{r} - \mathbf{R}|)d^3r = \int \mathbf{A}(\mathbf{r} + \mathbf{R}, t)\rho(|\mathbf{r}|)d^3r. \qquad (8.2)$$

179

In this relation $\rho(r)$ is the charge distribution for the electron and hence

$$\int \rho(r)d^3r = e. \tag{8.3}$$

If the motion of the electron is confined to a small region around the origin, then we set $\mathbf{R} = 0$ and we have

$$e\tilde{\mathbf{A}}(t) = \int \mathbf{A}(\mathbf{r},t)\rho(|\mathbf{r}|)d^3r. \tag{8.4}$$

Dipole Approximation for the Electromagnetic Field — We can diagonalize the Hamiltonian for a harmonically bound electron if we assume the validity of the dipole approximation. In this approximation we can expand $\mathbf{A}(\mathbf{r})$ in terms of a complete set of functions obtained for waves confined inside a large sphere of radius L, and thus we find

$$\mathbf{A}(\mathbf{r},t) = \text{Transverse part of} \left\{ \sum_{n=1}^{\infty} \sqrt{\frac{3}{L}} \mathbf{q}_n(t) \frac{\sin(k_n r)}{r} \right\}, \tag{8.5}$$

where $k_n = \frac{n\pi}{L}$. Note that there are three directions of polarization corresponding to the three components of $\mathbf{q}_n$. Similarly we expand $\mathbf{E}(\mathbf{r})$ as

$$\mathbf{E}(\mathbf{r},t) = -\text{Transverse part of} \left\{ \sum_{n=1}^{\infty} \sqrt{\frac{3}{L}} \mathbf{p}_n(t) \frac{\sin(k_n r)}{r} \right\}, \tag{8.6}$$

where $\mathbf{p}_n(t)$ is the canonical conjugate of $\mathbf{q}_n(t)$. This follows from the fact that $\left(-\frac{\mathbf{E}}{4\pi}\right)$ is the canonical conjugate of $\mathbf{A}$. Then $e\tilde{\mathbf{A}}$ can be obtained from [1]

$$e\tilde{\mathbf{A}}(t) = \sum \mathbf{q}_n(t) \sqrt{\frac{4}{3L}} \int \sin(k_n r)\,\rho(r)4\pi r dr = \sum \epsilon_n \mathbf{q}_n(t), \tag{8.7}$$

where

$$\epsilon_n = \delta_n \frac{n\pi c}{L} \sqrt{\frac{4e^2}{3L}}, \tag{8.8}$$

and δ_n is the form factor of the electron ($\delta_n = 1$ for a point particle). Substituting (8.5), (8.6) and (8.7) in (8.1) and carrying out the integrals we find the following expression for the Hamiltonian:

$$H = \frac{1}{2m_0}(\mathbf{P})^2 + V(\mathbf{R}) - \frac{1}{m_0}\mathbf{P}\cdot\sum_n \epsilon_n \mathbf{q}_n + \frac{1}{2m_0}\left(\sum_n \epsilon_n \mathbf{q}_n\right)^2$$

$$+ \frac{1}{2}\sum_n \left(\mathbf{p}_n^2 + k_n^2 \mathbf{q}_n^2\right). \tag{8.9}$$

This Hamiltonian is similar to the one describing the coupling between a particle and a bath consisting of harmonic oscillators, the only difference being that here the momentum of the particle is coupled to the amplitude of the field.

For the case of a harmonically bound electron, viz, when $V(\mathbf{R}) = \frac{1}{2}K\mathbf{R}^2$ the Hamiltonian is quadratic in the canonical variables and can be diagonalized by orthogonal transformation [1],[2].

Now let us consider the following canonical transformations

$$\mathbf{p}_n = \mathbf{p}'_n, \quad \mathbf{q}_n = \mathbf{q}'_n + \frac{\epsilon_n}{mk_n^2}\mathbf{P}', \quad \mathbf{P}_n = \mathbf{P}'_n, \tag{8.10}$$

and

$$\mathbf{R}_n = \mathbf{R}'_n + \sum_n \frac{\epsilon_n}{mk_n^2}\mathbf{p}'_n, \tag{8.11}$$

where m is the renormalized mass

$$m = m_0 + \sum_n \frac{\epsilon_n^2}{k_n^2}, \tag{8.12}$$

then the Hamiltonian becomes

$$H = \frac{1}{2m}\left(\mathbf{P}'\right)^2 + \frac{1}{2}K\left[\mathbf{R}' + \sum_n \left(\frac{\epsilon_n}{mk_n^2}\right)\mathbf{P}_n\right]^2 + \frac{1}{2m}\left(\sum_n \epsilon_n \mathbf{q}'_n\right)^2$$

$$+ \frac{1}{2}\sum_n \left(\mathbf{p}'^2_n + k_n^2\mathbf{q}'^2_n\right). \tag{8.13}$$

This is one of the several forms of Hamiltonians quadratic in both momenta and coordinates that we can find for realistic physical systems.

8.2   An Oscillator Coupled to a String of Finite Length

Another problem which can be reduced to a quadratic Hamiltonian is that of an oscillator coupled to a finite or infinite string (Sollfrey's model) [5]–[8]. First consider a string of length $2L$ fixed at both ends and right at its mid-point, $x = 0$, is coupled to an oscillator of natural frequency ν_0. We choose the units so that the density of the string and also its tension be equal to unity.

The coupled equations of motion for this system are:

$$\left(\frac{\partial^2}{\partial x^2} - \frac{\partial^2}{\partial t^2}\right)y(x,t) = e\delta(x)\left[y(0,t) - q(t)\right], \tag{8.14}$$

$$\left(\frac{d^2}{dt^2} + \nu_0^2\right)q(t) = \frac{e}{m}\left[y(0,t) - q(t)\right]. \tag{8.15}$$

We note that only the even solutions of (8.14) will be coupled to the oscillator, and we will only consider these solutions.

Let us take the solutions of the coupled set (8.14) and (8.15) to be of the form

$$y(x, t) = y(x, \omega) \exp(i\omega t), \tag{8.16}$$

and

$$q(t) = q(\omega) \exp(i\omega t). \tag{8.17}$$

By substituting these in Eqs. (8.14) and (8.15) and imposing the boundary conditions

$$y(\pm L, t) = 0, \tag{8.18}$$

we find the following eigenfunctions

$$y_n(x, \omega_n) = \frac{\sin\left[\omega_n(L - |x|)\right]}{\sqrt{N_{\omega_n}} \sin(\nu_n L)}, \tag{8.19}$$

with ν_n s being the eigenfrequencies of the uncoupled system,

$$\nu_n = \left(n - \frac{1}{2}\right)\left(\frac{\pi}{L}\right), \tag{8.20}$$

and

$$q_n(\omega_n) = \frac{-e\sin(\omega_n L)}{m\sqrt{N_{\omega_n}} \sin(\nu_n L)\left(\omega_n^2 - \nu_0'^2\right)}, \tag{8.21}$$

where the eigenvalues ω_n, $n = 0, 1, 2 \cdots$ are the roots of the transcendental equation

$$2\omega\left(\omega^2 - \nu_0'^2\right)\cos(\omega L) + e\left(\omega^2 - \nu_0^2\right)\sin(\omega L) = 0, \tag{8.22}$$

and where

$$\nu_0'^2 = \nu_0^2 + \frac{e}{m}. \tag{8.23}$$

We can calculate the roots, $\omega = \omega_n$, correct to the second order in e, from Eq. (8.22) and these approximate roots are given by

$$\omega_n = \nu_n + \frac{e}{2\nu_n L} - \frac{e^2}{4\nu_n^2 L^2} + \frac{e^2}{2m\nu_n L\left(\nu_n^2 - \nu_0^2\right)} + \mathcal{O}\left(e^3\right), \nu_n = \left(n - \frac{1}{2}\right)\frac{\pi}{L}$$

and

$$\omega_0 = \nu_0 + \frac{e}{2\nu_0 m} - \frac{e^2}{8\nu_0^3 m^2} - \frac{e^2 \tan(\nu_0 L)}{4\nu_0^2 m} + \mathcal{O}\left(e^3\right). \tag{8.24}$$

Since as we will see in Eq. (8.29 the Hamiltonian for this system is Hermitian and positive definite, when the eigenvalue equation (8.22) is regarded as a function of ω^2, all of the roots of the latter equation must be real.

Orthonormal Set of Eigenfunctions — The eigenfunctions $y(x, \omega)$ are normalized in such a way that in the presence of the coupling the orthogonality relation reads as

$$\int_{-L}^{L} y_i(x, \omega)y_j(x, \omega)dx + q_i(\omega)q_j(\omega) = \delta_{ij}, \tag{8.25}$$

whereas in the absence of coupling this relation takes the form

$$\int_{-L}^{L} y_i(x)y_j(x)dx + q_iq_j = \delta_{ij}. \tag{8.26}$$

In the latter case we have

$$y_n(x) = \frac{\cos(\nu_n x)}{\sqrt{L}}, \quad q_n = \frac{\delta_{\nu_n,\nu_0}}{\sqrt{m}}. \tag{8.27}$$

From Eqs. (8.19), (8.21) and (8.25) we can calculate the normalization constant N_{ω_n} and this constant turns out to be

$$N_{\omega_n} = L + \left\{ \frac{e\sin^2(\omega_n L)\left[\omega_n^4 + \left(\nu_0'^2 - 3\nu_0^2\right)\omega_n^2 + \nu_0^2\nu_0'^2\right]}{2\omega_n^2\left(\omega_n^2 - \nu_0'^2\right)^2} \right\}. \tag{8.28}$$

Hamiltonian for Vibrating String of Finite Length — The equations of motion (8.14),(8.15) can be derived from the Hamiltonian

$$H = \frac{1}{2}\left[\left(\frac{p^2}{m} + m\nu_0^2 q^2\right) + \int_{-L}^{L}\left(\pi^2(x,t) + \left(\frac{\partial y(x,t)}{\partial x}\right)^2\right)dx\right]$$
$$+ \frac{e}{2}(y(0,t) - q(t))^2, \tag{8.29}$$

where $\pi(x,t)$ is the momentum density conjugate to $y(x,t)$, i.e. y and π satisfy the Poisson bracket

$$\{\pi(x,t), y(x',t)\} = -\delta(x - x'). \tag{8.30}$$

We expand both $y(x,t)$ and $\pi(x,t)$ in terms of Fourier series:

$$y(x,t) = \sum_{\nu=1}^{\infty} y_\nu(x)q_\nu'(t), \quad q(t) = \frac{q_0'(t)}{\sqrt{m}}, \tag{8.31}$$

and

$$\pi(x,t) = \sum_{\nu=1}^{\infty} y_\nu(x)p_\nu'(t), \quad p(t) = \sqrt{m}p'(t). \tag{8.32}$$

By substituting these expansions in the Hamiltonian (8.29), we find

$$H = \frac{1}{2}\sum_{\nu}^{\infty}\left(p_\nu'^2 + \nu^2 q_\nu'^2\right) + \frac{e}{2}\left[\sum_{\nu=1}^{\infty}\frac{q_\nu'}{\sqrt{L}} - \frac{q_0'}{\sqrt{m}}\right]^2. \tag{8.33}$$

The quadratic Hamiltonian H (8.33) can be diagonalized. This is achieved with the help of the orthogonal matrix $T_{\omega\nu}$, and this matrix satisfies the following conditions:

$$\omega^2\delta_{\omega\omega'} = \sum_{\nu=0}^{\infty}\nu^2 T_{\omega\nu}T_{\omega'\nu} + e\left[\sum_{\nu=1}^{\infty}\frac{T_{\omega\nu}}{\sqrt{L}} - \frac{T_{\omega\nu_0}}{\sqrt{m}}\right] \times \left[\sum_{\nu=1}^{\infty}\frac{T_{\omega'\nu}}{\sqrt{L}} - \frac{T_{\omega'\nu_0}}{\sqrt{m}}\right], \tag{8.34}$$

and

$$\sum_{\omega=0}^{\infty} T_{\omega\nu} T_{\omega\nu'} = \delta_{\nu\nu'}. \tag{8.35}$$

To find the matrix elements of T we observe that they are the Fourier transform components of the coupled system eigenfunctions with respect to the uncoupled system eigenfunctions. By evaluating the corresponding integrals we find the following matrix elements:

$$T_{\omega_n \nu_k} = \frac{e\left(\omega_n^2 - \nu_0^2\right)\sin\left(\omega_n L\right)}{\sqrt{LN_{\omega_n}}\left(\omega_n^2 - \nu_0'^2\right)\left(\omega_n^2 - \nu_k^2\right)\sin\left(\nu_n L\right)}, \tag{8.36}$$

and

$$T_{\omega_n \nu_0} = -\frac{e\,\sin\left(\omega_n L\right)}{\sqrt{mN_{\omega_n}}\left(\omega_n^2 - \nu_0'^2\right)\sin\left(\nu_n L\right)}. \tag{8.37}$$

Time-Dependence of the Oscillator in Sollfrey's Model — Now we want to determine the motion of the oscillator as a function of time. Let us first consider the general case of initial value problem for this system, where $y(x,0)$, $\left(\frac{\partial y(x,t)}{\partial t}\right)_{t=0}$, $q(0)$ and $\dot{q}(0)$ are given, and the object is to find $y(x,t)$ and $q(t)$. For this purpose we expand the various functions in terms of the complete set of eigenfunctions $y_n(x,\omega_n)$ and $q_n(\omega_n)$ Eqs. (8.19)-(8.21):

$$y(x,t) = \sum_{n=0}^{\infty} y_n(x,\omega_n) B_n(t), \tag{8.38}$$

$$q(t) = \sum_{n=0}^{\infty} q_n(\omega_n) B_n(t), \tag{8.39}$$

$$\pi(x,t) = \sum_{n=0}^{\infty} y_n(x,\omega_n) A_n(t), \tag{8.40}$$

$$p(t) = \sum_{n=0}^{\infty} p_n(\omega_n) A_n(t). \tag{8.41}$$

Since $y_n(x,\omega_n)$ and q_n are the exact eigenfunctions of the system, the Hamiltonian when expressed in terms of the amplitudes $A_n(t)$ and $B_n(t)$, becomes diagonal

$$H = \frac{1}{2}\sum_{n=0}^{\infty}\left[A_n^2(t) + \omega_n^2 B_n^2(t)\right]. \tag{8.42}$$

From this Hamiltonian we find two differential equations for $A_n(t)$ and $B_n(t)$

$$\dot{B}_n(t) = A_n(t), \tag{8.43}$$

and

$$\ddot{B}_n(t) + \omega_n^2 B_n(t) = 0. \tag{8.44}$$

The solution of this last equation can be written as

$$B_n(t) = B_n(0)\cos(\omega_n t) + \dot{B}_n(0)\frac{\sin(\omega_n t)}{\omega_n}. \tag{8.45}$$

We now calculate $B_n(t)$ in terms of $y(x,t)$ and $q(t)$ by noting that

$$B_n(t) = \int_{-L}^{L} y(x,t)y_n(x,\omega_n)dx + q(t)q_n(\omega_n), \tag{8.46}$$

and from this relation we find the initial values $B_n(0)$ and $\dot{B}_n(0)$ to be

$$B_n(0) = \int_{-L}^{L} y(x,0)y_n(x,\omega_n)dx + q(0)q_n(\omega_n), \qquad \cdot(8.47)$$

and

$$\dot{B}_n(0) = \int_{-L}^{L}\left(\frac{\partial y(x,t)}{\partial t}\right)_{t=0} y_n(x,\omega_n)dx + \dot{q}(0)q_n(\omega_n). \tag{8.48}$$

Thus we obtain the following expression for $B_n(t)$

$$B_n(t) = \int_{-L}^{L} y_n(x,\omega_n)dx \left[y(x,0)\cos(\omega_n t) + \left(\frac{\partial y(x,t)}{\partial t}\right)_{t=0}\left(\frac{\sin(\omega_n t)}{\omega_n}\right)\right]$$

$$+ q_n\left[q(0)\cos(\omega_n t) + \dot{q}(0)\left(\frac{\sin(\omega_n t)}{\omega_n}\right)\right]. \tag{8.49}$$

By substituting for $B_n(t)$ in (8.38) and (8.39) we get the displacement of the string at any point x and at any time t, and also the displacement of the oscillator at any time. These are given by the following relations:

$$y(x,t) = \int_{-L}^{L}\left[y(x',0)\sum_{n=0}^{\infty} y_n(x,\omega_n)y_n(x')\cos(\omega_n t)\right.$$

$$+ \left(\frac{\partial y(x',t)}{\partial t}\right)_{t=0}\sum_{n=0}^{\infty} y_n(x,\omega_n)y_n(x')\left(\frac{\sin(\omega_n t)}{\omega_n}\right)\Bigg]dx'$$

$$+ q(0)\sum_{n=0}^{\infty} y_n(x,\omega_n)q_n\cos(\omega_n t) + \dot{q}(0)\sum_{n=0}^{\infty} y_n(x,\omega_n)q_n\left(\frac{\sin(\omega_n t)}{\omega_n}\right),$$

$$\tag{8.50}$$

and

$$q(t) = \int_{-L}^{L}\left[y(x,0)\sum_{n=0}^{\infty} y_n(x,\omega_n)q_n\cos(\omega_n t)\right.$$

$$+ \left(\frac{\partial y(x,t)}{\partial t}\right)_{t=0}\sum_{n=0}^{\infty} y_n(x,\omega_n)q_n\left(\frac{\sin(\omega_n t)}{\omega_n}\right)\Bigg]dx$$

$$+ \sum_{n=0}^{\infty} q_n^2\left[q_0\cos(\omega_n t) + \dot{q}(0)\left(\frac{\sin(\omega_n t)}{\omega_n}\right)\right], \tag{8.51}$$

If initially the string is at rest, i.e. $y(x,0) = 0$, and $\left(\frac{\partial y(x,t)}{\partial t}\right)_{t=0} = 0$ for all values of x, and the oscillator is displaced and released with zero initial velocity, then Eqs. (8.50) and (8.51) simplify:

$$y(x,t) = q(0) \sum_{n=0}^{\infty} y_n(x) q_n \cos(\omega_n t), \tag{8.52}$$

and

$$q(t) = q(0) \sum_{n=0}^{\infty} q_n^2 \cos(\omega_n t). \tag{8.53}$$

(Compare this equation with (9.21) of Ullersma's model). In order to show that $q(t)$ is exponentially damped we can follow the method discussed in Sec. 9.1. **Unruh and Zurek Model** — A model similar to the one we discussed in this section has been proposed by Unruh and Zurek [9]. In Unruh's model the system is described by the Lagrangian

$$L = L_{F+P} + L_I, \tag{8.54}$$

where

$$L_{F+P} = \int \left[\left(\frac{\partial y(x,t)}{\partial t}\right)^2 - \left(\frac{\partial y(x,t)}{\partial x}\right)^2 + \delta(x) \left\{ \left(\frac{dq(t)}{dt}\right)^2 - \nu_0^2 q^2 \right\} \right] dx, \tag{8.55}$$

is the Lagrangian for the field and the particle and

$$L_I = -e \int q(t) \delta(x) \left(\frac{\partial y(x,t)}{\partial t}\right) dx, \tag{8.56}$$

is the Lagrangian for the coupling between the two. The Lagrangian L generates the equations of motion similar to (8.14) and (8.15)

$$\frac{\partial^2 y(x,t)}{\partial t^2} - \frac{\partial^2 y(x,t)}{\partial x^2} = e \frac{dq(t)}{dt} \delta(x). \tag{8.57}$$

and

$$\frac{d^2 q(t)}{dt^2} + \nu_0^2 q(t) = -e \frac{\partial y(0,t)}{\partial t}, \tag{8.58}$$

Unlike the model of Sollfrey [6] this Lagrangian as well as the resulting equations of motion are not invariant under time reversal transformation. Thus the interaction Lagrangian L_I has a distinctive arrow of time.

8.3   An Oscillator Coupled to an Infinite String

In the case of an infinite string the equations of motion (8.14) and (8.15) remain unchanged, however the boundary conditions (8.18) must be lifted. We solve this problem by means of the Fourier integrals [6],[7]. Thus we write

$$y(x,t) = \frac{1}{\sqrt{\pi}} \int_0^\infty q'(\nu,t) \cos(\nu x) d\nu, \tag{8.59}$$

$$q'(\nu,t) = \frac{1}{\sqrt{\pi}} \int_{-\infty}^\infty y(x,t) \cos(\nu x) dx, \tag{8.60}$$

with a similar expansion for $\pi(x,t)$. The Hamiltonian for this case is analogous to Eq. (8.33) but now the summations are replaced by integrations

$$H = \frac{1}{2} \int_0^\infty \left[p'^2(\nu) + \nu^2 q'^2(\nu) \right] d\nu + \left(p_0^2 + \nu_0^2 q_0^2 \right)$$

$$+ \frac{e}{2} \left[\int_0^\infty \frac{q'(\nu)}{\sqrt{\pi}} d\nu - \frac{q_0'}{\sqrt{m}} \right]^2. \tag{8.61}$$

We can diagonalize H by means of the orthogonal transformation $T(\omega, \nu)$ and $T(\omega, \nu_0)$ where

$$T(\omega, \nu) = \frac{\omega}{\sqrt{F(\omega)}} \left[(\omega^2 - \nu_0'^2) \delta(\omega - \nu) + e(\omega^2 - \nu_0^2) \mathcal{P} \frac{1}{\pi(\omega^2 - \nu^2)} \right], \tag{8.62}$$

and

$$T(\omega, \nu_0) = \frac{-e\,\omega}{\sqrt{\pi m F(\omega)}}, \tag{8.63}$$

where

$$F(\omega) = \omega^2 (\omega^2 - \nu_0'^2)^2 + \frac{e^2}{4} (\omega^2 - \nu_0^2)^2. \tag{8.64}$$

The matrix elements $T(\omega, \nu)$ and $T(\omega, \nu_0)$ have the following properties:

$$\omega^2 \delta(\omega - \omega') = \int_0^\infty \nu^2 T(\omega, \nu) T(\omega', \nu) d\nu + \nu_0^2 T(\omega, \nu_0) T(\omega', \nu_0)$$

$$+ e \left[\int_0^\infty T(\omega, \nu) \frac{d\nu}{\sqrt{\pi}} - \frac{T(\omega, \nu_0)}{\sqrt{m}} \right]$$

$$\times \left[\int_0^\infty T(\omega', \nu) \frac{d\nu}{\sqrt{\pi}} - \frac{T(\omega', \nu_0)}{\sqrt{m}} \right], \tag{8.65}$$

$$\int_0^\infty T(\omega, \nu) T(\omega, \nu') d\omega = \delta(\nu - \nu'), \tag{8.66}$$

$$\int_0^\infty T(\omega, \nu) T(\omega, \nu_0) d\omega = 0, \tag{8.67}$$

and

$$\int_0^\infty T^2(\omega, \nu_0) d\omega = 1. \tag{8.68}$$

In the next chapter the motion of the central particle which in this case is a damped harmonic oscillator, will be discussed in detail.

Yurke's Model — A slightly different way of considering the coupling between a harmonic oscillator and an infinite string is the one suggested by Yurke [10],[11]. Here the equations are similar to the equations (8.14),(8.15), but we now write them in the form

$$\rho\frac{\partial^2 y(x,t)}{\partial t^2} - T\frac{\partial^2 y(x,t)}{\partial x^2} = 0 \quad x > 0, \tag{8.69}$$

$$m\frac{\partial^2 y(0,t)}{\partial t^2} - T\left(\frac{\partial y(x,t)}{\partial x}\right)_{x=0+} + m\Omega_0^2 y(0,t) = 0, \quad x = 0, \tag{8.70}$$

In these equations ρ is the density of the string, m is the mass of the oscillator and Ω_0 is its natural frequency. The infinite string is under the tension T.

Now the D'Alembert general solution of (8.69) is

$$y(x,t) = y_{in}\left(\frac{x}{v} + t\right) + y_{out}\left(-\frac{x}{v} + t\right), \tag{8.71}$$

where v is the velocity of propagation of the transverse wave in the string and is given by

$$v = \sqrt{\frac{T}{\rho}}. \tag{8.72}$$

By substituting (8.71) in (8.70) we find

$$m\frac{\partial^2 y(0,t)}{\partial t^2} + m\lambda\frac{\partial y(0,t)}{\partial t} + m\Omega_0^2 y(0,t) = 2m\lambda\frac{\partial y_{in}(0,t)}{\partial t}, \tag{8.73}$$

where

$$\lambda = \frac{\sqrt{\rho T}}{m}, \tag{8.74}$$

which is the damping constant has the dimension of time^{-1}. In the absence of any incoming wave $\frac{\partial y_{in}(0,t)}{\partial t}$ is zero and Eq. (8.73) represents the equation of motion of a damped harmonic oscillator. Equations of motion (8.69),(8.70) can be derived from the Lagrangian density

$$\mathcal{L} = \left\{\frac{1}{2}\rho\left(\frac{\partial y(x,t)}{\partial t}\right)^2 - \frac{1}{2}T\left(\frac{\partial y(x,t)}{\partial x}\right)^2\right\}\theta(x)$$

$$+ \delta(x)\left[m\left(\frac{\partial y(0,t)}{\partial t}\right)^2 - m\Omega_0^2 y^2(0,t)\right], \tag{8.75}$$

where $\theta(x)$ is the step function

$$\theta(x) = \begin{cases} 1 & \text{for } x > 0 \\ 0 & \text{for } x < 0 \end{cases}, \tag{8.76}$$

and where we have used the integral [12]

$$\int_0^x f(x)\delta(x)dx = \frac{1}{2}f(0). \tag{8.77}$$

The momentum density for the transverse motion of the string is found from the Lagrangian density, $\mathcal{L}$, to be

$$\pi(x,t) = \frac{\partial\mathcal{L}}{\partial\left(\frac{\partial y(x,t)}{\partial t}\right)} = [2m\delta(x) + \theta(x)\rho]\left(\frac{\partial y(x,t)}{\partial t}\right). \tag{8.78}$$

Using this expression for momentum density we can determine the Hamiltonian density of the system $\mathcal{H}$

$$\mathcal{H} = \frac{1}{2}\left[\rho\left(\frac{\partial y(x,t)}{\partial t}\right)^2 + \mathcal{T}\left(\frac{\partial y(x,t)}{\partial x}\right)^2\right]\theta(x)$$
$$+ \delta(x)\left[m\left(\frac{\partial y(0,t)}{\partial t}\right)^2 + m\Omega_0^2 y^2(0,t)\right], \tag{8.79}$$

where $\left(\frac{\partial y}{\partial t}\right)$ should be replaced by $\pi(x,t)$ using Eq. (8.78).

A Model Suggested by Lamb — The Lamb model [5],[13] is one of the earliest models in which the motion of a particle which is subject to the potential $V(x)$ is coupled to a field $y(x,t)$. The Lagrangian for this model is similar to Yurke's model and is given by

$$L = \frac{1}{2}m\dot{q}^2(t) - V(q) + \frac{1}{2}\int_0^\infty\left[\rho\left(\frac{\partial y(x,t)}{\partial t}\right)^2 - \mathcal{T}\left(\frac{\partial y(x,t)}{\partial x}\right)^2\right]dx. \tag{8.80}$$

Since there is no coupling between $y(x,t)$ and $q(t)$ in the Lagrangian, one imposes the following condition:

$$q(t) = y(0,t). \tag{8.81}$$

Thus the equation of motion of the particle is given by

$$m\ddot{q} + \frac{dV(q)}{dq} = f(t), \tag{8.82}$$

where $f(t)$ is the force exerted by the string on the particle and has to be determined from Eq. (8.81).

The equation for the wave motion of the string is given by the inhomogeneous wave equation

$$\frac{\partial^2 y(x,t)}{\partial t^2} - c^2\frac{\partial y(x,t)}{\partial x^2} = -\left(\frac{f(t)}{\rho}\right)\delta(x), \tag{8.83}$$

where $c = \sqrt{\frac{\mathcal{T}}{\rho}}$ is the speed of the wave. The retarded solution of (8.83) can be written as

$$y(x,t) = y_g(x,t) - \frac{1}{2\rho c} \int_{-\infty}^{t - \frac{|x|}{c}} f(t')\, dt', \tag{8.84}$$

where $y_g(x,t)$ is the solution of the homogeneous wave equation. By taking $x = 0$ in Eq. (8.84) and then differentiating with respect to t we find $f(t)$;

$$f(t) = 2\rho c \left[\frac{\partial y_g(0,t)}{\partial t} - \frac{\partial y(0,t)}{\partial t} \right]. \tag{8.85}$$

From Eqs. (8.81), (8.82) and (8.85) we obtain

$$m\ddot{q} + m\lambda\dot{q} + \frac{dV(q)}{dq} = F(t), \tag{8.86}$$

where λ has the same form as in Yurke's model, Eq. (8.74), and the particle is subject to the additional time-dependent force

$$F(t) = 2\sqrt{\rho\mathcal{T}} \left(\frac{\partial y_g(0,t)}{\partial t} \right). \tag{8.87}$$

References

[1] H.G. van Kampen, Dans. mat.-fys. Medd. 26, Nr.15 (1951).

[2] P. Ullersma, Physica 32, 27 (1966).

[3] H. Dekker, Physica A, 133, 1 (1985).

[4] A. Beruni, Found. Phys. 30, 121 (2000).

[5] H. Lamb, Proc. London, Math. Soc. 32, 208 (1900).

[6] W. Sollfrey and G. Goertzel, Phys. Rev. 83, 1038 (1951).

[7] W. Sollfrey, Ph.D. Dissertation, New York University (1950).

[8] A. Mouchet, Eur. J. Phys. 29, 1033 (2008).

[9] W.G. Unruh and W.H. Zurek, Phys. Rev. D 40, 1071 (1989).

[10] B. Yurke and O. Yurke, MSC report # 4240 (1980).

[11] B. Yurke, Am. J. Phys. 54, 1133 (1986).

[12] See for instance B. Friedman, *Principles and Techniques of Applied Mathematics*, (John Wiley & Sons, New York, N.Y. 1957) p. 154.

[13] G.W. Ford, J.T. Lewis and R.F. O'Connell, J. Stat. Phys. 53, 439 (1988).

Chapter 9

Damped Motion of the Central Particle

In the previous chapter we showed that the motion of a particle coupled to a field can be described by a Hamiltonian which is similar to that of an oscillator coupled to a heat bath. Now we want to study the motion of the particle (or the central oscillator) when all of the oscillators forming the bath are initially at the equilibrium position with zero velocity. All of the Hamiltonians which we have considered are quadratic functions of momenta and coordinates and can be diagonalized exactly by canonical transformations [1]–[8]. For the specific case of an oscillator either coupled to a finite or to an infinite string (Sollfrey's model), we discussed the technique of diagonalization in the last chapter. Here we consider a general linear coupling between the central particle and the heat bath or the field, where the coupling can depend on a number of parameters.

9.1 Diagonalization of the Quadratic Hamiltonian

Let us write a typical quadratic Hamiltonian as (see Eq. (7.52))

$$H = \frac{1}{2}\left(P^2 + \Omega_0^2 Q^2\right) + \sum_{n=1}^{N} \frac{1}{2}\left(p_n^2 + \omega_n^2 q_n^2\right) + \sum_{n=1}^{N} \epsilon_n q_n Q, \tag{9.1}$$

where P and Q are the momentum and the coordinate of the central oscillator. We have written H when there are N oscillators in the bath, but later we will take the limit as $N \to \infty$. Now we want to find the time evolution of both Q and P, and to this end we first diagonalize the Hamiltonian H. The canonical transformations that we need are of the forms

$$Q = \sum_{\nu=0}^{N} X_{0\nu} q'_\nu, \quad P = \sum_{\nu}^{N} X_{0\nu} p'_\nu, \tag{9.2}$$

191

and

$$q_n = \sum_{\nu=0}^{N} X_{n\nu} q'_\nu, \quad p_n = \sum_\nu X_{n\nu} p'_\nu. \tag{9.3}$$

By substituting these in Eq. (9.1) and using the orthogonality of the transformations we find the diagonal form of the Hamiltonian

$$H = \frac{1}{2} \sum_{\nu=0}^{N} \left(p'^2_\nu + s^2_\nu q'^2_\nu \right). \tag{9.4}$$

Here s_0, $s_1 \cdots s_N$ are the eigenfrequencies of the system that we would like to determine. Note that this Hamiltonian is the quite similar to the one we found for an oscillator coupled to a scalar field, Eq. (8.42). From the Hamiltonian (9.1) we derive the equations of motion for Q and q_n

$$\frac{d^2 Q(t)}{dt^2} + \Omega_0^2 Q(t) = \sum_{n=1}^{N} \epsilon_n q_n(t), \tag{9.5}$$

and

$$\frac{d^2 q_n(t)}{dt^2} + \omega_n^2 q_n(t) = \epsilon_n Q(t), \quad n = 1, 2, \cdots N. \tag{9.6}$$

By substituting

$$Q(t) = Q(s) e^{ist} \quad \text{and} \quad q_n(t) = q(s) e^{ist}, \tag{9.7}$$

in Eqs. (9.5) and (9.6) we get the set of homogeneous equations

$$\left(\Omega_0^2 - s^2 \right) Q(s) = \sum_{n=1}^{N} \epsilon_n q_n(s) \tag{9.8}$$

and

$$\left(\omega_n^2 - s^2 \right) q_n(s) = \epsilon_n Q(s). \tag{9.9}$$

Equations (9.8) and (9.9) will have nontrivial solutions if and only if

$$\left(\Omega_0^2 - s^2 \right) = \sum_{n=1}^{N} \frac{\epsilon_n^2}{\omega_n^2 - s^2}, \tag{9.10}$$

where s_0, $s_1 \cdots s_N$, the roots of (9.10), are the normal mode frequencies. These normal modes have the following properties: [9]:

(1) The normal mode frequencies are all real.

(2) The number of s_ν s is one more than N, i.e. $\nu = 0, 1, \cdots N$.

(3) Between any two successive s_ν s there is just one ω_n. Moreover $s_N^2 > \omega_n^2$ and $s_0^2 < \omega_1^2$.

(4) The smallest eigenvalue s_0^2 can be positive or negative depending on the sign of the quantity

$$\Omega_0^2 - \sum_{n=1}^{N} \frac{\epsilon_n^2}{\omega_n^2}. \tag{9.11}$$

If all s_ν^2 s are positive then there is no self-accelerating solution. Therefore we assume that the condition

$$\Omega_0^2 - \sum_{n=1}^{N} \frac{\epsilon_n^2}{\omega_n^2} \geq 0 \tag{9.12}$$

is satisfied.

Now we want to determine the orthogonal transformation which diagonalizes the Hamiltonian (9.1). From (9.4) and the Hamilton canonical equations we find $q_\nu'(t)$ and $p_\nu'(t)$ to be

$$q_\nu'(t) = q_\nu'(0) \cos(s_\nu t) + p_\nu'(0) \left[\frac{\sin(s_\nu t)}{s_\nu} \right], \tag{9.13}$$

$$p_\nu'(t) = -q_\nu'(0) s_\nu \sin(s_\nu t) + p_\nu'(0) \cos(s_\nu t). \tag{9.14}$$

Next we invert the linear transformations (9.2) and (9.3);

$$q_\nu'(t) = X_{0\nu} Q(t) + \sum_{n=1}^{N} X_{n\nu} q_n(t), \tag{9.15}$$

$$p_\nu'(t) = X_{0\nu} P(t) + \sum_{n=1}^{N} X_{n\nu} p_n(t). \tag{9.16}$$

Then from Eqs. (9.2),(9.3), (9.13),(9.14) and (9.15),(9.16) we find $Q(t)$ and $q_j(t)$;

$$Q(t) = \frac{dA(t)}{dt} Q(0) + A(t) P(0) + \sum_{n=1}^{N} \left\{ \frac{dA_n(t)}{dt} q_n(0) + A_n(t) p_n(0) \right\}, \tag{9.17}$$

$$q_j(t) = \frac{dA_j(t)}{dt} Q(0) + A_j(t) P(0) + \sum_{n=1}^{N} \left\{ \frac{dA_{nj}(t)}{dt} q_n(0) + A_{nj}(t) p_n(0) \right\}, \tag{9.18}$$

$$P(t) = \frac{dQ(t)}{dt}, \tag{9.19}$$

and

$$p_n(t) = \frac{dq_n(t)}{dt}. \tag{9.20}$$

In these equations $A(t)$, $A_n(t)$ and $A_{nj}(t)$ are used to denote the following sums:

$$A(t) = \sum_{\nu=0}^{N} X_{0\nu}^2 \left[\frac{\sin(s_\nu t)}{s_\nu} \right], \tag{9.21}$$

$$A_n(t) = \sum_{\nu=0}^{N} X_{0\nu} X_{n\nu} \left[\frac{\sin(s_\nu t)}{s_\nu} \right], \tag{9.22}$$

and

$$A_{nj}(t) = \sum_{\nu=0}^{N} X_{n\nu} X_{j\nu} \left[\frac{\sin(s_\nu t)}{s_\nu} \right]. \tag{9.23}$$

The initial values for $A(t)$, $A_n(t)$ and $A_{nj}(t)$ and their derivatives can be found from the definitions of these quantities and the orthogonality conditions for $X_{0\nu}$ and $X_{n\nu}$. These initial values are:

$$A(0) = 0, \quad \left(\frac{dA(t)}{dt} \right)_{t=0} = 1, \quad \left(\frac{d^2 A(t)}{dt^2} \right)_{t=0} = 0, \tag{9.24}$$

$$A_n(0) = 0, \quad \left(\frac{dA_n(t)}{dt} \right)_{t=0} = 0, \quad \left(\frac{d^2 A_n(t)}{dt^2} \right)_{t=0} = 0, \tag{9.25}$$

and

$$A_{nj}(0) = 0, \quad \left(\frac{dA_{nj}(t)}{dt} \right)_{t=0} = \delta_{nj}, \quad \left(\frac{d^2 A_{nj}(t)}{dt^2} \right)_{t=0} = 0, \tag{9.26}$$

and these agree with Eqs. (9.17)-(9.20).

Our next task is to determine the coefficients $X_{0\nu}$ and $X_{n\nu}$ of the transformation. From the eigenvector equation

$$X_{n\nu} = \frac{\epsilon_n}{s_\nu^2 - \omega_n^2} X_{0\nu}, \tag{9.27}$$

we find the normalization condition of the eigenvectors to be

$$\sum_{n=1}^{N} X_{n\nu}^2 + X_{0\nu}^2 = \left[1 + \sum_{n=1}^{N} \frac{\epsilon_n^2}{(s_\nu^2 - \omega_n^2)^2} \right] X_{0\nu}^2 = 1, \tag{9.28}$$

or if we define $G(z)$ by

$$G(z) = z - \Omega_0^2 - \sum_{n=1}^{N} \frac{\epsilon_n^2}{(z - \omega_n^2)}, \tag{9.29}$$

we can write the eigenvalue equation as

$$G(z) = 0, \quad z_\nu = s_\nu^2. \tag{9.30}$$

From the definition of $G(z)$, Eq. (9.29), we can see that the frequencies of the bath oscillators ω_n, $n = 1, 2, \cdots N$ are the poles of $G(z)$ whereas the normal mode frequencies s_ν, $\nu = 0, 1, 2, \cdots N$ are the zeros of $G(z)$. Therefore we can write $G(z)$ as

$$G(z) = \frac{\Pi_{\nu=0}^{N} (z - s_\nu^2)}{\Pi_{n=1}^{N} (z - \omega_n^2)}. \tag{9.31}$$

The normalization condition for the eigenvectors can be expressed as

$$X_{0\nu}^2 = \left[\left(\frac{dG(z)}{dz} \right)_{z=s_\nu^2} \right]^{-1}. \tag{9.32}$$

This last relation shows that $X_{0\nu}^2$ is the residue of the pole of $[G(z)]^{-1}$ at $z = s_\nu^2$.

Having found the complete solution for the quadratic Hamiltonian (9.1), we want to study the dissipative motion of the central particle with the canonical coordinates $Q(t)$ and $P(t)$. For this purpose we assume that all of the oscillators, $q_n(t)$, are initially at rest with zero velocity, i.e.

$$q_n(0) = 0, \quad p_n(0) = 0. \tag{9.33}$$

Using these conditions we can simplify Eq. (9.17),

$$Q(t) = \frac{dA(t)}{dt}Q(0) + A(t)P(0). \tag{9.34}$$

Thus once $A(t)$ is known, the position and momentum of the central particle can be completely determined. For finite N, we can solve Eq. (9.21) for $A(t)$ since the eigenvalues s_ν and the eigenvectors $X_{0\nu}, X_{n\nu}$ are all known. However it is simpler to take the limit of $N \to \infty$ for our model (this is definitely the case for an oscillator coupled to a string discussed in Chapter 8). Neglecting the detailed structure of the spectrum and assuming that it is continuous, we can replace the summation over ν by integration. We will do this in the following way (see also Sec. 18.6):

$$\frac{dA(t)}{dt} = \sum_{\nu=0}^{\infty} X_{0\nu}^2 \cos(s_\nu t) = \sum_{\nu=0}^{\infty} \cos(s_\nu t) \left[\left(\frac{dG(z)}{dz} \right)_{z=s_\nu^2} \right]^{-1}, \tag{9.35}$$

where ν now runs from zero to infinity. By applying Cauchy's theorem we can write (9.35) as a contour integral

$$\frac{dA(t)}{dt} = \frac{1}{2\pi i} \oint_C \frac{\cos(\sqrt{z}t)}{G(z)} dz, \tag{9.36}$$

where the contour C encircles the positive part of the real axis in the z-plane. In the limit where s_ν s are infinite, we can approximate $G(z)$ by

$$G(z) = z - \Omega_0^2 - \int_0^\infty \left(\frac{\gamma(\omega)}{z - \omega^2} \right) d\omega, \tag{9.37}$$

where γ is defined by

$$\gamma(\omega)d\omega = \sum_{\omega < \omega_n < \omega + \Delta\omega} \epsilon_n^2. \tag{9.38}$$

In this limit the condition for non-existence of self-accelerating solution, Eq. (9.12) becomes

$$\int_0^\infty \frac{\gamma(\omega)}{\omega^2} d\omega \leq \Omega_0^2. \tag{9.39}$$

In order to find an approximate expression for $G(z)$, Eq. (9.37), we observe that this function has a cut along the positive real axis, and therefore we write the contour integral as

$$\frac{1}{2\pi i}\oint_C \cos\left(\sqrt{x}t\right)\left\{\frac{1}{G_-(x)} - \frac{1}{G_+(x)}\right\}dx, \tag{9.40}$$

where $G_\pm$ is defined by

$$G_\pm(x) = x - \Omega_0^2 + \mathcal{P}\int_0^\infty \frac{\gamma(\omega)}{\omega^2 - x}d\omega \pm i\pi\frac{\gamma(\sqrt{x})}{2\sqrt{x}}. \tag{9.41}$$

In Eq. (9.41) $\mathcal{P}$ denotes the principal value of the integral. Changing the variable x to s where $s = \sqrt{x}$ we have

$$\frac{dA(t)}{dt} = \int_0^\infty \frac{\gamma(s)\cos(st)ds}{\{s^2 - \Omega_0^2 + \mathcal{P}\int_0^\infty \frac{\gamma(\omega)d\omega}{\omega^2 - s^2}\}^2 + \left(\frac{\pi^2}{4s^2}\right)\gamma^2(s)}. \tag{9.42}$$

Knowing $\dot{A}(t)$ and hence $A(t)$ enables us to calculate $Q(t)$ and $P(t)$ from Eqs. (9.21),(9.34) and the result shows that the motion of the central particle is damped. We can simplify the result when the following approximation is valid

$$\frac{\pi^2 s^2 \gamma(s)}{4s^2} \approx \frac{\pi^2 s^2}{4\Omega_0^2}\gamma(\Omega_0), \tag{9.43}$$

i.e. when $\frac{\gamma(s)}{s^2}$ does not change rapidly around $s = \Omega_0$. If this is the case we have

$$\Omega_1^2 = \Omega_0^2 - \mathcal{P}\int_0^\infty \frac{\gamma(\omega)}{(\omega^2 - \Omega_0^2)}d\omega \approx \Omega_0^2 - \int_0^\infty \frac{\gamma(\omega)}{\omega^2}d\omega. \tag{9.44}$$

By substituting Eqs. (9.43) and (9.44) in (9.42) we find

$$\frac{dA(t)}{dt} = \frac{2\lambda}{\pi}\int_0^\infty \frac{s^2\cos(st)ds}{(s^2 - \Omega_1^2)^2 + \lambda^2 s^2}, \tag{9.45}$$

where λ denotes the quantity

$$\lambda = \left(\frac{\pi}{2}\right)\frac{\gamma(\Omega_0)}{\Omega_0^2}. \tag{9.46}$$

Under these conditions $\dot{A}(t)$ can be calculated analytically:

$$\frac{dA(t)}{dt} = e^{-\frac{1}{2}\lambda t}\left[\cos(\omega t) - \frac{\lambda}{2\Omega}\sin(\omega t)\right], \quad \omega^2 = \Omega_1^2 - \frac{\lambda^2}{4}. \tag{9.47}$$

From (9.47) it follows that $A(t)$ is the solution of the differential equation

$$\frac{d^2 A(t)}{dt^2} + \lambda\frac{dA(t)}{dt} + \Omega_1^2 A(t) = 0, \tag{9.48}$$

and therefore both $Q(t)$ and $P(t)$ satisfy this differential equation, i.e.

$$\frac{d^2 Q(t)}{dt^2} + \lambda \frac{dQ(t)}{dt} + \Omega_1^2 Q(t) = 0, \tag{9.49}$$

$$\frac{d^2 P(t)}{dt^2} + \lambda \frac{dP(t)}{dt} + \Omega_1^2 P(t) = 0. \tag{9.50}$$

Here the equation of motion for $P(t)$ is not the same as the phenomenological Eq. (3.29) for $p(t)$, even though Eq.(9.49) for $Q(t)$ agrees with Eq. (3.27). In particular we note that in this derivation we have the canonical momentum $P(t)$ equal to the mechanical momentum $\dot{Q}(t)$ (note that we have set the mass of the central particle equal to unity). Thus we have obtained the same set of equations of motion as those generated by Dekker's Hamiltonian, by $H(x, p; m, \theta)$, Eq. (3.74) or other Hamiltonians qp-equivalent to these (Sec. 3.5).

Energy Flow between Two Systems — Let us consider two systems, one with the central oscillator $Q_I(t)$ and the other with the central oscillator $Q_{II}(t)$. The first one interacts with a group of oscillators labeled by odd integers and the second one is coupled to a group of oscillators labeled by even integers. In addition we assume that $Q_I(t)$ is coupled linearly with $Q_{II}(t)$ but except this there is no interaction between the two systems. Thus we have the following set of equations of motion for these two systems (see Eqs. (9.5) and (9.6))

$$\frac{d^2 Q_I}{dt^2} + \Omega_I^2 Q_I = B Q_{II} + \epsilon_I \sum_{n=3,5,7\cdots} q_k, \tag{9.51}$$

$$\frac{d^2 q_n}{dt^2} + \omega_n^2 q_n = \epsilon_I Q_I, \quad n = 3,\, 5,\, 7\cdots, \tag{9.52}$$

$$\frac{d^2 Q_{II}}{dt^2} + \Omega_{II}^2 Q_{II} = B Q_I + \epsilon_{II} \sum_{k=4,6,8\cdots} q_k \tag{9.53}$$

and

$$\frac{d^2 q_k}{dt^2} + \omega_k^2 q_k = \epsilon_{II} Q_{II}, \quad k = 4,\, 6,\, 8\cdots \tag{9.54}$$

Now by considering $B Q_{II}$ and $B Q_I$ as external forces acting on the oscillators Q_I and Q_{II} respectively, we can follow the same technique used earlier and average over the motion of the oscillators q_n and q_k to obtain the equations of motion for the central oscillators Q_I and Q_{II}:

$$\frac{d^2 Q_I}{dt^2} + \lambda_I \frac{dQ_I}{dt} + \Omega_I^2 Q_I = B Q_{II} + F_1(t), \tag{9.55}$$

and

$$\frac{d^2 Q_{II}}{dt^2} + \lambda_{II} \frac{dQ_{II}}{dt} + \Omega_{II}^2 Q_I = B Q_I + F_2(t). \tag{9.56}$$

In these equations $F_1(t)$ and $F_2(t)$ are fluctuating forces and λ_I and λ_{II} are defined by Eq. (9.46) with Ω_0 being replaced by Ω_I and Ω_{II} respectively. Let us note that the coupling between the two conservative systems $\{q_{2i}, Q_I\}$ and $\{q_{2i+1}, Q_{II}\}$ which is assumed to be BQ_IQ_{II} does not yield the forces BQ_{II} and BQ_I as are written in these equations. The problem is that the right-hand sides of these equations are obtained from a Hamiltonian for two dissipative systems, and the addition of the interaction term BQ_IQ_{II} which is for a conservative system, (9.51)–(9.54) is at best an approximation. Now the forces BQ_{II} and BQ_I act upon Q_I and Q_{II} oscillators, therefore the work done by the first oscillator on the second per unit time is given by

$$W = -B\left(Q_{II}\frac{dQ_I}{dt} - Q_I\frac{dQ_{II}}{dt}\right). \tag{9.57}$$

By differentiating this equation with respect to time and substituting for the second derivatives from Eqs. (9.55) and (9.56) we obtain

$$
\begin{aligned}
\frac{1}{B}\frac{dW}{dt} &= -\left(Q_{II}\frac{d^2Q_I}{dt^2} - Q_I\frac{d^2Q_{II}}{dt^2}\right) \\
&= \left(\lambda_I Q_{II}\frac{dQ_I}{dt} - \lambda_{II}Q_I\frac{dQ_{II}}{dt}\right) + \left(\Omega_I^2 - \Omega_{II}^2\right)Q_IQ_{II} - B\left(Q_{II}^2 - Q_I^2\right) \\
&\quad - \left(Q_{II}F_1(t) - Q_IF_2(t)\right) \\
&= \frac{1}{2}(\lambda_I + \lambda_{II})\left(Q_{II}\frac{dQ_I}{dt} - Q_I\frac{dQ_{II}}{dt}\right) + \frac{1}{2}(\lambda_{II} - \lambda_I)\frac{d}{dt}(Q_IQ_{II}) \\
&\quad + \left(\Omega_I^2 - \Omega_{II}^2\right)Q_IQ_{II} - B\left(Q_{II}^2 - Q_I^2\right) - \left(Q_{II}F_1(t) - Q_IF_2(t)\right). \tag{9.58}
\end{aligned}
$$

Now we want to replace the microscopic variable by the macroscopic variable, the latter being the time average of the former [6]. Consider a time interval $(t_2 - t_1)$ which is long compared to the interaction time but short relative to the macroscopic time, i.e. $(t_2 - t_1)$ is short with respect to the time it takes for the macroscopic properties to change appreciably. Let $t = \frac{1}{2}(t_2 - t_1)$ and define $\overline{W}$ by

$$\overline{W} = \frac{1}{t_2 - t_1}\int_{t_1}^{t_2} W(t)dt. \tag{9.59}$$

By taking the average, we find the energy flow from the first to the second oscillator to be

$$\frac{1}{B}\frac{d}{dt}\overline{W} = -\frac{1}{2}(\lambda_I + \lambda_{II})\overline{W} - B\overline{(Q_1^2 - Q_2^2)}. \tag{9.60}$$

From this result it follows that for the steady state, i.e. for $\frac{d}{dt}\overline{W} = 0$, we have

$$\overline{W} = \frac{2B}{\lambda_I + \lambda_{II}}\left(\overline{Q_I^2} - \overline{Q_{II}^2}\right). \tag{9.61}$$

Thus the energy flow between the two systems, each considered to be a thermostat, is proportional to the difference in their energies. If these thermostats are characterized by their temperatures, then as the right-hand side of Eq. (9.61) shows that the rate of flow is proportional to the difference between the temperatures, and this is the macroscopic law of heat conduction [8].

For similar but more realistic formulations of the problem of heat conduction in one-dimensional chains see [10]–[12].

References

[1] H.G. van Kampen, Dans. mat.-fys. Medd. 26, Nr.15 (1951).

[2] P. Ullersma, Physica 32, 27 (1966).

[3] P. Ullersma, Physica 32, 56 (1996).

[4] P. Ullersma, Physica, 32, 78 (1996).

[5] M. Razavy, Phys. Rev. A 41, 1211 (1990).

[6] See for instance L.A. Girifalco *Statistical Mechanics of Solids*, (Oxford University Press, Oxford, 2000) p. 38.

[7] Y. Kogure, J. Phys. Soc. Japan, 16, 14 (1961).

[8] Y. Kogure, J. Phys. Soc. Japan, 17, 36 (1962).

[9] G.W. Ford, J.T. Lewis and R.F. O'Connell, J. Stat. Phys. 53, 439 (1988).

[10] B. Hu, B. Li and H. Zhao, Phys. Rev. E 57, 2992 (1998).

[11] S. Lepri, R. Livi and A. Politi, Phys. Rep. 377, 1 (2003)

[12] B-S. Xie, Europhys. Lett. 69, 35 8 (2005).

Chapter 10

Classical Microscopic Models of Dissipation and Minimal Coupling Rule

In classical dynamics, for the motion of a charged particle in an external electromagnetic field, one can simply solve the equations of motion by adding the Lorentz force to the other forces acting on the particle. However for quantizing such a classical motion we need either the Lagrangian or the Hamiltonian, and in the presence of damping the canonical momentum is not simply related to the mechanical momentum and this creates problems both in connection with the minimal coupling rule and also in regard to the correct form of the momentum operator.

As we discussed in Chapter 5 this problem arises from the non-uniqueness of the Hamiltonians and the canonical momenta for the classical dissipative systems. A possible way of resolving this difficulty is to consider the full conservative system consisting of the subsystem of interest and the heat bath with which it is interacting. Then we can apply the minimal coupling prescription to the full system and eliminate the bath degrees of freedom to get the equation of motion for a charged damped system interacting with electromagnetic field.

Here we will study two microscopic models of damping in which the central particle of charge e is linearly coupled to a bath of neutral harmonic oscillators. The first is a model similar to the van Kampen's model [1], and Ullersma model [2] that we have discussed earlier in Chapter 9. The Hamiltonian for this model is given by [3],[4]

$$H = \left[\frac{1}{2}\mathbf{p}^2 + V(\mathbf{r})\right] + \left[\sum_j \left(\frac{1}{2m_j}\mathbf{p}_j^2 + \frac{1}{2}m_j\omega_j^2\mathbf{x}_j^2\right)\right] +$$

$$+ \left(\mathbf{r} \cdot \sum_j \epsilon_j\mathbf{x}_j + \mathbf{r}^2 \sum_j \frac{\epsilon_j^2}{2m_j\omega_j^2}\right). \tag{10.1}$$

In this relation $\mathbf{p}$ is the canonical momentum of the particle, and $\mathbf{x}_j$ and $\mathbf{p}_j$ are the coordinate and momentum of the jth oscillator with mass m_j and frequency ω_j, and ϵ_j s are the coupling coefficients. The last term in (10.1) is added so as to cancel the shift in the energy of the central particle in the corresponding quantum mechanical problem [5]-[7].

As we have seen earlier (Sec. 7.4) for the case of linear damping we choose ϵ_j, m_j and ω_j in such a way that

$$\sum_j \frac{\epsilon_j^2}{m_j\omega_j^2}\cos\left[\omega_j\left(t - t'\right)\right] = \lambda\delta\left(t - t'\right), \tag{10.2}$$

and this gives us a very simple expression for $K\left(t - t'\right)$.

Lorentz Force and Minimal Coupling Rule for Motion in Viscus Medium — When the central particle of charge e moves in an external electromagnetic field $(\mathbf{A}, \phi)$, then using the minimal coupling rule to incorporate the effect of this external field we get the equations of motion (note that the mass of the central particle is $m = 1$)

$$\dot{x}_\alpha = p_\alpha - \frac{e}{c}A_\alpha\left(\mathbf{r}, t\right), \tag{10.3}$$

$$\dot{p}_\alpha = -\frac{\partial V}{\partial x_\alpha} - e\frac{\partial \phi}{\partial x_\alpha} + \frac{e}{c}\mathbf{p}\cdot\left(\frac{\partial \mathbf{A}}{\partial x_\alpha}\right) - \frac{e^2}{c^2}\mathbf{A}\cdot\left(\frac{\partial \mathbf{A}}{\partial x_\alpha}\right) - \sum_j\left[\epsilon_j x_{j\alpha} + \frac{x_\alpha\epsilon_j^2}{m_j\omega_j^2}\right],$$

$$m_j\dot{x}_{j\alpha} = p_{j\alpha}, \tag{10.4}$$

and

$$\dot{p}_{j\alpha} = -m_j\omega_j^2 x_{j\alpha} - \epsilon_j x_\alpha, \qquad \alpha = 1, 2, 3. \tag{10.5}$$

In Eqs. (10.3)-(10.5) x_α, p_α are the components of $\mathbf{r}$ and $\mathbf{p}$ and $x_{j\alpha}, p_{j\alpha}$ are the components of $\mathbf{x}_j$ and $\mathbf{p}_j$ respectively. By eliminating $p_{j\alpha}$ between (10.4) and (10.5) we find the equation of motion for $x_{j\alpha}$;

$$\ddot{x}_{j\alpha} = -\omega_j^2 x_{j\alpha} - \frac{\epsilon_j}{m_j}x_\alpha. \tag{10.6}$$

This equation with the initial conditions $x_{j\alpha}(0) = 0$, $\dot{x}_{j\alpha}(0) = 0$ has the solution

$$x_{j\alpha}(t) = -\frac{\epsilon_j}{m_j\omega_j}\int_0^t x_\alpha\left(t'\right)\sin\left[\omega_j\left(t - t'\right)\right]dt'. \tag{10.7}$$

From Eqs. (10.3), (10.4) and (10.7) we get

$$\ddot{x}_\alpha = -\frac{\partial V}{\partial x_\alpha} - e\frac{\partial \phi}{\partial x_\alpha} - \frac{e}{c}\dot{A}_\alpha + \frac{e}{c}\mathbf{p}\cdot\left(\frac{\partial \mathbf{A}}{\partial x_\alpha}\right) - \frac{e^2}{c^2}\mathbf{A}\cdot\left(\frac{\partial \mathbf{A}}{\partial x_\alpha}\right)$$

$$- x_\alpha \sum_j \left[\frac{\epsilon_j^2}{m_j\omega_j^2}\right] + \sum_j \frac{\epsilon_j^2}{m_j\omega_j} \int_0^t \sin\left[\omega_j\left(t - t'\right)\right] x_\alpha\left(t'\right) dt'.$$

$$(10.8)$$

We now use (10.3) to replace $\mathbf{p}$ by $\dot{\mathbf{x}} + \frac{e}{c}\mathbf{A}$ and write the right hand side of (10.8) in terms of the Lorentz force. Integrating the last term by parts and using (10.2) we obtain

$$\ddot{x}_\alpha = -\frac{\partial V}{\partial x_\alpha} + e\left[\mathbf{E} + \frac{1}{c}\left(\dot{\mathbf{r}} \wedge \mathbf{B}\right)\right]_\alpha - \lambda\dot{x}_\alpha - \lambda x_\alpha\delta(t). \qquad (10.9)$$

This result shows that in the case of linear coupling to the heat bath we can recover the phenomenological damping force with the explicit inclusion of the Lorentz force. The last term in (10.9) comes from the particular type of coupling between the bath and the particle that we have used.

In the second model that we want to study, the interaction Hamiltonian

$$H_I = \left(\mathbf{r}\cdot\sum_j \epsilon_j\mathbf{x}_j + \mathbf{r}^2\sum_j \frac{\epsilon_j^2}{2m_j\omega_j^2}\right), \qquad (10.10)$$

is replaced by

$$H_I' = \mathbf{p}\cdot\sum_j \varepsilon_j\mathbf{x}_j. \qquad (10.11)$$

Obviously ϵ_j and ε_j have different dimensions, and for H_I' to be invariant under time-reversal invariance ε_j must change sign when t is changed to $-t$. This type of coupling has been studied by Vyatchnin [8], (see also [9],[11]).

Now we follow the same steps that led us to the equation of motion for $\ddot{x}_\alpha$ Eq. (10.8). Here we have the canonical equations

$$\dot{x}_\alpha = p_\alpha - \frac{e}{c}A_\alpha\left(\mathbf{r}, t\right) + \sum_j \varepsilon_j x_{j\alpha}, \qquad (10.12)$$

$$\dot{p}_\alpha = -\frac{\partial V}{\partial x_\alpha} - e\frac{\partial \phi}{\partial x_\alpha} + \frac{e}{c}\mathbf{p}\cdot\left(\frac{\partial \mathbf{A}}{\partial x_\alpha}\right) - \frac{e^2}{c^2}\mathbf{A}\cdot\left(\frac{\partial \mathbf{A}}{\partial x_\alpha}\right) + \frac{e}{c}\sum_j \varepsilon\mathbf{x}_j\cdot\left(\frac{\partial \mathbf{A}}{\partial x_\alpha}\right),$$

$$m\dot{x}_{j\alpha} = p_{j\alpha}, \qquad (10.13)$$

and

$$\dot{p}_{j\alpha} = -m_j\omega_j^2 x_{j\alpha} - \varepsilon_j\left(p_\alpha - \frac{e}{c}A_\alpha\right). \qquad (10.14)$$

Just as in the case of the first model, using the initial conditions $x_{j\alpha}(0) = \dot{x}_{j\alpha}(0) = 0$ we find the solution of the Eqs. (10.12) and (10.14) to be

$$x_{j\alpha}(t) = -\frac{\varepsilon_j}{\omega_j} \int_0^t \left[p_\alpha(t') - \frac{e}{c} A_\alpha(t') \right] \sin[\omega_j(t-t')]\, dt'. \qquad (10.15)$$

From Eqs. (10.12), (10.13) and (10.15) after some algebra it follows that

$$\ddot{x}_\alpha = -\frac{\partial V}{\partial x_\alpha} + e\left[\mathbf{E} + \frac{1}{c}(\dot{\mathbf{r}} \wedge \mathbf{B}) \right]_\alpha - \lambda' \dot{x}_\alpha, \qquad (10.16)$$

where λ' is now defined by a relation similar to (10.2)

$$\sum_j \varepsilon_j^2 \cos[\omega_j(t-t')] = \lambda' \delta(t-t'). \qquad (10.17)$$

Equation (10.16) is identical with the equation of motion of a charged particle moving in an external electromagnetic field, and there is no additional impulsive term that we found in (10.9).

References

[1] N.G. van Kampen, Dans. mat.-fys. Medd. 26, Nr. 15 (1951).

[2] P. Ullersma, Physica 32, 27 (1966).

[3] X.L. Li, G.W. Ford and R.F. O'Connell, Phys. Rev. A 41, 5287 (1990).

[4] G.W. Ford, J.T. Lewis and R.F. O'Connell, Phys. Rev. A 37, 4419 (1988).

[5] A.O. Caldeira and A. Leggett, Ann. Phys. (NY) 149, 374 (1983).

[6] G.W. Ford, J.T. Lewis and R.F. O'Connell, J. Stat. Phys. 53, 439 (1988).

[7] A. Pimpale and M. Razavy, Phys. Rev. A 36, 2739 (1987).

[8] S.P. Vyatchanin, Dok. Akad. Nauk. SSSR 286, 1379 (1986).

[9] F. Kheirandish and M. Anmooshahi, Int. J. Theo. Phys. 45, 33 (2005).

[10] H. Pahlavani and F. Keirandish, Chin. J. Phys. 48, 245 (2010).

[11] M. Daeimohamad, F. Kheirandish and M.R. Abolhassani, Int. J. Theor. Phys. 48, 693 (2009).

Chapter 11

Quantization of Dissipative Systems

In this and in the next five chapters we will review attempts to quantize classically damped systems and study the resulting wave equations and their spectra. As we will observe there are serious problems regarding both the mathematical aspects of the formulation and the physical interpretation of the results. We have at least three different options for quantizing such a motion. If at the classical level the frictional force is introduced phenomenologically, e.g. as a force proportional to a given power of velocity, then we can choose (a) - the canonical formalism, (b) - the Heisenberg equations, or (c) - the path integral method to formulate the quantum version of these dissipative systems.

Each one of these attempts for quantization leads to a variety of difficulties, ranging from non-uniqueness of the Hamiltonian, to the violation of the uncertainty principle in some formulations, and to the nonlinearity, and hence the violation of the superposition principle of the solutions of the wave equation in others. We will discuss these problems in detail in the following chapters.

Facing these complications, we can argue that the phenomenological, velocity-dependent forces are the result of isolating the motion of a particle or a subsystem from the classical solution of a many-body system (Chapters 7-9). The results found in this way are classical forces which depend on position, velocity and time, and therefore the classical nature of these forces may be the cause of some of the problems. To circumvent these difficulties, it has been suggested that one has to start with the quantization of a many-body conservative Hamiltonian, and then isolate the motion of a particle or a subsystem. If we follow this approach we find that in addition to the classical dissipative forces, in general, we have non-classical forces and non-Hermitian Hamiltonians. But even in this approach one finds ambiguities in the mathematical description and also problems associated with the interpretation of the results.

11.1 Early Attempts to Quantize the Damped Oscillator

From the early days of quantum mechanics, the question of the radiation damping and its effect on the stability of quantum states (or Bohr orbits) was the subject of critical discussions. For instance we know that in a seminar at E.T.H. on the topic of the Bohr atom, Max von Laue criticized the Bohr theory by saying "This is all nonsense! Maxwell's equations are valid under all circumstances. An electron in a circular orbit must emit radiation" [1].

The discovery of the matrix mechanics and later the Schrödinger equation renewed the interest in this problem and one of the first papers on the subject of radiation damping and the quantum mechanics of the damped oscillator was written by Seeger [2]. Seeger considered the problem of quantization of a damped harmonic oscillator using Rayleigh's dissipation function, Eq. (2.3), and the matrix mechanics of Born, Jordan and Heisenberg [3],[4].

Seeger's Approach to the Quantization of the Damped Oscillator — Starting with the Hamiltonian operator

$$H = \frac{1}{2m}p^2 + \frac{1}{2}m\Omega_0^2 x^2, \tag{11.1}$$

and the damping term

$$\mathcal{F} = \frac{m}{2}\lambda\dot{x}^2, \tag{11.2}$$

in the Heisenberg picture, we find the equations of motion for p and x to be

$$\dot{p} = -m\Omega_0^2 x - m\lambda\dot{x}, \quad \text{and} \quad m\dot{x} = p, \tag{11.3}$$

or

$$\ddot{x} + \lambda\dot{x} + \Omega_0^2 x = 0. \tag{11.4}$$

This equation must be satisfied by the matrix elements

$$x_{jk}(t) = x_{jk}(0)\exp\left(i\omega_{jk}t\right), \tag{11.5}$$

i.e.

$$\left\{-\omega_{jk}^2 + i\lambda\omega_{jk} + \Omega_0^2\right\}x_{jk}(0) = 0. \tag{11.6}$$

Therefore $x_{jk}(0)$ must be zero unless

$$\omega_{jk} = \pm a + i\frac{\lambda}{2}, \tag{11.7}$$

where

$$a = \sqrt{\Omega_0^2 - \frac{\lambda^2}{4}} \approx \Omega_0 - \frac{\lambda^2}{8\Omega_0}. \tag{11.8}$$

Here we are assuming that the motion is underdamped and that $2\Omega_0 \gg \lambda$. Let us now consider the commutation relation

$$\sum_j \left(p_{nj}x_{jn} - x_{nj}p_{jn}\right) = -i\hbar. \tag{11.9}$$

If

$$\alpha_{jk} = \text{Re} \left[\omega_{jk} \right], \tag{11.10}$$

then

$$\sum_j \alpha_{jk} |x_{jk}|^2 = -\frac{\hbar}{2m}. \tag{11.11}$$

We assume that there is no degeneracy. Then for each n there is at least one n' such that $x_{nn'} \neq 0$, but since $\omega_{nn'} = \pm a + i\frac{\lambda}{2}$, therefore there are at most two values n', say n'_1 and n'_2 for which $x_{nn'} \neq 0$. Using the Planck-Bohr relation we have

$$E_n - E_{n'_1} = \hbar a, \quad E_n - E_{n'_2} = -\hbar a, \tag{11.12}$$

or

$$E_n - E_{n'_1} = -\hbar a, \quad E_n - E_{n'_2} = \hbar a. \tag{11.13}$$

Next from the Hamiltonian (11.1) we calculate H_{nn}

$$H_{nn} = \frac{m}{2} e^{-\lambda t} \left(a^2 + \frac{\lambda^2}{4} + \Omega_0^2 \right) \sum_j |x_{nj}|^2, \tag{11.14}$$

which, in view of what we have seen earlier, can be written as

$$H_{nn} = \frac{m}{2} e^{-\lambda t} \left(a^2 + \frac{\lambda^2}{4} + \Omega_0^2 \right) \left\{ |x_{nn'_2}|^2 + |x_{nn'_1}|^2 \right\}. \tag{11.15}$$

On the other hand from the commutation relation (11.11) we have

$$\alpha_{nn'_1} \left\{ |x_{nn'_2}|^2 - |x_{nn'_1}|^2 \right\} = -\frac{\hbar}{2m}. \tag{11.16}$$

Just as in the problem of undamped harmonic oscillator we suppose that for the ground state n'_2 does not exist, hence

$$\alpha_{n_0 n'_0} |x_{n_0 n'_0}|^2 = \frac{\hbar}{2m}, \tag{11.17}$$

and

$$H_{n_0 n_0} = m\Omega_0^2 e^{-\lambda t} |x_{n_0 n'_0}|^2. \tag{11.18}$$

From (11.17) we obtain

$$|x_{n_0 n'_0}|^2 = \frac{\hbar}{2ma}, \tag{11.19}$$

and substituting this in (11.18) we find

$$H_{n_0 n_0} = \frac{\hbar \Omega_0^2}{2a} e^{-\lambda t} \approx \frac{\hbar \Omega_0}{2} \left(1 + \frac{\lambda^2}{8\Omega_0^2} \right) e^{-\lambda t}. \tag{11.20}$$

For higher levels we have

$$H_{jj} - H_{kk} = \frac{\hbar \Omega_0^2}{a} (j - k) e^{-\lambda t}, \tag{11.21}$$

and for the energy differences we obtain [2]

$$E_j - E_k = (j - k)a\hbar \approx \hbar\Omega_0(j - k)\left(1 - \frac{\lambda^2}{8\Omega_0^2}\right), \tag{11.22}$$

or

$$E_j - E_k = (H_{jj} - H_{kk})\left(\frac{a^2}{\Omega_0^2}\right)e^{\lambda t} \approx (H_{jj} - H_{kk})\left(1 - \frac{\lambda^2}{4\Omega_0^2}\right)e^{\lambda t}. \tag{11.23}$$

Thus Seeger concludes that the energies of the states of oscillator when $\lambda \ll 2\Omega_0$ remain constant over a considerable time.

Let us emphasize the following aspects of this formulation:

(a) The commutator $[p, x]$ stays constant in time.

(b) Equations for the matrix elements of x and p are identical.

(c) The Hamiltonian used is directly related to the energy, i.e. it is the energy operator.

(d) The change in the energy differences caused by damping is proportional to $\frac{\lambda^2}{8\Omega_0^2}$.

Incompatibility of the Canonical Commutation Relations and the Heisenberg's Equations of Motion — This method of quantizing a dissipative system has been criticized by Brittin [5]. Brittin has argued that for frictional forces depending on velocity, the commutation relation and the Heisenberg equations of motion are incompatible. Let us consider a one-dimensional motion under the force of friction Q. In Chapter 2 we have seen that the classical equation of motion can be derived from the Euler-Lagrange equation

$$\frac{d}{dt}\left(\frac{\partial L}{\partial \dot{x}}\right) - \frac{\partial L}{\partial x} = Q(x, \dot{x}). \tag{11.24}$$

By introducing the canonical momentum p by

$$p = \frac{\partial L}{\partial \dot{x}}, \tag{11.25}$$

we can write the Hamiltonian as

$$H(x, p) = p\dot{x} - L. \tag{11.26}$$

The equations of motion can also be obtained from (11.26) and the Hamilton canonical equations, which in this case are:

$$\dot{x} = \frac{\partial H}{\partial p}, \quad \text{and} \quad \dot{p} = -\frac{\partial H}{\partial x} + Q\left[x, \dot{x}(x, p)\right]. \tag{11.27}$$

In classical dynamics we can replace (11.27) by the definition of the time derivative of any dynamical variable $A(x, p, t)$ in terms of the Poisson bracket [6]

$$\frac{dA}{dt} = \{A, H\} + \frac{\partial A}{\partial p}Q(x, p), \tag{11.28}$$

where the Poisson bracket $\{A, H\}$ is defined by

$$\{A, H\} = \frac{\partial A}{\partial x}\frac{\partial H}{\partial p} - \frac{\partial A}{\partial p}\frac{\partial H}{\partial x}. \tag{11.29}$$

The transition to quantum mechanics can be made by associating the Poisson bracket with the commutator using the well-known rule of association [4],[7],[8]

$$\{A, H\} \rightarrow \frac{1}{i\hbar}[A, H] = \frac{1}{i\hbar}(AH - HA). \tag{11.30}$$

Hence for the time derivative of an operator A we find the Heisenberg equation

$$i\hbar\frac{dA}{dt} = [A, H] + i\hbar\frac{\partial A}{\partial p}Q(x, p). \tag{11.31}$$

Next we differentiate the commutator

$$[x, p] = i\hbar, \tag{11.32}$$

with respect to time to obtain

$$\left[\frac{dx}{dt}, p\right] + \left[x, \frac{dp}{dt}\right] = 0. \tag{11.33}$$

With the help of Eq. (11.31), taking A first to be x and then to be p, we rewrite (11.33) as

$$[[x, H], p] + [x, [p, H] + i\hbar Q] = 0. \tag{11.34}$$

From the last relation it follows that

$$[x, Q] = 0, \tag{11.35}$$

which in turn implies that Q can only be a function of x and not of $\dot{x}$ [5]. For a more detail discussion see Sec. 11.2 and also the paper by Srinivas and Shankara [9]. Thus Brittin's argument shows the inconsistency between the canonical equations of motion, the Hamiltonian and the commutation relation.

11.2 Tarasov's Conditions for a Self-Consistent Quantum-Mechanical Formulation of Dissipative Systems

In our study of the Hamiltonian and the Poisson bracket for the variational formulation of damped systems according Sedov, and Tarasov we found a set of canonical equations for q_i and p_i, Eqs. (3.272) and (3.273) of Sec. 3.13. Now from the Dirac's rule of association we find the quantum analogues of Eqs. (3.272) and (3.273) for the coordinate and momentum operators:

$$\frac{dQ_i}{dt} = \frac{i}{\hbar}\{H, Q_i\} - G_i(t, Q, P), \tag{11.36}$$

and

$$\frac{dP_i}{dt} = \frac{i}{\hbar}\{H,\ P_i\} + F_i(t, Q, P),\tag{11.37}$$

where H is the Hamiltonian operator for the energy of the system and Q_i and P_i are operators corresponding to the classical q_i and p_i. The two operators G_i and F_i are related to the operator D which is the quantum version of $D_0(A)$, the dissipation operator, Eq. (3.281) by

$$G_i(t, Q, P) = -D(Q_i),\quad\text{and}\quad F_i(t, Q, P) = D(P_i)\tag{11.38}$$

If we assume that for a system we can find the dissipation operator $D(A)$ satisfying the relation

$$D(A) = \frac{i}{\hbar}[A,\ W],\tag{11.39}$$

then we have

$$F_i(t, Q, P) = -\frac{i}{\hbar}[W,\ P_i],\quad\text{and}\quad G_i(t, Q, P) = \frac{i}{\hbar}[W,\ Q].\tag{11.40}$$

These relations imply that the quantum mechanical operator corresponding to the classical $\Omega_{ij}^{(s)}$ can be expressed by the following equations:

$$\Omega_{ij}^{(1)} = -\frac{i}{\hbar}\left([Q_i,\ G_j] - [Q_j,\ G_i]\right) = J(Q_i,\ Q_j,\ W)$$

$$= \frac{1}{\hbar^2}\left([\,[W,\ Q_i],\ Q_j] - [[W,\ Q_j],\ Q_i]\right),\quad i \neq j,\tag{11.41}$$

$$\Omega_{ij}^{(2)} = -\frac{i}{\hbar}\left([G_i,\ P_j] - [Q_i,\ F_j]\right) = J(Q_i,\ P_j,\ W)$$

$$= \frac{1}{\hbar^2}\left([\,[W,\ Q_i],\ P_j] - [\,[W,\ P_j],\ Q_i]\right),\tag{11.42}$$

$$\Omega_{ij}^{(3)} = -\frac{i}{\hbar}\left([P_j,\ F_i] - [P_i,\ F_j]\right) = J(P_i,\ P_j,\ W)$$

$$= \frac{1}{\hbar^2}\left([\,[W,\ P_i],\ P_j] - [\,[W,\ P_j],\ P_i]\right),\quad i \neq j,\tag{11.43}$$

where $J(A,\ B,\ C)$ is the Jacobian of the operators, $A,\ B,\ C$,

$$J(A,\ B,\ C) = \frac{1}{\hbar^2}\{[\,[A,\ B],\ C] + [\,[B,\ C],\ A] + [\,[C,\ A],\ B]\}.\tag{11.44}$$

Now consider a quantum system where the time evolution of an observable $A(t, Q, P)$ is defined by the commutator

$$\frac{dA}{dt} = \frac{\partial A}{\partial t} + \frac{i}{\hbar}[H - W,\ A],\tag{11.45}$$

then according to Tarasov this system is dissipative if at least one of the $\Omega_{ij}^{(s)}$ s is different from zero. From this definition of a quantum dissipative system Tarasov concludes that the usual requirement of consistency which is the absence of contradiction between the evolution equations and the commutations relations for dissipative systems should be replaced by a formalism involving anti-commutative non-Lie algebras [10], [11].

We note that the Leibnitz rule for differentiation does not apply to the quantum dissipative systems defined by Sedov's and Tarasov's variational formulation and the problem of compatibility of equations of motion and the commutation relations as mentioned earlier is not clearly addressed in this approach [11].

As an application of this formulation one can find the energy eigenvalues of a linearly damped harmonic oscillator where $\delta w = -\lambda p \delta q$. Using an approximate method of solving this problem Tarasov has found the eigenvalues to be the same as those found in Caldirola-Kanai Hamiltonian formulation which will be given later (see Eq. (12.28) [11]).

More About Consistency of the Equations of Motion, the Hamiltonian and the Commutation Relation — For a group of classical time-independent Hamiltonians constructed for the damped systems we can investigate the question of consistency in the following way:

Let us start with a separable Lagrangian, i.e. when L is a sum of two functions one of x and the other of $\dot{x}$. For a particle with unit mass we write L in the form of [12]

$$L\left(x, \dot{x}\right) = \dot{x} \int^{\dot{x}} \frac{1}{v^2} G(v) dv - V(x), \tag{11.46}$$

where $G(v)$ is an arbitrary function of v. As before we impose the following condition that $G(v)$ has to satisfy:

$$(1/v)(dG(v)/dv) \neq 0. \tag{11.47}$$

The canonical momentum conjugate to x is given by

$$p = \int^{\dot{x}} \frac{dv}{v^2} G(v) + \frac{G(\dot{x})}{\dot{x}} = \int^{\dot{x}} \frac{dv}{v} \frac{dG(v)}{dv}. \tag{11.48}$$

The Hamiltonian which is found from (11.46) but expressed in terms of x and p is

$$H(x, p) = G[\dot{x}(p)] + V(x), \tag{11.49}$$

where we have solved (11.48) for $\dot{x} = \dot{x}(p)$. Using the canonical commutation relation $[x, p] = i\hbar$ we can calculate $\frac{i}{\hbar}[H, x]$,

$$\frac{i}{\hbar}[H, x] = \frac{i}{\hbar}[G(\dot{x}), x] = \frac{d}{dp} G(\dot{x}) = \frac{d\dot{x}}{dp} \frac{dG(\dot{x})}{d\dot{x}} = \dot{x}. \tag{11.50}$$

This result follows from Eq. (11.48). Equation (11.50) is the correct Heisenberg equation for the time derivative of the position operator. Next we consider the time derivative of $\dot{x}$;

$$\ddot{x} = \frac{i}{\hbar}[H, \dot{x}] = \frac{i}{\hbar}[V, \dot{x}]. \tag{11.51}$$

To calculate the right hand side of (11.51) we first note that from the commutation relation, $[x, p] = i\hbar$, it follows that

$$\frac{i}{\hbar}[x, P(p)] = -\frac{dP(p)}{dp}, \tag{11.52}$$

for any function $P(p)$, therefore

$$\frac{i}{\hbar}[x, \dot{x}] = -\frac{d\dot{x}}{dp} = -\frac{\dot{x}}{\left(\frac{dG(\dot{x})}{d\dot{x}}\right)} = -R(\dot{x}), \tag{11.53}$$

where $R(\dot{x})$ is defined by (11.53).

Next we calculate the commutator

$$\frac{i}{\hbar}[x^n, \dot{x}] = \frac{i}{\hbar}\sum_{j=0}^{n-1} x^j[x, \dot{x}]x^{n-1-j} = -\sum_{j=0}^{n-1} x^j R(\dot{x})x^{n-1-j}. \tag{11.54}$$

In the classical limit the right hand side of (11.54) becomes

$$-nx^{n-1}R(\dot{x}) = -R(\dot{x})\frac{d}{dx}x^n, \tag{11.55}$$

and thus for this case we have the rule of association

$$\left[R(\dot{x})\frac{d}{dx}x^n\right]_{Q.M.} \rightarrow \sum_{j=0}^{n-1} x^j R(\dot{x})x^{n-1-j}. \tag{11.56}$$

Now we assume that $V(x)$ can be written as a power series in x, i.e.

$$V(x) = \sum_{n=0}^{\infty} a_n x^n, \tag{11.57}$$

then we can calculate (11.51),

$$\frac{i}{\hbar}[V(x), \dot{x}] = -\left(R(\dot{x})\frac{dV(x)}{dx}\right)_{Q.M.}. \tag{11.58}$$

Using (11.58) we find the Heisenberg equation of motion for x to be

$$\ddot{x} + \left(R(\dot{x})\frac{dV(x)}{dx}\right)_{Q.M.} = 0. \tag{11.59}$$

What we have shown here is that an equation of motion of the form (11.59) is compatible with the canonical commutation relation. But the mathematical consistency does not imply that the resulting quantum equation is physically meaningful.

As another example let us consider the simple case where $R(\dot{x}) = \dot{x}$ and thus the classical equation of motion becomes

$$\ddot{x} + \dot{x}\left(\frac{dV(x)}{dx}\right) = \ddot{x} + \frac{dV(x)}{dt} = 0. \tag{11.60}$$

For this case the Lagrangian and the canonical momentum are:

$$L = \frac{1}{2}\dot{x}\ln\dot{x}^2 - \dot{x} - V(x), \tag{11.61}$$

and

$$p = \frac{\partial L}{\partial \dot{x}} = \frac{1}{2}\ln\dot{x}^2. \tag{11.62}$$

The Hamiltonian obtained from (11.61) is given by

$$H = \pm e^p + V(x). \tag{11.63}$$

The quantum mechanical solution of this problem when $V(x) = x$ will be discussed later in this section where it will be shown that the result is unphysical.

Similar argument for obtaining the Heisenberg equation compatible with the commutator has been obtained for a Hamiltonian of the form

$$H(x,p) = R(p)S(x), \tag{11.64}$$

for conservative systems. Apparently the consistency requirement is closely related to the problem of the Hamiltonian operator giving the correct equation of motion in the phase space.

More on the Question of Consistency — In some cases the problem of consistency of the quantization procedure can be circumvented by changing the Lagrangian and the Hamiltonian for the system. As an example consider the following simple case where the equation of motion is given by [13]

$$m\ddot{x} + m\lambda\dot{x} = 0. \tag{11.65}$$

As we have already seen we can construct different Hamiltonians in addition to the one mentioned earlier i.e. $H = (p^2/2m)e^{-\lambda t}$. Let us take the mass of the particle to be equal to one and write as alternative Hamiltonian the function that we found in Eq. (3.33);

$$H = \mathcal{E}\exp\left(\frac{p}{p_0}\right) + \lambda p_0 x, \tag{11.66}$$

where $\mathcal{E}$ and p_0 are constants. Let p and x be the canonically conjugate operators satisfying the commutation relation $[p, x] = i\hbar$, and let us write the Heisenberg equations of motion as

$$\dot{x} = \frac{i}{\hbar}\left[\hat{H}, x\right] = \frac{\mathcal{E}}{p_0}\exp\left(\frac{p}{p_0}\right), \tag{11.67}$$

and

$$\dot{p} = \frac{i}{\hbar}\left[\hat{H}, p\right] = -\lambda p_0. \tag{11.68}$$

In these equations $\hat{H}$ represents the Hamiltonian operator corresponding to the classical function (11.66). By differentiating (11.67) with respect to t we find

$$\ddot{x} = \frac{\mathcal{E}}{2p_0^2}\left\{\dot{p}\exp\left(\frac{p}{p_0}\right) + \exp\left(\frac{p}{p_0}\right)\dot{p}\right\} = -\lambda\dot{x}, \tag{11.69}$$

where in (11.69) we have substituted for $\exp\left(p/p_0\right)$ and for $\dot{p}$ from Eqs. (11.67) and (11.68). Thus the canonical commutation relation and the equations of motion are compatible and there is no inconsistency. The Hamiltonian (11.66) also generates the correct equation of motion in phase space but it does not have the correct form as $\lambda \to 0$, and it is not the energy of the system. Therefore as a quantum-mechanical operator it is not acceptable. This can easily be seen by writing a "wave equation" for the operator (11.66) in momentum space

$$-i\hbar\lambda p_0\frac{\partial\psi}{\partial p} + \mathcal{E}\exp\left(\frac{p}{p_0}\right)\psi = E'\psi, \tag{11.70}$$

and solving it to find

$$\psi = \exp\left[\frac{-i}{\lambda\hbar p_0}\left\{\mathcal{E} - E'p_0\exp\left(\frac{p}{p_0}\right)\right\}\right]. \tag{11.71}$$

Again we note that ψ does not have a well-defined limit as $\lambda \to 0$.

11.3 Yang-Feldman Method of Quantization

In Chapters 2 and 3 we observed that the classical equations of motion in the coordinate space are the basis for construction of families of q-equivalent Lagrangian and Hamiltonian, and that these Hamiltonians (or Lagrangians) once quantized give different eigenfunctions and eigenvalues. We may ask whether it is possible to bypass the canonical formalism and to try to quantize the equation of motion (or its solution) directly. This method of quantization has been studied in quantum theory of fields and is known as the Yang-Feldman method [14],[15],[16].

The starting point here is the classical equation of motion which for a particle of mass $m = 1$ can be written as

$$\ddot{x}_i(t) = f_i\left(\mathbf{x}, \dot{\mathbf{x}}, t\right). \tag{11.72}$$

A formal solution of this classical equation with the initial values $x_i(0)$ and $\dot{x}_i(0)$ can be found with the aid of the Green function $G(t - t')$ defined by the differential equation

$$\frac{d^2}{dt^2}G(t - t') = \delta(t - t'),\qquad(11.73)$$

and has the special solution

$$G(t - t') = (t - t')\,\theta(t - t').\qquad(11.74)$$

From Eqs. (11.72) and (11.73) we find $x_i(t)$ to be

$$x_i(t) = x_i(0) + \dot{x}_i(0)t + \int_0^t (t - t')\,f_i(\mathbf{x'}, \dot{\mathbf{x}'}, t')\,dt',\qquad(11.75)$$

and this is the solution of the classical problem (11.72) subject to the initial conditions $x_i(0)$ and $\dot{x}_i(0)$. We can take Eq. (11.75) as the starting point for quantization in the Heisenberg picture. We also assume that for a given force $f_i(\mathbf{x'}, \dot{\mathbf{x}'}, t')$, we can construct a Hermitian operator using any one of the standard rules of association of classical function and quantum operators [17],[18]. But even in this, the Yang-Feldman method, the operator $\hat{p}_i(0) = \hat{\dot{x}}_i(0)$ which satisfies the commutation relation $[\hat{x}_i(0), \hat{p}_j(0)] = i\hbar\delta_{ij}$ can appear in the formal solution of the Heisenberg equation corresponding to (11.75) in different ways. **Non Uniqueness of Yang-Feldman Method of Quantization** — Let us demonstrate this by a simple example using two different q-equivalent Hamiltonians which we found in Eqs. (3.32),(3.33). First we note that if we choose $H_1 = (p^2/2)e^{-\lambda t}$ (we have set $m = 1$), then $p(t) = \dot{x}e^{\lambda t}$ and therefore as an operator

$$\hat{p}(0) = \hat{\dot{x}} = -i\hbar\frac{\partial}{\partial x},\qquad(11.76)$$

where the symbol $\hat{\dot{x}}$ indicates that we are using operators. Thus the Heisenberg equation corresponding to the classical equation (11.75) becomes

$$\hat{x}(t) = \hat{x} - i\hbar t\frac{\partial}{\partial x} - \lambda\int_0^t (t - t')\,\hat{\dot{x}}(t')\,dt',\qquad(11.77)$$

and this equation can be solved by iteration with the result that

$$\hat{x}(t) = \hat{x} - \frac{i\hbar}{\lambda}\left(1 - e^{-\lambda t}\right)\frac{\partial}{\partial x}.\qquad(11.78)$$

On the other hand for $\hat{H}_2 = e^p + \lambda x$ which is the same as H_2 in Eq. (3.33) but with $\mathcal{E} = p_0 = 1$ we have

$$\dot{x} = e^p \quad \text{and} \quad p = \ln\dot{x}.\qquad(11.79)$$

Using the operator identity

$$\exp\hat{A}\,\exp\hat{B} = \exp\hat{B}\,\exp\hat{A}\,\exp[\hat{A, B}],\qquad(11.80)$$

with $\hat{B} = \hat{p}$ and $\hat{A} = \frac{i}{\hbar}\hat{H}_2 t$, we get

$$\hat{\bar{x}}(t) = \exp\left(\frac{i}{\hbar}\hat{H}_2 t\right) e^{\hat{p}} \exp\left(-\frac{i}{\hbar}\hat{H}_2 t\right) = e^{\hat{p}} \exp\left(\frac{i}{\hbar}\left[\hat{H}_2, \hat{p}\right]\right)$$

$$= e^{\hat{p}} e^{-\lambda t} = \exp\left(-i\hbar\frac{\partial}{\partial x} - \lambda t\right). \tag{11.81}$$

By the direct integration of (11.81) we find

$$\hat{\bar{x}}(t) = \hat{x} + \frac{1}{\lambda}\left(1 - e^{-\lambda t}\right)\exp\left(-i\hbar\frac{\partial}{\partial x}\right), \tag{11.82}$$

and this is the solution of equation

$$\hat{\bar{x}}(t) = \hat{x} + t\exp\left(-i\hbar\frac{\partial}{\partial x}\right) - \lambda\int_0^t (t - t')\,\hat{\bar{x}}\,(t')\,dt', \tag{11.83}$$

which is different from (11.77). We can also show that the commutators $\left[\hat{\bar{x}}(t), \hat{\dot{\bar{x}}}(t)\right]$ for the two Hamiltonians are not the same. Thus from Eq. (11.78) and its time derivative we can calculate the commutator $\left[\hat{\bar{x}}(t), \hat{\dot{\bar{x}}}(t)\right]$ which turns out to be

$$\left[\hat{\bar{x}}(t), \hat{\dot{\bar{x}}}(t)\right] = \left[x - \frac{i\hbar}{\lambda}\left(1 - e^{-\lambda t}\right)\frac{\partial}{\partial x}, i\hbar e^{-\lambda t}\frac{\partial}{\partial x}\right] = i\hbar e^{-\lambda t}. \tag{11.84}$$

Similarly from (11.81) and (11.82) we find

$$\left[\hat{\bar{x}}(t), \hat{\dot{\bar{x}}}(t)\right] = \left[x, e^{p}e^{-\lambda t}\right] = i\hbar e^{p}e^{-\lambda t}. \tag{11.85}$$

Making use of the relation $\dot{p} = (i/\hbar)[p, H]$, Eq. (11.68), or its integral $p = \lambda t$ we find that $\left[\hat{\bar{x}}(t), \hat{\dot{\bar{x}}}(t)\right] = i\hbar$.

11.4 Heisenberg's Equations of Motion for Dekker's Formulation

The complex Hamiltonian of Dekker (Sec. 3.4) in operator form can be regarded as the generator for the Heisenberg equations of motion [19]-[21]. In this formalism we start with the operator H given by

$$\mathsf{H} = H_1 + i\Gamma = -i\omega\pi q - \frac{1}{4}\lambda\hbar, \tag{11.86}$$

where the operators q and π satisfy the commutation relation

$$[q, \pi] = i\hbar, \tag{11.87}$$

and this follows from the commutation relation $[x, p] = i\hbar$. Thus we have linear canonical transformations (3.61),(3.62). The time evolution of q and of π are given by

$$\dot{q} = -\frac{i}{\hbar}[q, \mathsf{H}] + \frac{1}{\hbar}[q, \Gamma]_+ + \mathcal{N}_q(t) \tag{11.88}$$

and

$$\dot{\pi} = -\frac{i}{\hbar}[\pi, \mathsf{H}] + \frac{1}{\hbar}[\pi, \Gamma]_+ + \mathcal{N}_\pi(t), \tag{11.89}$$

where $[\ ,\ ]$ is the usual commutator, $[\ ,\ ]_+$ is the anti-commutator and $\mathcal{N}_q$ and $\mathcal{N}_\pi$ are the noise operators with the property that [20]

$$\langle \mathcal{N}_q(t) \rangle = \langle \mathcal{N}_\pi(t) \rangle = 0. \tag{11.90}$$

The noise operator will be discussed later in Sec. 17.2. Here the average is taken with respect to the noise only.

We can rewrite $\mathsf{H} = H_1 + i\Gamma$ in terms of the original coordinate and momentum operators x and p (see Eq. (3.68));

$$H_1 = \frac{1}{2}p^2 + \frac{\lambda}{2}(px + xp) + \frac{1}{2}m\left(\omega^2 + \frac{\lambda^2}{4}\right)x^2, \tag{11.91}$$

$$\Gamma = -\frac{1}{4}\lambda\hbar. \tag{11.92}$$

For x and p operators we find that the equations of motion are:

$$\dot{x} = -\frac{i}{\hbar}[x, \mathsf{H}] + \mathcal{N}_x(t) = p + \mathcal{N}_x(t) \tag{11.93}$$

and

$$\dot{p} = -\frac{i}{\hbar}[p, \mathsf{H}] + \mathcal{N}_p(t) = -\lambda p - \left(\omega^2 + \frac{\lambda^2}{4}\right)x + \mathcal{N}_p(t). \tag{11.94}$$

Since both $\langle \mathcal{N}_p(t) \rangle_N$ and $\langle \mathcal{N}_x(t) \rangle_N$ are zero, we have the correct result for the Ehrenfest theorem. The properties of the noise operators in this formulation have been studied in detail by Dekker [20],[21]. This particular formulation of the damped oscillator will be further discussed in Sec. 14.3.

11.5 Quantization of the Bateman Hamiltonian

In Sec. 3.7 we discussed the Bateman dual Lagrangian and Hamiltonian for a damped-amplified harmonic oscillator. We also found a similar Hamiltonian, Eq. (3.310), resulting from the extended phase space formulation. Now we want to study the quantized version of these systems. First we note that the Bateman Hamiltonian (3.97) does not depend on time explicitly, and therefore is a constant of motion. By doubling the dimensions of the phase space, i.e. by introducing the y variable in Eq. (3.97) we obtain an effective isolated system. That is the new degrees of freedom represent a single equivalent (collective) degree of freedom for the bath which absorbs the energy dissipated by the oscillator [22].

We can obtain the eigenvalues of this system by introducing the creation and annihilation operators $A, B, A^\dagger$ and $B^\dagger$ and writing the Hamiltonian in terms of these operators.

Let us define these operators by:

$$A = \frac{1}{2\sqrt{m\omega\hbar}} \left[(p_x + p_y) - im\omega(x + y) \right], \qquad (11.95)$$

$$B = \frac{1}{2\sqrt{m\omega\hbar}} \left[(p_x - p_y) - im\omega(x - y) \right], \qquad (11.96)$$

$$A^\dagger = \frac{1}{2\sqrt{m\omega\hbar}} \left[(p_x + p_y) + im\omega(x + y) \right], \qquad (11.97)$$

and

$$B^\dagger = \frac{1}{2\sqrt{m\omega\hbar}} \left[(p_x - p_y) + im\omega(x - y) \right], \qquad (11.98)$$

where $\omega^2 = (\omega_0^2 - \lambda^2/4)$. These operators satisfy the commutation relations

$$\left[A, A^\dagger \right] = \left[B, B^\dagger \right] = 1, \qquad (11.99)$$

and

$$\left[A, B \right] = \left[A^\dagger, B^\dagger \right] = 0. \qquad (11.100)$$

By standard factorization method we can write the operator form of the Hamiltonian (3.97) as [21],[23]

$$H = H_0 + H_I = \hbar\omega \left(A^\dagger A - B^\dagger B \right) + \frac{i\Gamma}{2} \left(A^\dagger B^\dagger - AB \right), \qquad (11.101)$$

where

$$\Gamma = \hbar\lambda. \qquad (11.102)$$

We note that the first term in (11.101) represents the difference between the number of quanta of the two free harmonic oscillator Hamiltonians. Denoting the eigenstates of H by $|n_A, n_B\rangle$, where $n_A, n_B = 0, 1, 2 \cdots$ we observe that in the limit of $\Gamma = \hbar\lambda \to 0$ we obtain the harmonic oscillator states provided that $B|n_A, 0\rangle = 0$, i.e. when the B oscillator is kept in its ground state.

Now if we just retain these states, then the eigenstates of H merge into those of the simple undamped harmonic oscillator when $\lambda \to 0$. Therefore the state generated by $B^\dagger$ represents the sink where the energy dissipated by the damped harmonic oscillator flows. We thus conclude that the B oscillator represents the single mode reservoir or heat bath coupled to the A oscillator [22].

Next let us define spin-like operators ϕ_0, ϕ_x, ϕ_y and ϕ_z by the relations [21]

$$\phi_0 = \frac{1}{2} \left(A^\dagger A - B^\dagger B \right), \qquad (11.103)$$

$$\phi_x = \frac{1}{2}\left(A^\dagger B^\dagger + AB\right), \qquad (11.104)$$

$$\phi_y = \frac{i}{2}\left(A^\dagger B^\dagger - AB\right), \qquad (11.105)$$

and

$$\phi_z = \frac{1}{2}\left(A^\dagger A + BB^\dagger\right). \qquad (11.106)$$

These ϕ operators satisfy the commutation relations similar those of the spin operators

$$[\phi_x, \phi_y] = i\phi_z, \qquad (11.107)$$

$$[\phi_z, \phi_y] = i\phi_x, \qquad (11.108)$$

and

$$[\phi_x, \phi_z] = i\phi_y. \qquad (11.109)$$

Now we write the Hamiltonian (11.101) in terms of these operators:

$$H = H_0 + H_I = 2\hbar\omega\phi_0 + \Gamma\phi_y. \qquad (11.110)$$

This form of H is not diagonal, therefore let us try to express H in terms of ϕ_0 and ϕ_z. For this we observe that just as in the case of spin operators, we have

$$\phi_0^2 = \frac{1}{4} + \phi_z^2 - \left(\phi_x^2 + \phi_y^2\right), \qquad (11.111)$$

where ϕ_0 commutes with ϕ_x, ϕ_y and ϕ_z. Since we have replaced $A, A^\dagger, B$ and $B^\dagger$ by ϕ s, we also label the eigenstates of H_0 (or ϕ_0) by j and m, i.e. $|jm\rangle$ where

$$j = \frac{1}{2}(n_A - n_B), \quad \text{and} \quad m = \frac{1}{2}(n_A + n_B). \qquad (11.112)$$

Here m is an eigenstate of $(\phi_z - 1/2)$;

$$\phi_0|jm\rangle = j|jm\rangle, \quad \text{and} \quad \phi_z|jm\rangle = \left(m + \frac{1}{2}\right)|jm\rangle. \qquad (11.113)$$

In order to find the eigenvalues of H we note that since ϕ_0 and ϕ_y in (11.110) commute with each other, we can find their simultaneous eigenstates. To this end we introduce the operators $\phi_\pm$ which we define by

$$\phi_\pm = \phi_x \mp \phi_z, \qquad (11.114)$$

where ϕ_+ and ϕ_- are the raising and the lowering operators respectively. These two operators satisfy the commutation relation

$$[\phi_x, \phi_\pm] = \pm i\phi_\pm. \qquad (11.115)$$

Using Baker-Campbell-Hausdorff relation [24],[25]

$$e^A B e^{-A} = B + [A, B] + \frac{1}{2!}[A, [A, B]] + \frac{1}{3!}[A, [A, [A, B]]] + \cdots, \qquad (11.116)$$

we can express ϕ_y in terms of ϕ_z and ϕ_x;

$$\phi_y = \pm i \exp\left[\mp\frac{\pi}{2}\phi_x\right] \phi_z \exp\left[\pm\frac{\pi}{2}\phi_x\right]. \tag{11.117}$$

Before discussing the solution of the wave equation for this case we must pay special attention to the two essential problems of this formulation:

(1) The eigenstates $|\psi_{jm}^{\pm}\rangle$ of H, are not normalizable, i.e. their norm $\langle\psi_{jm}^{\pm}|\psi_{jm}^{\pm}\rangle$ is divergent [23].

(2) The states $|\psi_{jm}^{\pm}\rangle$ cannot be obtained from the eigenstates of H_0, viz, $|jm\rangle$ by a proper rotation in $SU(1,1)$. But these are linked together by a pseudo-rotation in $SL(2,\mathbb{C})$, and this corresponds to a non-unitary transformation in $SU(1,1)$ [22]. As Lindblad and Nagel have shown in any unitary irreducible representation of $SU(1,1)$, ϕ_y must have a purely continuous and real spectrum, and this is not the case for $|\psi_{jm}^{\pm}\rangle$, and therefore $|\psi_{jm}^{\pm}\rangle$ does not have a unitary irreducible representation [26]. So it is not surprising to find the following contradiction, that whereas ϕ_y is apparently a Hermitian operator, its spectrum is pure imaginary. The eigenfunctions $|\psi_{jm}^{\pm}\rangle$ do not belong to ordinary Hilbert space, since they are not normalizable. Thus if we calculate $\langle jm| \exp(2\nu\phi_x)|jm\rangle$, we find

$$\langle jm| \exp(2\nu\phi_x)|j,m\rangle = \frac{(\tan\nu)^{2m+1}}{(\sin\nu)^{2j+1}} \times \sum_l \binom{m-j}{l}\binom{2m-l}{m+j}(\cot\nu)^{2l}, \tag{11.118}$$

where the sum is over all integers l. Now if we set $\nu = \pm\frac{\pi}{2}$, the norm of $\langle jm| \exp(2\nu\phi_x)|jm\rangle$ becomes infinite.

To bypass this problem, let us define ϕ_y in a Hilbert space with a suitable inner product, so that $|\psi_{jm}^{\pm}\rangle$ would have a finite norm.

Normalizable Wave Function for Quantized Bateman Hamiltonian — To construct a normalizable wave function we first consider the effect of the time reversal operation $\mathcal{T}$ which is antiunitary on operators A and B:

$$\mathcal{T}(A,\, B) \to \left(-A^\dagger,\, -B^\dagger\right), \tag{11.119}$$

and introduce the conjugation operation by

$$\langle\psi_{jm}| \equiv [\mathcal{T}|\psi_{jm}\rangle]^\dagger, \tag{11.120}$$

with

$$\mathcal{T}|\psi_{jm}\rangle = |\psi_{j,\,-(m+1)}\rangle. \tag{11.121}$$

Equation (11.121) can be viewed as a rule for continuation of positive m s to negative m s. Thus the Hermitian conjugate of the ket

$$\phi_2|\psi_{jm}\rangle = \mu|\psi_{jm}\rangle, \qquad \mu = i\left(m+\frac{1}{2}\right), \tag{11.122}$$

is given by [22]

$$\langle\psi_{j\,-(m+1)}|\phi_y = \mu_T\langle\psi_{j\,-(m+1)}|, \tag{11.123}$$

where

$$\mu_T = -i\left[-(m+1) + \frac{1}{2}\right] = -\mu^* = \mu. \tag{11.124}$$

Now let us consider the time evolution of the vacuum state $(|n_A = 0, \ n_B = 0\rangle)$ which is the eigenstate found from $A|0\rangle = B|0\rangle = 0$, or the state with $j = 0$ and $m_0 = 0$. Then this state at time t is given by

$$|0(t)\rangle = \exp\left(-\frac{i}{\hbar}Ht\right)|0\rangle = \exp\left(-\frac{i}{\hbar}H_It\right)|0\rangle$$

$$= \exp\left[\frac{\lambda t}{2}\left(A^\dagger B^\dagger - AB\right)\right]|0\rangle. \tag{11.125}$$

The operator on the right-hand side of (11.125) acting on $|0\rangle$ can also be written as [27]

$$\exp\left[\frac{\lambda t}{2}\left(A^\dagger B^\dagger - AB\right)\right] = \exp\left[\tanh\frac{\lambda t}{2}\,A^\dagger B^\dagger\right]$$

$$\times \exp\left[-2\left(\ln\cosh\frac{\lambda t}{2}\right)\phi_z\right]$$

$$\times \exp\left[-\tanh\frac{\lambda t}{2}\,AB\right]. \tag{11.126}$$

From the definition of the vacuum state $|j = 0, \ m_0 = 0\rangle$ it follows that

$$AB|0\rangle = 0, \tag{11.127}$$

and consequently

$$\exp\left[\left(-\tanh\frac{\lambda t}{2}\right)AB\right]|0\rangle = |0\rangle. \tag{11.128}$$

Also from the definition of ϕ_z we find

$$\phi_z|0\rangle = \frac{1}{2}|0\rangle, \tag{11.129}$$

and therefore

$$\exp\left\{-2\left[\ln\left(\cosh\frac{\lambda t}{2}\right)\right]\phi_z\right\}|0\rangle = \left(\frac{1}{\cosh\frac{\lambda t}{2}}\right)|0\rangle. \tag{11.130}$$

By substituting (11.126) in (11.125) and simplifying using Eqs. (11.127)-(11.130) we obtain

$$|0(t)\rangle = \frac{1}{\cosh\frac{\lambda t}{2}}\exp\left[\left(\tanh\frac{\lambda t}{2}\right)\right]|0\rangle \tag{11.131}$$

From this result we conclude that the state $|0\rangle$ has unit norm

$$\langle 0(t)|0(t)\rangle = 1, \tag{11.132}$$

however the asymptotic state $\lim\langle 0(t)|$ as $t \to \infty$ is orthogonal to the initial state $|0\rangle$;

$$\lim_{t\to\infty} \langle 0(t)|0\rangle = \lim_{t\to\infty} \exp\left[-\ln\cosh\left(\frac{\lambda t}{2}\right)\right] \to 0. \qquad (11.133)$$

This last relation shows the instability of the vacuum under time evolution operator $\exp\left(-i\frac{tH}{\hbar}\right)$ in the limit of $t \to \infty$. Thus in this limit the time evolution generates a state outside the original Hilbert, i.e. those states that are unitarily equivalent to the representation of the canonical commutation relations. However as Celeghini *et al.* have shown if one gives up the condition of finiteness of degrees of freedom, then the removal of this condition allows for the coexistence of infinitely many unitarily inequivalent representation of the canonical commutation relations [22]. Then in such a system, i.e. a heat bath with infinite number of degrees of freedom, replacing a sink of one degree of freedom, one bypasses the mathematical problems associated with the quantization of the Bateman Hamiltonian which we have encountered [22].

Eigenfunctions and Eigenvalues of the Quantized Bateman Hamiltonian — Now let us define $|\psi_{jm}^{(\pm)}\rangle$ by

$$|\psi_{jm}^{(\pm)}\rangle = \exp\left[\mp\frac{\pi}{2}\phi_x\right]|jm\rangle, \qquad (11.134)$$

then from (11.117) and (11.134) we can show that $|\psi_{jm}^{(\pm)}\rangle$ is an eigenfunction of ϕ_y;

$$\phi_y|\psi_{jm}^{(\pm)}\rangle = \pm i\left(m+\frac{1}{2}\right)|\psi_{jm}^{(\pm)}\rangle. \qquad (11.135)$$

Suppose that m_0 is the smallest value that m can take, then

$$\phi_\pm|\psi_{jm_0}^{(\pm)}\rangle = 0. \qquad (11.136)$$

By multiplying (11.136) from the left by $\phi_\pm$ and then substituting for $\phi_\pm\phi_\mp$ from (11.114), for $\phi_+\phi_-$ we find

$$\left(\phi_x^2 - \phi_z^2 + [\phi_x,\phi_z]\right)|\psi_{jm_0}^{(\pm)}\rangle = \left(\phi_x^2 - \phi_z^2 + i\phi_y\right)|\psi_{jm_0}^{(\pm)}\rangle = 0, \qquad (11.137)$$

and a similar result for $\phi_-\phi_+$. Now we substitute for $\phi_x^2 - \phi_z^2$ from Eq. (11.111) in (11.137) so that only ϕ_0 and ϕ_y operate on $|\psi_{jm_0}^{(\pm)}\rangle$,

$$\left(\frac{1}{4} - \phi_0^2 - \phi_y^2 + i\phi_y\right)|\psi_{jm_0}^{(\pm)}\rangle = 0, \qquad (11.138)$$

and then we use the eigenvalue equations (11.113) and (11.135) in (11.138) with the result that $m_0^2 = j^2$, and consequently

$$m = |j|, \; |j| + \frac{1}{2}, \; |j| + 1, \cdots \qquad (11.139)$$

Thus the eigenvalue equation

$$H|\psi_{jm}^{(\pm)}\rangle = \left(2\hbar\omega\phi_0 + \Gamma\phi_y|\right)\psi_{jm}^{(\pm)}\rangle = i\hbar\frac{\partial}{\partial t}|\psi_{jm}^{(\pm)}\rangle \qquad (11.140)$$

can be integrated to yield

$$|\psi_{jm}^{(\pm)}\rangle = \exp\left[-2i\omega jt \pm \frac{\Gamma}{2}(2m+1)t \mp \frac{\pi}{2}\phi_x\right]|jm\rangle. \qquad (11.141)$$

As we mentioned earlier we want the B oscillator to be in the ground state, i.e. $n_B = 0$. Imposing this condition on the eigenvalues we get

$$2j = 2m = n_A = n. \qquad (11.142)$$

In this way we find that the eigenvalues of the Hamiltonian H are given by

$$\begin{aligned} H_n^{(\pm)} &= n\hbar\omega \pm \frac{i\Gamma}{2}(n+1) \\ &= n\hbar\omega \pm i\frac{\hbar\lambda}{2}(n+1). \end{aligned} \qquad (11.143)$$

We observe that the shift in the energy is proportional to $\left(\frac{\lambda^2}{8\omega_0^2}\right)$ just as in Seeger's method of quantization.

11.6 Quantization of Pseudo-Hermitian Hamiltonian for a Damped Harmonic Oscillator

Returning to the definitions of Hamiltonians H_+ and H_-, Eq. (3.109), that we found from the Bateman's Lagrangian we observe that these Hamiltonians when viewed as quantum mechanical operators have the property that

$$H_\pm^\dagger = H_\mp, \qquad (11.144)$$

i.e. the Hermitian conjugation property corresponds to the time reversal operation. In addition this property ensures the Hermiticity of the complete Hamiltonian [28]. The operators H_+ and H_- are pseudo-Hermitian, i.e. if η is the PT operator, then these operators transform as [28]–[30]

$$H_\pm^\dagger = \eta H_\pm \eta^{-1}. \qquad (11.145)$$

To show this property of $H_\pm$, let us note that under $\eta = PT$ transformation we have

$$\eta x_i \eta^{-1} = \sum_{j=1}^{2} g_{ij}x_j, \qquad (11.146)$$

$$\eta p_i \eta^{-1} = -\sum_{j=1}^{2} g_{ij} p_j, \tag{11.147}$$

or

$$\eta x \eta^{-1} = x^\dagger, \quad \text{and} \quad \eta p \eta^{-1} = -p^\dagger. \tag{11.148}$$

For a rigorous discussion of the non-Hermitian but PT symmetric Hamiltonians the reader is referred to references [72],[73].

Quantization of the Composite Hamiltonian — We observe that the Hamiltonians H_+ and H_-, for a particle of unit mass, are of the form

$$H = \frac{1}{2} \left(p^2 + w^2 x^2 \right)^\smile \tag{11.149}$$

where x and p are non-Hermitian and w is a complex number. We quantize the composite Hamiltonian (3.108) in the Heisenberg picture by writing H in terms of the creation and annihilation operators for the PT transformed operators;

$$a = \sqrt{\frac{w}{2\hbar}} \left(x + \frac{i}{w} p \right), \tag{11.150}$$

and

$$\tilde{a} = \eta^{-1} a^\dagger \eta = \sqrt{\frac{w}{2\hbar}} \left(x - \frac{i}{w} p \right). \tag{11.151}$$

The operator $\tilde{a}$ is the pseudo-Hermitian adjoint of a with respect to η, and these two operators have the commutator

$$[\tilde{a}, \, a] = -1, \tag{11.152}$$

a result that can be found from Eqs. (11.150) and (11.151). Next we introduce the operator N, which is the number operator in this formulation, by the following relation:

$$N = \tilde{a} a. \tag{11.153}$$

This operator is not self-adjoint, but under $\eta = PT$ transformation we find that $N^\dagger$ is transformed according to the relation

$$\eta^{-1} N^\dagger \eta = N. \tag{11.154}$$

From the operator N we obtain the following commutators:

$$[N, \, a] = -a, \tag{11.155}$$

and

$$[N, \, \tilde{a}] = \tilde{a} \tag{11.156}$$

Also from the structure of the total Hamiltonian $H = H_+ + H_-$ it follows that we have to construct a complete bi-dimensional eigenstates of the operators

$H_{\pm}$ and N which we denote by $\{|\psi_n\rangle, |\phi_n\rangle\}$. These states satisfy the following relations:

$$N|\psi_n\rangle = n|\psi_n\rangle, \tag{11.157}$$

$$N^{\dagger}|\phi_n\rangle = n^*|\phi_n\rangle, \tag{11.158}$$

$$\langle\phi_n|\psi_j\rangle = \delta_{jn} \tag{11.159}$$

and

$$\sum_n |\phi_n\rangle\langle\psi_n| = \sum_n |\psi_n\rangle\langle\phi_n| = 1. \tag{11.160}$$

Now from Eqs. (3.109), (11.150) and (11.151) we find that the total Hamiltonian can be written as

$$H = \hbar w\left(N + \frac{1}{2}\right) = \hbar\omega_{\pm}\left(N + \frac{1}{2}\right) = \hbar\left(\omega \pm i\frac{\lambda}{2}\right)\left(N + \frac{1}{2}\right), \tag{11.161}$$

i.e. the total Hamiltonian, in this representation, is diagonal. In addition we will show that the eigenvalues of N are real and positive. To show this we start with the commutation relation $[N, a] = -a$ and we obtain

$$N(a|\psi_n\rangle) = (n-1)a|\psi_n\rangle, \tag{11.162}$$

and thus we can write

$$a|\psi_n\rangle = c|\psi_{n-1}\rangle. \tag{11.163}$$

Similarly for the state $\langle\phi_n|$ we find

$$\langle\phi_n|\tilde{a} = d\langle\phi_{n-1}|. \tag{11.164}$$

where c and d are c-numbers. Now using the pseudo-Hermiticity of N, we can relate $\eta|\phi_n\rangle$ to $|\psi_n\rangle$. Thus from Eq. (11.154) we have [29]

$$N\eta|\phi_n\rangle = \eta|\phi_n\rangle, \tag{11.165}$$

and this result when compared with (11.157), gives us, up to a constant phase factor the relation

$$\eta|\phi_n\rangle = |\psi_n\rangle. \tag{11.166}$$

From this relation we can deduce the important result

$$\langle\phi_n|\tilde{a}|\psi_{n-1}\rangle = \langle\phi_{n-1}|a|\psi_n\rangle^*. \tag{11.167}$$

Equation (11.167) together with (11.163) and (11.164) give us $d = c^*$, and hence we conclude that n is positive

$$n = |c|^2. \tag{11.168}$$

As in the problem of quantized harmonic oscillator repeated applications of a would yield negative eigenvalues for N. Therefore there exists a state $|0\rangle$ such that $a|0\rangle = 0$, and this is the ground state of the system. We can find higher energy states by the repeated application of $\tilde{a}$.

Now regarding the wave functions for these states, we note that if the eigenfunctions of H_+ are $\{|\psi_n\rangle, |\phi_n\rangle\}$, then the eigenfunctions of H_- are $\{|\phi_n\rangle, |\psi_n\rangle\}$. Earlier, using the Feshbach and Tikochinsky method we found the eigenvalues of H a given by Eq. (11.143) (see also Eq. (12.28)). But here, according to (11.161), we have found the eigenvalues to be

$$H_n^{\pm} = \hbar \left(\omega \pm \frac{i}{2}\lambda \right) \left(n + \frac{1}{2} \right). \tag{11.169}$$

We note that the two methods yield similar results for large n, but in the latter the zero-point energy is present, and for this reason it may be regarded as a better candidate for the energy eigenvalues.

Quantization of the Hamiltonian which is Linear in x for a Damped Oscillator — In Sec. 3.7 we observed that classically it is possible to construct a Hamiltonian which is dependent linearly on displacement x, H_1 or H_2, Eq. (3.115) or (3.116) and is a constant of motion. Let us write the symmetrized form of the quantal Hamiltonian as

$$H = -\frac{i\omega_1}{2}(px + xp) + \frac{\omega_1}{\omega_1 - \omega_2} p^{1 - \frac{\omega_1}{\omega_2}}. \tag{11.170}$$

Setting $\hbar = 1$, the corresponding "wave equation" in momentum space for this case is

$$H_1 \left(x = i\frac{d}{dp}, p \right) \phi(p) = E\phi(p). \tag{11.171}$$

This equation is easy to integrate with the result that

$$\phi(p) \propto p^{\frac{E}{\omega_1}} \exp \left[-\frac{\omega_1}{\omega_1 - \omega_2} p^{1 - \frac{\omega_2}{\omega_1}} \right]. \tag{11.172}$$

This solution is not acceptable on physical grounds since the wave function depends on nonintegral and in general complex powers of p in the exponential. Other difficulties associated with quantizing H_1 are discussed in detail by Bender and collaborators [31].

11.7 Fermi's Nonlinear Equation for Quantized Radiation Reaction

One of the earliest attempts to include radiative reaction or damping force in the wave mechanics of charged particles was due to Fermi [32]–[35]. If we go back to the classical description we note that for a radiating electron moving in a conservative potential $V(r)$ the equation of motion is

$$m\frac{d^2\mathbf{r}}{dt^2} + \nabla V(r) = m\tau \frac{d^3\mathbf{r}}{dt^3}, \tag{11.173}$$

which is the three-dimensional form of Eq. (1.84) for the potential $V(r)$. Since the radiation reaction force on the right hand side of (11.173) is very small, we can find an iterative solution of (11.173) by writing

$$m\frac{d^2\mathbf{r}^{(0)}}{dt^2} + (\nabla V(r))_{r^{(0)}} = 0, \tag{11.174}$$

and

$$m\frac{d^2\mathbf{r}^{(1)}}{dt^2} + (\nabla V(r))_{r^{(1)}} = \frac{m\tau}{e}\frac{d^3}{dt^3}\left(e\mathbf{r}^{(0)}\right) = \frac{m\tau}{e}\frac{d^3\mathbf{d}(t)}{dt^3}, \tag{11.175}$$

where $\mathbf{d}(t)$ is the classical dipole moment of the electron and the subscripts (0) and (1) denote the order of the iteration. The right hand side of (11.175) is only a function of time, therefore we can write it as

$$\frac{m\tau}{e}\frac{d^3\mathbf{d}(t)}{dt^3} = \frac{m\tau}{e}\left[\nabla\left(\frac{d^3\mathbf{d}(t)}{dt^3}\cdot\mathbf{r}\right)\right]_{\mathbf{r}^{(0)}} = -(\nabla W)_{\mathbf{r}^{(0)}}, \tag{11.176}$$

where

$$W = -\mathbf{r}\cdot\frac{d^3}{dt^3}\left(\frac{m\tau}{e}\mathbf{d}(t)\right). \tag{11.177}$$

Thus the electron is subject to a conservative potential $V(r)$ and a time-dependent potential

$$\left\{-\left(\frac{m\tau}{e}\right)\left[\mathbf{r}\cdot\frac{d^3\mathbf{d}(t)}{dt^3}\right]\right\}. \tag{11.178}$$

In wave mechanics we replace the dipole moment by the dipole generated by the charge distribution given in terms of the wave function [8]

$$\rho(\mathbf{r},t) = e\psi^*(\mathbf{r},t)\psi(\mathbf{r},t). \tag{11.179}$$

Thus we find

$$\mathbf{d}(t) \to \mathbf{d}\left(\psi^*(\mathbf{r},t)\psi(\mathbf{r},t)\right) = \frac{\int \mathbf{r}\rho(\mathbf{r},t)d^3r}{\int \rho(\mathbf{r},t)d^3r}, \tag{11.180}$$

and the Schrödinger equation for the combined potential will be

$$i\hbar\frac{\partial\psi(\mathbf{r},t)}{\partial t} = H_0\psi(\mathbf{r},t) + W\left(\psi^*\psi\right)\psi(\mathbf{r},t), \tag{11.181}$$

where

$$H_0 = -\frac{\hbar^2}{2m}\nabla^2 + V(r). \tag{11.182}$$

Exactly Solvable Two-Level System — While the most general solution of the nonlinear equation (11.181) is not known, for some special cases we can find exact solutions. First we observe that for any eigenstate of the Hamiltonian H_0, the radiation reaction potential W vanishes. Therefore the stationary states of the atom will not be affected by this damping force. We will see that some of the other nonlinear equations describing the dissipation such as Schrödinger-Langevin equation also have this property [36],[37].

For a two-level system we can simplify (11.180) by noting that the two levels have energies E_1 and E_2 respectively and these are given by the solution of the Schrödinger equation

$$H\psi_i(\mathbf{r}) = E_i\psi_i(\mathbf{r}), \quad i = 1, 2. \tag{11.183}$$

The wave function for this two-level system is a linear combination of $\psi_1(\mathbf{r})$ and $\psi_2(\mathbf{r})$;

$$\psi(\mathbf{r}, t) = C_1(t)\exp\left(-\frac{iE_1 t}{\hbar}\right)\psi_1(\mathbf{r}) + C_2(t)\exp\left(-\frac{iE_2 t}{\hbar}\right)\psi_2(\mathbf{r}). \tag{11.184}$$

Now by substituting (11.183) and (11.184) in (11.181) we find the following coupled equations:

$$\frac{dC_1(t)}{dt} = \frac{1}{2}AC_1(t)C_2(t)C_2^*(t), \tag{11.185}$$

and

$$\frac{dC_2(t)}{dt} = -\frac{1}{2}AC_1^*(t)C_1(t)C_2(t), \tag{11.186}$$

where

$$A = \frac{4e^2}{3c^3\hbar}\left(\frac{E_2 - E_1}{\hbar}\right)^3\left|\int\psi_1^*(\mathbf{r})\mathbf{r}\psi_2(\mathbf{r})d^3 r\right|^2. \tag{11.187}$$

By changing the independent functions $C_1(t)$ and $C_2(t)$ in (11.185) and (11.186) to $|C_1(t)|^2$ and $|C_2(t)|^2$ we can rewrite these equations as

$$\frac{d\,|C_1(t)|^2}{dt} = A\,|C_1(t)|^2\,|C_2(t)|^2, \tag{11.188}$$

and

$$\frac{d\,|C_2(t)|^2}{dt} = -A\,|C_1(t)|^2\,|C_2(t)|^2. \tag{11.189}$$

From these two relations it follows that

$$\frac{d\,|C_1(t)|^2}{dt} + \frac{d\,|C_2(t)|^2}{dt} = 0. \tag{11.190}$$

Thus we can set

$$|C_1(t)|^2 + |C_2(t)|^2 = 1. \tag{11.191}$$

These coupled first order differential equations can be integrated with the result that [34]

$$|C_1(t)|^2 = \frac{|C_1(0)|^2\,e^{At}}{1 + |C_1(0)|^2\,(e^{At} - 1)}, \tag{11.192}$$

and

$$|C_2(t)|^2 = \frac{1 - |C_1(0)|^2}{1 + |C_1(0)|^2\,(e^{At} - 1)}. \tag{11.193}$$

If at $t = 0$ the electron is in one of these states, say level 2, i.e. if $|C_2(0)|^2 = 1$, then $|C_1(0)|^2 = 0$, and the initial state does not decay. However if we choose $|C_2(0)|^2 = 1 - \epsilon$, where ϵ is a small positive number, then the initial state will decay, and for $At \gg 1$ this decay is exponential, viz,

$$|C_2(t)|^2 \approx \exp(-At). \tag{11.194}$$

The above formulation by Fermi which results in a nonlinear wave equation containing the radiation reaction potential suffers from the wrong behavior for the spontaneous emission. An exact eigenstate ($\epsilon = 0$) cannot radiate spontaneously and when $\epsilon \neq 0$, it takes a long time for the atom to get de-excited [34].

11.8 Attempts to Quantize Systems with a Dissipative Force Quadratic in Velocity

As we have seen earlier (Sec. 2.3) the Lagrangian for the one-dimensional motion of a particle of mass m subject to a damping force $\frac{1}{2}m\gamma\dot{x}^2$ and the conservative force $-\frac{dV}{dx}$ is given by

$$L = \frac{1}{2}m\dot{x}^2 e^{\gamma x} - \int^x e^{\gamma y}\frac{dV}{dy}dy. \tag{11.195}$$

The corresponding Hamiltonian for this system is

$$H = \frac{1}{2m}p^2 e^{-\gamma x} + \int^x e^{\gamma y}\frac{dV}{dy}dy, \tag{11.196}$$

where the canonical momentum p is related to the velocity by

$$p = \frac{\partial L}{\partial \dot{x}} = m\dot{x}e^{\gamma x}. \tag{11.197}$$

The classical system is invariant under the time-reversal transformation $t \to -t$ as well as time translation $t \to t + t_0$, but the energy of the system is not conserved.

Problem of Ordering x and p Operators in the Hamiltonian — In quantizing this system just as the case of linear damping we have the violation of the uncertainty principle. In addition we also encounter the additional problem of ordering the factors of x and p appearing in the Hamiltonian H. To illustrate this point let us consider the case where the potential V in (11.196) is zero and the Hamiltonian has a single term:

$$H = \frac{p^2}{2m}\exp(-\gamma x). \tag{11.198}$$

Using the Dirac rule of association [7],[17],[18] and noting that p^3, $e^{-\gamma x}$, $pe^{-\frac{1}{3}\gamma x}p$ and $e^{-\frac{1}{3}\gamma x}pe^{-\frac{1}{3}\gamma x}$ are all Hermitian, we can write

$$\hat{H}_1 = \hat{O}_1 \left[\frac{p^2}{2m} e^{-\gamma x} \right] = \frac{1}{6im\gamma\hbar} \left[p^3, e^{-\gamma x} \right]$$

$$= \frac{1}{2m} e^{-\gamma x} \left[p^2 + i\hbar\gamma p - \frac{1}{3}\hbar^2\gamma^2 \right], \tag{11.199}$$

where $\hat{O}$ denotes a Hermitian ordering of the argument. But we can order H in other ways. For instance,

$$\hat{H}_2 = \hat{O}_2 \left[\frac{p^2}{2m} e^{-\gamma x} \right] = \frac{i}{2m\gamma\hbar} \left[e^{-\frac{1}{3}\gamma x}pe^{-\frac{1}{3}\gamma x}, pe^{-\frac{1}{3}\gamma x}p \right]$$

$$= \frac{1}{2m} e^{-\gamma x} \left[p^2 + i\hbar\gamma p - \frac{2}{9}\hbar^2\gamma^2 \right], \tag{11.200}$$

is another Hermitian Hamiltonian operator for the classical H, Eq. (11.198). From Eqs. (11.199) and (11.200) we find that

$$\hat{H}_2 = \hat{H}_1 + \frac{1}{18m}\hbar^2\gamma^2 e^{-\gamma x}, \tag{11.201}$$

and hence the commutator of $\hat{H}_2$ and $\hat{H}_1$ is not zero:

$$\left[\hat{H}_2, \hat{H}_1 \right] = \left(\frac{1}{18m^2} \right) (\hbar\gamma)^3 e^{-2\gamma x} (-\hbar\gamma + 2ip). \tag{11.202}$$

In this way we can construct an infinite number of operators and these, in general, do not commute with each other. However in the classical limit of $\hbar \to 0$ or when $\gamma \to 0$ all of the Hamiltonians constructed by different rules of association commute. As we discussed earlier the classical H is a constant of motion, but in quantum mechanics, if we choose one of the different Hamiltonians to be a constant of motion, the others will not have this essential property [38].

Other ordering rules such as symmetric rule [18]

$$e^{-\gamma x}p^2 \to \frac{1}{2} \left(p^2 e^{-\gamma x} + e^{-\gamma x}p^2 \right), \tag{11.203}$$

or Weyl's rule [39] or Born-Jordan's rule [18] can also be used for constructing Hermitian operators. In fact one can write analogues of these three forms as special cases of the general expression [40]

$$\hat{O}_a \left[p^2 e^{-\gamma x} \right] = \left[ae^{-\gamma x}p^2 + (1 - 2a)pe^{-\gamma x}p + ap^2 e^{-\gamma x} \right]. \tag{11.204}$$

Denoting any of these Hermitian Hamiltonians by $\hat{H}_a(x, p)$, we have the Heisenberg equations of motion

$$\dot{p} = \frac{1}{i\hbar} \left[p, \hat{H}_a \right] = \gamma \hat{H}_a, \tag{11.205}$$

and

$$\dot{x} = \frac{1}{i\hbar}\left[x, \hat{H}_a\right] = \frac{1}{2}\left[e^{-\gamma x}p + pe^{-\gamma x}\right]. \tag{11.206}$$

The expectation value of these operators amounts to what can be called as Ehrenfest's theorem, and the infinite set of the Hamiltonians constructed by the Dirac's rule of association, or by other rules all satisfy Ehrenfest's theorem. Therefore the classical limit or the correspondence principle will not be helpful in discriminating among the infinite set of the Hamiltonians.

11.9 Solution of the Wave Equation for Linear and Newtonian Damping Forces

Let us start with the assumption that one of the Hermitian Hamiltonians, say $\hat{H}_1$, is a constant of motion. Then the time-dependent Schrödinger equation

$$-\frac{\hbar^2}{2m}e^{-\gamma x}\left[\frac{\partial^2}{\partial x^2} - \gamma\frac{\partial}{\partial x} + \frac{1}{3}\gamma^2\right]\psi(x,t)$$
$$+ \left(\int^x V(y)e^{\gamma y}dy\right)\psi(x,t) = i\hbar\frac{\partial\psi(x,t)}{\partial t} \tag{11.207}$$

can be separated into space- and time-dependent parts. If we write

$$\psi(x,t) = \psi(x,\mathcal{E})\exp\left(\frac{i\mathcal{E}t}{\hbar}\right), \tag{11.208}$$

then $\psi(x,\mathcal{E})$ the solution of

$$\frac{-\hbar^2}{2m}e^{-\gamma x}\left[\frac{d^2}{dx^2} - \gamma\frac{d}{dx} + \frac{1}{3}\gamma^2\right]\psi(x,\mathcal{E})$$
$$+ \left(\int^x V(y)e^{\gamma y}dy\right)\psi(x,\mathcal{E}) = \mathcal{E}\psi(x,\mathcal{E}). \tag{11.209}$$

This equation is exactly solvable for $V(x) = 0$, $V(x) = Ax$ and for $V(x) = e^{-2\gamma x}$ [41], but these solutions are not physically acceptable. Another possible way of quantizing this type of motion is by the path integral method which essentially yields similar results [42].
Energy Operator as a First Integral for Some Dissipative Motions—
We have noted that the classical Hamiltonian found for a damped motion is not the same as the energy of the particle, but it is a first integral of motion. If the damping is proportional to a power of velocity, say $\dot{x}^\nu$, then we can find an energy first integral in the following way:
Let us write the equation of motion as

$$\left(m\ddot{x} + \frac{dV(x)}{dx} + \alpha\dot{x}^\nu\right) = 0, \tag{11.210}$$

and as in the case of a conservative system, we find the energy first integral by multiplying (11.210) by $\dot{x}$ and in this way we find the rate of change of energy of the particle plus environment;

$$\frac{dE}{dt} = \dot{x} \left(m\ddot{x} + \frac{dV(x)}{dx} + \alpha \dot{x}^\nu \right) = 0. \tag{11.211}$$

Thus the total energy which is conserved is given by

$$E = \frac{1}{2} m\dot{x}^2 + V(x) + \Delta E, \tag{11.212}$$

where

$$\Delta E = \alpha \int_0^x \dot{x}^2 dx. \tag{11.213}$$

Denoting the mechanical (kinematical) momentum by $p = m\dot{x}$, we can write Eq. (11.212) as a Hamiltonian

$$H = H_0 + \Delta H = \frac{p^2}{2m} + V(x) + \Delta H, \tag{11.214}$$

where

$$\Delta H = \alpha \int_0^x \dot{x}^\nu dx. \tag{11.215}$$

Next we formally quantize (11.215) by replacing the mechanical momentum p by the operator $-i\hbar\partial/\partial x$, i.e.

$$H = -\frac{\hbar^2}{2m} \frac{\partial^2}{\partial x^2} + V(x) + \alpha \left(-\frac{i\hbar}{m} \right)^\nu \left(\frac{\partial^{\nu-1}}{\partial x^{\nu-1}} \right). \tag{11.216}$$

This Hamiltonian is, in general, complex and non-Hermitian. The time-dependent Schrödinger equation which is obtained from H and is given by

$$i\hbar \frac{\partial \psi(x,t)}{\partial t} = \left[H_0 + \alpha \left(-\frac{i\hbar}{m} \right)^\nu \left(\frac{\partial^{\nu-1}}{\partial x^{\nu-1}} \right) \right] \psi(x,t), \tag{11.217}$$

can be separated by introducing $\psi(x, E)$ by

$$\psi(x,t) = \psi(x, E) \exp\left(-\frac{iEt}{\hbar} \right), \tag{11.218}$$

and substituting $\psi(x,t)$ in (11.217) with the result that

$$-\frac{\hbar^2}{2m} \frac{d^2\psi(x,E)}{dx^2} + [V(x) - E]\psi(x,E) + \alpha \left(\frac{-i\hbar}{m} \right)^\nu \frac{d^{\nu-1}\psi(x,E)}{dx^{\nu-1}} = 0. \tag{11.219}$$

Now let us consider the specific cases of linear and quadratic damping.
Wave Equation for Linear Damping in Velocity — For linear damping $\nu = 1$ and the wave equation is

$$-\frac{\hbar^2}{2m} \frac{d^2\psi(x,E)}{dx^2} + \left[V(x) - \frac{i\alpha\hbar}{m} \right] \psi(x,E) = E\psi(x,E), \tag{11.220}$$

i.e. the potential $\mathcal{V}(x) = V(x) - i\alpha\hbar/m$ is an optical potential with negative imaginary part (Chapter 20).

From the time-dependent wave equation (11.217) and its complex conjugate with $\nu = 1$ we get

$$\frac{\partial \rho(x,t)}{\partial t} + \frac{\partial}{\partial x} j_x(x,t) = -\frac{2\alpha}{m} \rho(x,t), \tag{11.221}$$

where $\rho(x,t) = |\psi(x,t)|^2$ is the probability density and

$$j_x(x,t) = \frac{-i\hbar}{2m} \left(\psi^* \frac{\partial \psi}{\partial x} - \psi \frac{\partial \psi^*}{\partial x} \right), \tag{11.222}$$

is the current density. By integrating (11.221) over all x and assuming that as $x \to \pm\infty$, $j_x(x,t)$ vanishes we obtain

$$\frac{d}{dt} \int_{-\infty}^{\infty} \rho(x,t)dx = -\frac{2\alpha}{m} \int_{-\infty}^{\infty} \rho(x,t)dx, \tag{11.223}$$

and by carrying out the time integration in (11.223) we find

$$\int_{-\infty}^{\infty} \rho(x,t)dx = \exp\left(-\frac{2\alpha t}{m}\right) \left(\int_{-\infty}^{\infty} \rho(x,0)dx \right). \tag{11.224}$$

Thus in this method of quantization all of the eigenstates including the ground state decay in time. The nonexistence of a stable ground state is an essential a defect of this model.

Wave Equation for Newtonian Frictional Force — If we choose $\nu = 2$, then the equation for $\psi(x,E)$ becomes [43]

$$\frac{\hbar^2}{2m} \frac{d^2\psi(x,E)}{dx^2} + \frac{\alpha\hbar^2}{m^2} \frac{d\psi(x,E)}{dx} + (E - V(x))\psi(x,E) = 0. \tag{11.225}$$

Here all of the coefficients in the differential equation are real and in contrast to the linear damping one can get normalizable eigenstates. For instance in the case of a harmonic oscillator potential $V(x) = \frac{1}{2}m\omega_0^2 x^2$, we find the solution of (11.225) for odd and even states to be [43]

$$\psi(x,E_j) = N_{2j+1} \exp\left(-\frac{\alpha x}{m}\right) \exp\left(-\frac{m\omega_0}{2\hbar}x^2\right) H_{2j+1}\left(\sqrt{\frac{m\omega_0}{\hbar}}x\right), \tag{11.226}$$

and

$$\psi(x,E_j) = N_{2j} \exp\left(-\frac{\alpha x}{m}\right) \exp\left(-\frac{m\omega_0}{2\hbar}x^2\right) H_{2j}\left(\sqrt{\frac{m\omega_0}{\hbar}}x\right), \tag{11.227}$$

where N_j s are the normalization constants.

The eigenvalues corresponding to these wave functions are given by

$$E_j = \left(j + \frac{1}{2}\right)\hbar\omega + \frac{\alpha^2\hbar^2}{2m^3}, \quad j = 0,1,2,\cdots. \tag{11.228}$$

We note that (a) - all of the eigenfunctions are stable but distorted slightly by the presence of the dissipative force, and (b) - the eigenfunctions, apart from a constant shift, are the same as those of the undamped oscillator.

The stability of the bound states in this case, or generally for any confining potential, $V(x)$, can be understood by an examination of the classical motion. The term $\alpha \dot{x}^2(t)$ in the energy equation shows a reduction in the energy when $\dot{x}(t) > 0$, but for $\dot{x}(t) < 0$ there will be a gain of energy for the particle. The solution of the classical equation

$$\ddot{x} + \omega_0^2 x + \frac{\alpha}{m}\dot{x}^2 = 0 \tag{11.229}$$

can be found exactly [44];

$$t = t_0 + \int_{x_0}^{x} \frac{dx}{\sqrt{Y(x)}}, \tag{11.230}$$

where

$$Y(x) = \omega_0 A^2 \exp\left(-\frac{2\alpha x}{m}\right) + \frac{m^2 \omega_0^2}{2\alpha^2}\left[1 - \frac{2\alpha x}{m} - \exp\left(-\frac{2\alpha x}{m}\right)\right], \tag{11.231}$$

and like the undamped oscillator it is periodic with a constant amplitude. Again we find that this model is unrealistic for the description of a dissipative motion. However if we replace $\frac{\alpha}{m}\dot{x}^2$ in (11.229) by $\frac{\alpha}{m}\dot{x}|\dot{x}|$, then $x(t)$ will exhibit damped oscillations similar to the linear damping.

11.10   Quantization of a Damped Hamiltonian Given by a Dynamical Matrix

In Sec. 3.14 we discussed an approach advanced by Enz in which the system is described by a Hamiltonian formulated in terms of a "dynamical matrix". There, we also introduced the concept of the generalized Poisson brackets Eq. (3.288). Such a Poisson bracket can be written in terms of its even and odd parts [45]

$$\{A, B\}_D = \{A, B\}_{D^a} - \{A, B\}_{D^s}, \tag{11.232}$$

where

$$\{A, B\}_{D^a} = -\{B, A\}_{D^a}, \quad \{A, B\}_{D^s} = \{B, A\}_{D^s}. \tag{11.233}$$

Since the classical observables are real, therefore D is real and we have

$$\{A, B\}_D^* = -\{B, A\}_D. \tag{11.234}$$

To obtain the quantized version of these classical systems, we can use the Dirac rule of association and write

$$\{x, A\}_{D^a} \rightarrow \frac{i}{\hbar}D^a D_0 [x, A]_-, \tag{11.235}$$

and

$$\{x,\ A\}_{D^s} \rightarrow \frac{1}{\hbar}D^s\,[x,\ A]_+\,, \tag{11.236}$$

where $[x,\ A]_-$ and $[x,\ A]_+$ denote the commutator and the anticommutator respectively. These two rules of association are unambiguous as long as D^a and D^s are independent of x. They have also the correct limit when $D = D_0$, since then the first one, Eq. (11.235), becomes the usual Dirac prescription, viz,

$$\{x,\ A\} \rightarrow -\frac{i}{\hbar}[x,\ A]_-, \tag{11.237}$$

while the second one, Eq. (11.236), vanishes identically.

From the quantum analogue of (3.290) and by replacing the classical Poisson brackets by commutators and anticommutators as we have seen in this section we obtain the following relations:

$$\dot{p} = \frac{i}{\hbar}D_{pq}[H,\ p]_- + \frac{1}{\hbar}D_{pp}[H,\ p]_+, \tag{11.238}$$

$$\dot{q} = \frac{i}{\hbar}D_{qp}[H,\ q]_- + \frac{1}{\hbar}D_{qq}[H,\ q]_+, \tag{11.239}$$

These lead to complicated operator equations.

Let us consider the simpler case of a one-dimensional motion of a particle subject to a frictional force given by (3.293). For this problem the equations of motion are

$$\dot{p} = \frac{i}{\hbar}[H,\ p]_- - \frac{\lambda}{\hbar}[H,\ p]_+, \tag{11.240}$$

$$\dot{q} = \frac{i}{\hbar}[H,\ q]_-. \tag{11.241}$$

We can integrate these last two equations formally and write $x(t) = (p(t), q(t))$ as

$$x(t) = \exp\left[\frac{1}{\hbar}(i - \lambda)Ht\right] x \exp\left[-\frac{1}{\hbar}(i + \lambda)Ht\right], \tag{11.242}$$

provided that

$$[\lambda H,\ q]_+ = 0. \tag{11.243}$$

But for a constant λ this condition cannot be satisfied, and therefore the operator approach does not work, hence we abandon the operator method in favor of the wave equation formulation [45].

Quantization Based on the Schrödinger Equation — In this method we start with a one-particle Schrödinger equation in n dimensions and write the Hamiltonian as

$$\langle \psi(q)|H|\psi(q)\rangle = \int_V \psi^*(q)H(q)\psi(q)d^n q$$

$$= \int_V \left[\frac{\hbar^2}{2m}|\nabla\psi(q)|^2 + V(q)|\psi(q)|^2|\right] d^n q. \tag{11.244}$$

Next we expand the wave function in terms of plane waves:

$$\psi(q) = \frac{1}{\sqrt{2\mathcal{V}}} \sum_k a_k e^{ikq} = \sum_k |k\rangle a_k, \qquad (11.245)$$

where $\mathcal{V}$ is a volume bounded by the closed surface $\mathcal{S}$, such that $\psi = 0$ on this surface. Then

$$\mathcal{H} = \langle \psi(q)|H|\psi(q)\rangle = \sum_{k,k'} H_{k,k'} a_k^* a_{k'}, \qquad (11.246)$$

where

$$H_{k,k'} = H_{k',k}^* = \langle k|H|k'\rangle = \frac{\hbar^2 k^2}{2m} \delta_{k,k'} + \tilde{V}_{k-k'}, \qquad (11.247)$$

and where $\tilde{V}$ is the Fourier transform of $V(q)$.

From the time-dependent Schrödinger equation, $i\hbar \frac{\partial \psi}{\partial q} = H\psi$, we find the time dependence of the coefficients of expansion a_k:

$$i\hbar \dot{a}_k = \sum_{k'} H_{k,k'} a_{k'} = \frac{\partial \mathcal{H}}{\partial a_k^*}. \qquad (11.248)$$

Combining this equation with the complex conjugate equation and introducing the notation $\dot{x}_k = (a_k, a_{k'}^*)$ we have

$$\dot{x}_k = \sum_{k'} D_{k,k'}^0 \frac{\partial \mathcal{H}}{\partial x_{k'}}. \qquad (11.249)$$

This is the canonical form of the Schrödinger equation analogue to (3.286) provided that we define the dynamical matrix $D_{k,k'}^0$ by

$$D_{k,k'}^0 = \begin{bmatrix} 0 & -\frac{i}{\hbar}\delta_{k,k'} \\ \frac{i}{\hbar}\delta_{k,k'} & 0 \end{bmatrix}. \qquad (11.250)$$

Now we introduce a generalized dynamical matrix by

$$D_{k,k'} = \begin{bmatrix} D_{a_k a_{k'}} & D_{a_k a_{k'}^*} \\ D_{a_k^* a_{k'}} & D_{a_k^* a_{k'}}^* \end{bmatrix}, \qquad (11.251)$$

and the generalized Poisson bracket by

$$\{A, B\}_D = \sum_{k,k'} \frac{\partial A}{\partial x_k} D_{k,k'} \frac{\partial B}{\partial x_{k'}}, \qquad (11.252)$$

which is the analogue of (3.288).

If we set $A = \psi(q)$ and $B = \psi^*(q')$, then using the expansion (11.245) and its complex conjugate we obtain the following brackets:

$$\{\psi(q),\ \psi^*(q')\}_D = \frac{1}{\mathcal{V}} \sum_{k,k'} D_{a_k a_{k'}^*} e^{i(kq - k'q')}, \qquad (11.253)$$

and

$$\{\psi(q),\ \psi(q')\}_D = \frac{1}{V}\sum_{k,k'} D_{a_k a_{k'}^*} e^{i(kq+k'q')}, \qquad (11.254)$$

These brackets satisfy the product rule given by

$$\{\psi(q),\ \psi^*(q')\,\psi(q')\}_D = \{\psi(q),\ \psi^*(q')\}_D\ \psi(q') \\ + \{\psi(q),\ \psi(q')\}_D\ \psi^*(q'). \qquad (11.255)$$

In order to obtain a reasonable generalization of the Schrödinger equation we must have

$$\{\psi(q),\ \mathcal{H}\}_D \propto \psi(q), \qquad (11.256)$$

and by combining this relation with (11.255) we find

$$\{\psi(q),\ \psi(q')\}_D = \{\psi^*(q),\ \psi^*(q')\}_D = 0. \qquad (11.257)$$

These relations together with (11.253) and (11.254) imply that that the diagonal elements of the dynamical matrix $D_{kk'}$ are zero, i.e.

$$D_{kk'} = \begin{bmatrix} 0 & D_{a_k a_{k'}^*} \\ D_{a_k^* a_{k'}^*} & 0 \end{bmatrix} \qquad (11.258)$$

and thus we find a generalized Schrödinger equation in Poisson-bracket form

$$\frac{\partial \psi}{\partial t} = \{\psi,\ \mathcal{H}\}_D = \sum_{kk'} |k\rangle D_{a_k a_{k'}^*} \langle k'|H|\psi\rangle. \qquad (11.259)$$

By writing the complex conjugate of (11.259) in terms of the conjugate of (11.255), and using (11.257) we obtain

$$D^*_{a_k a_{k'}} = D^*_{a_k^* a_{k'}}. \qquad (11.260)$$

In order to divide (11.260) into even and odd parts, we start with the plane wave expansion, Eqs. (11.253) and (11.254) and separate these into even and odd parts (similar to Eq. (11.233))

$$\{\psi(q),\ \psi^*(q')\}_{D^a} = -\{\psi^*(q'),\ \psi(q)\}_{D^a}, \qquad (11.261)$$

$$\{\psi(q),\ \psi^*(q')\}_{D^s} = \{\psi^*(q'),\ \psi(q)\}_{D^s}. \qquad (11.262)$$

In this way we obtain

$$D^a_{a_k a_{k'}^*} = -D^a_{a_k^* a_{k'}}, \qquad D^s_{a_k a_{k'}^*} = D^s_{a_k^* a_{k'}}. \qquad (11.263)$$

Thus the matrix $D^a_{kk'}$ is antisymmetric. From the analogue of Eq. (3.285) we find the matrix $D_{k,k'}$;

$$D_{kk'} = D^a_{kk'} - D^s_{kk'} = \begin{bmatrix} 0 & -\frac{i}{\hbar}\delta_{kk'} - D^s_{a_k a_{k'}^*} \\ \frac{i}{\hbar}\delta_{kk'} - D^s_{a_k a_{k'}} & 0 \end{bmatrix}. \qquad (11.264)$$

Substituting $D_{kk'}$ in (11.259) one obtains the mixed canonical-dissipative Schrödinger equation

$$i\hbar \frac{\partial}{\partial t}|\psi\rangle = (H + i\gamma H)|\psi\rangle, \tag{11.265}$$

where γ which is now an operator is given by

$$\gamma = -\hbar \sum_{kk'} |k\rangle\, D^s_{a_k a^*_{k'}}\, \langle k'| = \gamma^* \tag{11.266}$$

From (11.259) and its complex conjugate we can find $\frac{d}{dt}\langle \psi | H^N | \psi\rangle$, for $N = 0, 1, \cdots$

$$\frac{d}{dt}\langle \psi | H^N | \psi\rangle = -\sum_{k,k'} \left[\langle \psi | H^N | k\rangle\, D^s_{a_k a^*_{k'}}\, \langle k' | H | \psi\rangle \right]$$

$$+ \sum_{k,k'} \left[\langle \psi | H | k'\rangle\, D^s_{a^*_k a_{k'}}\, \langle k | H^N | \psi\rangle \right],$$

$$N = 0, 1, \cdots. \tag{11.267}$$

A Specific Form of Schrödinger Equation with Dissipation — By choosing a simple form for $D^s_{a_k a_{k'*}}$ we can obtain a set of equations for $\frac{d}{dt}\langle\psi|H|\psi\rangle$ from (11.267). Let us consider the case where

$$D^s_{a_k a^*_{k'}} = \frac{\nu}{\hbar}\delta_{kk'} + \frac{\mu}{\hbar} a_k a^*_{k'}, \tag{11.268}$$

where μ and ν are parameters. In order to satisfy (11.260) and (11.263) ν and μ must be real quantities, and they must be subject to the condition that the system be dissipative, i.e. losing energy (rather than gaining). By substituting (11.268) in (11.267) and setting $N = 0$ and $N = 1$ respectively we find

$$\frac{d}{dt}\langle\psi|\psi\rangle = -\frac{2}{\hbar}\left(\nu + \mu\langle\psi|\psi\rangle\right)\langle\psi|H|\psi\rangle, \tag{11.269}$$

and

$$\frac{d}{dt}\langle\psi|H|\psi\rangle = -\frac{2}{\hbar}[(\nu\langle\psi|H^2|\psi\rangle + \mu\langle\psi|H|\psi\rangle^2]$$

$$-\frac{2}{\hbar}(\nu + \mu)\langle\psi|H|\psi\rangle)^2 - \frac{2\nu}{\hbar}\sum_j |\langle\psi|H|j\rangle|^2, \tag{11.270}$$

where the states $|j\rangle$ and $|\psi\rangle$ form a complete set in Hilbert space

$$|\psi\rangle\langle\psi| + \sum_j |j\rangle\langle j| = 1. \tag{11.271}$$

Now from Eqs. (11.253), (11.254), (11.264) and (11.268) we obtain the following brackets

$$\{\psi(q),\ \psi^*(q')\}_{D_o} = -\frac{i}{\hbar}\delta(q - q'), \tag{11.272}$$

and

$$\{\psi(q),\ \psi^*(q')\}_{D^s} = -\frac{i}{\hbar}\delta(q - q') - \frac{\mu}{\hbar}\psi(q)\psi^*(q').\qquad (11.273)$$

If we want to describe the quantum analogue of a particle subject to friction with the probability equal to one to remain in a box of volume $\mathcal{V}$, then we have to impose the conservation of probability

$$\frac{d}{dt}\langle\psi|\psi\rangle = 0.\qquad (11.274)$$

To satisfy this condition from Eq. (11.269) we get

$$\mu = -\frac{\nu}{\langle\psi|\psi\rangle} = -\nu.\qquad (11.275)$$

This relation, when substituted in (11.270) yields the inequality

$$\frac{d}{dt}\langle\psi|H|\psi\rangle = -\frac{2\nu}{\hbar}\sum_j |\langle\psi|H|j\rangle|^2 \le 0,\qquad (11.276)$$

where the equality sign is correct when $|\psi\rangle$ is an eigenstate of the Hamiltonian H.

Some of the results that have been found by Enz can be found by other methods [46],[47]. In particular this approach provides a bridge between the classical description of dissipation which was the starting point of the present formulation, Sec. 3.14, and the Gisin model which we will consider in Sec. 18.11 [45].

11.11   Embedding a Damped Motion in a Volume-Preserving Dynamical System

We start this section with the proposition that any mechanical N-dimensional system can be embedded into a volume-preserving dynamical system of $2N$ dimensions. We have already seen such an enlargement of the phase space in Bateman's approach, Sec. 3.7 and also in symmetric phase space discussed in Sec. 3.15 , but now we want to consider a different formulation.

A volume preserving embedding enables one to construct a Hamiltonian and by the process of canonical quantization find the corresponding quantal description in the Heisenberg picture [48]–[50]. In the following discussion we will only consider a one-dimensional motion with the classical equations of motion

$$\dot{x} = p,\qquad (11.277)$$

$$\dot{p} = -\frac{\partial V(x)}{\partial x} - \lambda(x)p.\qquad (11.278)$$

This type of damped motion may be embedded in a 4-dimensional volume-preserving phase space (x, p, z, w) where in addition to (11.277) and (11.278) we have two additional first-order differential equations

$$\dot{z} = -wp \frac{d\lambda(x)}{dx}, \tag{11.279}$$

$$\dot{w} = w\lambda(x). \tag{11.280}$$

In this 4-dimensional phase space the time derivative of any dynamical quantity A is given by

$$\frac{dA}{dt} = \{A, \, H\}_1 + \{A, \, H'\}_2, \tag{11.281}$$

where for the damping problem defined by (11.277) and (11.278) the Hamiltonians H and H' are given by

$$H = \frac{1}{2}p^2 + V(x), \qquad H' = wp\lambda(x), \tag{11.282}$$

and the brackets $\{, \}_1$ and $\{, \}_2$ are defined by

$$\{A, \, B\}_1 = \frac{\partial A}{\partial x}\frac{\partial B}{\partial p} - \frac{\partial A}{\partial p}\frac{\partial B}{\partial x} + \frac{\partial A}{\partial z}\frac{\partial B}{\partial w} - \frac{\partial A}{\partial w}\frac{\partial B}{\partial z}, \tag{11.283}$$

and

$$\{A, \, B\}_2 = \frac{\partial A}{\partial x}\frac{\partial B}{\partial z} - \frac{\partial A}{\partial z}\frac{\partial B}{\partial x} + \frac{\partial A}{\partial w}\frac{\partial B}{\partial p} - \frac{\partial A}{\partial p}\frac{\partial B}{\partial w}. \tag{11.284}$$

By setting $B = H$ and $A = x$ and using (11.283) we find (11.277), and by setting $B = H$ and $A = p$ and using (11.283) and (11.284) we obtain (11.278). Therefore we can say that $H + H'$ is the generator of these equations of motion provided that the Poisson brackets are defined by (11.283) and (11.284).

Heisenberg's Equations for Motion in an Embedded Volume-Preserving Coordinate System — As a solvable example for this formulation let us assume that (a) - the potential $V(x)$ is given by a set of rectangular barriers

$$V(x) = \sum_j V_j \theta(x - x_j), \tag{11.285}$$

where $\theta(x)$ is the step function. and (b) - that $\lambda(x_j) = 0$ for the discrete set $\{x_j\}$. Mathematically we can write this condition as that of the intersection of the following two sets being empty [49]

$$\left\{ x : \frac{\partial V(x)}{\partial x} \neq 0 \right\} \cap \{x : \lambda(x) \neq 0\} = 0. \tag{11.286}$$

Let us point out two significant points that will result from $V(x)$ and $\lambda(x)$ satisfying (11.286).

(1) Since the energy dissipation depends on the integral of $\lambda(x)$ alone, therefore by requiring that $\lambda(x)$ to vanish in a set of measure zero, no physically significant effect will be lost.

(2) Any one-dimensional potential can be approximated by a sequence of rectangular potentials $V(x - x_j)$ according to (11.285). Going back to the canonical equations (11.277) and (11.278), we note that these equations can be integrated for two possible cases:

When $\frac{\partial V(x)}{\partial x} = 0$, then

$$p(x_2) - p(x_1) = - \int_{x_1}^{x_2} \lambda(x)dx, \tag{11.287}$$

and when $\lambda(x) = 0$, then

$$\frac{1}{2} \left[p^2(x_2) - p^2(x_1) \right] = V(x_1) - V(x_2). \tag{11.288}$$

From the formal identity of the classical and the operator equations of motion in the Heisenberg picture (with proper symmetrization), the solutions (11.287) and (11.288) can be used both for the classical and also for the time-dependent operators $p(t)$ and $x(t)$ [51].

In the quantum mechanical description one may choose an operator representation at $t = 0$ for the four dynamical variables (x, p, z, w) and these operators are given by

$$x(0) = \begin{bmatrix} x \\ x \end{bmatrix}, \quad p(0) = \begin{bmatrix} -i\frac{\partial}{\partial x} \\ -i\frac{\partial}{\partial w} \end{bmatrix}, \quad z(0) = \begin{bmatrix} z \\ -i\frac{\partial}{\partial x} \end{bmatrix}, \quad w(0) = \begin{bmatrix} -i\frac{\partial}{\partial z} \\ w \end{bmatrix}. \tag{11.289}$$

Having chosen these operators we can use them to construct the complete solution of the problem. With these initial conditions we can determine the time evolution of these operators from the Heisenberg equations of motion. That is for any dynamical operator A which does not depend explicitly on time, i.e. $\frac{\partial A}{\partial t} = 0$ we find $\frac{dA}{dt}$ from the commutator

$$\frac{dA}{dt} = -i\,[A,\,H]_1 - i\,[A,\,H']_2\,, \tag{11.290}$$

where we have set $\hbar = 1$. The states in the Heisenberg picture are described by the matrix $\begin{bmatrix} \psi_1(x) \\ \psi_2(x) \end{bmatrix}$. In simple classical problems where the equations of motion can be solved analytically, one may be able to obtain the operator solution directly, and then using states of the form $\begin{bmatrix} \psi_1(x) \\ 0 \end{bmatrix}$ avoid any reference to the embedding procedure [49]. For system whose description is given by (11.277) and (11.278) we can find the eigenstates of the constants of motion. Thus for the case when $\frac{dV(x)}{dx} = 0$, and $\lambda(x) \neq 0$, from these equations it follows that

$$C_1 = p + \int^x \lambda(y)dy, \tag{11.291}$$

is the first integral of motion. Without any loss of generality we use the $t = 0$ operator representation of $p(0)$, Eq. (11.289), and obtain the eigenvector for C_1,

$$C_1\phi_1^\pm(x) = \left(-i\frac{\partial}{\partial x} + \int^x \lambda(y)dy\right)\phi_1^\pm = \pm k\phi_1^\pm, \qquad (11.292)$$

or by solving (11.292) we have

$$\phi_1^\pm(x) = \exp\left[-i\left(\int^x dz \int^z \lambda(y)dy \mp kx\right)\right], \qquad (11.293)$$

and $\pm k$ are the eigenvalues of C_1. Now by operating on $\phi_1(x)$ with

$$p_0 = p(t = 0) = -i\frac{\partial}{\partial x}, \qquad (11.294)$$

we find

$$\phi_1^{-1}(x_2)p_0\phi_1(x_2) - \phi_1^{-1}(x_1)p_0\phi_1(x_1) = \int_{x_1}^{x_2}\lambda(y)dy. \qquad (11.295)$$

In the same way when $\lambda(x) = 0$, the constant of motion according (11.288) is the energy

$$C_2 = \frac{1}{2}p^2 + V(x), \qquad (11.296)$$

with the corresponding eigenvector $\phi_2(x)$. In this case the quantum relation corresponding to the consistency condition (11.287) is obtained by replacing $p(x_1)$ and $p(x_2)$ in (11.288) by the quantities

$$\phi_2^{-1}(x_1)p_0\phi_2(x_1), \quad \text{and} \quad \phi_2^{-1}(x_2)p_0\phi_2(x_2), \qquad (11.297)$$

respectively. Now we can construct stationary states for this system in the following way:

We require that the operators representing the observables, when acting on these states at a fixed time (e.g. $t = 0$), obey the consistency relations (11.287) and (11.288),(11.297). It should be pointed out that when we apply the consistency conditions, the Heisenberg operators p, x_1 and x_2 are, in general, defined at different times. Here using a single time, $t = 0$, is admissible since we want to study stationary states [49].

Transmission, Reflection and Capture for Two Identical Rectangular Barriers in the Presence of Dissipation — The solution of Heisenberg equations of motion for a general dissipative system is difficult and even numerical methods are unstable and the results are unreliable [52],[53]. For the simple system consisting of two identical rectangular barriers each of height V and located at $-\frac{1}{2}(a + b) < x < -\frac{1}{2}b$ and $\frac{1}{2}b < x < \frac{1}{2}(a + b)$ and with constant dissipation present in the space between them, i.e. $\lambda(x) = \lambda$, $-\frac{1}{2}b < x < \frac{1}{2}b$,

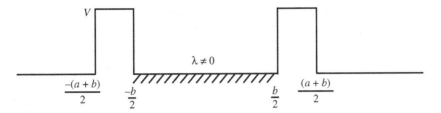

Figure 11.1: Two identical rectangular barriers each of height V, separated by a valley of width b, where there is a constant dissipative force λ (see Eqs. (11.285 and (11.286))[49].

we can solve the problem analytically [49]. This is achieved by applying the consistency conditions piecewise and then by joining the stationary states, viz, requiring the continuity of the function and its derivative in going from one part of the system to the other. In addition we have to impose boundary conditions at $x \to \pm\infty$, and to take into account the presence of several components resulting from reflection at the boundaries.

For problems where there are distinct regions of $V(x)$ and $\lambda(x)$ compatible with (11.286), the operator expectation values are piecewise constant and the Heisenberg states corresponding to stationary solutions contain all of the dynamical information. Denoting the incoming wave number by K, the eigenstates for the five regions are given as follows [49]:

$$\exp\left[iK\left(x + \frac{1}{2}(a+b)\right)\right] + \sum_{n=0} a_n \exp\left[-i(K - 2n\Lambda)\left(x + \frac{1}{2}(a+b)\right)\right],$$

$$x \leq -\frac{1}{2}(a+b), \tag{11.298}$$

$$\sum_{n=0} b_n \exp\left[-\sqrt{2V - (K - 2n\Lambda)^2}\left(x + \frac{1}{2}(a+b)\right)\right]$$

$$+ \sum_{n=0} c_n \exp\left[\sqrt{2V - (K - 2n\Lambda)^2}\left(x + \frac{1}{2}(a+b)\right)\right], \quad -\frac{1}{2}(a+b) \leq x \leq -\frac{1}{2}b,$$

$$\tag{11.299}$$

$$\sum_{n=0} d_n \exp\left[i\left\{(K - 2n\Lambda)\left(x + \frac{b}{2}\right) - \frac{\lambda}{2}\left(x + \frac{b}{2}\right)^2\right\}\right]$$

$$+ \sum_{n=0} e_n \exp\left[-i\left\{(K - 2n\Lambda - \Lambda)\left(x - \frac{b}{2}\right) - \frac{\lambda}{2}\left(x - \frac{b}{2}\right)^2\right\}\right],$$

$$-\frac{1}{2}b \leq x \leq \frac{1}{2}b, \tag{11.300}$$

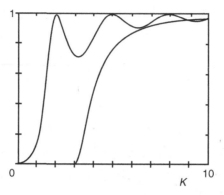

Figure 11.2: Transmission coefficient for tunneling through two identical barriers of Fig. 11.1 plotted as a function of the incoming wave number (momentum) K. The upper curve is for transmission in the absence of the damping force, whereas the lower curve is for the case when $\lambda \neq 0$. The main effect of dissipation, as can be seen from this graph, is the suppression of resonance tunneling. The parameters used for this calculation are $\Delta = 2$, $b = 1$ and $\lambda = 3$ [49].

$$\sum_{n=0} f_n \exp\left[-\sqrt{2V - (K - 2n\Lambda - \Lambda)^2}\left(x - \frac{1}{2}b\right)\right]$$
$$+ \sum_{n=0} g_n \exp\left[\sqrt{2V - (K - 2n\Lambda - \Lambda)^2}\left(x - \frac{1}{2}b\right)\right], \quad \frac{1}{2}b \leq x \leq \frac{1}{2}(a + b),$$

$$\tag{11.301}$$

$$\sum_{n=0} h_n \exp\left[i(K - 2n\Lambda - \Lambda)\left(x - \frac{1}{2}(a + b)\right)\right], \quad x \geq \frac{1}{2}(a + b). \tag{11.302}$$

In these relations $\Lambda = \lambda b$, and the sums in these these relations are finite sums, the largest n , $n = N$, being the one for which either $K = 2N\Lambda$ or $K - 2N\Lambda - \Lambda$ is positive whereas for $n = N + 1$ both become negative.

We apply the consistency conditions to the components having the same momentum at each boundary in order to determine the coefficients a_n, $b_n \cdots h_n$. We can make the problem much simpler by considering the barriers to have a very narrow width, $a \to 0$, and a very large height, $V \to \infty$ in such a way that $\Delta = aV$ has a nonzero finite value. Defining ε_n and δ_n by

$$\varepsilon_n = \frac{\Delta}{2i(K - n\Lambda)}, \quad \delta_n = 1 - \varepsilon_n, \tag{11.303}$$

we have for $n = 0$

$$a_0 = \frac{\varepsilon_0}{\delta_0}, \quad h_0 = \frac{1}{\delta_1 \delta_0} \exp\left[i\left(K - \frac{1}{2}\Lambda\right)b\right] \tag{11.304}$$

and for $n \geq 1$ we find

$$a_n = \frac{\varepsilon_{2n-1} \cdots \varepsilon_1}{\delta_{2n} \cdots \delta_0} \exp\left[2ni(K - n\Lambda)b\right], \tag{11.305}$$

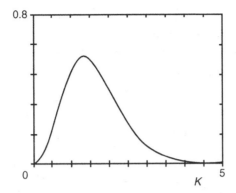

Figure 11.3: Momentum dependence of the probability of capture in the valley between the barriers. Here the parameters $\Delta = 2$, $b = 1$ and $\lambda = 3$ have been used [49].

and

$$h_n = \frac{\varepsilon_{2n} \cdots \varepsilon_1}{\delta_{2n+1} \cdots \delta_0} \exp\left\{ i \left[(2n+1)K - 2\left(n^2 + n\right)\Lambda - \frac{1}{2}\Lambda \right] b \right\}. \qquad (11.306)$$

The quantities $|a_n|^2$ and $|h_n|^2$ are the reflection and transmission probabilities at each momentum. For the case of capture the quantity $|d_N|^2$ (or $|e_N|^2$) is the probability of capture of the particle between the two barriers, where N denotes the integer part of $\frac{K}{2\Lambda}$. The result of the calculation of $|h_0|^2$ for $\Delta = 2$, $b = 1$ and $\lambda = 0$ is shown in Fig. 11.2 (the oscillating upper curve). However in the presence of damping with $\lambda = 3$, for the transmission coefficient one finds the lower curve of Fig. 11.2. We observe that the main effect of damping is a reduction in the transmission probability and the disappearance of resonant tunneling. The reason for suppression of resonances is that the dissipation causes the incident and the reflected waves between barriers to have different momenta, and not add coherently.

Finally in Fig. 11.3 the probability of capture, $|d_N(K)|^2$ (or $|a_N(K)|^2$) is shown. This probability is small for very small and also for very large values of K. It reaches its maximum for some intermediate value of K, and the position and the height of this maximum depends strongly on λ.

In this approach we have tried to find the solution of the damped system in the Heisenberg picture. It is important to remember that in this formulation there is no unitary evolution operator and therefore, in general, no Schrödinger picture.

11.12 Lagrangians and Hamiltonians for Velocity-Dependent Forces

We will now consider Lagrangians which depend quadratically on generalized velocities and that the coefficients of these terms are functions of the generalized coordinates. These types of interactions occur in nuclear, atomic and condensed matter physics as we will see shortly.

Noting that in quantum mechanics velocities are operators we write a typical Lagrangian as [54],[55]

$$L = \frac{1}{2} \sum_{i,j} \dot{q}_i g_{ij}(q) \dot{q}_j - V(q). \tag{11.307}$$

Here q_i and $\dot{q}_i$ ($i = 1$, 2, 3) stand for the generalized coordinates and their time derivatives respectively, $g_{ij}(q)$ and $V(q)$ are functions of coordinates q_i only, and $g_{ij}(q)$ is symmetric with respect to i and j which is the condition for L to be Hermitian. Using this Lagrangian we define the canonical momentum conjugate to q_i by;

$$p_i = \frac{\partial L}{\partial \dot{q}_i} = \frac{1}{2} \sum_j [g_{ij}(q), \dot{q}_j]_+ , \tag{11.308}$$

where $[\,,\,]_+$ denotes anti-commutator. Now if we assume the standard canonical relations between p_i and q_j, viz,

$$[p_i, q_j] = -i\hbar \delta_{ij}, \tag{11.309}$$

then form (11.308) and (11.309) it follows that

$$[\dot{q}_i, q_j] = -i\hbar f_{ij}(q), \tag{11.310}$$

where $f_{ij}(q)$ is a function of q_i and is symmetric with respect to the interchange of i and j. As we have seen earlier for a dissipative system which is quadratic in velocity and is a special case of (11.307), $f(q)$ tends to zero as $q \to \infty$. Thus we have a violation of the uncertainty principle, since the product $\Delta \dot{q} \Delta q$ also goes to zero in the limit of large q. Therefore let us assume that $f_{ij}(q)$ is such a set of functions that the uncertainty principle is not violated and in addition $\det g_{ij}(q) \neq 0$. Then from Eqs. (11.308) and (11.310) it follows that

$$\sum_j f_{ij}(q) g_{jk}(q) = \sum_j g_{ij}(q) f_{jk}(q) = \delta_{ik}. \tag{11.311}$$

Now when $f_{ij}(q)$ is not a c-number, $\delta \dot{q}_i$ is not, in general, a c-number, and in this case the operator

$$K = \frac{1}{2} \sum_i (p_i \dot{q}_i + \dot{q}_i p_i) - L, \tag{11.312}$$

is not the correct Hamiltonian and does not satisfy the Heisenberg equations of motion. On the other hand if we assume that $\delta \dot{q}_i$ is a c-number, then the Lagrangian and the Hamiltonian formalisms are not compatible with each other [54]. In order to get a dynamical description of the system which is free from the compatibility problem, we consider a set of canonically conjugate operators P_α and Q_α, $(\alpha = 1, 2, 3)$ with the following commutators

$$[P_\alpha, Q_\beta] = -i\hbar\delta_{\alpha,\beta}, \quad \left[\dot{Q}_\alpha, Q_\beta\right] = -i\hbar\kappa^{-1}\delta_{\alpha,\beta}, \tag{11.313}$$

that is $P_\alpha = \kappa\dot{Q}_\alpha$ with κ being a constant. We choose these new operators so that the Lagrangian assumes the simple form of

$$L\left(Q, \dot{Q}\right) = \frac{\kappa}{2}\sum_\alpha \left(\dot{Q}_\alpha\dot{Q}_\alpha\right) - V(Q). \tag{11.314}$$

Using this Lagrangian we find the Hamiltonian in the standard form;

$$H(P,Q) = \sum_\alpha P_\alpha\dot{Q}_\alpha - L\left(Q,\dot{Q}\right). \tag{11.315}$$

Once such a Hamiltonian is found, then the quantal $L\left(Q,\dot{Q}\right)$ and $H(P,Q)$ will be compatible. Let $F_1(p,Q)$ represent the generating function of this transformation which we write in the symmetrized form as

$$F_1(p,Q) = -\frac{1}{2}\sum_i [p_i, q_i(Q)]_+, \tag{11.316}$$

then from the generating function $F_1(p,Q)$ we find

$$q_i = -\frac{\partial F_1}{\partial p_i} = q_i(Q), \tag{11.317}$$

$$P_\alpha = -\frac{\partial F_1}{\partial Q_\alpha} = \frac{1}{2}\left[p_i, \frac{\partial q_i(Q)}{\partial Q_\alpha}\right]_+. \tag{11.318}$$

Now the well-known result of the canonical transformation relates $K(p,q)$ to $H(P,Q)$ by [56],[57]

$$\frac{dF_1}{dt} + \sum_\alpha \left(P_\alpha\dot{Q}_\alpha\right) - H(P,Q)$$

$$-\frac{1}{2}\sum_i [\dot{p}_i, q_i]_+ - K(p,q), \tag{11.319}$$

where

$$\dot{Q}_\alpha = \frac{1}{2}\sum_i \left[\dot{q}_i, \frac{\partial Q_\alpha}{\partial q_i}\right]_+. \tag{11.320}$$

We can write Eq. (11.319) as the following equation

$$\frac{dF_1}{dt} = -\frac{1}{2}\sum_i [\dot{p}_i, q_i]_+ - \frac{1}{4}\sum_i\sum_\alpha \left[p_i, \left[\dot{Q}_\alpha, \frac{\partial q_i}{\partial Q_\alpha}\right]_+\right]_+$$

$$= -\frac{1}{2}\sum_i [\dot{p}_i, q_i] - \sum_\alpha P_\alpha\dot{Q}_\alpha - Z, \tag{11.321}$$

248

Classical and Quantum Dissipative Systems

where

$$Z = \frac{1}{4}\sum_i\sum_\alpha \left[[p_i,\ \dot{Q}_\alpha],\ \frac{\partial q_i}{\partial Q_\alpha}\right] = -\frac{i\hbar}{4\kappa}\sum_i\sum_\alpha [[p_i,\ P_\alpha],\ [q_i,\ P_\alpha]].\quad (11.322)$$

It is important to note that while F_1 does not depend explicitly on time, its time derivative does not vanish. Also note that we have replaced the anti-commutator in (11.321) with commutator to get (11.322). Now by transforming (P,Q) back to (p,q) we find the Hamiltonian $H(p,q)$;

$$H(p,q) = K(p,q) - Z(p,q) = \sum_i [p_i,\ \dot{q}_i]_+ - L - Z(p,q).\quad (11.323)$$

By solving (11.317) for Q_α in terms of q_i we get

$$Q_\alpha = Q_\alpha(q),\qquad P_\alpha = \frac{1}{2}\left[p_i,\ \frac{\partial q_i}{\partial Q_\alpha}\right]_+,\qquad \alpha = 1,\ 2,\ 3.\quad (11.324)$$

Using these relations together with $[q_i,\ p_i] = i\hbar$, we find an identical commutation relation for Q_α and P_α, viz, $[Q_\alpha,\ P_\alpha] = i\hbar$. We can also solve (11.324) and find q_i and p_i in terms of Q_α and P_α [55];

$$q_i = q_i(Q),\qquad p_i = \frac{1}{2}\left[P_\alpha,\ \frac{\partial Q_\alpha}{\partial q_i}\right]_+,\qquad i = 1,\ 2,\ 3.\quad (11.325)$$

Now from Eqs. (11.322) and (11.324) we find the following expression for Z:

$$Z = \frac{1}{4\kappa}\sum_{i,j,\alpha}\left[p_i,\ \frac{\partial q_j}{\partial Q_\alpha}\right]\left[p_j,\ \frac{\partial q_i}{\partial Q_\alpha}\right],\quad (11.326)$$

where we have used the commutativity of $\frac{\partial q_i}{\partial Q_\alpha}$ and $\left[p_j,\ \frac{\partial q_k}{\partial Q_\beta}\right]$ to get (11.326).

When the Lagrangian (11.307) is given, then we can determine Z as a function of the matrix $g_{ij}(q)$ provided that its inverse matrix $f_{ij}(q)$ exists (see Eq. (11.311)). We now substitute $\dot{q}_i(Q)$ which in symmetrized form is given by

$$\dot{q}_i = \frac{1}{2}\sum_\alpha\left[\dot{Q}_\alpha,\ \frac{\partial q_i}{\partial Q_\alpha}\right]_+\quad (11.327)$$

in the Lagrangian (11.307). We then set

$$\sum_{i,j}\frac{\partial q_i}{\partial Q_\alpha}\frac{\partial q_j}{\partial Q_\beta}g_{ij}(q) = \delta_{\alpha,\beta},\quad (11.328)$$

and identify the resulting L with the L given by (11.314), and after some algebra we find

$$Z = \frac{-\hbar^2}{4\kappa}\sum_{jklm}\frac{\partial}{\partial q_i}\left(f_{ij}f_{kl}\frac{\partial g_{kl}(q)}{\partial q_j}\right) - \frac{\hbar^2}{4\kappa}\sum_{ij}\frac{\partial^2 f_{ij}(q)}{\partial q_i q_j}$$
$$-\frac{\hbar^2}{16\kappa}\sum_{ij}\sum_{klmn}\left(f_{ij}f_{kl}f_{mn}\frac{\partial g_{ij}(q)}{\partial q_m}\frac{\partial g_{kl}(q)}{\partial q_n}\right),\quad (11.329)$$

where $f_{ij}(q)$ is the inverse of $g_{ij}(q)$, Eq. (11.311). When $g_{ij}(q)$ is a diagonal matrix with elements $g_{ij}(q) = g(q)\delta_{i,j}$, then Z reduces

$$Z = \hbar^2 \left\{ \frac{n-1}{4\kappa} \sum_i \frac{\partial^2 f(q)}{\partial q_i^2} - \frac{n^2}{16\kappa} \frac{1}{f(q)} \sum_i \left(\frac{\partial f(q)}{\partial q_i} \right)^2 \right\}, \tag{11.330}$$

where n is the number of dimensions of the system and $f(q)$ is defined by $f(q) = (g(q))^{-1}$.

Application to the Problem of Newtonian Damping — From the foregoing discussion it is clear that if we start with the Lagrangian (11.195) considered as an operator and hence written in symmetrized form, then the Hamiltonian operator derived from this Lagrangian will have an additional term

$$H = \frac{1}{2m} p e^{-\gamma x} p + \int^x e^{\gamma y} \frac{dV(y)}{dy} dy - Z, \tag{11.331}$$

where according to (11.330) Z is given by

$$Z = \frac{\hbar^2 \gamma^2}{16m} e^{-\gamma x}. \tag{11.332}$$

This is the same kind of term that we found by considering the Dirac rule of association (see Eq. (11.201)).

The Classical Limit of the Schrödinger Equation with Velocity-Dependent Forces — There are indications, for example from the phase shift analysis, that the interaction between two nucleons at very short distances is velocity-dependent [58],[59]. Similar conclusions can be reached by considering potentials derived from the meson theory of nuclear forces [60]. In low and intermediate energies the two nucleon interaction can be described by the Schrödinger equation

$$-\frac{\hbar^2}{2m} \nabla^2 \psi - \frac{\hbar^2}{2m} \nabla \cdot [g(r)\nabla \psi] + V(r)\psi = i\hbar \frac{\partial \psi}{\partial t}, \tag{11.333}$$

where $\nabla \cdot g(r)\nabla$ is a short-range velocity-dependent force and $V(r)$ represents all other (static) forces.

The velocity-dependent potential used in (11.333) is phenomenological and can be regarded as the leading term of the expansion of a nonlocal interaction (Sec. 20.1). When viewed as the first term of expansion (11.333) does not describe a closed and conservative system. This is also reflected on the q-dependence of the commutation relation (11.310).

Similar Schrödinger equation with a space-dependent effective mass, is commonly used used in the band theory of solids [61] except that in this case $\left(\frac{m}{m^*} \right)$ are regarded as a tensor, i.e. the first two terms of (11.333) are replaced by

$$-\frac{\hbar^2}{2m} \sum_{i,j} \frac{\partial}{\partial x_i} \left(\frac{m}{m^*(x_k)} \right)_{i,j} \frac{\partial \psi}{\partial x_j}. \tag{11.334}$$

Now let us write (11.333) as

$$-\frac{\hbar^2}{2m}\nabla \cdot \left[\frac{1}{m^*(r)}\nabla\psi\right] + V(r)\psi = i\hbar\frac{\partial\psi}{\partial t}, \tag{11.335}$$

where $m^*(r) = [m/(1 - g(r))]$. To find the classical limit of (11.333) we take $\psi(\mathbf{r}, t)$ to be

$$\psi(\mathbf{r}, t) = \exp\left(\frac{iS(\mathbf{r}, t)}{\hbar}\right). \tag{11.336}$$

Substituting (11.336) in (11.333) and ignoring terms proportional to $\hbar$ we find

$$\frac{1}{2m^*(r)}(\nabla S)^2 + V(r) + \frac{\partial S}{\partial t} = 0, \tag{11.337}$$

which is the same as (4.41) provided that we choose m^* to be given by (4.38).

11.13 Quadratic Damping as an Externally Applied Force

In the next chapter we will see how the damping force can be assumed to be external and dependent only on time. For the Newtonian frictional force we can formulate the problem in the following way. Again for the sake of simplicity we assume a one-dimensional motion along the x-axis;

Let us study the classical Hamiltonian

$$H = \frac{p^2}{2m} + V(x) + \frac{1}{2}m\gamma\dot{\eta}^2\left[x - \eta(t)\right], \tag{11.338}$$

where x and p are the coordinate and the momentum of the particle respectively, and $\eta(t)$ is the solution of the classical equation

$$m\ddot{\eta} + \frac{1}{2}m\gamma\dot{\eta}^2 + \frac{dV(\eta)}{d\eta} = 0. \tag{11.339}$$

If we compare (11.339) with the equation of motion derived from H, Eq. (11.338), i.e. with

$$m\ddot{x} + \frac{dV(x)}{dx} + \frac{1}{2}m\gamma\dot{\eta}^2 = 0, \tag{11.340}$$

we find that $x = \eta(t)$ is the particular solution of (11.340). The Schrödinger equation for the quantized form of the Hamiltonian (11.338) is

$$-\frac{\hbar^2}{2m}\frac{\partial^2\psi}{\partial x^2} + \left[V(x) + \frac{1}{2}\gamma m\dot{\eta}^2(t)(x - \eta(t))\right]\psi = i\hbar\frac{\partial\psi}{dt}. \tag{11.341}$$

An examination of this Schrödinger equation shows that it shares certain features with the classical equation of motion, and some of these features are lost in the Hamiltonian formulation. For instance if we consider the special and simple case where $V(x) = 0$, then from (11.339) it is clear that the equation of motion is invariant under the displacement $\eta(t) \to \eta(t) + x_0$, whereas (11.340), with $V = 0$, is not invariant under the translation of the coordinate $x \to x + x_0$. On the other hand Eq. (11.341), again with $V = 0$, remains unchanged when we replace $\eta(t)$ by $\eta(t) + x_0$ and x by $x + x_0$.

The solution of Eq. (11.341) with $V = 0$ is given by

$$\psi(x,t) = \exp\left\{\frac{i}{\hbar}\left[m\dot{\eta}(t)(x - \eta(t)) + \frac{1}{2}\int^t m\dot{\eta}^2(t)dt\right]\right\}, \qquad (11.342)$$

which has the correct limit as $\gamma \to 0$. Since there is no potential energy, the energy of the particle is purely kinematical

$$-\frac{\hbar^2}{2m}\frac{\partial^2 \psi}{\partial x^2} = \frac{1}{2}m\dot{\eta}^2(t)\psi, \qquad (11.343)$$

and is a decreasing function of time.

We can also solve the wave equation for other potentials. For instance when $V(x) = \frac{1}{2}m\omega_0^2 x^2$, the solution of (11.341) is expressible as

$$\psi(x,t) = \phi_n(x - \eta(t))$$
$$\times \exp\left[\frac{i}{\hbar}\left(m\dot{\eta}(x - \eta(t)) - \epsilon_n t + \int^t \frac{1}{2}\left\{m\dot{\eta}(t)^2 - m\omega_0^2\eta^2(t)\right\}dt\right)\right], \qquad (11.344)$$

where ϕ_n s are the normalized harmonic oscillator wave functions and

$$\epsilon_n = \left(n + \frac{1}{2}\right)\hbar\omega_0, \qquad n = 0, 1, 2\cdots, \qquad (11.345)$$

are the harmonic oscillator eigenvalues. This result is similar to what we will find for damping linear in velocity in Sec. 12.7. In both cases the frequency eigenvalues are not affected by damping.

By substituting (11.344) in (11.341), we find the differential equation satisfied by ϕ_n

$$-\left(\frac{\hbar^2}{2m}\right)\frac{d^2\phi_n}{dy^2} + \frac{1}{2}m\omega_0^2 y^2 \phi_n = \epsilon_n \phi_n, \qquad (11.346)$$

where we have replaced the coordinate x by the "coherent state" coordinate (see Chapter 13)[62]-[64]

$$y = x - \eta(t). \qquad (11.347)$$

We observe that in the absence of damping, $\gamma \to 0$, both (11.344) and (11.346) reduce to the coherent state wave function for the harmonic oscillator.

From the wave function (11.344) we can calculate the mean energy, E_n, for this system when it is in its nth excited state

$$E_n = \int_{-\infty}^{\infty} \psi^*(x,t) \left[-\left(\frac{\hbar^2}{2m}\right) \frac{\partial^2}{\partial x^2} + \frac{1}{2} m\omega_0^2 x^2 \right] \psi(x,t) dx$$

$$= \epsilon_n + \frac{m}{2} \left(\dot{\eta}^2(t) + \omega_0^2 \eta^2(t) \right). \tag{11.348}$$

Using the same wave function we can also determine the expectation values of the coordinate and momentum of the particle. These are found to be

$$\langle x \rangle = \eta(t), \quad \text{and} \quad \langle p \rangle = m\dot{\eta}(t). \tag{11.349}$$

respectively.

11.14 Motion in a Viscous Field of Force Proportional to an Arbitrary Power of Velocity

In Sec. 3.10 we have seen that

$$H = p\left[C - \beta(2-a)x\right]^{\frac{1}{(2-a)}} = pf(x), \tag{11.350}$$

is the Hamiltonian for the motion of a particle in a dissipative force, $\beta \dot{x}^\nu$. We can formally "quantize" (11.350) by first symmetrizing H and then by constructing the Hermitian operator

$$H = -i\hbar f(x) \frac{\partial}{\partial x} - \frac{i}{2} \hbar \frac{df(x)}{dx}. \tag{11.351}$$

The "wave equation" for this time-independent Hamiltonian is

$$\frac{d\psi}{dx} + \left[\frac{1}{2} \frac{df(x)}{dx} - \frac{i\epsilon}{\hbar} f(x) \right] \psi = 0. \tag{11.352}$$

This first order differential equation can be integrated to yield

$$\psi(x) = \frac{1}{\sqrt{f(x)}} \exp\left[\frac{i\epsilon}{\hbar} \int^x \frac{dx'}{f(x')} \right]. \tag{11.353}$$

Again we observe that this solution is unphysical.

We can also start with the classical Hamiltonian given by Eq. (3.190) for $\dot{x}^\nu$ and write the Hamiltonian operator as

$$H = \frac{m}{2-\nu} \left[i\hbar \left(\frac{\nu-1}{m} \right) \frac{\partial}{\partial x} \right]^{\frac{\nu-2}{\nu-1}} + \alpha x. \tag{11.354}$$

For an arbitrary value of ν this method of canonical quantization is problematic. However for some special cases such as $\nu = \frac{n}{n+1}$, where n is an integer Geicke has shown that the Schrödinger equation leads to the correct classical limit according to the Ehrenfest theorem [65].

By a method similar to what we have seen in the previous section we can formulate a wave equation for a general damping force $F(\eta, \dot{\eta}, t)$ where η is the solution of the classical equation

$$m\ddot{\eta} + F(\eta, \dot{\eta}, t) + \frac{dV(\eta)}{d\eta} = 0. \tag{11.355}$$

Thus starting with the Scrödinger equation

$$-\left(\frac{\hbar^2}{2m}\right)\frac{\partial^2 \psi}{\partial x^2} + V(x)\psi + F[\eta(t), \dot{\eta}(t), t][x - \eta(t)]\psi = i\hbar\frac{\partial \psi}{\partial t}, \tag{11.356}$$

we observe that ψ will have the same functional dependence as (11.341). Again when $V(x) = 0$, we have the solution given by (11.342), but now $\eta(t)$ must satisfy the solution of the Newtonian equation of motion (11.355) with $V = 0$.

11.15 The Classical Limit and the Van Vleck Determinant

The formal connection between the classical equation of motion and the time-dependent Schrödinger equation can be established via Ehrenfest's theorem, and this is what we will use to discuss the time-dependence of the expectation values of x and p. But there is another way of relating the wave function and the classical probability of finding the particle in a given part of space. According to Van Vleck in the limit of $\hbar \to 0$, for a system where H is a constant of motion, we have [66]

$$|\psi(x)|^2 \, dx \to \left|\frac{\partial H}{\partial p}\right|^{-1} dx = \frac{dx}{|\dot{x}(\mathcal{E}, x)|}, \tag{11.357}$$

where the last term is the classical probability of finding the particle in the range x and $x + dx$, and $\mathcal{E}$ is the constant value of H.

In the case of a free particle moving in a viscous medium where the drag force is $\frac{1}{2}m\gamma\dot{x}^2$, we have the classical Hamiltonian given by (11.198) and therefore

$$\dot{x} = \sqrt{\frac{2\mathcal{E}}{m}}\exp\left(-\frac{\gamma x}{2}\right). \tag{11.358}$$

For the dissipative force $\beta\dot{x}^a$ from (11.350) we have

$$\dot{x} = \frac{\partial H}{\partial p} = f(x). \tag{11.359}$$

We have also the wave function $\psi(x)$ which is given by (11.353). Thus

$$|\psi(x)|^2 dx = \frac{dx}{|f(x)|} = \frac{dx}{|\dot{x}(x)|}. \tag{11.360}$$

The result shows that this wave function, though unphysical, has the correct classical limit.

11.16   Fractional Derivatives and the Wave Equation with Linear Damping

In Chapters 3 and 4 we observed that by introducing functional derivatives we can enlarge the coverage of the Hamiltonian and the Hamilton-Jacobi formulations to include certain dissipative systems. We can inquire whether the introduction of functional derivatives in these formulations will lead us to new ways of quantizing damped systems and whether the results are physically meaningful or not. Our starting point can be either the classical Hamiltonian, or the Hamilton-Jacobi equations, but the latter seems closer to the Schrödinger approach to quantization. We begin our inquiry by considering the wave equation which reduces to the Hamilton-Jacobi equation (4.50) as $\hbar \to 0$:

$$\left[-\frac{\hbar^2}{2m} \left(\frac{\partial^2}{\partial x^2} \right) - i\hbar x_{\frac{1}{2}} \left(\frac{\partial}{\partial x_{-\frac{1}{2}}} \right) + V(x) - \frac{i}{2}\lambda x_{\frac{1}{2}}^2 \right] \psi = i\hbar \frac{\partial \psi}{\partial t}, \tag{11.361}$$

where we have replaced the variable q by x. Next we want to show that the classical limit of (11.361) is indeed Eq. (4.50). To this end we write ψ as the product of two factors

$$\psi\left(x, x_{\frac{1}{2}}, x_{-\frac{1}{2}}, t\right) = A\left(x, x_{\frac{1}{2}}, x_{-\frac{1}{2}}, t\right) \exp\left[\frac{i}{\hbar} S\left(x, x_{\frac{1}{2}}, x_{-\frac{1}{2}}, t\right) \right]. \tag{11.362}$$

Since in Eq. (11.361) there are no fractional derivatives of the wave function we can perform the differentiation in the usual way and find the following equation [67],[68] :

$$-\frac{\hbar^2}{2m} \left(\frac{\partial^2 A}{\partial x^2} \right) - i\hbar x_{\frac{1}{2}} \left(\frac{\partial A}{\partial x_{-\frac{1}{2}}} \right) - i\frac{\hbar}{m} \frac{\partial A}{\partial x} \frac{\partial S}{\partial x} - \frac{i\hbar}{2m} A \frac{\partial^2 S}{\partial x^2} + x_{\frac{1}{2}} \frac{\partial S}{\partial x_{-\frac{1}{2}}}$$

$$+ \frac{1}{2m} A \left(\frac{\partial S}{\partial x} \right)^2 + AV(x) - \frac{i\lambda}{2} A x_{\frac{1}{2}}^2 = i\hbar \frac{\partial A}{\partial t} - A \frac{\partial S}{\partial t}. \tag{11.363}$$

In the limit of $\hbar \to 0$, this equation reduces to the classical Hamilton-Jacobi equation if S is the solution of

$$\frac{\partial S}{\partial t} + \frac{1}{2m} \left(\frac{\partial S}{\partial x} \right)^2 + x_{\frac{1}{2}} \left(\frac{\partial S}{\partial x_{-\frac{1}{2}}} \right) + V(x) - \frac{i\lambda}{2} x_{\frac{1}{2}}^2 = \frac{\hbar^2}{2mA} \left(\frac{\partial^2 A}{\partial x^2} \right), \tag{11.364}$$

provided that the equation of continuity

$$m\frac{\partial A}{\partial t} + mx_{\frac{1}{2}}\left(\frac{\partial A}{\partial x_{-\frac{1}{2}}}\right) + \frac{\partial A}{\partial x}\frac{\partial S}{\partial x} + \frac{1}{2}A\frac{\partial^2 S}{\partial x^2} = 0, \qquad (11.365)$$

is satisfied. The presence of the term $\left(-\frac{i}{2}\lambda x_{\frac{1}{2}}^2\right)$ in (11.364) shows that both S and consequently A are complex functions. We also observe that in (11.364), the potential has two components $V(x)$ which is real and $-\frac{i}{2}\lambda x_{\frac{1}{2}}^2$ which is pure imaginary, and that the imaginary part of the potential is negative for all values of $x_{\frac{1}{2}}$. Such a potential called "the optical potential" will be treated in detail in Chapter 20.

References

[1] See J. Mehta and H. Rechenberg, *The Historical Development of Quantum Theory*, Vol. 1, part 1 (Springer-Verlag, New York, N.Y. 1982) p. 201.

[2] R.J. Seegert, Proc. Nat. Acad. Sci. (USA) 18, 303 (1932).

[3] M. Born, and P. Jordan, Z. Phys. 34, 873 (1925).

[4] M. Razavy, *Heisenberg's Quantum Mechanics*, (World Scientific, Singapore, 2011).

[5] W.E. Brittin, Phys. Rev. 77, 396 (1950).

[6] H. Goldstein, *Classical Mechanics*, Second Edition (Addison-Wesley Reading, 1980).

[7] P.A.M. Dirac, *The Principles of Quantum Mechanics*, Fourth Edition (Oxford University Press, London, 1958).

[8] L.I. Schiff, *Quantum Mechanics*, Third Edition (McGraw-Hill, New York, N.Y. 1968).

[9] M.D. Srinivas and T.S. Shankara, Lettere Al Nuovo Cimento, 6, 321 (1973).

[10] V.E. Tasarov, Theor. Math. Phys. 100, 1100, (1994).

[11] V.E. Tasarov, Theor. Math. Phys. 110, 57, (1997).

[12] S. Okubo, Phys. Rev. D 22, 919 (1980).

[13] S. Okubo, Phys. Rev. A 23, 2776 (1981).

[14] C.N. Yang and D. Feldman, Phys. Rev. 79, 972 (1950).

[15] V. Dodonov, V.I. Man'ko and V.D. Skarzhinsky, in *Quantization and Group Methods in Physics*, Edited by A.A. Koma (Nova Science, Commack, 1988).

[16] V. Dodonov, V.I. Man'ko and V.D. Skarzhinsky, Phys. Lett. 82 A, 61 (1981).

[17] H.J. Groenewold, Physica 12, 405 (1946).

[18] J.R. Shewell, Am. J. Phys. 27, 16 (1959).

[19] H. Dekker, Z. Phys. B 12, 295 (1975).

[20] H. Dekker, Phys. Rev. A 16, 2126 (1977).

[21] H. Dekker, Phys. Rep. 1, 80 (1981).

[22] E. Celeghini, M. Rasetti and G. Vitiello, Ann. Phys. (NY), 215, 156 (1992).

[23] H. Feshbach and Y. Tikochinsky, Trans. N.Y. Acad. Sci. 38, 44 (1977).

[24] See for instance A.Z. Capri, *Nonrelativistic Quantum Mechanics*, Third Edition (World Scientific, Sigapore, 2002) p. 205.

[25] S. Gasiorowicz, *Quantum Mechanics*, Second Edition, (John Wiley & Sons, New York, N.Y. 1996) p. 438.

[26] G. Lindblad and B. Nagel, Ann. Inst. Henri Poincaré, Sec. A 13, 27 (1970).

[27] B. Yurke and M. Potasek, Phys. Rev. A 36, 3464 (1987).

[28] R. Banerjee and P. Mukherjee, J. Phys. A 35, 5591 (2002).

[29] C.M. Binder and S. Boettcher, Phys. Rev. Lett. 80, 5243 (1998).

[30] A. Mostafazadeh, J. Math. Phys. 43, 205 (2002).

[31] C.M. Bender, M. Gianfreda, N. Hassanpour and H.F. Jones, J. Math. Phys. 57, 084101 (2016).

[32] E. Fermi, Rend. Lincei. 5, 795 (1927).

[33] E. Fermi, *Collected Papers*, Vol. I (University of Chicago Press, 1962) p. 271.

[34] K. Wodkiewicz, in *Foundations of Radiation Theory and Quantum Electrodynamics* edited by A.O. Barut (Plenum, New York, N.Y. 1980) p. 109.

[35] K. Wodkiewicz and M.O. Scully, Phys. Rev. A 42, 5111 (1990).

[36] M.D. Kostin, J. Chem Phys. 57, 3589 (1972).

[37] M. Razavy, Lett. Nuovo Cimento 24, 293 (1979).

[38] M. Razavy, Phys. Rev. A 36, 482 (1987).

[39] H. Weyl, Z. Phys. 46, 1 (1927).

[40] P. Crehan, J. Phys. A 22, 811 (1989).

[41] F. Negro and A. Tartaglia, Phys. Rev. A 23, 1591 (1981)

[42] C. Stukens and D.H. Kobe, Phys. Rev. A 34, 3569 (1986).

[43] J. Geicke, J. Phys. A 22, 1017 (1989).

[44] E. Kamke, *Differentialgleichungen, Lösungsmethoden und Lösungen*, (Chelsea, New York, N.Y. 1948) p. 551.

[45] C.P. Enz, Found. Phys. 24, 1281 (1994).

[46] J.A. Holyst and L.A. Turski, Phys. Rev. A 45, 6180 (1992).

[47] N. Gisin, Hel. Phys. Acta, 54, 457 (1981).

[48] R.V. Mendes, Phys. Rev. D 26, 3446 (1982).

[49] S.M. Eleutério and R.V. Mendes, Nuovo Cimento, 96 B, 26 (1986).

[50] R.V. Mendes, in *Stochastic Processes in Quantum Theory and Statistical Physics, Lecture Notes in Physics, 173* (Springer, Berlin, 1982), p. 332.

[51] For a full discussion of this point as well as the question of symmtrization see M. Razavy, *Heisenberg's Quantum Mechanics*, (World Scientific, Singapore 2011).

[52] Conditions for resonance in the absence of dissipation for systems such as this is discussed in M. Razavy, *Quantum Theory of Tunneling*, (World Scientific, Singapore 2014) Chapter 3.

[53] M. Razavy, Phys. Rev. A 41, 6668 (1990).

[54] D. Kiang, K. Nakazawa and R. Sugano, Phys. Rev. 181, 1380 (1969).

[55] H.E. Lin, W.C. Lin and R. Sugano, Nucl. Phys. B 16, 431 (1970).

[56] L.D. Landau and E.M. Lifshitz, *Mechanics*, (Addison-Wesley, Reading, 1960.

[57] H. Goldstein, *Classical Mechanics*, Second Edition (Addison-Wesley Reading, 1980) Chapter 9.

[58] M. Razavy, G. Field and J.S. Levinger, Phys. Rev. 125, 269 (1962).

[59] M. Lacombe, B. Loiseau, J.M. Richard, R. Vinh Mau, J. Cote', P. Pires and R. de Tourreil, Phys. Rev. C 21, 861 (1980).

[60] A.E.S. Green, T. Sawada and D.S. Saxon, *The Nuclear Independent Particle Model*, (Academic Press, New York, N.Y. 1968).

[61] See for example, W.A. Harrison, *Solid State Theory*, (McGraw-Hill, New York, N.Y. 1970) p. 143.

[62] For the definition of coherent states in quantum theory the reader is reffered to the book by J.R. Klauder and B-S Skagerstam, *Coherent States: Application in Physics and Mathematical Physics*, (World Scientific, 1985).

[63] J-P. Gazeau, *Coherent States in Quantum Physics*, (Wiley-VCH, New York, 2009).

[64] C.F. Lo, Phys. Rev. A 43, 404 (1991).

[65] J. Geicke, Rev. Brasil Fisica, 21, 429 (1991).

[66] J.H. Van Vleck, Proc. Nat. Acad. Sci. U.S.A. 14, 178 (1928).

[67] F. Riewe, Phys. Rev. E 53, 1890 (1996).

[68] F. Riewe, Phys. Rev. E 56, 3581 (1997).

Chapter 12

Quantization of Explicitly Time-Dependent Hamiltonians

We have already seen that for a damping force linear in velocity we can derive the equation of motion from the Caldirola-Kanai Hamiltonian, Eq (3.24),

$$H = \frac{p^2}{2m} \exp(-\lambda t) + V(x) \exp(\lambda t). \tag{12.1}$$

Similarly in the case of a radiating electron when the equation of motion is given by (3.56) we find the explicitly time-dependent Hamiltonian (3.59). In this chapter we will study a number of problems that we encounter when we try to quantize these Hamiltonians. Just as the case of time-independent Hamiltonians for nonconservative systems, here we have many wave equations corresponding to the same classical problem.

12.1 Wave Equation for the Caldirola-Kanai Hamiltonian

The Caldirola-Kanai Hamiltonian for the potential $V(x)$, which is given by Eq. (3.24), leads to the correct equation of motion in configuration space but not in phase space (Chapter 3), and in addition the Hamiltonian H is not the energy function of the system. The operator $\hat{H}$ corresponding to the classical function H is found by using the conventional quantization rule and replacing the canonical momentum p by $-i\hbar \frac{\partial}{\partial x}$,

$$\hat{H} = -\frac{\hbar^2}{2m} \exp(-\lambda t) \frac{\partial^2}{\partial x^2} + V(x) \exp(\lambda t), \tag{12.2}$$

and thus the time-dependent Schrödinger equation for (12.2) is given by

$$i\hbar\frac{\partial\psi}{\partial t} = -\frac{\hbar^2}{2m}e^{-\lambda t}\frac{\partial^2\psi}{\partial x^2} + V(x)e^{\lambda t}\psi. \tag{12.3}$$

Solvable Examples — Now we want to study the problem of the damped harmonic oscillator to illustrate the unphysical nature of the wave function found for Cardirola-Kanai Hamiltonian. Equation (12.3) can be separated and solved only for few special forms of $V(x)$, e.g. when $V(x) = 0$ or for the damped harmonic oscillator [1]-[7]. In the latter case the wave equation can be transformed to the Schrödinger equation for the harmonic oscillator,

$$V(x) = \frac{1}{2}m\omega_0^2 x^2 = \frac{1}{2}m\left(\omega^2 + \frac{\lambda^2}{4}\right)x^2, \tag{12.4}$$

if we change the variable x to y where [1]

$$y = x\exp\left(\frac{\lambda t}{2}\right), \tag{12.5}$$

and thus change (12.3) to

$$i\hbar\left(\frac{\partial\psi}{\partial t} + \frac{\lambda}{2}y\frac{\partial\psi}{\partial y}\right) = -\frac{\hbar^2}{2m}\frac{\partial^2\psi}{\partial y^2} + \frac{1}{2}m\left(\omega^2 + \frac{\lambda^2}{4}\right)y^2\psi. \tag{12.6}$$

Now we write $\psi(y,t)$ as the product of two factors [2]

$$\psi(y,t) = \phi(y)\exp\left(-\frac{im\lambda}{4\hbar}y^2 - i\omega't\right), \tag{12.7}$$

and substitute this trial solution in (12.6) to find the eigenvalue equation for $\phi(y)$,

$$\frac{-\hbar^2}{2m}\frac{d^2\phi}{dy^2} + \frac{1}{2}m\omega^2 y^2\phi = \hbar\Omega\phi. \tag{12.8}$$

In this equation Ω which is a real number is related to ω' by

$$\omega' = \Omega + \frac{i}{4}\lambda = \left(n + \frac{1}{2}\right)\omega + \frac{i\lambda}{4}, \tag{12.9}$$

and $\phi(y)$ s are the harmonic oscillator wave functions. Substituting for $\phi(y)$, y and ω' in Eq. (12.7) we obtain the eigenfunction $\psi_n(x,t)$. The solution of Eq. (12.3) for the nth excited state of the damped harmonic oscillator in terms of the original variable x can be obtained from Eqs. (12.7) and (12.8);

$$\psi_n(x,t) = N_n\exp\left\{\left[-i\omega\left(n + \frac{1}{2}\right) + \frac{\lambda}{4}\right]t - \frac{m}{\hbar}\left(\frac{i\lambda}{4} + \frac{\omega}{2}\right)x^2 e^{\lambda t}\right\}$$

$$\times H_n\left[\left(\frac{m\omega}{\hbar}\right)^{\frac{1}{2}}xe^{\frac{\lambda t}{2}}\right], \tag{12.10}$$

where

$$N_n = \frac{1}{\sqrt{2^n n!}} \left(\frac{m\omega}{\pi\hbar}\right)^{\frac{1}{4}}, \qquad (12.11)$$

and H_n is the Hermite polynomial of order n.

The eigenvalues (12.9) found for the damped oscillator are different from those obtained from the dual Hamiltonian formulation of Bateman Eq. (11.142). They also differ from the set of eigenvalues found from the "loss-energy states" which will be defined below [8],[9].

Loss-Energy States and the Symmetry of the Hamiltonian — The loss-energy states are special states with certain symmetry property. To obtain these, first we note that the Hamiltonian (12.3) has the property that

$$\hat{H}\left(t + \frac{2\pi i}{\lambda}\right) = \hat{H}(t), \qquad (12.12)$$

therefore let us study the class of solutions of the wave equation for the Hamiltonian (12.2) with $V(x) = \frac{1}{2}m\omega_0^2 x^2$ which remains invariant under the transformation

$$\psi_n(x, t + iT) = \exp\left(\varepsilon_n T\right)\psi_n(x, t), \quad \text{Im } T = 0. \qquad (12.13)$$

Dodonov and Man'ko have called these states "loss-energy states" [8],[9] and these, in some respect, are similar to Bloch states (or Floquet solution) for a pure imaginary one-dimensional lattice [10],[11]. The simplest first integrals of motion for this problem are linear functions of x and p:

$$I(t) = \eta(t)x + \xi(t)p, \qquad (12.14)$$

and satisfy the relation

$$\left[i\hbar\frac{\partial}{\partial t} - \hat{H}, I\right] = 0. \qquad (12.15)$$

From this commutator and the time-dependent Hamiltonian for the damped harmonic oscillator, Eq. (12.2), with $V(x) = \frac{1}{2}m\omega_0^2 x^2$, we find $\dot{\eta}(t)$ and $\dot{\xi}(t)$ to be,

$$\dot{\eta}(t) = m\omega_0^2 e^{\lambda t}\xi(t), \qquad (12.16)$$

$$\dot{\xi}(t) = -\frac{1}{m}e^{-\lambda t}\eta(t), \qquad (12.17)$$

where $\xi(t)$ is a solution of $\ddot{\xi} + \lambda\dot{\xi} + \omega_0^2\xi = 0$ (see Eq. (4.19)). Next we search for the first integrals of motion with the property that

$$I\left(t + \frac{2i\pi}{\lambda}\right) = \alpha I(t), \qquad (12.18)$$

where α is a constant. If we choose $\xi(t)$ to be

$$\xi(t) = \exp\left[\left(-\frac{\lambda}{2} + i\omega\right)t\right], \quad \omega = \sqrt{\omega_0^2 - \frac{\lambda^2}{4}}, \qquad (12.19)$$

then we find the following two first integrals of motion [8];

$$I_- = e^{i\omega t}\left[e^{-\frac{\lambda t}{2}}p + m\left(\frac{\lambda}{2} - i\omega\right)e^{\frac{\lambda t}{2}}x\right],$$
(12.20)

and

$$I_+ = e^{-i\omega t}\left[e^{-\frac{\lambda t}{2}}p + m\left(\frac{\lambda}{2} + i\omega\right)e^{\frac{\lambda t}{2}}x\right],$$
(12.21)

with both I_- and I_+ having the time translation invariances given by

$$I_\pm\left(t + \frac{2\pi i}{\lambda}\right) = -\exp\left(\pm\frac{2\pi\omega}{\lambda}\right)I_\pm(t).$$
(12.22)

Using I_+ and I_- we can construct the first integral $K(t)$ with the property

$$K\left(t + \frac{2\pi i}{\lambda}\right) = K(t),$$
(12.23)

and this K is related to the Hamiltonian by

$$K(t) = \frac{1}{4m}\left(I_- I_+ + I_+ I_-\right) = H(t) + \frac{\lambda}{4}(xp + px).$$
(12.24)

The eigenstates of the operator $K(t)$, are the solutions of the eigenvalue equation

$$K(t)\psi_n(x,t) = \hbar\omega\left(n + \frac{1}{2}\right)\psi_n(x,t),$$
(12.25)

and are the loss-energy states

$$\psi_n(x,t) = \frac{1}{\sqrt{2^n n!}}\left(\frac{m\omega}{\pi\hbar}\right)^{\frac{1}{4}}$$
$$\times \exp\left[\frac{1}{2}\left(\frac{\lambda}{2} - i\omega\right)t - in\omega t - \frac{m}{2\hbar}\left(\omega + \frac{i\lambda}{2}\right)e^{\lambda t}x^2\right]$$
$$\times H_n\left(\sqrt{\frac{m\omega}{\hbar}}xe^{\frac{\lambda t}{2}}\right), \quad n = 0,1,2\cdots.$$
(12.26)

These wave functions are identical to the wave functions that we found earlier for Caldirola-Kanai Hamiltonian, and by construction, they have the symmetry property

$$\psi_n\left(x, t + \frac{2\pi i}{\lambda}\right) = \psi_n(x,t)\exp\left[\left(i\pi + \frac{2\omega\pi}{\lambda}\right)\left(n + \frac{1}{2}\right)\right].$$
(12.27)

By comparing (12.13) and (12.27) we have, for the loss-energy states, the eigenvalues

$$\varepsilon_n = \hbar\left(\omega + \frac{i\lambda}{2}\right)\left(n + \frac{1}{2}\right), \quad n = 0,1,2\cdots.$$
(12.28)

and these are similar, but not identical to the eigenvalues of the damped oscillator H^+ of the Bateman dual Hamiltonian, Eq. (11.143), and to the eigenvalues found earlier.

Let us examine some of the properties of the solution of the wave equation (12.8) when $\psi_n(x,t)$ is given by (12.26). Here we note that (a) - the expectation values $\langle n|x|n\rangle$ and $\langle n|p|n\rangle$ are both equal to zero and (b) - $\psi_n(x,t)$ is not an eigenfunction of $\hat{H}$ but if we use $\psi_n(x,t)$ to calculate the expectation value of $\hat{H}$ we get

$$\langle n|\hat{H}|n\rangle = \left(n + \frac{1}{2}\right)\hbar\frac{4\omega^2 + \lambda^2}{4\omega}, \tag{12.29}$$

is a constant in time.

From Eq. (12.6) and its complex conjugate we find that [2]

$$i\hbar\left(\frac{\partial}{\partial t} + \frac{\lambda}{2}y\frac{\partial}{\partial y}\right)\psi^*\psi = -\frac{\hbar^2}{2m}\frac{\partial}{\partial y}\left(\psi^*\frac{\partial\psi}{\partial y} - \psi\frac{\partial\psi^*}{\partial y}\right). \tag{12.30}$$

Now we multiply (12.30) by $(i\hbar)^{-1}\exp\left(\frac{\lambda t}{2}\right)$ and we write the result as

$$\frac{\partial}{\partial t}\left(\psi^*\psi e^{-\frac{\lambda t}{2}}\right) + \frac{\partial}{\partial y}\left[\frac{\hbar}{2im}\left(\psi^*\frac{\partial\psi}{\partial y} - \psi\frac{\partial\psi^*}{\partial y}\right)e^{-\frac{\lambda t}{2}} + \frac{\lambda}{2}y\psi^*\psi e^{-\frac{\lambda t}{2}}\right] = 0, \tag{12.31}$$

and this relation corresponds to the equation of continuity if we define the probability density $\rho(y,t)$ and the probability current $j(y,t)$ by

$$\rho(y,t) = \left(\psi^*\psi e^{-\frac{\lambda t}{2}}\right), \tag{12.32}$$

and

$$j(y,t) = \frac{\hbar}{2im}\left(\psi^*\frac{\partial\psi}{\partial y} - \psi\frac{\partial\psi^*}{\partial y}\right)e^{-\frac{\lambda t}{2}} + \frac{\lambda}{2}y\psi^*\psi e^{-\frac{\lambda t}{2}}, \tag{12.33}$$

respectively. The probability density, $\rho(y,t)$, defined by (12.32) satisfies the normalization condition

$$\int \rho(y,t)dy = \int \psi^*(y,t)\psi(y,t)e^{-\frac{\lambda t}{2}}dy = \int \psi^*(x,t)\psi(x,t)dx = 1. \tag{12.34}$$

Wave Packet Solution for the Damped Oscillator — An interesting solution of the Schrödinger equation (12.3) for the harmonic oscillator is of the form of a wave packet [5]. Let $\xi(t)$ represent the classical solution of the equation of the damped harmonic oscillator (4.19), then the wave packet

$$\psi(x,t) = \left(\frac{m\omega_0\left(\omega_0 + \frac{i\lambda}{2}\right)e^{\lambda t}}{\pi\omega^2}\right)^{\frac{1}{4}}\exp\left[-m\left(\frac{\omega_0 + i\frac{\lambda}{2}}{2\hbar}\right)e^\lambda(x - \xi(t))^2\right.$$
$$\left. + \frac{i}{\hbar}\left\{m\dot{\xi}(t)e^{\lambda t}(x - \xi(t)) + \frac{1}{2}\hbar\omega_0 t + \int^t L\left(\xi,\dot{\xi},t\right)dt\right\}\right] \tag{12.35}$$

is a solution of (12.3) with $V(x) = \frac{1}{2}m\omega_0^2 x^2$. In this equation $L\left(\xi, \dot{\xi}, t\right)$ represents the classical Lagrangian,

$$L = \frac{1}{2}\left(m\dot{\xi}^2 - m\omega_0^2\xi^2\right)e^{\lambda t}. \tag{12.36}$$

The probability density for the wave packet (12.35) is given by

$$|\psi(x,t)|^2 = \left(\frac{m\omega_0 e^{\lambda t}}{\pi}\right)^{\frac{1}{2}}\exp\left[-\frac{m\omega_0}{\hbar}e^{\lambda t}\left(x - \xi(t)\right)^2\right], \tag{12.37}$$

and thus the center of the wave packet (12.35) oscillates exactly as that of a damped harmonic oscillator, i.e. $\xi(t)$. From the wave packet (12.35) we can also calculate the expectation values of $\langle x \rangle$, $\langle p_{mech} \rangle$, $\langle p_{mech}^2 \rangle$ and $\langle x^2 \rangle$, where p_{mech} denotes the mechanical (or kinematical) momentum. These are given by

$$\langle x \rangle = \langle \psi(x,t)|x|\psi(x,t) \rangle = 0, \tag{12.38}$$

$$\langle p_{mech} \rangle = \left\langle \psi(x,t)\left|-i\hbar\frac{\partial}{\partial x}\right|\psi(x,t)\right\rangle = 0, \tag{12.39}$$

$$\langle x^2 \rangle = \langle \psi(x,t)|x^2|\psi(x,t) \rangle = \left(\frac{\hbar\omega^2}{2m\omega_0^3}\right)e^{-\lambda t}, \tag{12.40}$$

$$\langle p_{mech}^2 \rangle = \left\langle \psi(x,t)\left|-\hbar^2\frac{\partial^2}{\partial x^2}\right|\psi(x,t)\right\rangle = \left(\frac{\hbar\omega^2}{2\omega_0}\right)e^{-\lambda t}. \tag{12.41}$$

The total energy of the wave packet is obtained from the last two relations

$$\frac{1}{2m}\langle p_{mech}^2 \rangle + \frac{1}{2}\omega_0^2\langle x^2 \rangle = \frac{\hbar\omega^2}{\omega_0}e^{-\lambda t}, \tag{12.42}$$

and this total energy tends to zero as $t \to \infty$.

The canonical momentum space wave function corresponding to the wave packet Eq. (12.35) is found from the Fourier transform of $\psi(x,t)$;

$$\phi(p,t) = \left(\frac{m\omega_0}{\pi}e^{\lambda t}\right)^{\frac{1}{4}}$$

$$\times \exp\left[-\frac{e^{-\lambda t}}{2m\hbar\left(\omega_0 + i\frac{\lambda}{2}\right)}\left(p - \eta(t)\right)^2 - \frac{i}{\hbar}\left\{p\xi(t) - \int^t L dt + \frac{1}{2}\omega_0 t\right\}\right], \tag{12.43}$$

where $\eta(t) = m\dot{\xi}(t)e^{\lambda t}$. Again the probability density in momentum space is given by

$$|\phi(p,t)|^2 = \left(\frac{m\omega_0}{\pi}e^{\lambda t}\right)^{\frac{1}{2}}\exp\left[-\frac{m\omega_0}{\hbar}e^{\lambda t}\left(p - m\dot{\xi}(t)e^{\lambda t}\right)\right], \tag{12.44}$$

and from this last relation it follows that the center of this wave packet moves as $m\dot{\xi}(t)e^{\lambda t}$. We note that according to Eqs. (12.40) and (12.41) the expectation values of $\langle p^2_{mech} \rangle$ and $\langle x^2 \rangle$ both tend to zero as $t \to \infty$, and the wave function in the limit of $t \to \infty$ collapses to a δ-function. Thus after a long time both the position and the mechanical momentum of the particle can be measured precisely, and in this limit the uncertainty principle will be violated. We can reach this conclusion directly by starting with the commutation relation which is

$$[x, p] = i\hbar. \tag{12.45}$$

Now if we write this commutation relation in terms of the mechanical or kinematical momentum operator $m\dot{x}$, we have

$$\left[x, m\dot{x}e^{\lambda t}\right] = i\hbar, \tag{12.46}$$

or

$$[x, \dot{x}] = \frac{i\hbar}{m}e^{-\lambda t}. \tag{12.47}$$

From the commutation relation (12.47) we can derive the uncertainty relation exactly as it is done for conservative systems. Thus if Δx and $\Delta \dot{x}$ represent the uncertainties in the position and the velocity of the particle then

$$\Delta x \Delta \dot{x} \geq \frac{\hbar}{2m}e^{-\lambda t}. \tag{12.48}$$

This expression shows that as $t \to \infty$ one can, in principle, measure the position and the velocity of the particle with arbitrary accuracy, and this is not an acceptable result.

By studying the propagation of the wave packet associated with the motion of the particle Caldirola has shown that the presence of friction reduces the product of the uncertainties $\Delta x \Delta \dot{x}$ shown in (12.48) compared to the corresponding values when $\lambda = 0$, and he has concluded that the Heisenberg uncertainty relation remains valid [12]. Let us note that the problem is not only the violation of the uncertainty principle, but the fact that $H(x, p, t)$ (12.1) is not the energy of the particle and it does not give us the correct equation for the time evolution of the momentum operator.

Motion of a Particle in a Viscous Medium — In addition to the damped harmonic oscillator, analytic solutions can be found (a) - for the motion of a particle moving in a viscous medium, but in the absence of a conservative force, and (b) - when the external force is just a function of time t, e.g. a force like $-mg(t)$.

(a) For the first case the wave function in momentum space is given by [5]

$$\phi(p) = \left(\frac{a_0}{\pi}\right)^{\frac{1}{4}}$$

$$\times \exp\left[-\frac{a_0 m\lambda + i\hbar\left(1 - e^{-\lambda t}\right)}{2\hbar^2 m\lambda}\left(p - m\dot{\xi}\right)^2 - \frac{i}{\hbar}\left\{p\xi - \int^t Ldt\right\}\right],$$

$$\tag{12.49}$$

where L is the Lagrangian for a particle in viscous medium, a_0 is a constant and $\xi(t)$ is the solution of

$$m\ddot{\xi} + m\lambda\dot{\xi} = 0. \tag{12.50}$$

In this case for $t < \frac{1}{\lambda}$ the width of the wave packet grows as

$$\left| a_0 + \frac{i\hbar t}{m} \right|^{-1}, \tag{12.51}$$

which is like the increase in the width when there is no dissipation. However for larger values of t the width of the wave packet becomes constant.

(b) The same wave function $\phi(p)$ Eq. (12.49) can be used when the particle is subject to a force which is constant in space but is a function of time, $-mg(t)$, with the provision that both L and $\xi(t)$ should be replaced by the corresponding functions for this motion.

12.2   Quantization of a System with Variable Mass

In Sec. 3.6 we studied the classical Hamiltonian formulation for the motion of a particle with variable mass, and now we want to consider the possibility of quantizing such a motion. Since there we assumed mc^2 (with $c = 1$) to be the variable mass of the particle (in units of energy) with its conjugate θ, therefore in quantum theory m should be regarded as a dynamical variable i.e. a "mass operator" [13],[14]. If we work in a representation where m is diagonal then θ becomes a differential operator. However there are two points that we must keep in mind:

(1) According to a theorem by Bargman, different mass states in non-relativistic quantum theory cannot be superimposed (super-selection rule), but the theorem is not applicable to this problem [15].

(2) Since we want to have positive values of m, therefore its conjugate, $\theta = i\hbar\frac{\partial}{\partial m}$ will not be a self-adjoint operator in the domain $0 \leq m < \infty$ exactly as in the case of the time and the energy operators [16]–[18].

We write the quantized version of the classical Hamiltonian, Eq. (3.74) as

$$H = -\frac{\hbar^2}{2m}\frac{\partial^2}{\partial x^2} + \frac{1}{2}m\omega_0^2 x^2 + \frac{\lambda\hbar}{i}\left(m\frac{\partial}{\partial m} + \frac{1}{2} \right), \tag{12.52}$$

where we have used the symmetrized form of the operator $m\theta$;

$$m\theta \to \frac{1}{2}(m\theta + \theta m) = i\hbar\left(m\frac{\partial}{\partial m} + \frac{1}{2} \right), \tag{12.53}$$

then the Schrödinger equation for the Hamiltonian (12.52) is given by

$$H(x, m, t)\psi(x, m, t) = i\hbar\frac{\partial\psi(x, m, t)}{\partial t}. \tag{12.54}$$

In order to separate the variables we define a quantity with the dimension of length $L(m)$

$$L(m) = \left(\frac{\hbar}{m\omega_0}\right)^{\frac{1}{2}}, \tag{12.55}$$

and also introduce a quantity with the dimension dimensionless variable $y = x/L(m)$. Using these we write the wave equation as

$$H\psi = \left\{\frac{\hbar\omega_0}{2}\left(-\frac{\partial^2}{\partial y^2} + y^2 + \frac{\lambda}{i\omega_0}y\frac{\partial}{\partial y}\right) + \frac{\lambda\hbar}{i}\left(m\frac{\partial}{\partial m} + \frac{1}{2}\right)\right\}\psi = i\hbar\frac{\partial\psi}{\partial t}. \tag{12.56}$$

This differential equation is separable and has a solution of the form

$$\psi(y, m, t) = Y(y)M(m)\exp\left(-\frac{i\mathcal{E}t}{\hbar}\right), \tag{12.57}$$

where $Y(y)$ and $M(m)$ satisfy the two differential equations

$$\left(-\frac{d^2}{dy^2} + y^2 + \frac{\lambda}{i\omega_0}y\frac{d}{dy}\right)Y(y) = 2LY(y), \tag{12.58}$$

and

$$\left(m\frac{d}{dm} + \frac{1}{2}\right)M(m) = \frac{i}{2}\beta M(m). \tag{12.59}$$

In (12.57) $\mathcal{E}$ is related to the separation constant β by

$$\mathcal{E} = \hbar\left(\omega_0 L + \frac{\lambda}{2}\beta\right). \tag{12.60}$$

Equation (12.59) can be integrated to yield the result

$$M = \frac{1}{\sqrt{4\pi m}}\left(\frac{m}{m_0}\right)^{\frac{i\beta}{2}}, \tag{12.61}$$

with m_0 being the integration constant. We can eliminate the first order derivative in (12.58) by writing

$$Y(y) = \exp\left(-\frac{i\lambda}{2\omega}y\right)U(y), \tag{12.62}$$

and substituting for $Y(y)$ in (12.58) to find the equation for $U(y)$;

$$\left[-\frac{d^2}{dy^2} + \left(\frac{\omega^2}{\omega_0^2}\right)y^2\right]U(y) = \left(2L + \frac{\lambda}{2\omega_0}\right)U(y), \tag{12.63}$$

where $\omega^2 = \omega_0^2 - \frac{\lambda^2}{4}$. This equation can be solved in terms of the Hermite polynomials:

$$U(u) = N_n H_n(u)\exp\left(-\frac{u^2}{2}\right), \tag{12.64}$$

where the new variable u is given by

$$u = \left(\frac{\omega_0}{\omega}\right)^{\frac{-1}{2}} y = \left(\frac{m\omega}{\hbar}\right)^{\frac{1}{2}} x, \tag{12.65}$$

and N_n is the normalization constant. For the eigenvalues of Eq. (12.63) we have

$$2L + \frac{\lambda}{2i\omega_0} = 2\left(\frac{\omega}{\omega_0}\right)\left(n + \frac{1}{2}\right), \tag{12.66}$$

and thus the eigenvalues of H in (12.56) are

$$\mathcal{E}(n,\beta) = \left(n + \frac{1}{2}\right)\hbar\omega + \frac{\beta\hbar\lambda}{2} + \frac{i\hbar\lambda}{4}. \tag{12.67}$$

Since β is the separation constant and is arbitrary, the complete solution is found by superimposing different $\psi(x,m,t,\beta)$ s;

$$\psi(x,m,t) = \sum_n \int \frac{A_n(\beta)}{\sqrt{4\pi m}} \left(\frac{m}{m_0}\right)^{\frac{i\beta}{2}} \exp\left[-\frac{m}{2\hbar}\left(\omega + \frac{i\lambda}{2}\right)x^2\right]$$
$$\times N_n H_n \left[\left(\frac{m\omega}{\hbar}\right)^{\frac{1}{2}} x\right] \exp\left[-\frac{i\mathcal{E}(n,\beta)t}{\hbar}\right] d\beta. \tag{12.68}$$

Let $\tilde{A}_n(\xi)$ be the Fourier transform of $A_n(\beta)$, i.e.

$$\tilde{A}_n(\xi) = \frac{1}{\sqrt{2\pi}} \int A_n(\beta)e^{i\beta\xi}d\beta, \tag{12.69}$$

then for a given n we have

$$\frac{1}{\sqrt{m}} \int A_n(\beta) \left(\frac{m}{m_0}\right)^{\frac{i\beta}{2}} \exp\left(-i\frac{\beta\lambda t}{2}\right) d\beta = \frac{\tilde{A}_n}{\sqrt{2\pi m}} \left[\frac{1}{2}\ln\left(\frac{m}{m_0}\right) - \frac{\lambda t}{2}\right]. \tag{12.70}$$

This result shows that the center of each wave packet for fixed n moves about the classical value according to the relation $m = m_0 e^{\lambda t}$.

For the case when $A_n(\beta)$ is constant and is equal to $\frac{1}{\sqrt{m_0}}$ we find

$$\int \frac{L(m)}{\sqrt{mm_0}} dy\, dm \int \exp\left\{i\beta\left[\frac{1}{2}\ln\left(\frac{m}{m_0}\right) - \frac{\lambda t}{2}\right]\right\} d\beta$$
$$= 2\pi \left(\frac{\hbar}{m_0\omega_0}\right)^{\frac{1}{2}} \int \frac{1}{m}\delta\left[\frac{1}{2}\ln\left(\frac{m}{m_0}\right) - \frac{\lambda}{2}\right] dy\, dm$$
$$= 4\pi L(m_0) \int \delta\left(m - m_0 e^{\lambda t}\right) dy\, dm. \tag{12.71}$$

Thus by replacing m by $m_0 e^{\lambda t}$ in (12.68) we find the wave function for the damped oscillator we obtained earlier from the solution of the time-dependent Schrödinger equation (12.10).

12.3 Evolution Operator Method for the Variable Mass Harmonic Oscillator

An interesting method to solve the problem of harmonic oscillator with time-dependent mass can be given in terms of the evolution operator for the Schrödinger equation when the Hamiltonian is of the form

$$H(t) = a_1(t)J_+ + a_2(t)J_0 + a_3(t)J_-. \tag{12.72}$$

In this Hamiltonian J_+, J_0, and J_- are the generators of SU_2 group and satisfy the Lie algebra :

$$[J_+, \ J_-] = 2J_0, \tag{12.73}$$

$$[J_0, \ J_\pm] = \pm J_\pm, \tag{12.74}$$

and $a_1(t)$, $a_2(t)$ and $a_3(t)$ are arbitrary functions of time [19],[20]. The time-dependent Schrödinger equation

$$H(t)\psi(t) = i\hbar \frac{\partial}{\partial t}\psi(t), \tag{12.75}$$

can be integrated with the result that

$$\psi(t) = U(t,0)\psi(0), \tag{12.76}$$

where $\psi(0)$ is the initial wave function and $U(t,0)$ is the evolution operator which is the solution of the differential equation

$$H(t)U(t,0) = i\hbar \frac{\partial}{\partial t}U(t,0), \tag{12.77}$$

with the initial condition $U(0,0) = 1$.

Since J_+, J_0 and J_- form a closed Lie algebra, the evolution operator can be expressed as a product of the exponentials

$$U(t,0) = \exp[c_1(t)J_+] \ \exp[c_2(t)J_0] \ \exp[c_3(t)J_-], \tag{12.78}$$

where the functions $c_1(t)$, $c_2(t)$ and $c_3(t)$ are to be determined from the Schrödinger equation (12.75). Thus by substituting for $U(t,0)$ from (12.78) in (12.77) we obtain

$$\frac{\partial}{\partial t}U(t.0) = \left\{ \left[\dot{c}_1 - c_1\dot{c}_2 - c_1^2 \exp(-c_2)\dot{c}_3\right] J_+ + \left[\dot{c}_2 + 2c_1 \exp(-c_2)\dot{c}_3\right] J_0 \right\} U(t,0)$$

$$+ \left[\exp(-c_2)\dot{c}_3 J_-\right] U(t,0) = \frac{-i}{\hbar}H(t), \tag{12.79}$$

where dots denote time derivatives of c_i s. In this and the following equations we will suppress the time-dependence of $c_i(t)$ s and $a_i(t)$ s. Now by substituting for $H(t)$ from Eq. (12.72) in (12.79) and by matching the coefficients of J_+, J_0

and J_- on the two sides, we find the following set of first order differential equations for the coefficients c_1, c_2 and c_3:

$$i\hbar \left[\dot{c}_1 - c_1 \dot{c}_2 - c_1^2 \exp(-c_2)\dot{c}_3 \right] = a_1, \tag{12.80}$$

$$i\hbar \left[\dot{c}_2 + 2c_1 \exp(-c_2)\dot{c}_3 \right] = a_2, \tag{12.81}$$

$$i\hbar \exp(-c_2)\dot{c}_3 = a_3. \tag{12.82}$$

The last three equations can also be written as

$$i\hbar \dot{c}_1 = a_1 + a_2 c_1 - a_3 c_1^2, \tag{12.83}$$

$$i\hbar \dot{c}_2 = a_2 - 2a_3 c_1, \tag{12.84}$$

$$i\hbar \dot{c}_3 = a_3 \exp(c_2). \tag{12.85}$$

The initial conditions for these differential equations are found from $U(0,0) = 1$ and are given by

$$c_1(0) = c_2(0) = c_3(0) = 0. \tag{12.86}$$

Once we have found the solution of the Riccati equation (12.80) for $c_1(t)$, we can obtain the solutions of (12.81) and (12.82) by quadrature, i.e. from

$$i\hbar c_2(t) = \int_0^t [a_2(t') - 2a_3(t') c_1(t')] \, dt', \tag{12.87}$$

$$i\hbar c_3(t) = \int_0^t a_3(t') \exp[c_2(t')] \, dt'. \tag{12.88}$$

This completes the solution of the time-dependent Schrödinger equation (12.75).
Evolution Operator Method for the Variable Mass Oscillator — Let us consider the system with the Hamiltonian

$$H(t) = \frac{1}{2m(t)}p^2 + \frac{1}{2}m(t)\omega^2 x^2, \tag{12.89}$$

in which the spring constant $K = m(t)\omega^2$ is also time dependent. We can relate (12.89) to the H given by (12.72) if we define J_+, J_0, J_- and $a_1(t)$, $a_2(t)$ and $a_3(t)$ by the operators

$$J_+ = \frac{1}{2\hbar}x^2, \tag{12.90}$$

$$J_- = \frac{1}{2\hbar}p^2, \tag{12.91}$$

$$J_0 = \frac{i}{4\hbar}(xp + px), \tag{12.92}$$

and

$$a_1(t) = \hbar m(t)\omega^2, \tag{12.93}$$

$$a_2(t) = 0, \tag{12.94}$$

$$a_3(t) = \frac{\hbar}{m(t)}. \tag{12.95}$$

By integrating Eqs. (12.80)-(12.82) when $a_i(t)$ s are given by (12.93)-(12.95), we find $c_1(t)$, $c_2(t)$ and $c_3(t)$ to be

$$c_1(t) = im(t)\frac{\partial}{\partial t}[\ln u(t)], \tag{12.96}$$

$$c_2(t) = -2\ln\frac{u(t)}{u(0)}. \tag{12.97}$$

and

$$c_3(t) = -iu^2(0)\int_0^t \frac{dt'}{m(t')\,u^2(t')}. \tag{12.98}$$

In the last three equations $u(t)$ denotes the solution of the differential equation

$$\ddot{u} + \lambda(t)\dot{u} + \omega_0^2 u = 0, \tag{12.99}$$

where

$$\lambda(t) = \frac{\partial}{\partial t}[\ln m(t)]. \tag{12.100}$$

For the special case of $m(t) = m(0)e^{\lambda t}$ we have $\lambda(t) = \lambda$, and Eq. (12.99) becomes the classical equation of motion of a damped harmonic oscillator, Eq. (1.76). Since the time-dependent functions $c_1(t)$, $c_2(t)$ and $c_3(t)$ are all known then the operator $U(t,0)$ can be determined from (12.78), and the time-dependent wave function from (12.76).

For this special case $c_i(t)$ s can be found analytically for underdamped, critically damped and overdamped cases respectively [19]. For instance for the underdamped case we find

$$c_1(t) = -im\omega_0^2\frac{2\exp(\lambda t)}{\lambda + 2\omega\cot(\omega t)}, \tag{12.101}$$

$$c_2(t) = \lambda t - 2\ln\left[\left(\frac{\lambda}{2\omega}\right)\sin(\omega t) + \cos(\omega t)\right], \tag{12.102}$$

and

$$c_3(t) = -\frac{i}{m}\left(\frac{2}{\lambda + 2\omega\cot(\omega t)}\right). \tag{12.103}$$

where $\omega = \sqrt{\omega_0^2 - \frac{\lambda^2}{4}}$.

We will construct the coherent and the squeezed wave function for the Hamiltonian (12.89) in Secs. 13.3-13.5.

12.4 The Schrödinger-Langevin Equation for a Charged Particle Moving in an External Electromagnetic Field

In this section we want to derive a wave equation which describes the motion of a charged particle in the presence of damping while it is moving in an external electromagnetic field.

We start with the general form of the Schrödinger equation for such a motion [21]

$$ i\hbar \frac{\partial \psi}{\partial t} = \left[\frac{1}{2m} \sum_{k=1}^{3} \left(-i\hbar \frac{\partial}{\partial x_k} - \frac{e}{c} A_k \right)^2 + V + e\phi + V_L \right] \psi, \qquad (12.104) $$

where V_L represents the damping term to be specified later. We know that the classical motion for this particle is given by the equation

$$ m\ddot{x}_k = -\frac{\partial V}{\partial x_k} - e\frac{\partial \phi}{\partial x_k} - \frac{e}{c}\frac{\partial A_k}{\partial t} - \lambda \dot{x}_k + \frac{e}{c} \sum_{j=1}^{3} \left(\frac{\partial A_j}{\partial x_k} - \frac{\partial A_k}{\partial x_j} \right) \dot{x}_j, \quad k = 1,2,3. $$

$$ (12.105) $$

In the absence of damping, Pauli has shown that (12.104) and (12.105) are related to each other [22]. Here we follow Pauli's method to find the damping potential which we assume to be real, a functional of ψ and ψ^* and furthermore makes the Schrödinger equation (12.104), the quantum analogue of the classical equation (12.105). First we define the current **J** with components J_1, J_2 and J_3, where

$$ J_k = \frac{1}{2m} \left[\psi^* \left(-i\hbar \frac{\partial}{\partial x_k} - \frac{e}{c} A_k \right) \psi - \psi \left(-i\hbar \frac{\partial}{\partial x_k} + \frac{e}{c} A_k \right) \psi^* \right], \qquad (12.106) $$

and then from (12.104) and its complex conjugate we obtain the equation of continuity

$$ \frac{\partial}{\partial t} (\psi^* \psi) + \nabla \cdot \mathbf{J} = 0. \qquad (12.107) $$

Next we differentiate J_k with respect to time to obtain

$$ m \frac{\partial J_k}{\partial t} = \frac{1}{2} \left[(H\psi)^* \frac{\partial \psi}{\partial x_k} - \psi^* \frac{\partial}{\partial x_k} (H\psi) + H\psi \frac{\partial \psi^*}{\partial x_k} - \psi \frac{\partial}{\partial x_k} (H\psi^*) \right], \qquad (12.108) $$

where H is the Hamiltonian used in the Schrödinger equation (12.104). Simplifying (12.108) we get

$$ m \frac{\partial J_k}{\partial t} = -\sum_{j=1}^{3} \frac{\partial T_{kj}}{\partial x_j} + \left(-\frac{\partial V}{\partial x_k} - \frac{\partial V_L}{\partial x_k} - \frac{e}{c}\frac{\partial A_k}{\partial t} \right) \psi^* \psi + \frac{e}{c} \sum_{j=1}^{3} B_{kj} J_j, \qquad (12.109) $$

where in Eq. (12.109) T_{kj} is the symmetric tensor given by

$$
\begin{aligned}
T_{kj} = \frac{\hbar^2}{4m} &\left[\left(-\psi^* \frac{\partial^2 \psi}{\partial x_j \partial x_k} - \psi \frac{\partial^2 \psi^*}{\partial x_j \partial x_k} + \frac{\partial \psi}{\partial x_j} \frac{\partial \psi^*}{\partial x_k} + \frac{\partial \psi}{\partial x_k} \frac{\partial \psi^*}{\partial x_j} \right) \right. \\
&\left. + \frac{2ie}{\hbar c} \left\{ A_k \left(\psi^* \frac{\partial \psi}{\partial x_j} - \psi \frac{\partial \psi^*}{\partial x_j} \right) + A_j \left(\psi^* \frac{\partial \psi}{\partial x_k} - \psi \frac{\partial \psi^*}{\partial x_k} \right) \right\} \right] \\
&+ \frac{e^2}{mc^2} A_k A_j \psi^* \psi,
\end{aligned}
\tag{12.110}
$$

and B_{kj} is the antisymmetric tensor defined in terms of the curl**A**;

$$
B_{jk} = \frac{\partial A_j}{\partial x_k} - \frac{\partial A_k}{\partial x_j}.
\tag{12.111}
$$

Now we find the time derivative of the expectation value of x_k

$$
\frac{d}{dt} \langle x_k \rangle = \frac{d}{dt} \int \psi^* x_k \psi d^3 x = \int \left\{ \frac{\partial \psi^*}{\partial t} x_k \psi + \frac{\partial \psi}{\partial t} x_k \psi^* \right\} d^3 x,
\tag{12.112}
$$

where $d^3 x$ is the element of volume. Again by substituting for $\frac{\partial \psi^*}{\partial t}$ and $\frac{\partial \psi}{\partial t}$ from the Schrödinger equation we find

$$
\frac{d}{dt} \langle x_k \rangle = \int J_k d^3 x,
\tag{12.113}
$$

and from this we calculate $\frac{d^2 \langle x_k \rangle}{dt^2}$

$$
\frac{d^2}{dt^2} \langle x_k \rangle = \int \frac{\partial J_k}{\partial t} d^3 x.
\tag{12.114}
$$

By combining (12.109) and (12.114) we have

$$
\frac{d^2}{dt^2} \langle x_k \rangle = \int \left[-\left(\frac{\partial V}{\partial x_k} + \frac{\partial V_L}{\partial x_k} + \frac{e}{c} \frac{\partial A_k}{\partial t} \right) \psi^* \psi + \frac{e}{c} \sum_{j=1}^{3} B_{kj} J_j \right] d^3 x,
\tag{12.115}
$$

where we have used the divergence theorem,

$$
\int \sum_{j=1}^{3} \frac{\partial T_{jk}}{\partial x_j} d^3 x = 0,
\tag{12.116}
$$

to get the simplified equation (12.115). By comparing Eqs. (12.115) and (12.105) and noting that $\frac{d^2}{dt^2} \langle x_k \rangle$ is given by (12.115), we find that the two are similar in structure if the components of the dissipative force ∇V_L have the following expectation values:

$$
\int \psi^* \frac{\partial V_L}{\partial x_k} \psi d^3 x = \lambda \int \left[\frac{-i\hbar}{2} \left(\psi^* \frac{\partial \psi}{\partial x_k} - \psi \frac{\partial \psi^*}{\partial x_k} \right) - \frac{e}{c} A_k \psi^* \psi \right] d^3 x.
\tag{12.117}
$$

This equation is satisfied for an arbitrary wave function provided that V_L which is a real function is given by

$$V_L = \frac{\lambda \hbar}{2i} \ln \left(\frac{\psi}{\psi^*} \right) - \frac{\lambda e}{c} \sum_{k=1}^{3} \int A_k dx_k. \tag{12.118}$$

The nonlinear differential equation (12.104) with V_L given by (12.118) apart for a time-dependent term which comes from normalization of the wave function is called the Scrödinger-Langevin equation [23]-[25]. This equation was first found by Kostin who used the Heisenberg equation of motion

$$\dot{P} + \lambda P + \omega_0^2 Q = 0, \quad \dot{Q} = P, \tag{12.119}$$

for the position and momentum of a damped harmonic oscillator, Eqs. (9.49)-(9.50) to derive this equation [23]. As we will see the equations of motion for Q and P, Eqs. (12.119), have the same form in classical and in quantum mechanics, the quantal equations were derived for a harmonic oscillator interacting with a bath of oscillators composed of harmonic oscillators by Ford *et al.* [26]. Later we will see that we can derive the Schrödinger-Langevin equation directly from the modified form of the Hamilton-Jacobi equation.

The presence of the term

$$\left(-\frac{\lambda e}{c} \sum_k \int A_k dx_k \right), \tag{12.120}$$

in Eq. (12.118) shows that the minimal coupling scheme for the Schrödinger-Langevin equation does not work, i.e. there is an additional term arising from the coupling between the electromagnetic potential and the damping force. If we write the Schrödinger-Langevin equation as

$$i\hbar \frac{\partial \psi}{\partial t} = \frac{1}{2} \left[-i\hbar \nabla - \frac{e}{c} \mathbf{A}(\mathbf{r}, t) \right]^2 \psi + (V(\mathbf{r}) + e\phi(\mathbf{r}, t)) \psi$$
$$+ \frac{\lambda \hbar}{2i} \left[\ln \frac{\psi}{\psi^*} \right] \psi - \left(\frac{\lambda e}{c} \int^{\mathbf{r}} \mathbf{A}(\mathbf{r}', t) \cdot d\mathbf{r}' \right) \psi \tag{12.121}$$

then we can easily see that (12.121) remains invariant under the usual gauge transformation

$$\mathbf{A} \to \mathbf{A}' = \mathbf{A} + \nabla \chi, \tag{12.122}$$

$$\phi \to \phi' = \phi - \frac{1}{c} \frac{\partial \chi}{\partial t}, \tag{12.123}$$

and

$$\psi \to \psi' = \exp \left(\frac{ie\chi}{\hbar c} \right) \psi, \tag{12.124}$$

where $\chi(\mathbf{r}, t)$ is an arbitrary function of $\mathbf{r}$ and t.

In the presence of a magnetic field with the vector potential $\mathbf{A}(\mathbf{r}, t)$, the solution of (12.121) is related to the solution in the absence of the field by the relation

$$\psi(\mathbf{r}, t) = \psi(\mathbf{r}, t, \mathbf{A} = 0) \exp\left[\frac{ie}{\hbar c} \int_{-}^{\mathbf{r}} \mathbf{A}\left(\mathbf{r}', t\right) \cdot d\mathbf{r}'\right]. \tag{12.125}$$

12.5 Extensions of the Madelung Hydrodynamical Formulation of Wave Mechanics

Consider the one-dimensional time-dependent Schrödinger equation and let us write the wave function in terms of a variable amplitude $A(x, t)$ and a variable phase function $S(x, t)$, i.e.

$$\psi(x, t) = A(x, t) \exp\left[\frac{iS(x, t)}{\hbar}\right]. \tag{12.126}$$

Substituting this wave function in the Schrödinger equation and equating the real and imaginary parts we find

$$\frac{\partial S}{\partial t} + \frac{1}{2m}\left(\frac{\partial S}{\partial x}\right)^2 + V = \left(\frac{\hbar^2}{2mA}\right)\frac{\partial^2 A}{\partial x^2}, \tag{12.127}$$

and

$$m\frac{\partial A}{\partial t} + \left(\frac{\partial A}{\partial x}\right)\left(\frac{\partial S}{\partial x}\right) + \frac{1}{2}A\left(\frac{\partial^2 S}{\partial x^2}\right) = 0. \tag{12.128}$$

By taking the derivative of (12.127) with respect to x we obtain the equation of motion of the probability fluid [27]-[30], whereas Eq. (12.128) expresses the equation of continuity.

Madelung Hydrodynamical Equations with Linear Damping — In order to write Eqs. (12.127) and (12.128) in a more familiar form used in hydrodynamics we introduce the probability density for the particle position by

$$\rho(x, t) = |\psi(x, t)|^2 = A^2(x, t), \tag{12.129}$$

and the corresponding velocity field $v(x, t)$ by

$$v(x, t) = \frac{\partial S(x, t)}{\partial x}. \tag{12.130}$$

Then by replacing $A(x, t)$ and $S(x, t)$ in (12.127) and (12.128) in terms of $\rho(x, t)$ and $v(x, t)$ we get a system of two evolution equations:

$$\frac{\partial}{\partial t}\rho(x, t) + \frac{\partial}{\partial x}[\rho(x, t)v(x, t)] = 0 \tag{12.131}$$

and

$$\frac{\partial}{\partial t}v(x,t) + v(x,t)\frac{\partial}{\partial x}v(x,t) = -\frac{1}{m}\frac{dV(x)}{dx} - \frac{1}{\rho(x,t)}\frac{\partial}{\partial x}T_q(x,t), \qquad (12.132)$$

where $T_q(x,t)$ is given by

$$T_q(x,t) = \frac{\hbar^2}{2m}\left\{ \frac{1}{\rho(x,t)}\left[\frac{\partial}{\partial x}\rho(x,t)\right]^2 - \frac{\partial^2}{\partial x^2}\rho(x,t)\right\}. \qquad (12.133)$$

In the classical limit $T_q(x,t) \to 0$, and $\rho(x,t)$ and $v(x,t)$ tend to their classical counterparts $\rho_c(x,t)$ and $v_c(x,t)$. For the case of linear damping, the classical damping force is given by $F_f = -\lambda v_c$ and this force, in the classical equation of motion is added to the force $-\frac{1}{m}\frac{dV(x)}{dx}$. In hydrodynamical formulation, the total force $-\frac{1}{m}\frac{dV(x)}{dx} - \lambda v(x,t)$ replaces $-\frac{1}{m}\frac{dV(x)}{dx}$ in (12.132). Thus while the conservation of probability (12.131) retains its form, the evolution equation for $v(x,t)$ takes the form of [31]

$$\frac{\partial}{\partial t}v(x,t) + v(x,t)\frac{\partial}{\partial x}v(x,t) = -\frac{1}{m}\frac{dV(x)}{dx} - \lambda v(x,t) - \frac{1}{\rho(x,t)}\frac{\partial}{\partial x}T_q(x,t), $$
$$(12.134)$$

Now we ask whether it is possible to find a wave equation (or equations) which when formulated in hydrodynamical model satisfies both equations (12.131) and (12.134). To satisfy the former, from (12.129) we have

$$\psi(x,t) = \sqrt{\rho(x,t)}\exp\left[\frac{im}{\hbar}\int_{x_0}^{x}v(q,t)dq\right], \qquad (12.135)$$

where x_0 is an arbitrary fixed coordinate. The product $\rho(x,t)v(x,t)$ also calculated from (12.135) yields the following equation

$$\rho(x,t)v(x,t) = \frac{\hbar}{2im}\left[\psi^*(x,t)\frac{\partial}{\partial x}\psi(x,t) - [\psi(x,t)\frac{\partial}{\partial x}\psi^*(x,t)]\right] \qquad (12.136)$$

Combining the three equations (12.129), (12.131) and (12.136) we find that

$$i\hbar\left(\psi^*\frac{\partial\psi}{\partial t} + \psi\frac{\partial\psi^*}{\partial t}\right) = -\frac{\hbar^2}{2m}\left(\psi^*\frac{\partial^2\psi}{\partial t^2} + \psi\frac{\partial^2\psi^*}{\partial t^2}\right) \qquad (12.137)$$

This equation is equivalent to the following two partial differential equations:

$$i\hbar\frac{\partial}{\partial t}\psi(x,t) = -\frac{\hbar^2}{2m}\frac{\partial^2}{\partial x^2}\psi(x,t) + \mathcal{W}(x,t)\psi(x,t), \qquad (12.138)$$

$$i\hbar\frac{\partial}{\partial t}\psi^*(x,t) = \frac{\hbar^2}{2m}\frac{\partial^2}{\partial x^2}\psi^*(x,t) + \mathcal{W}(x,t)\psi^*(x,t), \qquad (12.139)$$

where $\mathcal{W}(x.t)$ is a real quantity to be determined later. Now from Eqs. (12.129), (12.135), (12.138) and (12.139) we calculate the time derivative of

$$\left(\psi^* \frac{\partial \psi}{\partial x} - \psi \frac{\partial \psi^*}{\partial x} \right) \tag{12.140}$$

i.e.

$$\frac{\hbar}{2im} \frac{\partial}{\partial t} \left(\psi^* \frac{\partial \psi}{\partial x} - \psi \frac{\partial \psi^*}{\partial x} \right) = \frac{1}{m} \rho \left(-\frac{dV(x)}{dx} \right)$$
$$+ \frac{\partial}{\partial x} \left\{ \frac{\hbar^2}{4m^2} \left[\frac{\partial^2 \rho}{\partial x^2} - \frac{1}{\rho} \left(\frac{\partial \rho}{\partial x} \right)^2 \right] - \rho v^2(x, t) \right\}. \tag{12.141}$$

The left-hand side of (12.141) according to (12.136) is the time derivative of $\rho(x, t)v(x, t)$. We can also calculate the time derivative of $\rho(x, t)v(x, t)$ from Eqs. (12.131) and (12.132). Thus we get

$$\frac{\partial}{\partial t} [\rho(x, t)v(x, t)] = -\lambda v(x, t) - \frac{1}{m} \frac{dV(x)}{dx} \rho(x, t)$$
$$- \frac{\partial}{\partial x} \left[T_q(x, t) + \rho(x, t)v^2(x, t) \right]. \tag{12.142}$$

Thus from (12.138) and (12.139) it follows that

$$-\frac{\partial}{\partial x} \mathcal{W} = -\lambda \frac{\hbar}{2i} \left[\frac{1}{\psi(x, t)} \frac{\partial}{\partial x} \psi(x, t) - \psi^*(x, t) \frac{\partial}{\partial x} \psi^*(x, t) \right] - \frac{dV(x)}{dx} \rho(x, t)$$
$$= -\lambda \frac{\hbar}{2i} \frac{\partial}{\partial x} \left[\ln \left(\frac{\psi(x, t)}{\psi^*(x, t)} \right) \right] - \frac{d}{dx} V(x), \tag{12.143}$$

or

$$\mathcal{W}(x, t) = \frac{\lambda \hbar}{2i} \ln \left(\frac{\psi(x, t)}{\psi^*(x, t)} \right) + V(x) + U(t), \tag{12.144}$$

where $U(t)$ is an arbitrary function of time. Equation (12.138) with $\mathcal{W}(x, t)$ given by (12.144) is the Schrödinger-Langevin equation. As we will see later, this equation can also be derived from the quantization of the modified Hamilton-Jacobi equation Sec. 12.6.

Other Methods of Generalizing the Madelung Hydrodynamical Equation for Dissipative Motion — Now let us study the following generalization of Eqs. (12.127) and (12.128) by replacing $\frac{\partial S}{\partial t}$ and $\frac{\partial S}{\partial t}$ by

$$\frac{\partial S}{\partial t} \rightarrow \frac{\partial S}{\partial t} + f(x, t), \tag{12.145}$$

and

$$\frac{\partial S}{\partial x} \rightarrow \frac{\partial S}{\partial x} + g(x, t), \tag{12.146}$$

where f and g are functions of x and t to be determined later. With these transformations Eqs. (12.127) and (12.128) are replaced by

$$\frac{\partial S}{\partial t} + f + \frac{1}{2m}\left(\frac{\partial S}{\partial x} + g\right)^2 + V = \left(\frac{\hbar^2}{2mA}\right)\frac{\partial^2 A}{\partial x^2}, \tag{12.147}$$

and

$$m\frac{\partial A}{\partial t} + \left(\frac{\partial A}{\partial x}\right)\left(\frac{\partial S}{\partial x} + g\right) + \frac{1}{2}A\left(\frac{\partial^2 S}{\partial x^2} + \frac{\partial g}{\partial x}\right) = 0, \tag{12.148}$$

and these equations form a generalization of Madelung equation. If we take the current density to be

$$J = \frac{A^2}{m}\left(\frac{\partial S}{\partial x} + g\right), \tag{12.149}$$

then the continuity equation becomes

$$\frac{\partial A^2}{\partial t} + \frac{\partial J}{\partial x} = 0, \tag{12.150}$$

and this follows from (12.148). We note that these equations reduce to the original forms (12.127) and (12.128) when $g = 0$. Thus in this extension of the Madelung formulation, the mechanical momentum mv is given by

$$mv = \frac{\partial S}{\partial x} + g. \tag{12.151}$$

Now let us take the classical limit of this generalization of the hydrodynamical model. As $\hbar \to 0$ we have

$$m\frac{dv}{dt} = -\frac{\partial V}{\partial x} + F_f, \tag{12.152}$$

where

$$F_f = \left(\frac{\partial g}{\partial t} - \frac{\partial f}{\partial x}\right). \tag{12.153}$$

Thus F_f in (12.153) represents an additional force which acts on the particle.

From Ehrenfest theorem it follows that the quantum mechanical expectation values satisfy an equation similar to (12.152);

$$m\frac{d}{dt}\langle v\rangle = -\left\langle\frac{\partial V}{\partial x}\right\rangle + \langle F_f\rangle, \tag{12.154}$$

where the symbol $\langle\ \rangle$ denotes the expectation value.

Some Solvable Cases of the Generalized Madelung Equation — The following three cases are of special interest:

(a) First we choose f and g in such a way that

$$\langle F_f\rangle = -\lambda m\frac{d}{dt}\langle x\rangle. \tag{12.155}$$

Next we derive the classical form of the Hamilton-Jacobi (H-J) equation which is the counterpart of the wave equation with the additional term $\langle F_f \rangle$. In the limit of $\hbar \to 0$, Eqs. (12.147) and (12.148) decouple and (12.147) becomes the H-J equation. For simplicity let us assume that $g = 0$, then we have

$$\frac{\partial S}{\partial t} + f + \frac{1}{2m} \left(\frac{\partial S}{\partial x} \right)^2 + V = 0. \tag{12.156}$$

We note that the only additional term in this equation is $f(x,t)$. Since we have set $g = 0$, therefore from (12.151), (12.153) and (12.155) it follows that

$$\frac{\partial f}{\partial x} = \lambda \frac{\partial S(x,t)}{\partial x}, \tag{12.157}$$

or

$$f(x,t) = \lambda S(x,t) + W(t), \tag{12.158}$$

where $W(t)$ is an arbitrary function of time. Thus the resulting equation is the same as the modified H-J equation that we found earlier, Eq. (4.5).

(b) For our second case we choose f and g to be:

$$f(x) = \lambda(1-a)\langle p \rangle (x - \langle x \rangle) - \frac{1}{2}\lambda^2 ma^2 (x - \langle x \rangle)^2, \tag{12.159}$$

and

$$g(x) = \lambda m a(x - \langle x \rangle), \tag{12.160}$$

where p is the momentum operator and a is an arbitrary parameter. Since $\langle g \rangle = 0$, we have $m\frac{d}{dt}\langle x \rangle = \langle p \rangle$, and thus the frictional force in this case is

$$F_f = -\lambda m \frac{d}{dt}\langle x \rangle + \lambda^2 ma^2 (x - \langle x \rangle). \tag{12.161}$$

Let us note that the expectation value of the last term in (12.161) is zero, and therefore (12.161) leads to (12.155).

(c) For the third possibility consider the case where $g = 0$ and

$$f(x,t) = \lambda \left(S - \langle S \rangle \right), \tag{12.162}$$

and thus a force of friction proportional to the velocity

$$F_f = -\lambda \frac{\partial S}{\partial x} = -\lambda m v, \tag{12.163}$$

acts on the particle. This choice gives us Kostin's nonlinear wave equation (or the Schrödinger-Langevin equation) and at the same time $\langle F_f \rangle$ satisfies (12.155) [23]. One can also generate nonlinear damping force proportional to v^n by choosing $g = 0$ and

$$f(x,t) = \gamma_n \int^x \left| \frac{\partial S}{\partial x} \right|^{n-1} \frac{\partial S}{\partial x} dx, \tag{12.164}$$

for which the damping force is

$$F_f = -\gamma_n |v|^{n-1} v. \tag{12.165}$$

Stability of the Stationary States — The extension of the Madelung formulation which leads to the Schrödinger-Langevin equation can also be used to study the stability of the stationary solutions of this equation. To this end we choose $\tilde{\rho}$ as the mass density and $\mathbf{v}$ as the velocity (see Eqs. (12.129) and (12.130)) to write the real and imaginary parts of the Schrödinger-Langevin equation as two equations of conservation of $\tilde{\rho}$ and momentum balance in hydrodynamics [32]-[34]

$$\frac{\partial \tilde{\rho}}{\partial t} + \nabla \cdot (\tilde{\rho} \mathbf{v}) = 0, \tag{12.166}$$

and

$$\frac{\partial (\tilde{\rho} \mathbf{v})}{\partial t} + \nabla \cdot (\tilde{\rho} \mathbf{v} \mathbf{v} + P_q) = \tilde{\rho} \left(-\nabla V + \mathbf{F}_f \right). \tag{12.167}$$

These are in fact the three-dimensional versions of (12.131) and (12.132). In these equations $\mathbf{v}\mathbf{v}$ denotes the tensor product of the two vectors [34], and P_q which is the quantum analogue of the pressure tensor is determined from the solution of the equation

$$\nabla \cdot P_q = \frac{\tilde{\rho}}{m} \nabla V_q, \tag{12.168}$$

where V_q which is called "the quantum potential" is given by [35]

$$V_q = -\frac{\hbar^2}{4m\tilde{\rho}} \left(\nabla^2 \tilde{\rho} - \frac{(\nabla \tilde{\rho})^2}{2\tilde{\rho}} \right). \tag{12.169}$$

The damping force $\mathbf{F}_f = -\nabla V_L$ corresponds to the potential

$$\left[-\frac{\lambda \hbar i}{2} \ln \left(\frac{\psi}{\psi^*} \right) \right], \tag{12.170}$$

in the Schrödinger-Langevin equation and is equal to $\mathbf{F}_f = -\lambda \mathbf{v}$.

For the stationary solutions of the Schrödinger equation in the absence of damping the pair of variables $(\tilde{\rho}, \mathbf{v})$ will satisfy the condition

$$\mathbf{v}_s = 0, \tag{12.171}$$

and

$$V_q(\tilde{\rho}_s) + V = E_s = \text{ constant}. \tag{12.172}$$

Now let us study the stability of the stationary solutions of Eqs. (12.166)-(12.167) when these solutions are given by (12.171) and (12.172). We note that

$$\int \tilde{\rho} \mathbf{v} \cdot \mathbf{F}_f \, d^3 r = -\lambda \int \tilde{\rho} \mathbf{v}^2 d^3 r \leq 0, \tag{12.173}$$

and we assume that both $\tilde{\rho}(\mathbf{r})$ and $\mathbf{v}$ are smooth functions of the coordinates and that the density $\tilde{\rho}(\mathbf{r})$ satisfies the asymptotic condition

$$\lim_{|\mathbf{r}|\to\infty} \tilde{\rho}(\mathbf{r})e^{|\mathbf{r}|} \to 0. \tag{12.174}$$

To simplify the writing of these equations we choose the units $\hbar = m = 1$, and in this set of units we examine the Liapunov functional, $Li(\tilde{\rho}, \mathbf{v})$, which for this problem is given by [32],[36],

$$Li(\tilde{\rho}, \mathbf{v}) = \int \tilde{\rho} \left[\frac{v^2}{2} + \frac{1}{2}\left(\frac{\nabla\tilde{\rho}}{2\tilde{\rho}}\right)^2 + V(r) - E_s \right] d^3r. \tag{12.175}$$

The variation of the functional $Li(\tilde{\rho}, \mathbf{v})$ can be written as

$$\delta Li(\tilde{\rho}, \mathbf{v}) = \int \left[\tilde{\rho}\mathbf{v} \cdot \delta\mathbf{v} + \left(\frac{v^2}{2} + V(r) + V_q - E_s\right)\delta\tilde{\rho} \right] d^3r$$

$$+ \oint \frac{1}{4}(\nabla \ln \tilde{\rho})\delta\tilde{\rho} \cdot d\mathbf{S}, \tag{12.176}$$

where we have used the divergence theorem and have written the last term in (12.176) as a surface integral over a sphere of radius R. As the asymptotic condition (12.174) shows this last term tends to zero as $R \to \infty$. Thus for the stationary solutions according to Eqs. (12.171) and (12.172) $\delta Li(\tilde{\rho}, \mathbf{v})$ must be zero.

If we calculate the time derivative of Li using the equations for $\tilde{\rho}$ and $\mathbf{v}$ we find

$$\frac{dLi(\tilde{\rho}, \mathbf{v})}{dt} = - \oint \left[\tilde{\rho}\mathbf{v}\left(\frac{v^2}{2} + V(r) + V_q - E_s\right) + \frac{1}{4}\nabla(\ln\tilde{\rho})\nabla \cdot (\tilde{\rho}\mathbf{v}) \right] \cdot d\mathbf{S}$$

$$+ \int \tilde{\rho}\mathbf{v} \cdot \mathbf{F}_f \, d^3r = -\lambda \int \tilde{\rho}\mathbf{v}^2 d^3r \leq 0. \tag{12.177}$$

The inequality in (12.177) follows from the fact that the surface integral in (12.177) vanishes as $R \to \infty$. Thus $\frac{dLi}{dt}$ is a negative semi-definite quantity.

Next we calculate the second variation of $Li(\tilde{\rho}, \mathbf{v})$. By calculating this second variation between $(\delta\tilde{\rho}, \delta\mathbf{v})$ and $(\delta'\tilde{\rho}, \delta'\mathbf{v})$ we find

$$\int [\delta\mathbf{v}, \ \delta\tilde{\rho}, \ \nabla\delta\tilde{\rho}] \begin{bmatrix} \tilde{\rho} & \mathbf{v} & \mathbf{0} \\ \mathbf{v} & \frac{(\nabla\tilde{\rho})^2}{4\tilde{\rho}^3} & -\frac{\nabla\tilde{\rho}}{4\tilde{\rho}^2} \\ \mathbf{0} & -\frac{\nabla\tilde{\rho}}{4\tilde{\rho}^2} & \frac{1}{4\tilde{\rho}} \end{bmatrix} \begin{bmatrix} \delta'\mathbf{v} \\ \delta'\tilde{\rho} \\ \nabla\delta'\tilde{\rho} \end{bmatrix} d^3r. \tag{12.178}$$

In this equation we have used the notation $\mathbf{0} = (0,0,0)$ and $\mathbf{v}$ for the velocity vector with three components (v_x, v_y, v_z). Therefore the integrand in (12.178) is the product of three matrices $[1 \times 7][7 \times 7][7 \times 1]$. The minors of the square matrix in (12.178) are all non-negative for the stationary solution; therefore the linearized system is marginally stable [32].

Let us emphasize that the continuity of $(\tilde{\rho}, \mathbf{v})$ and the asymptotic condition (12.174) are satisfied by the stationary solutions of the harmonic oscillator and Coulomb potentials. In particular the stability condition for the harmonic oscillator has been discussed in detail by Ván and Fülöp [32]. For a damped system the Liapunov functional expresses the difference between the energy of a given state and the energy of the stationary state, and can be expressed in terms of $(\tilde{\rho}, \mathbf{v})$ as we had in (12.175) or in terms of the wave function [32],[33].

12.6   Quantization of a Modified Hamilton-Jacobi Equation for Damped Systems

A different way of arriving at the Schrödinger-Lagevin equation equation is to quantize a modified form of the Hamilton-Jacobi equation [37]-[39]. Here we start with the time-dependent Hamiltonian of Cardirola-Kanai, Eq. (3.24), and we use the minimal coupling rule to account for the interaction of the charged particle with the electromagnetic field. The Hamiltonian in this case is

$$H = \frac{1}{2m} \left[\mathbf{p} - \frac{e}{c} \mathbf{A}(\mathbf{r}, t) \right]^2 e^{-\lambda t} + [V(\mathbf{r}) + e\phi(\mathbf{r}, t)] e^{\lambda t}, \qquad (12.179)$$

and from this H we can find that the standard Hamilton-Jacobi equation for this Hamiltonian is

$$I_1(\zeta) = \frac{1}{2m} \left[\nabla\zeta - \frac{e}{c} \mathbf{A}(\mathbf{r}, t) \right]^2 + [V(\mathbf{r}) + e\phi(\mathbf{r}, t)] e^{2\lambda t} + e^{\lambda t} \frac{\partial \zeta}{\partial t} = 0. \quad (12.180)$$

Schrödinger's Method of Quantization — The passage from this Hamilton-Jacobi to the wave equation can be achieved by using a method originally proposed by Schrödinger [40]. In this method we write the real function $\zeta(\mathbf{r}, t)$ in terms of a complex function $u(\mathbf{r}, t)$ where

$$\zeta(\mathbf{r}, t) = \frac{\hbar}{2i} \ln \left(\frac{u(\mathbf{r}, t)}{u^*(\mathbf{r}, t)} \right), \qquad (12.181)$$

and then find the extremum of the functional

$$uu^* I_1 (u, u^*), \qquad (12.182)$$

using the variational principle. The functional $I_1(\zeta)$ is given by (12.180) with ζ being replaced by u and u^*, Eq. (12.181). For the present problem we get the wave equation

$$i\hbar \frac{\partial u(\mathbf{r}, t)}{\partial t} = \frac{1}{2m} \left[-i\hbar\nabla - \frac{e}{c} \mathbf{A}(\mathbf{r}, t) \right]^2 e^{-\lambda t} u(\mathbf{r}, t) + [V(\mathbf{r}) + \phi(\mathbf{r}, t)] e^{\lambda t} u(\mathbf{r}, t).$$
$$(12.183)$$

This result can also be obtained by using the well-known quantization rule, i.e. by replacing $\mathbf{p}$ by $-i\hbar\nabla$ in the Hamiltonian (12.179). But as we have seen in Sec. 12.1 the rule of association $\mathbf{p} \rightarrow -i\hbar\nabla$ gives us unacceptable result for the velocity-position uncertainty. Therefore let us make the following transformation and replace ζ by S where

$$S(\mathbf{r}, t) = \zeta(\mathbf{r}, t)e^{-\lambda t}, \qquad (12.184)$$

and derive the following equation for the Jacobi function $S(\mathbf{r}, t)$ [38],[39]

$$I_2(S) = \frac{1}{2m}\left[\nabla S - \frac{e}{c}\mathbf{A}(\mathbf{r}, t)e^{-\lambda t}\right]^2 + (V(\mathbf{r}) + e\phi(\mathbf{r}, t)) + \frac{\partial S}{\partial t} + \lambda S = 0. \quad (12.185)$$

This Hamilton-Jacobi equation for a time independent electromagnetic potentials $\phi(\mathbf{r})$ and $\mathbf{A}(\mathbf{r})$ will remain invariant under time translation $t \rightarrow t+t_0$, whereas (12.180) will not.

Again we want to use the Schrödinger method of quantization and for this we express S in terms of ψ and its complex conjugate

$$S = -\frac{i\hbar}{2}\ln\left(\frac{\psi}{\psi^*}\right), \qquad (12.186)$$

and try to find the extremum of the functional

$$\psi\psi^* I_2(\psi, \psi^*) = \frac{\hbar^2}{2m}\left(\nabla + \frac{e\mathbf{A}}{i\hbar c}\right)\psi \cdot \left(\nabla + \frac{e\mathbf{A}}{i\hbar c}\right)\psi^*$$
$$+ \left[V + e\phi + \frac{\lambda\hbar}{2i}\ln\left(\frac{\psi}{\psi^*}\right)\right]\psi\psi^* + \frac{\hbar}{2i}\left[\psi^*\frac{\partial\psi}{\partial t} - \psi\frac{\partial\psi^*}{\partial t}\right],$$

$$(12.187)$$

subject to the normalization condition

$$\int \psi^*(\mathbf{r}, t)\psi(\mathbf{r}, t)d^3r = 1. \qquad (12.188)$$

The normalization of the wave function in the case of a dissipative system should be imposed as a constraint on the variational problem. Thus we try to find the extremum of the functional

$$J(\psi, \psi^*) = \int \psi^*\psi I_2(\psi, \psi^*)\,d^3rdt + \eta\int \psi^*\psi d^3rdt, \qquad (12.189)$$

where η is a complex Lagrange's multiplier. The Euler-Lagrange equation for the functional in (12.189) gives us two equations for ψ and ψ^*. For the wave function ψ we have the nonlinear equation

$$i\hbar\frac{\partial\psi(\mathbf{r}, t)}{\partial t} = \frac{1}{2m}\left[-i\hbar\nabla - \frac{e}{c}\mathbf{A}(\mathbf{r}, t)e^{-\lambda t}\right]^2\psi(\mathbf{r}, t)$$
$$+ \frac{\lambda\hbar}{2i}\left[\ln\left(\frac{\psi}{\psi^*}\right)\right]\psi + [V(\mathbf{r}) + \phi(\mathbf{r}, t)]\psi(\mathbf{r}, t) + \left(\eta - \frac{\lambda\hbar}{2i}\right)\psi,$$

$$(12.190)$$

and we have the complex conjugate of this equation for ψ^*. If we impose the normalization condition (12.188) and assume the boundary condition

$$\psi(\mathbf{r}, t) \to 0, \quad \text{as} \quad r \to \infty, \tag{12.191}$$

then from (12.190) and its complex conjugate it follows that

$$\eta - \eta^* - \frac{\lambda\hbar}{i} = 0, \quad \text{or} \quad \text{Im} \, \eta = -\frac{1}{2}\lambda\hbar. \tag{12.192}$$

In addition to the conservative force $-\nabla V(\mathbf{r})$ and the electromagnetic force there can also be a random force $\mathbf{F}(t)$ acting on the particle. This force arises from the interaction of the particle with the heat bath and is only a function of time. To include this random force we add a term $-\mathbf{r} \cdot \mathbf{F}(t)$ to the equation (12.190) which now takes the form

$$i\hbar\frac{\partial\psi(\mathbf{r}, t)}{\partial t} = \frac{1}{2m}\left[-i\hbar\nabla - \frac{e}{c}\mathbf{A}(\mathbf{r}, t)e^{-\lambda t}\right]^2 \psi(\mathbf{r}, t)$$
$$+ \frac{\lambda\hbar}{2i}\left[\ln\left(\frac{\psi}{\psi^*}\right)\right]\psi + [V(\mathbf{r}) + e\phi(\mathbf{r}, t) - \mathbf{r}\cdot\mathbf{F}(t)]\,\psi(\mathbf{r}, t). \tag{12.193}$$

For this wave equation the law of conservation of probability can be derived the same way as for the ordinary Schrödinger and the result is

$$\frac{\partial}{\partial}\,(\psi^*\psi) + \nabla\cdot\mathbf{J} = 0, \tag{12.194}$$

where

$$\mathbf{J} = \frac{i\hbar}{2m}\,(\psi\nabla\psi^* - \psi^*\nabla\psi) - \frac{e}{mc}\psi^*\mathbf{A}e^{-\lambda t}\psi. \tag{12.195}$$

From Eq. (12.190) we can deduce that the gauge transformation is

$$\mathbf{A}' = \mathbf{A} + \nabla\chi(\mathbf{r}, t)e^{-\lambda t}, \quad \phi' = \phi - \frac{1}{c}e^{-\lambda t}\frac{\partial\chi}{\partial t}, \tag{12.196}$$

and this reduces to the standard form as λ goes to zero. The wave function in this new gauge is related to the wave function in the old gauge by the relation

$$\psi' = \psi\exp\left[\frac{ie}{\hbar c}\chi(\mathbf{r}, t)e^{-\lambda t}\right]. \tag{12.197}$$

Wagner's Derivation of the Schrödinger-Langevin Equation — This derivation of the Schrödinger-Langevin equation from the modified form of the Hamilton-Jacobi equation poses some difficulties. In a very careful analysis of the Schrödinger quantization method applied to the linear dissipative systems Wagner [41] has shown that the choice of a complex Lagrange's multiplier as given in Eq. (12.192) will result in having an equation for ψ^* which is not the same as that of taking the complex conjugate of the differential equation for ψ. This is a serious problem in the above derivation. However Wagner [41] has demonstrated that the same method of quantization can be used for the functional $I(S)$ rather than $I_2(S)$ of Eq. (12.185) where

$$I(S) = e^{\lambda t} \left[\frac{1}{2m} (\nabla S)^2 + V(\mathbf{r}) + \frac{\partial S}{\partial t} + \lambda S \right] = 0. \tag{12.198}$$

Here for simplicity we have set $\mathbf{A}$ and ϕ in $I(S)$ equal to zero. Now the functional $\psi^* \psi I(\psi, \psi^*)$ becomes

$$\psi\psi^* I(\psi, \psi^*) = e^{\lambda t} \left[\frac{\hbar^2}{2m} (\nabla\psi \cdot \nabla\psi^*) + \left\{ V + \frac{\lambda\hbar}{2i} \ln\left(\frac{\psi}{\psi^*}\right) \right\} \psi\psi^* \right]$$

$$+ \frac{\hbar}{2i} \left(\psi^* \frac{\partial\psi}{\partial t} - \psi \frac{\partial\psi^*}{\partial t} \right) e^{\lambda t}. \tag{12.199}$$

The Euler-Lagrange equation for ψ in this formulation is given by

$$\frac{i\hbar}{2} \left[e^{\lambda t} \frac{\partial\psi}{\partial t} + \frac{\partial}{\partial t} \left(e^{\lambda t} \psi \right) \right] = \left(-\frac{\hbar^2}{2m} \nabla^2 \psi + V\psi \right) e^{\lambda t}$$

$$+ \frac{i\hbar\lambda}{2} \psi \left[1 - \ln\left(\frac{\psi}{\psi^*}\right) \right] e^{\lambda t}, \tag{12.200}$$

with its complex conjugate for ψ^*. By multiplying (12.200) by $e^{-\lambda t}$ we find the Schrödinger-Langevin equation (12.193) without the electromagnetic potentials. Thus by introducing the multiplying factor $e^{\lambda t}$ in $I(S)$ we can resolve the difficulty associated with the complex Lagrange's multiplier.

The second problem with the Schrödinger-Langevin equation is the way that the minimal coupling rule modifies this equation. A comparison between (12.121) and (12.193) shows that the magnetic potential appears with a factor $e^{-\lambda t}$ in the latter equation but not in the former, and in neither of these equations the minimal coupling prescription is equivalent to the addition of the Lorentz force to the equation of motion.

Let us examine the conditions under which Eq. (12.121) was derived. We started with the Heisenberg equations of motion for the operator Q and P, Eqs. (12.119), however those equations were obtained for a central harmonic oscillator interacting with a heat bath. Therefore Eqs. (12.119) were valid when the particle was subject to a harmonic force or a random force given by the potential $(-QF(t))$ (in three-dimensional problem by $(-\mathbf{r} \cdot \mathbf{F}(t))$). When we try to include arbitrary conservative and or electromagnetic forces in the Schrödinger-Langevin equation then there is no guarantee that the resulting equation yields acceptable results.

In principle we can add a purely time-dependent potential $U(t)$ in the Schrödinger-Langevin equation and this addition will only affect the phase of the wave function. Let us write (12.193) without the electromagnetic potentials, but with the potential $U(t)$ as

$$i\hbar \frac{\partial \psi}{\partial t} = -\frac{\hbar^2}{2m}\nabla^2 \psi + [V(\mathbf{r}) - \mathbf{r} \cdot \mathbf{F} + U(t)]\,\psi + \frac{\lambda \hbar}{2i}\left[\ln\left(\frac{\psi}{\psi^*}\right)\right]\psi. \quad (12.201)$$

Now we determine $U(t)$ by requiring that the expectation value of the total energy $\langle E(t)\rangle$ be equal to the sum of the expectation values of kinetic and potential energies, i.e.

$$\langle E(t)\rangle = \left\langle \frac{1}{2}m\dot{\mathbf{r}}^2 \right\rangle + \langle V(\mathbf{r})\rangle - \langle \mathbf{r} \cdot \mathbf{F}(t)\rangle. \quad (12.202)$$

Next we calculate $\langle E(t)\rangle$ and $\langle \frac{1}{2}m\dot{\mathbf{r}}^2 \rangle$,

$$\langle E(t)\rangle = i\hbar \int \psi^* \frac{\partial \psi}{\partial t} d^3 r, \quad (12.203)$$

$$\left\langle \frac{1}{2}m\dot{\mathbf{r}}^2 \right\rangle = \frac{\hbar^2}{2m} \int \left(\nabla \psi^* \cdot \nabla \psi\right) d^3 r. \quad (12.204)$$

Substituting (12.203) and (12.204) in (12.202) and making use of the wave equation (12.201) we find $U(t)$;

$$U(t) = -\frac{\hbar \lambda}{2i} \int |\psi|^2 \ln\left(\frac{\psi}{\psi^*}\right) d^3 r. \quad (12.205)$$

This completes our derivation of the Schrödinger-Langevin equation from the modified Hamilton-Jacobi equation. The fact that the superposition principle does not hold is a drawback of this nonlinear equation. However the type of logarithmic nonlinearity $-\frac{\lambda \hbar}{2i}\left[\ln\left(\frac{\psi}{\psi^*}\right)\right]\psi$ occurring in the equation has the following important property:

Motion of Two Non-Interacting Particles Moving in a Viscous Medium — In Sec. 4.3 we discussed the separability of the Jacobi function $S(\mathbf{r}_1, \mathbf{r}_2, t)$ which can be written as the sum of two terms $S_1(\mathbf{r}_1, t)$ and $S_2(\mathbf{r}_2, t)$, Eq. (4.36). For any linear wave equation, derived from Caldirola-Kanai Hamiltonian, this separability is well-known. However in the case of a nonlinear wave equation such as Schrödinger-Langevin equation this property must be verified. In the case of a nonlinear wave equation the lack of any correlation between the motion of the two particles implies that the total wave function must factorize:

$$\psi_{12}(\mathbf{r}_1, \mathbf{r}_2, t) = \psi_1(\mathbf{r}_1, t)\psi_2(\mathbf{r}_2, t). \quad (12.206)$$

To show this we write the Schrödinger-Langevin equation for two noninteracting particles moving in a dissipative environment

$$-\frac{\hbar^2}{2}\left(\frac{1}{m_1}\nabla_1^2 + \frac{1}{m_2}\nabla_2^2\right)\psi_{12} + [V_1(\mathbf{r}_1) + V_2(\mathbf{r}_2)]\,\psi_{12}$$

$$+ \frac{\lambda \hbar}{2i}\left[\ln\left(\frac{\psi_{12}}{\psi_{12}^*}\right)\right]\psi_{12} = i\hbar \frac{\partial \psi_{12}}{\partial t}, \quad (12.207)$$

then we observe that the solution of this equation with logarithmic nonlinearity can be factorized as a product $\psi_{12} = \psi_1(\mathbf{r}_1, t)\psi_{12}(\mathbf{r}_2, t)$ where each factor satisfies the nonlinear equation

$$-\frac{\hbar^2}{2m_i}\nabla_i^2\psi_i + V_i(\mathbf{r}_i)\psi_i + \frac{\lambda\hbar}{2i}\left[\ln\left(\frac{\psi_i}{\psi_i^*}\right)\right]\psi_i = i\hbar\frac{\partial}{\partial t}\psi_i, \quad i = 1, 2. \tag{12.208}$$

12.7 Exactly Solvable Cases of the Schrödinger-Langevin Equation

Having discussed some of the merits and deficiencies of the nonlinear Schrödinger-Langevin equation we now want to study few solvable examples. But before considering specific cases, we observe that because of the nonlinearity of this equation, and the inapplicability of the superposition principle, we cannot find the general solution and also we cannot ascertain the uniqueness of the solution that we have found. For the linear partial differential equations the method of separation of variables is an important technique. In the present case we note that the time-dependent part can be factored out if we assume that the wave function is a product of the form $\psi(\mathbf{r}, t) = \phi(\mathbf{r})T(t)$. Thus if we write

$$\psi(\mathbf{r}, t) = \phi(\mathbf{r})\exp\left[\frac{i\mathcal{E}}{\lambda\hbar}\left(e^{-\lambda t} - 1\right) - \frac{i}{\hbar}e^{-\lambda t}\int_0^t e^{\lambda t'}U(t')\,dt'\right], \tag{12.209}$$

and substitute $\psi(\mathbf{r}, t)$ in the Schrödinger-Langevin equation

$$i\hbar\frac{\partial\psi}{\partial t} = -\frac{\hbar^2}{2m}\nabla^2\psi + [V(\mathbf{r}) + U(t)]\psi + \frac{\lambda\hbar}{2i}\left[\ln\left(\frac{\psi}{\psi^*}\right)\right]\psi, \tag{12.210}$$

we find that $\phi(\mathbf{r})$ satisfies the nonlinear partial differential equation

$$-\frac{\hbar^2}{2m}\nabla^2\phi + V(\mathbf{r})\phi + \frac{\lambda\hbar}{2i}\left[\ln\left(\frac{\phi}{\phi^*}\right)\right]\phi = \mathcal{E}\phi. \tag{12.211}$$

The factors in (12.209) are chosen in such a way that in the limit $\lambda \to 0$, the solution goes over to that of the Schrödinger equation. However a separable solution of the type given by (12.209) may not be an interesting one.

Now let us consider the following two cases where the wave function can be obtained analytically:

(1) The first case is the motion of a particle in a viscous medium subject to the action of the random force $\mathbf{F}(t)$, but in the absence of the conservative force $(-\nabla V)$. We can solve the resulting equation by writing

$$\psi(\mathbf{r}, t) = \exp\left[\frac{iS(\mathbf{r}, t)}{\hbar}\right]\exp\left[-\frac{i}{\hbar}e^{-\lambda t}\int_0^t e^{\lambda t'}U(t')\,dt'\right], \tag{12.212}$$

and substituting it in (12.201) to find the following equation for S

$$\frac{\partial S}{\partial t} = \frac{1}{2m}\left[i\hbar\nabla^2 S - (\nabla S)^2\right] + \mathbf{r}\cdot\mathbf{F}(t) - \frac{\lambda}{2}\left(S + S^*\right). \tag{12.213}$$

In the classical limit of $\hbar \to 0$ and with the assumption that S is real, this equation becomes the Hamilton-Jacobi equation for the time-dependent force $\mathbf{F}(t)$

$$\frac{\partial S}{\partial t} = -\frac{1}{2m}(\nabla S)^2 + \mathbf{r}\cdot\mathbf{F}(t) - \lambda S. \tag{12.214}$$

This equation can be solved to yield the principal function S:

$$S = \mathbf{r}\cdot\mathbf{p}(t) + \theta(t), \tag{12.215}$$

where $\mathbf{p}(t)$ and $\theta(t)$ are given by

$$\mathbf{p}(t) = \mathbf{p}_0 e^{-\lambda t} + e^{-\lambda t}\int_0^t \mathbf{F}\left(t'\right)e^{\lambda t'}\,dt', \tag{12.216}$$

and

$$\theta(t) = \theta(0)e^{-\lambda t} - \frac{1}{2m}\int_0^t |\mathbf{p}\left(t'\right)|^2 e^{\lambda t'}\,dt'. \tag{12.217}$$

Since S is a linear function of $\mathbf{r}$, therefore $\nabla^2 S = 0$ and hence S is an exact solution of (12.213). Equation (12.216) shows how the momentum of the particle changes in time. We also note that $\mathbf{p}(t)$ is an eigenvalue of

$$-i\hbar\nabla\psi(\mathbf{r}, t) = \mathbf{p}(t)\psi(\mathbf{r}, t), \tag{12.218}$$

where $\psi(\mathbf{r}, t)$ is given by (12.212).

Spreading of a Gaussian Wave Packet — We can also find a solution for the Schrödinger-Langevin equation in the viscous medium when $V = 0$ which is the analogue of the spreading of a Gaussian wave packet [42]. Let us consider the one-dimensional motion of the wave packet

$$\Psi(x, t) = Ne^{ik(t)x}\exp\left[-\frac{(x - \xi(t))^2}{2a}\right], \tag{12.219}$$

where the time-dependent wave number $k(t)$ is given by

$$k(t) = k_0 e^{-\lambda t}, \tag{12.220}$$

and $\xi(t)$ is a solution of $\ddot{\xi}(t) + \lambda\dot{\xi}(t) = 0$, i.e.

$$\xi(t) = \xi_0 + \frac{\hbar k_0}{m\lambda}\left(1 - e^{-\lambda t}\right). \tag{12.221}$$

When $\lambda \to 0$, the complex number $a = a_R + ia_I$ is given by

$$a = 2\left(\Delta x_0\right)^2 + \frac{i\hbar t}{m}, \tag{12.222}$$

with Δx_0 being the initial width of the wave packet. Now if we substitute (12.219) in the Schrödinger-Langevin equation we obtain two coupled first order differential equations for a_R and a_I;

$$\frac{da_R}{dt} = \frac{2\lambda a_R a_I^2}{|a|^2}, \qquad (12.223)$$

and

$$\frac{da_I}{dt} = \frac{\hbar}{m} - \lambda a_I \frac{(a_R^2 - a_I^2)}{|a|^2}. \qquad (12.224)$$

The asymptotic solutions of these equations for large t are:

$$a_I \to \frac{\hbar}{m\lambda}, \quad |a| \approx a_R \to \frac{2\hbar}{m}\sqrt{\frac{t}{\lambda}}. \qquad (12.225)$$

Thus in a viscous medium the width of the wave packet grows as $t^{\frac{1}{2}}$, whereas for $\lambda = 0$, according to (12.222) the width grows linearly as a function of time. This result can be interpreted in the following way: Because the velocity component at the leading edge of the packet is greater than the packet's group velocity, the wave packet spreads. Now in the viscous medium the leading edge experiences a larger force of friction than the center of the packet. A larger viscous force on the leading edge and a smaller one at the trailing edge causes a retardation in the spreading of the wave packet and hence a slower rate for growth of $|a(t)|$.

(2)- The Schrödinger-Langevin equation for the problem of damped harmonic oscillator can be written as

$$i\hbar\frac{\partial\psi(x,t)}{\partial t} = -\frac{\hbar^2}{2m}\frac{\partial^2\psi(x,t)}{\partial x^2} + \frac{1}{2}m\omega_0^2 x^2 \psi(x,t) + \frac{\lambda\hbar}{2i}\left(\ln\frac{\psi}{\psi^*}\right)\psi(x,t). \quad (12.226)$$

As we have noted before, the classical action S_J, Eq. (4.18), is given by a complex quadratic function of x which depends on $(x - \xi(t))$ as well as $mx\dot{\xi}(t)$. Let us write the wave function as

$$\psi_n(x,t) = \phi_n[x - \xi(t)]\exp\left\{-\frac{iE_n}{\lambda\hbar} + i\hbar\left[mx\dot{\xi}(t) + C(t)\right]\right\}. \qquad (12.227)$$

In this relation $\phi_n(x)$ is the harmonic oscillator wave function

$$\phi_n(x) = \left(\frac{m\omega_0}{\hbar\pi}\right)^{\frac{1}{4}}\frac{1}{\sqrt{2^n n!}}H_n\left[\left(\frac{m\omega_0}{\hbar}\right)^{\frac{1}{2}}x\right]\exp\left(-\frac{m\omega_0}{2\hbar}x^2\right), \qquad (12.228)$$

and E_n the eigenvalue of $\phi_n(x)$,

$$E_n = \hbar\omega_0\left(n + \frac{1}{2}\right). \qquad (12.229)$$

The time-dependent terms $\xi(t)$ and $C(t)$ are given by (4.19) and (4.20) respectively [39],[43],[44].

12.8   Harmonically Bound Radiating Electron and the Schrödinger-Langevin Equation

In Sec. 1.6 we have seen that the equation of motion for the non-relativistic radiating electron is given by Eq. (1.84). Loinger [45] has shown that the quantal position operator of the electron, X, satisfies an equation which is formally analogous to the classical Dirac-Lorentz equation, i.e.

$$\frac{d^2 X}{dt^2} + \nu^2 X - \tau \left(\frac{d^3 X}{dt^3} \right) = \frac{1}{m} F(t), \tag{12.230}$$

where $F(t)$ is the external force, m is the mass of the electron and τ is the constant defined by equation (1.85). As in the classical problem we can transform (12.230) by the Bopp [46] transformation, Sec. 1.6, i.e.

$$x = \frac{(a\tau)^{\frac{1}{2}}}{\nu} \left(\dot{X} - \frac{1}{a} \ddot{X} \right), \tag{12.231}$$

to find a quadratic equation for the operator x:

$$\ddot{x} + \lambda \dot{x} + \omega_0^2 x = \frac{1}{\sqrt{a\tau} m \nu} \frac{dF}{dt}. \tag{12.232}$$

In this elation λ is the positive root of (1.93) and ω_0^2 is given by

$$\omega_0^2 = \frac{\nu^2}{a\nu}. \tag{12.233}$$

Let us emphasize the fact that x in (12.231) is not simply the position operator, but a generalized coordinate which is a linear combination of velocity and acceleration of the particle. If we assume that $m[x, \dot{x}] = i\hbar$ then as we have seen earlier a Heisenberg-Langevin equation like (12.232) is related to the Schrödinger-Langevin equation

$$i\hbar \frac{\partial \psi}{\partial t} = -\frac{\hbar^2}{2m} \frac{\partial^2 \psi}{\partial x^2} + \frac{1}{2} m\omega_0^2 x^2 \psi + \frac{\lambda \hbar}{2i} \left(\ln \frac{\psi}{\psi^*} \right) \psi - \frac{x \dot{F}(t)}{m\nu \sqrt{a\tau}}. \tag{12.234}$$

A solution of (12.234) which is of the form of a wave packet is given by [47],[48]

$$\psi_n(x, t) = \phi_n(x - \xi(t)) \exp \left[-\frac{i E_n}{\hbar \lambda} + \frac{i}{\hbar} \left(m x \dot{\xi}(t) + C(t) \right) \right], \tag{12.235}$$

where in this case $C(t)$ is given by (4.20) and $\xi(t)$ is the solution of the inhomogeneous differential equation

$$\ddot{\xi}(t) + \lambda \dot{\xi}(t) + \omega_0^2 \xi(t) = \frac{\dot{F}(t)}{m\nu \sqrt{a\tau}}. \tag{12.236}$$

A direct way of quantizing the classical problem of radiating electron can be formulated in the following way:

We start by constructing a Hamiltonian of Ostrogradsky type. Since the equation of motion is of the third order, therefore the corresponding phase space is six-dimensional. By introducing a boson, an anti-boson and an additional "ghost" operator according to the Feshbach and Tikochinsky method of Sec. 11.5 we can determine the eigenvalues of the harmonically bound electron.

We can also use the fourth order equation, (3.56), for a radiating electron and write the Schrödinger equation for the Hermitian Hamiltonian operator obtained from (3.59) which we write as

$$ i\hbar \frac{\partial}{\partial t} \left(e^{iS} \psi \right) = \hat{H} \left(e^{iS} \psi \right), \tag{12.237} $$

or

$$ i\hbar \frac{\partial \psi}{\partial t} = \exp(-iS) \sum_i \left(-i\hbar \frac{\partial}{\partial q_i} Q_i \right) \psi - \frac{8}{m\tau^2} \exp(-iS) \exp\left(\frac{4t}{\tau} \right) $$
$$ \times \sum_i \left(-\hbar^2 \frac{\partial^2 \psi}{\partial Q_i^2} \right) + \exp(-iS) \exp\left(-\frac{4t}{\tau} \right) V(q,t)\psi + \hbar \frac{\partial S}{\partial t}\psi, \tag{12.238} $$

where S is the Hermitian operator

$$ S = -\frac{it}{\tau} \sum_i \left(q_i \frac{\partial}{\partial q_i} + \frac{\partial}{\partial q_i} q_i + Q_i \frac{\partial}{\partial Q_i} + \frac{\partial}{\partial Q_i} Q_i \right). \tag{12.239} $$

We can find the continuity equation for the wave equation (12.238) by defining

$$ \rho(q_i, t) = \int |\psi(q_i, Q_i, t)|^2 \, dQ_1 dQ_2 dQ_3, \tag{12.240} $$

and

$$ j_k(q_i, t) = \int |\psi(q_i, Q_i, t)|^2 Q_k dQ_1 dQ_2 dQ_3, \tag{12.241} $$

as the probability and current density respectively. Then from (12.238) and its complex conjugate we obtain

$$ \frac{\partial \rho}{\partial t} + \nabla_{\mathbf{q}} \cdot \mathbf{j} = -\frac{8i\hbar}{m\tau^2} \exp\left(\frac{4t}{\tau} \right) \int_{S(Q_i)} (\psi^* \nabla_{\mathbf{Q}} \psi - \psi \nabla_{\mathbf{Q}} \psi^*) \cdot d\mathbf{S}(Q_i), \tag{12.242} $$

where $\mathbf{S}$ is a closed surface in Q space. By requiring that ψ and $\nabla \psi$ vanish at the boundary of Q space, Eq. (12.242) reduces to the standard form of the continuity equation.

The conditions that ψ and $\nabla \psi$ tend to zero as $Q \to \infty$ means that the probability of finding a particle with infinite velocity is zero [49].

12.9   Other Phenomenological Nonlinear Potentials for Dissipative Systems

In addition to the Schrödinger-Langevin equation that we obtained from the quantization of a classical damped systems, we can find other equations in a similar way [5].

Let us start with the classical equations of motion with a damping term proportional to the momentum of the particle,

$$\frac{dP}{dt} + \lambda P + \frac{dV(x)}{dx} = 0, \quad P = m\frac{dX}{dt}, \tag{12.243}$$

where P is the mechanical momentum of the particle. The total energy of the particle, E, which is given by $E = \frac{P^2}{2m} + V(x)$ decreases in time, and from (12.243) we find the rate of energy loss to be

$$\frac{dE}{dt} = -\frac{\lambda}{m}P^2. \tag{12.244}$$

The quantal equations corresponding to (12.243) according to Ehrenfest's theorem are

$$\left\langle \frac{dp}{dt} \right\rangle + \lambda \langle p \rangle + \left\langle \frac{dV(x)}{dx} \right\rangle = 0, \quad \langle p \rangle = m\left\langle \frac{dx}{dt} \right\rangle, \tag{12.245}$$

where x and p are the position and momentum operators and where the expectation values of these operators are defined by the relation

$$\langle O \rangle = \int_{-\infty}^{\infty} \psi^*(x,t) O(x,p) \psi(x,t) dx. \tag{12.246}$$

In Eq. (12.246) $\psi(x,t)$ is a wave packet solution of the time-dependent Schrödinger equation

$$i\hbar\frac{\partial \psi(x,t)}{\partial t} = H\psi(x,t). \tag{12.247}$$

We assume that the Hamiltonian can be written as the sum of kinetic energy and the potential energy plus an additional term $\lambda W(x,t)$ which accounts for the dissipative force, i.e.

$$H = \frac{p^2}{2m} + V(x) + \lambda W(x,t). \tag{12.248}$$

Now we want to determine $W(x,t)$ in such a way that by making use of the Heisenberg equation of motion $\frac{dp}{dt} = -\frac{i}{\hbar}[p,H]$ and the definition of the expectation value (12.246), we obtain (12.245). Following these steps we find that

$$\left\langle \frac{\partial W(x,t)}{\partial x} \right\rangle = \langle p \rangle. \tag{12.249}$$

But this relation does not yield a unique solution for $W(x)$, and as we will see we can construct a family of $W(x,t)$ s with the property (12.249).

The Schrödinger-Langevin Equation — The simplest solution of (12.249) is

$$\int \psi^* \frac{\partial W(x,t)}{\partial x} \psi dx = -\frac{i\hbar}{2} \int \left[\psi^* \frac{\partial \psi}{\partial x} - \frac{\partial \psi^*}{\partial x} \psi \right] dx, \qquad (12.250)$$

from which we obtain

$$W(x,t) = -\frac{i\hbar}{2} \left[\ln \left(\frac{\psi}{\psi^*} \right) - \left\langle \ln \left(\frac{\psi}{\psi^*} \right) \right\rangle \right]. \qquad (12.251)$$

The second term which is a function of time is added so as to make the expectation value of W equal to zero. This choice of $W(x,t)$ gives us the nonlinear Schrödinger-Langevin equation which we have discussed in Sec. 12.6.

Hasse Wave Equation — Next let us examine some other possibilities for $W(x,t)$ [5],[12]. We want W to be Hermitian and from (12.249) we conclude that W must be a general quadratic function of x, p, $\langle x \rangle$ and $\langle p \rangle$. Consider the Hasse dissipative potential

$$W_a(x,t) = a \left[\frac{1}{2}(xp + px) - \langle x \rangle p \right] + (1-a)\langle p \rangle (x - \langle x \rangle) \qquad (12.252)$$

$$= (x - \langle x \rangle) \left[ap + (1-a)\langle p \rangle \right] - \frac{i}{2}a\hbar, \qquad (12.253)$$

which depends on the real parameter a, and satisfies (12.249). This parameter, a, can be a constant or can be a function of time. The mechanical momentum of the system

$$m\dot{x} = m\frac{\partial H}{\partial p} = p + m\lambda a(x - \langle x \rangle), \qquad (12.254)$$

also depends on a. However for this damping potential the expectation value of W_a is given by

$$\langle W_a(x,t) \rangle = a \left[\frac{1}{2} \langle xp + px \rangle - \langle x \rangle \langle p \rangle \right] \approx 0, \qquad (12.255)$$

whereas for $W(x,t)$ this expectation value is exactly zero.

An Exactly Solvable Problem — As a specific example of the damping potential W_a let us consider the damped harmonic oscillator, $V(x) = \frac{1}{2}m\omega^2 x^2$. Here the wave function also depends on a;

$$\psi_n(x,t) = N_n \exp \left[-i\omega_a \left(n + \frac{1}{2} \right) t - \left(\frac{m}{2\hbar} \right) (\omega_a + ia\lambda) x^2 \right]$$

$$\times H_n \left(\sqrt{\frac{m\omega_a}{\hbar}} x \right), \qquad (12.256)$$

where N_n is the normalization constant

$$N_n = \left(\frac{m\omega_a}{\pi\hbar}\right)^{\frac{1}{4}} \frac{1}{\sqrt{2^n n!}}, \tag{12.257}$$

and ω_a is given by [5]

$$\omega_a = \begin{cases} \omega_0 & \text{for } a = 0 \\ \left(\omega_0^2 - \frac{\lambda^2}{4}\right)^{\frac{1}{2}} & \text{for } a = \pm\frac{1}{2} \end{cases}. \tag{12.258}$$

In addition to the damped harmonic oscillator the problem of motion of a particle when $V(x) = 0$ is also exactly solvable [5].

12.10 Scattering in the Presence of Frictional Forces

In the theory of heavy-ion collisions, e.g. collision of ^{84}Kr with ^{209}Bi, certain features of the scattering can be explained if we assume the presence of a viscous medium around the heavy nuclei (See Chapter 2). Dissipation of the relative kinetic energy and relative angular momentum and the trapping of the nuclei are among the phenomena which can be attributed to a simple damping force [50]. This force must be nonzero for a finite separation between the target and the projectile i.e. $\lambda \to \lambda(\mathbf{r})$ and $\lambda(\mathbf{r}) \to 0$ as $|\mathbf{r}| \to \infty$, or it must act for a finite duration, i.e. $\lambda \to \lambda(t), \lambda(t) \to 0$ as $t \to \pm\infty$. For the latter case we can study the problem within the framework of the time-dependent scattering theory.

Let us consider a Hamiltonian of the form

$$H = \frac{1}{2m}\mathbf{p}^2 \exp\left[-\Lambda(t)\right] + V(r)\exp\left[\Lambda(t)\right], \tag{12.259}$$

where

$$\Lambda(t) = \int_{-\infty}^{t} \lambda\left(t'\right) dt'. \tag{12.260}$$

Here it is assumed that $\lambda(t)$ is non-negative and that it goes to zero as $t \to \pm\infty$ fast enough so that $\Lambda(t)$ is well-defined for all values of t. The equation of motion for the position of the projectile found from (12.259) is explicitly time-dependent;

$$\ddot{\mathbf{r}} + \lambda(t)\dot{\mathbf{r}} + \frac{1}{m}\nabla V(r) = 0. \tag{12.261}$$

Using the Hamiltonian (12.259) we find the modified Hamilton-Jacobi equation

$$\frac{1}{2m}(\nabla S)^2 + \frac{\partial S}{\partial t} + V(r) + \lambda(t)S - U(t) = 0, \tag{12.262}$$

where $U(t)$ is the time-dependent potential introduced in Eq. (12.201). This classical equation can be quantized exactly as before with the result that

$$-\frac{\hbar^2}{2m}\nabla^2\psi + \frac{\hbar}{i}\frac{\partial\psi}{\partial t} + [V(r) - U(t)]\psi + \frac{\hbar\lambda(t)}{2i}\ln\left(\frac{\psi}{\psi^*}\right)\psi = 0, \qquad (12.263)$$

provided that $U(t)$ is given by

$$U(t) = -\frac{\hbar\lambda(t)}{2i}\int \psi^* \ln\left(\frac{\psi}{\psi^*}\right)\psi d^3r. \qquad (12.264)$$

As is well-known in the theory of time-dependent scattering, we use the "adiabatic switching", i.e. we replace $V(r)$ by $V(r)e^{-\varepsilon|t|}$, where ε is a small positive number, to get the correct asymptotic forms for the incoming and outgoing waves as $t \to -\infty$ and $t \to +\infty$ respectively [51]. The solution of (12.263) for large $|t|$ is

$$\psi(\mathbf{r}, t) \to \exp\left\{\frac{i}{\hbar}\left[\mathbf{r}\cdot\mathbf{q}\, e^{-\Lambda(t)} + e^{-\Lambda(t)}\Phi(t)\right]\right\}, \qquad (12.265)$$

where

$$\Phi(t) = B + \int_{-\infty}^{t}\left(U\left(t'\right)e^{\Lambda\left(t'\right)} - \frac{\mathbf{q}^2}{2m}e^{-\Lambda\left(t'\right)}\right)dt', \qquad (12.266)$$

and $\mathbf{q}$ and B are constants.

When $t \to -\infty$, (12.265) reduces to

$$\psi(\mathbf{r}, t) \to \exp\left[\frac{i}{\hbar}(\mathbf{q}\cdot\mathbf{r})\right], \qquad (12.267)$$

and this shows that $\mathbf{q}$ is the initial momentum of the particle. Using the asymptotic forms of the wave function (12.265) as $t \to \pm\infty$ we can calculate the kinetic energy per unit volume as $t \to -\infty$ and as $t \to +\infty$:

$$\lim_{t\to-\infty}\left(\frac{\hbar^2}{2m}\nabla\psi^* \cdot \nabla\psi\right) = \frac{q^2}{2m}, \qquad (12.268)$$

and

$$\lim_{t\to+\infty}\left(\frac{\hbar^2}{2m}\nabla\psi^* \cdot \nabla\psi\right) = \frac{q^2}{2m}\exp[-2\Lambda(\infty)]. \qquad (12.269)$$

Thus the energy per unit volume dissipated in collision is

$$\frac{q^2}{2m}\left(1 - e^{-2\Lambda(\infty)}\right). \qquad (12.270)$$

Since Eq. (12.263) is nonlinear, the complete solution of the scattering problem can only be found numerically. Among the numerical technique which can be applied to this problem are the method of quasi-linearization of Bellman and Kalaba [38],[52] or the method of Immele *et al.* [42] for the one-dimensional motion of a wave packet.

12.11 Application of the Noether Theorem: Linear and Nonlinear Wave Equations for Dissipative Systems

We can apply Noether's theorem to compare the conserved quantities for the two types of the wave equation which we have derived from a single Lagrangian for the damped motion linear in velocity [41],[53].

Let us recall that from the Lagrangian (3.11) we obtained the linear equation, Eq. (12.3), and from the same Lagrangian and the corresponding Hamiltonian, via the modified Hamilton-Jacobi equation we found the Schrödinger-Langevin equation (12.193). The Lagrangian densities for the linear and for the Schrödinger-Langevin wave equations are given by

$$\mathcal{L}_L = i\hbar\psi^*\dot{\psi} - \left(\frac{\hbar^2}{2m}\right)\nabla\psi^* \cdot \nabla\psi e^{-\lambda t} - V(\mathbf{r})e^{\lambda t}\psi^*\psi, \qquad (12.271)$$

and

$$\mathcal{L}_S = e^{\lambda t}\left[i\hbar\psi^*\dot{\psi} - \left(\frac{\hbar^2}{2m}\right)\nabla\psi^* \cdot \nabla\psi - V(\mathbf{r})\psi^*\psi\right]$$
$$+ e^{\lambda t}\left[\frac{i\hbar\lambda}{2}\psi^*\psi\left\{\ln\left(\frac{\psi}{\psi^*}\right) + 1\right\}\right], \qquad (12.272)$$

where the subscripts L and S refer to the linear and the Schrödinger-Langevin equations respectively. From these Lagrangians it is easy to deduce that:

(a) The momentum densities for the two cases are different

$$\pi_L(\mathbf{r}, t) = \frac{\partial\mathcal{L}_L}{\partial\dot{\psi}} = i\hbar\psi^*(\mathbf{r}, t), \qquad (12.273)$$

and

$$\pi_S(\mathbf{r}, t) = \frac{\partial\mathcal{L}_S}{\partial\dot{\psi}} = i\hbar e^{\lambda t}\psi^*(\mathbf{r}, t). \qquad (12.274)$$

(b) The corresponding Hamiltonian densities for the two wave fields obtained from the corresponding Lagrangians are

$$\mathcal{H}_L = \pi_L\dot{\psi} - \mathcal{L}_L = \left(\frac{\hbar^2}{2m}\right)\nabla\psi^* \cdot \nabla\psi e^{-\lambda t} + V(\mathbf{r})e^{\lambda t}\psi^*\psi, \qquad (12.275)$$

and

$$\mathcal{H}_S = e^{\lambda t}\left[\left(\frac{\hbar^2}{2m}\right)\nabla\psi^* \cdot \nabla\psi + V(\mathbf{r})\psi^*\psi - \frac{i\hbar\lambda}{2}\psi^*\psi\left\{\ln\left(\frac{\psi}{\psi^*}\right) + 1\right\}\right].$$
$$(12.276)$$

Both of the Lagrangian densities $\mathcal{L}_L$ and $\mathcal{L}_S$ reduce to the Lagrangian $\mathcal{L}$ of the Schrödinger field when λ tends to zero. The same is true of the Hamiltonian densities $\mathcal{H}_L$ and $\mathcal{H}_S$, i.e. they both tend to $\mathcal{H}$ as $\lambda \to 0$.

For simplicity we only consider the motion of a free particle in a viscous medium. In this case we set $V(\mathbf{r}) = 0$ in our equations for $\mathcal{H}_L$ and $\mathcal{H}_S$.

First let us study the conservation laws for the ψ_L field i.e. the linear wave equation [41]. These are given by the following relations:

(1) Conservation of the energy which for $V(\mathbf{r}) = 0$ is just the conservation of the kinetic energy

$$\frac{d}{dt} \int \left[\left(\frac{\hbar^2}{2m}\right) \nabla \psi^* \cdot \nabla \psi\right] d^3r = \frac{d}{dt}\left(Ee^{2\lambda t}\right) = 0. \tag{12.277}$$

(2) Conservation of linear momentum

$$\frac{d}{dt} \int \left(\frac{\hbar}{2i}\right) \{\psi^* \nabla \psi - \psi \nabla \psi^*\} d^3r = \frac{d}{dt}\left(\mathbf{P}e^{\lambda t}\right) = 0. \tag{12.278}$$

(3) Conservation of angular momentum

$$\frac{d}{dt} \int \left(\frac{\hbar}{2i}\right) \mathbf{r} \wedge \{\psi^* \nabla \psi - \psi \nabla \psi^*\} d^3r = \frac{d}{dt}\left(\mathbf{L}e^{\lambda t}\right) = 0. \tag{12.279}$$

(4) Motion of the center of the wave packet (or the center of mass for a system of particles)

$$\frac{d}{dt} \int \left[\left(\frac{\hbar}{2i}\right) [\psi^* \nabla \psi - \psi \nabla \psi^*] \left(\frac{1 - e^{-\lambda t}}{\lambda}\right) - m\mathbf{r}\psi^* \psi\right] d^3r$$
$$= \frac{d}{dt}\left(\mathbf{P}\frac{e^{\lambda t} - 1}{\lambda} - m\mathbf{R}\right) = 0. \tag{12.280}$$

For the Schrödinger-Langevin equation we have the following conservation laws:

(1) Conservation of the (kinetic) energy

$$\frac{d}{dt} \int \left[\left(\frac{\hbar^2}{2m}\right) \nabla \psi^* \cdot \nabla \psi - \frac{i\hbar\lambda}{2}\psi^* \psi \ln\left(\frac{\psi}{\psi^*}\right)\right] e^{\lambda t} d^3r$$
$$= \frac{d}{dt}\left[\left\{E - \frac{i\hbar\lambda}{2} \int \psi^* \psi \ln\left(\frac{\psi}{\psi^*}\right)\right\} e^{\lambda t} d^3r\right] = 0. \tag{12.281}$$

(2) Conservation of linear momentum

$$\frac{d}{dt} \int \left(\frac{\hbar}{2i}\right) \{\psi^* \nabla \psi - \psi \nabla \psi^*\} e^{\lambda t} d^3r = \frac{d}{dt}\left(\mathbf{P}e^{\lambda t}\right) = 0. \tag{12.282}$$

(3) Conservation of angular momentum

$$\frac{d}{dt} \int \left(\frac{\hbar}{2i}\right) \mathbf{r} \wedge [\psi^* \nabla \psi - \psi \nabla \psi^*] e^{\lambda t} d^3r = \frac{d}{dt}\left(\mathbf{L}e^{\lambda t}\right) = 0. \tag{12.283}$$

(4) Conservation of the motion of the center of the wave packet (or the center of mass for a system of particles)

$$\frac{d}{dt}\int\left[\left(\frac{\hbar}{2i}\right)\{\psi^*\nabla\psi - \psi\nabla\psi^*\}\left(\frac{e^{\lambda t}-1}{\lambda}\right) - m\mathbf{r}\psi^*\psi\right]d^3r$$
$$= \frac{d}{dt}\left(\mathbf{P}\frac{e^{\lambda t}-1}{\lambda} - m\mathbf{R}\right) = 0. \tag{12.284}$$

In addition to these we have the invariance of the Lagrangian densities $\mathcal{L}_L$ and $\mathcal{L}_S$ with respect to a global change of phase of the wave function,

$$\psi(\mathbf{r},t) \to e^{i\alpha}\psi(\mathbf{r},t) \quad \text{for the } L \text{ field}, \tag{12.285}$$

and

$$\psi(\mathbf{r},t) \to \exp\left(i\alpha e^{-\lambda t}\right)\psi(\mathbf{r},t) \quad \text{for the } S \text{ field}. \tag{12.286}$$

For both fields, this invariance implies the conservation of probability [41]

$$\frac{d}{dt}\int\rho(\mathbf{r},t)d^3r = 0. \tag{12.287}$$

12.12 Wave Equation for Impulsive Forces Acting at Certain Intervals

In Chapter 3 we studied a Hamiltonian formulation of a classical system expressed by first order linear difference equation (1.153). The difference equation results from simplifying assumptions regarding the nature of impulsive forces, i.e. using them in the form of constraints at times $t = t_n$, on the dynamics of otherwise free motion of a particle (or a system).

To find the quantum equivalent of Eq. (1.153) we note that the Hamiltonian (3.322) is of the Kanai-Cardirola type, and that the commutation relation

$$[x,p] = [x,P]\exp\left\{-\int_0^t f\left(t'\right)dt'\right\} = i\hbar\exp\left\{-\int_0^t f\left(t'\right)dt'\right\} \tag{12.288}$$

violates the position-velocity uncertainty unless $\rho = 1$. Therefore as we have seen in this chapter the direct quantization of (3.322) leads to unacceptable results. However we can use Kostin's method and start with the Schrödinger equation

$$i\hbar\frac{\partial\psi}{\partial t} = -\frac{\hbar^2}{2m}\left(\frac{\partial^2\psi}{\partial x^2}\right) + (V_L(x,t) - xg(t))\,\psi, \tag{12.289}$$

and follow the same steps to find that

$$\langle p\rangle = \frac{\hbar}{2i}\int_{-\infty}^{\infty}\left(\psi^*\frac{\partial\psi}{\partial x} - \psi\frac{\partial\psi^*}{\partial x}\right)dx, \tag{12.290}$$

and

$$\left\langle \frac{\partial V_L(x,t)}{\partial x} \right\rangle = f(t)\langle p \rangle. \tag{12.291}$$

From Eqs. (12.290) and (12.291) we obtain $V_L(x,t)$ to be

$$V_L(x,t) = \frac{\hbar f(t)}{2i} \ln \left(\frac{\psi(x,t)}{\psi^*(x,t)} \right) + W(t). \tag{12.292}$$

By imposing the condition that

$$\langle E(t) \rangle = \frac{1}{2m} \langle p^2 \rangle - g(t)\langle x \rangle, \tag{12.293}$$

we find $W(t)$;

$$W(t) = -\frac{\hbar}{2i} f(t) \frac{\int_{-\infty}^{\infty} |\psi(x,t)|^2 \ln \left(\frac{\psi(x,t)}{\psi^*(x,t)} \right) dx}{\int_{-\infty}^{\infty} |\psi(x,t)|^2 dx}. \tag{12.294}$$

Thus Eq. (12.289) is the Schrödinger-Langevin equation for the classical Hamiltonian (3.322) [54].

12.13 Classical Limit for the Time-Dependent Problems

In the classical limit when $\hbar \to 0$, the Schrödinger equation as well as the wave equations for linear and nonlinear damping reduce to the corresponding Hamilton-Jacobi equations. Now we want to study a different formulation of the same problem based on the concept of distribution function of the positions and momenta of an ensemble of identical particles [55]-[57]. Consider an ensemble of identical particles whose initial positions ξ_0 are distributed with a distribution function $\mathcal{F}(\xi_0)$ with all of the initial momenta being equal to p_0. Starting from the classical Hamilton-Jacobi equation one can show that the probability density $P(x,t)$ for this ensemble is given by

$$P(x,t) = \frac{\partial^2 S}{\partial p_0 \partial x} \left| \mathcal{F} \left(\frac{\partial S}{\partial p_0} \right) \right|^2, \tag{12.295}$$

where $S(x,p_0,t)$ is the Hamilton principal function. The fact that the above ensemble is defined for a fixed p_0 is problematic. Therefore we choose a Gaussian form for the distribution function $\mathcal{F}(\xi_0)$ and calculate the wave packet for the free particle $\psi_F(x,t)$ and its Fourier transform $\phi_F(p,t)$. From these we obtain one of the possible quantum mechanical distribution functions for this problem which is in the form of the product [58]

$$D(x,p,t=0) = |\psi_F(x,0)|^2 \, |\phi_F(p,0)|^2. \tag{12.296}$$

This distribution function does not have a fixed momentum, p_0, but a Gaussian distribution about p_0 [59].

Now we start with the following definition of the classical limit:

Let us consider an ensemble with a given initial distribution of positions and momenta and then using Liouville theorem we determine $D(x, p, t)$ by replacing x and p from the solution of the classical equations of motion, i.e.

$$D(x, p, t) = D\left[x(t = 0), p(t = 0)\right], \qquad (12.297)$$

where $x(t = 0)$ and $p(t = 0)$ are the initial coordinates of the particle written in terms of $x(t)$ and $p(t)$. This gives us the distribution function at the time t.

Let us compare two ensembles, one classical and the other quantum-mechanical, and let us assume that the initial distribution of position and momentum are the same for both ensembles. Then the classical limit of the quantum-mechanical ensembles at a given time t is defined as that of the classical ensemble at t. This definition can be used with any distribution function such as the one given by (12.296) or by others like the Wigner distribution function [60].

Existence of the Classical Limit for the Damped Harmonic Oscillator — Next we apply this idea to the problem of the damped harmonic oscillator where the wave function for the ground state is given by Eq. (12.227)

$$\psi_0(x, t) = \left(\frac{m\omega_0}{\hbar\pi}\right)^{\frac{1}{4}} \exp\left[-\frac{m\omega_0}{2\hbar} (x - \xi(t))^2\right] \times$$

$$\times \exp\left[-\frac{i\omega_0}{2\lambda} + i\hbar\left(mx\dot{\xi} + C(t)\right)\right]. \qquad (12.298)$$

Thus the initial quantum mechanical distribution function at $t = 0$ is given by

$$P(x, t = 0) = |\psi_0(x, t = 0)|^2 = \left(\frac{m\omega_0}{\hbar\pi}\right)^{\frac{1}{2}} \exp\left[-\frac{m\omega_0}{\hbar} (x - \xi_0)^2\right], \qquad (12.299)$$

and at a later time t, $P(x, t)$ becomes

$$P(x, t) = \left(\frac{m\omega_0}{\hbar\pi}\right)^{\frac{1}{2}} \exp\left[-\frac{m\omega_0}{\hbar} (x - \xi(t))^2\right]. \qquad (12.300)$$

The corresponding classical phase space density function $D(x, p, t = 0)$ is the product of two Gaussian functions one for x and the other for p

$$D(x, p, t = 0) = \frac{1}{\pi} \exp\left\{-\left(\frac{m\omega_0}{\hbar}\right)(x - \xi_0)^2\right\} \exp\left\{-\left(\frac{1}{\omega_0 m\hbar}\right)(p - p_0)^2\right\}. \qquad (12.301)$$

The classical distribution function at time t can be found from the solution of the classical equation of motion. Thus by solving the differential equation for $x = \xi(t)$, Eq. (4.19), with the initial conditions $x(0)$ and $p(0) = m\dot{x}(0)$ and then inverting the result we find $x(0)$ and $p(0)$ as functions of x and p;

$$x(0) = e^{-\frac{\lambda}{2}t} \left\{x\left(\cos\omega t - \frac{\lambda}{2\omega} \sin\omega t\right) - \frac{p}{m\omega} \sin\omega t\right\} \qquad (12.302)$$

and

$$p(0) = e^{-\frac{\lambda}{2}t} \left\{ p \left(\cos \omega t + \frac{\lambda}{2\omega} \sin \omega t \right) + m\omega_0 x \sin \omega t \right\}.$$ (12.303)

Using these classical results we determine $D(x, p, t)$;

$$D(x, p, t) = D\left(x(0), p(0)\right) = \frac{1}{\pi} \exp\left\{-\left(\frac{m\omega_0}{\hbar}\right)(x(0) - \xi_0)^2\right\}$$

$$\times \exp\left\{-\left(\frac{1}{\omega_0 m\hbar}\right)(p(0) - p_0)^2\right\},$$ (12.304)

where $x(0)$ and $p(0)$ are given by (12.302) and (12.303) respectively. Now if we calculate the integral

$$P_c(x, t) = \int D(x, p, t) dp,$$ (12.305)

and if this $P_c(x, t)$ is identical to the quantum mechanical result (12.300), we say that $D(x, p, t)$ is the classical limit of the quantum distribution function. When $\lambda = 0$, i.e. no damping, then $\xi(t) = \xi_0 \cos \omega_0 t$, and (12.305) will be the same as (12.300). But with $\lambda \neq 0$ the classical limit in this sense does not exist.

Let us note two points about the question of finding the classical limit of quantum damped motions. First we know that there are many distribution functions [58] that we can construct for a given quantum mechanical state. For instance we can replace (12.296) by the Wigner distribution function

$$D_W(x, p) = \frac{1}{\pi\hbar} \int \psi^*(x + y)\psi(x - y) \exp\left(\frac{2ipy}{\hbar}\right) dy.$$ (12.306)

The second point is the suggestion of Korsch *et al.* that one should replace the quantum operators by their expectation values in coherent state representation [61]. This approach has been studied for non-Hermitian PT symmetric systems. Details of this method can be found in ref. [61].

References

[1] E.H. Kerner, Can. J. Phys. 36, 371 (1958).

[2] F. Bopp, Zeit. Angw. Phys. 14, 699 (1962).

[3] W.K.H. Stevens, Pro. Phys. Soc. Lond. 72, 1072 (1958).

[4] For a systematic and carful analysis of the problem of quantized motion of a radiating electron see B-G. Englert, Ann. Phys. (NY) 129, 1 (1980).

[5] R.W. Hasse, J. Math. Phys. 16, 2005 (1975).

[6] R.W. Hasse, Rep. Prog. Phys. 41, 1027 (1978).

[7] I.K. Edwards, Am. J. Phys. 47, 153 (1979).

[8] V.V. Dodonov and V.I. Man'ko, Nuovo Cimento, 44B, 265 (1978).

[9] V.V. Dodonov and V.I. Man'ko, Phys. Rev. A 20, 550 (1979).

[10] F. Bloch, Z. Phys. 52, 555 (1928).

[11] Floquet solution of differential equations with periodic coefficients is discussed in many books, e.g. E.T. Whittaker and G.N. Watson, *A Course of Modern Analysis*, Fourth Edition, (Cambridge University Press, London, 1948) p. 412.

[12] P. Cardirola, Lett. Nuovo Cimento, 20, 589 (1977).

[13] D.M. Greenberger J. Math. Phys. 20, 762 (1979).

[14] D.M. Greenberger J. Math. Phys. 20, 771 (1979).

[15] See for instance, F.A. Kaempfer, *Concepts in Quantum Mechanics*, (Acadmic Press, New York, N.Y. 1965).

[16] M. Razavy, Nuovo Cimento, 63B, 271 (1969).

[17] F. Englemann and E. Fick, Nuovo Cimento, 12, 63 (1959).

[18] E. Fick and F. Englemann, Z. Phys. 175, 271 (1964).

[19] C.M. Cheng and P.C.W. Fung, J. Phys. A 21, 4115 (1988).

[20] C.F. Lo, Phys. Rev. A 43, 404 (1991).

[21] A. Pimpale and M. Razavy, Phys. Rev. A 36, 2739 (1987).

[22] W. Pauli, *General Principles of Quantum Mechanics*, translated by P. Achuthan and K. Vankatesan (Springer-Verlag, Berlin, 1980) pp. 28-30.

[23] M.D. Kostin, J. Chem. Phys. 57, 3589 (1972).

[24] M.D. Kostin, J. Stat. Phys. 12, 145 (1975).

[25] H. Dekker, Phys. Rep. 80, 1 (1981).

[26] G.W. Ford, M. Kac and P. Mazur, J. Math. Phys. 6, 504 (1965).

[27] W. Stoker and K. Albrecht, Ann. Phys. (NY) 117, 436 (1979).

[28] E. Madelung, Z. Phys. 40, 322 (1926).

[29] J. Mehta and H. Rechenberg, *The Historical Development of Quantum Theory*, Vol. 5, Part 2 (Springer-Verlag, New York, N.Y. 1987) p. 856.

[30] L.A. Turski, Acta Phys. Polonica B 26, 1311 (1995).

[31] R.J. Wysocki, Phys. Rev. A 61, 022104 (2000).

[32] P. Ván and T. Fülöp, Phys. Lett. A 323, 374 (2004).

[33] S. Bhattacharya, S. Dutta and S. Roy, J. Mod. Phys. 2, 231 (2011).

[34] See for example: Z.U.A. Warsi, *Fluid Dynamics*, (CRC Press, Boca Raton, 1992) p. 38.

[35] P.R. Holland, *The Quantum Theory of Motion*, (Cambridge University Press, Cambridge, 1993).

[36] L. Meirovitch, *Methods of Analytical Dynamics*, (McGraw-Hill, New York, N.Y. 1970) Sec. 6.7.

[37] M. Razavy, Z. Physik B 26, 201 (1977).

[38] M. Razavy, Can. J. Phys. 56, 311 (1978).

[39] M. Razavy, Can. J. Phys. 56, 1372 (1978).

[40] E. Scrödinger, Ann. Phys. (Leipzig), 79, 361 (1926).

[41] H-J Wagner, Z. Physik, B 95, 261 (1994).

[42] J.D. Immele, K-K. Kan and J.J. Griffin, Nucl. Phys. A241, 47 (1975).

[43] J. Messer, Acta Physica Aust. 50, 75 (1997).

[44] B. K. Skagerstam, J. Math. Phys. 18, 308 (1977).

[45] A. Loinger, Nuovo Cimento, 2, 511 (1955).

[46] F. Bopp, Zeit. Angw. Phys. 14, 699 (1962).

[47] I.R. Senitzky, Phys. Rev. 95, 1115 (1954).

[48] M. Razavy, Lett. Nuovo Cimento, 24, 293 (1979).

[49] G. Valentini, Nuovo Cimento, XIX, 1280 (1961).

[50] W. Nörenberg and H.A. Weidenmüller, *Introduction to the Theory of Heavy-Ion Collisions*, (Springer-Verlag, New York, N.Y. 1980) Sec. 2.6.

[51] See, for example, C.J. Joachain, *Quantum Collision Theory*, (North-Holland, Amsterdam, 1975) p. 304.

[52] R.E. Bellman and R.E. Kalaba, *Quasilinearization and Nonlinear Boundary Value Problems*, (Rand Corp. Santa Monica, 1965).

[53] K. Yasue, Ann. Phys. (NY) 114, 479 (1978).

[54] M. Razavy, Hadronic J. 17, 515 (1994).

[55] L.S. Brown, Am. J. Phys. 40, 371 (1972).

[56] D. Home, *Conceptual Foundations of Quantum Physics*, (Springer, New York, 1997) Sec. 3.2.1.

[57] J.R. Klauder and B-S. Skagerstam, *Coherent States: Applications in Physics and Mathematical Physics*, (World Scientific, Singapore, 1985) p. 401.

[58] L. Cohen, J. Math. Phys. 7, 781 (1966).

[59] D. Home and S. Sengupta, Am. J. Phys. 51, 265 (1983).

[60] E.P. Wigner, Phys. Rev. 40, 749 (1932).

[61] E-V Graefe, M. Höning and H.J. Korsch, J. Phys. A 43, 075306 (2010).

Chapter 13

Coherent State Formulation of Damped Systems

In the preceding chapter we studied the time-dependent formulation of the quantized damped systems, and our approach was based on the Schrödinger equation. In this chapter we will start with a special group of problems which can be described as "linear oscillators with time-dependent frequencies" where we will consider the application of Heisenberg's equations to solve these problems. Later in this chapter, we try to use this approach and discuss the coherent state theory of the damped motion for a single particle or a collection of particles interacting with a central particle modelled after Ullersma's Hamiltonian.

13.1 First Integral of Motion and Quantum Description of an Oscillator with Variable Frequency

For a system given by a Hamiltonian with a time-dependent frequency and having the general form of

$$H = \frac{1}{2m}p^2 + \frac{1}{2}m\Omega^2(t)x^2, \tag{13.1}$$

we can find a first integral of motion $I(x, p, t)$ from the Heisenberg equation defining the time derivative of any operator I. Thus I is a first integral if

$$\frac{dI}{dt} = \frac{\partial I}{\partial t} + \frac{1}{i}[I,\,H] = 0, \qquad \hbar = 1. \tag{13.2}$$

Let us try to find a first integral, I, which is quadratic function of x and p, and is a Hermitian operator [1]-[5]. Of course there is no reason to limit the first integrals to be quadratic functions of x and p, particularly for two- or three-dimensional motions [6]-[8]. But here we want to consider the simplest forms of $I(x,p,t)$ that we can find.

The most general form of I quadratic in x and p is given by

$$I(x,p,t) = \frac{1}{2}\left[\mathcal{A}(t)x^2 + \mathcal{B}p^2 + \mathcal{C}(t)(px+xp)\right]. \qquad (13.3)$$

the Hermiticity of I is evident from the symmetrized form of (13.3). By substituting (13.1) and (13.3) in (13.2) we obtain

$$\frac{1}{2}\left[\left(\dot{\mathcal{A}}(t) - 2m\Omega^2(t)\mathcal{C}\right)x^2 + \left(\dot{\mathcal{B}}(t) + \frac{2}{m}\mathcal{C}(t)\right)p^2\right]$$
$$+ \frac{1}{2}\left[\dot{\mathcal{C}}(t) + \frac{1}{m}\mathcal{A}(t) - m\Omega^2(t)\mathcal{B}(t)\right](px+xp) = 0. \qquad (13.4)$$

This relation must be satisfied by all matrix elements of $I(x,p,t)$, therefore the following relations must be true:

$$\dot{\mathcal{A}}(t) = 2m\Omega^2(t)\mathcal{C}(t), \qquad (13.5)$$

$$\dot{\mathcal{B}}(t) = -\frac{2}{m}\mathcal{C}(t), \qquad (13.6)$$

and

$$\dot{\mathcal{C}}(t) = m\Omega^2(t)\mathcal{B}(t) - \frac{1}{m}\mathcal{A}(t). \qquad (13.7)$$

Now let us introduce the function $\sigma(t)$ by

$$\sigma^2(t) = \mathcal{B}, \qquad (13.8)$$

and express $\mathcal{A}$ and $\mathcal{B}$ in terms of $\sigma(t)$. By substituting for these two functions in (13.5) we obtain the following differential equation for $\sigma(t)$

$$\sigma(t)\frac{d}{dt}\left[m^2\left(\ddot{\sigma}(t) + \Omega^2(t)\sigma(t)\right)\right] + 3m^2\left[\dot{\sigma}(t)\left(\ddot{\sigma}(t) + \Omega^2(t)\sigma(t)\right)\right] = 0. \qquad (13.9)$$

This equation can be integrated with the result that

$$m^2\ddot{\sigma}(t) + m^2\Omega^2(t)\sigma(t) - \frac{c}{\sigma^3(t)} = 0, \qquad (13.10)$$

where c is the constant of integration. By choosing

$$\rho(t) = \left(\frac{c}{m^2}\right)\sigma(t), \qquad (13.11)$$

we can write (13.10) in the standard form of [2],[3]

$$\ddot{\rho}(t) + \Omega^2(t)\rho(t) - \frac{1}{\rho^3(t)} = 0, \qquad (13.12)$$

Since we have found $\mathcal{A}$, $\mathcal{B}$ and $\mathcal{C}$ in terms of ρ, we can also write $I(x, p, t)$ in terms of $\rho(t)$;

$$I(x, p, t) = \frac{1}{2} \left\{ \frac{1}{\rho^2(t)} x^2 + [\rho(t)p - m\dot{\rho}x]^2 \right\}. \tag{13.13}$$

Now let us investigate the quantization of this classical motion using the first integral $I(x, p, t)$. For this purpose we will use a representation in which $I(x, p, t)$ is diagonal. Let $|n\rangle$ be a state such that

$$I(x, p, t)|n\rangle = n|n\rangle, \tag{13.14}$$

and

$$\langle n'|I(x, p, t)|n\rangle = \delta_{n,n'}. \tag{13.15}$$

While the operator $I(x, p, t)$ depends explicitly on time, its eigenvalues, $\{n\}$ are time-independent. To show this we first find the derivative of (13.14) with respect to t;

$$\frac{\partial I}{\partial t}|n\rangle + I\frac{\partial}{\partial t}|n\rangle = \frac{\partial n}{\partial t}|n\rangle + n\frac{\partial}{\partial t}|n\rangle. \tag{13.16}$$

Now from (13.2) we have

$$i\frac{\partial I}{\partial t}|n\rangle + IH|n\rangle - nH|n\rangle = 0. \tag{13.17}$$

From this relation we find the matrix elements of $\frac{\partial I}{\partial t}$;

$$i\left\langle n' \left| \frac{\partial I}{\partial t} \right| n \right\rangle + (n' - n)\langle n'|H|n\rangle = 0, \tag{13.18}$$

and this result implies that

$$\left\langle n \left| \frac{\partial I}{\partial t} \right| n \right\rangle = 0. \tag{13.19}$$

By calculating the scalar product of Eq. (13.16) with $\langle n|$ we find that

$$\frac{\partial n}{\partial t} = 0, \tag{13.20}$$

and therefore the eigenvalues of $I(x, p, t)$ do not depend on time.
Matrix Elements of the First Integral $I(x, p, t)$ for a Solvable Case —
For calculating the matrix elements of $I(x, p, t)$ we introduce the lowering and the raising operators

$$a(t) = \frac{1}{\sqrt{2\hbar}} \left[\frac{x}{\rho(t)} + i(p\rho - x\dot{\rho}) \right], \tag{13.21}$$

and

$$a^\dagger(t) = \frac{1}{\sqrt{2\hbar}} \left[\frac{x}{\rho(t)} - i(p\rho - x\dot{\rho}) \right]. \tag{13.22}$$

The operators $a^\dagger(t)$ and $a(t)$ satisfy the commutation relations

$$[a(t),\, a^\dagger(t)] = 1. \tag{13.23}$$

As a solvable example let us consider the special case [9]

$$\Omega(t) = (\gamma_0 t)^{-1}, \tag{13.24}$$

where γ_0 is a constant. We observe that for this $\Omega(t)$, Eq. 13.12 can be solved analytically

$$\rho(t) = \left[Ct^{\left(\gamma_0 + \frac{i\gamma}{\gamma_0}\right)} + Dt^{\left(\gamma_0 - \frac{i\gamma}{\gamma_0}\right)} + 2\sqrt{\left(CD + \frac{\gamma_0^2}{\gamma^2}\right)t}\,\right]^{\frac{1}{2}}. \tag{13.25}$$

Here for physically acceptable solution as $t \to \infty$ we set $C = D = 0$, and thus we get

$$\rho(t) = \sqrt{\left(\frac{2\gamma_0}{\gamma}t\right)}. \tag{13.26}$$

Diagonal and Off-Diagonal Matrix Elements of H for a General $\rho(t)$ — The first integral $I(x, p, t)$, Eq. (13.1), can be written in terms of a and $a^\dagger$ (13.21) and (13.22).

$$I = \left(a^\dagger a + \frac{1}{2}\right), \tag{13.27}$$

from which it follows that the normalized eigenstates of I and $a^\dagger a$ are the same as those of the harmonic oscillator. Using the standard lowering and raising relations

$$a|n\rangle = \sqrt{n}\,|n-1\rangle, \tag{13.28}$$
$$a^\dagger|n\rangle = \sqrt{n+1}\,|n+1\rangle, \tag{13.29}$$

we find the eigenvalues of I to be

$$i_n = \left(n + \frac{1}{2}\right), \quad n = 0,\, 1,\, 2\cdots, \quad \hbar = 1. \tag{13.30}$$

Diagonal and Off-Diagonal Matrix Elements of the Time-Dependent Hamiltonian — In the representation in which $I(x, p, t)$ is diagonal, $H(x, p, t)$ is a matrix with diagonal as well as non-diagonal matrix elements. To find these elements we write the Hamiltonian (13.1) in terms of a and $a^\dagger$ Eqs. (13.21)-(13.22). In this way, for the diagonal elements we find

$$\langle n|H|n\rangle = \frac{1}{4m}\left(m^2\dot{\rho}^2(t) + \Omega^2(t)\rho^2(t) + \frac{1}{\rho^2(t)}\right)\langle n|\left\{a,\, a^\dagger\right\}_+|n\rangle$$
$$= \frac{1}{2m}\left(m^2\dot{\rho}^2(t) + \Omega^2(t)\rho^2(t) + \frac{1}{\rho^2(t)}\right)\left(n + \frac{1}{2}\right), \quad \hbar = 1,$$

$$\tag{13.31}$$

and for the off-diagonal elements of H we get:

$$
\langle n'|H|n \rangle = \frac{1}{4} \left\{ m \left[\dot{\rho}^2(t) - \ddot{\rho}(t)\rho(t) \right] - 2i \left(\frac{\dot{\rho}(t)}{\rho(t)} \right) \right\} \sqrt{n(n-1)} \, \delta_{n'+2,n}
$$

$$
+ \frac{1}{4} \left\{ m \left[\dot{\rho}^2(t) - \ddot{\rho}(t)\rho(t) \right] + 2i \left(\frac{\dot{\rho}(t)}{\rho(t)} \right) \right\} \sqrt{(n+1)(n+2)} \, \delta_{n',n+2},
$$

$$
\hbar = 1. \tag{13.32}
$$

13.2 Wave Function and the Coherent State Representation for the Time-Dependent Harmonic Oscillator

According to Eq. (13.30), i_n is the eigenvalue of the operator $I(x,p,t)$. If we denote the corresponding eigenfunction by $|i_n,t\rangle$, then we have

$$
I(t)|i_n,t\rangle = i_n|i_n,t\rangle, \quad \frac{\partial i_n}{\partial t} = 0. \tag{13.33}
$$

The eigenstate $|i_n,t\rangle$, in general, does not satisfy the Schrödinger equation. However if we modify the phases of the states $|i_n,t\rangle$ by writing

$$
|i_n,t\rangle_S = e^{i\alpha_n(t)}|i_n,t\rangle, \tag{13.34}
$$

and if we choose $\alpha_n(t)$ in such a way that

$$
\hbar \frac{d\alpha_n(t)}{dt} = \left\langle i_n,t \left| i\hbar \frac{\partial}{\partial t} - H(t) \right| i_n,t \right\rangle, \tag{13.35}
$$

then $|i_n,t\rangle_S$ will satisfy the Schrödinger equation. In Eq. (13.35) we have written down $\hbar$ explicitly. By integrating (13.35), we find that the phase $\alpha_n(t)$ can be expressed as the following integral [4]

$$
\alpha_n(t) = -\left(n + \frac{1}{2} \right) \int_0^t \frac{dt'}{\rho^2(t')}. \tag{13.36}
$$

We can write the general solution of the Schrödinger equation as a linear superposition of the eigenstates of $I(x,p,t)$ or H^*;

$$
|\psi,t\rangle_S = \sum_n c_n e^{i\alpha_n(t)}|i_n,t\rangle, \tag{13.37}
$$

where c_n s are the coefficients of expansion and $|\psi,t\rangle_S$ indicates that the states evolve in time according to the time-dependent Schrödinger equation.

The Schrödinger equation for the Hamiltonian H^*, Eq. (3.174) is

$$
H^*|\psi_n\rangle = E_n|\psi_n\rangle, \tag{13.38}
$$

where

$$E_n = \hbar\omega\left(n + \frac{1}{2}\right) = \hbar\sqrt{\omega_0^2 - \frac{\lambda^2}{4}}\left(n + \frac{1}{2}\right), \tag{13.39}$$

and

$$\begin{aligned}
\psi_n(x,t) &= \frac{1}{\sqrt{2^n n!}}\left(\frac{m\omega}{\hbar}\right)^{\frac{1}{4}}\exp\left(-\frac{m\omega Q^2}{2\hbar}\right)H_n\left(\sqrt{\frac{m\omega}{\hbar}}Q\right) \\
&\times \exp\left\{\left[\frac{\lambda - 2i\omega}{4} - \frac{in\lambda}{2}\right]t - \frac{im\lambda}{4\hbar}Q^2\right\}
\end{aligned} \tag{13.40}$$

This wave function, apart from a phase, is that of a stationary harmonic oscillator, however ω_0 has been replaced by the shifted frequency ω. The same change of frequency is also present in the eigenvalues E_n. Thus it seems that the problem of a damped oscillator can be reduced to that of a stationary problem. We can inquire whether in this case, for large quantum numbers, the quantum mechanical probability will oscillate about the classical probability, as it does for the the undamped oscillator or not?. The normalized classical probability can be obtained from the velocity of the particle $v(x)$ [10]

$$P_{cl}(x) = \frac{1}{\pi|v(x)|}. \tag{13.41}$$

For calculating $v(x)$ we consider the parametric equations for $x(t)$ and $v(t)$:

$$\begin{cases} x(t) = a_n e^{-\frac{\lambda t}{2}}\sin(\omega t) \\ v(t) = a_n\left(-\frac{\lambda}{2}e^{-\frac{\lambda t}{2}}\sin(\omega t) + \omega e^{-\frac{\lambda}{2}}\cos(\omega t)\right) \end{cases}. \tag{13.42}$$

In these relations a_n is the amplitude of the oscillation which is given by

$$a_n = \sqrt{\frac{2E_n}{m\omega^2}}. \tag{13.43}$$

Now we want to compare the probabilities of the classical and the quantum oscillators in the regime of large quantum numbers. To this end we equate the energies in the two cases;

$$E_n = \frac{1}{2}m\omega^2 a_n^2 = \left(n + \frac{1}{2}\right)\hbar\omega, \tag{13.44}$$

and thus we find that the amplitude is given by

$$a_n = \sqrt{\frac{(2n+1)\hbar}{m\omega}}. \tag{13.45}$$

Let us emphasize that in this approach the wave function (13.40) is stationary and E_n is constant. For large quantum numbers, $P_q(Q) = |\psi_n(Q,t)|^2$, which is the probability density oscillates about the classical probability $P_{cl}^d(Q) =$

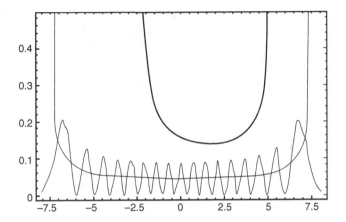

Figure 13.1: The classical probability $P_{cl}^d(Q)$ Eq. (13.41) for a damped harmonic oscillator for $\lambda = 0.3$, $\omega_0 = \frac{\pi}{5}$ and the amplitude $a_n = \sqrt{\frac{2n+1}{\omega_0}}$ shown by thin line. The quantum probability $|\psi(Q,t)|^2$ for the harmonic oscillator is also shown for $n = 16$. The probability for the damped oscillator calculated directly from the classical equations of motion, Eq. (13.42), is the U-shaped curve drawn with thick line.

$\frac{1}{\pi\sqrt{a_n^2-Q^2}}$. In Fig. 13.1 the probability $P_{cl}(Q)$ is shown by a thin line, and $P_q(Q)$ which is a rapidly oscillating function about $P_{cl}(Q)$ by thick line. For this calculation the parameters $\omega_0 = \frac{\pi}{5}$, $\lambda = 0.3$, $n = 16$ and $\hbar = m = 1$ have been used.

But how these compare with the classical probability obtained from Eq. (13.42)? Using the same set of parameters we find $P_{cl}^d(x)$ to be the U-shaped curve shown by the thick line in Fig. 13.1. Thus we conclude that the correct classical probability for a damped oscillator differs from the probability for the undamped oscillator in the following ways:

(a) Unlike the undamped probability, $P_{cl}(x)$, the probability for the damped motion $P_{cl}^d(x)$ is not symmetric about $x = 0$, and its minimum is not at zero, but is shifted to larger x.

(b) The width of $P_{cl}^d(x)$ relative to $P_{cl}(x)$ is shorter, i.e. the turning points are closer to each other. Thus we conclude that this formulation of the damped oscillation is not free of problems. It is worth mentioning that even for conservative systems, the quantization of the first integrals of motion, except for the Hamiltonian leads to unacceptable results [5],[11],[12].

13.3 First Integrals of Motion and the Creation and Annihilation Operators

There are a number of ways of obtaining a wave packet solution for a linear Schrödinger equation describing a damped harmonic oscillator. Among these the coherent state formulation which is closely related to the classical motion is of special interest [2],[13]. Here we will first determine the two first integrals of motion from the Hamiltonian operator for a linearly damped system, and then from these construct creation and annihilation operators. Having obtained the annihilation operator, we can find the coherent state of the damped motion.

We start with the one-dimensional wave equation $i\frac{\partial \psi}{\partial t} = \hat{H}\psi$, where

$$\hat{H} = \frac{1}{2}\left[-\frac{\partial^2}{\partial x^2}e^{-\lambda t} + \omega_0^2 x^2 e^{\lambda t} \right] - f e^{\lambda t}x, \tag{13.46}$$

and where for the sake of simplicity we have used the units where $\hbar = m = 1$. In this expression for $\hat{H}$, f is a constant external force which acts on the particle.

As we have seen earlier the Hamiltonian (13.46) has a symmetry property expressed by Eq. (12.12) [14]. There are also two independent first integrals of motion linear in x and $-i\frac{\partial}{\partial x}$ for this system similar to $I_\pm$ that was introduced earlier

$$\hat{I}(t) = a(t)x - i\xi(t)\frac{\partial}{\partial x} + \delta(t), \tag{13.47}$$

where $\xi(t)$ is a solution of (4.19), and $a(t)$ and $\delta(t)$ will be defined later. According to the definition of the first integral, $\hat{I}(t)$, satisfies the equation

$$\left[\left(i\frac{\partial}{\partial t} - \hat{H} \right), \, \hat{I}(t) \right] = 0. \tag{13.48}$$

From (13.47) and (13.48) it follows that

$$\dot{a}(t) = \omega_0^2 e^{\lambda t}\xi(t), \tag{13.49}$$

$$\dot{\xi}(t) = -e^{-\lambda t}a(t), \tag{13.50}$$

and

$$\dot{\delta} = -f e^{\lambda t}\xi(t). \tag{13.51}$$

Thus $\hat{I}(t)$ can be written as

$$\hat{I}(t) = -i\xi(t)\frac{\partial}{\partial x} - \dot{\xi}(t)e^{\lambda t}x - f\int^t e^{\lambda t'}\xi(t')\,dt'. \tag{13.52}$$

Since Eq. (4.19) admits two linearly independent solutions $\xi_\pm(t)$, therefore there are two independent first integrals of motion $\hat{I}_\pm(t)$.

We choose the two independent solutions of (4.19) to be

$$\xi_\pm = \frac{1}{\sqrt{\omega}}\exp\left[\left(\pm i\omega - \frac{\lambda}{2} \right)t \right], \quad \omega = \sqrt{\omega_0^2 - \frac{\lambda^2}{4}}, \tag{13.53}$$

so that

$$e^{\lambda t}\left(\dot{\xi}_+\xi_- - \dot{\xi}_-\xi_+ \right) = 2i. \tag{13.54}$$

We can now define two mutually Hermitian conjugate first integrals of motion $\hat{A}(t)$ and $\hat{A}^\dagger(t)$ satisfying the commutation relation

$$\left[\hat{A}(t), \hat{A}^\dagger(t)\right] = 1, \tag{13.55}$$

where $\hat{A}(t)$ and $\hat{A}^\dagger(t)$ are given by

$$\hat{A}(t) = \frac{i}{\sqrt{2}}\left[-i\xi_+(t)\frac{\partial}{\partial x} - \dot{\xi}_+(t)e^{\lambda t}x\right] + \frac{\delta(t)}{\sqrt{2}}, \tag{13.56}$$

and

$$\hat{A}^\dagger(t) = -\frac{i}{\sqrt{2}}\left[i\xi_-(t)\frac{\partial}{\partial x} - \dot{\xi}_-(t)e^{\lambda t}x\right] + \frac{\delta^*(t)}{\sqrt{2}}, \tag{13.57}$$

with

$$\delta(t) = -\frac{if}{\sqrt{\omega}}\frac{\exp\left[\left(i\omega + \frac{\lambda}{2}\right)t\right]}{\left(i\omega + \frac{\lambda}{2}\right)}. \tag{13.58}$$

By writing $\hat{A}(t)$ in the expanded form

$$\hat{A}(t) = \frac{1}{\sqrt{2\omega}}\left[\exp\left\{\left(i\omega - \frac{\lambda}{2}\right)t\right\}\frac{\partial}{\partial x} - i\left(i\omega - \frac{\lambda}{2}\right)\exp\left\{\left(i\omega + \frac{\lambda}{2}\right)t\right\}x\right]$$
$$- \frac{if}{\sqrt{2\omega}}\frac{\exp\left[\left(i\omega + \frac{\lambda}{2}\right)t\right]}{\left(i\omega + \frac{\lambda}{2}\right)}, \tag{13.59}$$

we observe that $\hat{A}(t)$ and $\hat{A}^\dagger(t)$ satisfy the symmetry properties

$$\hat{A}\left(t + \frac{2i\pi}{\lambda}\right) = -e^{-\frac{2\pi\omega}{\lambda}}\hat{A}(t), \tag{13.60}$$

and

$$\hat{A}^\dagger\left(t + \frac{2i\pi}{\lambda}\right) = -e^{\frac{2\pi\omega}{\lambda}}\hat{A}^\dagger(t). \tag{13.61}$$

We also observe that the eigenfunctions of the operator $\hat{A}^\dagger(t)\hat{A}(t)$ are [15]

$$\psi_n(x,t) = \left(\frac{1}{2^n n!\sqrt{\pi\hbar\xi_-\xi_+}}\right)^{\frac{1}{2}}\left(\frac{\xi_+}{\sqrt{\xi_-\xi_+}}\right)^{n+\frac{1}{2}}$$
$$\times H_n\left(\frac{x}{\sqrt{2\xi_-\xi_+}}\right)\exp\left[\frac{ie^{\lambda t}\dot{\xi}_- x^2}{2\xi_-}\right]. \tag{13.62}$$

Wave Function for the Coherent State — The coherent state wave function which we denote by $\psi(\alpha, x, t)$ is the eigenstate of $\hat{A}(t)$, i.e.

$$\hat{A}(t)\psi(\alpha, x, t) = \alpha\psi(\alpha, x, t), \tag{13.63}$$

and at the same time it satisfies the equation [14]

$$\hat{A}^{\dagger}(t)\psi(\alpha, x, t) = \frac{\partial}{\partial \alpha}\left[\exp\left(\frac{|\alpha|^2}{2}\right)\psi(\alpha, x, t)\right]\exp\left(-\frac{|\alpha|^2}{2}\right). \tag{13.64}$$

The wave function $\psi(\alpha, x, t)$ which is the solution of (13.63),(13.64) is given by

$$\psi(\alpha, x, t) = \left(\frac{\omega}{\pi}\right)^{\frac{1}{4}}\exp\left[-\frac{i}{2}\left(\omega + \frac{i\lambda}{2}\right)t - \frac{1}{2}\left(\omega + \frac{i\lambda}{2}\right)e^{\lambda t}x^2\right.$$
$$+ \sqrt{2\omega}\alpha x e^{\left(\frac{\lambda}{2} - i\omega\right)t} - \frac{1}{2}\alpha^2 e^{-2i\omega t} - \frac{1}{2}|\alpha|^2 - \sqrt{\omega}x e^{\left(\frac{\lambda}{2} - i\omega\right)t}\delta(t)$$
$$\left. - \alpha\sqrt{2\omega}e^{\left(\frac{\lambda}{2} - i\omega\right)t}z(t) - \frac{\omega}{2}z^2(t)e^{\lambda t} + i\phi(t)\right], \tag{13.65}$$

where $\phi(t)$ and $z(t)$ are defined by

$$\phi(t) = \frac{\omega}{2}\ \mathrm{Re}\int^t e^{-2i\omega t}\delta^2(t)dt, \tag{13.66}$$

and

$$z(t) = -\mathrm{Re}\left[\frac{\delta(t)}{\sqrt{\omega}}e^{-\left(\frac{\lambda}{2} + i\omega\right)t}\right]. \tag{13.67}$$

Returning to the symmetry property of the Hamiltonian expressed by (12.13) we observe that the wave function $u(\alpha, x, t)$ defined by

$$u(\alpha, x, t) = \psi(\alpha, x, t)\exp\left(\frac{1}{2}|\alpha|^2\right) \tag{13.68}$$

has a symmetry property similar to $\hat{H}$

$$u\left(\alpha, x, t + \frac{2\pi i}{\lambda}\right) = u\left(-\alpha e^{\frac{2\pi\omega}{\lambda}}, x, t\right)\exp\left(\frac{i\pi}{2} + \frac{\pi\omega}{\lambda}\right). \tag{13.69}$$

An important feature of the wave packet (13.65) is that the product of the widths $\langle x^2 \rangle$ and $\langle p^2 \rangle$ is independent of time,

$$\langle x^2 \rangle = \int_{\infty}^{\infty}\psi^* x^2 \psi dx = \frac{e^{-\lambda t}}{2\omega}, \tag{13.70}$$

and

$$\langle p^2 \rangle = \int_{\infty}^{\infty}\psi^*\left(-\frac{\partial^2}{\partial x^2}\right)\psi dx = \frac{\omega_0^2 e^{\lambda t}}{2\omega}, \tag{13.71}$$

therefore

$$\langle x^2 \rangle\langle p^2 \rangle = \frac{\omega_0^2}{4\omega^2} \geq \frac{1}{4}. \tag{13.72}$$

Thus in this solution of the damped harmonic oscillator the uncertainty relation remains valid for all times.

13.4 Coherent State for the Central Oscillator

For a single harmonic oscillator let us define the coherent state as an eigenstate of the annihilation operator with the eigenvalue α

$$a|\alpha\rangle = \alpha|\alpha\rangle, \tag{13.73}$$

where $\alpha = u + iv$ is a complex number. We can write this state as an infinite series in terms of the number states as

$$|\alpha\rangle = \exp\left(-\frac{1}{2}|\alpha|^2\right) \sum_{n=0}^{\infty} \frac{\alpha^n}{\sqrt{n!}}|n\rangle. \tag{13.74}$$

If a and $a^\dagger$ denote annihilation and creation operators for a harmonic oscillator, i.e.

$$a = \frac{1}{\sqrt{2\hbar\omega_0}}(\omega_0 q + ip), \tag{13.75}$$

$$a^\dagger = \frac{1}{\sqrt{2\hbar\omega_0}}(\omega_0 q - ip), \tag{13.76}$$

then

$$[a, a^\dagger] = 1, \tag{13.77}$$

and

$$|n\rangle = \frac{(a^\dagger)^n}{\sqrt{n!}}|0\rangle. \tag{13.78}$$

From Eqs. (13.74) and (13.78) it follows that

$$|\alpha\rangle = \exp\left(-\frac{1}{2}|\alpha|^2\right) \exp\left(aa^\dagger\right)|0\rangle. \tag{13.79}$$

The transformation from $|0\rangle$ to $|\alpha\rangle$ as is shown in (13.79) can be generated by the unitary operator $D(\alpha)$:

$$D(\alpha) = \exp\left(\alpha a^\dagger - a^\dagger a\right). \tag{13.80}$$

This follows from Baker-Campbell-Hausdorff identity for two operators A and B, when $[A, B]$ is a c-number, i.e. from

$$\exp A \exp B = \exp\left(A + B + \frac{1}{2}[A, B]\right). \tag{13.81}$$

Since $|n\rangle$ is an eigenstate of the Hamiltonian for the harmonic oscillator with the eigenvalue $\hbar\omega_0\left(n + \frac{1}{2}\right)$, we can find the time-dependence of $|\alpha\rangle$ by noting that

$$|\alpha, t\rangle = U(t)|\alpha, t = 0\rangle = \exp\left(-\frac{iH_0 t}{\hbar}\right)|\alpha\rangle = \exp\left(-\frac{1}{2}|\alpha|^2\right) \sum_{n=0}^{\infty} \frac{1}{\sqrt{n!}}(\alpha)^n$$

$$\times \exp\left[-i\omega_0 t\left(n + \frac{1}{2}\right)\right]|n\rangle = \exp\left(-\frac{i}{2}\omega_0 t\right)|\alpha(t)\rangle, \tag{13.82}$$

where

$$\alpha(t) = \alpha \exp(-i\omega_0 t). \tag{13.83}$$

Now let us consider a general Hamiltonian which is given in terms of annihilation and creation operators, and that it is arranged in normal order, i.e. all a s are to the right of $a^\dagger$ s:

$$H = \sum_{j,k} h_{j,k}(t) \left(a^\dagger\right)^j (a)^k. \tag{13.84}$$

Here $h_{jk}(t)$ are coefficients arising from the normal ordering and may be time-dependent. Next we write the Schrödinger equation with the Hamiltonian (13.84) as

$$i\hbar \frac{\partial}{\partial t} |\Psi, t\rangle = H |\Psi, t\rangle, \tag{13.85}$$

and also write the formal solution of (13.85) in terms of the operator $U(t)$ to be determined later

$$|\Psi, t\rangle = U(t)|\Psi, 0\rangle \tag{13.86}$$

Obviously the initial condition for $U(t)$ is $U(0) = 1$, and the time evolution of $U(t)$ follows from Eq. (13.85);

$$i\hbar \frac{\partial}{\partial t} U(t) = HU(t). \tag{13.87}$$

Now denoting the normal ordered form of $U(t)$ by $U^{(n)}$, we can express $a^j U^{(n)}(t)$ as [16],[17]

$$a^j U^{(n)}(t) = \mathcal{N} \left\{ \left(\alpha + \frac{\partial}{\partial \alpha^*}\right)^j \bar{U}^{(n)}(\alpha, \alpha^*, t)\right\} \tag{13.88}$$

where

$$\bar{U}^{(n)}(\alpha, \alpha^*, t) = \langle \alpha | U\left(a^\dagger, a, t\right)| \alpha\rangle, \tag{13.89}$$

and where $|\alpha\rangle$ is a coherent state and $\mathcal{N}$ denotes the normal ordering operator. The operator $\mathcal{N}$ replaces α^* by $a^\dagger$ and α by a, with all $a^\dagger$ s to the left of all a s. Thus by substituting (13.84) in (13.85), and with the help of (13.88) we find the c-number equation

$$i\hbar \frac{\partial}{\partial t} \bar{U}^{(n)}(t) = \sum_{j,k} h_{j,k}(t) (\alpha^*)^j \left[\alpha + \frac{\partial}{\partial \alpha^*}\right]^k \bar{U}^{(n)}(t). \tag{13.90}$$

Once the solution of this equation is found we can determine $|\Psi, t\rangle$ from

$$|\Psi, t\rangle = \mathcal{N} \left[\bar{U}^{(n)}(\alpha, \alpha^*, t)\right] |\Psi, 0\rangle. \tag{13.91}$$

13.5 Coherent States for Harmonic Oscillator with Time-Dependent Frequency

We have seen that the motion of a linearly damped oscillator, in classical or quantum mechanics, subject to certain conditions, can be transformed to that of a harmonic oscillator with time-dependent frequency. There are a number of other important applications of the latter problem, among them, the well-known example of semiclassical theory of radiation, which can be reduced to the problem of an oscillator with time-dependent frequency [4]. In this case, the problem is that of an atom or a molecule undergoing radiation transition, and the Hamiltonian operator describes the interaction of the atom or the molecule with the classical radiation field. Here we assume that the Hamiltonian tends to constant operators in sufficiently remote past and future, and in these limits, the operators have known complete sets of eigenvalues and eigenstates [4].

Earlier we defined the coherent state for an oscillator when H was a constant of motion. We assume the validity of Eqs. (13.73)-(13.80) at time $t = 0$. But at later times, $|n\rangle$ will no longer remain an eigenstate of the Hamiltonian, rather, it can be regarded as an eigenstate of the first integral of motion, H^* as we noted in Eq. (13.38). By allowing for $|\alpha, t\rangle_S$, Eq. (13.40) to evolve in time we find

$$|\alpha, t\rangle_S = \exp\left(-\frac{1}{2}|\alpha^2|\right) \sum_n \frac{\alpha^n}{\sqrt{n!}} |n, t\rangle_S, \qquad (13.92)$$

where

$$|n, t\rangle = e^{-i\alpha_n(t)} e^{-\frac{iHt}{\hbar}} |n, 0\rangle, \qquad (13.93)$$

and $\alpha_n(t)$ is given by (13.36). When $\omega(t) = \omega_0$ is independent of time then $\alpha_n(t) = -\left(n + \frac{1}{2}\right)\omega_0 t$ and (13.93) reduces to

$$|n, t\rangle = |n, 0\rangle. \qquad (13.94)$$

Coherent State Wave Function for Variable Mass Oscillator — In Sec. 12.3 we found the solution of the time-dependent Schrödinger equation using time evolution operator. Now we want to apply the same method to find the time development of a coherent state. Let us assume that initially the oscillator in the coherent state which we denote by $|\alpha\rangle$ [62],[63]

$$|\psi(0)\rangle = |\alpha\rangle. \qquad (13.95)$$

At any later time this wave function becomes $|\psi(t)\rangle$, where according to (12.76)

$$|\psi(t)\rangle = U(t, 0)|\alpha\rangle, \qquad (13.96)$$

Let us define the coherent state operator $\hat{\alpha}$ at time $t = 0$ by

$$\hat{\alpha} = \frac{1}{\sqrt{2m(0)\hbar\omega}}(m(0)\omega x + ip). \qquad (13.97)$$

This operator is proportional to $\hat{I}(t = 0)$ where $\hat{I}(t)$ is given by (13.47) and has the eigenvalue α and an eigenvector $|\alpha\rangle$:

$$\hat{a}|\alpha\rangle = \alpha|\alpha\rangle. \tag{13.98}$$

The time dependence of $\hat{a}$ can be found from

$$\hat{A}(t) = U(t,0)\hat{a}U^\dagger(t,0), \tag{13.99}$$

and the eigenvalue equation for the operator $\hat{A}(t)$ is given by (13.63).

Some of the Properties of the Coherent States — As we have seen, by construction, the state $|\alpha, t\rangle$ is the eigenstate of the lowering operator $a(t)$ with the eigenvalue $\alpha \exp[2i\alpha_0(t)]$, where $\alpha_0(t)$ is given by (13.36). Other states can be found from the ground state using the unitary operator $D(\alpha)$. That is we first construct the state $|\alpha, 0\rangle_S$ from the ground state by operating with $D(\alpha)$ and then letting $|\alpha, 0\rangle_S$ evolve in time.

Now we inquire whether these coherent states are states of minimum uncertainty or not. We can calculate x, x^2, p and p^2 in terms of $a(t)$ and $a^\dagger(t)$, Eqs. (13.21),(13.22), and then find $(\Delta x)^2$ and $(\Delta p)^2$ using the coherent states $|\alpha, t\rangle_S$. This calculation yields the following results:

$$(\Delta x)^2 = \frac{\hbar}{2}\rho^2(t) \tag{13.100}$$

and

$$(\Delta p)^2 = \frac{\hbar}{2}\left(\dot{\rho}^2(t) + \frac{1}{\rho^2}\right). \tag{13.101}$$

Thus the uncertainty relation in this case becomes

$$\Delta x \Delta p = \frac{\hbar}{2}\sqrt{\rho^2(t)\dot{\rho}^2(t) + 1}. \tag{13.102}$$

This relation shows that the coherent state defined by (13.92) is not a state of minimum uncertainty and that it spreads in time [20]. Next let us determine the motion of the center of the wave packet $|\alpha, t\rangle_S$ which is defined by

$$\bar{x}(t) =_S \langle \alpha, t|x|\alpha, t\rangle_S. \tag{13.103}$$

To this end we introduce a time-dependent frequency $\Phi(t)$ by

$$\Phi(t) = -2\alpha_0(t) = \int_0^t \frac{dt'}{\rho^2(t')}. \tag{13.104}$$

Using the wave packet (13.92) we calculate the expectation value of x, Eq. (13.103) to find that

$$\bar{x} = \sqrt{2\hbar|\alpha|^2\rho^2(t)}\sin[\Phi(t) + \delta], \tag{13.105}$$

where δ is the phase of α. The solution of the equation of motion for the time-dependent harmonic oscillator

$$\ddot{x}_{cl} + \omega^2(t)x_{cl} = 0, \tag{13.106}$$

can be written in terms of $\rho(t)$ and $\Phi(t)$ the two functions appearing in the coherent state description,

$$x_{cl}(t) = \sqrt{2I\rho^2(t)}\sin[\Phi(t) + \delta]. \tag{13.107}$$

In this relation $I(x, p, t)$ is the classical counterpart of the operator $I(x, p, t)$, Eq. (13.13), but with the classical variable, $p \to p_{cl}$ and $x \to x_{cl}$. Therefore (13.105) is exactly the same as classical oscillator with the invariant $\hbar|\alpha|^2 = s$ $\langle \alpha, t|I|\alpha, t \rangle_S - \frac{1}{2}\hbar$ [20].

The Central Oscillator Coupled to a Large Number of Oscillators — As an example of the application of coherent state we consider the quantization of Ullersma's model [16]. Writing P, Q and p_n s and q_n s, (see Eq. (9.1)), in terms of creation and annihilation operators and arranging them in normal order with the help of commutation relations, we find

$$H = \hbar\Omega_0 a^\dagger a + \sum_{n=1}^{\infty} \hbar\omega_n \left(b_n^\dagger b_n + \frac{1}{2} \right)$$

$$+ \sum_{n=1}^{\infty} \frac{\hbar\epsilon_n}{\sqrt{2\omega_n\Omega_0}} \left(a^\dagger b_k + ab_k^\dagger \right) + c\text{-number terms}, \tag{13.108}$$

where a and $a^\dagger$ are annihilation and creation operators for the central particle, and b_n and $b_n^\dagger$ are similar operators for the heat bath. In order to find the energy exchanged between these two parts, we introduce the function $\bar{U}^{(n)}$ as in Eq. (13.89) which now will depend on the coherent state α for a and β_k for b. Furthermore we define a function $G(\alpha, \alpha^*, \{\beta_k\}, \{\beta_k^*\}, t)$ by

$$\bar{U}^{(n)}(\alpha, \alpha^*, \{\beta_k\}, \{\beta_k^*\}, t) = \exp[G(\alpha, \alpha^*, \{\beta_k\}, \{\beta_k^*\}, t)], \tag{13.109}$$

where

$$G = A(t)\alpha\alpha^* + \sum_{k=1}^{\infty} B_k(t)\beta_k^*\beta_k + \sum_{k=1}^{\infty} C_k(t)(\alpha^*\beta_k + \beta_k^*\alpha). \tag{13.110}$$

By substituting (13.109) and (13.110) in (13.90) we find that the coefficients $A(t)$, $B_n(t)$ and $C_n(t)$ satisfy the following equations

$$i\frac{d\mathcal{A}}{dt} = \Omega_0 \mathcal{A} + \sum_{k=1}^{\infty} \frac{\epsilon_k}{\sqrt{2\omega_k\Omega_0}} C_k, \tag{13.111}$$

$$i\frac{d\mathcal{B}_k}{dt} = \omega_k \mathcal{B} + \frac{\epsilon_k}{\sqrt{2\omega_k\Omega_0}} C_k, \tag{13.112}$$

and

$$i\frac{dC_k}{dt} = \omega_k C_k + \frac{\epsilon_k}{\sqrt{2\omega_k \Omega_0}} B_k,$$

(13.113)

where

$$\mathcal{A}(t) = A(t) + 1, \quad \text{and} \quad \mathcal{B} = B(t) + 1$$

(13.114)

The initial condition $U(t=0) = 1$ requires that

$$A(0) = B_n(0) = C_n(0) = 0.$$

(13.115)

Since we are interested in the motion of the central particle, we only need to solve Eqs. (13.111) and (13.113). The solutions of equations similar to these have been discussed before for the classical systems (Eqs. (9.5) and (9.6)). Later we will be considering the same set of equations in connection with the quantum theory of natural line width (Sec. 18.9). By assuming that the number of oscillators is large enough so that they can be considered a dense set we replace the summation over n by integration. To this end we define a function $\gamma(\omega)d\omega$ by

$$\gamma(\omega)d\omega = \sum_{\omega < \omega_n < \omega + \delta\omega} \frac{\epsilon_n^2}{\sqrt{2\omega_n \Omega_0}},$$

(13.116)

and then we use the method given in Sec. 9.1 to find that

$$A(t) = e^{-i\Omega_1 t} e^{-\frac{\lambda t}{2}} - 1.$$

(13.117)

where

$$\Omega_1^2 = \Omega_0^2 - \mathcal{P}\int \frac{\gamma(\omega)}{(\omega^2 - \Omega_0^2)},$$

(13.118)

and

$$\lambda = \frac{\pi}{2}\frac{\gamma(\Omega_0)}{\Omega_0^2}.$$

(13.119)

Having found the time-dependent coefficients $A(t), B(t)$ and $C(t)$, we can write the wave function in terms of coherent sates. First we note that at $t = 0$, when there is no coupling between the central oscillator and the oscillators forming the heat bath, the many particle wave function will be formed from the products of single particle wave functions, i.e.

$$|\Psi, 0\rangle = |\alpha, \{\beta_n\}\rangle = |\alpha\rangle |\{\beta_n\}\rangle.$$

(13.120)

After a time t the total wave function in the coherent state representation becomes

$$|\Psi, t\rangle = U(t)|\Psi, 0\rangle = \exp\left[A\alpha a^\dagger + \int h(\omega)B(\omega, t)\beta(\omega)b^\dagger(\omega)d\omega \right.$$

$$\left. + \int h(\omega)C(\omega, t)\left(\beta(\omega)a^\dagger + \alpha b^\dagger(\omega)\right)d\omega \right] |\alpha\rangle |\{\beta_n\}\rangle.$$

(13.121)

If we assume that initially all of the oscillators in the heat bath are in their ground state, then from (13.121) it follows that only terms depending on a and $a^\dagger$ survive

$$|\Psi,t\rangle = \exp\left(A\alpha a^\dagger\right)\exp\left[\int h(\omega)C(\omega,t)\alpha b^\dagger(\omega)d\omega\right]|\alpha\rangle|0\rangle. \tag{13.122}$$

Using Eq. (13.79) we can write (13.122) as

$$|\Psi,t\rangle = \exp\left(-\frac{1}{2}|\alpha|^2\right)\left[\exp\left(\frac{1}{2}|\alpha\mathcal{A}(t)|^2\right)\right]\exp\left[\int h(\omega)C(\omega,t)\alpha b^\dagger(\omega)d\omega\right]|\chi(t)\rangle|0\rangle, \tag{13.123}$$

where

$$|\chi(t)\rangle = |\alpha\mathcal{A}(t)\rangle = \exp\left(-\frac{1}{2}|\alpha\mathcal{A}(t)|^2\right)\exp\left[\mathcal{A}(t)\alpha a^\dagger\right]|0\rangle, \tag{13.124}$$

and $\chi(t)$ is a non-normalized coherent state. The function $\chi(t)$ is an eigenvalue of the state $|\Psi\rangle$ for any time t;

$$a|\Psi,t\rangle = \chi(t)|\Psi,t\rangle = \alpha\mathcal{A}(t)|\Psi,t\rangle$$
$$= e^{-i\Omega t}e^{-\frac{\lambda t}{2}}\alpha|\Psi,t\rangle. \tag{13.125}$$

The expectation values of p, q and H can easily be determined with the coherent state wave function $|\Psi,t\rangle$. For instance by writing q as (see Eqs. (13.75) and (13.76))

$$q = \sqrt{\frac{\hbar}{2\omega}}\left(a + a^\dagger\right), \tag{13.126}$$

we find

$$\langle\Psi,t|q|\Psi,t\rangle = \left[\frac{2\hbar|\alpha|^2}{\Omega_0}\right]^{\frac{1}{2}}e^{-\frac{\lambda t}{2}}\sin(\Omega t + \phi), \tag{13.127}$$

where we have denoted the phase of α by ϕ, i.e. $\alpha = |\alpha|e^{-i\phi}$.

References

[1] H.R. Lewis, Phys. Rev. Lett. 18, 635 (1967).

[2] M. Razavy, Phys. Rev. 44, 2384 (1991).

[3] H.R. Lewis and P.G.L. Leach, J. Math. Phys. 23, 2371 (1982).

[4] H.R. Lewis and W.B. Riesenfeld, J. Math. Phys. 10, 1458 (1969).

[5] M. Razavy, *Heisenberg's Quantum Mechanics*, (World Scientific, Singapore, 2011).

[6] M. Razavy, Phys. Lett. A 118, 387 (1986).

[7] J. Haitarinta, J. Math. Phys. 25, 1883 (1984).

[8] J. Haitarinta, Phys. Lett. A 246, 97 (1998).

[9] J.M. Cerveró and J. Villarroel, J. Phys. 17, 1777 (1984).

[10] See, for example, M.A. Morrison *Understanding Quantum Mechanics: A User's Manual*, (Prentice-Hall, Englewood Cliffs, 1990) Chapter 9.

[11] F.J. Kennedy and E.H. Kerner, Am. J. Phys. 33, 463 (1965).

[12] F. Calogero and A. Degasperis, Am. J. Phys. 72, 1202 (2004).

[13] V.V. Dodonov and V.I. Man'ko, Nuovo Cimento, 44 B, 265 (1978).

[14] V.V. Dodonov and V.I. Man'ko, Phys. Rev. A 20, 550 (1979).

[15] S.P. Kim, J. Phys. A 36, 12089 (2003).

[16] I.A. Pedrosa and B. Baseia, Phys. Rev. D 30, 765 (1984).

[17] W.H. Louisell and J.H. Marburger, IEEE J. Quantum Electronics, QE-3 383 (1967).

[18] J.R. Klauder and B-S Skagerstam, *Coherent States: Applications in Physics and Mathematical Physics*, (World Scientific, Singapore, 1985).

[19] J-P. Gazeau, *Coherent States in Quantum Physics*, (Wiley-VCH, New York, 2009).

[20] J. Hartley and J.R. Ray, Phys. Rev. D 25, 385 (1982).

Chapter 14

Density Matrix and the Wigner Distribution Function

In Chapters 2, 3 and 4 we discussed the analytical dynamics of simple damped systems from the point of view of Lagrangian, Hamiltonian and Hamilton-Jacobi formulations. Since the quantization of dissipative motion based on variational method are, as we have seen, problematic, there have been attempts to circumvent some of the difficulties by the direct use of the classical (Newtonian) equations of motion. The Yang-Feldman method discussed in Sec. 11.3 is one way of quantizing the second law of motion directly. In this chapter we want to study another powerful method of studying dissipative system in quantum theory which is based on the idea of the density matrix (or the density operator) and its connection to the classical Liouville theorem [1]-[3].

14.1 Classical Distribution Function for Nonconservative Motions

According to the Liouville theorem, for a conservative system the distribution function or the density in phase space which here we denote by $w_c(x, p, t)$ satisfies the equation [4],[5]

$$\frac{dw_c}{dt} = \frac{\partial w_c}{\partial t} + \{w_c, H\} = 0, \tag{14.1}$$

where $\{w_c, H\}$ is the Poisson bracket,

$$\{w_c, H\} = \sum_j \left(\frac{\partial w_c}{\partial x_j} \frac{\partial H}{\partial p_j} - \frac{\partial w_c}{\partial p_j} \frac{\partial H}{\partial x_j} \right). \tag{14.2}$$

323

Different ways of extending the Liouville equation, (14.1), to the non-conservative system have been suggested by Gerlich [6], by Steep [7] and by Bolivar [8]-[11].

We will study a generalization of the Liouville theorem for the simple case of a one-dimensional system where the equations of motion are given by

$$\dot{x} = X_1(x, p, t), \tag{14.3}$$

and

$$\dot{p} = X_2(x, p, t). \tag{14.4}$$

Since (14.1) is an equation expressing the conservation of $w_c(x, p, t)$ as a function of time, we generalize the Liouville theorem so that the new form also expresses a conservation law, i.e.

$$\frac{\partial w_c}{\partial t} + \text{div}\,(w_c \mathbf{X}) = 0, \tag{14.5}$$

where

$$\text{div}(\mathbf{X}) = \frac{\partial X_1}{\partial x} + \frac{\partial X_2}{\partial p}. \tag{14.6}$$

Equation (14.5) has the correct limit when damping becomes negligible. Thus for the motion of a single particle in the absence of dissipation $X_1 = p/m$, and $X_2 = X_2(x, t)$, therefore div$(\mathbf{X}) = 0$, and

$$\frac{dw_c}{dt} = \frac{\partial w_c}{\partial t} + \frac{p}{m}\frac{\partial w_c}{\partial x} + X_2(x, t)\frac{\partial w_c}{\partial p} = 0, \tag{14.7}$$

which is the same as Eq. (14.1).

The function $w_c(x, p, t)$ is the analogue of the Wigner distribution function $W(x, p, t)$ which will be introduced later [12]. To establish the connection between $w_c(x, p, t)$ and the quantum mechanical wave function, we introduce a new function $\chi\left(x + \frac{1}{2}\delta\xi, x - \frac{1}{2}\delta\xi\right)$ by [8],[9],[13]

$$\chi\left(x + \frac{1}{2}\delta\xi, x - \frac{1}{2}\delta\xi\right) = \int_{-\infty}^{\infty} w_c(x, p, t) \exp\left(\frac{ip\delta\xi}{\ell}\right) dp, \tag{14.8}$$

where ℓ is a constant having the dimension of action and $\delta\xi$ is a small but nonzero quantity. First let us obtain the equation that χ satisfies when the system is conservative and the Hamiltonian is given by $H = \frac{1}{2m}p^2 + V(x)$ and therefore $X_2(x) = -dV(x)/dx$. Substituting for X_2 in (14.7), multiplying the resulting equation by $\exp\left(\frac{ip\delta\xi}{\ell}\right)$ and integrating over all p we find

$$-\frac{\partial\chi}{\partial t} + \frac{i\ell}{m}\frac{\partial^2\chi}{\partial x\partial(\delta\xi)} - \frac{i}{\ell}\delta V(x)\chi = 0, \tag{14.9}$$

where we have used the condition

$$\left|w_c(x, p, t)\exp\left(\frac{ip\delta\xi}{\ell}\right)\right|_{p=-\infty}^{p=\infty} = 0, \tag{14.10}$$

which is expected from a probability distribution function. In Eq. (14.9) $\delta V(x)$ stands for the expression

$$\delta V(x) = \frac{\partial V(x)}{\partial x}\delta\xi = V\left(x + \frac{1}{2}\delta\xi\right) - V\left(x - \frac{1}{2}\delta\xi\right). \tag{14.11}$$

Now if we change the variables to

$$y = x + \frac{1}{2}\delta\xi, \quad \text{and} \quad y' = x - \frac{1}{2}\delta\xi, \tag{14.12}$$

then we can write (14.9) as

$$\left[\frac{\ell^2}{2m}\left(\frac{\partial^2}{\partial y^2} - \frac{\partial^2}{\partial y'^2}\right) - \{V(y) - V(y')\}\right]\chi(y, y', t) = -i\ell\frac{\partial}{\partial t}\chi(y, y', t). \tag{14.13}$$

Thus for a conservative system, if we assume that $\chi(y, y', t)$ can be written as

$$\chi(y, y', t) = \psi(y, t)\psi^*(y', t), \tag{14.14}$$

then $\psi^*(y', t)$ and $\psi(y, t)$ each satisfy the Schrödinger equation if we set $\ell = \hbar$.

Classical Distribution Function and Quantization of Damped System — In this method of quantization our starting point is classical distribution function $w_c(x, p, t)$. rather than the classical canonical formulation of the motion. If we apply this approach to to quantize conservative systems we obtain the Schrödinger equation. But using the present method has the added advantage that we can circumvent problems associated with trying to quantize classical motions given in terms of the generalized (non-Cartesian) coordinates [13]. The same method can be applied to quantize non-conservative systems.

Let us consider the details of the formulation for a one-dimensional non-conservative system. Here we choose

$$X_1 = \frac{p}{m}, \quad \text{and} \quad X_2 = F_f(x, p, t) - \frac{\partial V(x, t)}{\partial x}, \tag{14.15}$$

then the equation for w_c takes the form

$$\frac{\partial w_c}{\partial t} + \frac{p}{m}\frac{\partial w_c}{\partial x} - \frac{\partial V(x, t)}{\partial x}\frac{\partial w_c}{\partial p} = \frac{\partial}{\partial p}[F_f(x, p, t)w_c]. \tag{14.16}$$

Following the same steps outlined earlier we find the following equation for $\chi(y, y', t)$

$$i\hbar\frac{\partial}{\partial t}\chi(y, y', t) = -\frac{\hbar^2}{2m}\left(\frac{\partial^2}{\partial y^2} - \frac{\partial^2}{\partial y'^2}\right)\chi(y, y', t) + \{V(y) - V(y')\}$$
$$\times \chi(y, y', t) - i\hbar I(y, y', t)\chi(y, y', t), \tag{14.17}$$

where

$$\hbar I(y, y', t)\chi(y, y', t) = \int\frac{\partial}{\partial p}[F_f(x, p, , t)w_c]\exp\left(\frac{ip\delta\xi}{\hbar}\right)dp, \tag{14.18}$$

and where y and y' are given by (14.12).

In general the operator $I(y, y', t)$ cannot be obtained in closed form, and even if $I(y, y', t) \chi(y, y', t)$ can be found analytically, the function $\chi(y, y', t)$ will not be separable, i.e, it cannot be written as $\psi^*(y', t) \psi(y, t)$. However for some simple cases such as linear damping where, $F_f = -\lambda p$, one can find $I(y', y, t)$ as a differential operator. For this type of damping the last term in (14.17) takes the form [10]

$$i\hbar I(y, y', t) \chi(y, y', t) = -\frac{i\hbar\lambda}{2}(y - y')\left(\frac{\partial}{\partial y} - \frac{\partial}{\partial y'}\right) \chi(y, y', t). \qquad (14.19)$$

Thus for this case $\chi(y, y', t)$ satisfies a partial differential equation of second order in y and y', and first order in time.

We note that in this approach we can bypass the construction of the classical Lagrangian and the Hamiltonian for the damped system, but like other cases the result is not unique and one can find different wave equations even for the simple case of linear damping [8].

The initial condition needed for the solution of the differential equation for $\chi(y, y', t)$ can be obtained from the initial condition imposed on $w_c(x, p, t)$. Since $w_c(x, p, t)$ is the classical distribution function, for the motion of a single particle at $t = 0$, this function is given by

$$w_c(x, p, 0) = \delta(x - x_0)\delta(p - p_0). \qquad (14.20)$$

But this initial form of w_c is not suitable for finding the quantum mechanical distribution function $\chi(y, y', t)$, since at $t = 0$ the exact values of x and p are given in violation of the uncertainty principle. Thus instead of (14.20) we can choose a Gaussian form for w_c;

$$w_c(x, p, 0) = \frac{1}{\pi\ell} \exp\left[-\frac{\varepsilon}{\ell}(x - x_0)^2\right] \exp\left[-\frac{1}{\varepsilon\ell}(p - p_0)^2\right], \qquad (14.21)$$

where ε is an arbitrary positive constant having the dimension of (mass/time). By substituting $w_c(x, p, 0)$ in (14.8) and then setting $\ell = \hbar$, we find the initial value of χ to be

$$\chi(y, y', 0) = \sqrt{\frac{\varepsilon}{\pi\hbar}} \exp\left[-\frac{\varepsilon}{2\hbar}(y - y_0)^2\right] \exp\left[-\frac{\varepsilon}{2\hbar}(y' - y_0')^2\right]. \qquad (14.22)$$

Here we have assumed that the initial form of $\chi(y, y', 0)$ is separable, but this does not guarantee that $\chi(y, y', t)$ will remain as a product of two factors $\psi^*(y', t) \psi(y, t)$ at some later time.

14.2 The Density Matrix

If a state of the system is given by the wave function $\psi(t)$, then we define the density matrix by the outer product

$$\rho(t) = |\psi(t)\rangle \langle\psi(t)|. \qquad (14.23)$$

This is the case when we are considering a pure state [1]. However if the system is not known precisely, but is known to be in a state $|\psi_n(t)\rangle$ with the probability p_n, then we define $\rho(t)$ in the Schrödinger picture by

$$\rho(t) = \sum_n p_n |\psi_n(t)\rangle \langle \psi_n(t)|, \tag{14.24}$$

with

$$\sum_n p_n = 1. \tag{14.25}$$

In either case we have the important result that

$$\text{Trace } \rho(t) = 1. \tag{14.26}$$

This follows from the fact that trace of $\rho(t)$ means summing over the diagonal elements of $\rho(t)$, i.e.

$$\text{Trace } \rho(t) = \sum_j \sum_n p_n \langle \psi_j|\psi_n\rangle \langle \psi_n|\psi_j\rangle = \sum_n p_n = 1. \tag{14.27}$$

In addition to its trace being equal to one, the density matrix has the following important properties:

(1) The mean value of an operator $\mathcal{O}$ can be written as

$$\langle \psi_n|\mathcal{O}|\psi_n\rangle = \text{Trace}(\mathcal{O}\rho(t)). \tag{14.28}$$

(2) The operator $\rho(t)$ is positive semi-definite, i.e. for any state A

$$\langle A|\rho(t)|A\rangle = \sum_n p_n |\langle A|\psi_n(t)\rangle|^2 \geq 0. \tag{14.29}$$

(3) For a pure state

$$\rho^2(t) = |\psi(t)\rangle\langle\psi(t)|\psi(t)\rangle\langle\psi(t)| = \rho(t), \tag{14.30}$$

but for mixed states we have

$$\text{Trace } \rho^2(t) = \sum_{j,n} p_j p_n |\langle\psi_j|\psi_n\rangle|^2 \leq 1. \tag{14.31}$$

This last inequality follows from the fact that $|\langle\psi_j|\psi_n\rangle|^2 \leq 1$, and $\sum_j p_j = 1$, therefore $\sum_n p_n |\langle\psi_j|\psi_n\rangle| \leq 1$ for any j, and we have the inequality

$$\text{Trace } \rho^2(t) \leq \sum_n p_n = 1. \tag{14.32}$$

The time evolution of the density matrix for an open system can be obtained in the Heisenberg picture by noting that $\rho(t)$ is defined by Eq. (14.23), and $|\psi(t)\rangle$ is the solution of the Schrödinger equation. If at $t = t_0$ the state of the system is given by $|\psi(t_0)\rangle$, then

$$|\psi(t)\rangle = U(t, t_0)|\psi(t_0)\rangle, \tag{14.33}$$

where $U(t, t_0)$ is the unitary time-evolution operator which satisfies the differential equation

$$i\hbar\frac{\partial}{\partial t}U(t, t_0) = H(t)U(t, t_0), \tag{14.34}$$

and is subject to the initial condition

$$U(t_0, t_0) = 1, \tag{14.35}$$

1 being the identity operator. For a time-dependent Hamiltonian of a non-conservative system the formal solution of (14.34) with the boundary condition (14.35) is expressible in terms of an integral equation

$$U(t, t_0) = 1 - \frac{i}{\hbar}\int_{t_0}^{t} H\left(t'\right)U\left(t', t_0\right)dt'. \tag{14.36}$$

This integral equation can be solved by iteration with the formal solution given by

$$U(t, t_0) = 1 + \sum_{n=1}^{\infty}\left(\frac{-i}{\hbar}\right)^n \int_{t_0}^{t} H(t_n)dt_n \int_{t_0}^{t_n} H(t_{n-1})dt_{n-1}\cdots$$

$$\times \int_{t_0}^{t_2} H(t_1)dt_1. \tag{14.37}$$

In general the operators $H(t_1), H(t_2)\cdots$ do not commute for different values of time, so that the indicated order must be maintained. For this reason the operators in (14.37) are arranged in order of increasing time as we go from right to left. Equation (14.37) can also be written as [14],[15]

$$U(t, t_0) = 1 + \sum_{n=1}^{\infty}\left(\frac{-i}{\hbar}\right)^n \frac{1}{n!}\int_{t_0}^{t} dt_n \int_{t_0}^{t} dt_{n-1}$$

$$\times \cdots dt_1 \int_{t_0}^{t} T\left[H(t_n)\cdots H(t_1)\right]dt_1, \tag{14.38}$$

where T is the time-ordering symbol, i.e. we arrange the n factors $H(t_i)$ in the order of increasing time from right to left. We can also write (14.38) formally as an exponential operator

$$U(t, t_0) = T\left\{\exp\left[-\frac{i}{\hbar}\int_{t_0}^{t} H(t_1)dt_1\right]\right\}. \tag{14.39}$$

From the definition of $\rho(t)$ in terms of the wave function Eq. (14.24), we have

$$\rho(t) = \sum_n U(t, t_0)|\psi_n(t_0)\rangle\langle\psi_n(t_0)|U^\dagger(t, t_0), \tag{14.40}$$

or

$$\rho(t) = U(t, t_0)\rho(t_0)U^\dagger(t, t_0). \tag{14.41}$$

By differentiating this equation using (14.34) we obtain the following equation for $\partial\rho/\partial t$

$$i\hbar\frac{\partial\rho(t)}{\partial t} = [H, \, \rho(t)]. \tag{14.42}$$

and this is the quantum Liouville equation.

14.3 Phase Space Quantization of Dekker's Hamiltonian

As an example of the density matrix formulation of the motion of a damped oscillator we want to study the complex Hamiltonian of Dekker (Sec. 3.4). Our starting point will be the classical density function $\rho(q, \pi, t)$ (or $w_c(x, p, t)$ of Sec. 14.1) which satisfies the continuity equation

$$\frac{\partial\rho}{\partial t} + \frac{\partial}{\partial q}(\dot{q}\rho) + \frac{\partial}{\partial\pi}(\rho\dot{\pi}) = 0. \tag{14.43}$$

For $\dot{q}$ and $\dot{\pi}$ we can substitute from the canonical relations

$$\dot{q} = \frac{\partial H}{\partial\pi} = \{q, H\}, \quad \dot{\pi} = -\frac{\partial H}{\partial q} = \{\pi, H\}, \tag{14.44}$$

where $\{q, H\}$ and $\{\pi, H\}$ are the Poisson brackets. This gives us the equation of continuity in terms of the Hamiltonian of the system. Since for any function F of the dynamical variables q and π we have [4]

$$\{q, F\} = \frac{\partial}{\partial\pi}F, \quad \text{and} \quad \{\pi, F\} = -\frac{\partial}{\partial q}F, \tag{14.45}$$

therefore we can write the equation for $\partial\rho/\partial t$ in terms of the double Poisson brackets;

$$\frac{\partial\rho}{\partial t} = \{\pi, \{q, H\}\rho\} - \{\rho\{H, \pi\}, q\}. \tag{14.46}$$

The quantal analogue of (14.46) can be found by using the Dirac rule of association between the classical Poisson bracket and the quantum mechanical commutator [16];

$$\{u, v\} \to \frac{1}{i\hbar}[u, v]. \tag{14.47}$$

In Sec. 11.8 we discussed the ambiguities resulting from this rule of association, but this ambiguity is not the only source of difficulty in making transition from classical to quantum mechanics, since here we have the additional problem of the non-Hermiticity of the Hamiltonian. Now assuming the validity of this rule of association, even for non-Hermitian Hamiltonians, we obtain the master equation [17],[18]

$$\frac{\partial \hat{\rho}}{\partial t} = -\frac{1}{\hbar^2} \left[\pi, [q, \mathsf{H}] \, \hat{\rho}\right] + \frac{1}{\hbar^2} \left[\hat{\rho} \left[\mathsf{H}, \pi\right], q\right], \tag{14.48}$$

where now $\hat{\rho}, \pi, q$ and H are all operators. The Hamiltonian H is the sum of two terms $\mathsf{H} = H_1 + i\Gamma$, where H_1 and Γ are Hermitian operators, $H_1^\dagger = H_1$. Substituting for H in (14.48) and simplifying the result we find the operator equation for $(\partial \hat{\rho}/\partial t)$

$$\frac{\partial \hat{\rho}}{\partial t} = -\frac{i}{\hbar} \left[H_1, \hat{\rho}\right] - \frac{i}{\hbar^2} \left[\pi, [q, \Gamma] \, \hat{\rho}\right] - \frac{i}{\hbar^2} \left[\hat{\rho} \left[\Gamma, \pi\right], q\right], \tag{14.49}$$

where H_1 and Γ are given by

$$\mathsf{H} = H_1 + i\Gamma = -i\omega\pi q - \frac{\lambda}{2}\pi q. \tag{14.50}$$

Using (14.50), Eq. (14.49) simplifies to

$$\frac{\partial \hat{\rho}}{\partial t} = -\frac{\omega}{\hbar} \left[\pi q, \hat{\rho}\right] + \frac{i\lambda}{2\hbar} \left([\pi, q\hat{\rho}] + [\hat{\rho}\pi, q]\right). \tag{14.51}$$

This equation can also be written in terms of creation and annihilation operators,

$$a^\dagger = -i\sqrt{\frac{2}{\hbar}}\pi, \quad a = \frac{1}{\sqrt{2\hbar}}q. \tag{14.52}$$

Noting that the Hamiltonian is

$$\mathsf{H} = \hbar\left(\omega - \frac{i\lambda}{2}\right)a^\dagger a, \tag{14.53}$$

then the master equation becomes

$$\frac{\partial \hat{\rho}}{\partial t} = -i\omega \left[a^\dagger a, \hat{\rho}\right] - \frac{\lambda}{2}\left([a^\dagger, a\hat{\rho}] + [\hat{\rho}a^\dagger, a]\right). \tag{14.54}$$

We can now write (14.54) in terms of the original operators defined by the quantum analogues of (3.61) and (3.67):

$$q = \frac{1}{\sqrt{\omega}}\left[p + \left(\frac{\lambda}{2} - i\omega\right)x\right], \tag{14.55}$$

and

$$\pi = \frac{i}{2\sqrt{\omega}}\left[p + \left(\frac{\lambda}{2} + i\omega\right)x\right], \tag{14.56}$$

Substituting for q and π in (14.51) the master equation takes the form

$$\frac{\partial \hat{\rho}}{\partial t} = -\frac{i}{\hbar}[H_1, \hat{\rho}] - \frac{\lambda}{4\hbar\omega}[p, [p, \hat{\rho}]] - \frac{\lambda^2}{8\hbar\omega}[x, [p, \hat{\rho}]]$$

$$- \frac{\lambda^2}{8\hbar\omega}[p, [x, \hat{\rho}]] - \frac{\lambda\omega_0^2}{4\hbar\omega}[x, [x, \hat{\rho}]] - \frac{i\lambda}{2\hbar}([x, \hat{\rho}p] - [p, \hat{\rho}x]), \qquad (14.57)$$

where

$$H_1 = \frac{1}{2}p^2 + \frac{\lambda}{4}(px + xp) + \frac{1}{2}\omega_0^2 x^2. \qquad (14.58)$$

We have observed that if the state of the system can be described by a single wave function, then the density matrix satisfies Eq. (14.42). But for the damped harmonic oscillator with the Hamiltonian (14.58), $(\partial\hat{\rho}/\partial t)$ in Eq. (14.57) has additional terms proportional to λ and λ^2, therefore one can conclude that for such a damped system the master equation cannot describe a system with a single wave function [17].

14.4 Density Operator and the Fokker-Planck Equation for Dekker's Hamiltonian

We have seen that the density matrix, $\hat{\rho}(t)$, in the Schrödinger picture satisfies Eq. (14.54) which we can write as

$$i\hbar\frac{\partial\hat{\rho}(t)}{\partial t} = \hbar\omega\left[a^\dagger a, \hat{\rho}(t)\right] + [V(t), \hat{\rho}(t)]. \qquad (14.59)$$

In this equation $V(t)$ is defined by

$$V(t) = -\frac{i\hbar\lambda}{2}\left(a^\dagger a\hat{\rho}(t) + \hat{\rho}(t)a^\dagger a - 2a\hat{\rho}(t)a^\dagger\right). \qquad (14.60)$$

We can transform $\hat{\rho}(t)$ and write it in the interaction picture as the operator $\hat{\rho}_I(t)$ [2];

$$\hat{\rho}(t) = U_0^\dagger(t)\hat{\rho}_I(t)U_0(t), \qquad (14.61)$$

where $U_0(t)$ is given by

$$U_0(t) = \exp\left[-i\omega t(a^\dagger a)\right]. \qquad (14.62)$$

By substituting (14.61) and (14.62) in (14.59) and simplifying we obtain the following equation for the operator $\hat{\rho}_I(t)$;

$$i\hbar\frac{\partial\hat{\rho}_I(t)}{\partial t} = [V_I(t), \hat{\rho}_I(t)], \qquad (14.63)$$

with the commutator $[V_I(t), \hat{\rho}_I]$ given by

$$[V_I(t), \hat{\rho}_I(t)] = U_0^\dagger V(t)U_0(t) = \frac{i\hbar\lambda}{2}\left(a^\dagger a\hat{\rho}_I(t) + \hat{\rho}_I(t)a^\dagger a - 2a\hat{\rho}_I(t)a^\dagger\right). \quad (14.64)$$

Coherent State Representation of the Density Operator — We can transform (14.63) to a Fokker-Planck equation, [19] using the coherent state representation for $\hat{\rho}_I(t)$, i.e. we write it as [20],[21]

$$\hat{\rho}_I(t) = \int P(\alpha, t) \, |\alpha\rangle\langle\alpha| \, d^2\alpha, \qquad (14.65)$$

where $|\alpha\rangle$ s are the coherent states,

$$a|\alpha\rangle = \alpha|\alpha\rangle. \qquad (14.66)$$

Using the following correspondences:

$$a\hat{\rho}_I(t) \longleftrightarrow \alpha P(\alpha, t), \qquad (14.67)$$

$$a^\dagger \hat{\rho}_I(t) \longleftrightarrow \left(\alpha^* - \frac{\partial}{\partial\alpha}\right) P(\alpha, t), \qquad (14.68)$$

$$\hat{\rho}_I(t) a^\dagger \longleftrightarrow P(\alpha, t)\alpha^*, \qquad (14.69)$$

and

$$\hat{\rho}_I(t) a \longleftrightarrow \left(\alpha - \frac{\partial}{\partial\alpha^*}\right) P(\alpha, t), \qquad (14.70)$$

we obtain the Fokker-Planck equation for $P(\alpha, t)$ which is defined by (14.65)

$$\frac{\partial P(\alpha, t)}{\partial t} = \frac{\lambda}{2} \left(\frac{\partial}{\partial\alpha}\alpha + \frac{\partial}{\partial\alpha^*}\alpha^*\right) P(\alpha, t). \qquad (14.71)$$

In addition to Eq. (14.71) there are other representations which do not have classical analogues. Here it is easier to work with $Q(\alpha, \alpha^*)$ defined in terms of the density matrix by [1]

$$Q(\alpha, \alpha^*) = \langle\alpha|\hat{\rho}_I|\alpha\rangle. \qquad (14.72)$$

From the normalization of the density operator, i.e.

$$\text{Trace}\{\hat{\rho}_I\} = \text{Trace} \int |\alpha\rangle\langle\alpha|\hat{\rho}_I \, d^2\alpha = \int \langle\alpha|\hat{\rho}_I|\alpha\rangle d^2\alpha = 1, \qquad (14.73)$$

we find the normalization of $Q(\alpha, \alpha^*)$

$$\int Q(\alpha, \alpha^*) \, d^2\alpha = 1. \qquad (14.74)$$

We can express the averages of anti-normally ordered products of creation and annihilation operators as an integral over $Q(\alpha, \alpha^*)$. Thus

$$\langle a^r a^{\dagger s}\rangle = \text{Trace}\left[a^r a^{\dagger s}\hat{\rho}_I\right] = \int \alpha^r \alpha^{*s} Q(\alpha, \alpha^*) \, d^2\alpha. \qquad (14.75)$$

Similarly, relations like Eqs. (14.67)-(14.70) for $P(\alpha, \alpha^*)$, can be obtained from the definition of $Q(\alpha, \alpha^*)$;

$$a\hat{\rho}_I \longleftrightarrow \left(\alpha + \frac{\partial}{\partial \alpha^*}\right) Q(\alpha, \alpha^*), \tag{14.76}$$

$$a^\dagger \hat{\rho}_I \longleftrightarrow \alpha^* Q(\alpha, \alpha^*), \tag{14.77}$$

$$\hat{\rho}_I a \longleftrightarrow \alpha Q(\alpha, \alpha^*), \tag{14.78}$$

and

$$\hat{\rho}_I a^\dagger \longleftrightarrow \left(\alpha^* + \frac{\partial}{\partial \alpha}\right) Q(\alpha, \alpha^*). \tag{14.79}$$

From these relations and Eq. (14.63) we find the partial differential equation satisfied by $Q(\alpha, \alpha^*, t)$.

$$\frac{\partial Q(\alpha, \alpha^*, t)}{\partial t} = \left[\frac{\lambda}{2}\left(\frac{\partial}{\partial \alpha}\alpha + \frac{\partial}{\partial \alpha^*}\alpha^*\right) + \lambda \frac{\partial^2}{\partial \alpha \partial \alpha^*}\right] Q(\alpha, \alpha^*, t). \tag{14.80}$$

14.5 Squeezed State of a Damped Harmonic Oscillator

In modern optics, in addition to the coherent state which is the minimum uncertainty state, the concept of squeezed state plays an important role [33]. The squeezed state may be regarded as a generalization of the idea of the coherent state. We know that there are fluctuations in conjugate variables (e.g. x and p) of an oscillator or a field and that these fluctuations impose a limit on the accuracy of the measurements of the conjugate variables. Now by reducing the fluctuations of one of the variables and increasing the fluctuations of the other, we can obtain fluctuations for the first variable less than those in vacuum. Thus for the undamped harmonic oscillator the uncertainty area in the generalized coordinate phase space (x, p) for a coherent state is a circle whereas for a squeezed state this uncertainty is an ellipse. In this section we will study the squeezed state of a damped oscillator. For a review of the squeezed state of the harmonic oscillator the reader is referred to the standard books on quantum optics for instance ref. [33].

Bogolyubov's Transformation of Creation and Annihilation Operators — Consider the creation and annihilation operators $\hat{A}^\dagger(t)$ and $\hat{A}(t)$ defined by Eqs. (13.56),(13.57) and let us a new set of operators $\hat{A}^\dagger_{r\theta}(t)$ and $\hat{A}_{r\theta}(t)$ related to $\hat{A}^\dagger(t)$ and $\hat{A}(t)$ by the Bogolyubov transformation [24],[34]

$$\hat{A}_{r\theta}(t) = \mu^* \hat{A}(t) - \nu^* \hat{A}^\dagger(t), \tag{14.81}$$

and

$$\hat{A}^\dagger_{r\theta}(t) = \mu \hat{A}^\dagger(t) - \nu \hat{A}^\dagger(t). \tag{14.82}$$

Here μ and ν are complex numbers, and thus the transformation depends on four parameters. By requiring that this transformation be unitary, i.e.

$$\left[\hat{A}^\dagger_{r\theta}(t),\ \hat{A}_{r\theta}(t)\right] = 1, \tag{14.83}$$

and using the fact that $\left[\hat{A}^\dagger(t),\ \hat{A}(t)\right] = 1$, and that the other commutators are zero, we find the following constraint on the parameters μ and ν of the transformation

$$|\mu|^2 - |\nu|^2 = 1. \tag{14.84}$$

Of the four parameters in the transformation, one is an overall phase which can be absorbed in the wave function and the other is determined by the constraint given by (14.84). Thus we are left with a family of transformations depending on two parameters which we write as

$$\mu = \cosh r, \quad \text{and} \quad \nu = e^{i\theta}\sinh r. \tag{14.85}$$

The generator of this unitary transformation is $U(\zeta)$ and this operator is defined by

$$U(\zeta) = \exp\left[\frac{1}{2}\left(\zeta\hat{A}^{\dagger 2}(t) - \zeta^*\hat{A}^2(t)\right)\right], \tag{14.86}$$

where

$$\zeta = -e^{i\theta}r. \tag{14.87}$$

The action of the operator $U(\zeta)$ is to change $\hat{A}^\dagger(t)$ and $\hat{A}(t)$ to $\hat{A}^\dagger_{r\theta}(t)$ and $\hat{A}_{r\theta}(t)$;

$$\hat{A}_{r\theta}(t) = U(\zeta)\hat{A}(t)U^\dagger(\zeta), \tag{14.88}$$

and

$$\hat{A}^\dagger_{r\theta}(t) = U(\zeta)\hat{A}^\dagger(t)U^\dagger(\zeta). \tag{14.89}$$

Noting that

$$U^\dagger(\zeta) = U^{-1}(\zeta) = U(-\zeta), \tag{14.90}$$

and using Baker-Campbell-Hausdorff formula Eq. (11.116) we have [33]

$$\hat{A}_{r\theta}(t) = U(\zeta)\hat{A}(t)U^\dagger(\zeta)$$
$$= \hat{A}(t) - \left[\frac{1}{2}\zeta^*\hat{A}^2(t) - \frac{1}{2}\zeta\hat{A}^{\dagger 2}(t),\ \hat{A}(t)\right] + \cdots = \hat{A}(t)\left(1 + \frac{r^2}{2!} + \cdots\right)$$
$$+ e^{i\theta}\hat{A}^\dagger(t)\left(r + \frac{r^3}{3!} + \cdots\right) = \hat{A}(t)\cosh r + \hat{A}^\dagger(t)e^{i\theta}\sinh r. \tag{14.91}$$

Thus we conclude that $U(\zeta)$ is the unitary operator which transforms $\hat{A}(t)$ to $\hat{A}_{r,\theta}(t)$, and $\hat{A}^\dagger(t)$ to $\hat{A}^\dagger_{r,\theta}(t)$ according to the Bogolyubov transformation Eqs. (14.81),(14.82).

We can determine the wave function for the squeezed state (r, θ) in a representation where the generalized number operator [22],[23]

$$N_{r,\theta} = \hat{A}^{\dagger}_{r\theta}(t)\hat{A}_{r\theta}(t), \tag{14.92}$$

is diagonal and has the eigenvalue n [35]

$$N_{r,\theta}\psi(t, r, \theta) = n\psi(t, r, \theta). \tag{14.93}$$

We note that the integer n in this case does not represent the number of quanta, since in the Bogolyubov transformation the eigenvalues of $\hat{A}^{\dagger}(t)\hat{A}(t)$ are not the same as those of $\hat{A}^{\dagger}_{r\theta}(t)\hat{A}_{r\theta}(t)$, and the vacuum of the coherent state is different from the vacuum defined by $\hat{A}_{r\theta}(t)|0\rangle = 0|0\rangle$ [24],[34].

It should be pointed out that in this approach the canonical momentum p which satisfies the equation $\ddot{p} - \lambda\dot{p} + \omega_0^2 p = 0$ has been used and that is why $\langle \Delta p^2 \rangle$ grows in time as $e^{\lambda t}$. But as we have seen in models of Dekker and Ullersma the momentum operator satisfies the time-reversed equation of the previous equation, i.e. $\ddot{p} + \lambda\dot{p} + \omega_0^2 p = 0$. In Sec. 14.6 we will observe that an approximate solution of the time-dependent squeezed states based on the density matrix formulation gives us exponential decays for x as well as p. Solving this eigenvalue equation in coordinate space with $\hat{A}_{r,\theta}(t)$ and $\hat{A}^{\dagger}_{r,\theta}(t)$ given by (14.81),(14.82) and $\hat{A}(t)$ and $\hat{A}^{\dagger}(t)$ defined by (13.56),(13.57) and by setting $\delta = 0$, we find the wave function to be

$$\psi(x, t; r, \theta) = \frac{1}{\sqrt{2^n n!}} \left(\frac{\alpha_{r,\theta}}{\sqrt{\pi}}\right)^{\frac{1}{2}} \exp\left[-i\Theta_{r,\theta}(t)\left(n + \frac{1}{2}\right)\right]$$
$$\times H_n\left(\alpha_{r,\theta}x\right) \exp\left(-B_{r,\theta}x^2\right), \tag{14.94}$$

where

$$\alpha_{r,\theta} = \sqrt{\frac{\omega e^{\lambda t}}{\hbar}} \left[\cosh 2r + \sinh 2r \cos(2\omega t + \theta)\right]^{-\frac{1}{2}}, \tag{14.95}$$

$$B_{r,\theta} = \frac{\omega e^{\lambda t}}{2\hbar} \left[\frac{e^{i\omega t} \cosh r - e^{-i(\omega t + \theta)} \sinh r}{e^{i\omega t} \cosh r + e^{-i(\omega t + \theta)} \sinh r} + i\frac{\lambda}{2\omega}\right], \tag{14.96}$$

and

$$\tan \Theta_{r,\theta} = \frac{\sin \omega t - \tanh r \sin(\omega t + \theta)}{\cos \omega t + \tanh r \cos(\omega t + \theta)}. \tag{14.97}$$

In the limit of $r \to 0$, and $\theta \to 0$, this wave function reduces to the one given by Eq. (12.10).

The uncertainty relation for the squeezed states can be obtained from the wave function $\psi(x, t; r, \theta)$, Eq. (14.94);

$$\langle \Delta x \rangle^2 \langle \Delta p \rangle^2 = \int_{-\infty}^{\infty} x^2 \, |\psi_n(x, t; r, \theta)|^2 \, dx$$

$$\times \int_{\infty}^{\infty} \psi_n(x, t; r, \theta) \left(-\frac{\hbar^2}{2} \frac{\partial^2}{\partial x^2} \right) \psi_n^*(x, t; r, \theta) dx$$

$$= \frac{\hbar^2}{4 \cos^2 \left(\frac{\delta_\lambda}{2} \right)} \left(n + \frac{1}{2} \right)^2 [\cosh 2r + \sinh 2r \cos(2\omega t + \theta)]^2$$

$$\times [\cosh 2r - \sinh 2r \cos(2\omega t + \theta + \delta_\lambda)], \qquad (14.98)$$

where

$$\cos \delta_\lambda = \frac{1 - \frac{\lambda^2}{4\omega^2}}{1 + \frac{\lambda^2}{4\omega^2}}, \quad 0 \le \delta_\lambda \le \pi. \qquad (14.99)$$

The uncertainty $(\Delta x)(\Delta p)$ found from (14.98) has a simple form if we choose the ground state $n = 0$, and assume $r = 0$ for the squeezing parameter. Then $(\Delta x)(\Delta p)$ becomes only a function of δ_λ [35];

$$(\Delta x)_{00\theta} (\Delta p)_{00\theta} = \frac{\hbar}{2} \sec \left(\frac{\delta_\lambda}{2} \right). \qquad (14.100)$$

From (14.100) we find that the uncertainty for $r = 0$ is independent of θ and is given by

$$(\Delta x)_{00\theta} (\Delta p)_{00\theta} = \frac{\hbar}{2} \left(\frac{\omega_0}{\omega} \right) \ge \frac{\hbar}{2}, \qquad (14.101)$$

which is the same as the inequality (13.72). We can also calculate exactly the time averaged uncertainty for the ground state from Eq. (14.98):

$$(\Delta x)_{00\theta} (\Delta p)_{00\theta} = \frac{\hbar}{2} \sec \left(\frac{\delta_\lambda}{2} \right) \left(\cosh^2 r - \frac{\cos \delta_\lambda}{2} \sinh^2 r \right)$$

$$\ge \frac{\hbar}{2} \sec \left(\frac{\delta_\lambda}{2} \right). \qquad (14.102)$$

Squeezed States Uncertainties Derived Using Evolution Operator — In Sec. 14.6 we will discuss the variances of the distributions for the squeezed state of a damped harmonic oscillator Eqs. (14.130),(14.134) from the results that we have already obtained for the coherent state. Now we will find the same results from the evolution operator formulation of Sec. 12.3. From Eqs. (12.78) and (13.99), it follows that A is related to $\hat{\alpha}$ and $\hat{\alpha}^\dagger$ by a Bogolyubov transformation

$$\hat{A} = \mu^* \hat{\alpha} - \nu^* \hat{\alpha}^\dagger y \qquad (14.103)$$

where $|\mu|^2 - |\nu|^2 = 1$, and μ^* and ν^* are given by

$$\mu^* = \frac{1}{2} \exp \left(-\frac{1}{2} c_2 \right) \left[1 + c_1 c_3 + \exp(c_2) - \frac{c_1}{m\omega_0} - m\omega_0 c_3 \right] \qquad (14.104)$$

$$\nu^* = \frac{1}{2} \exp\left(-\frac{1}{2}c_2\right) \left[1 - c_1 c_3 - \exp(c_2) + \frac{c_1}{m\omega_0} - m\omega_0 c_3\right]. \qquad (14.105)$$

As Eqs. (13.99) and (14.103) show, the coherent state starts at the time $t = 0$ and evolves in a way which is reminiscent of a squeezed state. Thus we define the squeezed state with the minimum uncertainty by the two operators Y_1 and Y_2 similar to the operators X_1 and X_2 that we will introduce later (Eqs. (14.122) and (14.124)):

$$Y_1 = \frac{1}{2}\left(\hat{a} + \hat{a}^\dagger\right), \qquad (14.106)$$

and

$$Y_2 = \frac{1}{2i}\left(\hat{a} - \hat{a}^\dagger\right). \qquad (14.107)$$

The expectation values of these operators with respect to the wave function $|\psi(t)\rangle$, Eq. (13.96), can be found from $U(t,0)$, and are given in terms of c_1, c_2 and c_3

$$\langle Y_1 \rangle = \frac{1}{2} \exp\left(-\frac{1}{2}c_2\right) \left[(1 + m\omega_0 c_3)\alpha + (1 - m\omega_0 c_3)\alpha^*\right], \qquad (14.108)$$

and

$$\langle Y_2 \rangle = \frac{1}{2i} \exp\left(-\frac{1}{2}c_2\right)$$
$$\times \left\{\left[c_1 c_3 + \exp(c_2) + \frac{c_1}{m\omega_0}\right]\alpha - \left[c_1 c_3 + \exp(c_2) - \frac{c_1}{m\omega_0}\right]\alpha^*\right\}. \qquad (14.109)$$

By calculating the fluctuations ΔY_1^2 and ΔY_2^2, using (14.106),(14.107) we find the following results

$$\Delta Y_1^2 = \frac{1}{4} \exp\left(-c_2\right) \left(1 - m^2 \omega_0^2 c_3^2\right), \qquad (14.110)$$

and

$$\Delta Y_2^2 = \frac{1}{4i} \exp\left(-c_2\right) \left\{\left[c_1 c_3 + e^{c_2}\right]^2 - \frac{c_1^2}{m^2 \omega_0}\right\}. \qquad (14.111)$$

Now from Eqs. (14.110) and (14.111) we obtain the asymptotic forms of ΔY_1 and ΔY_2, and these are:

$$(\Delta Y_1)^2 \propto \exp\left(-c_2\right), \qquad (14.112)$$

and

$$(\Delta Y_2)^2 \propto \exp\left(c_2\right). \qquad (14.113)$$

The two operators Y_1 and Y_2 are proportional to x and p respectively, as can be deduced from Eqs. (13.97),(14.106) and (14.107), i.e.

$$x = \left(\frac{2\hbar}{m\omega}\right) Y_1, \qquad (14.114)$$

and

$$p = (2\hbar m \omega) \, Y_2. \tag{14.115}$$

Then from these equations and Eqs. (14.108),(14.109), we get the product of the uncertainties $\Delta x \Delta p$ to be

$$\Delta x \Delta p = \frac{1}{2}\hbar \left(\frac{\sin(\omega t)}{\omega} \right)^2 \left[\left(\omega_0^2 + \frac{1}{4}\lambda^2 - \omega^2 \cot^2(\omega t) \right)^2 + 4\omega_0^2 \omega^2 \cot^2(\omega t) \right]^{\frac{1}{2}}. \tag{14.116}$$

These are similar to Eqs. (13.70) and (13.71) which we have found earlier. Let us emphasize the following points regarding the uncertainties Δx and Δp.

(1) The uncertainty Δp is not the uncertainty in the mechanical momentum $(m\Delta \dot{x})$, but it is the uncertainty in the canonical momentum.

(2) By examining the explicit forms of $c_1(t)$, $c_2(t)$ and $c_3(t)$, Eqs. (12.101)-(12.103) for the harmonic oscillator we find that while $\Delta x \Delta p$ has a minimum equal to $\frac{\hbar}{2}$, when $t \to \infty$, this value cannot be reached, since both $c_1(t)$ and $c_2(t)$ become infinite at times when $\lambda + \cot(\omega t) = 0$. As we can see from (14.116) the product $\Delta p \Delta x$ is an oscillatory function of time with a period of $\left(\frac{\pi}{\omega} \right)$, and its maximum depends on the coefficient of friction λ. For the parameters $\omega_0 = 1$ and $\lambda = 0.25$, we find that the maximum amplitude of the oscillation is $0.5158\hbar$.

14.6   A Different Formulation of the Problem of Time-Dependence of the Squeezed State

As an interesting application of (14.80) we find the time evolution of the squeezed state $|\chi, \zeta\rangle$ for the damped harmonic oscillator where unlike the squeezed states of Caldirola-Kanai Hamiltonian both (Δx^2) and (Δp^2) decay in time. Consider the combined action of the squeezing operator $U(\zeta)$, and the displacement operator $D(\chi)$ defined by

$$D(\chi) = \exp\left[-\frac{1}{2}|\chi|^2 \right] e^{\chi a^\dagger(t)} e^{\chi^* a(t)} \tag{14.117}$$

on the vacuum $|0\rangle$. If we choose $\theta = \pi$, U becomes a function of r, and we denote the resulting state by $|\chi, r\rangle$,

$$|\chi, r\rangle = D(\chi)U(r)|0\rangle. \tag{14.118}$$

From the Baker-Campbell-Hausdorff theorem, (11.116), we find the effect of the operator $U(r)$ on the annihilation (creation) operator a $(a^\dagger)$

$$U^\dagger(r)a(t)U(r) = a(t)\cosh r + a^\dagger(t)\sinh r. \tag{14.119}$$

Similarly, we have for the $D(\chi)$ operator

$$D^\dagger(\chi)a(t)D(\chi) = a(t) + \chi. \tag{14.120}$$

We observe that $a(t)$ which is now dependent on r and will be denoted by $a_r(t)$ is a complex operator satisfying the commutation relation $\left[a_r^\dagger(t), a_r(t)\right] = 1$, we can write it in terms of the real and imaginary operators X_1 and X_2:

$$a_r(t) = X_1 + iX_2, \tag{14.121}$$

where X_1 and X_2 satisfy the commutation relation

$$[X_1, X_2] = \frac{i}{2}. \tag{14.122}$$

The variances in X_i s are defined by

$$\left(\Delta X_i^2\right) = \left\langle X_i^2 \right\rangle - \left\langle X_i \right\rangle^2, \quad i = 1, 2. \tag{14.123}$$

Thus from (14.122) it follows that the squeezed states are states of minimum uncertainty, i.e.

$$\Delta X_1 \Delta X_2 = \frac{1}{4}, \tag{14.124}$$

but the uncertainties in X_1 and X_2 are given by

$$\Delta X_1 = \frac{1}{2}e^{-r}, \quad \text{and} \quad \Delta X_2 = \frac{1}{2}e^r. \tag{14.125}$$

Compared with the coherent state where $\Delta X_1 = \Delta X_2 = \frac{1}{2}$, here we have reduced fluctuations in one and increased fluctuations in the other, so that the product $\Delta X_1 \Delta X_2$ remains unchanged. Using these properties of the squeezed state, $|\chi, r\rangle$, the representation of $Q(\alpha, \alpha^*, 0)$ has the following analytical form: [20]-[23]

$$Q(\alpha, \alpha^*, 0) = \frac{1}{\pi \cosh r} \exp\left[-\frac{1}{2}(\mathbf{z} - \langle\mathbf{z}\rangle)^T \sigma^{-1} (\mathbf{z} - \langle\mathbf{z}\rangle)\right], \tag{14.126}$$

with

$$\mathbf{z} = \begin{bmatrix} \alpha \\ \alpha^* \end{bmatrix}, \tag{14.127}$$

and

$$\langle\mathbf{z}\rangle^T = \left[\langle a \rangle, \langle a^\dagger \rangle\right] = [\chi, \chi^*]. \tag{14.128}$$

Here T denotes the transpose and σ is defined by the following 2×2 matrix:

$$\sigma = \frac{1}{2} \begin{bmatrix} -\sinh 2r & \cosh 2r + 1 \\ \cosh 2r + 1 & -\sinh 2r \end{bmatrix}. \tag{14.129}$$

When damping is present then we solve Eq. (14.80) with the initial condition (14.126), but now $\mathbf{z}$ in (14.127) is replaced by [20]

$$\mathbf{z} = e^{-\frac{\lambda t}{2}} \begin{bmatrix} \chi \\ \chi^* \end{bmatrix} \tag{14.130}$$

and by its transpose $\mathbf{z}^T$. Then σ becomes

$$\sigma = \frac{1}{2} \begin{bmatrix} -\sinh 2r & \cosh 2r + 1 \\ \cosh 2r + 1 & -\sinh 2r \end{bmatrix} e^{-\lambda t} + \begin{bmatrix} 0 & 1 \\ 1 & 0 \end{bmatrix} \left(1 - e^{-\lambda t}\right). \qquad (14.131)$$

With these changes we can write the solution of the Fokker-Planck equation (14.80) as

$$Q\left(\alpha, \alpha^*, t\right) = \frac{1}{\pi\sqrt{\det\sigma}} \exp\left[-\frac{1}{2}\left(\mathbf{z} - \langle\mathbf{z}(t)\rangle\right)^T \sigma^{-1}\left(\mathbf{z} - \langle\mathbf{z}(t)\rangle\right)\right]. \qquad (14.132)$$

The variances of the distribution in this case are:

$$(\Delta X_1^2)(t) = \frac{1}{4}\left[\left(e^{-2r} - 1\right)e^{-\lambda t} + 1\right], \qquad (14.133)$$

and

$$(\Delta X_2^2)(t) = \frac{1}{4}\left[\left(e^{2r} - 1\right)e^{-\lambda t} + 1\right]. \qquad (14.134)$$

These relations show that if $r > 0$, then $(\Delta X_1^2)(t) < 1/4$, and $(\Delta X_2^2)(t) > 1/4$, that is the fluctuations for $X_1(t)$ are decreased and for $X_2(t)$ are amplified.

14.7 Density Matrix Formulation of a Solvable Model

In Chapter 8 we observed that for a harmonic oscillator coupled linearly to a bath of oscillators the classical equations of motion can be solved exactly. Now we want to consider the damped motion of a more general nonlinear system [25],[26].

Again we start with the total Hamiltonian

$$H_T = H_S + H_B + H_I, \qquad (14.135)$$

where H_S, H_B and H_I are the Hamiltonians for the system, the bath and the interaction respectively. Here we take the dissipative system S to be a particle and this particle can be in any number of the sites ν, where ν is an integer $-\infty < \nu < +\infty$. We denote a given state of the particle by $|\nu\rangle$ and assume that the collection of these states form an orthonormal set;

$$\langle\nu|\nu'\rangle = \delta_{\nu\nu'}. \qquad (14.136)$$

We also assume that the Hamiltonian H_S is given by

$$H_S = \Omega S = \frac{1}{2}\Omega\left(s + s^{-1}\right), \qquad \Omega \geq 0 \qquad (14.137)$$

where we have set $\hbar = 1$ and where Ω is a constant and s and s^{-1} are the raising and lowering operators

$$s|\nu\rangle = |\nu + 1\rangle, \quad s^{-1}|\nu\rangle = |\nu - 1\rangle. \tag{14.138}$$

Clearly S is a Hermitian operator, and its eigenfunctions are $|\eta\rangle$ s where

$$|\eta\rangle = \frac{1}{\sqrt{2\pi}} \sum_{\nu=-\infty}^{\infty} e^{i\eta\nu} |\nu\rangle, \quad 0 \le \eta < 2\pi \tag{14.139}$$

From Eqs. (14.137) and (14.139) we can calculate the matrix elements of H in η-representation and these matrix elements are given by

$$\langle\eta|H_S|\eta'\rangle = \Omega \cos\eta \, \delta(\eta - \eta'). \tag{14.140}$$

The bath consists of a collection of harmonic oscillators for which the Hamiltonian is

$$H_B = \sum_n \omega_n a_n^\dagger a_n, \quad \omega_n > 0. \tag{14.141}$$

We also assume that the interaction Hamiltonian is of the form

$$H_I = \frac{1}{2}\Omega \left(s + s^{-1}\right) \sum_n \epsilon_n \left(a_n^\dagger + a_n\right), \tag{14.142}$$

with the coupling constants ϵ_n.

The total Hamiltonian H_T commutes with H_S, and therefore it is diagonal in η-representation

$$\langle\eta|H_T|\eta'\rangle = \delta(\eta - \eta')\{\Omega \cos\eta + H_\eta\}, \tag{14.143}$$

where in (14.143) H_η is defined as

$$H_\eta = \sum_n H_{\eta,n} = \sum_n \left[\omega_n a_n^\dagger a_n + \cos\eta \, \epsilon_n \left(a_n^\dagger + a_n\right)\right]. \tag{14.144}$$

We can write $H_{\eta,n}$ in the form

$$H_{\eta,n} = \omega_n \left(a_n^\dagger + u_n\right)(a_n + u_n) - \omega_n u_n^2, \tag{14.145}$$

with

$$u_n = \frac{\epsilon_n}{\omega_n} \cos\eta. \tag{14.146}$$

Having found the matrix elements of H_T, we want to find the master equation giving us the time development of the density matrix $\hat{\rho}_S(t)$. We note that at $t = 0$, the density matrix of the total system is of the form of a product

$$\hat{\rho}_T(0) = \hat{\rho}_S(0) \otimes \hat{\rho}_B, \tag{14.147}$$

where $\hat{\rho}_S(0)$ is the initial state of S, and $\hat{\rho}_B(0)$ is the initial state of the bath. This density matrix $\hat{\rho}_T(0)$ evolves in time, and at some later time $\hat{\rho}_T(t)$ is given by

$$\hat{\rho}_T(t) = \exp\left(-iH_T t\right) \hat{\rho}_T(0) \exp\left(iH_T t\right). \qquad (14.148)$$

The time evolution of the density matrix of S alone can be obtained by averaging $\hat{\rho}_T(t)$ over the bath;

$$\hat{\rho}_S(t) = \text{Trace} \exp\left(-iH_T t\right) \hat{\rho}_S(0) \otimes \hat{\rho}_B \exp\left(iH_T t\right). \qquad (14.149)$$

Next we calculate the matrix elements of $\hat{\rho}_T(t)$ with the help of (14.143) and (14.144);

$$\langle \eta \,|\hat{\rho}_T(t)|\, \eta' \rangle = \langle \eta \,|\hat{\rho}_T(0)|\, \eta' \rangle \, e^{-i\Omega t\left(\cos\eta - \cos\eta'\right)} \otimes e^{-iH_\eta t} \hat{\rho}_B(0) e^{iH_{\eta'}t}, \quad (14.150)$$

where the last factor in (14.150) can be determined by noting that

$$e^{-iH_\eta t} \hat{\rho}_B(0) e^{iH_{\eta'}t} \rightarrow \prod_n \left\{ \sum_{N_n} \langle N_n \,|e^{-iH_{\eta,n}t} e^{itH_{\eta',n}}|\, N_n \rangle \right\}. \qquad (14.151)$$

Now we substitute for $H_{\eta,n}$ and $H_{\eta',n}$ from (14.145) to obtain

$$e^{-iH_{\eta,n}t} e^{iH_{\eta',n}t} = \exp\left[i\omega_n t \left(u_n^2 - u_n'^2\right)\right] \exp\left[-i\omega_n t \left(a_n^\dagger + u_n\right)\left(a_n + u_n\right)\right]$$
$$\times \exp\left[i\omega_n t \left(a_n^\dagger + u_n'\right)\left(a_n + u_n'\right)\right]. \qquad (14.152)$$

By taking the trace of this operator over the bath states we have

$$\text{Trace } e^{-iH_{\eta,n}t} e^{iH_{\eta',n}t} = \exp\left[i\left(\omega_n t - \sin\omega_n t\right)\left(u_n^2 - u_n'^2\right)\right]$$
$$\times \text{Trace} \exp\left\{ u_n' \left[a_n \left(e^{i\omega_n t} - 1\right) - a_n^\dagger \left(e^{-i\omega_n t} - 1\right)\right] \right.$$
$$\left. + u_n \left[a_n \left(1 - e^{i\omega_n t}\right) - a_n^\dagger \left(1 - e^{-i\omega_n t}\right)\right] \right\}. \qquad (14.153)$$

If we write the trace as a sum over the eigenstates of $a_n^\dagger a_n$, we have the final result for $\hat{\rho}_S$ which is

$$\langle \eta \,|\hat{\rho}_S(t)|\, \eta' \rangle = \langle \eta \,|\hat{\rho}_S(0)|\, \eta' \rangle \, e^{-i\Omega t\left(\cos\eta - \cos\eta'\right)}$$
$$\times \exp\left[iF(t)\left(\cos^2\eta - \cos^2\eta'\right) - G(t)\left(\cos\eta - \cos\eta'\right)^2\right], \qquad (14.154)$$

where

$$F(t) = \sum_n \left(\frac{\epsilon_n^2}{\omega_n}\right)\left(t - \frac{\sin\omega_n t}{\omega_n}\right), \qquad (14.155)$$

and

$$G(t) = \sum_n \left(\frac{\epsilon_n^2}{\omega_n^2}\right)\left(1 - \cos\omega_n t\right). \qquad (14.156)$$

By differentiating (14.154) with respect to time we obtain an equation similar to Eqs. (14.63) and (14.64)

$$\frac{\partial \hat{\rho}_S(t)}{\partial t} = -i\Omega \left[S, \, \hat{\rho}_S(t) \right] + i\dot{F}(t) \left[S^2, \, \hat{\rho}_S(t) \right]$$
$$- \dot{G}(t) \left\{ S^2 \hat{\rho}_S(t) - 2S\hat{\rho}_S(t)S + \hat{\rho}_S(t)S^2 \right\}, \tag{14.157}$$

but here dF/dt and dG/dt are time-dependent functions. Let us suppose that the ω_n s are dense on the segment $0 < \omega_n < \infty$, and let us define as in the case of Ullersma's model a strength function $\gamma(\omega)$ by

$$\gamma(\omega)d\omega = \sum_{\omega < \omega_n < \omega + \Delta \omega} \epsilon^2, \tag{14.158}$$

and this strength function must satisfy the integrability condition, viz,

$$\int_0^\infty \frac{\gamma(\omega)}{\omega} d\omega = \text{finite}. \tag{14.159}$$

In terms of $\gamma(\omega)$ we can write $F(t)$ and $G(t)$ as

$$F(t) = \int_0^\infty \frac{\gamma(\omega)}{\omega} \left(t - \frac{\sin(\omega t)}{\omega} \right) d\omega, \tag{14.160}$$

and

$$G(t) = \int_0^\infty \frac{\gamma(\omega)}{\omega^2} \left(1 - \cos(\omega t) \right) d\omega. \tag{14.161}$$

For instance if we choose

$$\gamma(\omega) = \frac{\omega}{\omega^2 + \nu^2}, \tag{14.162}$$

then we can calculate $F(t)$ and dG/dt analytically:

$$F(t) = \frac{\pi}{2\nu} \left(t - \frac{1 - e^{-\nu t}}{\nu} \right), \tag{14.163}$$

and

$$\frac{dG}{dt} = \frac{1}{2\nu} \left\{ \left[e^{\frac{-\nu t}{2}} Ei\left(\frac{\nu t}{2} \right) - e^{\frac{\nu t}{2}} Ei\left(-\frac{\nu t}{2} \right) \right] \right\}, \tag{14.164}$$

where $Ei(z)$ is the exponential integral function [27]. For $\nu t \gg 2$, both $F(t)$ and $G(t)$ are approximately linear functions of t (for the latter see Fig. (14.1)), i.e. the coefficient $\frac{dF}{dt}$ in Eq. (14.157), becomes time-independent, whereas $\frac{dG}{dt}$ goes slowly to zero.

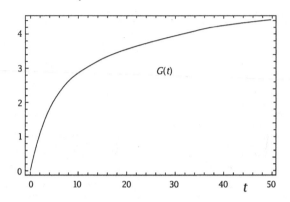

Figure 14.1: The function $G(t)$ calculated from (14.164), is shown as a function of time.

14.8 Wigner Distribution Function for the Damped Oscillator

The coherent state representation of $\hat{\rho}$ was just one of a number of ways of converting the operator equation for the density to a c-number equation. There are other ways of transforming the generalized Liouville equation like Eq. (14.57) to a c-number equation. For instance we can start with Eq. (14.43) and write a master equation for the Wigner distribution function, $W(p, x, t)$, in terms of the original coordinate x and momentum p [12]. This distribution function W is related to $\hat{\rho}$ by

$$W(p, x, t) = \frac{1}{2\pi\hbar} \int_{-\infty}^{\infty} e^{\frac{ipy}{\hbar}} \left\langle x - \frac{1}{2}y \,|\hat{\rho}|\, x + \frac{1}{2}y \right\rangle dy. \tag{14.165}$$

We know that in the absence of damping, $\lambda = 0$, the Wigner distribution function for the harmonic oscillator is given by the solution of the partial differential equation

$$\frac{\partial W}{\partial t} = -\frac{\partial}{\partial x}(pW) + \omega_0^2 \frac{\partial}{\partial p}(xW), \quad \lambda = 0, \tag{14.166}$$

and we want to find a similar equation in the presence of damping. Let I_1 denote the contribution of the second term on the left-hand side of Eq. (40.13), i.e.

$$\left\{ -\left(\frac{\lambda}{4\hbar\omega}\right)[p, [p, \hat{\rho}]] \right\} \tag{14.167}$$

to the distribution function, i.e.

$$I_1 = -\frac{\lambda}{4\hbar\omega} \frac{1}{2\pi\hbar} \int_{-\infty}^{\infty} e^{\frac{ipy}{\hbar}} \left\langle x - \frac{1}{2}y \,|p\,[p, \hat{\rho}] - [p, \hat{\rho}]\,p|\, x + \frac{1}{2}y \right\rangle dy. \tag{14.168}$$

We can write I_1, using a complete set of states,

$$\int |r\rangle\langle r|\, dr = 1, \tag{14.169}$$

as

$$I_1 = -\frac{\lambda}{8\pi\hbar^2\omega} \int_{-\infty}^{\infty} \exp\left(\frac{ipy}{\hbar}\right) dy \int_{-\infty}^{\infty} ds \int_{-\infty}^{\infty} dr$$

$$\times \left[\left\langle x - \frac{1}{2}y|p|r\right\rangle \left\{ \langle r|p|s\rangle \left\langle s|\hat{\rho}|x + \frac{1}{2}y\right\rangle - \langle r|\hat{\rho}|s\rangle \left\langle s|p|x + \frac{1}{2}y\right\rangle \right\} \right.$$

$$- \left\{ \left\langle x - \frac{1}{2}y|p|s\right\rangle \langle s|\hat{\rho}|r\rangle - \left\langle x - \frac{1}{2}y\,|\hat{\rho}|\, s\right\rangle \langle s|p|r\rangle \right\}$$

$$\left. \times \left\langle r\,|p|\, x + \frac{1}{2}y\right\rangle\right]. \tag{14.170}$$

We simplify (14.170) by noting that

$$\langle r'\,|p|\, r\rangle = -i\hbar\frac{\partial}{\partial r'}\delta\left(r' - r\right). \tag{14.171}$$

Thus substituting for the expectation value of p in (14.170) and using partial integration we obtain

$$I_1 = \frac{\lambda}{8\pi\omega}\frac{\partial}{\partial x}\int_{-\infty}^{\infty} \exp\left(\frac{ipy}{\hbar}\right) dy \int_{-\infty}^{\infty}\left[\delta\left(s - x + \frac{1}{2}y\right)\frac{\partial}{\partial s}\left\langle s\,|\hat{\rho}|\left(x + \frac{1}{2}y\right)\right\rangle\right.$$

$$\left. + \delta\left(s - x - \frac{1}{2}y\right)\frac{\partial}{\partial s}\left\langle\left(x - \frac{1}{2}y\right)|\hat{\rho}|\, s\right\rangle\right]ds. \tag{14.172}$$

This expression for I_1 reduces to

$$I_1 = \frac{\lambda\hbar}{4\omega}\frac{1}{2\pi\hbar}\frac{\partial^2}{\partial x^2}\int_{-\infty}^{\infty} e^{\frac{ipy}{\hbar}}\left\langle x - \frac{1}{2}y\,|\hat{\rho}|\, x + \frac{1}{2}y\right\rangle dy = \frac{\hbar\lambda}{4\omega}\frac{\partial^2 W}{\partial x^2}. \tag{14.173}$$

The contribution of other terms in (14.57) to the equation for distribution function can be calculated in a similar way. The final result for the distribution function for Dekker's Hamiltonian is given by the partial differential equation for $W(p, x, t)$ [17]:

$$\frac{\partial W}{\partial t} = -\frac{\partial}{\partial x}\,(pW) + \omega_0^2\frac{\partial}{\partial p}\,(xW) + \lambda\frac{\partial}{\partial p}(pW)$$

$$+ \frac{\lambda\hbar}{4\omega}\frac{\partial^2 W}{\partial x^2} - \frac{\hbar\lambda^2}{4\omega}\frac{\partial^2 W}{\partial x\partial p} + \frac{\hbar\lambda\omega_0^2}{4\omega}\frac{\partial^2 W}{\partial p^2}. \tag{14.174}$$

This equation can be solved approximately (or numerically) for $W(x, p, t)$. For the initial condition we assume that at $t = 0$ there is no coupling between the oscillator and the damping force, therefore

$$W(x, p, 0) = \frac{1}{2\pi\hbar} \int_{-\infty}^{\infty} \psi_n^* \left(x + \frac{y}{2} \right) \psi_n \left(x - \frac{y}{2} \right) \exp\left(\frac{ipy}{\hbar} \right) dy, \qquad (14.175)$$

where $\psi_n(x)$ is the normalized harmonic oscillator wave function for the nth excited state. The integral in (14.175) can be obtained analytically [28]:

$$W(x, p, 0) = \frac{(-1)^n}{\pi\hbar} L_n \left(r^2 \right) \exp\left(-\frac{1}{2}r^2 \right), \qquad (14.176)$$

where r^2 in this equation represents the dimensionless quantity

$$r^2 = 2 \left(\frac{m\omega_0}{\hbar^2} x^2 + \frac{1}{m\omega_0\hbar} p^2 \right), \qquad (14.177)$$

and L_n is the Laguerre polynomial of order n. For the ground state of the oscillator (14.176) takes the simple form of

$$W_0(x, p, 0) = \frac{1}{\pi\hbar} \exp\left(-\frac{1}{2}r^2 \right), \qquad (14.178)$$

which is always non-negative, whereas for the first excited state we find

$$W_1(x, p, 0) = \frac{1}{\pi\hbar} \left(r^2 - 1 \right) \exp\left(-\frac{1}{2}r^2 \right), \qquad (14.179)$$

which is negative around the origin but becomes positive for larger r.

Here we have discussed the density matrix and the Wigner distribution function starting with the Heisenberg equations for the damped harmonic oscillator. We can change our starting point and use the model where a harmonically bound particle is coupled to a field or to a bath of oscillators and for the motion of this central particle derive both $\hat{\rho}$ and W. Once these are known for the whole system, then one can try to isolate the motion of the central particle. This formulation of the damped motion of the particle will be studied in the next section as well as in Sec. 17.4.

14.9   Density Operator for a Particle Coupled to a Heat Bath

When the central particle is linearly coupled to a heat bath of N degrees of freedom and at the same time is subject to an external force, then the Hamiltonian will be quadratic in coordinates and momenta. We can write this Hamiltonian in the compact form of

$$H = \frac{1}{2} z^T B z + C(t) z, \qquad (14.180)$$

where z^T and $C(t)$ are row matrices with $2N + 2$ elements

$$z^T = [P, \, p_1, \cdots p_N, Q, \, q_1, \cdots q_N], \qquad (14.181)$$

and

$$C(t) = [C_1(t), C_2(t), \cdots C_{2N}(t)]. \qquad (14.182)$$

In these equations T denotes the transpose of the matrix and B is a symmetric matrix $B^T = B$.

The equations of motion derived from (14.180) can be written in the matrix form:

$$\frac{dz}{dt} = -ABz - AC(t). \qquad (14.183)$$

Here A is a $(2N + 2) \times (2N + 2)$ antisymmetric matrix with constant matrix elements

$$A = \begin{bmatrix} 0 & I_{N+1} \\ -I_{N+1} & 0 \end{bmatrix}, \quad A^2 = -I_{2N+2}, \qquad (14.184)$$

where I_{N+1} in an $(N+1) \times (N+1)$ unit matrix. Since B is symmetric and A is antisymmetric, therefore Trace $(AB) = 0$.

As an example consider the Hamiltonian given by (9.1) where

$$B = \begin{bmatrix} K_{N+1} & 0 \\ 0 & -I_{N+1} \end{bmatrix}, \qquad (14.185)$$

and

$$K_{N+1} = \frac{1}{2} \begin{bmatrix} 2\Omega_0^2 & \epsilon_1, & \epsilon_2 & \cdots & \epsilon_N \\ \epsilon_1 & 2\omega_1^2 & 0 & \cdots & 0 \\ \epsilon_2 & 0 & 2\omega_2^2 & \cdots & 0 \\ \cdots & \cdots & \cdots & \cdots \\ \epsilon_N & 0 & 0 & \cdots & 2\omega_N^2 \end{bmatrix}. \qquad (14.186)$$

Now let us consider the first order differential equation (14.183) with B given by (14.185). This represents a central particle which interacts with a heat bath. From what we have seen earlier for a single harmonic oscillator, the Wigner distribution function is of the form given by (14.166) but now it is for $N + 1$ degrees of freedom. Therefore we can write this distribution function as

$$\frac{\partial W}{\partial t} = \sum_{j=1}^{2N+2} \frac{\partial}{\partial z_j} \left[\sum_k \{(AB)_{jk} z_k + A_{jk} C_k\} W \right], \qquad (14.187)$$

but the corresponding density matrix will not be positive definite. A way to address this problem is to add the terms

$$\sum_{jk} D_{jk} \frac{\partial^2 W}{\partial z_j \partial z_k}, \qquad (14.188)$$

to the right hand side of (14.187) and make it like a Fokker-Planck equation. The coefficients D_{jk} (diffusion coefficients) are independent of the coordinates

z_j, but can depend on time. These coefficients are not arbitrary, but are subject to certain conditions, and these conditions ensure that $W(z,t)$ remains a positive distribution function for all times [29].

We can also solve (14.187) for the whole system and then average over the coordinates and momenta of the bath. For this we assume that initially the Wigner distribution function is given as a product

$$W(z,t) = W(P,Q)W(p_1....p_N; q_1.....q_N) \qquad (14.189)$$

and then solve the partial differential equation (14.187) with this initial condition. Averaging $W(z,t)$ over the bath degrees of freedom gives us $W(P,Q,t)$ for the motion of the central particle [29]-[32].

References

[1] C.W. Gardiner, *Quantum Noise*, (Springer-Verlag, Berlin 1991) Chapter 2.

[2] W.H. Louisell, *Quantum Statistical Properties of Radiation*, (John Wiley & Sons, New York, N.Y. 1973).

[3] H.-P. Breuer and F. Petruccione, *The Theory of Open Quantum Systems*, (Oxford University Press, Oxford, 2002) Chapter 3.

[4] H. Goldstein, *Classical Mechanics*, Second Edition, (Addison-Wesley Reading, 1980).

[5] J.B. Marion, *Classical Dynamics of Particles and Systems*, Second Edition, (Academic Press, New York, N.Y. 1970) Sec. 7.14.

[6] G. Gerlich, Physica A 69, 4586 (1973).

[7] W.-H. Steep, Physica A 95, 181 (1979).

[8] A.O. Bolivar, Phys. Rev. A 58, 4330 (1998).

[9] A.O. Bolivar, Physica A 301, 210 (2001).

[10] A.O. Bolivar, Can. J. Phys. 81, 663 (2003).

[11] A.O. Bolivar, *Quantum-Classical Correspondence: Dynamical Quantization and the Classical Limit*, (Springer, Berlin, 2004).

[12] E. Wigner, Phys. Rev. 40, 749 (1932).

[13] L.S.F. Olavo, Physica A 262, 197 (1999).

[14] F.J. Dyson, Phys. Rev. 75, 486 (1949).

[15] J.J. Sakurai, *Modern Quantum Mechanics*, (Addison-Wesley, Menlo Park, California, 1985) p. 325.

[16] P.A.M. Dirac, *The Principles of Quantum Mechanics*, Fourth Edition, (Oxford University Press, Oxford, 1958).

[17] H. Dekker, Physica A 95, 311 (1979).

[18] H.Dekker, Phys. Rep. 80, 1 (1981).

[19] H. Risken, *The Fokker-Planck Equation: Methods of Solutions and Applications*, Second Edition, (Springer-Verlag, Berlin, 1996).

[20] G.J. Milburn and D.F. Walls, Am. J. Phys. 51, 1134 (1983).

[21] D.F. Walls and G.I. Millburn, *Quantum Optics*, (Springer-Verlag, Berlin,1995) Chapter 6.

[22] H.P. Yuen, Phys. Rev. A 13, 2226 (1976).

[23] R.W. Henry and S.C. Glotzer, Am. J. Phys. 56, 318 (1988).

[24] N.N. Bogolyubov, J. Phys. (USSR), 11, 77 (1947).

[25] N.G. van Kampen, J. Stat. Phys. 78, 299 (1995).

[26] N.G. van Kampen in *25 Years of Non-Equilibrium Statistical Mechanics*, (Springer-Verlag, Berlin, 1995).

[27] I.S. Gradshetyn and I.M. Ryzhik, *Tables of Integrals, Series and Products*, (Academic Press, New York, N.Y. 1965) p. 925.

[28] Y.S. Kim and M.E. Noz, *Phase Space Picture of Quantum Mechanics*, (World Scientific, Singapore, 1991) Sec. 3.4.

[29] V.V. Dodonov and O.V. Man'ko, Theor. Math. Phys. Vol. 65, 1033 (1986).

[30] V.V. Dodonov and V.I. Man'ko, in *Classical and Quantum Effects in Electrodynamics*, Edited by A.A. Komar (Nova Science, Commack, 1986) p. 53.

[31] A detailed account of the quantum Liouville equation, of Wigner distribution function and of the Fokker-Planck equation for a particle coupled linearly to a heat bath is given by V.V. Dodonov, O.V. Man'ko and V.I. Man'ko, in J. Russian Laser Res. 16, 1 (1995).

[32] V.V. Dodonov, and V.I. Man'ko, in *Quantum Field Theory, Quantum Mechanics and Quantum Optics*, Preceeding of the Lebedev Physics Institute Vol. 187, Edited by V.V. Dodonov, and V.I. Man'ko (Nova Scientific, Commack, 1989) p. 209.

[33] J-S. Peng and G-X Li, *Introduction to Modern Optics*, (World Scientific, Singapore, 1998) p. 160.

[34] A.S. Davydov, *Quantum Mechanics*, translated by D. ter Harr, (Pergamon Press, 1965) p. 610.

[35] S.P. Kim, J. Phys. A 36, 12089 (2003).

Chapter 15

Path Integral Formulation of a Damped Harmonic Oscillator

An elegant way of quantizing the dissipative systems which is specially suited for the damped harmonic oscillator is by means of the Feynman formulation where one can obtain the propagator and from the propagator determine the time development of the wave function [1], [2]. This same method of path integration can be applied to a central particle under the action of the potential $V(Q)$ but coupled linearly to a bath of harmonic oscillators. We will discuss the latter problem in detail later.

For the construction of the propagator we can use either the Lagrangian or the Hamiltonian, but here like other methods of quantization (Dirac or Schrödinger) the Feynman approach will not lead to a unique result. The non-uniqueness arises from the fact that we can use an infinite set of classically equivalent Lagrangians as the starting point. But even if we decide on a single Lagrangian L, the ordering of $x(t)$ and $\dot{x}(t)$ operators will not be a trivial task in the case of dissipative systems where, as we have noted, there are terms like $x^n(t)\dot{x}^j(t)$ in the Lagrangian. Once it was thought that the path integral method can be used to determine a unique way of ordering the operators $x(t)$ and $\dot{x}(t)$ [3]. But later it was discovered that this is not the case and in the path integral formulation like other methods there is no unique way of ordering x and $\dot{x}$ operators [4],[5].

In addition to the ambiguity due to the ordering of x and $\dot{x}$ factors, there is also the problem of choosing the proper classical Lagrangian for the path integral quantization. By "proper Lagrangian" we mean those Lagrangians that are quadratic functions of velocity, since for equivalent Lagrangians like $\frac{1}{6p_0}m^2\dot{x}^3 e^{2\lambda t}$, Eq. (2.166), the Feynman method of construction of propagator does not work [1].

Among the many possible ways of constructing the propagator for the damped harmonic oscillator we focus our attention on two simple cases. In the first one we start with the time-dependent Lagrangian for linear damping, find the propagator and using this propagator calculate the time-dependent wave function. For the second example we determine the propagator for the classical Lagrangian assuming an optical harmonic potential, Eq. (2.192).

15.1   Propagator for the Damped Harmonic Oscillator

There are different ways of constructing the propagator from the classical Lagrangian for one degree of freedom when the damping is linear in velocity, i.e.

$$L(x, \dot{x}, t) = \left[\frac{1}{2} m \dot{x}^2 - V(x) \right] e^{\lambda t}. \tag{15.1}$$

We can start with the definition of the kernel $G(x'', x', t'', t')$ in the phase space [6]-[14]

$$G(x'', x', t'', t') = \int \exp \left\{ \frac{i}{\hbar} \int_{t'}^{t''} [p\dot{x} - H(p, x, t)] \, dt \right\} \mathcal{D}[x] \mathcal{D}[p]$$

$$= \lim_{N \to \infty} \int_{-\infty}^{\infty} \cdots \int_{-\infty}^{\infty} \frac{dp'}{\sqrt{2\pi\hbar}} \prod_{k=1}^{N-1} \frac{dp_k dx_k}{\sqrt{2\pi\hbar}}$$

$$\times \exp \left\{ \frac{i}{\hbar} \sum_{k=0}^{N-1} [p_k(x_{k+1} - x_k) - \varepsilon H(p_k, x_k, t)] \right\}, \tag{15.2}$$

and carry out the functional integration in (15.2) to find $G(x'', x', t'', t')$. In Eq. (15.2) $H(p, x, t)$ is the Caldirola-Kanai Hamiltonian found from (15.1), $\varepsilon = \frac{t''-t'}{N}$ is the small time increment and $x'' = x(t'')$ and $x' = x(t')$ are the initial and final coordinates respectively. By substituting for $H(p, x, t)$ in (15.2) and integrating over all momenta in the phase space we get [6],[7]

$$G(x'', x'; t'', t') = \int \exp \left\{ \frac{i}{\hbar} \int_{t'}^{t''} L(x, \dot{x}, t) \, dt \right\} \mathcal{D}[x]$$

$$= \lim_{N \to \infty} \left(\frac{m e^{\lambda t'}}{2\pi i \hbar \varepsilon} \right)^{\frac{1}{2}} \prod_{k=1}^{N-1} \left(\frac{m e^{\lambda t_k}}{2\pi i \hbar \varepsilon} \right)^{\frac{1}{2}} \times$$

$$\times \int_{-\infty}^{\infty} \cdots \int_{-\infty}^{\infty} \exp \left\{ \frac{i\varepsilon}{\hbar} \sum_{k=0}^{N-1} \left[\frac{m}{2} \left(\frac{x_{k+1} - x_k}{\varepsilon} \right)^2 - V(x_k) \right] e^{\lambda t_k} \right\} \prod_{k=1}^{N-1} dx_k, \tag{15.3}$$

where we have used the Gaussian integral [8]–citeDittrich

$$\frac{1}{\sqrt{2\pi\hbar}} \int_{-\infty}^{\infty} \exp\left\{\frac{i}{\hbar}\left[p_k\left(x_{k+1} - x_k\right) - \frac{\varepsilon p_k^2}{2m}e^{-\lambda t_k}\right]\right\} dp_k$$

$$= \left(\frac{me^{\lambda t_k}}{i\varepsilon}\right) \exp\left[\frac{im\varepsilon e^{\lambda t_k}}{2\hbar}\left(\frac{x_{k+1} - x_k}{\varepsilon}\right)^2\right], \tag{15.4}$$

to carry out the integration over momenta.

Propagator for the Damped Harmonic Oscillator — Now let us consider the damped harmonic oscillator with $V(x) = \frac{1}{2}m\omega_0^2 x^2$ for which (15.3) can be written as

$$G\left(x'', x'; t'', t'\right) = \lim_{N\to\infty} \left[\prod_{k=1}^{N-1}\left(\frac{me^{\lambda t_k}}{2\pi i\hbar}\right)^{\frac{1}{2}}\right]$$

$$\times \int_{-\infty}^{\infty} \cdots \int_{-\infty}^{\infty} \exp\left\{\frac{im\varepsilon}{2\hbar}\left[\left(\frac{1}{\varepsilon^2}\right)\sum_{k=0}^{N-1} e^{\lambda t_k}\left(x_{k+1} - x_k\right)^2 - \sum_{k=0}^{N-1} e^{\lambda t_k}\omega_0^2 x_k^2\right]\right\}$$

$$\times \prod_{k=1}^{N-1} dx_k. \tag{15.5}$$

It is convenient to change the variable x_k to s_k where

$$s_k = \left(\frac{m}{2\hbar\varepsilon}\right)^{\frac{1}{2}} \exp\left(\frac{\lambda t_k}{2}\right) x_k, \tag{15.6}$$

and then $G\left(x'', x', t'', t'\right)$ can be rewritten as

$$G\left(s'', s'; t'', t'\right) = \lim_{N\to\infty} \frac{1}{(i\pi)^{\frac{N}{2}}}\left(\frac{me^{\lambda t''}}{2\hbar\varepsilon}\right)^{\frac{1}{2}}$$

$$\times \exp\left\{i\left[\left(s''^2 + s'^2\right) - \varepsilon^2\omega_0^2 s'^2\right]\right\}$$

$$\times \int_{-\infty}^{\infty} \cdots \int_{\infty}^{\infty} \exp\left[i\left(\sum_{k=1}^{N-1}\left(1 + e^{\lambda\varepsilon} - \omega_0^2\varepsilon^2\right)s_k^2 - 2\sum_{k=0}^{N-1} e^{\frac{\lambda\varepsilon}{2}} s_k s_{k+1}\right)\right]$$

$$\times \prod_{k=1}^{N-1} ds_k. \tag{15.7}$$

We can transform this multiple integral to a Gaussian integral [8],[9]

$$\int_{-\infty}^{\infty} \cdots \int_{\infty}^{\infty} \exp\left[i\left(s^T A s + 2b^T s\right)\right] \prod_{k=1}^{N-1} ds =$$

$$= (i\pi)^{\frac{N}{2}} \frac{1}{\sqrt{\det A}} \exp\left(-ib^T A^{-1} b\right), \tag{15.8}$$

where the matrix A is given by

$$A = \begin{bmatrix} a_1 & -d & 0 & 0\cdots\cdots 0 & 0 & 0 & 0 \\ -d & a_2 & -d & 0\cdots\cdots 0 & 0 & 0 & 0 \\ 0 & -d & a_3 & -d\cdots\cdots 0 & 0 & 0 & 0 \\ \cdots & \cdots & \cdots & \cdots & \cdots & \cdots & \cdots \\ 0 & 0 & 0 & 0\cdots\cdots -d & a_{N-3} & -d & 0 \\ 0 & 0 & 0 & 0\cdots\cdots 0 & -d & a_{N-2} & -d \\ 0 & 0 & 0 & 0\cdots\cdots 0 & 0 & -d & a_{N-1} \end{bmatrix}. \tag{15.9}$$

The matrix elements a_k and d in (15.9 are functions of ε;

$$a_k = 1 + e^{\lambda \varepsilon} - \omega_k^2 \varepsilon^2, \quad \text{and} \quad d = \exp\left(\frac{\lambda \varepsilon}{2}\right), \tag{15.10}$$

while the column matrix b in Eq. (15.9) is composed of the following elements

$$b_1 = -s' \exp\left(\frac{\lambda \varepsilon}{2}\right) = -\sqrt{\frac{m}{2\hbar\varepsilon}}\, x' \exp\left(\frac{\lambda t'}{2}\right), \tag{15.11}$$

$$b_k = 0, \quad k = 2, 3 \cdots N - 2, \tag{15.12}$$

and

$$b_{N-1} = -s'' \exp\left(\frac{\lambda \varepsilon}{2}\right) = -\sqrt{\frac{m}{2\hbar\varepsilon}}\, x'' \exp\left[\frac{\lambda}{2}\left(t'' + \varepsilon\right)\right]. \tag{15.13}$$

Now if we substitute (15.8) and (15.9) in (15.7) and note that as $\varepsilon \to 0$, $\exp\left(-i\varepsilon^2 \omega_0^2 s'^2\right) \to 1$, we obtain

$$G\left(s'', s'; t'', t'\right) = \lim_{\varepsilon \to 0} \left(\frac{m e^{\lambda t''}}{2\pi i \hbar\varepsilon \det A}\right)^{\frac{1}{2}} \exp\left[iB\left(s'', s'; \varepsilon\right)\right], \tag{15.14}$$

where

$$B\left(s'', s'; \varepsilon\right) = \left(s''^2 + s'^2\right) - b^T(\varepsilon) A^{-1}(\varepsilon) b(\varepsilon). \tag{15.15}$$

Next we find the limit of $\varepsilon \to 0$ of (15.14). First let us find the limit $\varepsilon \det A$ as ε goes to zero. To this end we consider the following determinants:

$$q_{N-1} = \varepsilon a_{N-1},$$

$$q_{N-2} = \varepsilon \begin{vmatrix} a_{N-2} & -d \\ -d & a_{N-1} \end{vmatrix},$$

$$q_{N-3} = \varepsilon \begin{vmatrix} a_{N-3} & -d & 0 \\ -d & a_{N-2} & -d \\ 0 & -d & a_{N-1} \end{vmatrix},$$

$$\cdots\cdots\cdots\cdots\cdots$$

$$q_1 = \varepsilon \det A \tag{15.16}$$

and

$$p_1 = \varepsilon a_1,$$

$$p_2 = \varepsilon \begin{vmatrix} a_1 & -d \\ -d & a_2 \end{vmatrix},$$

$$p_3 = \varepsilon \begin{vmatrix} a_1 & -d & 0 \\ -1 & a_2 & -d \\ 0 & -d & a_3 \end{vmatrix},$$

. . . .

$$p_{N-1} = \varepsilon \det A. \tag{15.17}$$

We can verify that q_k and p_k satisfy the finite difference equations

$$\frac{1}{\varepsilon^2}(q_{n+1} - 2q_n + q_{n-1}) = -\omega_0^2 q_n - \frac{\lambda}{\varepsilon}(q_{n+1} - q_n), \tag{15.18}$$

and

$$\frac{1}{\varepsilon^2}(p_{n+1} - 2p_n + p_{n-1}) = -\omega_0^2 p_n + \frac{\lambda}{\varepsilon}(p_n - p_{n-1}), \tag{15.19}$$

respectively. In the limit of $\varepsilon \to 0$, Eqs. (15.18) and (15.19) become differential equations

$$\ddot{q} + \lambda\dot{q} + \omega_0^2 q = 0, \tag{15.20}$$

and

$$\ddot{p} - \lambda\dot{p} + \omega_0^2 p = 0. \tag{15.21}$$

These equations are subject to the boundary conditions that at the final time t'',

$$q(t'') = 0, \quad \dot{q}(t'') = -1, \tag{15.22}$$

whereas at the initial time t'

$$p(t') = 0, \quad \dot{p}(t') = 1. \tag{15.23}$$

The solutions of (15.20) and (15.21) subject to these conditions have simple forms of

$$q(t) = \frac{1}{\omega} \exp\left[-\frac{\lambda}{2}(t - t'')\right] \sin\omega(t'' - t), \tag{15.24}$$

and

$$p(t) = \frac{1}{\omega} \exp\left[-\frac{\lambda}{2}(t' - t)\right] \sin\omega(t - t'), \tag{15.25}$$

where $\omega = \sqrt{\omega_0^2 - \frac{\lambda^2}{4}}$. We note that while $q(t)$ and $p(t)$ are the solutions of canonical equations for coordinate and momentum found from Caldirola-Kanai Hamiltonian, the initial conditions are different. We also note that both $q(t)$ and $p(t)$ have the same dimension.

From the definitions of q_k and p_k we can also find the following relation

$$q_{k+1}p_k - d^2 q_{k+2}p_{k-1} = q_k p_{k-1} - d^2 q_{k+1}p_{k-2}. \tag{15.26}$$

Using this together with the difference equations for q_k and p_k we obtain p_k in terms of q_k s and p_{k-1}, i.e.

$$p_k = \frac{\varepsilon q_1 q_{k+2}}{q_{k+1}q_{k+2}} + d^2 \frac{q_{k+2}}{q_{k+1}} p_{k-1}. \tag{15.27}$$

By continuing this iteration we can write p_k in terms of q_k s only

$$p_k = \varepsilon q_1 q_{k+2} \sum_{j=1}^{k+1} \frac{d^{2(k-j+1)}}{q_j q_{j+1}}. \tag{15.28}$$

Thus $b^T A^{-1} b$ in (15.15) is expressible in terms of q_k s [6]

$$b^T A^{-1} b = \sum_{k=1}^{N-1} \frac{\left(\sum_{j=k}^{N-1} b_j d^j q_{j+1} \right)^2}{q_k q_{k+1} d^{2k}}. \tag{15.29}$$

After a lengthy calculation we obtain the following result;

$$b^T A^{-1} b = \left(\frac{mq_2}{2\hbar \varepsilon q_1} \right) e^{\lambda t'} x^2(t') + \left(\frac{m}{\hbar q_1} \right) e^{\lambda t''} x(t') x(t'')$$

$$+ \frac{m}{2\hbar} \left(\sum_{k=1}^{N-1} \frac{d^{2(N-k)}}{q_k q_{k+1}} \right) e^{\lambda t''} x^2(t''). \tag{15.30}$$

Substituting for $\varepsilon \det A$, $B(s'', s'; \varepsilon)$ and $b^T(\varepsilon)A^{-1}(\varepsilon)b(\varepsilon)$ in the expression for the propagator (15.14), taking the limit as $\varepsilon \to 0$ and simplifying the result we obtain

$$G(x'', x'; t', t'') = \left(\frac{me^{\lambda t''}}{2\pi i \hbar q(t')} \right)^{\frac{1}{2}}$$

$$\times \exp \left[\left(\frac{me^{\lambda t''}}{2i\hbar q(t')} \right) \left[e^{-\lambda(t''-t')} \dot{q}(t') x'^2 + 2x'x'' + (\lambda q(t') - \dot{p}(t'')) x''^2 \right] \right]. \tag{15.31}$$

We can rewrite (15.31) by substituting for $q(t)$ and $p(t)$ from (15.24) and (15.25)

$$G(x'', x'; t', t'') = \left(\frac{m\omega e^{\frac{\lambda}{2}(t''+t')}}{2\pi i \hbar \sin \omega T} \right)^{\frac{1}{2}} \exp \left[-\frac{\lambda}{2i\omega} \left(X'^2 - X''^2 \right) \right]$$

$$\times \exp \left[\left(\frac{im\omega}{2\hbar \sin \omega T} \right) \left\{ \left(X'^2 + X''^2 \right) \cos \omega T - 2X'X'' \right\} \right], \tag{15.32}$$

where

$$X(t) = \left(\frac{me^{\lambda t}}{2\hbar}\right)^{\frac{1}{2}} x(t) \quad \text{and} \quad T = t'' - t'. \tag{15.33}$$

For $\lambda = 0$, the propagator $G\left(x'', x'; t', t''\right)$ reduces to the well-known form of the harmonic oscillator propagator, and only then G will become a function of $T = t'' - t'$.

The time-dependent wave function for this propagator can be obtained from the initial wave function. Suppose that at $t = 0$ we have

$$\psi_n(X, 0) = \psi_n(X, \omega) \exp\left(-\frac{i\lambda X^2}{2\omega}\right), \tag{15.34}$$

where $\psi_n(X, \omega)$ is the normalized nth eigenfunction of the harmonic oscillator with shifted frequency ω, and $\exp\left(-\frac{i\lambda X^2}{2\omega}\right)$ is a space-dependent phase factor. Using Eq. (15.34) we can calculate the wave function at a later time t from the integral

$$\psi_n(X, t) = \int G\left(X, X'; t, 0\right) \psi_n\left(X', 0\right) dX'. \tag{15.35}$$

By substituting for $G\left(x'', x'; t', t''\right)$ from (15.31) and for $\psi_n\left(X', 0\right)$ from (15.34) in (15.35) and carrying out the integration over X' we find

$$\psi_n(x, t) = N_n \exp\left(\frac{i\mathcal{E}_n t}{\hbar}\right) \exp\left[-\frac{m\omega}{2\hbar}\left(1 + \frac{i\lambda}{2\omega}\right) e^{\lambda t} x^2\right]$$
$$\times H_n\left[\sqrt{\frac{m\omega}{\hbar}} e^{\frac{\lambda t}{2}} x\right], \tag{15.36}$$

where

$$\mathcal{E}_n = \left[\left(n + \frac{1}{2}\right)\omega + \frac{i\lambda}{4}\right]\hbar, \tag{15.37}$$

is the eigenvalue for the nth excited state, and

$$N_n = \left(\frac{m\omega}{\pi\hbar}\right)^{\frac{1}{4}} \left(\frac{1}{2^n n!}\right)^{\frac{1}{2}}. \tag{15.38}$$

In Eq. (15.36) we have expressed the wave function in terms of the original coordinate x. This result is identical with the wave function obtained from Caldirola-Kanai Hamiltonian, Eq. (12.10) [7],[11]-[14].

The damped motion of a harmonic oscillator in a uniform magnetic field is another problem for which the wave function as well as the kernel G can be obtained exactly , provided that the magnetic field increases with time as $H(t) = H_0 e^{\lambda t}$ [15].

15.2 Path Integral Quantization of a Harmonic Oscillator with Complex Spring Constant

In the classical formulation of the damped harmonic oscillator Sec. 4.2 we observed that we can use complex time formulation to obtain a very simple Lagrangian as well as the classical action for this motion [16]. As we have seen in Sec. 4.2 the action for the complex time τ is

$$S\left(x_2, x_2; \tau_2, \tau_2\right) = \frac{m\omega}{2\sin\omega(\tau_2 - \tau_2)} \left[\left(x_2^2 + x_1^2\right)\cos\omega(\tau_2 - \tau_1) - 2x_1x_2\right], \quad (15.39)$$

where $\tau = (1 - i\alpha)t$ and this τ can also be written as $\tau = \left(1 - \frac{i\lambda}{2\omega}\right)t$, with λ being the damping constant. We can change the complex time τ to the real time t by writing

$$\beta = (i + \alpha)\omega, \quad \text{or} \quad \omega\tau = -i\beta t, \quad (15.40)$$

and then express S in terms of $T = t_2 - t_1$,

$$S\left(x_2, x_2; T\right) = -\frac{m\omega}{2i\sinh\beta T} \left[\left(x_2^2 + x_1^2\right)\cosh\beta T - 2x_1x_2\right]. \quad (15.41)$$

Since the action is quadratic in the coordinates x_2 and x_1, we can write the propagator as [8]

$$G(2, 1) = \int_1^2 \exp\left[\frac{i}{\hbar}S(2, 1)\right] \mathcal{D}[x], \quad (15.42)$$

where 2 and 1 refer to the space-time coordinates (x_2, t_2) and (x_1, t_1). Now the action (15.39) is identical with the action for a harmonic oscillator, therefore using the method of Gaussian integrals we find $G(2, 1)$ to be [8]

$$G(2, 1) \doteq \frac{\alpha}{[2\pi\sinh\beta T]^{\frac{1}{2}}}$$

$$\times \exp\left[\frac{\alpha^2}{2\sinh\beta T}\left\{\left(x_2^2 + x_1^2\right)\cosh\beta(\tau_2 - \tau_1) - 2x_1x_2\right\}\right].$$

$$(15.43)$$

From the summation formula [17]

$$\sum_{k=0}^{\infty} \left(\frac{\lambda}{2}\right)^k \frac{1}{k!} H_k(x)H_k(y) \exp\left[-\frac{1}{2}\left(x^2 + y^2\right)\right]$$

$$= \frac{1}{\sqrt{1 - \lambda^2}} \exp\left[-\frac{1 + \lambda^2}{2(1 - \lambda)^2}\left(x^2 + y^2\right) + \frac{2\lambda}{1 - \lambda^2}xy\right], \quad (15.44)$$

it follows that $G(2,1)$ can be written as a product of the harmonic oscillator wave functions

$$G(2,1) = \sqrt{\frac{m\omega}{\pi\hbar}} \sum_{n=0}^{\infty} \frac{1}{2^n n!} H_n\left(\sqrt{\frac{m\omega}{\hbar}}x_2\right) H_n\left(\sqrt{\frac{m\omega}{\hbar}}x_1\right)$$

$$\times \exp\left[-\frac{m\omega}{2\hbar}\left(x_1^2 + x_2^2\right)\right] \exp\left[-\beta\left(n + \frac{1}{2}\right)T\right]. \tag{15.45}$$

This expression for the propagator can be used to verify that

$$G(3,1) = \int G(3,2)G(2,1)dx_2, \tag{15.46}$$

i.e. the amplitudes for the events occurring in succession in time multiply. **Normalized Wave Function Found from the Propagator** — Now we want to examine the evolution of a displaced wave packet in time with the help of the propagator $G(2,1)$. We choose the initial wave packet to be

$$\Psi_n(x_1,0) = \left(\frac{m\omega}{\pi\hbar}\right)^{\frac{1}{4}} \frac{1}{\sqrt{2^n n!}} H_n\left[\sqrt{\frac{m\omega}{\hbar}}(x_1 + \xi_0)\right] \exp\left[-\frac{m\omega}{2\hbar}(x_1 + \xi_0)^2\right], \tag{15.47}$$

where ξ_0 is a constant. The wave packet at a later time, t, can be obtained from the propagator G and the initial wave packet $\Psi_n(x_1,t)$;

$$\Phi_n(x,t) = \int_{-\infty}^{\infty} G(x,t;x_1,0)\Psi_n(x_1,0)dx_1, \tag{15.48}$$

where from the definition of $\Phi_n(x,t)$ it follows that

$$\Phi_n(x,0) = \Psi_n(x,0). \tag{15.49}$$

By substituting for G and Ψ_n from (15.45) and (15.47) in (15.48) and carrying out the integration over x_1 we find

$$\Phi_n(x,t) = \left(\frac{m\omega}{\hbar\pi}\right)^{\frac{1}{4}} H_n\left[\sqrt{\frac{m\omega}{\hbar}}\left\{x + \xi_0 \cosh\left(i + \frac{\lambda}{2\omega}\right)\omega t\right\}\right]$$

$$\times \exp\left[-\beta\left(n + \frac{1}{2}\right)t - \frac{m\omega}{2\hbar}\left(x + \xi_0 \cos\omega t e^{\frac{-\lambda t}{2}}\right)^2 + \frac{m\omega}{4\hbar}\xi_0^2\left(e^{-\lambda t} - 1\right)\right]$$

$$\times \frac{1}{\sqrt{2^n n!}} \exp\left[\frac{im\omega}{\hbar}\left\{\frac{1}{4}\xi_0^2 \sin(2\omega t)e^{-\lambda t} + x\xi_0 \sin(\omega t)e^{-\frac{\lambda t}{2}}\right\}\right], \tag{15.50}$$

where in this equation we have replaced α by $\frac{\lambda}{2\omega}$. This wave function is not normalized except at the time $t = 0$. If we want a properly normalized wave packet, we change the propagator by a time-dependent factor. Let $\Gamma(x,t;x_1,0)$ be the new propagator

$$\Gamma_n(x,t;x_1,0) = \frac{1}{D_n(t)}G(x,t;x_1,0), \tag{15.51}$$

and let us use the same initial wave packet, then the wave packet at a later time is given by

$$\Psi_n(x,t) = \int_{-\infty}^{\infty} \Gamma_n(x,t;x_1,0)\Psi_n(x_1,0)dx_1 = \frac{\Phi_n(x,t)}{D_n(t)}. \tag{15.52}$$

Since we want $\Psi_n(x,t)$ to be normalized, from (15.52) we obtain

$$\langle \Psi_n(x,t)|\Psi_n(x,t)\rangle = \frac{\langle \Phi_n|\Phi_n\rangle}{|D_n(t)|^2} = 1. \tag{15.53}$$

From Eqs. (15.50) and (15.53) we find $D_n(t)$ to be

$$D_n(t) = \exp\left[-\left(n+\frac{1}{2}\right)\frac{\lambda t}{2} - \frac{m\omega}{4\hbar}\xi_0^2\left(1-e^{-\lambda t}\right)\right]$$

$$\times \left[L_n\left\{-\frac{2m\omega}{\hbar}\xi_0^2 \sinh^2\left(\frac{\lambda t}{2}\right)\right\}\right]^{\frac{1}{2}}, \tag{15.54}$$

where we have used the following integral [18]:

$$\int_{-\infty}^{\infty} e^{-x^2} H_n(x+y)H_n(x+z)dx = 2^n\sqrt{\pi}n!L_n(-2yz). \tag{15.55}$$

The wave packet at the time t is given by (15.52), and the probability density at this time is

$$|\Psi_n(x,t)|^2 = \sqrt{\frac{m\omega}{\pi\hbar}}\frac{1}{2^n n!}\exp\left[-\frac{m\omega}{\hbar}\left(x+\xi_0\cos(\omega t)e^{\frac{-\lambda t}{2}}\right)^2\right]$$

$$\times \frac{\left|H_n\left[\sqrt{\frac{m\omega}{\hbar}}\left(x+\xi_0\cosh\left[\left(i\omega+\frac{\lambda}{2}\right)t\right]\right)\right]\right|^2}{L_n\left[-\frac{2m\omega}{\hbar}\xi_0^2\sinh^2\left(\frac{\lambda t}{2}\right)\right]}. \tag{15.56}$$

From this result we can conclude that:

(1) For $n=0$, $H_0=1$ and $L_0=1$, thus we have the damped oscillations of a Gaussian wave packet with the damping constant λ. Asymptotically for large t (15.56) becomes

$$|\Psi_0(x,t)|^2 \to \sqrt{\frac{m\omega}{\hbar\pi}}\exp\left(-\frac{m\omega x^2}{\hbar}\right). \tag{15.57}$$

(2) For all values of n, $|\Psi_n(x,t)|^2$ decays to the ground state, $n=0$, and the lifetime of the excited state is proportional to $\frac{2}{\lambda}$.

(3) We can use the Feynman-Kac formula [5] and relate the energy of the ground state to the asymptotic form of the propagator Γ_n. This formula is expressible as

$$E_0 = -\hbar \lim_{t\to\infty}\frac{1}{it}\ln\left\{\int_{-\infty}^{\infty}\Gamma_0(x,t;x,0)dx\right\}. \tag{15.58}$$

If we substitute for $\Gamma_0(x, t; x, 0)$ from (15.51), we find

$$\int_{-\infty}^{\infty} \Gamma_0(x, t; x, 0)dx = \frac{1}{D_0(t)} \int_{-\infty}^{\infty} G(x, t; x, 0)dx$$

$$= \frac{\exp\left[-\frac{1}{2}\beta t - \frac{m\omega}{4\hbar}\xi_0^2 \left(1 - e^{-\lambda t}\right)\right]}{(1 - e^{-\beta t})}. \tag{15.59}$$

By substituting (15.59) in (15.58) and taking the limit as $t \to \infty$, we find $E_0 = \frac{1}{2}\hbar\omega$, as expected.

15.3 Other Formulations of Classical Actions and the Corresponding Propagators for the Damped Harmonic Oscillator

In addition to the propagator that we found earlier from Caldirola-Kanai Hamiltonian we can also use a form of the classical action that we discussed in Chapter 4. The starting point of this formulation is the Hamiltonian

$$H = \frac{p^2}{2m} + \frac{m}{2}\omega_0^2 x^2 + \lambda m\dot{\xi}(t)(x - \xi(t)), \tag{15.60}$$

and the action (4.28) which we now write as

$$S(x, t) = \frac{m}{2}\omega_0 \left[x - \xi(t)\right]^2 \cot \omega_0 t + mx\dot{\xi}(t) + g(t). \tag{15.61}$$

In addition we have replaced α by ω_0 where

$$\omega_0^2 = \left(1 + \alpha^2\right)\omega^2 = \left(\omega^2 + \frac{\lambda^2}{4}\right). \tag{15.62}$$

The time-dependent function $g(t)$ will be determined later. We can write this action for two space-time points (x_2, t_2) and (x_1, t_1) as

$$S(2, 1) = \frac{m\omega_0}{2\sin\omega_0 T} \left[\left\{(x_1 - \xi_1)^2 + (x_2 - \xi_2)^2\right\} \cos\omega_0 T\right.$$

$$\left. - 2(x_1 - \xi_1)(x_2 - \xi_2)\right] + m\left(x_2\dot{\xi}_2 - x_1\dot{\xi}_1\right) + g(t_2) - g(t_1). \tag{15.63}$$

In this relation we have used the following notations: $T = t_2 - t_1$, $\xi_1 = \xi(t_1)$ and $\xi_2 = \xi(t_2)$.

We observe that $S(2, 1)$ is a quadratic function of x_1 and x_2, therefore the exact propagator can be written as [10]

$$G(2, 1) = F(t_2, t_1) \exp\left[\frac{i}{\hbar}S(2, 1)\right], \tag{15.64}$$

where $F(t_2, t_1)$ is defined by the functional integral

$$F(t_2, t_1) = \int_0^0 \exp\left[\frac{i}{\hbar} S(2, 1)\right] \mathcal{D}[x]. \tag{15.65}$$

For a complete determination of $G(2, 1)$ we can either carry out the functional integration in (15.65) or alternatively try to find $F(t_2, t_1)$ by considering the time evolution of a wave packet. In the latter method for the wave packet of the nth excited state we choose $\Psi_n(x_1, t)$ to be

$$\Psi_n(x_1, 0) = \left(\frac{1}{2^n n!} \frac{\alpha_0}{\pi^{\frac{1}{2}}}\right)^{\frac{1}{2}} H_n\left[\alpha_0(x_1 - \xi_1)\right] \exp\left[-\frac{\alpha_0^2}{2}(x - \xi_1)^2\right] \exp\left(\frac{im}{\hbar} x_1 \dot\xi_1\right), \tag{15.66}$$

where we have denoted $\frac{m\omega_0}{\hbar}$ by α_0^2. The last factor in (15.66), $\exp\left(\frac{im}{\hbar} x_1 \dot\xi_1\right)$, has been included in $\Psi_n(x_1, 0)$, so that the corresponding factor in the kernel can be cancelled. The wave function at t_2 is given by

$$\Psi_n(x_2, t_2) = \int_{-\infty}^{\infty} G(x_2, t_2; x_1, 0) \Psi_n(x_1, 0) dx_1. \tag{15.67}$$

Substituting for G and Ψ from (15.64) and (15.66) in (15.67) and integrating over x_1 we obtain

$$\Psi_n(x_2, t_2) = \frac{\alpha_0^{\frac{1}{2}}}{(2^n n!)^{\frac{1}{2}} \pi^{\frac{1}{4}}} H_n\left[\alpha_0(x_2 - \xi_2)\right] \exp\left[-\frac{\alpha_0^2}{2}(x_2 - \xi_2)^2\right]$$

$$\times \exp\left\{-i\omega_0\left(n + \frac{1}{2}\right) t_2 + \frac{i}{\hbar} m x_2 \dot\xi_2 + \frac{i}{\hbar}[g(t_2) - g(t_0)]\right\}, \tag{15.68}$$

provided that

$$F(t_2) = \frac{\alpha_0}{\sqrt{2\pi i \sin(\omega_0 t_2)}}. \tag{15.69}$$

This is consistent with the form of the initial wave packet (15.66), and $\Psi_n(x_2, t_2)$ is normalized for all values of t_2.

The time-dependent function $g(t)$ which has not been defined so far can be determined from the energy of the particle. For the nth excited state the energy $E_n(t)$ is

$$E_n(t) = \left\langle \Psi_n \left| i\hbar \frac{\partial}{\partial t} \right| \Psi_n \right\rangle = \left(n + \frac{1}{2}\right) \hbar\omega_0 - m\xi(t)\ddot\xi(t) - \dot g(t). \tag{15.70}$$

This energy should be equal to the quantized energy of the oscillator, i.e. $\left(n + \frac{1}{2}\right) \hbar\omega_0$ plus the classical energy of the displaced oscillator. Thus from $\partial\Psi(x, t)/\partial t$ we find

$$-\left(m\xi(t)\ddot\xi(t) + \dot g(t)\right) = \frac{1}{2} m\dot\xi^2(t) + \frac{1}{2} m\omega_0^2 \xi^2(t), \tag{15.71}$$

where $\xi(t)$ is the solution of Eq. (4.19). From Eq. (15.71) we find $g(t)$ to be

$$g(t) = m \int \left(\frac{1}{2}\omega_0^2 \xi^2(t) - \frac{1}{2}\dot{\xi}^2(t) + \lambda \xi(t)\dot{\xi}(t) \right) dt. \tag{15.72}$$

A Different Form of Classical Action — A different Hamiltonian for the linearly damped oscillator is obtained from L', Eq, (2.184) [20],[21];

$$H\left(x', p'\right) = \frac{1}{2m}p'^2 + \frac{1}{2}m\omega_0^2 x'^2 + \frac{1}{2}\lambda x' p'. \tag{15.73}$$

This is the same first integral of motion that we found earlier from group theoretical consideration, Eq. (3.174). The Hamilton-Jacobi equation is obtained from (15.73) is given by

$$\frac{1}{2m}\left(\frac{\partial S'}{\partial x'}\right)^2 + \frac{1}{2}m\omega_0^2 + \frac{\lambda}{2}x'\frac{\partial S'}{\partial x'} + \frac{\partial S'}{\partial t} = 0. \tag{15.74}$$

The solution of this equation gives us the action

$$S\left(x', x'_0, t\right) = \frac{m}{4}\left(\alpha x_0'^2 - 2\beta x'_0 x' + \gamma x'^2\right), \tag{15.75}$$

where the three coefficients α, β and γ are functions of time and are given by the following relations;

$$\alpha = 2\omega \cot \omega t + \lambda, \tag{15.76}$$

$$\beta = \left(\frac{2\omega}{\sin \omega t}\right)\cosh\left(\frac{\lambda t}{2}\right), \tag{15.77}$$

and

$$\gamma = 2\omega \cot \omega t - \lambda. \tag{15.78}$$

For a quadratic Hamiltonian, Eq. (15.73) the propagator $G\left(x', t; x'_0, 0\right)$ can be written in closed form [5]

$$G\left(x', t; x'_0, 0\right) = \left[\frac{i}{2\pi\hbar}\left(\frac{\partial^2 S'}{\partial x'_0 \partial x'}\right)\right]^{\frac{1}{2}} \exp\left(\frac{i}{\hbar}S'\right), \tag{15.79}$$

where S' is the classical action given by (15.75). Using this propagator we find the wave function for the nth state of the system from the relation

$$\Psi_n\left(x', t\right) = \int_{-\infty}^{\infty} G\left(x'.t; x'_0, 0\right)\Psi_n\left(x'_0, 0\right) dx'_0. \tag{15.80}$$

Now let us assume that at $t = 0$, λ is zero and therefore initially we have an undamped oscillator with the wave function

$$\Psi_n\left(x'_0, 0\right) = N_0 H_n\left(\alpha_0 x'_0\right)\exp\left(-\frac{1}{2}\alpha_0^2 x_0^2\right) \tag{15.81}$$

where

$$N_0 = \left(\frac{\alpha_0}{\sqrt{\pi}2^n n!}\right)^{\frac{1}{2}} \quad \text{and} \quad \alpha_0 = \sqrt{\frac{m\omega}{\hbar}}. \tag{15.82}$$

From this initial wave function we obtain the wave function at a later time from (15.80) [21];

$$\Psi_n(x',t) = N_0\sqrt{A}\exp\left[-i\left(n+\frac{1}{2}\right)\cot^{-1}\left(\frac{\alpha}{2\omega}\right)\right]\exp\left(-Bx'^2\right)H_n(Ax'), \tag{15.83}$$

where

$$A = \frac{\beta}{\sqrt{\alpha^2 + 4\omega^2}}, \tag{15.84}$$

and

$$B = \frac{im}{4\hbar}\left\{\frac{\beta^2(\alpha - 2\omega)}{\alpha^2 + 4\omega^2} - i\gamma\right\}. \tag{15.85}$$

In getting (15.83) we have used the following integral [18]

$$\int_{-\infty}^{\infty} e^{-(x-y)^2} H_n(ax)dx = \left[\pi\left(1-a^2\right)^n\right]^{\frac{1}{2}} H_n\left[\frac{ay}{\sqrt{1-a^2}}\right], \tag{15.86}$$

in addition to the identity [22]

$$\left(\frac{a-ib}{a+ib}\right)^{\frac{n}{2}} = \exp\left\{-in\cot^{-1}\left(\frac{a}{b}\right)\right\}, \tag{15.87}$$

which is found from the well-known relation

$$\ln\left(\frac{z-i}{a+ib}\right) = -2i\cot z. \tag{15.88}$$

The Mechanical Energy of the Damped Oscillator — Now let us calculate the way that mechanical energy of the oscillator is changing as a function of time. To this end we calculate the canonical momenta

$$p' = \frac{\partial L'}{\partial x'} = m\left(\dot{x}' - \frac{1}{2}\lambda x'\right), \tag{15.89}$$

and from this relation we find the velocity operator

$$\dot{x}' = \frac{1}{m}p' + \frac{1}{2}\lambda x'. \tag{15.90}$$

Next we find the energy operator for the oscillator which is the sum of kinetic and potential energies and after symmetrization, the sum of these energies is given by

$$\begin{aligned}
\hat{E} &= \frac{1}{2}m\left(\dot{x}'^2 + \omega^2 x'^2\right) \\
&= \frac{1}{2m}p'^2 + \frac{1}{2}\left(\omega^2 + \frac{1}{4}\lambda^2\right)x'^2 + \frac{1}{2}\lambda p'x' \\
&= -\frac{\hbar^2}{2m}\frac{\partial^2}{\partial x'^2} + \frac{m}{2}\left(\omega^2 + \frac{1}{4}\lambda^2\right)x'^2 - \frac{i\lambda\hbar}{4}\left(x'\frac{\partial}{\partial x'} + \frac{\partial}{\partial x'}x'\right).
\end{aligned} \tag{15.91}$$

Let us emphasize that the replacement of $\hat{p}'$ by $-i\hbar \frac{\partial}{\partial \hat{x}'}$ is not permissible when the motion is dissipative, as will be seen in the solutions of many-body models simulating dissipation. Therefore $\hat{E}$ is an operator which represents the energy of the system only when $\lambda \to 0$. Just to indicate the problems with this expression for $\hat{E}$, we first find the expectation values of $\frac{\partial^2}{\partial \hat{x}'^2}$, $\hat{x}'^2$, $\hat{x}' \frac{\partial}{\partial \hat{x}'}$ and $\frac{\partial}{\partial \hat{x}'} \hat{x}'$, using the wave function (15.83). The results show that (a) - the off-diagonal elements $\langle \Psi_n(x',t)|\hat{E}|\Psi_j(x',t)\rangle$ are not zero, and (b) - the diagonal elements $\langle \psi_n(x',t)|\hat{E}|\psi_n(x',t)\rangle$ are given by

$$\langle \Psi_n(x',t)|\hat{E}|\Psi_n(x',t)\rangle = \frac{1}{2}\left(n+\frac{1}{2}\right)\hbar\omega\left[\left(1+C^2\right)A^2 - \frac{\lambda}{\omega}C + \frac{1}{A^2}\left(1+\frac{\lambda^2}{4\omega^2}\right)\right],$$
$$(15.92)$$

where A is given by (15.84) and C is related to A by

$$C = \left(\frac{1}{2\omega}\right)\left(\alpha - \frac{\gamma}{A^2}\right). \qquad (15.93)$$

Now the sum of all four terms in $\hat{E}$, in addition to the diagonal elements have nonzero off-diagonal elements for $n = j\pm2$ [20]. These off-diagonal elements vanish as $\lambda \to 0$, and in this limit the diagonal elements reduce to the simple form of $(n+\frac{1}{2})\hbar\omega$. However if for $\lambda \neq 0$, we calculate α, β and γ and from these parameters determine A and C, then in the limit of $t \to \infty$, $\langle \Psi_n(x',t)|\hat{E}|\Psi_n(x',t)\rangle$ diverges. Therefore the Lagrangian (2.184), though classically acceptable leads to unacceptable result when quantized. That this unphysical behaviour of $\hat{E}$ is not the result of the method of quantization can be demonstrated by replacing the path integral technique by other methods of quantization and arrive at the same conclusion[21].

15.4 Path Integral Formulation of a System Coupled to a Heat Bath

Here again we start with the Lagrangian corresponding to the Hamiltonian (9.1), i.e.

$$L = \frac{1}{2}\dot{Q}^2 - V(Q) + \sum_n \left[\frac{1}{2}\left(\dot{q}_n^2 - \omega_n^2 q_n^2\right) - \epsilon_n q_n Q\right]. \qquad (15.94)$$

As in our earlier discussions we want to eliminate the bath degrees of freedom and find an effective Lagrangian for the motion of the central particle [23]. To this end we must choose the initial and final states for the coordinates of the oscillator. Let us consider the paths with a given initial and final coordinates where [24]

$$Q(t=0) = Q_i, \quad \text{and} \quad Q(t=T) = Q_f. \qquad (15.95)$$

Thus the ground state of the nth oscillator initially is a Gaussian centered about $q_{n_i} = \epsilon_n Q_i / \omega_n^2$ and finally is at $q_{n_f} = \epsilon_n Q_f / \omega_n^2$. Now we calculate the integral

$$\langle Q_f, q_{n_f} | Q_i, q_{n_i} \rangle = \oint_{Q_i}^{Q_f} \mathcal{D}[Q] \prod_n \left[\int \phi_0^* \left(x_{n_f} - q_{n_f} \right) dx_{n_f} \right.$$

$$\left. \times \int \phi_0 \left(x_{n_i} - q_{n_i} \right) dx_{n_i} \oint_{x_{n_i}}^{x_{n_f}} \exp \left\{ \frac{i}{\hbar} \int_0^T L(Q, q_n) dt \right\} \right] \mathcal{D}[q_n].$$

$$(15.96)$$

The integrals over the bath coordinates form a series of Gaussian integrals, and these can be done analytically. We can also write (15.96) as

$$\oint_{Q_i}^{x_{Q_f}} \exp \left\{ \frac{i}{\hbar} \int_0^T L(Q, q_n) dt \right\} \mathcal{D}[Q] \prod_n C_{q_n}, \qquad (15.97)$$

where suppressing the index n in C_{q_n} we have

$$C_q = \int \left(\frac{\omega}{\pi \hbar} \right)^{\frac{1}{4}} \exp \left[-\frac{\omega}{2\hbar} \left(x_f - q_f \right)^2 \right] dx_f$$

$$\times \int \left(\frac{\omega}{\pi \hbar} \right)^{\frac{1}{4}} \exp \left[-\frac{\omega}{2\hbar} \left(x_i - q_i \right)^2 \right] dx_i$$

$$\times \oint_{x_i}^{x_f} \exp \left\{ \frac{i}{\hbar} \int_0^T \left(\frac{1}{2} \dot{q}^2 - \omega^2 q^2 - \epsilon q Q \right) dt \right\} \mathcal{D}[q]. \qquad (15.98)$$

Now let us change the variables to the dimensionless quantities

$$\zeta(t) = \frac{\epsilon}{\sqrt{2\hbar\omega^3}} Q(t), \quad \zeta_i = \sqrt{\frac{\omega}{2\hbar}} q_i, \quad \zeta_f = \sqrt{\frac{\omega}{2\hbar}} q_f, \qquad (15.99)$$

and

$$\beta(T) = \omega \int_0^T \zeta(t) e^{-i\omega t} dt. \qquad (15.100)$$

Using these we can calculate C_q [8];

$$C_q = e^{\frac{-i\omega T}{2}} \exp \left[-\frac{1}{2} \left(\zeta_i^2 + \zeta_f^2 - 2\zeta_i \zeta_f e^{-i\omega T} \right) \right]$$

$$\times \exp \left[-i \left(\zeta_i \beta + \zeta_f \beta^* e^{-i\omega T} \right) + \omega^2 \int_0^T \int_0^t \zeta(t) \zeta(s) e^{-i\omega(t-s)} ds dt \right].$$

$$(15.101)$$

We can simplify C_q by integrating $\beta(T)$ and the double integral in (15.101) by parts;

$$C_q = e^{\frac{-i\omega T}{2}} \exp \left[i\omega \int_0^T \zeta^2(t) dt - \int_0^T \int_0^t \dot{\zeta}(t) \dot{\zeta}(s) e^{-i\omega(t-s)} ds dt \right]. \qquad (15.102)$$

By substituting for $\zeta(t)$ in terms of $Q(t)$ we obtain

$$C_q = \exp\left[\frac{i}{\hbar}\int_0^T \left\{\frac{\epsilon^2}{2\omega^2}Q^2(t) - \frac{\hbar\omega}{2}\right\}dt\right]$$

$$\times \exp\left[-\frac{1}{\hbar}\int_0^T\int_0^t \dot{Q}(t)\dot{Q}(s)\frac{\epsilon^2}{2\omega^3}e^{-i\omega(t-s)}\,ds\,dt\right]. \tag{15.103}$$

This will be the contribution of a typical oscillator (e.g. the nth) in the bath. Summing over all of the bath oscillators we find the following results:

(a) The last integral in (15.96) will be modified in that the potential term in $L(Q, q_n)$ will be replaced by the effective potential

$$V_{eff}(Q) = V(Q) - \sum_n \frac{\epsilon_n^2}{2\omega_n^2}Q^2 + \sum_n \frac{\hbar\omega_n}{2}. \tag{15.104}$$

(b) There will be a time-retarded self interaction with the kernel

$$\Delta(t - s) = \sum_n \frac{\epsilon_n^2}{2\omega_n^3}e^{-i\omega_n(t-s)}. \tag{15.105}$$

Having eliminated the bath degrees of freedom we extend the time integration to be from $-\infty$ to ∞ and thus (15.96) becomes

$$\langle Q_f, q_{n_f} | Q_i, q_{n_i} \rangle =$$

$$= \oint \exp\left\{\frac{i}{\hbar}\int_{-\infty}^{\infty}\left[\frac{1}{2}\dot{Q}^2 - V_{eff}(Q) + i\int_{-\infty}^{t}\dot{Q}(t)\dot{Q}(s)\Delta(t-s)\,ds\right]dt\right\}\mathcal{D}[Q]. \tag{15.106}$$

The effective Lagrangian, L, found in (15.106) is a complex quantity, and thus there are two degrees of freedom for the motion of the central particle, Re Q and Im Q. This L is given by

$$L = \frac{1}{2}\dot{Q}^2(t) - V_{eff}(Q) + i\int_{-\infty}^{\infty}\dot{Q}(t)\dot{Q}(s)\Delta(t-s)\,ds. \tag{15.107}$$

For certain types of potential $V_{eff}(Q)$, when the particle is tunnelling through a potential barrier, it is easier to work with imaginary time, i.e. to write (15.106) as [24],[25]

$$\oint \exp\left[-\frac{1}{\hbar}\int_{\infty}^{\infty}\left\{\frac{1}{2}\dot{Q}^2 + V(Q) + \frac{1}{2}\int_{-\infty}^{\infty}\dot{Q}(\sigma)\dot{Q}(\tau)U(\tau-\sigma)\,d\sigma\right\}d\tau\right]\mathcal{D}[Q], \tag{15.108}$$

where

$$U(\tau - \sigma) = \sum_k \frac{\epsilon_n^2}{2\omega_n^3}\exp\left[-\omega_n|\tau - \sigma|\right]. \tag{15.109}$$

This formulation has been used extensively in the problem of macroscopic quantum tunnelling [25],[26] (see also Sec. 7.5).

References

[1] R.P. Feynman, Rev. Mod. Phys. 20, 367 (1948).

[2] An extensive account of the path integral method as applied to the damped systems is given in G.-T. Ingold's article in *Coherent Evolution in Noisy Environments*, Edited by A. Buchleitner and H. Hornberger, (Springer-Verlag, Berlin, 2002) p.1.

[3] E.H. Kerner and W.G. Sutcliffe, J. Math. Phys. 11, 391 (1970).

[4] L. Cohen, J. Math. Phys. 11, 3296 (1970).

[5] L.S. Schulman, *Techniques and Applications of Path Integration*, (John Wiley & Sons, New York, N.Y. 1981).

[6] B.K. Cheng, J. Phys. A 17, 2475 (1984).

[7] C-I Um, K-H Yeon and T.F. George, Phys. Rep. 362, 64 (2002).

[8] R.P. Feynman and A.R. Hibbs, *Quantum Mechanics and Path Integrals*, (McGraw-Hill, New York, N.Y. 1965).

[9] J. Zinn-Justin, *Path Integrals in Quantum Mechanics*, (Oxford University Press, Oxford, 2005) Chapter 2.

[10] T. Dittrich, P. Hänggi, G.-L. Ingold, B. Kramer, G. Schön and W. Zwerger, *Quantum Transport and Dissipation*, (Wiley-VCH, Weinheim, 1998). Chapter 4.

[11] E.H. Kerner, Can. J. Phys. 36, 371 (1958).

[12] R.W. Hasse, J. Math. Phys. 16, 2005 (1975). A 110, 347 (1985).

[13] B.K. Cheng, Phys. Lett. A 110, 347 (1985).

[14] G.J. Papadopoulos, J. Phys. A 7, 209 (1974).

[15] A.D. Jannussis, G.N. Brodimas and A. Streclas, Phys. Lett. A 74, 6 (1979).

[16] M. Razavy, Hadronic J. 10, 7 (1987).

[17] E.R. Hanson, *A Table of Series and Products*, (Prentice-Hall, Englewood Cliffs, 1975). p. 329.

[18] I.S. Gradshetyn and I.M. Ryzhik, *Table of Integrals, Series, and Products*, (Academic Press, New York, N.Y. 1965) p. 838.

[19] See for instance W. Dittrich and M. Reuter, *Classical and Quantum Dynamics*, (Springer-Verlag, Berlin, 1992) Chapter 16.

[20] S. Pal, Phys. Rev. A 39, 3825 (1989).

[21] S. Srivastava, Vishwamittar and I.S. Minhar, J. Math. Phys. 32, 1510 (1991).

[22] I.S. Gradshetyn and I.M. Ryzhik, *Table of Integrals, Series, and Products,* (Academic Press, New York, N.Y. 1965) p. 47.

[23] R.P. Feynman and F.L. Vernon, Jr. Ann. Phys. (NY) 24, 118 (1963).

[24] J.P. Sethna, Phys. Rev. B 24, 698 (1981).

[25] A.O. Caldeira and A. Leggett, Ann. Phys. (NY) 149, 374 (1983).

[26] M.Razavy, *Quantum Theory of Tunneling*, Second Edition, (World Scientific, Singapore, 2014).

Chapter 16

Quantization of the Motion of an Infinite Chain

In the next two chapters we want to study the problem of quantization of a conservative many-body system and the subsequent separation of the motion of the central particle from the motion of the rest of the particles. As we have seen earlier, Sec. 14.3, the motion of the central particle cannot be described exactly by a single particle wave function. In this chapter we study the quantum dynamics of a uniform infinite chain (Schrödinger chain) Sec. 7.1 [1], and that of a non-uniform chain. We can also solve the problem of a finite chain of oscillators [2], but the motion of infinite chain, because of the translational symmetry, is a simpler one to study.

16.1   Quantum Mechanics of a Uniform Chain

Our starting point is the Hamiltonian for the Schrödinger chain, which we write as

$$H = \sum_{j=-\infty}^{\infty} \frac{1}{2m} p_j^2 + \frac{1}{2} \sum_{j=-\infty}^{\infty} \sum_{k=-\infty}^{\infty} \frac{1}{4} m\nu^2 (x_j - x_k)^2 \delta_{j,k+1}. \tag{16.1}$$

The time-dependent Schrödinger equation for this Hamiltonian gives us the wave function Ψ which depends on the coordinates x_k s and the time:

$$i\hbar \frac{\partial \Psi}{\partial t} = -\frac{\hbar^2}{2m} \sum_{k=-\infty}^{\infty} \frac{\partial^2 \Psi}{\partial x_k^2} + \frac{1}{8} m\nu^2 \sum_{j,k=-\infty}^{\infty} (x_k - x_j)^2 \delta_{j,k+1} \Psi. \tag{16.2}$$

Let us ignore the correlations between the particles in the chain and consider a specific and simple solution of (16.2) which is in the form of the product of single particle wave functions;

$$\Psi(x_1, x_2 \cdots ; \xi_1(t), \xi_2(t) \cdots) = \prod_{k=-\infty}^{\infty} \phi_k[x_k - \xi_1(t)]$$

$$\times \exp\left[\frac{i}{\hbar}\left\{mx_k\dot{\xi}_k(t) + C_k(t)\right\}\right], \qquad (16.3)$$

where $\phi_k[x_k - \xi_1(t)]$ is the normalized single particle wave function, $\xi_k(t)$ is the classical solution of the displacement of the kth particle in the chain, $C_k(t)$ a function of time which is defined by

$$C_k(t) = \int^t \left[\frac{1}{8}m\nu^2 \left\{2\xi_k^2(t) - \xi_{k+1}^2(t) - \xi_{k-1}^2(t)\right\} - \frac{1}{2}\dot{\xi}_k^2(t) - \mathcal{E}_k\right] dt, \qquad (16.4)$$

and $\mathcal{E}_k$ is a constant to be determined later. Substituting (16.3) in (16.2) we find the differential equation for $\phi_k[x_k - \xi_k(t)]$,

$$\sum_{k=-\infty}^{\infty}\left[-\frac{\hbar^2}{2m}\frac{\partial^2\phi_k}{\partial y_k^2} + \frac{1}{2}m\dot{\xi}_k^2\phi_k + \left(m\ddot{\xi}_k x_k + \dot{C}_k\right)\phi_k\right]$$

$$\times \exp\left[\frac{i}{\hbar}\left(mx_k\dot{\xi}_k + C_k\right)\right]\prod_{j\neq k}\phi_j(y_j)\exp\left[\frac{i}{\hbar}\left(mx_j\dot{\xi}_j + C_j\right)\right]$$

$$+ \frac{1}{8}m\nu^2 \sum_{k=-\infty}^{\infty}\left[(x_{k+1} - x_k)^2 + (x_k - x_{k-1})^2\right]\prod_{j=-\infty}^{\infty}\phi_j(y_j)$$

$$\times \exp\left[\frac{i}{\hbar}\left(mx_j\dot{\xi}_j + C_j\right)\right] = 0, \qquad (16.5)$$

where ϕ_k is a function of the variable y_k;

$$y_k = x_k - \xi_k(t). \qquad (16.6)$$

Note that in (16.5) $\xi_k, \dot{\xi}_k$ and C_k are all functions of time. If we multiply (16.5) by

$$\prod_{j\neq k}\phi_j(y_j)\exp\left[-\frac{i}{\hbar}\left(mx_j\dot{\xi}_j + C_j\right)\right], \qquad (16.7)$$

and integrate over all y_j s except y_k and use the relation

$$\int_{-\infty}^{\infty}\phi_{k+1}^*(y_{k+1})x_{k+1}\phi_{k+1}(y_{k+1})dy_{k+1} = \xi_{k+1}(t), \qquad (16.8)$$

we find that ϕ_k apart from a time-dependent phase factor satisfies the wave equation for a harmonic oscillator

$$-\frac{\hbar^2}{2m}\frac{d^2\phi_k}{dy_k^2} + \left(\frac{1}{4}m\nu^2 y_k^2 - \mathcal{E}_k\right)\phi_k = 0. \qquad (16.9)$$

The fact that $\phi_k(y)$ is the wave function for a harmonic oscillator justifies using Eq. (16.8) for this problem.

The eigenvalues and the eigenfunctions of (16.9) are given by

$$\mathcal{E}_k^{(n_k)} = \left(n_k + \frac{1}{2}\right) \frac{\hbar \nu}{\sqrt{2}}, \tag{16.10}$$

and

$$\phi_k^{(n_k)}(y_k) = \left(\frac{\alpha}{\pi^{\frac{1}{2}} 2^{n_k} n_k!}\right)^{\frac{1}{2}} H_{n_k}(\alpha y_k) \exp\left(-\frac{1}{2}\alpha^2 y_k^2\right), \tag{16.11}$$

respectively. Here the constant α is given by

$$\alpha^2 = \frac{m\nu}{2^{\frac{1}{2}}\hbar}. \tag{16.12}$$

We can also formulate this problem in the following way: We write the Hamiltonian for the kth particle as

$$H_k = \frac{1}{2m}p_k^2 + \frac{1}{8}m\nu^2 \left[(x_k - \xi_{k-1})^2 + (x_k - \xi_{k+1})^2\right], \tag{16.13}$$

and this form of H_k implies that the kth particle is subject to the action of two time-dependent classical forces, these forces are exerted by $(k-1)$th and $(k+1)$th particles on the kth. But for the motion of the kth particle the Hamiltonian (16.13) is not unique. The same equation of motion can be obtained from H_k', where

$$H_k' = H_k - \frac{1}{8}m\nu^2 \left(\xi_{k-1}^2(t) + \xi_{k+1}^2(t)\right). \tag{16.14}$$

In quantum mechanics the wave functions corresponding to the two Hamiltonians H_k and H_k' differ from each other by a time-dependent phase factor.

The single particle wave function for the kth particle is

$$i\hbar\frac{\partial \psi_k}{\partial t} = H_k \psi_k, \tag{16.15}$$

where H_k is defined by (16.13), and for this case the solution of (16.15) is given by

$$\psi_k^{(n_k)}(x_k, t) = \phi_k^{(n_k)}(x_k - \xi_k(t)) \exp\left\{\frac{i}{\hbar}\left(mx_k\dot{\xi}_k + C_k(t)\right)\right\}. \tag{16.16}$$

Now using this wave function we can determine the mean energy associated with the motion

$$\langle E_k \rangle = \int_{-\infty}^{\infty} \psi_k^{*(n_k)} \left(i\hbar\frac{\partial}{\partial t}\right) \psi_k^{(n_k)} dx_k$$

$$= \frac{m}{2}\dot{\xi}_k^2 + \frac{1}{8}m\nu^2 \left[(\xi_k - \xi_{k-1})^2 + (\xi_k - \xi_{k+1})^2\right] + \mathcal{E}_k^{(n_k)}. \tag{16.17}$$

From this relation we obtain the energy of the kth particle which turns out to be

$$e_k = \langle E_k \rangle - \frac{1}{8} m \nu^2 \left(\xi_{k-1}^2 + \xi_{k+1}^2 \right),$$ (16.18)

and thus the total energy of the system is

$$\langle E \rangle = \sum_{k=-\infty}^{\infty} \langle E_k \rangle = \frac{m}{2} \sum_{k=-\infty}^{\infty} \left[\dot{\xi}_j^2 + \left(\sum_{j=-\infty}^{\infty} \frac{1}{4} \nu^2 (\xi_j - \xi_k)^2 \delta_{j,k+1} \right) \right]$$

$$+ \sum_{j=-\infty}^{\infty} \mathcal{E}_j^{(n_j)}.$$ (16.19)

If on the classical equation of motion, $\xi_k(t)$, we impose the initial conditions

$$\xi_k(0) = \dot{\xi}_k(0) = 0, \quad k \neq 0, \quad \xi_0(0) = A, \quad \dot{\xi}_0(0) = 0,$$ (16.20)

then as we have seen in Sec. 7.1 the total energy associated with the classical motion is $\frac{1}{4} m \nu^2 A^2$. Thus

$$\langle E \rangle = \frac{1}{4} m \nu^2 A^2 + \sum_{j=-\infty}^{\infty} \mathcal{E}_j^{(n_j)}.$$ (16.21)

We observe that the total momentum of the system is also a constant of motion

$$\langle P \rangle = \sum_{k=-\infty}^{\infty} \int_{-\infty}^{\infty} \psi_k^*(x_k) \left(-i\hbar \frac{\partial}{\partial x_k} \right) \psi_k(x_k) dx_k$$

$$= \sum_{k=-\infty}^{\infty} m \dot{\xi}_k(t) = \frac{1}{2} m \nu A \sum_{k=-\infty}^{\infty} (J_{2k-1} - J_{2k+1}) = 0.$$ (16.22)

16.2   Ground State of the Central Particle

The energy of the ground state according to (16.21) is equal to

$$\langle E \rangle = \frac{1}{4} m \nu^2 A^2 + \sum_{j=-\infty}^{\infty} \left(\frac{\hbar \nu}{2^{\frac{3}{2}}} \right),$$ (16.23)

which is a divergent quantity. The sum $\sum_{j=-\infty}^{\infty} \left(\frac{\hbar \nu}{2^{\frac{3}{2}}} \right)$ may be regarded as an unobservable quantity and omitted, for some problems, but not for others [3]–[7]. The question as to whether the zero-point energy can have observable consequences or not has been addressed for systems of many degrees of freedom and fields [8],[9]. In particular the Casimir effect that we will be considering in Sec. 19.1 is an important example of observable forces arising from vacuum fluctuations.

The wave function of the system in its ground state is

$$\Psi_G = \prod_{k=-\infty}^{\infty} \left(\frac{m\nu}{2^{\frac{1}{2}} \hbar \pi} \right)^{\frac{1}{4}} \exp \left\{ \left(-\frac{m\nu}{2^{\frac{3}{2}} \hbar} \right) (x_k - \xi_k)^2 + \frac{i}{\hbar} \left(m x_k \dot{\xi}_k + C_k \right) \right\}.$$

(16.24)

For the central particle, $k = 0$, the wave function $\phi_0(x_0, t)$ is

$$\phi_0(x_0, t) = \left(\frac{m\nu}{2^{\frac{1}{2}} \hbar \pi} \right)^{\frac{1}{4}} \exp \left\{ \left(-\frac{m\nu}{2^{\frac{3}{2}} \hbar} \right) (x_0 - AJ_0(\nu t))^2 \right\}$$

$$= \sum_{N=0}^{\infty} \frac{\alpha^N A^N J_0^N(\nu t)}{(2^N N!)^{\frac{1}{2}}} \exp \left[-\frac{1}{4} \alpha^2 A^2 J_0^2(\nu t) \right] u_N(x_0), \qquad (16.25)$$

where we have expanded the wave function using the generating function of the Hermite polynomials. In this relation $u_N(x_0)$ is the Nth eigenfunction for the harmonic oscillator and α is given by (16.12). The expansion in (16.25) shows that a large number of stationary states $u_N(x_0)$ contribute to the time-dependent wave function. However the most important contributions at the time t come from the values of N close to $N_0(t)$, where

$$N_0(t) \approx \frac{1}{2} \alpha^2 A^2 J_0^2(\nu t). \qquad (16.26)$$

We can use the same argument for the expansion of $\phi_k(x_k, t)$, noting that in this case we have

$$N_k(t) \approx \frac{1}{2} \alpha^2 A^2 J_{2k}^2(\nu t). \qquad (16.27)$$

This means that the probability of finding the N_kth excited state in $\phi_k(x_k, t)$ follows a Poisson distribution with the mean value given by (16.27), and that it only becomes appreciable after a time $t = \frac{2k}{\nu}$.

The classical motion of the chain is invariant under the time reversal transformation, since the total Hamiltonian of the chain is conservative and the solution $\xi_k(t) = AJ_{2k}(\nu t)$ remains unchanged if we change t to $-t$. The single particle Hamiltonian has a time-dependent effective potential

$$V(x_k, t) = \frac{1}{8} m\nu^2 \left[\xi_{k-1}^2 + \xi_{k+1}^2 - 2x_k \left(\xi_{k-1} + \xi_{k+1} \right) \right], \qquad (16.28)$$

which is also invariant under this transformation.

The time-reversed wave function for the kth particle which we will denote by $\bar{\psi}_k(x_k, t)$ is given by

$$\bar{\psi}_k(x_k, t) = \psi_k^*(x_k, -t) = \psi_k(x_k, t). \qquad (16.29)$$

We observe that in this case, in contrast with the model of a particle coupled to the heat bath (e.g. Ullersma's model), for each oscillator including the central one, the time-reversal transformed motion leads to a physically realizable "reversed" motion. But the direction of the flow of energy remains unchanged as $t \to -t$, and therefore the system is irreversible [10].

Similar results can be obtained when the initial conditions are:

$$\xi_k(0) = 0, \quad k = 0, \pm 1, \pm 2 \cdots \tag{16.30}$$

and

$$\dot{\xi}_0(0) = \frac{2}{\nu}B, \quad \dot{\xi}_k(0) = 0, \quad k \neq 0. \tag{16.31}$$

In this case the classical solution is $\xi_k(t) = BJ_{2k+1}(\nu t)$, where B is a constant. Here the Hamiltonian has the following invariance:

$$H(-x_k, p_k, -t) = H(x_k, p_k, t), \tag{16.32}$$

and the time-reversed wave function is related to the original wave function by the following relation [10]

$$\bar{\psi}_k(x_k, t) = \psi_k^*(x_k, -t) = (-1)^{(n_k)} \psi_k(-x_k, t). \tag{16.33}$$

16.3   Wave Equation for a Non-Uniform Chain

The problem of non-uniform chain is similar to that of the Schrödinger chain. Following the method that we discussed in Sec. 16.1 we write the wave equation for the conservative Hamiltonian describing the motion of the chain, Eq. (7.23), as

$$H = \sum_{j=1}^{\infty} \left\{ \frac{p_j^2}{2m_j} + \frac{1}{4} \sum_{k=1}^{\infty} K_{kj}(x_k - x_j)^2 \right\}, \tag{16.34}$$

where

$$K_{kj} = K_k \delta_{k+1,j} + K_j \delta_{j+1,k}, \tag{16.35}$$

is a symmetric matrix.

The Hamilton canonical equations for (16.34) generate an equation for x_j which is identical to Eq. (7.34) for $\xi_k(t)$. From the Hamiltonian we obtain the time-dependent Schrödinger equation which is given by

$$\sum_{j=1}^{\infty} \left\{ -\frac{\hbar^2}{2m_j} \frac{\partial^2}{\partial x_j^2} + \frac{1}{4} \left[K_j(x_{j+1} - x_j)^2 + K_{j-1}(x_{j-1} - x_j)^2 \right] \right\} \Psi = i\hbar \frac{\partial \Psi}{\partial t}. \tag{16.36}$$

As in the case of a uniform chain, we introduce a new set of variables y_1, $y_2, \cdots y_j, \cdots$ by

$$y_j = x_j - \xi_j(t), \tag{16.37}$$

where $\xi_j(t)$ s are defined as the solutions of Eq. (7.34) with the initial conditions (7.35) and (7.36). Next we write the total wave function as

$$\Psi(x_1, x_2 \cdots, t) = \Phi(y) \exp\left[-\frac{i}{\hbar}C(y,t)\right], \tag{16.38}$$

and choose $C(y,t)$ to be

$$
C(y,t) = \mathcal{E}t
$$
$$
+ \int \left\{ \sum_{j=1}^{\infty} \frac{1}{2}m_j\dot{\xi}_j^2 + \frac{1}{4}\sum_{j,k=1}^{\infty}\left[K_{kj}(\xi_k - \xi_j)^2 + \frac{1}{2}K_{kj}(\xi_k - \xi_j)(y_k - y_j)\right]\right\} dt. \tag{16.39}
$$

By substituting (16.38) and (16.39) in (16.36), we find that Φ satisfies the Schrödinger equation

$$\sum_{j=1}^{\infty}\left(-\frac{\hbar^2}{2m_j}\frac{\partial^2\Phi}{\partial y_j^2}\right) + \frac{1}{4}\sum_{k,j=1}^{\infty}K_{kj}(y_k - y_j)^2\,\Phi = \mathcal{E}\Phi, \tag{16.40}$$

provided that $\xi_k(t)$ is a solution of Eq. (7.34).

In the many-body wave function (16.38) there are two kinds of correlations between the motions of the particles. The first is the classical correlations between different mass points as can be seen from Eq. (7.34) and this enters in the expression for Φ as well as the phase $C(y,t)$. The second one is the quantum correlation as implied by the solution of (16.40). Ignoring the correlations in (16.40), we write $\Phi(y_1, y_2, \cdots)$ as a product of wave functions for each particle [11];

$$\Phi(y_1, y_2, ...) = \prod_{i=1}^{\infty}\phi_{ni}(y_i). \tag{16.41}$$

Following the method that we used earlier for a uniform chain we find the single particle wave equation to be

$$H_k\phi_{k_n}(y_k) = -\frac{\hbar^2}{2m_k}\frac{d^2}{dy_k^2}\phi_{n_k}(y_k) + \frac{1}{2}\left[(K_k + K_{k-1})y_k^2\right]\phi_{n_k}(y_k) = \mathcal{E}_{n_k}\phi_{k_n}(y_k), \tag{16.42}$$

which is the equation of motion for a simple harmonic oscillator (K_k and K_{k-1} are defined by Eq. (16.35)). Thus $\phi_{n_k}(y_k)$ can be written as

$$\phi_{k_n}(y_k) = \left(\frac{\alpha_k}{\pi^{\frac{1}{2}}2^{n_k}n_k!}\right)^{\frac{1}{2}}H_{n_k}(\alpha_k y_k)\exp\left(-\frac{1}{2}\alpha_k^2 y_k^2\right), \tag{16.43}$$

where

$$\alpha_k^2 = \frac{1}{\hbar}[m_k(K_{k+1} + K_k)]^{\frac{1}{2}}. \tag{16.44}$$

To calculate the single particle phase factor we note that

$$
\int \phi^2_{n_{k+1}}(y_{k+1})\phi^2_{n_{k-1}}(y_{k-1})dy_{k+1}dy_{k-1}
$$

$$
\times \exp\left[-\frac{i}{\hbar}\left\{\beta_k K_k(y_{k+1}-y_k)-\beta_{k-1}K_{k-1}(y_{k-1}-y_k)\right\}\right]
$$

$$
= L_{n_{k+1}}\left[\frac{1}{2}\left(\frac{\beta_k K_k}{\hbar\alpha_{k+1}}\right)^2\right]L_{n_{k-1}}\left[\frac{1}{2}\left(\frac{\beta_{k-1}K_{k-1}}{\hbar\alpha_k}\right)^2\right]
$$

$$
\times \exp\left[\frac{i}{\hbar}y_k(\beta_k K_k-\beta_{k-1}K_{k-1})\right],
$$

(16.45)

where

$$
\beta_k(t)=\int^t\left[\xi_{k+1}(t)-\xi_k(t)\right]dt,
$$

(16.46)

and $L_{n_{k+1}}$ s are Laguerre polynomials. We can now determine the single particle phase factor by calculating the expectation value of $\exp\left[-\frac{i}{\hbar}C(y_k,t)\right]$;

$$
\exp\left(-\frac{i}{\hbar}C_k\right)=\frac{1}{N_k(t)}\int\cdots\int\left(\prod_{j\neq k}^{\infty}\phi^*_{n_j}\right)\exp\left(-\frac{i}{\hbar}C(y_k,t)\right)\left(\prod_{j\neq k}^{\infty}\phi_{n_j}\right)\prod_{j\neq k}^{\infty}dy_j
$$

$$
= \exp\left[-\frac{i}{\hbar}\left(\mathcal{E}_{n_k}t+\int\left\{\frac{1}{2}m_k\dot{\xi}_k^2+\frac{1}{2}\sum_{j\neq k}^{\infty'}K_{kj}(\xi_k-\xi_j)^2\right\}dt-y_k m_k\dot{\xi}_k\right)\right],
$$

(16.47)

where

$$
N_k(t)=L_{n_{k+1}}\left[\frac{1}{2}\left(\frac{\beta_k K_k}{\hbar\alpha_{k+1}}\right)^2\right]L_{n_{k-1}}\left[\frac{1}{2}\left(\frac{\beta_{k-1}K_{k-1}}{\hbar\alpha_k}\right)^2\right].
$$

(16.48)

With this choice of $N_k(t)$, the single particle wave function

$$
\psi_k(y_k,t)=\phi_{n_k}(y_k)\exp\left[-\frac{i}{\hbar}C_k(y_k,t)\right]
$$

(16.49)

becomes normalized.

16.4   Connection with Other Phenomenological Frictional Forces

We have seen that the single particle wave function can be defined by the solution of Eq. (16.42) and Eq. (16.49). From these two relations we can derive the time-dependent Schrödinger equation for $\psi_k(y_k, t)$

$$H_k\psi_k - i\hbar\frac{\partial\psi_k}{\partial t} = \left[y_k m_k\ddot{\xi}_k - \frac{1}{2}\sum_{j\neq k}^{\infty} K_{jk}(\xi_k - \xi_j)^2\right]\psi_k. \qquad (16.50)$$

This equation is still invariant under time-reversal transformation, i.e. $\psi_k(y_k, t)$ satisfies the same equation as $\psi_k^*(y_k, -t)$. It is convenient to work with the wave function $u_k(y_k, t)$ which we define as

$$u_k(y_k, t) = \psi_k(y_k, t)\exp\left[\frac{i}{2\hbar}\int^t \sum_{j\neq k}^{\infty} K_{jk}(\xi_k - \xi_j)^2 dt\right]. \qquad (16.51)$$

Now if in (16.50) we replace $\ddot{\xi}_k(t)$ in terms of $\dot{\xi}_k(t)$ and $\xi_k(t)$, Eq. (7.51), we find that $u_k(y_k, t)$ is a solution of

$$i\hbar\frac{\partial u_k(y_k, t)}{\partial t} = H_k u_k(y_k, t) + \left[\alpha\gamma(k)m_k\dot{\xi}_k - f_k(\xi_k)\right]y_k u_k(y_k, t), \qquad (16.52)$$

where α which is defined by (7.31) is the damping constant. We can also express $u_k(y_k, t)$ in terms of $\phi_k(y_k)$ using Eqs. (16.49) and (16.51);

$$u_k(y_k, t) = \phi_k(y_k)\exp\left[-\frac{i}{\hbar}\left(\mathcal{E}_{n_k}t - m_k y_k\dot{\xi}_k - \int^t \frac{1}{2}m_k\dot{\xi}_k^2 dt\right)\right]. \qquad (16.53)$$

The expectation value of the momentum of the kth particle is given by

$$\langle p_k\rangle = \int_{-\infty}^{\infty} u_k^*\left(-i\hbar\frac{\partial}{\partial y_k}\right)u_k(y_k)\,dy_k = m_k\dot{\xi}_k. \qquad (16.54)$$

Substituting for $\dot{\xi}_k$ in Eq. (16.52) we obtain

$$i\hbar\frac{\partial u_k(y_k, t)}{\partial t} = H_k u_k(y_k, t) + \left[\alpha\langle p_k\rangle - f_k(\xi_k)\right]y_k u_k(y_k, t), \qquad (16.55)$$

and this is a special case of phenomenological wave equations studied by Hasse [12] and by Stoker [13].

16.5 Fokker-Planck Equation for the Probability Density

From the wave function $\psi_k(x_k, \xi_k)$ we can find the probability density, ρ_k, of finding the kth particle between x_k and $x_k + dx_k$;

$$\frac{\partial\rho_k}{\partial t}dx_k = |\psi_k(x_k, \xi_k)|^2\,dx_k = |\phi_k(x_k - \xi_k)|^2\,dx_k. \qquad (16.56)$$

This probability density satisfies the Fokker-Planck equation

$$\frac{\partial \rho_k}{\partial t} + \frac{\partial}{\partial x_k}(v_k \rho_k) - \frac{\hbar}{2m_k}\frac{\partial^2 \rho_k}{\partial x_k^2} = 0, \qquad (16.57)$$

provided that the velocity v_k is given by

$$v_k = \dot{\xi}_k + \frac{\hbar}{2m_k}\frac{\partial}{\partial x_k}(\ln \rho_k). \qquad (16.58)$$

We can also relate the current density, $j_k(x_k, t)$, to the probability density $\rho_k(x_k, t)$ by noting that

$$j_k(x_k, t) = -\frac{i\hbar}{2m_k}\left(\psi_k^* \frac{\partial \psi_k}{\partial y_k} - \psi_k \frac{\partial \psi_k^*}{\partial y_k}\right), \qquad (16.59)$$

and substituting for ψ_k and ψ_k^* from (16.49) to obtain

$$j_k(x_k, t) = \dot{\xi}(t)\rho_k(x_k, t). \qquad (16.60)$$

The coefficients of expansion of the wave function $\psi_k(x_k, \xi_k)$ also satisfy a master equation originally derived by Bloch [14]. Let us consider the motion of the kth particle in the chain ($k \neq 1$). Initially this particle is in its ground state, and at some later time, t, the wave function becomes

$$\psi_k(x_k, \xi_k) = \left(\frac{\alpha_k}{\pi^{\frac{1}{2}}}\right)^{\frac{1}{2}} \exp\left[-\frac{1}{2}\alpha_k^2(x_k - \xi_k)^2\right] \exp\left(-\frac{i}{\hbar}C_k(x_k, t)\right). \qquad (16.61)$$

We expand $\psi_k(x_k, \xi_k)$ in terms of the complete set of harmonic oscillator states;

$$\psi(x_k, \xi_k) = \sum_{n=0}^{\infty} A_n(\xi_k)\phi_{n_k}(x_k), \qquad (16.62)$$

where

$$A_n(\xi_k) = \frac{(\alpha_k \xi_k)^n}{(2^n n!)^{\frac{1}{2}}} \exp\left[-\frac{1}{4}(\alpha_k \xi_k)^2\right]. \qquad (16.63)$$

Thus $A_n(\xi_k)$ is the time-dependent probability amplitude for the kth particle to be in the state n_k. Since for $\alpha t > 1$, ξ_k decreases exponentially, i.e.

$$\xi_k = \left(\frac{m_1}{m_k}\right)^{\frac{1}{2}} e^{-\alpha t}, \quad t > \frac{1}{\alpha}, \qquad (16.64)$$

therefore from Eqs. (16.63) and (16.64) it follows that for $t > \frac{1}{\alpha}$, the probability amplitude $A_n(\xi_k)$ satisfies the master equation

$$\frac{1}{2\alpha}\frac{d}{dt}|A_n|^2 = (n+1)|A_{n+1}|^2 - n|A_n|^2. \qquad (16.65)$$

This equation has also been studied in detail by Bauer and Jensen [15].

References

[1] E. Schrödinger, Ann. Phys. (Leipzig), 44, 916 (1914).

[2] D.H. Zanette, Am. J. Phys. 62, 404 (1994).

[3] A historical account of the problem of zero-point energy for a single oscillator or a collection of oscillators can be found in J. Mehra and H. Rechenberg's article in Found. Phys. 29, 91 (1999).

[4] The history of zero-point energy in early quantum theory is given in P.W. Milonni and M.-L. Shih, Am. J. Phys. 59, 68 (1991).

[5] For a lucid and informative account of the properties of the vacuum the reader is referred to T.H. Boyer's article in Sci. Am. 253, 70 (1985).

[6] Various interpretation of the divergent term appearing in the total energy, and whether the zero-point energy is observable or not, can be found in the very interesting paper of I.J.R. Aitchison, Contemp. Phys. 26, 333 (1985).

[7] C.P. Enz, in *Physical Reality and Mathematical Description*, Edited by C.P. Enz and J. Mehra, (D. Reidel Pub. Comp. Dordrecht, Holland, 1974) pp. 124-132.

[8] M. Kimball, *The Casimir Effect: Physical Manifestations of Zero-Point Energy*, (World Scientific, Singapore, 2001).

[9] F.Y. Khalili, Phys.-Uspekhi, 46, 293 (2003).

[10] M. Razavy, Can. J. Phys. 57, 1731 (1979).

[11] M. Razavy, Can J. Phys. 58, 1019 (1980).

[12] R.W. Hasse, Rep. Prog. Phys. 41, 1027 (1978).

[13] W. Stoker and K. Albrecht, Ann. Phys. (NY), 117, 436 (1979).

[14] F. Bloch, Z. Phys. 29, 58 (1928).

[15] H. Bauer and J.H.D. Hansen, Z. Phys. 124, 580 (1948).

Chapter 17

The Heisenberg Equations of Motion for a Particle Coupled to a Heat Bath

The classical damped motion of a central particle coupled to a field or to a large number of oscillators forming a heat bath was discussed in Chapter 9. The corresponding quantum problem can be formulated in the same way. When the central particle is harmonically bound and the Hamiltonian is quadratic in the coordinates and momenta of the oscillators as well as the coupling, then the Heisenbergs equations have similar dependence on the initial conditions as the classical equations.

17.1 Heisenberg Equations for a Damped Harmonic Oscillator

Let us consider an oscillator of unit mass density which interacts with a scalar field, a string of length $2L$, a model similar to the one we discussed in Sec. 8.2, but with a different type of coupling (Harris's model).

The Lagrangian for this system is given by [1],[2]

$$L_q = \int_{-L}^{L} \mathcal{L}\left(Q, \dot{Q}, \frac{\partial y}{\partial t}, \frac{\partial y}{\partial x}\right) dx, \tag{17.1}$$

where the Lagrangian density $\mathcal{L}$ has the following form:

$$\mathcal{L}\left(Q, \dot{Q}, \frac{\partial y}{\partial t}, \frac{\partial y}{\partial x}\right)$$
$$= \frac{1}{2}\left[\left(\frac{\partial y}{\partial t}\right)^2 - \left(\frac{\partial y}{\partial x}\right)^2\right] + \frac{1}{2}\delta(x)\left[\dot{Q}^2 - \omega_0^2 Q^2 - 2\epsilon Q\left(\frac{\partial y}{\partial t}\right)\right]. \tag{17.2}$$

We note that in this model $\mathcal{L}$ is not invariant under time reversal transformation.

To simplify the Lagrangian we expand $y(x,t)$ in a Fourier series in the interval $-L < x < L$, and assume that the string is fixed at both ends, i.e. $y(-L,t) = y(L,t) = 0$ (see also Sec. 8.2). Thus

$$y(x,t) = \sum_k \frac{1}{\sqrt{L}} q_k(t) \cos(kx), \quad k = \left(n + \frac{1}{2}\right)\frac{\pi}{L}, \quad n = 0,1,2\cdots. \quad (17.3)$$

The terms involving $\sin(kx)$ do not couple to the motion of the oscillator, and therefore are not included in the sum. By substituting (17.3) in (17.2) and simplifying the result we find the Lagrangian

$$L_q\left(Q,\dot{Q},q_k,\dot{q}_k\right) = \frac{1}{2}\left(\dot{Q}^2 - \omega_0^2 Q^2\right) + \frac{1}{2}\sum_k(\dot{q}_k^2 - k^2 q_k^2) - \frac{\epsilon}{\sqrt{L}}Q\sum_k \dot{q}_k. \quad (17.4)$$

The Hamiltonian obtained from the above Lagrangian has the simple form of

$$H = \frac{1}{2}\left(P^2 + \omega_0^2 Q^2\right) + \frac{1}{2}\sum_k\left[\left(p_k + \frac{\epsilon Q}{\sqrt{L}}\right)^2 + k^2 q_k^2\right]. \quad (17.5)$$

Now we regard P, Q, p_k and q_k as quantum mechanical operators obeying the canonical commutation relations

$$[Q,P] = [q_k, p_k] = i\hbar. \quad (17.6)$$

Using these together with the Heisenberg equations of motion for P and Q, i.e.

$$\dot{Q} = -\frac{i}{\hbar}[Q,H], \quad \dot{P} = -\frac{i}{\hbar}[P,H], \quad (17.7)$$

and similar equations for $\dot{q}_k$ and $\dot{p}_k$ we find the operator equations for Q and q_k

$$\ddot{Q} + \omega_0^2 Q = -\frac{\epsilon}{\sqrt{L}}\sum_k \dot{q}_k, \quad (17.8)$$

and

$$\ddot{q}_k + k^2 q_k = \frac{\epsilon}{\sqrt{L}}\dot{Q}. \quad (17.9)$$

These are linear coupled equations for Q and q_k which we can solve exactly. We first solve (17.9) to obtain $q(t)$

$$q_k(t) = \frac{\epsilon}{\sqrt{L}}\int_0^t \cos\left[k\left(t - t'\right)\right]q\left(t'\right)dt' + \left(q_k(0)\cos(kt) + p_k(0)\frac{\sin(kt)}{k}\right). \quad (17.10)$$

Noting that

$$\frac{1}{L}\sum_k \cos\left[k\left(t - t'\right)\right] = \delta\left(t - t'\right), \quad (17.11)$$

we find from (17.10) that

$$\frac{\epsilon}{\sqrt{L}} \sum_k q_k = \lambda Q + \frac{\epsilon}{\sqrt{L}} \sum_k \left(q_k(0) \cos(kt) + p_k(0) \frac{\sin(kt)}{k} \right), \qquad (17.12)$$

where $\lambda = \frac{\epsilon^2}{2}$, L and ϵ having the dimensions of time and time$^{-\frac{1}{2}}$ respectively.

By differentiating (17.12) and substituting for $\sum_k \dot{q}_k$ in (17.8) we find the equation of motion for the oscillator [1]

$$\ddot{Q} + \lambda \dot{Q} + \omega_0^2 Q = \frac{\epsilon}{\sqrt{L}} \sum_k \left[q_k(0) k \sin(kt) - p_k(0) \cos(kt) \right], \qquad (17.13)$$

$$\ddot{P} + \lambda \dot{P} + \omega_0^2 Q = \frac{\epsilon}{\sqrt{L}} \sum_k \left[q_k(0) k^2 \sin(kt) - k p_k(0) \cos(kt) \right], \qquad (17.14)$$

From the commutation relation

$$[q_k(0), p_k(0)] = i\hbar, \qquad (17.15)$$

it follows that unlike the classical case we cannot set both $q_k(0)$ and $p_k(0)$ equal to zero.

Solutions of the Equations of Motion — Since the equations of motion are linear in the operators Q, P, q_k and p_k we can write $Q(t)$ and $P(t)$ in terms of the initial operators [1]. These are the analogues of the equations (9.17) and (9.20) of Ullersma's model

$$Q(t) = Q(0) f_1(t) + P(0) f_2(t) + \frac{\epsilon}{\sqrt{L}} \sum_k \left\{ q_k(0) f_3(k,t) - p_k(0) f_4(k,t) \right\}, \quad (17.16)$$

and

$$P(t) = Q(0) \dot{f}_1(t) + P(0) \dot{f}_2(t) + \frac{\epsilon}{\sqrt{L}} \sum_k \left\{ q_k(0) \dot{f}_3(k,t) - p_k(0) \dot{f}_4(k,t) \right\}. \quad (17.17)$$

The time-dependent coefficients $f_1(t)$, $f_2(t)$, $f_3(k,t)$ and $f_4(k,t)$ will be defined later.

If we assume that initially all of the oscillators are in the ground state (or the string has no initial displacement, and no initial velocity), then we can average Q and P over the ground states of $q_k(0)$ and $p_k(0)$. Since

$$\langle 0|q_k(0)|0 \rangle = \langle 0|p_k(0)|0 \rangle = 0, \qquad (17.18)$$

therefore by averaging we get

$$\frac{d}{dt} \langle 0|P(t)|0 \rangle + \lambda \langle 0|P(t)|0 \rangle + \omega_0^2 \langle 0|Q(t)|0 \rangle = 0, \qquad (17.19)$$

and

$$\frac{d}{dt} \langle 0|Q(t)|0 \rangle = \langle 0|P(t)|0 \rangle, \qquad (17.20)$$

where $\langle 0|Q(t)|0\rangle$ and $\langle 0|P(t)|0\rangle$ denote the averages of these operators taken over the ground state of the heat bath.

Equations (17.19) and (17.20) can be solved to yield the following results:

$$\langle 0|Q(t)|0\rangle = \langle 0|Q(0)|0\rangle f_1(t) + \langle 0|P(0)|0\rangle f_2(t), \tag{17.21}$$

and

$$\langle 0|P(t)|0\rangle = \langle 0|Q(0)|0\rangle \dot{f_1}(t) + \langle 0|P(0)|0\rangle \dot{f_2}(t), \tag{17.22}$$

where

$$f_1(t) = e^{\frac{-\lambda t}{2}} \left(\cos(\omega t) + \frac{\lambda}{2\omega} \sin(\omega t) \right), \tag{17.23}$$

and

$$f_2(t) = \frac{1}{\omega} e^{\frac{-\lambda t}{2}} \sin(\omega t), \tag{17.24}$$

i.e. $f_1(t)$ and $f_2(t)$ are the solutions of the equation of motion of a damped harmonic oscillator with the boundary conditions

$$f_1(0) = 1, \quad \dot{f_1}(0) = 0, \tag{17.25}$$

and

$$f_2(0) = 0, \quad \dot{f_2}(0) = 1, \tag{17.26}$$

with $\omega = \left(\omega_0^2 - \frac{\lambda^2}{4} \right)^{\frac{1}{2}}$.

From Eqs. (17.21) and (17.22) it follows that

$$[\langle 0|Q(t)|0\rangle, \langle 0|P(t)|0\rangle] = [\langle 0|Q(0)|0\rangle, \langle 0|P(0)|0\rangle]$$
$$\times \left(f_1(t)\dot{f_2}(t) - \dot{f_1}(t)f_2(t) \right) = [\langle 0|Q(0)|0\rangle, \langle 0|P(0)|0\rangle] e^{-\lambda t}. \tag{17.27}$$

This relation shows that by averaging over $q_k(0)$ and $p_k(0)$, the commutator $[\langle 0|Q(t)|0\rangle, \langle 0|P(t)|0\rangle]$ will decrease as a function of time. From the solution of (17.13) we find $f_3(k,t)$ and $f_4(k,t)$ to be

$$f_3(k,t) = \frac{k^2 \exp(s_+ t)}{(s_+ - s_-)(s_+^2 + k^2)} + \frac{ik \exp(ikt)}{2(k + is_+)(k + is_-)}$$
$$+ \frac{k^2 \exp(s_- t)}{(s_- - s_+)(s_-^2 + k^2)} - \frac{ik \exp(-ikt)}{2(k - is_-)(k - is_+)}, \tag{17.28}$$

and

$$f_4(k,t) = \frac{s_+ \exp(s_+ t)}{(s_+ - s_-)(s_+^2 + k^2)} + \frac{\exp(ikt)}{2(k + is_+)(k + is_-)}$$
$$+ \frac{s_- \exp(s_- t)}{(s_- - s_+)(s_-^2 + k^2)} + \frac{\exp(-ikt)}{2(k - is_-)(k - is_+)}. \tag{17.29}$$

In these equations $s_\pm = -\frac{\lambda}{2} \pm i\omega$ are the roots of the quadratic equation

$$s^2 + \lambda s + \omega_0^2 = 0. \tag{17.30}$$

If we calculate the commutator of $Q(t)$, and $P(t)$ rather than $[\langle 0|Q(t)|0\rangle, \langle 0|P(t)|0\rangle]$ we find that

$$[Q(t), P(t)] = [Q(0), P(0)] \left(f_1(t)\dot{f}_2(t) - \dot{f}_1(t)f_2(t) \right)$$

$$- \frac{\epsilon^2}{L} \sum_k [q_k(0), p_k(0)] \left(f_3(k,t)\dot{f}_4(k,t) - \dot{f}_3(k,t)f_4(k,t) \right). \tag{17.31}$$

But we have already shown that $(f_1(t)\dot{f}_2(t) - \dot{f}_1(t)f_2(t))$ is equal to $e^{-\lambda t}$, therefore we need to calculate the last parenthesis in (17.31) with the help of Eqs. (17.28) and (17.29);

$$\frac{\epsilon^2}{L} \sum_k \left(f_3(k,t)\dot{f}_4(k,t) - \dot{f}_3(k,t)f_4(k,t) \right) = e^{-\lambda t} - 1. \tag{17.32}$$

Substituting these in (17.31) we obtain the standard commutator for $Q(t)$ and $P(t)$;

$$[Q(t), P(t)] = i\hbar. \tag{17.33}$$

That is once all the terms contributing to $Q(t)$ and $P(t)$ are taken into account, then the commutator of $Q(t)$ and $P(t)$ becomes independent of time.

It is important to note that by omitting the sum in Eq. (17.31) we have lost the constancy of $[Q(t), P(t)]$ in time. We will see in Sec. 17.2 that this sum is an operator that for general system may be regarded as the noise operator. **Averages of P^2, Q^2 and $Q(t)P(t) + P(t)Q(t)$** — In addition to the time dependence of the expectation values of $Q(t)$ and $P(t)$ given by (17.16) and (17.17) we can calculate the time dependence of $Q^2(t)$, $P^2(t)$ and $Q(t)P(t) + P(t)Q(t)$. Again we want these expectation values over the ground state of the oscillators $q_k(0)$ and $p_k(t)$. These are given by [1]

$$\langle 0|Q^2(t)|0\rangle = \langle 0|Q^2(0)|0\rangle f_1^2(t) + \langle 0|P^2(0)|0\rangle f_2^2(t)$$

$$+ \langle 0|Q(0)P(0) + P(0)Q(0)|0\rangle f_1(t)f_2(t) + \frac{\lambda}{2\pi} g_1(t), \tag{17.34}$$

$$\langle 0|P^2(t)|0\rangle = \langle 0|Q^2(0)|0\rangle \dot{f}_1^2(t) + \langle 0|P^2(0)|0\rangle \dot{f}_2^2(t)$$

$$+ \langle 0|Q(0)P(0) + P(0)Q(0)|0\rangle \dot{f}_1(t)\dot{f}_2(t) + \frac{\lambda}{2\pi} g_2(t), \tag{17.35}$$

and

$$\langle 0|Q(t)P(t) + P(t)Q(t)|0\rangle = \frac{d}{dt} \langle 0|Q^2(t)|0\rangle, \tag{17.36}$$

where

$$g_1(t) = \hbar \int_0^\infty k \left[\frac{1}{k^2} f_3^2(k,t) + f_4^2(k,t) \right] dk, \tag{17.37}$$

and

$$g_2(t) = \hbar \int_0^\infty k \left[\frac{1}{k^2} \dot{f}_3^2(k,t) + \dot{f}_4^2(k,t) \right] dk. \tag{17.38}$$

In arriving at the last two relations we have used the fact that the summation over k can be replaced by integration

$$\sum_k \rightarrow \frac{L}{2\pi} \int_{-\infty}^{+\infty} dk. \tag{17.39}$$

The function $g_2(t)$ is logarithmically divergent, and thus $\langle P^2 \rangle$ is infinite, a defect of the present model. This can be remedied by making ϵ a function of k in such a way that the high frequency oscillators decouple from the motion of the particle. This has been achieved in Ullersma's model by assuming that ϵ_n is a decreasing function of n. The same approach can be used for the quantization of a system composed of a particle coupled to a string of finite or of infinite length (Sollfrey's model, Chapter 8), when the coupling is between Q and q_n rather than between Q and p_n.

17.2   Heisenberg-Langevin Equations for a Damped Harmonic Oscillator: Quantized Noise Operator

In Secs. 11.4, 14.3 and 17.1 we found the Heisenberg equations of motion for a damped harmonic oscillator. Also from the motion of a harmonically bound particle coupled to a scalar field we obtained similar equations (17.19) and (17.20). In both cases the resulting operator equations had the general form of

$$\frac{dQ(t)}{dt} = \omega P(t) - \frac{\lambda}{2} Q(t) + \phi(t), \tag{17.40}$$

$$\frac{dP(t)}{dt} = -\omega Q(t) - \frac{\lambda}{2} Q(t) + \pi(t). \tag{17.41}$$

In writing these relations we have assumed that the particle has a unit mass, and that $\omega = \sqrt{\omega_0^2 - \frac{\lambda^2}{4}}$ is the shifted frequency of the oscillator. The two added terms $\phi(t)$ and $\pi(t)$ are operators depending on time and representing the noise. For the motion of a harmonically bound particle, these operators are defined by the right-hand sides of Eqs. (17.15) and (17.14). We have already shown that by including $\phi(t)$ and $\pi(t)$ in this way, we get the commutation relation (17.33).

If these noise terms are omitted, then the coupled equations (17.40) and (17.41) are exactly solvable. Denoting the solutions in the absence of noise by $Q_1(t)$ and $P_1(t)$, we can write;

$$Q_1(t) = (Q_0 \cos(\omega t) + P_0 \sin(\omega t)) e^{\frac{-i\lambda t}{2}}, \qquad (17.42)$$

$$P_1(t) = (-Q_0 \sin(\omega t) + P_0 \cos(\omega t)) e^{\frac{-i\lambda t}{2}}, \qquad (17.43)$$

where Q_0 and P_0 are the operators at $t = 0$. Setting $\hbar = 1$, the commutation relation at $t = 0$ is

$$[Q(0), \ P(0)] = i, \qquad (17.44)$$

then, at a later time from (17.42)-(17.45), we get

$$[Q_1(t), \ P_1(t)] = ie^{-\lambda t}, \qquad (17.45)$$

and therefore we have a violation of the uncertainty principle as we have seen earlier Sec.

With the inclusion of quantum noise one can inquire as to whether it is possible to choose the non-commuting noise operators $\phi(t)$ and $\pi(t)$ such that the canonical commutation relations hold for all time [3] ? As an answer to this question Lax showed that if one assumes the commutator

$$[\phi(t), \ \pi(t)] = ib\delta \left(t - t' \right), \qquad (17.46)$$

for the quantum noise operators, then a suitable choice of the parameter b exactly compensates for the decay of the commutator $[Q_1(t), \ P_1(t)]$ as given by (17.45) [4]. A deficiency of this proposal of Lax is the fact one needs a full range of energies, $-\infty < k < \infty$ to construct white noise, and thus negative energies are introduced into the theory, and this is not physically acceptable [5],[6]. As an alternative, let us study a general approach advanced by Streater in which one assumes a positive energy Gaussian noise [6].

In this formulation we introduce two operators $A(t)$ and $a(t)$ by

$$A(t) = \frac{1}{\sqrt{2}} (P(t) - iQ(t)), \qquad (17.47)$$

and

$$a(t) = \frac{1}{\sqrt{2}} (\pi(t) - i\phi(t)). \qquad (17.48)$$

Using these operators we can write Eqs. (17.40) and (17.41) as two differential equations for $A(k)$ and $A^\dagger$

$$\frac{d}{dt} A(t) = \left(-i\omega - \frac{\lambda}{2} \right) A(t) + a(t), \qquad (17.49)$$

$$\frac{d}{dt} A^\dagger(t) = \left(i\omega - \frac{\lambda}{2} \right) A^\dagger(t) + a(t). \qquad (17.50)$$

By solving (17.49) as an initial value problem we get

$$A(t) = \exp\left[\left(-i\omega - \frac{\lambda}{2}\right)t\right] A(0) + \int_0^t \exp\left[\left(-i\omega - \frac{\lambda}{2}\right)(t-s)\right] a(s)ds,$$

(17.51)

and the corresponding relation for $A^\dagger(t)$.

Let us assume that $a(t)$ and $a^\dagger$ have Fourier decompositions given by

$$a(t) = \int_0^\infty \rho(k)e^{-ikt}a(k)dk,$$

(17.52)

and

$$a^\dagger(t) = \int_0^\infty \rho(k)e^{ikt}a^\dagger(k)dk,$$

(17.53)

where $a(k)$ s and $a_k^\dagger$ s with $k \geq 0$ are oscillators, so that

$$\left[a(k),\, a^\dagger(k')\right] = \delta(k - k'),$$

(17.54)

and

$$\left[a^\dagger(k),\, a^\dagger(k')\right] = [a(k),\, a(k')] = 0.$$

(17.55)

Furthermore we take $\rho(k)$ to be non-negative. The fact that the integration in (17.52) is over positive values of k only means that we do not need to add quanta of negative energy to $A^\dagger(t)$. Next we want to determine the most general form of $\rho(k)$ such that

$$\left[A(t),\, A^\dagger(t')\right] = 1,$$

(17.56)

for all $t \geq 0$. Noting that $\left[A(0),\, A^\dagger(0)\right] = 1$, from Eqs. (17.51)–(17.55) we obtain

$$\left[A(t),\, A^\dagger(t')\right] = 1 = e^{-\lambda t} + \int_0^\infty \rho^2(k) \int_0^t ds \int_0^t ds'$$
$$\times \exp\left[\left(-i\omega - \frac{\lambda}{2}\right)(t-s) - iks + \left(i\omega - \frac{\lambda}{2}\right)(t-s') + iks'\right].$$

(17.57)

Now defining $G(k)$ by

$$G(k) = \left[\frac{\lambda^2}{4} + (\omega - k)^2\right]^{-1},$$

(17.58)

we can write(17.57) as

$$e^{-\lambda t}\left\{1 + \int_0^\infty \rho^2(k)G(k)\left[e^{\lambda t} + 1 - 2e^{\frac{\lambda t}{2}}\cos[(\omega - k)t]\right]dk\right\} = 1.$$

(17.59)

As $t \to \infty$, Eq. (17.59) reduces to

$$\int_0^\infty \rho^2(k)G(k)dk = 1.$$

(17.60)

From Eqs. (17.59) and (17.60) it follows that for all $t \geq 0$ we have

$$1 + e^{\lambda t} + 1 - 2e^{\frac{\lambda t}{2}} \int_0^\infty \rho^2(k) G(k) \cos[(\omega - k)t] dk = e^{\lambda t}. \qquad (17.61)$$

We can write (17.61) as an integral equation for unknown function $\rho(k)$

$$\int_0^\infty \rho^2(k) \left[\frac{\lambda^2}{4} + (\omega - k)^2 \right]^{-1} \cos[(\omega - k)t] \, dk = e^{\frac{-\lambda t}{2}}. \qquad (17.62)$$

We can change this last equation into the following equation for $\rho(k)$

$$\int_0^\infty [\rho^2(k + \omega) + \rho^2(\omega - k)] \left[\frac{\lambda^2}{4} + (\omega - k)^2 \right]^{-1} \cos(kt) \, dk = e^{\frac{-\lambda t}{2}}, \qquad (17.63)$$

and then using the uniqueness of the cosine transform we obtain

$$\rho^2(k + \omega) + \rho^2(\omega - k) = \frac{\lambda}{\pi}. \qquad (17.64)$$

If $\rho(k)$ could be nonzero for negative values of k, then a possible solution is given by

$$\rho(k) = \frac{\lambda}{2\pi}, \qquad (17.65)$$

and this is the solution found by Lax [4]. We are interested in the solutions with positive energy, i.e. we want $\rho(k) = 0$ for negative values of k. The general solution with $\rho(k)$ satisfying this condition obtained by Streater is as follows: Let $\sigma(k)$, $0 \leq k \leq \omega$ be any function satisfying $0 \leq \sigma(k) \leq \frac{\lambda}{\pi}$, then $\rho(k)$ is given by [6]

$$\rho^2(k) = \begin{cases} 0 & k < 0 \\ \sigma(\omega - k) & 0 \leq k < \omega \\ \frac{\lambda}{\pi} - \sigma(k - \omega) & \omega \leq k < 2\omega \\ \frac{\lambda}{\pi} & k \geq 2\omega \end{cases}. \qquad (17.66)$$

An analysis very similar to the Streater's treatment but with the removal of the condition that $\rho(k < 0) \equiv 0$, has been considered by Hasegawa *et al.* (see also [7]). In this analysis it is assumed that the expansion (17.49) and (17.50) are replaced by

$$\frac{d}{dt} A(t) = \left(-i\omega - \frac{\lambda}{2} \right) A(t) + \xi w(t), \qquad (17.67)$$

$$\frac{d}{dt} A^\dagger(t) = \left(i\omega - \frac{\lambda}{2} \right) A^\dagger(t) + \xi w(t)^*, \qquad (17.68)$$

where ξ is a constant and

$$w(t) = a(t) + a^\dagger(t) = \int_0^\infty a(k)\rho(k)e^{-ikt} dk + \int_0^\infty a^\dagger(t)\rho^*(k)e^{ikt} dk. \qquad (17.69)$$

This choice of $w(t)$ implies the commutation relation

$$[w(s), \ w(s')] = \int_0^\infty |\rho(k)|^2 \exp\left[-ik(s-s')\right] dk - \int_0^\infty |\rho(k)|^2 \exp\left[ik(s-s')\right] dk$$

$$= \int_{-\infty}^\infty \text{sgn}(k)|\rho(k)|^2 \exp\left[-ik(s-s')\right] \qquad (17.70)$$

where $\text{sgn}(k)$ is defined by Eq. (2.18) and where we have assumed the symmetry $|\rho(k)|^2 = |\rho(-k)|^2$. The commutator (17.70) replaces the commutator

$$[a(s), \ a(s')] = \int_0^\infty |\rho(k)|^2 \exp\left[-ik(s-s')\right] dk, \qquad (17.71)$$

in Streater's formulation. The solution of (17.51) for $\mathcal{A}(t)$ is given by

$$\mathcal{A}(t) = \exp\left[\left(-i\omega - \frac{\lambda}{2}\right)t\right] \mathcal{A}(0) + \xi \int_0^t w(s) \exp\left[\left(-i\omega - \frac{\lambda}{2}\right)(t-s)\right] w(s) ds,$$

$$(17.72)$$

and the complex conjugate of this relation for $\mathcal{A}^\dagger(t)$. Now following the steps taken from Eq. (17.57) to reach (fk1.19), but replacing $A(t)$ and $A^\dagger(t)$ by $\mathcal{A}(t)$ and $\mathcal{A}^\dagger(t)$ we find that the equation corresponding to (17.62) in Streater's approach can be written as

$$\left(\frac{\xi^2}{2}\right) \int_{-\infty}^\infty \left[\text{sgn}(k+\omega)|\rho(k+\omega)|^2 + \text{sgn}(-k+\omega)|\rho(-k+\omega)|^2\right]$$

$$\times \frac{\cos(kt)}{\left(k^2 + \frac{\lambda^2}{4}\right)} dk = \exp\left(-\frac{\lambda t}{2}\right), \qquad t \geq 0. \qquad (17.73)$$

Now inverting the Fourier cosine transform in (17.73) we find the functional equation for $\rho^2(k)$ to be of the form

$$\text{sgn}(k+\omega)|\rho(k+\omega)|^2 + \text{sgn}(-k+\omega)|\rho(-k+\omega)|^2 = \frac{\lambda}{\pi\xi^2}. \qquad (17.74)$$

Here $\text{sgn}(k)|\rho(k)|^2$ is not necessarily positive, nor it ha to be identically equal to zero on the negative k-axis [8].

On physical grounds we expect that $|\rho(k)|^2$ be expressible as a product of a constant factor depending on ω and λ multiplied by a universal function of k, the latter representing the spectrum of the heat bath [8]. If such a parameter dependence is assumed, then the constant may be absorbed into the coupling constant ξ (17.73). Then the left-hand side of (17.74) becomes only a function of k independent of ω and λ. In this case the acceptable solution of the functional equation (17.74) is given by

$$\text{sgn}(k)\rho^2(k) = Bk, \qquad (17.75)$$

with B a constant. The details of finding this solution is discussed in the paper of Hasegawa *et al.* [8]. Thus we find $\rho^2(k)$ to be

$$\rho^2(k) = \frac{\lambda}{\pi\omega}|k|, \qquad (17.76)$$

17.3   Quantization of the Motion of an Oscillator Coupled to a String

Now we will examine the interesting case of an oscillator coupled to an infinite string with the Hamiltonian given by (8.61) (Sollfrey's model). To quantize this system we replace $y(x,t)$, $\pi(x,t)$, q'_0 and p'_0 by operators satisfying the canonical commutation relations [9],[10]

$$[y(x,t), \pi(x',t)] = i\hbar\delta(x - x'), \qquad (17.77)$$

$$[q'_0, p'_0] = i\hbar, \qquad (17.78)$$

with all of the other commutators being equal to zero. We now define the creation and annihilation operators by

$$q'(\nu) = \left(\frac{\hbar}{2\nu}\right)^{\frac{1}{2}} \left(a^\dagger(\nu) + a(\nu)\right), \qquad (17.79)$$

$$p'(\nu) = i\left(\frac{\hbar\nu}{2}\right)^{\frac{1}{2}} \left(a^\dagger(\nu) - a(\nu)\right), \qquad (17.80)$$

$$q'_0 = \left(\frac{\hbar}{2\nu_0}\right)^{\frac{1}{2}} \left(a_0^\dagger + a_0\right), \qquad (17.81)$$

and

$$p'_0 = i\left(\frac{\hbar\nu_0}{2}\right)^{\frac{1}{2}} \left(a_0^\dagger - a_0\right), \qquad (17.82)$$

where these operators satisfy the commutation relations

$$\left[a(\nu), a^\dagger(\nu')\right] = \delta(\nu - \nu'), \qquad (17.83)$$

and

$$\left[a_0, a_0^\dagger\right] = 1. \qquad (17.84)$$

We can write the Hamiltonian in terms of $a^\dagger(\nu)$, $a(\nu)$, a_0 and $a_0^\dagger$ operators, and in this representation the Hamiltonian will not be diagonal. To find a Hamiltonian which is diagonal in the number operator, we make use of the transformations that we introduced earlier for the classical system and define $Q(\omega)$ and $P(\omega)$ operators by

$$Q(\omega) = \int_0^\infty T(\omega, \nu)q'(\nu)d\nu + T(\omega, \nu_0)q'_0, \qquad (17.85)$$

and

$$P(\omega) = \int_0^\infty T(\omega, \nu)p'(\nu)d\nu + T(\omega, \nu_0)p'_0. \qquad (17.86)$$

These operators in turn can be replaced by creation and annihilation operators

$$Q(\omega) = \left(\frac{\hbar}{2\omega}\right)^{\frac{1}{2}} \left(A^\dagger(\omega) + A(\omega)\right), \tag{17.87}$$

and

$$P(\omega) = i \left(\frac{\hbar\omega}{2}\right)^{\frac{1}{2}} \left(A^\dagger(\omega) - A(\omega)\right). \tag{17.88}$$

From the Hamiltonian (8.61) and the transformations $T(\omega,\nu)$ and $T(\omega,\nu_0)$, Eqs. (8.62) and (8.63) it follows that H can be written as

$$H = \int_0^\infty \hbar\omega A^\dagger(\omega)A(\omega)d\omega, \tag{17.89}$$

where we have omitted the zero point energy arising from ordering of the operators. The question of the physical significance of the zero point energy and whether we can arbitrarily omit it from the Hamiltonian has been discussed in Sec. 16.1 and we will return to it later in this section and also in Sec. 19.1.

The ground state energy of the uncoupled system, $|\Phi_0'\rangle$, can be found by noting that this state is defined by

$$a(\nu)|\Phi_0'\rangle = a_0|\Phi_0'\rangle = 0, \tag{17.90}$$

whereas the ground state for the coupled system is defined by

$$A(\omega)|\Phi_0\rangle = 0. \tag{17.91}$$

Ground State Expectation Value of H — Taking the expectation value of the Hamiltonian (17.89) with $|\Phi_0\rangle$ using (17.90) and then returning to the representation where $a^\dagger(\nu), a_0^\dagger, a(\nu)$ and a_0 have been used, we obtain

$$\langle H\rangle_{\Phi_0} = \frac{\hbar}{4}\int_0^\infty \left[\int_0^\infty T^2(\omega,\nu)\frac{(\omega-\nu)^2}{\nu}d\nu + T^2(\omega,\nu_0)\frac{(\omega-\nu_0)^2}{\nu_0}\right]d\omega. \tag{17.92}$$

By substituting $T(\omega,\nu)$ and $T(\omega,\nu_0)$ from Eqs. (8.62) and (8.63) we find

$$\langle H\rangle_{\Phi_0} = \frac{\hbar}{4}\int_0^\infty \left[\frac{e^2}{\pi^2}\left(\omega^2 - \nu_0^2\right)^2 \int_0^\infty \frac{d\nu}{\nu(\nu+\omega)^2} + \frac{e^2}{\pi m}\frac{(\omega-\nu_0)^2}{\nu_0}\right]\frac{\omega^2 d\omega}{F(\omega)}. \tag{17.93}$$

This last expression shows that the integration over ν leads to a logarithmic divergence of $\langle H\rangle_{\Phi_0}$ at low frequencies.

Next let us assume that the system at $t = 0$ is in the state $\Phi_1' = a_0^\dagger\Phi_0'$ (primes denote states of the uncoupled system), we want to determine the probability that it will remain in this state after a time t. This probability is given by

$$\mathcal{P}(t) = |W(t)|^2, \tag{17.94}$$

where the probability amplitude is

$$W(t) = \left\langle a_0^\dagger(t)\Phi_0', \; a_0^\dagger(0)\Phi_0' \right\rangle. \tag{17.95}$$

To evaluate $W(t)$ we note that $A(\omega)$ and $A^\dagger(\omega)$ have simple time dependence which follows from the Hamiltonian (17.89);

$$A(\omega, t) = A(\omega)e^{-i\omega t} \quad \text{and} \quad A^\dagger(\omega, t) = A(\omega)e^{i\omega t}, \tag{17.96}$$

Therefore starting with $A(\omega)$ and $A^\dagger(\omega)$ we expand both of these operators in terms of q_0' and p_0' using Eqs. (17.85)-(17.88), and then replace q_0' and p_0' by $a_0^\dagger$ and a_0 and let these operators act on Φ_0. This gives us the following expression for $W(t)$

$$W(t) = \frac{1}{4} \int_0^\infty \int_0^\infty T(\omega, \nu_0) T(\omega', \nu_0) \left\langle \Phi_0' \right| \left[(\omega + \nu_0) A(\omega, t) + (\nu_0 - \omega) A^\dagger(\omega, t) \right]$$

$$\times \left[(\omega' + \nu_0) A^\dagger(\omega, 0) + (\nu_0 - \omega') A(\omega', 0) \right] \left| \Phi_0' \right\rangle \frac{d\omega' d\omega}{\nu_0 \sqrt{\omega \omega'}}. \tag{17.97}$$

Simplifying (17.97) we find

$$W(t) = \frac{1}{4} \int_0^\infty \int_0^\infty \frac{T^2(\omega, \nu_0) T^2(\omega', \nu_0)}{\nu_0 \omega \omega'} \left[(\nu_0 + \omega)^2 e^{-i\omega t} - (\nu_0 - \omega)^2 e^{i\omega t} \right]$$

$$\times \omega' d\omega' d\omega. \tag{17.98}$$

The integration over ω' can be carried out using Eq. (8.68) and then we are left with a simple expression for the probability amplitude $W(t)$,

$$W(t) = \frac{e^2}{4\pi m \nu_0} \int_0^\infty \left[(\nu_0 + \omega)^2 e^{-i\omega t} - (\nu_0 - \omega)^2 e^{i\omega t} \right] \frac{\omega d\omega}{F(\omega)}$$

$$= -\frac{e^2}{4\pi m \nu_0} \int_{-\infty}^\infty \frac{\omega(\omega - \nu_0)^2}{F(\omega)} e^{i\omega t} d\omega. \tag{17.99}$$

In the formulation of this problem we have used the system where ν_0, e and ω are measured in units of inverse time, m is measured in the units of time, and $W(t)$ is dimensionless as it should be.

Since as $\omega \to \infty$, $F(\omega) \to \omega^6$, it follows that at $t = 0$, $W(t)$ and its first derivative are finite, however the second derivative of $W(t)$ is infinite. This corresponds to the problem of a displaced string with a corner which starts its motion with infinite acceleration, or alternatively, it is impossible to maintain a corner in a string since the motion will immediately smooth it out [9] Next let us examine the denominator, $F(\omega)$, in (17.99) can be written as the product of two cubic factors

$$F(\omega) = \left(\omega^3 - \frac{ie}{2}\omega^2 - \nu_0'^2 \omega + \frac{1}{2}ie\nu_0^2 \right) \left(\omega^3 + \frac{ie}{2}\omega^2 - \nu_0'^2 \omega - \frac{1}{2}ie\nu_0^2 \right). \tag{17.100}$$

If we calculate the roots of the first cubic factor in (17.100), then the roots of the second cubic factor will be the negatives of the roots of the first one.

Determination of the Roots of $F(\omega)$ — We want to evaluate $W(t)$ using contour integration in the complex ω-plane. Let us denote the roots of the first cubic polynomial by $\alpha + i\lambda$, $-\alpha + i\lambda$ and $i\gamma$. By substituting $\omega = i\gamma$ in the first cubic polynomial in (8.64) we find a cubic equation for γ;

$$\gamma^3 - \frac{1}{2}e\gamma^2 + \nu_0'^2\gamma - \frac{1}{2}e\nu_0^2 = 0. \tag{17.101}$$

This equation has a real positive root and the pure imaginary root is $i\gamma$ and not $-i\gamma$. Next if we substitute $\omega = \alpha + i\lambda$ in the first factor of $F(\omega)$, Eq. (8.64) and separate the real and imaginary parts we obtain

$$\alpha^3 - 3\alpha\lambda^2 - \nu_0'^2\alpha + e\alpha\lambda = 0, \tag{17.102}$$

and

$$3\alpha^2\lambda - \lambda^3 - \nu_0'^2\lambda - \frac{1}{2}e\alpha^2 + \frac{1}{2}e\lambda^2 + \frac{1}{2}e\nu_0^2 = 0. \tag{17.103}$$

The solution of the first equation gives us either $\alpha = 0$ which corresponds to the root $i\gamma$ or α satisfying the relation

$$\alpha^2 = 3\lambda^2 - e\lambda + \nu_0'^2. \tag{17.104}$$

If we substitute this result in (17.103) and substitute for ν_0^2 from Eq. (3.73) we find a cubic equation for λ

$$8\lambda^3 - 4e\lambda^2 + \left(2\nu_0'^2 + \frac{1}{2}e^2\right)\lambda - \frac{1}{2}\frac{e^2}{m} = 0. \tag{17.105}$$

This equation also has a real positive root and two complex roots. Now the second factor in $F(\omega)$ has roots $\alpha - i\lambda$, $-\alpha - i\lambda$ and $-i\gamma$.

For positive values of t the integral in (17.99) can be evaluated by contour integration. Closing the contour by a large semi-circle in the upper-half of the complex ω-plane, and noting that the integral along the semicircle vanishes, we find that the contribution to the integral comes from the residue at the poles in the upper-half plane. Having found the roots of $F(\omega)$ we can evaluate the integral and find $W(t)$;

$$W(t) = -\frac{2\pi i e^2}{4\pi m\nu_0}\left[\frac{(\alpha + i\lambda - \nu_0)^2 e^{i\alpha t - \lambda t}}{8i\alpha\lambda\left[(\alpha + i\lambda)^2 + \gamma^2\right]}\right.$$
$$\left. - \frac{(\alpha - i\lambda + \nu_0)^2 e^{-i\alpha t - \lambda t}}{8i\alpha\lambda\left[(\alpha - i\lambda)^2 + \gamma^2\right]} + \frac{(i\gamma - \nu_0)^2 e^{-\gamma t}}{2\left[(\alpha + i\lambda)^2 + \gamma^2\right]\left[(\alpha - i\lambda)^2 + \gamma^2\right]}\right]. \tag{17.106}$$

Approximate Determination of the Roots of $F(\omega)$ when the Coupling e Is Small — When e is small, then from the solutions of (17.101), (17.104) and (17.105) we can obtain α, λ and γ each as a power series in e. Thus to the fourth power of e we get

$$\alpha = \nu_0' - \frac{e^2}{8m\nu_0'} + \frac{3}{32}\left(\frac{e^4}{m^2\nu_0'^2}\right), \tag{17.107}$$

$$\lambda = \frac{e^2}{4m\nu_0'^2} - \frac{e^4}{16m\nu_0'^4}, \tag{17.108}$$

and

$$\gamma = \frac{1}{2}e - \frac{e^2}{2m\nu_0'^2} - \frac{e^4}{8m\nu_0'^4}. \tag{17.109}$$

From Eqs. (17.107)–(17.109) it is clear that in Eq. (17.106) that α and γ are of the order of unity and e respectively, whereas the second term λ is of the order e^2. Therefore for small e we find

$$W(t) \approx \exp(-i\alpha t - \lambda t), \tag{17.110}$$

and we conclude that for the case of infinite string, the probability $\mathcal{P}(t)$ for the system to be in its initial state is given by

$$\mathcal{P}(t) \approx \exp(-2\lambda t) \approx \exp\left(-\frac{e^2}{2m\nu_0^2}t\right), \tag{17.111}$$

i.e. the decay is exponential and the decay rate is proportional to e^2.

Perturbation Theory Applied to the Calculation of the Damping Constant — So far in this section we have studied the exact solution of the problem of an oscillator which is coupled to a string of infinite length. Now we want to calculate the damping constant for the motion of the oscillator using the perturbation theory. The standard quantum theory of natural line breadth gives an exponential decay of the state Φ_1' with the decay constant

$$\lambda = \frac{2\pi}{\hbar}\rho_E \left|H'\right|^2_{E=E'}. \tag{17.112}$$

In this relation H' is the matrix element of the perturbation operator between the state Φ_1' and the states it is directly coupled to, provided that these states have the same unperturbed energy as that of Φ_1'. The density of the final state in this case is $\hbar^{-1}$ [9]. From the total Hamiltonian (8.61) we have the perturbation part which we write in terms of the operators $a(\nu)$, $a^\dagger(\nu)$, a_0 and $a_0^\dagger$

$$H' = \frac{1}{4}\hbar e \left[\frac{1}{\sqrt{\pi}}\int_0^\infty \left(a(\nu) + a^\dagger(\nu)\right)\frac{d\nu}{\sqrt{\nu}} - \frac{1}{\sqrt{m\nu_0}}\left(a_0 + a_0^\dagger\right)\right]^2. \tag{17.113}$$

This perturbation causes transition to the states where there is a quantum in the string and none in the oscillator. We will denote the part of H' which induces this transition by H_1';

$$H_1' = -\frac{1}{2}\hbar e \frac{1}{\sqrt{m\pi\nu_0}}\int_0^\infty a^\dagger(\nu)a_0\frac{d\nu}{\nu}. \tag{17.114}$$

From this operator we obtain the energy conserving matrix element $\langle H_1' \rangle$. The absolute value squared of this is

$$|\langle H_1' \rangle|^2 = \frac{\hbar^2}{4\pi m \nu_0^2}. \tag{17.115}$$

Substituting this in (17.112) we find

$$\lambda = \frac{e^2}{2m\nu_0^2}, \tag{17.116}$$

and this is the first term of the expansion of λ in powers of e (17.108).

Decay Constant in Sollfrey's Model for Large Coupling Constants — The motion of an oscillator attached to a string of infinite length in Sollfry's model can be overdamped if the coupling constant e is large, or if e has a value that makes the motion critically damped. For these cases we can calculate the transition amplitude $W(t)$ the same way as we did for underdamped oscillations. For calculating $W(t)$ we need to know the roots of the polynomial $F(\omega)$. For approximate determination of these roots, we consider three possible cases depending on the mass of the oscillator m, and its frequency when it is coupled to the string, ν_0', and the coupling constant e between the oscillator and the string. In all cases as e increases, α first increases and reaches a maximum value, and then decreases to zero. This maximum or critical value of α is less than ν_0', as can be deduced from (17.107). Now let us consider possible motions of the particle as a function of the dimensionless parameter $\eta = m^2 \nu_0^2$. Depending on the value of this parameter, i.e. whether it is less than or greater than one, we have two distinct possibilities.

(a) If $\eta < 1$, then α remains zero for all values of e larger than the critical value. As a consequence, for these values of α all the roots will be imaginary and the system will be overdamped. . For e very large, and $\eta < 1$ we denote the three roots for the equation $F(\omega) = 0$ by $i\gamma_1$, $i\gamma_2$ and $i\gamma_3$ where

$$\gamma_1 = \frac{1}{m}\left(1 - \sqrt{1-\eta} + \frac{4}{em} + \frac{12}{em\sqrt{1-\eta}}\right), \tag{17.117}$$

$$\gamma_2 = \frac{1}{m}\left(1 + \sqrt{1-\eta} + \frac{4}{em} - \frac{12}{em\sqrt{1-\eta}}\right), \tag{17.118}$$

and

$$\gamma_3 = \left(\frac{1}{2}e - \frac{2}{m} - \frac{8}{em^2}\right). \tag{17.119}$$

Using these approximate roots, by contour integration, we can find the following asymptotic form for $W(t)$;

$$W(t) \approx -\frac{i}{4\sqrt{\eta}\sqrt{1-\eta}}\left\{\left[\sqrt{\eta} - i(1 - \sqrt{1-\eta})\right]^2 \exp\left[-(1 - \sqrt{1-\eta})\frac{t}{m}\right]\right.$$
$$\left. - \left[\sqrt{\eta} - i(1 + \sqrt{1-\eta})\right]^2 \exp\left[-(1 + \sqrt{1-\eta})\frac{t}{m}\right]\right\} \quad \eta < 1. \tag{17.120}$$

This expression for $W(t)$ shows us that the first term lies between 0 and m^{-1}, while the second term lies between m^{-1} and $2m^{-1}$. Therefore for large t the second term of (17.120) becomes negligible compared to the first. So when the coupling is very strong the probability that the system remains in its initial state decreases exponentially. In addition the phase of the wave function remains constant in time rather than having an oscillatory behaviour, as in the case of small e.

(b) Now let us consider the other possibiblity, i.e. when $\eta > 1$. In this case, i.e. $\eta > 0$ the roots of $F(\omega) = 0$ are given by

$$\alpha = \frac{1}{m}\left(\sqrt{\eta - 1} + \frac{12}{em\sqrt{\eta - 1}}\right) \tag{17.121}$$

$$\lambda = \frac{1}{m}\left(1 + \frac{4}{em}\right), \tag{17.122}$$

and

$$\gamma = \frac{1}{2}e - \frac{2}{m} - \frac{8}{em^2}. \tag{17.123}$$

As e increases further, for certain range of its variations, α will be zero and the motion of the oscillator will be overdamped. By increasing e beyond this range, we have an oscillatory motion of the particle. In the limit of $e \to \infty$, $W(t)$ will assume the simple form of

$$W(t) \approx -\frac{1}{4\sqrt{\eta}\sqrt{\eta - 1}} e^{-\frac{t}{m}} \left\{ \left(\sqrt{\eta} - \sqrt{\eta - 1} - i\right)^2 \exp\left[i\sqrt{\nu_0^2 - \frac{1}{m^2}}\, t\right] \right.$$
$$\left. - \left(\sqrt{\eta} - \sqrt{\eta - 1} + i\right)^2 \exp\left[-i\sqrt{\nu_0^2 - \frac{1}{m^2}}\, t\right] \right\}, \quad \eta > 1. $$

$$\tag{17.124}$$

Again we observe that for large e the damping constant is m^{-1}. Also we note that if the particle is very heavy, the system oscillates with a frequency close to the fundamental frequency. In this situation, i.e. the strong coupling between the oscillator and the string, there are rapid surges of energy between these two parts for the time intervals of the order e^{-1}, and then the system settles down to a rather stable behaviour given by (17.124).

17.4 Quantized Motion of a Spring Attached to a Finite String

Having discussed the motion of a spring coupled to a string of infinite length, we want to find the corresponding solution when the spring is attached at midpoint to a string of length $2L$. In Sec. 8.3 we have obtained the characteristic frequencies of such a coupled system, Eq. (8.22), and also the Hamiltonian for this system. Here our aim is to relate the creation and annihilation operators and the ground states for both the uncoupled and for the coupled systems. Let us denote these pair of operators for the two cases by $(a_\nu^\dagger, a_\nu)$ and $(A_\omega^\dagger, A_\omega)$ respectively. If the ground states for the two cases are Φ_0' and Φ_0 for the uncoupled and coupled systems respectively, then we have

$$a_\nu|\Phi_0'\rangle = a_0|\Phi_0'\rangle = 0, \qquad (17.125)$$

and

$$A_\omega|\Phi_0\rangle = 0. \qquad (17.126)$$

The operators A_ω and $A_\omega^\dagger$ depend linearly on a_ν and $a_\nu^\dagger$. The rule for the transformation between these pairs can be found from Eqs. (17.77)–(17.89) with the result that

$$A_\omega = \sum_{\nu=0}^{\infty} \frac{1}{2\sqrt{\omega\nu}} T_{\omega\nu} \left[(\omega+\nu)a_\nu + (\omega-\nu)a_\nu^\dagger \right], \qquad (17.127)$$

and

$$A_\omega^\dagger = \sum_{\nu=0}^{\infty} \frac{1}{2\sqrt{\omega\nu}} T_{\omega\nu} \left[(\omega+\nu)a_\nu^\dagger + (\omega-\nu)a_\nu \right]. \qquad (17.128)$$

Using these relations and noting that H is diagonal in the number operator $A_\omega^\dagger A_\omega$, i.e.

$$H = \hbar \sum_{\omega=0}^{\infty} \omega A_\omega^\dagger A_\omega, \qquad (17.129)$$

we obtain the expectation value of H with the ground state of the uncoupled system $|\Phi_0'\rangle$ to be

$$\langle \Phi_0' |H| \Phi_0' \rangle = \frac{1}{4}\hbar \frac{1}{\sqrt{\nu\nu'}} T_{\omega\nu} T_{\omega\nu'} \left[(\omega-\nu)(\omega-\nu') \langle \Phi_0' | a_\nu a_\nu' | \Phi_0' \rangle \right]. \qquad (17.130)$$

Since $|\Phi_0'\rangle$ is defined by (17.125)

$$\langle \Phi_0' | a_\nu a_\nu' | \Phi_0' \rangle = \delta_{\nu\nu'}, \qquad (17.131)$$

and thus

$$\langle \Phi_0' |H| \Phi_0' \rangle = \frac{1}{4}\hbar \sum_{\omega=0}^{\infty} \sum_{\nu=0}^{\infty} \frac{1}{\nu} T_{\omega\nu}^2 \left(\omega^2 - 2\omega\nu + \nu^2 \right). \qquad (17.132)$$

To simplify this expression we note that the orthogonality conditions for $T_{\omega\nu}$ elements enable us to carry out the summation over ν and ω for the terms involving $2\omega\nu$ and ν^2 in (17.132). For the first term on the right-hand side of (17.132), i.e. $\frac{\hbar}{4}\sum_\omega T_{\omega\nu}^2 \omega^2$, we can use Eq. (8.34) to sum over ω. The final result can be written as

$$\langle \Phi_0' |H| \Phi_0' \rangle = \frac{1}{4}\hbar \left[2\sum_{\nu=0}^{\infty} \nu - 2\sum_{0}^{\infty} \omega + e \sum_{\nu=1}^{\infty} \left(\frac{1}{\nu L} \right) + \frac{e}{\nu_0 m} \right]. \qquad (17.133)$$

This result shows us that the first two terms in (17.133) are quadratically divergent, whereas the third term is logarithmically divergent. However from the approximate form of the roots, ω_n given by Eq. (8.24), it is clear that $\langle \Phi_0' | H | \Phi_0' \rangle$ is finite. This finite result is a consequence of reordering of the Hamiltonian by subtracting the zero-point energy of the coupled system. If we subtract the zero-point energy of the uncoupled Hamiltonian, then $\langle \Phi_0' | H | \Phi_0' \rangle$ diverges, i.e. the third sum in (17.133) would remain. Therefore we conclude that the state $| \Phi_0' \rangle$ has a logarthmically infinite energy [10]. We can also determine the magnitude of the rate of change of the initial wave function. In the Schrödinger representation, this quantity is proportional to the expectation value of the square of the Hamiltonian in the initial state. For the vacuum state we find the following result:

$$\langle \Phi_0' | H^2 | \Phi_0' \rangle = |\langle \Phi_0' | H | \Phi_0' \rangle|^2 + \frac{e^2}{8} \left[\sum_{\nu=1}^{\infty} \left(\frac{1}{\nu L} \right) + \frac{1}{\nu_0 m} \right]^2. \qquad (17.134)$$

This matrix element diverges like the square of a logarithm, so that the rate of change of $| \Phi_0' \rangle$ is infinite. The divergence of (17.134) indicates that the initial state $| \Phi_0 \rangle$ cannot be prepared, since it would jump discontinuously into some other states. To prepare such a state, $| \Phi_0 \rangle$, one needs to have an apparatus with infinite band width. The infinite band width is also required if we want to ascertain the absence of quanta from string at all frequencies. Then we either measure or in some way detect their absence, but neither of these alternatives is achievable [9].

17.5   Senitzky's Model

Other models similar to Harris's model have been proposed by Senitzky [3] and by Weber [11],[12]. In Senitzky's model again the central particle is coupled to a heat bath, but the detailed form of the bath is not specified. The total Hamiltonian of the system is of the form

$$H = \frac{P^2}{2M} + \frac{1}{2} M \omega^2 Q^2 + H_B + \epsilon \, \Gamma(t) P, \qquad (17.135)$$

where H_B is the Hamiltonian of the bath, ϵ is the coupling constant and $\Gamma(t)$ is the coordinate of loss mechanism which couples the bath to the oscillator. The Heisenberg equations of motion for P and Q obtained from (17.135) are

$$\frac{dP}{dt} = -M \omega^2 Q, \qquad (17.136)$$

and

$$\frac{dQ}{dt} = \frac{1}{M} P + \epsilon \Gamma(t). \qquad (17.137)$$

Comparing these with Eqs. (17.7) we find that $\Gamma(t)$ plays the same role as $\frac{1}{\sqrt{L}} \sum_k q_k$ in Harris's model.

In Senitzky's approach it is assumed that the interaction is turned on at $t = 0$ and that in the absence of coupling, the operator $\Gamma^{(0)}(t)$ has the matrix elements

$$\Gamma_{ij}^{(0)}(t) = \Gamma_{ij}^{(0)}(0) \exp\left[\frac{i(E_i - E_j)t}{\hbar}\right] \equiv \tilde{\Gamma}_{ij} e^{i\omega_{ij}t}, \qquad (17.138)$$

where E_i s are the energy states of the uncoupled loss mechanism. Now the Heisenberg equation for $\Gamma(t)$ found from the Hamiltonian (17.135) is

$$i\hbar\frac{d\Gamma(t)}{dt} = [\Gamma(t), H_B(t)], \qquad (17.139)$$

where from (17.135) we can determine H_B;

$$H_B(t) = H_B^{(0)} + \frac{\epsilon}{i\hbar}\int_0^t [H_B(t_1), \Gamma(t_1)] P(t_1) dt_1. \qquad (17.140)$$

By substituting (17.140) in (17.139) we obtain

$$\frac{d\Gamma(t)}{dt} = \frac{1}{i\hbar}\left[\Gamma(t), H_B^{(0)}\right] + \frac{\epsilon}{\hbar^2}\int_0^t [\Gamma(t), [\Gamma(t_1), H_B(t_1)] P(t_1)] dt_1, \qquad (17.141)$$

and the integral of this first order integro-differential equation is [3]

$$\Gamma(t) = \Gamma^{(0)}(t) + \frac{\epsilon}{\hbar^2}\int_0^t dt_1 \int_0^{t_1} \exp\left[\frac{i}{\hbar}H_B^{(0)}(t - t_1)\right]$$

$$\times [\Gamma(t_1), [\Gamma(t_2), H_B(t_2)] P(t_2)] \exp\left[-\frac{i}{\hbar}H_B^{(0)}(t - t_1)\right] dt_2. \qquad (17.142)$$

The next step is to integrate Eqs. (17.136) and (17.137);

$$P(t) = P^{(0)}(t) - M\omega\epsilon \int_0^t \Gamma(t_1) \sin[\omega(t - t_1)] dt_1, \qquad (17.143)$$

and

$$Q(t) = Q^{(0)}(t) + \epsilon \int_0^t \Gamma(t_1) \cos[\omega(t - t_1)] dt_1, \qquad (17.144)$$

where $P^{(0)}(t)$ and $Q^{(0)}(t)$ are the solutions of the loss-free oscillator given in terms of the initial operators $P^{(0)}(0)$ and $Q^{(0)}(0)$. By substituting for $\Gamma(t)$ from (17.142) in (17.143) we find an integral equation for $P(t)$

$$P(t) = P^{(0)}(t) - M\omega\epsilon \int_0^t \Gamma^{(0)}(t_1) \sin[\omega(t - t_1)] dt_1$$

$$- \frac{M\omega\epsilon^2}{\hbar^2}\int_0^t dt_1 \int_0^{t_1} dt_2 \int_0^{t_2} \sin[\omega(t - t_1)]$$

$$\times \exp\left[\frac{i}{\hbar}H_B^{(0)}(t_1 - t_2)\right] [\Gamma(t_2), [\Gamma(t_3), H_B(t_3)] P(t_3)]$$

$$\times \exp\left[-\frac{i}{\hbar}H_B^{(0)}(t_1 - t_2)\right] dt_3. \qquad (17.145)$$

The first term on the right hand side of (17.145) represents the effect of the fluctuations of the unperturbed loss mechanism on the central oscillator. If we find the average of $P(t)$ over a state of the bath which may be described by a temperature T and for which the density matrix is diagonal

$$\rho_{nm} = \frac{\delta_{mn} \exp\left(-\frac{E_n}{kT}\right)}{\sum_j \exp\left(-\frac{E_n}{kT}\right)}, \qquad (17.146)$$

then the first term will not contribute. Thus the last term in (17.145) contains the interaction between the oscillator and the field and is responsible for the damping. Equation (17.145) is exact, and one can use it without specifying the details of the heat bath. This is clearly an advantage of the model, but then (17.145) can only be solved approximately. The details of the rather lengthy calculation and justifying the method of approximation used are given in Senitzky's work [3]. The final result is a simple differential equation for the momentum operator $P(t)$ in the form

$$\frac{d^2 P}{dt^2} + \lambda \frac{dP}{dt} + \omega^2 P = -M\omega^2 \epsilon \Gamma^{(0)}(t). \qquad (17.147)$$

Here the damping constant λ is given by

$$\lambda = \frac{\pi \omega \epsilon^2 M \left(1 - e^{\frac{-\hbar\omega}{kT}}\right) \int_0^\infty \rho(E + \hbar\omega)\rho(E)\tilde{\Gamma}^2(E + \hbar\omega, E)e^{\frac{-E}{kT}} dE}{\int_0^\infty \rho(E)e^{\frac{-E}{kT}} dE}. \qquad (17.148)$$

In this approximate calculation of $P(t)$ we have assumed that the energy levels of the loss mechanism are closely spaced, and therefore we have replaced the summation by integration

$$\sum_{j,k} \rightarrow \int_0^\infty \rho(E_j) dE_j \int_0^\infty \rho(E_k) dE_k, \qquad (17.149)$$

where $\rho(E)$ is the density of states in energy space.

17.6 Density Matrix for the Motion of a Particle Coupled to a Field

As we have seen earlier, the density matrix formulation provides an alternative way of studying the quantal problem of linearly damped harmonic oscillator. The same approach can be used for the quantum description of a particle coupled to a heat bath or to a field. For instance we can start with the Harris model with the Hamiltonian (17.5) and introduce the set of coordinates q_α and momenta p_α by

$$q_\alpha = Q, q_1, q_2 \cdots q_N \cdots, \quad p_\alpha = P, p_1, p_2 \cdots p_N \cdots, \qquad (17.150)$$

and express H as a function of q_α and p_α. This H can be used to define the density matrix ρ by the operator equation

$$\frac{\partial \rho}{\partial t} = \frac{i}{\hbar} [\rho, H].$$ (17.151)

For the present case it is more convenient to study the properties of the Wigner distribution function $W(q_\alpha, p_\alpha, t)$ which is directly related to ρ by the Fourier transform

$$W(q_\alpha, p_\alpha, t) = \lim_{N \to \infty} \int \rho \left(q_\alpha + \frac{x_\alpha}{2} ; q_\alpha - \frac{x_\alpha}{2} \right) \exp \left(-\frac{i}{\hbar} \sum_\beta p_\beta x_\beta \right) \frac{d^N x_\alpha}{(2\pi\hbar)^N},$$ (17.152)

where in (17.152) q_α and p_α are c numbers. From Eqs. (17.151) and (17.152) we obtain a partial differential equation for $W(q_\alpha, p_\alpha, t)$

$$-i\hbar \frac{\partial W(q_\alpha, p_\alpha, t)}{\partial t} = H \left[q_\alpha + \frac{\hbar}{2i} \frac{\partial}{\partial p_\alpha}, p_\alpha - \frac{\hbar}{2i} \frac{\partial}{\partial q_\alpha} \right] W(q_\alpha, p_\alpha, t)$$
$$- H \left[q_\alpha - \frac{\hbar}{2i} \frac{\partial}{\partial p_\alpha}, p_\alpha + \frac{\hbar}{2i} \frac{\partial}{\partial q_\alpha} \right] W(q_\alpha, p_\alpha, t).$$ (17.153)

Using the Hamiltonian operator Eq. (17.5) we find

$$\frac{\partial W}{\partial t} + \dot{Q} \frac{\partial W}{\partial Q} + \dot{P} \frac{\partial W}{\partial P} + \sum_k \left(\dot{q}_k \frac{\partial W}{\partial q_k} + \dot{p}_k \frac{\partial W}{\partial p_k} \right) = 0,$$ (17.154)

where

$$\dot{Q} = \frac{\partial H}{\partial P} = P, \quad \dot{P} = -\frac{\partial H}{\partial Q} = -\omega_0^2 Q - \frac{\epsilon}{\sqrt{L}} \sum \left[p_k + \frac{\epsilon}{\sqrt{L}} Q \right],$$ (17.155)

and

$$\dot{q}_k = \frac{\partial H}{\partial p_k} = p_k + \frac{\epsilon}{\sqrt{L}} Q, \quad \dot{p}_k = -\frac{\partial H}{\partial q_k} = -k^2 q_k,$$ (17.156)

i.e. we have the same equation for W as the classical distribution function. This is not surprising, since for potentials of the general form $V(Q) = AQ + BQ^2$, the distribution function does not depend on $\hbar$ [13].

In order to determine the time evolution of (17.154) we can proceed in the following way [1]:

Let us assume that an acceptable initial distribution function is given by $W(q_{\alpha 0}, p_{\alpha 0}, 0) = W_0(q_{\alpha 0}, p_{\alpha 0})$ and then we integrate Eqs. (17.155),(17.156) to obtain

$$q_\alpha(t) = q_\alpha(q_{\alpha 0}, p_{\alpha 0}, t),$$ (17.157)

and

$$p_\alpha(t) = p_\alpha(q_{\alpha 0}, p_{\alpha 0}, t),$$ (17.158)

where $q_{\alpha 0}$ and $p_{\alpha 0}$ are the initial values of $q_\alpha(t)$ and $p_\alpha(t)$. Solving (17.158) for $q_{\alpha 0}$ and $p_{\alpha 0}$ we have

$$q_{\alpha 0} = q_{\alpha 0}\left(q_\alpha(t),\, p_\alpha(t), t\right), \quad p_{\alpha 0} = p_{\alpha 0}\left(q_\alpha(t), p_\alpha(t), t\right). \tag{17.159}$$

Substituting these in $W_0(q_{\alpha 0}, p_{\alpha 0})$ we find

$$W\left(q_\alpha(t), p_\alpha(t), t\right) = W_0\left[q_{\alpha 0}\left(q_\alpha(t), p_\alpha(t), t\right),\, p_{\alpha 0}\left(q_\alpha(t), p_\alpha(t), t\right)\right]. \tag{17.160}$$

We can express this result in terms of the functions $f_1(t), f_2(t), f_3(k,t)$ and $f_4(k,t)$ defined by (17.23),(17.24) and (17.28),(17.29). To this end we replace t by $-t$ in the argument of these functions and interchange q_α with $q_{\alpha 0}$ and p_α with $p_{\alpha 0}$. For instance for Q_0 we get

$$Q_0 = Q f_1(-t) + P f_2(-t) + \frac{\epsilon}{\sqrt{L}} \sum_k \left[q_k f_3(k, -t) - p_k f_4(k, -t)\right], \tag{17.161}$$

and we find similar expressions for P_0, q_{k0} and p_{k0}.

Since we are interested in the motion of the central particle we want to find the reduced distribution function which depends only on Q, P and t. This can be achieved by integrating over all q_k s and p_k s. The N-fold integrals as $N \to \infty$ can be carried out if we choose the initial distribution function to be Gaussian, i.e.

$$W_0 = C \exp\left[-\left(A_1 P^2 + A_2 Q^2 + A_3 PQ + A_4 Q + A_5 P\right) - A_6 \sum_k \left(p_k^2 + k^2 q_k^2\right)\right], \tag{17.162}$$

where A_i s are constants. By integrating over q_k and p_k we find the reduced distribution function to be .

$$W(Q, P, t) = \mathcal{N} \exp\left[-a_1^2 (\Delta P)^2 - a_2^2 (\Delta Q)^2 - a_3^2 \Delta P \Delta Q\right]. \tag{17.163}$$

In this relation the quantities ΔP and ΔQ are defined by

$$\Delta P = P - \langle P(t)\rangle, \quad \Delta Q = Q - \langle Q(t)\rangle, \tag{17.164}$$

where a_1, a_2, a_3 and $\mathcal{N}$ are generally functions of time. By normalizing $W(Q, P, t)$ we get

$$\mathcal{N} = \frac{1}{2\pi}\left(4 a_1^2 a_2^2 - a_3^4\right)^{\frac{1}{2}}. \tag{17.165}$$

Similarly we find

$$\langle \Delta P^2(t)\rangle = \frac{\partial \ln(\mathcal{N})}{\partial a_1^2}, \tag{17.166}$$

$$\langle \Delta Q^2(t)\rangle = \frac{\partial \ln(\mathcal{N})}{\partial a_2^2}, \tag{17.167}$$

and

$$\langle \Delta Q(t)\Delta P(t) + \Delta P(t)\Delta Q(t)\rangle = 2\frac{\partial \ln(\mathcal{N})}{\partial a_2^3}. \tag{17.168}$$

Now if we substitute for $\mathcal{N}$ from (17.165) in (17.166),(17.168) we obtain the following results:

$$\langle \Delta P^2 \rangle \langle \Delta Q^2 \rangle - \frac{1}{4} \langle \Delta P \Delta Q + \Delta Q \Delta P \rangle = \frac{1}{(2\pi\mathcal{N})^2}, \qquad (17.169)$$

$$a_1^2 = \frac{(2\pi\mathcal{N})^2}{2} \langle \Delta Q^2(t) \rangle, \qquad (17.170)$$

$$a_2^2 = \frac{(2\pi\mathcal{N})^2}{2} \langle \Delta P^2(t) \rangle, \qquad (17.171)$$

and

$$a_3^2 = \frac{(2\pi\mathcal{N})^2}{2} \langle \Delta P(t) \Delta Q(t) + \Delta Q(t) \Delta P(t) \rangle. \qquad (17.172)$$

Thus the reduced distribution function $W(Q, P, t)$, Eq. (17.163), is completely determined.

17.7 Commutation Relations for the Motion for the Central Particle

Before considering the general case where the central particle is moving in a potential $V(x)$, let us consider the commutator in Ullersma's model.

Commutation Relations in Ullersma's Model — In the quantum version of Ullersma's model we regard $Q(t), P(t), Q(0), P(0) \cdots q_j(0), p_j(0)$ as operators. Using these we can calculate the commutator

$$\left[\dot{Q}(t), Q(t) \right] = [P(t), Q(t)]$$

$$= -i\hbar \sum_{\nu=0}^{N} \sum_{\mu=0}^{N} X_{0\nu} X_{0\mu} \left[\cos(s_\nu t) \cos(s_\mu t) + \frac{s_\mu}{s_\nu} \sin(s_\nu t) \sin(s_\mu t) \right]$$

$$\times \left\{ X_{0\nu} X_{0\mu} + \sum_{n=1}^{N} X_{n\nu} X_{n\mu} \right\}. \qquad (17.173)$$

Then noting that

$$\left\{ X_{0\nu} X_{0\mu} + \sum_{n=1}^{N} X_{n\nu} X_{n\mu} \right\} = \delta_{\mu\nu}, \qquad (17.174)$$

we find that the commutator of $P(t)$ and $Q(t)$ is the same as in Eq. (17.33), whereas if we average over all of the oscillators forming the bath then we obtain (17.27).

We have already discussed the classical equations of motion for a particle which is subject to a potential $V(Q)$ and at the same time is coupled linearly to a heat bath. The classical and the quantum Hamiltonians for the system are given by Eq. (8.9).

The Heisenberg equation of motion averaged over the ground state of the bath oscillators is a solution of

$$\frac{d}{dt}\langle 0|P(t)|0\rangle + \left\langle 0\left|\frac{\partial V(Q)}{\partial Q}\right|0\right\rangle - \int_0^t K(t-t')\langle 0|Q(t')|0\rangle\,dt', \qquad (17.175)$$

From Eqs. (17.175) and $M\dot{Q}(t) = P(t)$ we have

$$\frac{d}{dt}\left[\langle 0|P(t)|0\rangle, \langle 0|Q(t)|0\rangle\right] = \left[\frac{d}{dt}\langle 0|P(t)|0\rangle, \langle 0|Q(t)|0\rangle\right]$$

$$= \int_0^t K(t-t')\left[\langle 0|Q(t)|0\rangle, \langle 0|Q(t')|0\rangle\right]dt'. \qquad (17.176)$$

Thus the equal time commutator of $\langle 0|Q(t)|0\rangle$ and $\langle 0|P(t)|0\rangle$ is not a constant of motion, rather, it is given by [14]

$$[\langle 0|P(t)|0\rangle, \langle 0|Q(t)|0\rangle] = -i\hbar$$

$$+ \int_0^t dt' \int_0^{t'} K(t-t'')\left[\langle 0|Q(t'')|0\rangle, \langle 0|Q(t')|0\rangle\right]dt''.$$

$$(17.177)$$

In the classical formulation of this problem discussed in Chapter 7 we observed that when (a) - the coupling to the bath is nonlinear or (b) - the potential $V(Q)$ is neither quadratic nor linear in Q, the Poisson bracket of $P(t)$ and $Q(t)$ will depend on $Q(0), P(0)$ as well as t. Therefore in quantum mechanics the right hand side of (17.177) will depend not only on time, but in general, it is a q-number. Thus an essential point in Kostin's derivation of the Scrödinger-Langevin equation which is the validity of the commutation relation (17.27) is missing when the potential $V(Q)$ is not harmonic or linear.

17.8 Wave Equation for the Motion of the Central Particle

If we start with the Hamiltonian operator

$$H = \frac{P^2}{2M} + V(Q) + \sum_{n=1}^{\infty}\frac{1}{2}\left(\frac{p_n^2}{m_n} + m_n\omega_n^2 q_n^2\right) + \sum_n^{\infty} M\epsilon_n q_n Q, \qquad (17.178)$$

and eliminate $q_n(t)$ and $p_n(t)$ from the equation of motion of $Q(t)$, we find as in the classical case

$$\frac{d^2 Q(t)}{dt^2} + \frac{1}{M}\frac{\partial V(Q)}{\partial Q} = -\sum_{n=1}^{\infty}\left[q_n(0)\cos(\omega_n t) + \frac{1}{\omega_n}\dot{q}_n(0)\sin(\omega_n t)\right]$$

$$+ \int_0^t K(t-t')\,Q(t')\,dt', \qquad (17.179)$$

where

$$K\left(t-t'\right) = M \sum_{n=1}^{\infty} \frac{\epsilon_n^2}{m_n \omega_n} \sin\left[\omega_n \left(t - t'\right)\right]. \tag{17.180}$$

By assuming, as we did earlier, that the oscillators are initially in the ground state, i.e.

$$\langle 0 | q_n(0) | 0 \rangle = \langle 0 | \dot{q}_n(0) | 0 \rangle = 0, \tag{17.181}$$

and by denoting the expectation value of $Q(t)$ over the initial states of these oscillator by $\bar{Q}(t)$, we have

$$\frac{1}{M} \frac{d\bar{P}(t)}{dt} + \frac{\partial V(\bar{Q})}{\partial \bar{Q}} = \int_0^t K\left(t - t'\right) \bar{Q}\left(t'\right) dt', \quad \bar{P}(t) = M \frac{d\bar{Q}(t)}{dt}. \tag{17.182}$$

For the special case of $V(\bar{Q}) = \frac{1}{2} M \Omega_0^2 \bar{Q}^2$, we can solve the operator equation in terms of two independent functions $A_1(t)$ and $A_2(t)$;

$$\bar{Q}(t) = A_1(t)\bar{Q}(0) + \frac{1}{M} A_2(t)\bar{P}(0). \tag{17.183}$$

These functions are defined as the independent solutions of the integro-differential equations

$$\frac{d^2 A_i(t)}{dt^2} + \Omega_0^2 A_i(t) = \int_0^t K\left(t - t'\right) A_i\left(t'\right) dt', \quad i = 1, 2, \tag{17.184}$$

with the boundary conditions

$$A_1(0) = 1, \quad \dot{A}_1(0) = 0, \quad A_2(0) = 0, \quad \dot{A}_2(0) = 1. \tag{17.185}$$

Let us investigate the motion of the central particle. At the time $t = 0$, the Hamiltonian for this particle which is not coupled to the bath of oscillators is given by

$$H_P(0) = \frac{1}{2M} \bar{P}^2(0) + \frac{1}{2} M \Omega_0^2 \bar{Q}^2(0). \tag{17.186}$$

But at any other time t this Hamiltonian is

$$\begin{aligned} H_P(t) &= \frac{1}{2M} \bar{P}^2(t) + \frac{1}{2} M \Omega_0^2 \bar{Q}^2(t) \\ &= \frac{1}{2} M G_1(t)\bar{Q}^2(0) + \frac{1}{2M} G_2(t)\bar{P}^2(0) + \frac{1}{2} G_3(t) \left[\bar{Q}(0)\bar{P}(0) + \bar{P}(0)\bar{Q}(0)\right], \end{aligned} \tag{17.187}$$

where $G_1(t), G_2(t)$ and $G_3(t)$ are defined by

$$G_1(t) = \dot{A}_1^2(t) + \Omega_0^2 A_1^2(t), \tag{17.188}$$

$$G_2(t) = \dot{A}_2^2(t) + \Omega_0^2 A_2^2(t), \tag{17.189}$$

and

$$G_3(t) = \dot{A}_1(t)\dot{A}_2(t) + \Omega_0^2 A_1(t)A_2(t). \tag{17.190}$$

Since the Hamiltonian for the central particle is $H_P(t)$, the wave equation for the central particle is

$$i\hbar\frac{\partial\psi}{\partial t} = \frac{1}{2}M\bar{Q}^2 G_1(t)\psi - \frac{\hbar^2}{2M}G_2(t)\left(\frac{\partial^2\psi}{\partial\bar{Q}^2}\right)$$
$$- \frac{1}{2}i\hbar G_3(t)\left[2\bar{Q}\frac{\partial}{\partial\bar{Q}} + 1\right]\psi. \tag{17.191}$$

We can eliminate the term $\frac{\partial\psi}{\partial\bar{Q}}$ in (17.191) by writing

$$\psi\left(\bar{Q},t\right) = \exp\left[-i\frac{MG_3(t)}{2\hbar G_2(t)}\bar{Q}^2\right]\phi\left(\bar{Q},t\right), \tag{17.192}$$

and substituting for $\psi\left(\bar{Q},t\right)$ in (17.191) to get

$$i\hbar\frac{\partial\phi}{\partial t} = -\frac{\hbar^2}{2M(t)}\left(\frac{\partial^2\phi}{\partial\bar{Q}^2}\right) + \frac{M(t)}{2}\Omega^2(t)\bar{Q}^2\phi, \tag{17.193}$$

where

$$M(t) = \frac{M}{G_2(t)}, \tag{17.194}$$

and

$$\Omega^2(t) = G_2(t)\left\{G_1(t) - \left(\frac{G_3^2(t)}{G_2(t)}\right) - \frac{d}{dt}\left(\frac{G_3(t)}{G_2(t)}\right)\right\}. \tag{17.195}$$

Wave equations for harmonic oscillators of the type where the mass and the frequency are time-dependent have been studied extensively, e.g. in ref. [15]. **Stability of the Ground State** — A very important point regarding any dissipative system is the question of the stability of its ground state. As we have already seen, in a number of models, the effect of dissipation is to cause transitions to the lower energy states. But once the system has reached its ground state, there cannot be any further decay [16]. Considering the ground state of (17.193), we note that this equation is invariant under the transformation $\bar{Q} \rightarrow -\bar{Q}$, and that the ground state, $\phi_0(\bar{Q},t)$, is an even function of $\bar{Q}$,

$$\phi_0(\bar{Q},t) = \exp\left(C_2(t)\bar{Q}^2 + C_0(t)\right). \tag{17.196}$$

By substituting (17.196) in (17.193) and equating different powers of $\bar{Q}$ on the two sides we find that the differential equations for $C_0(t)$ and $C_2(t)$ are given by:

$$i\hbar\frac{dC_2}{dt} = -\left(\frac{2\hbar^2}{M}\right)G_2(t)C_2^2(t) + \frac{1}{2}M(t)\Omega^2(t), \tag{17.197}$$

and

$$ i\hbar \frac{dC_0}{dt} = - \left(\frac{\hbar^2}{M} \right) G_2(t) C_2(t). \tag{17.198} $$

We note that at $t = 0$, the oscillator is not coupled to the bath and therefore the initial values for $C_0(t)$ and $C_2(t)$ are:

$$ C_0(t = 0) = 0, \quad \text{and} \quad C_2(t = 0) = -\frac{M\Omega_0}{2\hbar}. \tag{17.199} $$

As $t \to \infty$, the real part of $C_2(t)$ should tend to a finite nonzero value, otherwise the particle comes to rest at $Q = 0$, a result which is incompatible with the uncertainty principle. The finiteness of Re $C_2(t)$ as $t \to \infty$ imposes certain conditions on functions $G_1(t)$, $G_2(t)$ and $G_3(t)$, and hence on $A_i(t)$ s and $K(t - t')$. To study the asymptotic form of $C_2(t)$ we first consider the normalization condition on the wave function $\psi(Q, t)$ which relates Re C_2 to Re C_0

$$ \int_{-\infty}^{\infty} \psi^*(Q,t) \psi(Q,t) dQ = \sqrt{\frac{\pi}{-(C_2 + C_2^*)}} \exp\left[-(C_0 + C_0^*)\right] = 1 \tag{17.200} $$

To study the stability of the solution of (17.197) we transform it to the normal form of Riccati differential equation by changing $C_2(t)$ to $D(t)$, [17]

$$ C_2(t) = \left(\frac{im}{2\hbar} \right) \left(\frac{D(t) + i\omega}{G_2(t)} \right), \tag{17.201} $$

and then substituting for $C_2(t)$ in (17.197) to get

$$ \frac{dD(t)}{dt} = -D^2(t) + \left(\frac{\dot{G}_2}{G_2} - 2i\omega \right) D(t) + \left(\frac{i\omega \dot{G}_2}{G_2} - G_2 G_4 \right) + \omega^2, \tag{17.202} $$

with the initial condition

$$ D(t = 0) = -i \left(\omega + \frac{2\hbar G_2(0)}{m} \right), \tag{17.203} $$

and with the function $G_4(t)$ defined by

$$ G_4(t) = G_1(t) - \frac{G_3^2(t)}{G_2(t)} - \frac{d}{dt}\left(\frac{G_3(t)}{G_2(t)} \right). \tag{17.204} $$

Here $D(t)$ and $G_2(t)$ have the dimensions of time^{-1}, and length^{-2} respectively. The formal solution of (17.202) is given by

$$ D(t) = \int^t \left(\omega^2 + \frac{i\omega \dot{G}_2(t)}{G_2(t)} - G_2(t) G_4(t) \right) \exp\left[f(t) - f(t')\right] dt' $$

$$ - \int^t D^2(t') \exp\left[f(t) - f(t')\right] dt', \tag{17.205} $$

where

$$f(t) = \int^t \left\{ \frac{\dot{G}_2(t')}{G_2(t')} - 2i\omega \right\} dt' = \ln\left(\frac{G_2(t)}{G_2(0)}\right) - 2i\omega t. \tag{17.206}$$

The solution of (17.205) can be obtained by iteration if we write this equation in following form [18]:

$$D_{n+1}(t) = \int^t \left(\omega^2 + \frac{i\omega \dot{G}_2(t)}{G_2(t)} - G_2(t)G_4(t) \right) \left(\frac{G_2(t)}{G_2(t')}\right) \exp\left[f(t) - f(t')\right]$$

$$- \int^t D_n^2(t') \left(\frac{G_2(t)}{G_2(t')}\right) \exp\left[f(t) - f(t')\right] dt', \tag{17.207}$$

with the condition that $D_{-1}(t) = 0$.

Since both of the integrands in (17.207) are complex functions of time, we do not have a simple test on these integrands to decide whether $D(t)$ is bounded as $t \to \infty$ or not. This would have been possible for real integrands in an equation like (17.207) [17]. But the iterative solution of (17.207) will show us the divergence or boundedness of $|D_{n+1}(t)|$ in the limit of large t for a given $K(t - t')$ in Eqs. (9.3.3) and (17.182).

Energy Transfer between the Particle and the Heat Bath — Let us now consider the question of the transfer of energy between the central particle and the heat bath. From the Hamiltonian of the central particle we can calculate the energy eigenvalues as a function of time. Thus by calculating the diagonal elements of $H_P(t)$ we find

$$\langle n|H_P(t)|n\rangle = \frac{1}{2}\left[\dot{A}_1^2(t) + \Omega_0^2 A_1^2(t)\right] \langle n|\bar{Q}^2(0)|n\rangle$$

$$+ \frac{1}{2}\left[\dot{A}_2^2(t) + \Omega_0^2 A_2^2(t)\right] \langle n|\bar{P}^2(0)|n\rangle. \tag{17.208}$$

We note that the last term in Eq. (17.187) has a vanishing expectation value

$$\langle n|\bar{Q}(0)\bar{P}(0) + \bar{P}(0)\bar{Q}(0)|n\rangle = 0, \tag{17.209}$$

and does not contribute to $\langle n|H_P(t)|n\rangle$. Using the well-known relations

$$\langle n|\bar{Q}^2(0)|n\rangle = \frac{\hbar}{M\Omega_0}\left(n + \frac{1}{2}\right), \tag{17.210}$$

and

$$\langle n|\bar{P}^2(0)|n\rangle = \hbar M\Omega_0\left(n + \frac{1}{2}\right), \tag{17.211}$$

we have

$$\frac{\langle n|H_P(t)|n\rangle}{\langle n|H_P(0)|n\rangle} = \frac{1}{2\Omega_0^2}\left[\left(\dot{A}_1^2(t) + \Omega_0^2\dot{A}_2^2(t)\right) + \Omega_0^2\left(A_1^2(t) + \Omega_0^2 A_2^2(t)\right)\right]. \tag{17.212}$$

In a similar way we can determine the Hamiltonian for the interaction between the particle and the bath in terms of $\bar{Q}(t)$ and therefore in terms of $\bar{Q}(0)$ and $\bar{P}(0)$;

$$H_I(t) = \sum_n \epsilon_n \bar{q}_n(t) \bar{Q}(t) = -\int_0^t K\left(t - t'\right) \bar{Q}(t) \bar{Q}\left(t'\right) dt'. \qquad (17.213)$$

Again by substituting for $\bar{Q}(t)$ and $\bar{Q}(t')$ in terms of $\bar{Q}(0)$ and $\bar{P}(0)$ we can write $H_I(t)$ in terms of $A_1(t)$ and $A_2(t)$ and their time derivatives. Then we find the matrix elements of $H_I(t)$ to be

$$\frac{\langle n|H_I(t)|n\rangle}{\langle n|H_P(0)|n\rangle} = -\frac{1}{2\Omega_0^2} \int_0^t K\left(t - t'\right) \left[A_1(t)A_1\left(t'\right) + \Omega_0^2 A_2(t)A_2\left(t'\right)\right] dt'. \qquad (17.214)$$

Finally the energy transferred to the bath of oscillators can be obtained by noting that the initial energy of the system is given by $\langle n|H_P(0)|n\rangle$. Denoting the energy of the bath (apart from the zero point energies of the oscillators) by $\langle n|H_B(0)|n\rangle$, we have

$$\langle n|H_B(t)|n\rangle = \langle n|H_P(0)|n\rangle - \langle n|H_P(t)|n\rangle - \langle n|H_I(t)|n\rangle. \qquad (17.215)$$

For the special case where $A_1(t)$ and $A_2(t)$ are given by Eqs. (17.184) and

$$K\left(t - t'\right) = 2\lambda \frac{d}{dt'} \delta\left(t' - t\right), \qquad (17.216)$$

we have calculated $\langle n|H_P(t)|n\rangle$, $\langle n|H_I(t)|n\rangle$ and $\langle n|H_B(t)|n\rangle$ all in units of $\langle n|H_P(0)|n\rangle$. Note that here we have used the integral

$$\int_0^t 2\lambda \frac{d}{dt} \delta\left(t - t'\right) A_i\left(t'\right) dt' = -\lambda \left(\frac{dA_i}{dt}\right). \qquad (17.217)$$

since t is the end point of integration [19]. The results are shown in Figs. (17.1), (17.2) and (17.3). The rate of loss of energy by the central particle which is

$$\frac{d}{dt} \langle n|H_P(t)|n\rangle, \qquad (17.218)$$

is also shown in Fig. (17.4). We observe that the rate of the energy flow is finite, unlike the situation in some other models such as the Wigner-Weisskopf model (Sec. 18.5).

For this simple case $\bar{Q}$ asymptotically goes to zero, and therefore there is a natural cut-off in the coupling between the bath and the central oscillator.

The operator equation (17.182) for any potential which is polynomial in $\bar{Q}$ can be solved numerically. Thus we write (17.182) as

$$\frac{d^2 \bar{Q}(t)}{dt^2} = \frac{1}{M} f(\bar{Q}) + \int_0^t K\left(t - t'\right) \bar{Q}\left(t'\right) dt' \qquad (17.219)$$

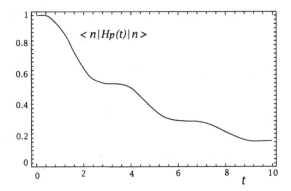

Figure 17.1: The matrix element $\langle n|H_P(t)|n\rangle$ of the energy of the central particle shown as a function of time. Since the energy is given in units of $\langle n|H_P(0)|n\rangle$, i.e. in terms of the initial energy of the system, this matrix element is independent of n.

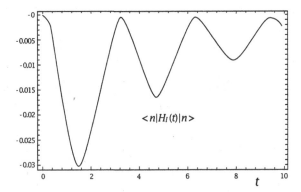

Figure 17.2: The interaction energy $\langle n|H_I(t)|n\rangle$ plotted as a function of time and expressed in units of the initial energy of the central particle.

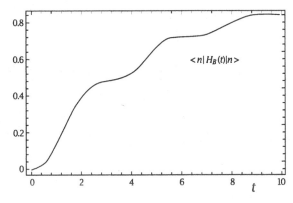

Figure 17.3: The energy transferred to the bath in units of the initial energy of the central oscillator.

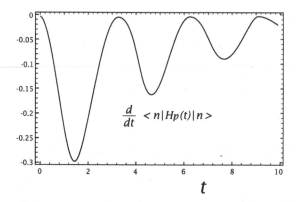

Figure 17.4: The rate of flow of energy from the particle to the bath.

where $f(\bar{Q})$ is a polynomial in $\bar{Q}$. Then we expand $\bar{Q}(t)$ as an infinite series in $\bar{Q}(0)$ and $\bar{P}(0)$;

$$\bar{Q}(t) = A_0(t) + \sum_{n=1}^{\infty} \left[A_n(t)(\bar{Q}(0))^n + B_n(t)(\bar{P}(0))^n \right]$$

$$+ \sum_{n=0}^{\infty} \sum_{m=1}^{\infty} C_{n,m}(t) \left[(\bar{P}(0))^m (\bar{Q}(0))^{n-m+1} + ((\bar{Q}(0))^{n-m+1} \bar{P}(0))^m \right].$$

$$(17.220)$$

By substituting (17.220) in (17.219) and ordering the operators on the right-hand side of (17.219) using the commutator

$$\left[\bar{Q}(0)\bar{P}(0) - \bar{P}(0)\bar{Q}(0) \right] = i\hbar, \qquad (17.221)$$

is such a way that the right-hand side is an infinite sum of Hermitian operators,

$$(\bar{P}(0))^m (\bar{Q}(0))^{n-m} + (\bar{Q}(0))^{n-m} (\bar{P}(0))^m, \qquad (17.222)$$

we find that the functions $A_0(t)$, $A_n(t)$, $B_n(t)$ and $C_{m,n}(t)$ all satisfy coupled second-order differential equations. Since these equations are nonlinear, the result of numerical integration is accurate only for relatively short times [20].

17.9   Motion of the Center-of-Mass in a Viscous Medium

In problems such as the collision or trapping of heavy ions models have been proposed where two particles interact while moving in a viscous medium [21]-[24].

In a model where dissipation arises from the linear coupling of the motion of the two particles with the heat bath, we can separate the motion of the center of mass from the relative coordinates. For the sake of simplicity we consider the one-dimensional motion and assume that the interaction between the two particles is given by the potential $V(|Q_1 - Q_2|)$. Thus the Hamiltonian operator is

$$H = \frac{P_1^2}{2M_1} + \frac{P_2^2}{2M_2} + V(|Q_1 - Q_2|) + \sum_n \frac{1}{2} \left[\frac{p_n^2}{m_n} + m_n \omega_n^2 q_n^2 \right]$$

$$+ \sum_n \epsilon_n q_n (M_1 Q_1 + M_2 Q_2), \qquad (17.223)$$

where M_1 and M_2 are the masses of the two particles. Introducing the center-of-mass and the relative coordinates by

$$(M_1 + M_2)Q = M_1 Q_1 + M_2 Q_2, \quad q = Q_1 - Q_2, \qquad (17.224)$$

and their conjugate momenta by P and p, we can write (17.223) as

$$H = \frac{P^2}{2M} + \frac{p^2}{2m} + V(q) + \sum_n \frac{1}{2} \left[\frac{p_n^2}{m_n} + m_n \omega_n^2 q_n^2 \right]$$

$$+ \sum_n M \epsilon_n q_n Q, \qquad (17.225)$$

where $M = M_1 + M_2$ is the total mass and

$$m = \left(\frac{M_1 M_2}{M_1 + M_2} \right), \qquad (17.226)$$

is the reduced mass. It can be seen from (17.225) that the relative motion is not coupled to the oscillators, whereas the center-of-mass is.

The equation of motion for the center of mass averaged over the ground states of the oscillators q_n is

$$\frac{d^2 \bar{Q}(t)}{dt^2} = \int_0^t K(t - t') \bar{Q}(t') dt', \qquad (17.227)$$

and for $\bar{q}$ we have

$$m \frac{d^2 \bar{q}}{dt^2} = -\frac{\partial V(\bar{q})}{\partial \bar{q}}. \qquad (17.228)$$

Since the motion of the center-of-mass operator $\bar{Q}$ is independent of $V(q)$, we can find the wave function for the motion by writing $\bar{Q}(t)$ in terms of $\bar{Q}(0)$ and $\bar{P}(0)$, Eq. (17.184), where

$$\frac{d^2 A_i(t)}{dt^2} = \int_0^t K(t - t') A_i(t') dt', \quad i = 1, 2, \qquad (17.229)$$

with the same initial conditions as those given in (17.185).

The Hamiltonian for the center-of-mass motion is given by

$$H_{cm} = \frac{1}{2M}\bar{P}^2(t) + \mathcal{F}(t)\bar{Q}, \quad \bar{P} = M\frac{d\bar{Q}(t)}{dt}, \tag{17.230}$$

where

$$\mathcal{F}(t) = M\sum_n \epsilon_n q_n(t), \tag{17.231}$$

is the time-dependent driving force. Thus we have the Schrödinger equation

$$i\hbar\left(\frac{\partial\Psi}{\partial t}\right) = \frac{1}{2}M\dot{A}_1^2\bar{Q}^2\Psi - \frac{i\hbar}{2M}\dot{A}_1\dot{A}_2\Psi$$

$$- \frac{\hbar^2}{2M}\dot{A}_2^2\left[\frac{\partial\Psi}{\partial\bar{Q}^2} + \left(\frac{2iM\dot{A}_1}{\hbar\dot{A}_2}\right)\bar{Q}\left(\frac{\partial\Psi}{\partial\bar{Q}}\right)\right] + \mathcal{F}(t)\bar{Q}\Psi. \tag{17.232}$$

Again if we eliminate the first order derivative $\frac{\partial\Psi}{\partial\bar{Q}}$ in (17.232) by changing Ψ to Φ, where

$$\Psi = \Phi\exp\left[-\frac{iM\dot{A}_1\bar{Q}^2}{2\hbar\dot{A}_2}\right], \tag{17.233}$$

we obtain a partial differential equation for Φ,

$$i\hbar\left(\frac{\partial\Phi}{\partial t}\right) = -\frac{\hbar^2}{2M}\dot{A}_2^2\left(\frac{\partial\Phi}{\partial\bar{Q}^2}\right) - \frac{M}{2}\left[\frac{d}{dt}\left(\frac{\dot{A}_1}{\dot{A}_2}\right)\bar{Q}^2\Phi\right] + \mathcal{F}(t)\bar{Q}\Phi. \tag{17.234}$$

For the special case where

$$K(t-t') = 2\lambda\frac{d}{dt'}\delta(t'-t), \tag{17.235}$$

Eq. (17.229) can be solved exactly;

$$A_1(t) = 1, \quad A_2(t) = \frac{1-e^{-\lambda t}}{\lambda}. \tag{17.236}$$

Thus for this form of the kernel $K(t-t')$, the equation of of motion of the center-of-mass becomes;

$$i\hbar\frac{\partial\Psi}{\partial t} = -\frac{\hbar^2}{2M}e^{-\lambda t}\frac{\partial^2\Psi}{\partial\bar{Q}^2} + \mathcal{F}(t)\bar{Q}\Psi, \tag{17.237}$$

i.e. the wave function in this case is identical with the one that we found by quantizing Caldirola-Kanai Hamiltonian. This simple result follows from the special form of the coupling that we have assumed. If we replace

$$\sum_n \epsilon_n q_n(M_1 Q_1 + M_2 Q_2) \tag{17.238}$$

by

$$\sum_n q_n(\epsilon_n M_1 Q_1 + \varepsilon_n M_2 Q_2) \tag{17.239}$$

then the damping force acting on the two particles will be different.

17.10 Invariance Under Galilean Transformation

In the microscopic derivation of dissipative forces from a many-particle system, we have seen that the time-reversal invariance and the constancy of the commutation relation $[\bar{P}(t), \bar{Q}(t)]$ are violated in the process of eliminating the degrees of freedom of the oscillators forming the bath. Now let us consider the question of the Galilean invariance of a dissipative system derived from a many-particle system which is assumed to be invariant under this transformation.

Let us start with a modified version of the Hamiltonian (17.178) which we write as

$$H = \frac{P^2}{2M} + \sum_j \frac{p_j^2}{2m_j} + \sum_j \frac{1}{2} m_j \omega_j^2 \left(Q - q_j\right)^2. \tag{17.240}$$

Here we assume that the frequencies ω_j and masses m_j satisfy the relation

$$\sum_{j=1}^{\infty} m_j \omega_j^2 = M\Omega^2, \tag{17.241}$$

and therefore the force $\frac{1}{2} M\Omega^2 Q^2$ on the particle is a finite. The equations of motion in this case are

$$M\ddot{Q} + \sum_j m_j \omega_j^2 \left(Q - q_j\right) = 0, \tag{17.242}$$

and

$$m_j \ddot{q}_j - m_j \omega_j^2 \left(Q - q_j\right) = 0. \tag{17.243}$$

These equations remain invariant under the Galilean transformation

$$P' = P + Mv, \quad Q' = Q + vt, \quad q_j' = q_j + vt. \tag{17.244}$$

As before by eliminating q_j between (17.242) and (17.243) we find the equation of motion for Q to be

$$M\ddot{Q} + M\Omega^2 Q - \sum_j m_j \omega_j^2 \left\{ q_j(0) \cos(\omega_j t) + \frac{1}{\omega_j} \dot{q}_j(0) \sin(\omega_j t) \right\}$$
$$+ \int_0^t \Gamma\left(t - t'\right) Q\left(t'\right) dt' = 0,$$

$$\tag{17.245}$$

where

$$\Gamma\left(t - t'\right) = \sum_j m_j \omega_j^3 \sin\left[\omega_j \left(t - t'\right)\right]. \tag{17.246}$$

Let us consider the equation of motion for Q' i.e. the motion of the central particle referred to a frame S' which is moving with constant velocity with respect to the initial frame of reference S;

$$M\ddot{Q}' + M\Omega^2 Q' - \sum_j m_j \omega_j^2 \left\{ q_j(0)\cos(\omega_j t) + \frac{1}{\omega_j}[\dot{q}_j(0) + v]\sin(\omega_j t) \right\}$$

$$+ \int_0^t \Gamma(t - t')\, Q'(t')\, dt' = 0,$$

$$(17.247)$$

If we average Q' over the initial distributions of $q_j(0)$ and $\dot{q}_j(0)$, the presence of the additional term proportional to v destroys the Galilean invariance of the original equations (17.242) and (17.243).

17.11   Velocity Coupling and Coordinate Coupling

In Sec. 8.2 we briefly mentioned two types of classical coupling between the central particle and the bath oscillators; either we can couple the coordinate of the central particle Q to the coordinates of the oscillators of the bath q_n s or we can couple its momentum P to q_n s . Here we want to show that these two models of interaction are related to each other. This connection can be made either at the classical level or at the quantum level.

Let us start with the momentum coupling model for which the total Hamiltonian operator is given by [38]

$$H_P = \frac{1}{2M}\left[P + \sum_j m_j \omega_j q_j\right]^2 + V(Q) + \sum_j \left[\frac{p_j^2}{2m_j} + \frac{1}{2}m_j\omega_j^2 q_j^2\right]. \quad (17.248)$$

Next consider the unitary transformation

$$U_1 = \exp\left(-\frac{i}{\hbar}Q\sum_j m_j\omega_j q_j\right), \quad (17.249)$$

which transforms P, Q, p_j and q_j to

$$P \to U_1^\dagger P U_1 = P - \sum_j m_j\omega_j q_j, \quad Q \to Q, \quad (17.250)$$

$$p_j \to U_1^\dagger p_j U_1 = p_j - m_j\omega_j Q, \quad q_j \to q_j, \quad (17.251)$$

and thus H_P transforms to H_I,

$$H_P \to H_I = U_1^\dagger H_P U_1 = \frac{P^2}{2M} + V(Q) + \sum_j \left[\frac{(p_j - m_j\omega_j Q)^2}{2m_j} + \frac{1}{2}m_j\omega_j^2 q_j^2\right],$$

$$(17.252)$$

This Hamiltonian is not yet of the form where Q couples to q_j s. Therefore we make a second unitary transformation

$$U_2 = \exp\left[\frac{i\pi}{2\hbar} \sum_j \left(\frac{p_j^2}{2m_j\omega_j} + \frac{1}{2}m_j\omega_j q_j^2\right)\right].$$ (17.253)

Under this transformation we change q_j s, p_j s and H_I;

$$q_j \to U_2^\dagger q_j U_2 = -\frac{p_j}{m_j\omega_j}, \quad p_j \to m_j\omega_j q_j,$$ (17.254)

$$H_I \to H_Q = U_2^\dagger H_I U_2 = \frac{P^2}{2M} + V(Q) + \sum_j \left[\frac{p_j^2}{2m_j} + \frac{1}{2}m_j\omega_j^2\left(q_j - Q\right)^2\right].$$ (17.255)

Thus after the second transformation we get a Hamiltonian where Q is coupled to all q_j s.

References

[1] E.G. Harris, Phys. Rev. A 42, 3685 (1990).

[2] W.G. Unruh and W.H. Zurek, Phys. Rev. D 40, 1071 (1989).

[3] I.R. Senitzky, Phys. Rev. 119, 670 (1960).

[4] M. Lax, Phys. Rev. 145, 111 (1965).

[5] H. Haken, *Laser Theory. Handbuch der Physik, Vol. XXV/2c* (Springer, Berlin, 1970) p. 43.

[6] R.F. Streater, J. Phys. A, 15, 1477 (1982).

[7] L. Accardi, Rev. Math. Phys. 02, 127 (1990).

[8] H. Hasegawa, J.R. Klauder and M. Lakshmanan, J. Phys. A 18, L123 (1985).

[9] W. Sollfrey, Ph.D. Dissertation, New York University (1950)

[10] W. Sollfrey and G. Goertzel, Phys. Rev. 83, 1038 (1951).

[11] J. Weber, Phys. Rev. 101, 1619 (1956).

[12] J. Weber, Phys. Rev. 90, 977 (1953).

[13] C.W. Gardiner, *Quantum Noise*, (Springer-Verlag, Berlin 1991).

[14] M. Razavy, Phys. Rev. A 41, 1211 (1990).

[15] V.V. Dodonov and V.I Man'ko in *Invariants and the Evolution of Non-stationary Quantum Systems*, Proceedings of the Lebedev Physics Institute Vol. 183, Edited by M.A. Markov. (Nova Scientific, Commack, 1989) p. 103.

[16] M. Razavy, Can. J. Phys. 69, 1235 (1991).

[17] R. Bellman, *Stability Theory of Differential Equations*, (Dover Publications, New Yok, N.Y. 1969) p. 130.

[18] R. Bellman, *Methods of Nonlinear Analysis*, Vol. 2. (Academic Press, New Yok, N.Y. 1973).

[19] B. Friedman *Principles and Techniques of Applied Mathematics*, (John Wiley & Sons, New York, N.Y. 1957) p. 154.

[20] M. Razavy, Phys. Rev. A41, 6668 (1990).

[21] W. Nöremberg and H.A. Weidemüler, *Introduction to the Theory of Heavy-Ion Collision*, (Springer-Verlag, Berlin, 1976).

[22] R.W. Hasse, J. Math. Phys. 16, 2005 (1976).

[23] R.W. Hasse, Rep. Prog. Phys. 41, 1027 (1978).

[24] K. Albrecht and R.W. Hasse, Physica D 4, 244 (1982).

Chapter 18

Quantum Mechanical Models of Dissipative Systems

In our studies of the quantum dissipative systems, up to this point, we relied on the quantization of classically damped motions. We observed that this classical damping can be introduced phenomenologically in the equation of motion, or can be derived from a conservative many-particle system. But there are damped or decaying systems in quantum mechanics with no classical counterparts. Perhaps the most striking example of a decaying system with no classical analogue is the problem of the escape of a particle originally trapped behind a barrier by the mechanism of quantum tunneling. Here we will first consider quantum mechanical limitations on initial decay rates and also the decaying rate after a very long time. Then we study few examples were the starting point is quantum mechanical, and these systems may or may not correspond to standard classical damped motions.

18.1 General Properties of the Decay Modes of Quantum Mechanical Systems

Without specifying a decaying system completely, from the general principles of quantum theory we can deduce some important results about the decay rate of a given system at very early stages of decay and also after a very long time [1]-[4]. These limitations on the form and the rate of decay arise from the fact that for a given unstable system the Hamiltonian is bounded below and has a minimum energy eigenvalue. In addition, the time-energy uncertainty adds specific constraints on the form of survival probability $S(t)$ and predicts certain forms of time-dependence for the rate of decay. The survival probability is defined as the probability of finding the system in its initial state after a time t [1].

Asymptotic Form of the Decay of an Unstable State after a Very Long Time — Let H_0 be the Hamiltonian of a system S_1 (e.g. a particle) when S_1 is not coupled to the system S_2 (e.g. the heat bath), and let $|\phi(0)\rangle|$ denotes the initial state of S_1. If we denote the Hamiltonian for the whole system by H and its set of eigenfunctions by $|\phi(E)\rangle$, then we can write the eigenvalue equation

$$H|\phi(E)\rangle = E|\phi(E)\rangle. \tag{18.1}$$

Now if the operators (H, A) represent the observables of the system $S_1 + S_2$ of which H is a member, then we can write the following relations

$$A|\phi(E)\rangle = a|\phi(E)\rangle, \tag{18.2}$$

and

$$\int |\phi(E, a)\rangle\langle\phi(E, a)| \, dE \, da = I, \tag{18.3}$$

where the second one, Eq. (18.3), is the completeness relation and I is the unit operator. By coupling S_1 to S_2, the initial state $|\phi(t = 0)\rangle = |\phi(0)\rangle$ will no longer be an eigenstate of H, and this initial state will decay in time. Setting $\hbar = 1$, we write the probability amplitude for the decay of $|\phi(0)\rangle$ as

$$C_0(t) = \int_{E_{min}}^{\infty} e^{-iEt} dE \int |\langle\phi(E, a)|\phi(0)\rangle|^2 da. \tag{18.4}$$

Now by introducing a function $\omega(E)$ by the relation

$$\omega(E) = \int |\langle\phi(E, a)|\phi(0)\rangle|^2 da, \tag{18.5}$$

we can express $C_0(E)$ as the Fourier transform of $\omega(E)$

$$C_0(t) = \int_{E_{min}}^{\infty} e^{-iEt} \omega(E) dE = \int_{-\infty}^{\infty} e^{-iEt} \tilde{\omega}(E) dE, \tag{18.6}$$

where $\tilde{\omega}(E)$ is defined by

$$\tilde{\omega}(E) = \begin{cases} 0 & \text{for } E < E_{min} \\ \omega(E) & \text{for } -\infty < E_{min} \leq E \end{cases}, \tag{18.7}$$

Since $\tilde{\omega}(E)$ is zero for $E < E_{min}$, therefore from Paley and Wiener theorem it follows that [5]

$$\int_{-\infty}^{\infty} \frac{|\ln |C_0(t)||^2}{1 + t^2} dt < \infty. \tag{18.8}$$

This integral converges as $t \to \infty$ if and only if $C_0(t)$ for large t behaves as

$$|\ln |C_0(t)|| \to Bt^{2-p}, \quad t \to \infty, \quad p > 1. \tag{18.9}$$

Since $|C_0(t)|$ goes to zero as t goes to infinity, therefore $\ln |C_0(t)|$ is negative and we find the following condition on $S(t) = |C_0(t)|^2$;

$$S(t) = ||C_0(t)|^2| \rightarrow \exp[-Ct^q], \quad \text{as} \quad t \rightarrow \infty, \quad q < 1, \tag{18.10}$$

where C is a positive constant.

Equation (18.10) shows that as $t \rightarrow \infty$ the decay is not exponential, and that the probability of finding the system in its initial state, $S(t)$, tends to zero slower than an exponential [6]-[8].

Decay Rate of an Unstable State at very Early Stages of Decay — Now let us study the initial decay rate of an unstable system. We can make use of the asymptotic properties of $\omega(E)$ function introduced earlier to find the decay rate, or alternatively we can apply the time-energy uncertainty and find bounds on the survival probability $S(t)$ [1], [9],[10]. Here we consider the latter method and for this we start with the formulation of uncertainty according to Mandelstam and Tamm [11],[12]. First we will review this form of uncertainty and then we will use it to derive bounds on variations of $S(t)$. The Heisenberg equation of motion for the mean value of an operator $A(t)$ can be written as

$$\frac{d}{dt}\langle A \rangle = \left\langle \frac{\partial A}{\partial t} \right\rangle + \frac{1}{i\hbar}\langle [A, H] \rangle. \tag{18.11}$$

Let us consider a system with the total Hamiltonian H which is time independent and let us assume that A also does not depend explicitly on time, then by setting $A_1 = A$ and $A_2 = H$ in (18.11) and substitute in the uncertainty relation

$$\Delta A_1 \Delta A_2 \geq \frac{1}{2}|\langle k| [A_1, A_2] |k\rangle|, \tag{18.12}$$

where $|k\rangle$ represents a wave packet, we find

$$\Delta A \Delta E \geq \frac{1}{2}|\langle k| [A_1, H] |k\rangle|. \tag{18.13}$$

The quantity $\Delta E = \Delta H$ is the uncertainty in the energy. By combining (18.11) and (18.13) we get the desired form uncertainty relation

$$\Delta \tau_A \Delta E \geq \frac{1}{2}\hbar. \tag{18.14}$$

In this relation $\Delta \tau_A$ which has the dimension of time is defined by

$$\Delta \tau_A = \left| \frac{d\langle k|A|k\rangle}{dt} \right|^{-1} \Delta A. \tag{18.15}$$

The uncertainty $\Delta \tau_A$ introduced by (18.15) is the time characteristic of the evolution of the observable A [11]-[14].

To apply this form of time-energy uncertainty relation to the problem of decay rate of an unstable state, we again consider the initial state $|\phi(0)\rangle$ which is not an eigenstate of the Hamiltonian of the system H, but is expressible as an integral over the continuous eigensates of H. After a time t, $|\phi(t)\rangle$ is given by

$$|\phi(t)\rangle = \exp\left(-i\frac{Ht}{\hbar}\right)|\phi(0)\rangle, \tag{18.16}$$

and therefore

$$S(t) = |\langle \phi(0)|\phi(t)\rangle|^2. \tag{18.17}$$

From (18.16) and (18.17) we can easily deduce that $S(t)$ is an even function of time. Now we choose the operator $A(t)$ in Eq. (18.15) to be

$$A(t) = |\phi(t)\rangle\langle\phi(t)|. \tag{18.18}$$

By substituting $A(t)$ in (18.15) and using the definition (18.17) we find [9]

$$\frac{[S(t)|\,(1-S(t))]^{\frac{1}{2}}}{\left|\frac{dS(t)}{dt}\right|} \geq \frac{\hbar}{2\Delta E}. \tag{18.19}$$

A rearrangement of this inequality shows us that

$$\left|\frac{dS(t)}{dt}\right| \leq \frac{2\Delta E}{\hbar}\left[|S(t)|\,(1-S(t))\right]^{\frac{1}{2}}. \tag{18.20}$$

The right-hand side of (18.20) takes its maximum value for $S(t_h) = \frac{1}{2}$ when

$$\left|\left|\frac{dS(t)}{dt}\right|\right|_{t=t_h} \leq \frac{\Delta E}{\hbar}. \tag{18.21}$$

This time, t_h, which is half-life of a decaying system has the property that only for $t = t_h$ (18.21) can become an equality. An important conclusion that we can draw from (18.21) is that no unstable quantum system can decay completely within a time $\frac{\hbar}{\Delta E}$ [9],[10]. Now we integrate (18.20) and find t as a function of S;

$$t \geq \frac{\hbar}{\Delta E}\cos^{-1}\sqrt{S(t)}, \tag{18.22}$$

where the initial condition $S(0) = 1$ has been used to determine the constant of integration. If we solve (18.22) for $S(t)$ we obtain

$$S(t) \geq \cos^2\left(\frac{\Delta t}{\hbar}\right), \qquad 0 \leq t \leq \frac{\pi}{2\Delta E}. \tag{18.23}$$

From (18.23) we obtain the following characteristics of quantum decay for very short and after a very long time:

$$\left|\frac{dS(t)}{dt}\right|_{t=0} = 0, \tag{18.24}$$

$$\frac{d}{dt}\cos^{-1}S(t) \leq \frac{2E}{\hbar}, \quad \text{as} \quad t \to 0, \tag{18.25}$$

and

$$\frac{d}{dt}\sqrt{S(t)} \leq \frac{\Delta E}{\hbar}, \quad \text{as} \quad t \to \infty. \tag{18.26}$$

As Eq. (18.24) shows $S(t)$ cannot be a purely exponential function of time since if $S(t) = S(0)e^{-\Gamma t}$ then we have

$$\left(\frac{dS(t)}{dt}\right)_{t=0} = -\Gamma S(0) < 0. \tag{18.27}$$

Initial Decay Rate for a Particle Coupled to a Heat Bath — For a large group of problems we encounter coupled systems with equations of motion of the form

$$\frac{dC_0(t)}{dt} = \int \frac{d^3k}{(2\pi)^3}\exp[i\Omega(\mathbf{k})]V(\mathbf{k})C_{\mathbf{k}}(t), \tag{18.28}$$

$$\frac{dC_{\mathbf{k}}(t)}{dt} = \exp[i\Omega(\mathbf{k})]V(\mathbf{k})C_0(t), \tag{18.29}$$

with the initial conditions

$$C_0(t=0) = 1, \quad C_{\mathbf{k}}(t=0) = 0. \tag{18.30}$$

Mathematically these equations have the same form as the equations of motion of an oscillator coupled to a field or an electron coupled to the electromagnetic field (see Eqs. (18.305) and (18.306)).

Now let us find the formal solutions of (18.28) and (18.29). By integrating (18.29) and substituting the resulting $C_{\mathbf{k}}(t)$ in (18.28) we find an integro-differential equation for $C_0(t)$ [15];

$$i\frac{dC_0(t)}{dt} = -\int_0^t g(t - t')C_0(t')dt', \tag{18.31}$$

where

$$g(t - t') = \int \frac{d^3k}{(2\pi)^3}|V(\mathbf{k})|^2\exp[i\Omega(\mathbf{k})(t - t')]. \tag{18.32}$$

We can also write (18.31) as

$$i\frac{dC_0(t)}{dt} = -\int_0^t g(t')C_0(t - t')dt'. \tag{18.33}$$

Equations (18.32) and (18.33) can be used to determine both the early time behaviour and also the non-exponential nature of the decay law [15],[16]. Here we will consider the onset of the 'decay, i.e. the limit of the survival probability $|C_0(t)|^2$ for very small t. For this we introduce a function of time $f(t)$ by

$$C_0(t) = \exp[-f(t)], \tag{18.34}$$

and substitute this in (18.33) to find an integro-differential equation for $f(t)$;

$$\frac{df(t)}{dt} = \int_0^t g(t') \exp\left[f(t) - f(t - t')\right] dt'.\tag{18.35}$$

Now from the initial conditions (18.30) and Eqs. (18.34) and (18.35) it follows that

$$f(0) = 0, \quad \left(\frac{df(t)}{dt}\right)_{t=0} = 0.\tag{18.36}$$

Using these boundary conditions we can write expansions for $f(t)$ and $g(t)$ for short times as a power series in t:

$$f(t) = \mathcal{A}t^2 + \mathcal{B}t^3 + \mathcal{C}t^4 + \cdots,\tag{18.37}$$

and

$$g(t) = \int \frac{d^3k}{(2\pi)^3}|V(\mathbf{k})|^2 + it \int \frac{d^3k}{(2\pi)^3}|V(\mathbf{k})|^2\Omega(\mathbf{k})$$
$$-\frac{1}{2}t^2 \int \frac{d^3k}{(2\pi)^3}|V(\mathbf{k})|^2\Omega(\mathbf{k})^2 + \cdots.\tag{18.38}$$

By substituting (18.37) and (18.38) in (18.35) and equating equal powers of t we find the coefficients of expansion (18.37) to be

$$\mathcal{A} = \frac{1}{2}\int \frac{d^3k}{(2\pi)^3}|V(\mathbf{k})|^2,\tag{18.39}$$

$$\mathcal{B} = \frac{1}{2}\int \frac{d^3k}{(2\pi)^3}|V(\mathbf{k})|^2\Omega(\mathbf{k}),\tag{18.40}$$

and

$$\mathcal{C} = \frac{1}{2}\int \frac{d^3k}{(2\pi)^3}|V(\mathbf{k})|^2\Omega^2(\mathbf{k}) + \frac{1}{2}\mathcal{A}^2.\tag{18.41}$$

Therefore for early stages of decay, $S(t) = |C_0(t)|^2$ has the form

$$S(t) = |C_0(t)|^2 = \left[1 - \frac{1}{2}(\sqrt{\mathcal{A}}\,t)^2 + \frac{1}{24}(\sqrt{\mathcal{A}}\,t)^4 + \cdots\right]$$
$$\times \exp\left[-(\mathcal{A}^2 - \mathcal{C})\frac{t^4}{4}\right],\tag{18.42}$$

or

$$S(t) \approx \cos^2\left(\sqrt{\mathcal{A}}\,t\right)\exp\left[-(\mathcal{A}^2 - \mathcal{C})\frac{t^4}{4}\right].\tag{18.43}$$

18.2 Forced Vibration of a Chain of Oscillators with Damping

Let us consider a linear chain of oscillators, each coupled to its neighbor. Suppose that at some initial time t_0 a pulse $g(t)$ of some arbitrary form is applied at one end. We want to show that the exact solution of the wave equation implies that this classical pulse propagates down the chain and is only modulated by quantum mechanical wave function as it propagates. Thus if at the other end of the chain a load $f(t)$ is applied, matched exactly by the arriving pulse, so that the energy is completely absorbed by this load as it emerges, then the chain returns exactly to its original quantum state [17].

Let us start with the Schrödinger equation [17]

$$-\frac{\hbar^2}{2m}\frac{\partial^2\psi(x,t)}{\partial x^2} + \frac{1}{2}Kx^2\psi(x,t) - g(t)x\psi(x,t)$$
$$- i\hbar\lambda x\psi(x,t)\int \psi^*(y,t)\frac{\partial\psi(y,t)}{\partial y}dy = i\hbar\frac{\partial\psi(x,t)}{\partial t}, \qquad (18.44)$$

where $g(t)$ is the external force and

$$i\hbar\lambda\int \psi^*(y,t)\frac{\partial\psi(y,t)}{\partial y}dy = -\lambda\langle p\rangle \qquad (18.45)$$

is the damping force. We observe that the effect of damping is zero if $\psi(x,t)$ is an eigenstate of the free lattice. This suggests, as in the case of Schrödinger-Langevine equation, that we change x to z, where

$$z = x - \xi(t), \qquad (18.46)$$

and assume a solution of the form

$$\psi(x,t) = u(z)\exp\left[-\frac{i}{\hbar}\Gamma(x,t)\right]. \qquad (18.47)$$

Here $u(z)$ is the solution of the Schrödinger equation for a free undamped oscillator

$$-\frac{\hbar^2}{2m}\frac{d^2u(z)}{dz^2} + \frac{1}{2}Kz^2u(z) = Eu(z). \qquad (18.48)$$

By substituting (18.47) in (18.44) and making use of (18.48) we find that $\Gamma(x,t)$ must satisfy the relation

$$\Gamma(x,t) = Et + \int\left[\frac{1}{2}m\dot{\xi}^2 - \frac{1}{2}K\xi^2 + \left(K\xi - g(t) + m\lambda\dot{\xi}\right)x\right]dt, \qquad (18.49)$$

where ξ is a solution of the classical equation of motion

$$m\ddot{\xi}(t) = -K\xi(t) + g(t) - m\lambda\dot{\xi}(t). \qquad (18.50)$$

By calculating $|\psi(x,t)|^2$ we find that

$$|\psi(x,t)|^2 = |u(z)|^2 = |u(x - \xi(t)|^2; \qquad (18.51)$$

therefore the center of the wave packet oscillates exactly as that of a damped and forced harmonic oscillator, i.e. $\xi(t)$.

In this model the expectation value of the energy of the system is given by

$$\langle E \rangle = i\hbar \int \psi^*(x,t) \frac{\partial \psi(x,t)}{\partial t} dx = E + \frac{m}{2}\dot{\xi}^2 + \frac{1}{2}K\xi^2 - g(t)\xi + m\lambda\xi\dot{\xi}, \quad (18.52)$$

which is the sum of the quantum energy and the classical energy associated with the oscillatory motion of $\xi(t)$. There are different ways of generalizing this model:

(a) We can change it to a form that can be used for dissipative forces proportional to any power of velocity.

(b) We can modify it so that it can be applied to N interacting particles with pair interaction.

First let us consider the following modification of (18.44), where we replace the term

$$-i\hbar\lambda x \int \psi^*(y,t) \frac{\partial \psi(y,t)}{\partial y} dy \qquad (18.53)$$

by

$$\gamma x \left[\int \psi^*(y,t) \left(-i\hbar\frac{\partial \psi(y,t)}{\partial y} \right) dy \right]^2. \qquad (18.54)$$

The new $\Gamma(x,t)$ and $\xi(t)$ must satisfy the following relations

$$\Gamma(x,t) = Et + \int \left[\frac{1}{2}m\dot{\xi}^2 - \frac{1}{2}K\xi^2 + \left(K\xi - g(t) + m\gamma\dot{\xi}^2 \right) x \right] dt \qquad (18.55)$$

and

$$m\ddot{\xi}(t) = -K\xi(t) + g(t) - m\gamma\dot{\xi}^2(t). \qquad (18.56)$$

Similar results can be found for dissipative forces proportional to $\dot{\xi}^\alpha, (\alpha > 0)$.

The second possible generalization is to the case where there are N interacting particles with $3N$ degrees of freedom. Here the simple harmonic pair interaction is of the form

$$V(x_1, x_2, ...x_{3N}) = \frac{1}{4} \sum_{j,k=1}^{3N} K_{jk}(x_k - x_j)^2, \qquad (18.57)$$

where both j and k run over all of the allowed values. For an externally applied force we assume the potential W to be

$$W(x_1, x_2, ...x_{3N}, t) = -\sum_{j=1}^{3N} g_j(t)x_j. \qquad (18.58)$$

With these two potentials the Schrödinger equation becomes

$$-\frac{\hbar^2}{2m}\sum_j \left(\frac{\partial^2 \psi}{\partial x_j^2}\right) + V(x_1, x_2, ...x_{3N})\,\psi(x_1....x_{3N}, t) - \sum_j g_j(t)x_j\psi$$

$$-i\hbar\lambda\sum_j\left(\int \psi^*\frac{\partial \psi}{\partial y_j}dy_j\right)\psi(x_1....x_{3N}, t) = i\hbar\frac{\partial}{\partial t}\psi(x_1....x_{3N}, t).$$

$$(18.59)$$

Next we change the variable x_k to z_k where

$$z_k = x_k - \xi_k(t), \tag{18.60}$$

and again seek a solution of the form

$$\psi(x_1....x_{3N}, t) = u(z_1, z_2, ...z_{3N})\exp\left[-\frac{i}{\hbar}\Gamma(x_1, x_2, ...x_{3N}, t)\right]. \tag{18.61}$$

Here $u(z_1, z_2, ...z_{3N})$ is the solution of the Schrödinger equation for the lattice with no damping force

$$\left[-\frac{\hbar^2}{2m}\sum_j\left(\frac{\partial^2}{\partial z_j^2}\right) + V(z_1, z_2, ...z_{3N})\right]u(z_1, ...z_{3N}) = Eu(z_1, ...z_{3N}).$$

$$(18.62)$$

Just as before, Eq. (18.61) will satisfy (18.59) provided that $\Gamma(x_1, x_2, ...x_{3N}, t)$ is given by

$$\Gamma(x_1, x_2, ...x_{3N}, t) = Et + \int\left[\frac{m}{2}\sum_{j=1}^{3N}\dot{\xi}_j^2 - \frac{1}{2}\sum_{j,k=1}^{3N}K_{jk}\left(\xi_k(t) - \xi_j(t)\right)^2\right.$$

$$\left.+\frac{1}{2}\sum_{j,k=1}^{3N}K_{jk}\left(x_k - x_j\right)\left(\xi_k(t) - \xi_j(t)\right) - \sum_{j=1}^{3N}g_j(t)x_j + m\lambda\sum_{j=1}^{3N}\dot{\xi}_j x_j\right]dt,$$

$$(18.63)$$

and where $\xi_j(t)$ satisfies the classical equation of motion

$$m\ddot{\xi}_j = -\sum_k K_{jk}\left[\xi_j(t) - \xi_k(t)\right] + g_j(t) - m\lambda\dot{\xi}_j. \tag{18.64}$$

From Eq. (18.61) it follows that the centers at $x_j = \xi_j(t)$ of the quantum wave packets follow the classical motion of the lattice.

The expression for the energy can be found in the same way as we obtained Eq. (18.52). The result shows that the total energy is the sum of quantum energy of the lattice, E, plus the energy of the classical motion

$$\langle E \rangle = E + \sum_j \left(\frac{m}{2} \right) \dot{\xi}_j^2 + \frac{1}{4} \sum_{jk} K_{jk} \left[\xi_j(t) - \xi_k(t) \right]^2$$
$$- \sum_j g_j(t) \xi_j(t) + m\lambda \sum_j \xi_j(t) \dot{\xi}_j(t). \tag{18.65}$$

18.3 A Spin System Coupled to a Tuned Circuit Coupled to a Transmission Line

We will start with a two-state system which can make radiative transition and is coupled to a high-frequency field. For the correct dynamical description of such a system one has to include damping effects due to the reaction of the system to its own radiation field. This damping becomes important for high-Q rf structures and also for systems with large dc susceptibilities [18].

Spin-Boson System — As a quantum-mechanical model of this problem we can consider the so called "spin-boson" Hamiltonian [20]-[21]

$$H = KS_z + \sum_j \left(\frac{p_j^2}{2m_j} + \frac{1}{2} m_j \omega_j^2 q_j^2 \right) + S_x \sum_j \epsilon_n q_n, \tag{18.66}$$

where K is the coupling constant and S_x and S_z are the components of the spin of the system. Various approximate methods have been used to determine the time evolution of the spin when it is coupled to the heat bath which in this case is assumed to be a large collection of harmonic oscillators [20]-[22].

A similar problem which is mathematically equivalent to this is the motion of a particle in a symmetric double-well potential which is coupled to a dissipative environment [23],[24]. When the two lowest energy levels are close to each other but far from from other levels, this system can be described by a spin-boson model more or less the same as (18.66)

As an alternative description one can study a model where the particle with spin is coupled to a damped oscillator [25]-[27]. The damped oscillator may describe a cavity mode, or lattice mode or possibly a spin wave [27].

For a simple solvable model we consider the problem of a mode of a resonant cavity coupled to the resistive losses arising from the walls of the cavity. This approach requires a detailed analysis of the shape and the material that the cavity is made of. A simpler model can be found if we describe the cavity by its field configurations, frequency and Q factor. Thus rather than solving Maxwell's equation inside the walls and matching the solutions at the boundaries, we consider a cavity with perfectly reflecting walls which is filled with a dielectric field of finite conductivity σ (similar to the problem of maser [26]). Choosing a gauge which makes the electric potential ϕ equal to zero, the wave equation for **A** becomes

$$\nabla^2 \mathbf{A} - \frac{4\pi\sigma\mu}{c^2}\frac{\partial \mathbf{A}}{\partial t} - \frac{\mu\epsilon}{c^2}\frac{\partial^2 \mathbf{A}}{\partial t^2} = 0. \tag{18.67}$$

This wave equation for **A** in a general case for a lossy medium can be derived from a Lagrangian density or a Hamiltonian density [28], but here we will deal with a simpler case.

Normal Modes of the Electromagnetic Wave Inside the Cavity — It is convenient to express **A** in rectangular coordinates as

$$\mathbf{A} = A_1(t)\left[\mathbf{i}\, k_2 \cos(k_1 x)\sin(k_2 y) - \mathbf{j}\, k_1 \sin(k_1 x)\cos(k_2 y)\right]\sin(k_3 z), \tag{18.68}$$

where **i** and **j** are the unit vectors in x and y directions and A_1 is a function of time only. Then the electric field **E** becomes

$$\mathbf{E} = -\frac{\mu}{c}\frac{dA_1}{dt}\left[\mathbf{i}\, k_2 \cos(k_1 x)\sin(k_2 y) - \mathbf{j}\, k_1 \sin(k_1 x)\cos(k_2 y)\right]\sin(k_3 x), \tag{18.69}$$

and that the tangential components of **E** vanish on the planes

$$x = 0,\ \frac{\pi}{k_1},\ \frac{2\pi}{k_1}, \cdots, \tag{18.70}$$

$$y = 0,\ \frac{\pi}{k_2},\ \frac{2\pi}{k_2}, \cdots, \tag{18.71}$$

and

$$z = 0,\ \frac{\pi}{k_3},\ \frac{2\pi}{k_3}, \cdots. \tag{18.72}$$

Thus we can consider a mode which can exist in a cavity bounded by a selection of these planes. From Eqs. (18.67) and (18.68) we find the differential equation for $A_1(t)$;

$$\left(k_1^2 + k_2^2 + k_3^2\right)A_1 + \frac{4\pi\sigma\mu}{c^2}\frac{dA_1}{dt} + \frac{\mu\epsilon}{c^2}\frac{d^2 A_1}{dt^2} = 0, \tag{18.73}$$

i.e. the time-dependent part of the vector potential for a given mode satisfies the equation for a damped oscillator. As we have seen earlier, this equation for a linearly damped oscillator can be found from the time-dependent Hamiltonian (3.25). To simplify (18.73) we introduce three constants L, R and C by

$$L = \alpha\left(\frac{\mu\epsilon}{c^2}\right), \quad R = \alpha\left(\frac{4\pi\sigma\mu}{c^2}\right), \quad \text{and} \quad \frac{1}{C} = \alpha\left(k_1^2 + k_2^2 + k_3^2\right), \tag{18.74}$$

where α is a constant with the dimension of (cm^{-1}). Setting $A_1(t) = Q(t)$, we write (18.73) as

$$L\frac{d^2 Q}{dt^2} + R\frac{dQ}{dt} + \frac{Q}{C} = 0, \tag{18.75}$$

i.e. the equation for the charge in an *LRC* circuit. While classically this equation is rigorously valid for Q, in quantum mechanics, there is always a time-dependent term on the right-hand side of (18.75) which arises from the fluctuations in the field. We will come back to this point later.

According to (3.25) the simplest Hamiltonian for (18.75) is of the form

$$H_e = \exp\left(-\frac{R}{L}t\right)\frac{P^2}{2\mathcal{A}L} + \exp\left(\frac{R}{L}t\right)\frac{\mathcal{A}Q^2}{2C}, \tag{18.76}$$

where P is the momentum conjugate to Q, and the constant $\mathcal{A}$ is determined by assuming that that when the damping is switched off, the total energy in the mode can be determined from the energy density of the electromagnetic field from the well-known expression [29]

$$\frac{1}{8\pi}\int\left(\epsilon\mathbf{E}^2 + \mu\mathbf{H}^2\right)dV. \tag{18.77}$$

This condition gives us the constant $\mathcal{A}$;

$$\mathcal{A} = \frac{\mu V\left(k_1^2 + k_2^2\right)}{32\pi\alpha}. \tag{18.78}$$

If the damping is switched on at time t_0, then we replace $\mathcal{A}$ by

$$\mathcal{A} = \frac{\mu V\left(k_1^2 + k_2^2\right)}{32\pi\alpha}\exp\left(-\frac{Rt_0}{L}\right). \tag{18.79}$$

The Spin System — Now let us suppose that the system that we are considering is a cavity which contains an assembly of particles with spin, and in addition, there is an external constant magnetic field $\mathbf{H}_0$ which is present inside the cavity. If these particles with spin are noninteracting and have identical g factors, then the contribution to the total energy of the particles is given by the Hamiltonian [30]

$$H_m = \frac{\mu ge}{2mc}\mathbf{H}_0 \cdot \mathbf{S}, \quad \mathbf{S} = \sum_j \mathbf{s}_j, \tag{18.80}$$

where m is the mass and e is the charge of each particle. There is also the additional energy arising from the interaction of these particles with the radiation field inside the cavity. This energy is given by

$$H_1 = \sigma_j\left(\frac{\mu ge}{2mc}\right)(\mathbf{s}_j \cdot \text{curl}\mathbf{A}). \tag{18.81}$$

Here again we note that the thermal noise has been omitted from the total Hamiltonian. Furthermore, while H_e, Eq. (18.76), might describe the Hamiltonian for the radiation in cavity alone (ignoring noise) and H_m is also correct for the spin system alone, however it is questionable whether these two Hamiltonians can be added together and also added to the interaction term H_1 [27]. Whereas the addition of two conservative Hamiltonians is legitimate, it is not clear as to what should be done when one part describes a nonconservative system and the other part a conservative system.

Returning to the total Hamiltonian, we can simplify the problem further if we assume that the spins are so close to one another that the vector potential remains constant over them. With this assumption H_1 can be written as

$$
\begin{aligned}
H_1 = \frac{\mu g e}{2mc} Q \{ & S_x k_1 k_3 \sin(k_1 x) \cos(k_2 y) \cos(k_3 z) \\
& + S_y k_2 k_3 \cos(k_1 x) \sin(k_2 y) \cos(k_3 z) \\
& - S_z (k_1^2 + k_2^2) \cos(k_1 x) \cos(k_2 y) \sin(k_3 z) \}.
\end{aligned}
\tag{18.82}
$$

Time-Dependent Wave Equation for the Particles with Spin — From the total Hamiltonian H

$$
H = H_e + H_m + H_1.
\tag{18.83}
$$

we have the Schrödinger equation

$$
i\hbar \frac{d}{dt} |t\rangle = H|t\rangle,
\tag{18.84}
$$

and thus knowing the state of the system at the initial time t_0, we can find the state at a later time t. To simplify the problem, we assume that inside the cavity the spins are placed in the region where the external constant electromagnetic field $\mathbf{H}_0$ is perpendicular to the oscillating field curl $\mathbf{A}$. Thus if $\mathcal{H}_0$ is along the z axis the spin system can be placed at a point where $\sin(k_2 y) = \sin(k_3 z) = 0$. When this is the case, H_1 reduces to

$$
H_1 = \frac{\mu e g Q}{2mc} S_x k_1 k_3 \sin(k_1 x) = \frac{\mu e g a Q}{2mc} S_x,
\tag{18.85}
$$

where $a = k_1 k_3 \sin(k_1 x)$. With these simplifications the total Hamiltonian takes the form

$$
\begin{aligned}
H = \hbar \omega_0 S_z & + \left[\exp\left(\frac{-Rt}{L} \right) \frac{P^2}{2AL} + \exp\left(\frac{Rt}{L} \right) \frac{AQ^2}{2C} \right] \\
& + \frac{\mu a g e}{2mc} Q S_x,
\end{aligned}
\tag{18.86}
$$

where we have introduced ω_0 by the relation [30]

$$
\hbar \omega_0 = \frac{\mu g e}{2mc} H_0.
\tag{18.87}
$$

Next we define $|T_1, t\rangle$ as

$$
|t\rangle = \exp\{ -i\omega_0 S_z (t - t_0) \} |T_1, t\rangle,
\tag{18.88}
$$

then the transformed ket satisfies the Scrödinger equation

$$
\frac{d}{dt} |T_1, t\rangle = H' |T_1, t\rangle,
\tag{18.89}
$$

with

$$H' = \exp[i\omega_0 S_z(t - t_0)]H \exp[-i\omega_0 S_z(t - t_0)] - \hbar\omega_0 S_z$$
$$= H_e + \frac{\mu age}{4mc}Q\left\{\exp[i\omega_0 S_z(t - t_0)](S_+ + S_-)\exp[-i\omega_0 S_z(t - t_0)]\right\},$$
$$(18.90)$$

where $S_\pm$ is defined by

$$S_\pm = S_x \pm iS_y. \tag{18.91}$$

Using the relations

$$S_\pm S_z = (S_z \mp 1)S_\pm, \tag{18.92}$$

and

$$S_\pm S_z^2 = (S_z \mp 1)^2 S_\pm, \tag{18.93}$$

we find the following equations:

$$S_\pm \exp[-i\omega_0 S_z(t - t_0)] = \exp[-i\omega_0(S_z \mp 1)(t - t_0)]S_\pm, \tag{18.94}$$

and

$$\left\{\exp[i\omega_0 S_z(t - t_0)]S_\pm \exp[-i\omega_0 S_z(t - t_0)]\right\} = \exp[\pm i\omega_0(t - t_0)]S_\pm. \tag{18.95}$$

Thus the Hamiltonian H' can be written as the sum of two parts

$$H' = \left[\exp\left(\frac{-Rt}{L}\right)\frac{P^2}{2AL} + \exp\left(\frac{Rt}{L}\right)\frac{AQ^2}{2C}\right]$$
$$+ \frac{\mu ageQ}{4mc}\left\{\exp[i\omega_0(t - t_0)]S_+ + \exp[-i\omega_0(t - t_0)]S_-\right\}, \tag{18.96}$$

where with this transformation we have eliminated the Zeeman energy, but have found a more complicated term for the interaction.

Next we try to solve the first part of the Hamiltonian H' which is related to the cavity mode. To this end we introduce two sets of creation and annihilation operators (Q_+, Q_-) and (q_+, q_-) in the following way:

$$2P = \left(2AL\hbar\omega e^{\frac{Rt_0}{L}}\right)^{\frac{1}{2}}(Q_+ + Q_-), \tag{18.97}$$

$$2iQ = \left(\frac{2C}{A}\hbar\omega e^{\frac{-Rt_0}{L}}\right)^{\frac{1}{2}}(Q_+ - Q_-), \tag{18.98}$$

$$Q_+ = i\cosh\theta\, q_+ + \sinh\theta\, q_-, \tag{18.99}$$

and

$$Q_- = \sinh\theta\, q_+ - i\cosh\theta\, q_-, \tag{18.100}$$

The constants ω and θ are related to L, C and R by

$$\omega = \frac{1}{\sqrt{LC}} \qquad \tanh(2\theta) = -\frac{R}{2\omega L}, \tag{18.101}$$

and thus

$$\cosh 2\theta = \frac{\omega}{\Omega}, \tag{18.102}$$

where

$$\Omega = \left(\frac{1}{LC} - \frac{R^2}{4L^2} \right)^{\frac{1}{2}}. \tag{18.103}$$

The operators Q_-, Q_+, q_- and q_+ satisfy the commutation relations

$$[Q_-, Q_+] = [q_-, q_+] = 1. \tag{18.104}$$

In order to eliminate the part H_e in the Hamiltonian, we define a ket $|T_2, t\rangle$ by

$$|T_1, t\rangle = e^{iS} \exp\left\{ -\frac{1}{2} i\Omega(q_+q_- + q_-q_+)(t - t_0) \right\} |T_2, t\rangle. \tag{18.105}$$

In this relation the operator S is given by

$$
\begin{aligned}
S &= \frac{R(t - t_0)}{4\hbar L} (PQ + QP) \\
&= -\frac{R(t - t_0)}{4\hbar L} \left[\frac{R}{2\Omega L} (q_+q_- + q_-q_+) - \frac{i\omega}{\Omega} \left(q_+^2 - q_-^2 \right) \right].
\end{aligned}
\tag{18.106}
$$

By substituting $|T_1, t\rangle$ from (18.105) in Eq. (18.89) we find that $|T_2, t\rangle$ satisfies another Schrödinger-type equation

$$i\hbar \frac{d}{dt} |T_2, t\rangle = H'' |T_2, t\rangle, \tag{18.107}$$

where now H'' is the following Hamiltonian

$$
\begin{aligned}
H'' &= \frac{\mu a g e}{2mc} \left(\exp[i\omega_0(t - t_0)]S_+ + \exp[-i\omega_0(t - t_0)]S_- \right) \left(\frac{C}{2A\hbar\omega} \right)^{\frac{1}{2}} \\
&\quad \times \exp\left(-\frac{Rt}{2L} \right) \{ q_+(\cosh\theta + i\sinh\theta) \exp[i\Omega(t - t_0)] \\
&\quad + q_-(\cosh\theta - i\sinh\theta) \exp[-i\Omega(t - t_0)] \}.
\end{aligned}
\tag{18.108}
$$

It is convenient to introduce the new variable x by

$$x = \exp\left[-\frac{R(t - t_0)}{2L} \right], \tag{18.109}$$

and new creation and annihilation operators by

$$\bar{q}_\pm = \exp(i\phi)q_\pm, \tag{18.110}$$

where

$$e^{-i\phi}(\cosh\theta + i\sinh\theta) = \sqrt{\cosh 2\theta}. \tag{18.111}$$

In addition let

$$Q_0 = \frac{\omega L}{R}, \tag{18.112}$$

denote the Q factor of the cavity mode. Then using these relations we find that the ket $|T_2, t\rangle$ satisfies the differential equation

$$i\hbar \frac{d}{dt}|T_2, t\rangle = K \left[\left(x^{\frac{-2iQ_0\omega_0}{\omega}} S_+ + x^{\frac{2iQ_0\omega_0}{\omega}} S_- \right) \right.$$
$$\left. \times \left(x^{\frac{-2iQ_0\Omega}{\omega}} \bar{q}_+ + x^{\frac{2iQ_0\Omega}{\omega}} \bar{q}_- \right) \right] |T_2, t\rangle, \tag{18.113}$$

where

$$K = -\frac{2age}{mc} Q_0 \left[\frac{\pi\mu\hbar}{(k_1^2 + k_2^2 + k_3^2)(k_1^2 + k_2^2) V\Omega} \right]^{\frac{1}{2}}. \tag{18.114}$$

By integrating the differential equation (18.113) we can determine the time dependence of $|T_2, t\rangle$ and hence the ket $|t\rangle$.

To simplify Eq. (18.113) further, we assume that the external d.c. magnetic field $\mathbf{H}_0$ is adjusted in such a way that $\omega_0 = \Omega$, i.e. the spins and cavity are resonant to the same frequency (ω_0 and Ω are defined by (18.87) and (18.103) respectively). When this is the case, then terms in $(x^{\frac{\pm 2iQ_0\Omega}{\omega}})$ can be neglected and (18.113) reduces to [26]

$$i\hbar \frac{d}{dx}|T_2, t\rangle = K \left(S_+ \bar{q}_- + S_- \bar{q}_+ \right) |T_2, t\rangle. \tag{18.115}$$

In Eq. (18.113) after carrying out the product of the two parentheses we find two terms $S_+ \bar{q}_-$ and $S_- \bar{q}_+$, or the terms independent of x, plus two other terms $x^{\frac{\pm 2iQ_0\Omega}{\omega}}$ which are dependent on x. We ignore the contribution of the last two terms. Here we have also used Eq. (18.109) to replace the derivative with respect to time by the derivative with respect to x. By integrating this equation we find

$$|T_2, t\rangle = \exp \left\{ K \left(S_+ \bar{q}_- + S_- \bar{q}_+ \right) (x - 1) \right\} |t_0\rangle, \tag{18.116}$$

where $|t_0\rangle$ is the initial state of the system. Finally the ket $|t\rangle$ describing the time evolution of the wave function found from $|T_1, t\rangle$ and $|T_2, t\rangle$ is given by

$$|t\rangle = \exp[-i\omega_0 S_z(t - t_0)] \exp(iS) \exp \left[-\frac{i}{2}\Omega(q_+q_- + q_-q_+)(t - t_0) \right]$$
$$\times \exp \left[-\frac{iK}{\hbar} \left(S_+ \bar{q}_- + S_- \bar{q}_+ \right) (x - 1) \right] |t_0\rangle. \tag{18.117}$$

Time-Dependence of the Expectation Value of $\langle t|S_z|t\rangle$ — Even with a number of assumptions that we have already made in order to simplify the expression for the time evolution of S_z, still the solution for the general case is not known. Therefore let us consider just a single particle which at the time t_0 is in the quantum state $M = \frac{1}{2}$ (M is the eigenvalue of the S_z operator).

Therefore the state of the system can be specified by two quantum numbers $\pm$ for the spin and n for the electromagnetic field inside the cavity, that is the the eigenvalue of the operator $\bar{q}_+\bar{q}_- + \bar{q}_-\bar{q}_+$. Thus the ket $|t_0\rangle$ is specified by two quantum numbers

$$|t_0\rangle = |n, +\rangle, \tag{18.118}$$

where the plus sign denotes the state $M = \frac{1}{2}$. Next let X represent the operator

$$X = -\frac{K}{\hbar}\left(S_+\bar{q}_- + S_-\bar{q}_+\right)(x - 1), \tag{18.119}$$

then from Eqs. (18.117) and (18.119) it follows that the expectation value of S_z at time t is given by

$$\langle t|S_z|t\rangle = \langle t_0|e^{-iX}S_z e^{iX}|t_0\rangle$$
$$= \langle n, +|e^{-iX}S_z e^{iX}|n, +\rangle. \tag{18.120}$$

In calculating the expectation value of $\langle t|S_z|t\rangle$ we observe that only the last factor in (18.117) does not commute with S_z and since both S_z and X commute with

$$\mathbf{S}^2 = S_x^2 + S_y^2 + S_z^2, \tag{18.121}$$

therefore the total spin remains a good quantum number. Now let us find the eigenvalue of the operator X^2

$$X^2|n, +\rangle = \frac{K^2}{\hbar^2}(x - 1)^2\left[\bar{q}_-\bar{q}_+ S_+ S_-\right]|n, +\rangle$$
$$= \frac{K^2}{\hbar^2}(x - 1)^2(n + 1)|n, +\rangle = b^2|n, +\rangle, \tag{18.122}$$

where

$$b^2 = \frac{K^2}{\hbar^2}(x - 1)^2(n + 1). \tag{18.123}$$

Thus we have

$$e^{iX}|n, +\rangle = \frac{1}{2}\left\{(e^{iX} + e^{-iX}) + (e^{iX} - e^{-iX})|n, +\rangle\right.$$
$$= \left(\cos b + i\frac{\sin b}{b}X\right)\bigg\}|n, +\rangle$$
$$= \cos b\,|n, +\rangle - i\sin b\,|n + 1, -\rangle$$
$$= \cos\left[\frac{K}{\hbar}(x - 1)\sqrt{n + 1}\right]|n, +\rangle - i\sin\left[\frac{K}{\hbar}(x - 1)\sqrt{n + 1}\right]|n + 1, -\rangle. \tag{18.124}$$

Similarly for $e^{iX}|n, -\rangle$ we find

$$e^{iX}|n, -\rangle = \cos\left[\frac{K}{\hbar}(x - 1)\sqrt{n}\right]|n, -\rangle - i\sin\left[\frac{K}{\hbar}(x - 1)\sqrt{n}\right]|n - 1, +\rangle. \tag{18.125}$$

Using these expressions, we find the time dependence of the expectation values of S_z;

$$\langle n, + \left| e^{-iX} S_z e^{iX} \right| n, + \rangle = \frac{1}{2} \cos\left[\frac{2K}{\hbar}(x-1)\sqrt{n+1} \right], \tag{18.126}$$

and

$$\langle n, - \left| e^{-iX} S_z e^{iX} \right| n, - \rangle = -\frac{1}{2} \cos\left[\frac{2K}{\hbar}(x-1)\sqrt{n} \right]. \tag{18.127}$$

From the definition of Eq. (18.109) it is clear that as t increases from t_0 to infinity x goes from 1 to 0. Thus depending on the magnitude of K and n, $\langle S_z \rangle$ changes sinusoidally with time. If initially the spin is $M = \frac{1}{2}$ and there is energy in the cavity, then this energy can be gradually dissipated in the wall losses and therefore there is a possibility that the spin will be simulated to emit. In this way the spin, by giving its energy to the cavity, makes a transition from the initial spin of the particle, $M = \frac{1}{2}$, to the final, $M = -\frac{1}{2}$ state. Now after the spin has made the transition to $M = -\frac{1}{2}$ state, if there is enough energy in the cavity, the spin may reabsorb some energy and return to $M = \frac{1}{2}$. This process may be repeated many times provided that adequate energy is available. We also note that because the cavity energy is dissipated (by wall losses) therefore the probability of the spin making the transition is reduced, and if the Q factor is small, the dissipation will be large and the spin has little chance of making the transition. It is interesting to note that as $t \to \infty$ the spin will not assume its lowest state. The reason for this is that in our formulation of the problem we have not added the fluctuating field (or noise) inside the cavity to the equation of motion for the damped oscillator (18.75). If we include the noise to the field, then $\langle S_z \rangle$ will tend to a constant value, and this value is determined by the temperature of the cavity as $t \to \infty$. An approximate solution for the case when S is large can be found in a paper by Stevens and Josephson [26].

Eigenvalues of H_e Operator — In our discussion of the dissipative electromagnetic field inside the cavity we described the cavity mode by the quantum number n. However for the determination of the constant $\mathcal{A}$ from the electromagnetic field inside the cavity we assumed that at $t = 0$ the damping is switched on (i.e. $R = 0$ for $t < 0$), and this is done without any change in the energy. In other words at $t = 0$ the expectation value of H_e is

$$\langle n|H_e|n \rangle = \left\langle n \left| \exp\left(-\frac{Rt_0}{L} \right) \frac{P^2}{2\mathcal{A}L} + \exp\left(\frac{Rt_0}{L} \right) \frac{\mathcal{A}Q^2}{2C} \right| n \right\rangle. \tag{18.128}$$

Using Eqs. (18.85),(18.86) we can express the diagonal matrix elements of H_e, Eq. (18.76), in terms of the corresponding matrix elements of Q_+ and Q_-

operators as

$$\frac{1}{2}\hbar\omega\langle n|Q_+Q_- + Q_-Q_+|n\rangle = \frac{1}{2}\hbar\omega\langle n\,|(q_+q_- + q_-q_+)\cosh 2\theta$$
$$+ i\left(q_+^2 - q_-^2\right)\sinh 2\theta|\,n\rangle$$
$$= \frac{1}{2}(2n + 1)\hbar\omega\cosh 2\theta$$
$$= \frac{(2n + 1)(\hbar\omega)^2}{2\hbar\Omega}. \tag{18.129}$$

This follows from the fact that n is an eigenstate of $(q_+q_- + q_-q_+)$, but is not an eigenstate of energy at t_0. Denoting the eigenstate of the energy, i.e. the eigenstate of the operator $Q_+Q_- + Q_-Q_+$ by $|N\rangle$, we find that expressing the general case of $|N\rangle$ in terms of $|n\rangle$ is difficult. Therefore we will consider the simple case where

$$(Q_+Q_- + Q_-Q_+)|N\rangle = (2N + 1)|N\rangle, \quad \text{with } N = 0. \tag{18.130}$$

and expand $|N\rangle$ in terms of $|n\rangle$;

$$|N = 0\rangle = \sum_{n=0}^{\infty} |n\rangle\langle n|N = 0\rangle. \tag{18.131}$$

Since Q is the annihilation operator we have

$$Q_-|N = 0\rangle = 0 = (\sinh\theta\, q_+ - i\cosh\theta\, q_-)\sum_{n=0}^{\infty} |n\rangle\langle n|N = 0\rangle. \tag{18.132}$$

Thus we find the following relation

$$\sum_{n=0}^{\infty} |n\rangle\left\{\langle n - 1|N = 0\rangle\sinh\theta\,\sqrt{n} - \langle n + 1|N = 0\rangle i\cosh\theta\,\sqrt{n+1}\right\} = 0.$$
$$\tag{18.133}$$

or

$$\langle n + 1|N = 0\rangle = -i\tanh\theta\frac{\sqrt{n}}{\sqrt{n+1}}\langle n - 1|N = 0\rangle. \tag{18.134}$$

From this result it follows that $|N = 0\rangle$ state of energy can be written as

$$|N = 0\rangle = \frac{1}{\sqrt{\cosh\theta}}\exp\left[-\frac{i}{2}\tanh\theta\, q_+^2\right]|n = 0\rangle. \tag{18.135}$$

(See a similar derivation given in Sec. 11.5).
More About the Coupling to a Spin System — Now let us consider a different model where the tuned circuit is coupled to a line and there is a spin system which is located in the coil of inductance. Added to this arrangement is a magnetic field. The current in the circuit is $\frac{dP}{dt}$ which according to Eq. (7.117)

is equal to $\left(-\frac{Q-Q_1}{L}\right)$. We also assume that the field set up at the position of spin is orthogonal to $\mathbf{H}_0$. Then as before we write the total Hamiltonian as the sum of three terms, Eq. (18.86)

$$H = H_e + \frac{\mu g e}{2mc} H_0 S_z + k(Q - Q_1) S_x. \qquad (18.136)$$

Let us note that this Hamiltonian is not complete since there are magnetic fields generated by the currents passing through the line and the spin may be coupled to to these. We will ignore this additional contribution to the Hamiltonian. Now if we are interested in the coupling to one of the modes, and if, at $t = 0$, the energy of this mode is well above the equilibrium noise energy, then it is reasonable to assume a coupling between the spin and the "smoothed current", i.e. to $\underline{C}\frac{d^2\tilde{Q}}{dt^2}$. Then from the Eq. (7.137) it follows that $\tilde{Q}$ satisfies the differential equation

$$\underline{L}\frac{d^2\tilde{Q}}{dt^2} + R\frac{d\tilde{Q}}{dt} + \frac{1}{\underline{C}}\tilde{Q} = 0, \qquad (18.137)$$

or

$$\underline{C}\frac{d^2\tilde{Q}}{dt^2} = -\frac{1}{\underline{L}}\left(\tilde{Q} + Re^{-2\gamma t}\tilde{P}\right), \qquad (18.138)$$

where $\gamma = \frac{R}{L}$. Therefore the Hamiltonian (18.136) can be written as

$$H = \hbar\omega_0 S_z + \left(e^{-2\gamma t}\frac{\tilde{P}^2}{2\underline{C}} + e^{2\gamma t}\frac{\tilde{Q}^2}{2\underline{L}}\right) + kS_x\left(\tilde{Q} + Re^{-2\gamma t}\tilde{P}\right). \qquad (18.139)$$

The problem of coupling of a quantum mechanical motion, in this case the spin of a particle, to a classical damped system, which here is the electromagnetic field, can be extended to other systems, and this is the topic that we will consider in the next section.

18.4 Coupling of a Quantum System to a Dissipative Classical Motion

In the last section we studied one particular example where an exact quantum mechanical formulation of the problem was not feasible, let alone its solution. Now, as in the case of time evolution of a spin in a cavity, we assume the principle of additivity of the classical and quantum Hamiltonians, i.e.

$$\hat{H} = \hat{H}_q + H_{cl} + \hat{H}_{int}, \qquad (18.140)$$

where hat denotes a quantum mechanical operator, $\hat{H}_q$ and H_{cl} stand for quantal and classical Hamiltonians, and $\hat{H}_{int}$ represents the coupling between the two. In this case we have q number of operators as well as a number of classical variables. These operators must close under Lie algebra with respect to the

Hamiltonian $\hat{H}$ of the system, i.e. we assume that there exists a set of relations of the type

$$\left[\hat{H}(t),\ \hat{O}_i(t)\right] = i\hbar \sum_{j=1}^{q} g_{ji}\hat{O}_j(t), \quad i = 1,\ 2 \cdots q, \qquad (18.141)$$

where g_{ji} are the elements of a $q \times q$ matrix G.

From the generalized Ehrenfest theorem we find that [31]

$$\frac{d\langle\hat{Q}_i(t)\rangle}{dt} = \frac{i}{\hbar}\left\langle\left[\hat{H},\ \hat{O}_i(t)\right]\right\rangle, \quad i = 1,\ 2 \cdots q. \qquad (18.142)$$

By substituting (18.141) in (18.142) we find a linear set of first order differential equations for the time evolution of the expectation values

$$\frac{d\langle\hat{Q}_i(t)\rangle}{dt} = -\sum_{j=1}^{q} g_{ji}(t)\langle\hat{O}_j(t)\rangle, \quad i = 1,\ 2 \cdots q. \qquad (18.143)$$

For the previous problem of the time evolution of spin in a cavity in the presence of damped electromagnetic field, we chose a damped Hamiltonian of Caldirola-Kanai type and coupled it to the equation of motion. Now we consider a different approach, where the total Hamiltonian, (classical plus quantum mechanical) remains constant in time, but the damping is introduced via Rayleigh's dissipative function, for the classical part of the system by modifying the canonical equations of motion. Thus from the Lagrange equation (2.1) we get

$$\frac{dx_j}{dt} = \frac{\partial}{\partial p_j}\langle\hat{H}\rangle, \qquad (18.144)$$

and

$$\frac{dp_j}{dt} = -\left(\frac{\partial}{\partial x_j}\langle\hat{H}\rangle + \lambda p_j\right). \qquad (18.145)$$

where x_j s are the coordinates and p_j s are their corresponding momenta.

The Case of a Quantum Harmonic Oscillator Coupled to a Classically Damped Harmonic Oscillator — In order to illustrate the present method we will consider the simple case of a quantum mechanical oscillator of mass m which is coupled to a classically damped oscillator of mass M for which the equations of motion are given by (18.144) and (18.145) [31]. The conserved Hamiltonian in this case can be written as

$$\hat{H} = \frac{1}{2}\left(\frac{\hat{p}_q^2}{m} + m\omega_0^2\hat{x}_q^2 + \frac{p_{cl}^2}{M} + M\omega^2 x_{cl}^2\right) + \gamma x_{cl}\hat{x}_q, \qquad (18.146)$$

where $\hat{p}_q^2$, $\hat{x}_q^2$ and $\hat{x}_q$ are quantum operators and p_{cl} and x_{cl} are classical variables. Later on we will see that the condition $\frac{\gamma^2}{mM} \leq \omega_0^2\omega^2$ guarantees a result which is free of divergent. Again, here we have coupled the classical displacement to the quantum position operator, but this type of coupling may be questionable. Setting $\hbar = 1$, for the sake of convenience, we introduce the following dimensionless quantum operators and classical functions:

$$\hat{x} = \sqrt{m\omega_0}\,\hat{x}_q, \tag{18.147}$$

$$\hat{p} = \frac{\hat{p}_q}{\sqrt{m\omega_0}}, \tag{18.148}$$

$$s = \sqrt{M\omega}\,x_{cl}, \tag{18.149}$$

and

$$p_s = \frac{p_{cl}}{\sqrt{M\omega}}. \tag{18.150}$$

In this simple model, the quantum mechanical operator $O_i(t)$ in Eq. (18.142) can be one of the following operators: $\hat{x}$, $\hat{p}$, $\hat{x}^2$, $\hat{p}^2$ and $\hat{S} = \hat{x}\hat{p} + \hat{p}\hat{x}$. the last one arising from differentiating $\hat{x}^2$ or $\hat{p}^2$ with respect to time and symmetrizing the result. Thus from Eq. (18.143) we find the following relations for the expectation values of the quantum operators:

$$\frac{d}{d\tau}\langle\hat{x}\rangle = \langle\hat{p}\rangle, \tag{18.151}$$

$$\frac{d}{d\tau}\langle\hat{p}\rangle = -\left(\langle\hat{x}\rangle + \chi s\right), \tag{18.152}$$

$$\frac{d}{d\tau}\langle\hat{x}^2\rangle = \langle\hat{S}\rangle, \tag{18.153}$$

$$\frac{d}{d\tau}\langle\hat{p}^2\rangle = -\left(\langle\hat{S}\rangle + 2\chi s\,\langle\hat{p}\rangle\right), \tag{18.154}$$

and

$$\frac{d}{d\tau}\langle\hat{S}\rangle = 2\left(\langle\hat{p}^2\rangle - \langle\hat{x}^2\rangle - \chi s\langle\hat{x}\rangle\right), \tag{18.155}$$

where we have introduced a dimensionless time variable, $\tau = \omega_0 t$, and a dimensionless constant

$$\chi = \frac{\gamma}{\sqrt{mM\omega\omega_0^3}}, \tag{18.156}$$

For the classical motion we have

$$\frac{ds}{d\tau} = \Omega p_s, \tag{18.157}$$

and

$$\frac{dp_s}{d\tau} = -\left(\chi\langle\hat{x}\rangle + \Omega s + \frac{\lambda}{\omega_0}p_s\right), \tag{18.158}$$

with $\Omega = \frac{\omega}{\omega_0}$. Now we observe that Eqs. (18.151), (18.152), (18.157) and (18.158) form an autonomous set of differential equations by themselves, and that these equations are linear and homogeneous. Assuming a trial solution of the form $A \exp(r\tau)$ for these, with r a constant, we find that r must be the root of the fourth degree equation

$$r^4 + \frac{\lambda}{\omega_0} r^3 + \left(\Omega^2 + 1 \right) r^2 + \frac{\lambda}{\omega_0} r + \Omega \left(\Omega - \chi^2 \right) = 0. \tag{18.159}$$

Denoting the roots of (18.159) by r_1, r_2, r_3 and r_4 we can write $\langle \hat{x} \rangle$, $\langle \hat{p} \rangle$, s and p_s as linear functions of $\exp(r_k \tau)$, i.e.

$$\langle \hat{x}(\tau) \rangle = \sum_{k=1}^{4} A_k \exp(r_k \tau), \tag{18.160}$$

$$\langle \hat{p}(\tau) \rangle = \sum_{k=1}^{4} r_k A_k \exp(r_k \tau), \tag{18.161}$$

$$s(\tau) = -\frac{1}{\chi} \left\{ \sum_{k=1}^{4} \left(r_k^2 + 1 \right) A_k \exp(r_k \tau) \right\}, \tag{18.162}$$

and

$$p_s(\tau) = -\frac{1}{\Omega\chi} \left\{ \sum_{k=1}^{4} \left(r_k^2 + 1 \right) r_k A_k \exp(r_k \tau) \right\}. \tag{18.163}$$

In these relations the amplitude A_k is given by

$$A_k = B_k \left\{ \langle \hat{x} \rangle(0) r_k \left[r_k \left(\frac{\lambda}{\omega_0} + r_k \right) - \Omega^2 \right] + s(0)\chi \left(\frac{\lambda}{\omega_0} + r_k \right) \right.$$
$$\left. - \langle \hat{p}(0) \rangle \left[r_k \left(\frac{\lambda}{\omega_0} + r_k \right) + \Omega^2 \right] + p_s(0)\chi\Omega \right\} \tag{18.164}$$

and B_k is the following cubic polynomial in r_k:

$$B_k = - \left[4r_k^3 + 3r_k^2 + 2 \left(\Omega^2 + 1 \right) r_k + \frac{\lambda}{\omega_0} \right]^{-1}. \tag{18.165}$$

Depending on the frequencies, the damping constant and the coupling between the two parts, we can have the following possibilities for the roots:
(1) Two pairs of complex conjugate roots.
(2) Two real roots together with a pair of complex conjugate ones.
(3) Four real roots.
All the roots are pure imaginary only if $\chi = 0$, i.e. when the two parts are uncoupled. For the physically acceptable solution, that is for negative real roots and and negative real parts of the complex roots the condition $\Omega \geq \chi^2$ or

$$\omega^2 \geq \frac{\gamma^2}{mM\omega_0^2}, \tag{18.166}$$

must be satisfied, and this is essentially the same condition, i.e. Eq. (9.12), that we found for a classical oscillator coupled to a heat bath. The time evolution of

$\langle \hat{x} \rangle$, $\langle \hat{p} \rangle$ and $\langle \hat{H}_q \rangle$ for small and moderate damping and satisfying the condition (18.166) are given in ref. [31].

An important result of this approach is that the uncertainty principle is satisfied for all times and is independent of the value of the damping constant λ. The uncertainty can be determined from the solution of Eqs. (18.151)–(18.155)

$$(\Delta x)^2(\tau)(\Delta p)^2(\tau) = (\Delta x)^2(0)(\Delta p)^2(0) - \left[\frac{1}{2} \langle \hat{S} \rangle(0) - \langle \hat{x} \rangle(0) \langle \hat{p} \rangle(0) \right]^2$$

$$+ \left\{ \frac{1}{2} \left[-(\Delta x)^2(0) + (\Delta p)^2(0) \right] \sin(2\tau) \right.$$

$$\left. + \left(\frac{1}{2} \langle \hat{S} \rangle(0) - \langle \hat{x} \rangle(0) \langle \hat{p} \rangle(0) \right) \cos(2\tau) \right\}^2 , \qquad (18.167)$$

where
$$(\Delta x)^2 = \langle \hat{x}^2 \rangle - \langle \hat{x} \rangle^2, \qquad (18.168)$$

and
$$(\Delta p)^2 = \langle \hat{p}^2 \rangle - \langle \hat{p} \rangle^2. \qquad (18.169)$$

Now if the Heisenberg uncertainty principle is satisfied at $\tau = 0$, then as (18.167) shows it will remain valid for all times;

$$(\Delta x)^2(\tau)(\Delta p)^2(\tau) \geq (\Delta x)^2(0)(\Delta p)^2(0) \geq \frac{1}{4}. \qquad (18.170)$$

We can also calculate $\langle H_q \rangle(\tau)$ from the expectation values of $\langle \hat{x}^2 \rangle$, $\langle \hat{p}^2 \rangle$, $\langle \hat{x} \rangle^2$ and $\langle \hat{p} \rangle^2$;

$$\langle \hat{H}_q \rangle = \frac{1}{2} \omega_0 \langle \hat{x}^2(\tau) + \hat{p}^2(\tau) \rangle$$

$$= \frac{1}{2} \omega_0 \left\{ (\Delta x)^2(0) + (\Delta p)^2(0) \right\}$$

$$+ \frac{\omega_0}{2} \sum_{k=1}^{4} \sum_{j=1}^{4} \left\{ A_k A_j (r_k r_j + 1) \exp\left[(r_k + r_j)\tau \right] \right\}. \qquad (18.171)$$

It is interesting to note that when the condition (18.166) for the existence of physically acceptable solution is satisfied then as $\tau \to \infty$, we find that the classical and quantal solutions decouple, and that the asymptotic value of $\langle \hat{H}_q \rangle$ becomes

$$\langle \hat{H}_q \rangle(\infty) = \frac{1}{2} \omega_0 \left[(\Delta x)^2(0) + (\Delta p)^2(0) \right]. \qquad (18.172)$$

In this model of dissipative system the amount of energy dissipated in the quantum oscillator is

$$\Delta E_q = \langle \hat{H}_q \rangle(\infty) - \langle \hat{H}_q \rangle(0)$$

$$. = -\frac{1}{2} \omega_0 \left\{ [\langle \hat{x} \rangle(0)]^2 + [\langle \hat{p} \rangle(0)]^2 \right\} \leq 0 \qquad (18.173)$$

The detailed calculation of the time evolution of $\langle \hat{x} \rangle(\tau)$, $\langle \hat{p} \rangle(\tau)$ and $\langle \hat{H}_q \rangle(\tau)$ for small and moderate damping have been given by Kowalski *et al.* [31].

A Model for the Damped Two-Level System — Earlier, in this chapter, we discussed the problem of a particle with spin interacting with a damped electromagnetic field inside a cavity. Now we want to study a different version of the same problem where a two-level system is coupled to a classically damped oscillator [31]. The Hamiltonian for the system written in terms of the dimensionless quantities introduced earlier for the quantum oscillator coupled to a classical oscillator is of the form

$$\hat{H} = E_1 \hat{a}_1^\dagger \hat{a}_1 + E_2 \hat{a}_2^\dagger \hat{a}_2 + \frac{1}{2}\omega \left(p_s^2 + s^2 \right)$$
$$+ \gamma s \left(\epsilon \hat{a}_1^\dagger \hat{a}_2 + \epsilon^* \hat{a}_2^\dagger \hat{a}_1 \right), \tag{18.174}$$

where γ and ϵ are dimensionless constants, and we assume that $E_2 > E_1$. The operators $\hat{a}_i^\dagger$ and $\hat{a}_i$, $(i=1, 2)$, are the standard creation and annihilation operators (they can represent bosonic or fermionic operators). Now we introduce four operators $\hat{O}_1$, $\hat{O}_2$, $\hat{O}_3$ and $\hat{O}_4$ by

$$\hat{O}_1 = \hat{a}_1^\dagger \hat{a}_1, \tag{18.175}$$

$$\hat{O}_2 = \hat{a}_2^\dagger \hat{a}_2, \tag{18.176}$$

$$\hat{O}_3 = i \left(\epsilon \hat{a}_1^\dagger \hat{a}_2 - \epsilon^* \hat{a}_2^\dagger \hat{a}_1 \right), \tag{18.177}$$

and

$$\hat{O}_4 = \left(\epsilon \hat{a}_1^\dagger \hat{a}_2 + \epsilon^* \hat{a}_2^\dagger \hat{a}_1 \right). \tag{18.178}$$

By applying the generalized Ehrenfest theorem (18.142)) we find the following set of coupled equations:

$$\frac{d}{dt}\langle \hat{O}_1 \rangle = -\gamma s \langle \hat{O}_3 \rangle, \tag{18.179}$$

$$\frac{d}{dt}\langle \hat{O}_2 \rangle = \gamma s \langle \hat{O}_3 \rangle, \tag{18.180}$$

$$\frac{d}{dt}\langle \hat{O}_3 \rangle = -2\gamma s |\epsilon|^2 \left(\langle \hat{O}_2 \rangle - \langle \hat{O}_1 \rangle \right) + \omega_0 \langle \hat{O}_4 \rangle, \tag{18.181}$$

and

$$\frac{d}{dt}\langle \hat{O}_4 \rangle = -\omega_0 \langle \hat{O}_3 \rangle, \tag{18.182}$$

where $\omega_0 = E_2 - E_1$. These equations are coupled to the classical equations of motion

$$\frac{ds}{dt} = \omega p_s, \tag{18.183}$$

and

$$\frac{dp_s}{dt} = -\left(\omega s + \gamma \langle \hat{O}_4 \rangle + \lambda p_s \right) \tag{18.184}$$

We can reduce the number of equations by defining the difference between the number operators $\hat{O}_1$ and $\hat{O}_2$

$$\Delta\hat{N} = \hat{O}_2 - \hat{O}_1. \tag{18.185}$$

Denoting the mean value of $\Delta\hat{N}$ by ΔN, changing t to τ, where $\tau = \omega_0 t$, and introducing

$$\alpha = \frac{2\gamma}{\omega_0}, \quad \text{and} \quad \Omega = \frac{\omega}{\omega_0}, \tag{18.186}$$

we can write Eqs. (18.175)-(18.178) as

$$\frac{d\Delta N}{d\tau} = \alpha s\langle \hat{O}_3 \rangle, \tag{18.187}$$

$$\frac{d}{d\tau}\langle \hat{O}_3 \rangle = -\alpha s\Delta N + \langle \hat{O}_4 \rangle, \tag{18.188}$$

and

$$\frac{d}{d\tau}\langle \hat{O}_4 \rangle = -\langle \hat{O}_3 \rangle. \tag{18.189}$$

To these we add Eqs. (18.183) and (18.184), but now written in terms of τ

$$\frac{ds}{d\tau} = \Omega p_s, \tag{18.190}$$

$$\frac{dp_s}{d\tau} = -\left(\Omega s + \frac{1}{2}\alpha\langle \hat{O}_4 \rangle + \frac{\lambda}{\omega_0}p_s\right). \tag{18.191}$$

Details of solutions of these five coupled equations have been discussed by Kowalski *et al.* From the numerical solutions for a set of parameters these authors have reached the following conclusions:

(1) The fixed points in the solution of the nonlinear algebraic equation for r in the trial solution $\exp(r\tau)$, similar to Eq. (18.159) are independent of the dissipation coefficient λ, but are dependent on the initial conditions and also on the parameters α and Ω, Eq. (18.186), in the differential equations.

(2) The equilibrium values of $(p_s)_f$ and $\langle \hat{O}_3 \rangle_f$ are zero.

(3) An important result that follows from the numerical calculation is that the flux of particles is directed from the excited state towards the ground state [31].

18.5 The Wigner-Weisskopf Model

One of the earliest attempts to understand of the mechanism for the decay of a quantum unstable state was advanced by Wigner and Weisskopf [32]. This model describes an unstable system which consists of a central particle with wave function $\psi(\mathbf{r}, t)$, and this particle interacts with a group of motionless objects with the amplitude $\chi_n(t), n = 1, 2, \cdots$ at the origin [32]-[36]. In the following formulation of this problem we use the units where $\hbar = 1$ and $m = \frac{1}{2}$. Then the Schrödinger equations for the coupled systems take the simple form of

$$i\frac{\partial \psi(\mathbf{r}, t)}{\partial t} = -\nabla^2 \psi(\mathbf{r}, t) + \sum_n g_n \delta(\mathbf{r}) \chi_n(t) \tag{18.192}$$

and

$$i\frac{d\chi_n}{dt} = \mathcal{E}_n \chi_n + g_n \psi(0, t), \quad n = 1, 2 \cdots, \tag{18.193}$$

where $\psi(0, t)$ is defined as the finite part of $\psi(\mathbf{r}, t)$ at the origin [34]. Since the interaction takes place at $r = 0$, therefore only S wave will be affected by the coupling. In this model one can also introduce a form factor $\Lambda(\mathbf{r})$ and write the interaction term as $\sum_n g_n \Lambda(\mathbf{r}) \chi_n(t)$ in Eq. (18.192) and $g_n \int \Lambda(\mathbf{r}) \psi(\mathbf{r}, t) d^3 r$ in (18.193) [35].

We can use the time Fourier transform and write the solution for the wave number k as

$$\begin{cases} \psi_k(r) = \frac{1}{r} \sin(kr + \eta(k)) \\ \chi_{nk}, \quad n = 1, 2, \cdots \end{cases} \tag{18.194}$$

Substituting (18.194) in (18.192) and equating the coefficients of $\delta(\mathbf{r})$ on the two sides yields the condition

$$4\pi \sin \eta(k) + \sum_n g_n \chi_{nk} = 0. \tag{18.195}$$

The same substitution in (18.193) gives us

$$\left(k^2 - \mathcal{E}_n\right) \chi_{nk} = g_n k \cos \eta(k), \tag{18.196}$$

where the finite part of $\psi(0, r)$ from (18.194) is $k \cos \eta(k)$. Denoting the eigenvalues by $\mathcal{E} = \mathcal{E}_n$, we note that if all $\mathcal{E}_n$ s are positive then we have no bound state, however if some $\mathcal{E}_n$ s are negative then we have bound states. Here we assume that all $\mathcal{E}_n$ s are positive. The completeness relation for this system is

$$\langle k'|k \rangle = \int \psi^* (\mathbf{r}, k') \psi (\mathbf{r}, k) d^3 r + \sum_n \chi_{nk'} \chi_{nk} = 2\pi^2 \delta (k - k'). \tag{18.197}$$

By eliminating χ_{nk} between (18.195) and (18.196) we find

$$\tan \eta(k) = -k \sum_n \frac{g_n^2}{4\pi \left(k^2 - \mathcal{E}_n\right)}. \tag{18.198}$$

Equation (18.198) shows that there are resonances at $k^2 = \mathcal{E}_n$ when $\mathcal{E}_n > 0$.

Now let us impose the initial conditions where one of the motionless particles, say $\chi_0(t = 0)$ has the maximum amplitude of one, while all the others have zero amplitudes, i.e.

$$|\Psi_0\rangle = \begin{cases} \psi(\mathbf{r}, 0) = 0 \\ \chi_0 = 1, \quad \text{and} \quad \chi_i = 0, \quad i \neq 0 \end{cases}. \tag{18.199}$$

We expand this initial state $|\Psi_0\rangle$ in terms of the complete set of states

$$|\Psi_0\rangle = \int_0^\infty |k\rangle \, \rho_0(k) dk, \tag{18.200}$$

where $\rho_0(k)$ can be determined from Eqs. (18.197)-(18.199)

$$\rho_0(k) = \frac{1}{2\pi^2} \langle k|\Psi_0\rangle = \frac{g_0}{2\pi^2} \left(\frac{k \cos \eta(k)}{k^2 - \mathcal{E}_0}\right). \tag{18.201}$$

While we do have the orthogonality of states, these states are not physically realizable. The reason for this is that the mean energy of the state $|\Psi_0\rangle$ is infinite

$$\langle \Psi_0 \,|k^2|\, \Psi_0\rangle = \frac{1}{2\pi^2} \int_0^\infty \frac{k^4 \cos^2 \eta(k)}{k^2 - \mathcal{E}_0} dk \to \infty. \tag{18.202}$$

This problem can be remedied either by introducing an appropriate form factor $\Lambda(\mathbf{r})$ that we mentioned earlier, or by replacing the interaction by a nonlocal one in time, i.e. by considering the following Lagrangian for the system [36]

$$L = \sum_n \chi_n^*(t) \left(i\frac{d}{dt} - \mathcal{E}_n\right) \chi_n(t) + \int \left[i\psi^* \frac{\partial \psi}{\partial t} - \nabla \psi^* \cdot \nabla \psi \right.$$

$$\left. - \Lambda(\mathbf{r}) \sum_n \int_{-\infty}^\infty K(t - t') \{\chi_n(t') \psi^*(\mathbf{r}, t') + \chi_n^*(t') \psi(\mathbf{r}, t')\} dt'\right] d^3r, \tag{18.203}$$

where $K(t - t')$ is the solution of the differential equation

$$\left[P\left(\frac{d^2}{dt^2}\right) - \beta^2\right] K(t - t') = -\beta^2 \delta(t - t'), \tag{18.204}$$

and P is a polynomial of its argument. The boundary conditions for (18.204) is that the function $K(t - t')$ should tend to zero as $t - t'$ goes to infinity. This Lagrangian generates the equations of motion (18.192) and (18.193) provided that

$$\Lambda(\mathbf{r}) \to \delta(\mathbf{r}), \quad \text{and} \quad K(t - t') \to \delta(t - t'). \tag{18.205}$$

Thus while $|\Psi_0\rangle$ is not a realizable state, we can find realizable states arbitrarily close to $|\Psi_0\rangle$ with finite mean energy.

Now let us study the way that $|\Psi_0\rangle$ decays in time. For this we need to examine the overlap between the initial state of the system $|\Psi_0(0)\rangle$ and the state of the system at the time t, i.e. $|\Psi_0(t)\rangle$. We note that $|\Psi_0(0)\rangle$ is expressible as a superposition of the energy eigenstates, Eq. (18.200), and hence

$$|\Psi_0(t)\rangle = \int_0^\infty |k\rangle \, \rho_0(k) \exp\left(-ik^2 t\right) dk. \tag{18.206}$$

By calculating $\langle \Psi_0(0)|\Psi_0(t)\rangle$ we find that

$$\langle \Psi_0(0)|\Psi_0(t)\rangle = 2\pi^2 \int_0^\infty |\rho_0(k)|^2 \exp\left(-ik^2 t\right) dk, \qquad (18.207)$$

and by substituting for ρ_0 from (18.201) we have

$$\langle \Psi_0(0)|\Psi_0(t)\rangle = \frac{g_0^2}{2\pi^2} \int_0^\infty \frac{k^2 \cos^2 \eta(k) e^{-ik^2 t} dk}{(k^2 - \mathcal{E}_0)^2}$$

$$= \frac{g_0^2}{2\pi^2} \int_0^\infty \frac{k^2 e^{-ik^2 t} dk}{\left[(k^2 - \mathcal{E}_0)^2 + k^2 g_0^4\right]}. \qquad (18.208)$$

The integral in (18.208) can be evaluated in terms of the error function. Let us assume that g_0^2 is not very large, then the integrand in (18.208) has four poles, one in each quadrant. If the pole in the first quadrant is located at $k = \kappa$, then the other poles are at $k = -\kappa$ and at $k = \pm\kappa^*$. When this is the case the integral in (18.208) can be calculated analytically:

$$\langle \Psi_0(0)|\Psi_0(t)\rangle = \frac{ig_0^2}{4\pi\left(\kappa^2 - \kappa^{*2}\right)} \left[\kappa e^{-i\kappa^2 t} \mathrm{erfn}\left\{\exp\left(\frac{-i\pi}{4}\right)\kappa\sqrt{t}\right\}\right.$$

$$\left. - \kappa^* e^{-i\kappa^{*2} t} \mathrm{erfn}\left\{\exp\left(\frac{-i\pi}{4}\right)\kappa^*\sqrt{t}\right\} + 2\kappa^* e^{-i\kappa^{*2} t}\right], \qquad (18.209)$$

where the error function, $\mathrm{erfn}(z)$, is defined by

$$\mathrm{erfn}(z) = \frac{2}{\sqrt{\pi}} \int_z^\infty e^{-x^2} dx. \qquad (18.210)$$

When g_0^2 is small and t is not too large, the largest term in (18.209) which dominates the decay is the last term, i.e.

$$\langle \Psi_0(0)|\Psi_0(t)\rangle \approx \frac{ig_0^2 \kappa^* e^{-i\kappa^{*2} t}}{2\pi\left(\kappa^2 - \kappa^{*2}\right)}, \qquad (18.211)$$

and from this relation it follows that $|\langle \Psi_0(0)|\Psi_0(t)\rangle|^2$ decreases exponentially in time. This exponential decay is for the initial state $\chi_0(0)$ which decays into $\psi(\mathbf{r}, t)$ and $\chi_i(t), (i \neq 0)$. Thus $\chi_0(t)$ plays the role of the central particle which by its absorption by the particle $\psi(\mathbf{r}, t)$ and re-emission into other $\chi_i(t)$ s loses energy to the other parts of the system.

From Eq. (18.209) we also deduce that the initial rate of decay $|\langle \Psi_0(0)|\Psi_0(t)\rangle|^2$ is infinite at $t = 0$ and hence this probability as a function of t has a cusp at $t = 0$. This result should be compared with the result obtained for the model discussed in Sec. 17.4.

The Wigner-Weisskopf model can also be derived using an approximate solution of the two-level system, such as the one which will be discussed in Sec. 18.9. Details of this derivation and the approximation used can be found in ref. [37].

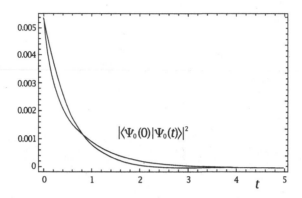

Figure 18.1: A plot of $|\langle\Psi_0(0)|\Psi_0(t)\rangle|^2$, Eq. (18.209), as a function of time t shows that the decay is very close to an exponential decay. For comparison the exponential decay curve which is higher in the range $0 < t < 1$ is also shown.

18.6 Equation of Motion for a Harmonically Bound Radiating Electron

In this and the following two sections (Secs. 18.7,18.8) we want to study three different formulations of the problem of nonrelativistic radiating electron. Following the works of Ford and O'Connell, we start with the Heisenberg description, where both the motion of the charged particle and the electromagnetic field which the electron is coupled to, are quantized [38]–[42]. Then by eliminating the degrees of freedom of the field we obtain an equation which is formally similar to the classical Abraham-Lorentz-Dirac equation discussed in Chapter 1. The second formulation is based on the Euclidean path integral method of quantization, and the calculation of the partition function. In this way we obtain the natural spectrum of the one-dimensional harmonic oscillator (Sec. 18.7). Finally in Sec. 18.8 we show that the expectation value of the velocity operator satisfies a second order equation, again similar to the Abraham-Lorentz-Dirac equation. In this approach the motion of the charged particle is quantized whereas the electromagnetic field is not.

From the Heisenberg equations of motion for a harmonically bound electron coupled to electromagnetic field we can derive an operator equation for the position of the electron which is the quantum analogue of the Abraham-Lorentz equation [38]–[43].

We start with the quantum mechanical Hamiltonian for the one-dimensional motion of the electron coupled to the radiation field in the dipole approximation

$$H = \frac{1}{2m_0}\left[p_x - \frac{e}{c}A_x\right]^2 + \frac{1}{2}Kx^2 + \frac{1}{2}\sum_{\mathbf{k},\sigma}\left(p_{\mathbf{k},\sigma}^2 + \omega_k^2 q_{\mathbf{k},\sigma}\right)^2, \qquad (18.212)$$

where we have written the Hamiltonian for the radiation field in terms of $p_{\mathbf{k},\sigma}$ and $q_{\mathbf{k},\sigma}$, and these are the components of $\mathbf{p_k}$ and $\mathbf{q_k}$ (see Sec. 8.1). In Eq. (18.212) $\omega_k = ck$, and the x-component of the quantized vector potential A_x is given by [44]-[46]

$$A_x = \sum_{\mathbf{k},\sigma} \left(\frac{2\pi\hbar c}{kV} \right)^{\frac{1}{2}} \left(\delta_k^* a_{\mathbf{k},\sigma} \mathbf{e}_{\mathbf{k},\sigma} \cdot \hat{\mathbf{x}} + \delta_k a_{\mathbf{k},\sigma}^\dagger \mathbf{e}_{\mathbf{k},\sigma}^* \cdot \hat{\mathbf{x}} \right). \tag{18.213}$$

Here δ_k is the electron form factor, $\mathbf{e}_{\mathbf{k},\sigma}$ is the polarization unit vector, $\hat{\mathbf{x}}$ is the unit vector in the direction $\mathbf{x}$ and V is the volume. The creation and annihilation operators $a_{\mathbf{k},\sigma}^\dagger$ and $a_{\mathbf{k},\sigma}$ for the electromagnetic field are related to $q_{\mathbf{k},\sigma}$ and $p_{\mathbf{k},\sigma}$ in (18.212) by the relations

$$a_{\mathbf{k},\sigma} = \frac{1}{\sqrt{2\omega}} \left(\omega_k q_{\mathbf{k},\,\sigma} + ip_{\mathbf{k},\sigma} \right), \tag{18.214}$$

$$a_{\mathbf{k},\sigma}^\dagger = \frac{1}{\sqrt{2\omega}} \left(\omega_k q_{\mathbf{k},\,\sigma} - ip_{\mathbf{k},\sigma} \right), \tag{18.215}$$

and these operators satisfy the commutation relations

$$\left[a_{\mathbf{k},\sigma}, a_{\mathbf{k}',\,\sigma'}^\dagger \right] = \delta_{\mathbf{k},\,\mathbf{k}'} \delta_{\sigma,\sigma'}, \tag{18.216}$$

and

$$\left[a_{\mathbf{k},\sigma}^\dagger, a_{\mathbf{k}',\sigma'}^\dagger \right] = [a_{\mathbf{k},\sigma}, a_{\mathbf{k}',\sigma'}] = 0. \tag{18.217}$$

Substituting for $p_{\mathbf{k},\sigma}$ and $q_{\mathbf{k},\sigma}$ in terms of $a_{\mathbf{k},\sigma}^\dagger$ and $a_{\mathbf{k},\sigma}$ we can write H as

$$H = \frac{1}{2m_0} \left[p_x - \frac{e}{c} A_x \right]^2 + \frac{1}{2} K x^2 + \frac{1}{2} \sum_{\mathbf{k},\sigma} \hbar k c\, a_{\mathbf{k},\sigma}^\dagger a_{\mathbf{k},\sigma}, \tag{18.218}$$

where we have omitted the zero-point energy term. From this Hamiltonian operator we find the Heisenberg equation of motion for x, p_x and $a_{\mathbf{k},\sigma}$;

$$\dot{x} = \frac{1}{m_0} \left(p_x - \frac{e}{c} A_x \right), \quad \dot{p}_x = -Kx, \tag{18.219}$$

and

$$\dot{a}_{\mathbf{k},\sigma} = -icka_{\mathbf{k},\sigma} + i \left(\frac{2\pi e^2}{\hbar ckV} \right)^{\frac{1}{2}} (\delta_k \hat{\mathbf{x}} \cdot \mathbf{e}_{\mathbf{k},\sigma})\, \dot{x}, \tag{18.220}$$

with a similar relation for $\dot{a}_{\mathbf{k},\sigma}^\dagger$. Now we integrate (18.220) and assume that the coupling is turned on at $t = -\infty$, i.e.

$$a_{\mathbf{k},\sigma}(t) = a_{\mathbf{k},\sigma}^f(t) + i \left(\frac{2\pi e^2}{\hbar ckV} \right)^{\frac{1}{2}} \delta_k \hat{\mathbf{x}} \cdot \mathbf{e}_{\mathbf{k},\sigma} \int_{-\infty}^t e^{-i\omega(t-t')} \dot{x}(t')\, dt'. \tag{18.221}$$

The operator $a^f_{\mathbf{k},\sigma}(t)$ is the free-field Heisenberg annihilation operator. We can also find a similar result for $a^\dagger_{\mathbf{k},\sigma}(t)$. Next we substitute for $a^\dagger_{\mathbf{k},\sigma}(t)$ and $a_{\mathbf{k},\sigma}(t)$ in A_x Eq. (18.213) and eliminate p_x between the two equations in (18.219) to obtain the Heisenberg equation for the position operator x;

$$m_0\ddot{x} + \int_{-\infty}^{t} J\left(t - t'\right) \dot{x}\left(t'\right) dt' + Kx = F(t), \qquad (18.222)$$

where

$$J\left(t - t'\right) = \left(\frac{8\pi e^2}{3V}\right) \sum_{\mathbf{k}} |\delta_k|^2 \cos\left[ck\left(t - t'\right)\right], \qquad (18.223)$$

and

$$F(t) = -\frac{e}{c}\left(\frac{\partial A^f_x(t)}{\partial t}\right). \qquad (18.224)$$

Once the electron form factor δ_k is given, then $J\left(t - t'\right)$ can be determined and the integro-differential equation (18.222) can be solved. For instance if we choose the electron form factor to be [38]

$$|\delta_k|^2 = \frac{\Omega^2}{\Omega^2 + c^2 k^2}, \qquad (18.225)$$

where Ω is a large cutoff frequency, then with this choice of $|\delta_k|^2$ we have

$$J\left(t - t'\right) = \left(\frac{8\pi e^2}{3V}\right) \sum_{\mathbf{k}} \frac{\Omega^2}{\Omega^2 + c^2 k^2} \cos\left[ck\left(t - t'\right)\right]. \qquad (18.226)$$

Now we replace the summation over $\mathbf{k}$ by integration

$$J\left(T\right) = \left(\frac{8\pi e^2}{3(2\pi)^3}\right) \int_0^\infty \frac{4\pi k^2 \Omega^2}{\Omega^2 + c^2 k^2} \cos\left(ckT\right) dk, \qquad (18.227)$$

and in this way we can evaluate the integral in (18.227) to find $J(T)$ analytically

$$J\left(T\right) = m\Omega^2 \tau \left[2\delta(T) - \Omega \exp(-\Omega T)\right], \qquad (18.228)$$

where $\tau = \frac{2e^2}{3mc^2}$. Here m is the renormalized mass of the electron (see e.g. Eq. (8.12) of the van Kampen's model). For the present model this renormalized mass is given by [38],[39]

$$m = m_0 + \left(\frac{2e^2\Omega}{3c^3}\right) m = \tau\Omega m. \qquad (18.229)$$

Substituting for $J\left(t - t'\right)$ from (18.228) in (18.222) we find the equation of motion of the electron

$$m_0\ddot{x}(t) + m\Omega^2\tau\dot{x}(t) - m\Omega^3\tau \int_{-\infty}^{t} \exp\left[-\Omega\left(t - t'\right)\right] \dot{x}\left(t'\right) dt' + Kx(t) = F(t). \qquad (18.230)$$

Now we multiply this equation by $e^{-\Omega t}\frac{d}{dt}e^{\Omega t}$ to get an operator differential equation for x;

$$m\frac{d^2x}{dt^2} + K\left(x + \frac{1}{\Omega}\dot{x}\right) = m\left(\tau - \frac{1}{\Omega}\right)\frac{d^3x}{dt^3} + F(t) + \frac{1}{\Omega}\dot{F}(t). \qquad (18.231)$$

This is one of the possible quantum mechanical analogues of the classical Abraham-Lorentz equation [40].

Motion of a Dirac Electron in Quantized Electromagnetic Field — The main result that we have obtained in this section is an equation for the nonrelativistic position operator of the electron when the field is approximated by the dipole contribution only. We want to inquire about the possibility of solving the problem of motion of an electron satisfying the quantal Dirac equation (i.e. a charged particle of spin $\frac{1}{2}$) which interacts with a quantized electromagnetic field and without using the dipole approximation [47],[48]. Following Rosen, we discuss the general formulation of this motion but without discussing his solution for one-dimensional version of the problem. Here we assume that the longitudinal part of the field is already in the mass term of the Dirac equation [47]. Furthermore we want to consider the case where the whole system, i.e. the electron plus the field are confined in a large cubic box of side L. Then we can expand the field in a set of plane polarized waves with proper frequencies and with quantized amplitudes [44]-[46]. We label these modes of oscillation of the field by ν. The total Hamiltonian for this system can be written as

$$H = c\boldsymbol{\alpha}\cdot\mathbf{p} + mc^2\beta - e\boldsymbol{\alpha}\cdot\sum_\nu \mathbf{A}_\nu + H_0, \qquad (18.232)$$

where $\boldsymbol{\alpha}$ and β are Dirac matrices [44],[45], $\mathbf{p}$ is the momentum of the electron and H_0 is the Hamiltonian of the field,

$$H_0 = \frac{1}{2}\sum_\nu \left(P_\nu^2 + \omega_\nu^2 Q_\nu^2 - \hbar\omega_\nu\right), \qquad (18.233)$$

and

$$\mathbf{A}_\nu = c\sqrt{\frac{\pi}{L}}\boldsymbol{\epsilon}_\nu\left[Q_\nu\left(e^{\mathbf{k}_\nu\cdot\mathbf{r}} + e^{-\mathbf{k}_\nu\cdot\mathbf{r}}\right) + \left(\frac{i}{\omega_\nu}\right)P_\nu\left(e^{\mathbf{k}_\nu\cdot\mathbf{r}} - e^{-\mathbf{k}_\nu\cdot\mathbf{r}}\right)\right], \qquad (18.234)$$

is the quantized vector potential for the degrees of freedom (denoted by ν) at the position of the electron $\mathbf{r}$. Here k_ν is $\frac{2\pi}{L}$ times a vector with integer components (n_1, n_2, n_3) and these components can take both positive and negative values. The angular frequency $\omega_\nu = c|\mathbf{k}_\nu|$ and ϵ_ν is the polarization vector which is orthogonal to $\mathbf{k}_\nu$. The zero-point energy of the electromagnetic field in the Hamiltonian for this field, H_0, has been subtracted. In Eq. (18.234) P_ν and Q_ν are pairs of conjugate operators satisfying the usual commutation relations for coordinate and momenta.

The time-independent wave equation for the stationary state with energy E is given by

$$H\left(-i\hbar\nabla,\ \mathbf{r};\ \left\{-i\hbar\frac{\partial}{\partial Q_\nu}\right\},\{Q_\nu\}\right)\psi\left(\mathbf{r},\{Q_\nu\}\right)=E\psi\left(\mathbf{r},\{Q_\nu\}\right),\quad(18.235)$$

where $\{Q_\nu\}$ denotes the set containing all Q_ν s. Next we expand $\psi\left(\mathbf{r},\{Q_\nu\}\right)$ in terms of the product of harmonic oscillator wave functions

$$\psi\left(\mathbf{r},\{Q_\nu\}\right)=\sum_{\{j\}}a_{j_1}a_{j_2}\cdots a_{j_\nu}\cdots\mathcal{U}_{j_1j_2\cdots j_\nu\cdots}\exp\left[i\left(\frac{1}{\hbar}\mathbf{p}'-j_1\mathbf{k}_1-j_2\mathbf{k}_2\cdots\right)\cdot\mathbf{r}\right],$$

$$(18.236)$$

where

$$\mathcal{U}_{j_1j_2\cdots j_\nu\cdots}=u_{j_1}(Q_1)u_{j_2}(Q_2)\cdots u_{j_\nu}(Q_\nu)\cdots,\quad(18.237)$$

and u_j is the solution of the unperturbed mode of oscillation

$$\frac{1}{2}\left[P^2+\omega^2Q^2-\hbar\omega\right]u_\nu(Q)=n\hbar\omega u_\nu(Q).\quad(18.238)$$

In Eq. (18.236) $\mathbf{p}'$ is a constant vector representing the total momentum of the system and $a_{j_1}a_{j_2}\cdots a_{j_\nu}\cdots$ are the constant coefficients of expansion. The summation $\sum_{\{j\}}$ indicates that the sum is over all of the states of all the degrees of freedom. By substituting (18.235) and (18.236) and making use of the relations

$$\left[Q+\frac{i}{\omega}P\right]u_n=\sqrt{\frac{2n\hbar}{\omega}}\ u_{n-1},\quad(18.239)$$

$$\left[Q-\frac{i}{\omega}P\right]u_n=\sqrt{\frac{2(n+1)\hbar}{\omega}}\ u_{n+1},\quad(18.240)$$

and after rearrangement of terms we find the following relation

$$\left[c\mathbf{p}'\cdot\boldsymbol{\alpha}+mc^2\beta+\hbar\sum_\nu n_\nu\left(\omega_\nu-c\mathbf{k}_\nu\cdot\boldsymbol{\alpha}\right)-E\right]a_{n_1n_2\cdots n_\nu\cdots}$$

$$-ce\sqrt{\frac{\pi}{L}}\sum_\nu\boldsymbol{\epsilon}_\nu\cdot\boldsymbol{\alpha}\left[\sqrt{\frac{2(n_\nu+1)\hbar}{\omega_\nu}}a_{n_1n_2\cdots n_\nu+1\cdots}+\sqrt{\frac{2n_\nu\hbar}{\omega_\nu}}a_{n_1n_2\cdots n_\nu-1\cdots}\right]$$

$$=0$$

$$(18.241)$$

We can transform the operator H in which the electron coordinates are diagonal to one in which the components of the total momentum are diagonal. Thus we define a function ϕ which is independent of $\mathbf{r}$ by

$$\phi=\sum_{\{n_\nu\}}a_{n_1n_2\cdots n_\nu\cdots}\,\mathcal{U}_{n_1n_2\cdots n_\nu\cdots},\quad(18.242)$$

and introduce an operator H' by

$$H' = c\mathbf{p}' \cdot \boldsymbol{\alpha} + mc^2\beta + \frac{1}{2}\sum_\nu \left(P_\nu^2 + \omega_\nu^2 Q_\nu^2 - \hbar\omega_\nu\right)\left[1 - \frac{c\mathbf{k}_\nu \cdot \boldsymbol{\alpha}}{\omega_\nu}\right]$$

$$- 2ce\sqrt{\frac{\pi}{L}}\sum_\nu Q_\nu \boldsymbol{\epsilon}_\nu \cdot \boldsymbol{\alpha}, \tag{18.243}$$

then

$$H'\phi = E\phi, \tag{18.244}$$

gives us the Schrödinger equation in which the electron coordinates have been eliminated. For a one-dimensional motion of the electron Rosen has shown that the Schrödinger equation (18.244) can be transformed to a system consisting of two sets of coupled oscillators which can be diagonalized [47]. In this case, similar to Sollfrey's model, all of the eigenfrequencies of the system can be determined by finding the roots of a transcendental equation. The main interest in studying this model is to investigate the question of divergence of the ground state energy of the system caused by the coupling between the electron and the electromagnetic field. However if we consider the rate of energy loss of an electron of momentum $\mathbf{p}'$ to the ground state of the quantized electromagnetic field, then the problem becomes similar to Sollfrey's model.

18.7 Path Integral Approach to the Spectrum of Harmonically Bound Oscillator

A method analogous to the Feynman formulation of the problem of polarons can be applied to obtain the natural broadening of the spectral lines of the harmonic oscillator and show that the intensity distribution can be well approximated by a series of Lorentz profiles [49]-[53]. We briefly studied the corresponding classical solution of this problem (see Sec. 1.10), and we will find a similar result for the spectrum of a charged quantum oscillator in the next section for a two-level system.

Let us start with the van Kampen model Eq. (8.1) and make a simpler model based on the following approximations:

(a) We retain only the dipole interaction.

(b) We neglect the self-interaction term $\mathbf{A}^2(\mathbf{r})$ of the field. That is, in this simplified form the interaction between the electron of mass m_0 and the electromagnetic field is given by the coupling $-\frac{e}{m_0}\mathbf{p} \cdot \mathbf{A}$, with $\mathbf{A}$ denoting the magnetic potential at the center of the oscillating electron, i.e. $\mathbf{A}(0)$. The Hamiltonian of this model can be written in the simplified form of [56]

$$H = \frac{1}{2m_0}\mathbf{P}^2 + \frac{1}{2}m_0\omega_0^2\mathbf{R}^2 + \sum_{\mathbf{k},\sigma}\left(\frac{1}{2}p_{\mathbf{k},\sigma}^2 + \frac{1}{2}\omega_\mathbf{k}^2 q_{\mathbf{k},\sigma}^2\right)$$

$$- \mathbf{P}\cdot\sum_{\mathbf{k},\sigma}\left(\frac{4\pi}{\mathcal{V}}\right)^{\frac{1}{2}}\frac{e}{m_0}\mathbf{e}_{\mathbf{k},\sigma}q_{\mathbf{k},\sigma}. \tag{18.245}$$

One way of formulating and solving this problem is the method of Euclidean path integral where one averages over the degrees of freedom of the electromagnetic field variables [51]–[55]. An interesting example of the application of this technique can also be found in the work of Caldeira and Leggett [57]. The result of this averaging is that the electron, in addition to the harmonic force, is subjected to an effective nonlocal potential. Once the Euclidean action for the simplified Hamiltonian (18.245) is found, the spectrum of the system can be obtained from the inverse Laplace transform of the partition function. Following Feynman, we write the action as the sum of the actions of the oscillator and that of the radiation field

$$S = S_{osc} + S_{rad}. \tag{18.246}$$

We then calculate the propagator of the electron, i.e. the amplitude for transition from a position $\mathbf{R}_0$ to the position $\mathbf{R}_\beta$, and this is given by integrating over all paths joining $\mathbf{R}_0$ to $\mathbf{R}_\beta$;

$$G(\mathbf{R}_0, 0; \mathbf{R}_\beta, \beta) = \int_{x_0}^{x_\beta} \exp\left[-\frac{1}{\hbar} S_{osc}\right] \exp\left[-\frac{1}{\hbar} S_{rad}\right] \mathcal{D}[R] \tag{18.247}$$

The radiation part of the action is the sum of contributions of the field oscillators

$$f_{\mathbf{k},\sigma} = \left(\frac{4\pi}{\mathcal{V}}\right)^{\frac{1}{2}} e\omega_0 \mathbf{e}_{\mathbf{k},\sigma} \cdot \hat{\mathbf{q}}, \tag{18.248}$$

and is given by [55]

$$S_{rad}^{\mathbf{k},\sigma}(\mathbf{R}) = -\frac{1}{2\omega^2} \int_0^\beta \int_0^\beta \frac{\omega^2 \cosh\left[\omega(|t-s| - \frac{\beta}{2})\right]}{2\sinh\left(\frac{1}{2}\omega\beta\right)} f_{\mathbf{k},\sigma}(t) f_{\mathbf{k},\sigma}(s) dt ds. \tag{18.249}$$

By summing over $\mathbf{k}$ and σ we find the total action, oscillator plus field, to be given by [49]-[55]

$$S(\mathbf{R}) = \int_0^\beta \frac{1}{2} m \left[\left(\frac{d\mathbf{R}(t)}{dt}\right)^2 + \omega_0^2 \mathbf{R}^2(t)\right] dt$$
$$+ \frac{\alpha\hbar\omega^2}{3\pi c^2} \int_0^\beta \int_0^\beta \left(\frac{\pi^2}{2\beta^2}\right) \frac{[\mathbf{R}(t) - \mathbf{R}(s)]^2}{\sin^2\left[\frac{\pi}{\beta}(t-s)\right]} dt ds, \tag{18.250}$$

where β is proportional to the inverse temperature, $\alpha = \frac{e^2}{\hbar c}$ is the fine structure constant and $m\omega^2$ is the spring constant. Note that in (18.250) m_0 has been replaced by m, the renormalized mass [55]. We observe that the interaction between the electron and the electromagnetic field is nonlocal.

Having found $S(\mathbf{r})$, we can find the partition function for the oscillator. Now the partition function $Z_\alpha^{(d)}(\beta)$ for the oscillator when it is coupled to the radiation field is equal to the partition function of the whole system divided by the partition function of the radiation field in the absence of the oscillator. The latter is obtained from the infinite product

$$Z_{rad}(\beta) = \prod_k \left[2 \sinh \left(\frac{\beta \omega_k}{2} \right) \right]^{-2}. \tag{18.251}$$

Thus the partition function for the oscillator can be determined from

$$Z_\alpha^{(d)}(\beta) = \oint \exp \left[-\frac{1}{\hbar} S[R] \right] \mathcal{D}[R] \tag{18.252}$$

From Eq. (18.250) we find the partition function for the reduced system to be

$$Z_\alpha^d(\beta) = \left[\frac{\rho}{2\pi} \exp[-(2 \cos \phi) \rho \ln \rho] \left| \Gamma \left(\rho e^{i\phi} \right) \right|^2 \right]^d, \tag{18.253}$$

where d is the number of dimensions, Γ denotes the gamma-function and the two dimensionless quantities ρ and ϕ are given by

$$\rho = \frac{\beta \omega_0}{2\pi} \quad \text{and} \quad \cos \phi = \frac{\alpha}{3} \left(\frac{\hbar \omega_0}{mc^2} \right). \tag{18.254}$$

To find the spectrum of the oscillator interacting with its own radiation field, $f_\alpha^{(d)}(\omega)$, we find the inverse Laplace transform of $Z_\alpha^{(d)}$

$$f_\alpha^{(d)}(\omega) = \frac{1}{2\pi} e^{a\omega} \int_{-\infty}^{+\infty} Z_\alpha^{(d)}(a + iy) e^{i\omega y} dy, \quad a > 0, \tag{18.255}$$

which is independent of a. Now we will consider two cases, the first case is when the electron is not coupled to its own radiation field, and the second case when it is coupled.

For a free d-dimensional oscillator, we set $\alpha = 0$, or $\phi = \frac{\pi}{2}$ and then $Z_\alpha^{(d)}(\beta)$ reduces to

$$Z_0^{(d)}(\beta) = \left[2 \sinh \left(\frac{\beta \omega_0}{2} \right) \right]^{-d}, \tag{18.256}$$

and the corresponding $f_0^{(d)}(\omega)$ found from (18.255) becomes

$$f_0^{(d)}(\omega) = \sum_{n=0}^{\infty} \binom{n+d-1}{d-1} \delta \left[\omega - \left(n + \frac{1}{2} d \right) \omega_0 \right], \tag{18.257}$$

i.e. there are $\binom{n+d-1}{d-1}$ states with the frequency $\omega = \left(n + \frac{1}{2} \right) \omega_0$ related to the frequency interval of length ω_0.

Returning to the motion of the oscillator coupled to the field, a detailed analysis shows that $f_\alpha^{(d)}(\omega)$ satisfies the following conditions [55]:

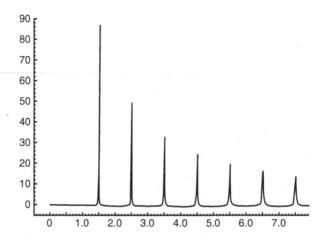

Figure 18.2: The natural spectrum of the one-dimensional harmonic oscillator $\omega f_\alpha^{(1)}$ plotted as a function of ω for $\frac{\phi}{\pi} = 0.499$. Here the δ-peak of the ground state is not shown [55].

(a) $f_\alpha^{(d)}(\omega) \geq 0$,   for $\omega \geq 0$.

(b) $f_\alpha^{(d)}(\omega) = \delta\left(\omega - \frac{1}{2}d\omega_\alpha\right) + \frac{d\pi}{3\omega_0}\cos(\phi)\theta\left(\omega - \frac{1}{2}d\omega_\alpha\right) + r_\alpha^{(d)}(\omega)$, where $r_\alpha^{(d)}(\omega)$ is a continuous function which is zero for $\omega \leq \frac{1}{2}d\omega_\alpha$.

(c) $\int_0^\Omega f_\alpha^{(d)}(\omega)d\omega \sim \frac{\left(\frac{\Omega}{\omega_0}\right)^d}{d!}$ as $\Omega \to \infty$.

While it is possible to find $f_\alpha^{(d)}(\omega)$ analytically, we can obtain an accurate result for this function by the numerical integration of Eq. (18.255). Thus we find that $f_\alpha^{(d)}(\omega)$ can be approximated very well by a sum of Lorentzian distributions each of the form

$$L_n(\Gamma_n, \omega_n, \omega) = \left(\frac{\Gamma_n}{2\pi}\right)\frac{1}{\left(\omega - \omega_n^{(d)}\right)^2 + \frac{1}{4}\Gamma^2}, \tag{18.258}$$

where

$$\omega_n^{(d)} = \left(n + \frac{1}{2}\right)\omega_\alpha, \tag{18.259}$$

and

$$\Gamma_n = 2n\omega_0\cos\phi = n\omega_0\frac{2\alpha}{3}\left(\frac{\hbar\omega_0}{m_0c^2}\right), \tag{18.260}$$

and these are truncated at $\omega = \frac{1}{2}d\,\omega_\alpha$. The angular frequency ω_α is also related to ω_0 by

$$\omega_\alpha = \frac{2}{\pi}(\cos\phi + \phi\sin\phi)\omega_0 = \left[1 - \frac{\alpha^2}{18}\left(\frac{\hbar\omega_0}{m_0c^2}\right)^2 + \cdots\right]\omega_0. \tag{18.261}$$

The result of the calculation of the natural spectrum for one-dimensional oscillator is shown in Fig. 18.2. This figure is a plot of dimensionless quantity

$\omega f_{\alpha}^{(1)}(\omega)$ versus ω when the angle ϕ is fixed at 0.499π, and where the δ-peak of the ground state is omitted. The approximate analytical result for $f_{\alpha}^{(1)}(\omega)$ is discussed in detail in [55].

18.8 More about the Quantum-Mechanical Formulation of the Abraham-Lorentz-Dirac Equation

In Chapter 1 we studied the Abraham-Lorentz-Dirac (ALD) equation and the problems associated with its solution, and in the last two sections we derived the quantal analogue of this equation for an harmonically bound charged particle. In this section we want to derive a quantized version of the same equation, i.e. the equation of motion of a charged particle interacting with its own electromagnetic field and an external electromagnetic field. For this we can describe the charged particle by means of a quantized Schrödinger field $\psi(\mathbf{r}, t)$ interacting with a quantized electromagnetic field and eliminate the field degrees of freedom. In this approach we find that in the nonrelativistic limit and in the dipole approximation, there is a complete correspondence between the result found from quantum electrodynamics and the Abraham-Lorentz-Dirac (ALD) equation. That is the time evolution of the expectation value of the position of the electron satisfies an equation similar to the classical ALD equation. In particular if one considers an harmonically bound electron, then the above expectation value is exactly the same as Dirac's classical equation [59]. A similar work by Loinger also shows that for a nonrelativistic point electron the position operator of the particle satisfies an equation which is formally analogous to the classical ALD equation [60].

Schrödinger Equation for Charged Particle Coupled to a Classical Radiation Field — A different and simpler approach is to assume that the electron obeys the Schrödinger equation and is coupled to a classical (nonquantized) radiation field [62]. With this assumption the equation of motion for the electron is given by

$$i\hbar \frac{\partial \psi}{\partial t} = H\psi = \left[\frac{1}{2m_0} \left(\mathbf{p} - \frac{e}{c} \mathbf{A}(\mathbf{r}, t) \right)^2 + e\phi(\mathbf{r}, t) + V(\mathbf{r}, t) \right] \psi, \qquad (18.262)$$

where m_0 is the bare mass of the electron and $\mathbf{A}(\mathbf{r}, t)$ and $\phi(\mathbf{r}, t)$ are the classical vector and scalar potentials respectively and $V(\mathbf{r}, t)$ is an external potential. Defining the charge density $\rho(\mathbf{r}, t)$ and the current density $\mathbf{J}(\mathbf{r}, t)$ for the electron by

$$\rho(\mathbf{r}, t) = e\psi^*(\mathbf{r}, t)\psi(\mathbf{r}, t), \qquad (18.263)$$

$$\mathbf{J}(\mathbf{r}, t) = \text{Re} \left[\frac{e}{m_0} \psi^*(\mathbf{r}, t) \left(\mathbf{p} - \frac{e}{c} \mathbf{A} \right) \psi(\mathbf{r}, t) \right], \qquad (18.264)$$

we can find $\phi(\mathbf{r}, t)$ and $\mathbf{A}(\mathbf{r}, t)$ in the Coulomb gauge from the well-known relations [63]

$$\phi(\mathbf{r}, t) = \phi_s(\mathbf{r}, t) = \int \frac{\rho(\mathbf{r}', t)}{|\mathbf{r} - \mathbf{r}'|} d^3 r', \tag{18.265}$$

and

$$\mathbf{A}(\mathbf{r}, t) = \mathbf{A}_s(\mathbf{r}, t) = \frac{1}{c} \int \frac{\mathbf{J}_t \left[\mathbf{r}', t - \frac{|\mathbf{r} - \mathbf{r}'|}{c}\right]}{|\mathbf{r} - \mathbf{r}'|} d^3 r', \tag{18.266}$$

where the subscript s indicates self-fields. These expressions are for the inhomogeneous wave equations and do not contain the free field solutions for $\phi(\mathbf{r}, t)$ and $\mathbf{A}(\mathbf{r}, t)$. In Eq. (18.266) $\mathbf{J}_t$ is the transverse part of the current density, i.e. $\mathbf{J} = \mathbf{J}_t + \mathbf{J}_l$, with $\mathbf{J}_l$ being the longitudinal part. By expanding $\mathbf{A}(\mathbf{r}, t)$ in (18.266) in powers of c^{-1}, we find

$$\mathbf{A}_s(\mathbf{r}, t) = \frac{1}{c} \int \frac{\mathbf{J}_t(\mathbf{r}', t)}{|\mathbf{r} - \mathbf{r}'|} d^3 r' - \frac{2}{3c^2} \frac{d}{dt} \int \mathbf{J}(\mathbf{r}', t) d^3 r' + \mathcal{O}\left(\frac{1}{c^3}\right), \tag{18.267}$$

where the fact that the charged particle is localized with a small volume guarantees the rapid convergence of the series given by (18.267).

By writing the vector potential $\mathbf{A}_s(\mathbf{r}, t)$ in terms of the expectation value of the velocity operator

$$\mathbf{v} = \frac{1}{m_0}\left(\mathbf{p} - \frac{e}{c}\mathbf{A}\right), \tag{18.268}$$

we have

$$\mathbf{A}_s(\mathbf{r}, t) = \frac{1}{c} \int \frac{\mathbf{J}_t(\mathbf{r}', t)}{|\mathbf{r} - \mathbf{r}'|} d^3 r' - \frac{2e}{3c^2} \frac{d}{dt} \langle \mathbf{v} \rangle, \tag{18.269}$$

where terms of the order c^{-3} and higher have been ignored. The last term in (18.269) is found from

$$\int \mathbf{J}(\mathbf{r}, t) d^3 r = \int \psi^*(\mathbf{r}, t) \mathbf{v} \psi(\mathbf{r}, t) d^3 r = e\langle \mathbf{v} \rangle, \tag{18.270}$$

Next we write Heisenberg's equation for the motion of the electron

$$m_0 \frac{d\mathbf{v}}{dt} = e\left(-\nabla\phi_s - \frac{1}{c}\frac{\partial \mathbf{A}}{\partial t}\right) - \nabla V(\mathbf{r}, t). \tag{18.271}$$

In this equation the magnetic force has been neglected since, in general, it is smaller than the electric field by a factor $\frac{v}{c}$. The expectation value of (18.271) with the wave function $\psi(\mathbf{r}, t)$ takes the form

$$m_0 \frac{d\langle \mathbf{v} \rangle}{dt} = e\langle -\nabla\phi_s \rangle - \frac{e}{c}\frac{\partial\langle \mathbf{A}_s \rangle}{\partial t} + \langle \mathbf{F}_{\text{ext}} \rangle, \tag{18.272}$$

where $\mathbf{F}_{\text{ext}} = -\nabla V$.

By calculating the partial derivative $\frac{\partial}{\partial t}\langle \mathbf{A}_s \rangle$ from Eq. (18.269) we obtain

$$m_0 \frac{d\langle \mathbf{v} \rangle}{dt} = e\langle -\nabla\phi_s \rangle + \langle \mathbf{F}_J \rangle + \frac{2e^2}{3c^3}\frac{d^2\langle \mathbf{v} \rangle}{dt^2} + \langle \mathbf{F}_{\text{ext}} \rangle. \tag{18.273}$$

Now let us examine the two terms $\langle \mathbf{F}_J \rangle$ and $\langle \nabla \phi_s \rangle$ in this equation. These are given by

$$\langle \mathbf{F}_J \rangle = \frac{e}{c^2} \int \int \frac{|\psi(\mathbf{r}',t)|^2}{|\mathbf{r}-\mathbf{r}'|} \frac{\partial}{\partial t}(\mathbf{J}-\mathbf{J}_l) d^3r d^3r', \tag{18.274}$$

and

$$-e\langle \nabla \phi_s \rangle = e^2 \int \int |\psi(\mathbf{r},t)|^2 \frac{(\mathbf{r}-\mathbf{r}')}{|\mathbf{r}-\mathbf{r}'|^3} |\psi(\mathbf{r}',t)|^2 d^3r d^3r'. \tag{18.275}$$

Since the integrand in (18.275) is an odd function of $(\mathbf{r}-\mathbf{r}')$, therefore the result of integration yields

$$-e\langle \nabla \phi_s \rangle = 0. \tag{18.276}$$

The force $\langle \mathbf{F}_J \rangle$ defined by (18.274) is composed of two parts, $\langle \mathbf{F}_{J1} \rangle$ and $\langle \mathbf{F}_{J2} \rangle$, where $\langle \mathbf{F}_{J1} \rangle$ comes from the contribution of $\mathbf{J}$ in (18.274) and $\langle \mathbf{F}_{J2} \rangle$ is the result found from integration over $\mathbf{J}_l$. Thus

$$\langle \mathbf{F}_{J1} \rangle = -\frac{e}{c^2} \int \int \frac{|\psi(\mathbf{r}',t)|^2}{|\mathbf{r}-\mathbf{r}'|} \frac{\partial \mathbf{J}}{\partial t} d^3r d^3r'$$

$$= -\frac{e^2}{2\pi^2 c^2} \int \left\{ \langle \exp(-i\mathbf{k}\cdot\mathbf{r}) \rangle \frac{d}{dt} \langle \exp(-i\mathbf{k}\cdot\mathbf{r})\mathbf{v} \rangle \right\} \frac{d^3k}{k^2}, \tag{18.277}$$

where we have used the identity

$$\frac{1}{|\mathbf{r}-\mathbf{r}'|} = \frac{1}{2\pi^2} \int \exp\left[i\mathbf{k}\cdot(\mathbf{r}-\mathbf{r}')\right] \frac{d^3k}{k^2}. \tag{18.278}$$

Noting that here the charge of the electron is distributed over a volume determined by the charge density (18.263), and that the dimensions of this volume are not smaller than the Compton wavelength of the electron, we can introduce a cut-off frequency k_{max} defined by

$$0 < k < k_{max} = \frac{m_0 c}{\hbar}, \tag{18.279}$$

for the integration over k in (18.278). Here $\hbar k$ is the momentum of the photon which acts on the extended charge of the electron. Therefore in the range of the k integration, $\mathbf{k}\cdot\mathbf{r} \ll 1$ and we can use the dipole approximation $\exp(\mathbf{k}\cdot\mathbf{r}) \approx 1$. In this approximation $\mathbf{F}_{j1}$ takes the simple form

$$\langle \mathbf{F}_{J1} \rangle = -\delta m \frac{d\langle \mathbf{v} \rangle}{dt}, \tag{18.280}$$

where δm is defined by

$$\delta m = \frac{e^2}{2\pi^2 c^2} \int_0^{k_{max}} dk \int \int \sin\vartheta \, d\vartheta \, d\varphi = \frac{2m_0}{\pi} \frac{e^2}{\hbar c}. \tag{18.281}$$

The second term of the force, i.e. $\langle \mathbf{F}_{J2} \rangle$ is given by

$$
\langle \mathbf{F}_{J2} \rangle = \frac{e}{c^2} \int \int \frac{|\psi(\mathbf{r}',t)|^2}{|\mathbf{r}-\mathbf{r}'|} \frac{\partial \mathbf{J}_l(\mathbf{r},t)}{\partial t} d^3 r d^3 r'
$$

$$
= \frac{e}{2\pi^2 c^2} \int \frac{d^3 k}{k^2} \int |\psi(\mathbf{r}',t)|^2 \exp(-i\mathbf{k}\cdot\mathbf{r}') d^3 r'
$$

$$
\times \frac{d}{dt} \int \exp(i\mathbf{k}\cdot\mathbf{r}) \mathbf{J}_l(\mathbf{r},t) d^3 r, \tag{18.282}
$$

where again we have replaced $|\mathbf{r}-\mathbf{r}'|^{-1}$ by its integral representation Eq. (18.278). Making use of the identity

$$
\langle \mathbf{J}_l \rangle = \frac{1}{(2\pi)^3} \int \frac{\mathbf{k}}{k} \left[\frac{\mathbf{k}\cdot\mathbf{J}(k)\exp(i\mathbf{k}\cdot\mathbf{r})}{k} \right] d^3 k
$$

$$
= -\frac{1}{4\pi} \int \nabla \left\{ \nabla \cdot \left[\frac{\mathbf{J}(\mathbf{r}',t)}{|\mathbf{r}-\mathbf{r}'|} \right] \right\} d^3 r', \tag{18.283}
$$

we find the following expression for $\langle \mathbf{F}_{J2} \rangle$:

$$
\langle \mathbf{F}_{J2} \rangle = -\frac{e}{8\pi^3 c^2} \int \frac{\exp(-i\mathbf{k}\cdot\mathbf{r})}{k^2} d^3 k \int \exp(i\mathbf{k}\cdot\mathbf{r}) d^3 r
$$

$$
\times \int \nabla \left\{ \nabla \cdot \left[\frac{\partial_t \mathbf{J}(\mathbf{r}',t)}{|\mathbf{r}-\mathbf{r}'|} \right] \right\} d^3 r', \tag{18.284}
$$

where ∂_t denotes the time derivative. To simplify this last expression for $\mathbf{F}_{J2}$ we note that the jth component of the vector in the curly bracket in (18.284) can be written as a sum of three terms

$$
\nabla_j \left\{ \nabla \cdot \left[\frac{\partial_t \mathbf{J}(\mathbf{r}',t)}{|\mathbf{r}-\mathbf{r}'|} \right] \right\} = \frac{3\left(x_j-x_j'\right)^2}{|\mathbf{r}-\mathbf{r}'|^5} \partial_t J_j(\mathbf{r}',t) - \frac{1}{|\mathbf{r}-\mathbf{r}'|^3} \partial_t J_j(\mathbf{r}',t)
$$

$$
+ \sum_{i\neq j} \frac{3\left(x_j-x_j'\right)\left(x_i-x_i'\right)}{|\mathbf{r}-\mathbf{r}'|^5} \partial_t J_j(\mathbf{r}',t), \tag{18.285}
$$

Since the integration in (18.284) does not depend on how we choose the direction of the axis of the coordinates, therefore in Eq. (18.285) we can set

$$
(x-x')^2 = (y-y')^2 = (z-z')^2 = \frac{1}{3}(\mathbf{r}-\mathbf{r}')^2, \tag{18.286}
$$

with the result that first and second terms in in the right-hand side of (18.285) cancel each other. By changing the coordinate $\mathbf{r}$ to $\mathbf{r}_1 = \mathbf{r} - \mathbf{r}'$, we can write the ith component of $\langle \mathbf{F}_{J2} \rangle$ as

$$
\langle \mathbf{F}_{J2} \rangle_i = -\frac{3e^2}{8\pi^3 c^2} \int \langle \exp(-i\mathbf{k}\cdot\mathbf{r}) \rangle \frac{d^3 k}{k^2}
$$

$$
\times \int \exp(i\mathbf{k}\cdot\mathbf{r}_1) \left\{ \sum_{j\neq i} \frac{x_{1i}x_{1j}}{|\mathbf{r}_1|^5} \frac{d}{dt} \langle \exp(i\mathbf{k}\cdot\mathbf{r})v_j \rangle \right\} d^3 r_1 \tag{18.287}
$$

In the dipole approximation that we are using, $\langle \mathbf{F}_{J2} \rangle_i$ can evaluated exactly

$$\langle \mathbf{F}_{J2} \rangle_i = -\frac{3e^2}{8\pi^3 c^2} \int \frac{d^3 k}{k^2} \int \left[\sum_{j \neq i} \left\{ \frac{x_{1i} x_{1j}}{|\mathbf{r}_1|^5} \right\} \frac{d\langle v_j \rangle}{dt} \right] d^3 r_1 \approx 0. \tag{18.288}$$

This result follows from the antisymmetry of the integrand in (18.288) with respect to x_{1i}. Now if we substitute for $\nabla \phi_s$ and for $\mathbf{F}_J = \mathbf{F}_{J1} + \mathbf{F}_{J2}$ in (18.273) we obtain

$$(m_0 + \delta m)\frac{d\langle \mathbf{v} \rangle}{dt} = \langle \mathbf{F}_{\text{ext}} \rangle + \frac{2e^2}{3c^3}\frac{d^2\langle \mathbf{v} \rangle}{dt^2} \tag{18.289}$$

where $m = m_0 + \delta m$ is the renormalized mass of the electron.

18.9 Quantum Theory of Line Width

In Sec. 1.10 we studied the classical theory of line width resulting from the exponential nature of damping of the motion of electron, and in Sec. 18.5 we found a similar result from the decay of an unstable system in the Wigner-Weisskopf model. Now the quantum theory of radiation offers a perturbative approach to the problem of the line width for the general case of a bound electron which is interacting with the electromagnetic field [64]. Here we want to study the same problem with using a simple solvable model of a two-level system where the two levels are coupled together by the transverse electromagnetic field $\mathbf{E}_\perp(\mathbf{r}, t)$ [65],[66]. The total Hamiltonian of the system is given by

$$H = H_0 + H', \tag{18.290}$$

where

$$H_0 = \sum_{\alpha=i,f} \int \left[\frac{1}{2m} \nabla \psi_\alpha^*(\mathbf{r}, t) \cdot \nabla \psi_\alpha(\mathbf{r}, t) + V(\mathbf{r})\psi_\alpha^*(\mathbf{r}, t)\,\psi_\alpha(\mathbf{r}, t) \right] d^3 r$$

$$+ \frac{1}{8\pi} \int \left\{ \mathbf{E}_\perp^2(\mathbf{r}, t) + [\nabla \wedge \mathbf{A}(\mathbf{r}, t)]^2 \right\} d^3 r, \tag{18.291}$$

and

$$H' = \frac{ie}{2m} \int \mathbf{A}^+(\mathbf{r}, t) \cdot [\psi_i^*(\mathbf{r}, t)\nabla \psi_f(\mathbf{r}, t) - \nabla \psi_i^*(\mathbf{r}, t)\psi_f(\mathbf{r}, t)] \, d^3 r$$

$$+ \frac{ie}{2m} \int \mathbf{A}^-(\mathbf{r}, t) \cdot [\psi_f^*(\mathbf{r}, t)\nabla \psi_i(\mathbf{r}, t) - \nabla \psi_f^*(\mathbf{r}, t)\psi_i(\mathbf{r}, t)] \, d^3 r. \tag{18.292}$$

In these equations we have set $\hbar = c = 1$, i and f refer to the initial and the final states of the system respectively, and $\mathbf{A}^+$ and $\mathbf{A}^-$ are the positive and negative frequency parts of the magnetic potential $\mathbf{A}$. The binding potential $V(\mathbf{r})$ can be short-range or long-range. Denoting the eigenfunctions of the Schrödinger equation with the potential $V(\mathbf{r})$ by $u_\alpha(\mathbf{r})$, i.e.

$$\left[-\frac{1}{2m}\nabla^2 + V(\mathbf{r}) \right] u_\alpha(\mathbf{r}) = E_\alpha u_\alpha(\mathbf{r}), \quad \alpha = i, f, \tag{18.293}$$

we can expand ψ_α and ψ_α^* as

$$\psi_\alpha(\mathbf{r}, t) = b_\alpha(t) u_\alpha(\mathbf{r}), \tag{18.294}$$

and

$$\psi_\alpha^*(\mathbf{r}, t) = b_\alpha^\dagger(t) u_\alpha^*(\mathbf{r}). \tag{18.295}$$

Here $b_\alpha^\dagger(t)$ and $b_\alpha(t)$ denote the creation and annihilation operators for the electron satisfying the anti-commutation relation

$$\left[b_\alpha(t),\ b_{\alpha'}^\dagger(t) \right]_+ = \delta_{\alpha,\alpha'}. \tag{18.296}$$

By choosing the periodic boundary conditions and unit normalization volume we can expand $\mathbf{A}^\pm(\mathbf{r}, t)$ in terms of plane waves:

$$\mathbf{A}^\pm(\mathbf{r}, t) = \sum_{\mathbf{k},\sigma} \left(\frac{2\pi}{k} \right)^{\frac{1}{2}} \mathbf{e}_\sigma(\mathbf{k}) \left(\begin{array}{c} a_{\mathbf{k},\sigma}(t) \\ a_{\mathbf{k},\sigma}^\dagger(t) \end{array} \right) \exp\left(\pm i \mathbf{k}.\mathbf{r} \right), \tag{18.297}$$

where the photon energy is $k = |\mathbf{k}|$, and $\mathbf{e}_\sigma(\mathbf{k})$ is the unit polarization vector which is orthogonal to $\mathbf{k}$. The sum over $\mathbf{k}$ extends over all plane waves in the unit normalization volume and the sum over σ is for the two allowed directions of polarization. In Eq. (18.297) the operators $a_{\mathbf{k},\sigma}^\dagger(t)$ and $a_{\mathbf{k},\sigma}(t)$ satisfy the commutation relation

$$\left[a_{\mathbf{k},\sigma}(t),\ a_{\mathbf{k}',\sigma'}^\dagger(t) \right] = \delta_{\mathbf{k},\mathbf{k}'}\delta_{\sigma,\sigma'}. \tag{18.298}$$

By substituting from (18.293) and (18.297) in H, Eqs. (18.290)-(18.292) and carrying out the integration over $\mathbf{r}$, we find H in the second quantized form:

$$H = E_f b_f^\dagger b_f + E_i b_i^\dagger b_i + \sum_k k \left[a_{\mathbf{k}}^\dagger a_{\mathbf{k}} + \frac{1}{2} \right]$$

$$+ e \sum_k \left(\frac{1}{2k} \right)^{\frac{1}{2}} \left[\beta(k) b_i^\dagger b_f a_{\mathbf{k}} + \beta(k)^* b_f^\dagger b_i a_{\mathbf{k}}^\dagger \right], \tag{18.299}$$

where we have suppressed the sum over σ. In the above equation the coefficient $\beta(k)$ is defined by

$$\beta(k) = \frac{2i}{m} \sqrt{\pi} \int e^{-i\mathbf{k}\cdot\mathbf{r}} u_i(\mathbf{r}) \mathbf{e}_\sigma(\mathbf{k}).\nabla u_f(\mathbf{r}) d^3 r. \tag{18.300}$$

Since $u_i(\mathbf{r})$ is the bound state wave function, the integral in (18.300) is convergent and $\beta(k)$ goes to zero as $k \to \infty$.

We assume that at $t = 0$, the particle is in the state i and no photon is present. Thus we can write the wave function as

$$|\Psi(0)\rangle = |0, 1_i\rangle, \qquad (18.301)$$

where 0 refers to the number of photons, and 1_i shows that the electron is in the state i. Because of the structure of the Hamiltonian, Eq. (18.299), after a time t we can have a photon, and an electron. The total wave function now takes the form

$$|\Psi(t)\rangle = C_0(t)\,|0, 1_i\rangle + \sum_{\mathbf{k}} C_{\mathbf{k}}(t)\,|1_{\mathbf{k}}, 1_f\rangle. \qquad (18.302)$$

Here $|1_{\mathbf{k}}, 1_f\rangle$ indicates a state which includes the electron in the state f plus one photon with the energy k and polarization $\mathbf{e}_\sigma(\mathbf{k})$. The coefficients $C_0(t)$ and $C_{\mathbf{k}}(t)$ measure the probability amplitudes for no photon and a single photon states respectively.

By comparing (18.301) and (18.302), we find the initial conditions for $C_0(t)$ and $C_{\mathbf{k}}(t)$ to be

$$C_0(t = 0) = 1, \quad C_{\mathbf{k}}(t = 0) = 0. \qquad (18.303)$$

To determine these amplitudes we write the time-dependent Schrödinger equation for this case as

$$i\frac{\partial}{\partial t}\,|\Psi(t)\rangle = H\,|\Psi(t)\rangle, \qquad (18.304)$$

and then by substituting for H and $|\Psi(t)\rangle$ using Eqs. (18.299) and (18.302) and equating the coefficients of the state vectors $|0, 1_i\rangle$ and $|1_k, 1_f\rangle$ we obtain

$$i\frac{d}{dt}C_0(t) = E_i C_0(t) + \sum_{\mathbf{k}} \frac{e\beta(k)}{\sqrt{2k}} C_{\mathbf{k}}(t), \qquad (18.305)$$

and

$$i\frac{d}{dt}C_{\mathbf{k}}(t) = (E_f + k)\,C_{\mathbf{k}}(t) + \frac{e\beta^*(k)}{\sqrt{2k}} C_0(t). \qquad (18.306)$$

We can find the normal modes for this coupled system by expressing $C_0(t)$ and $C_{\mathbf{k}}(t)$ in terms of their Fourier transforms [67]:

$$C_{\mathbf{q}}(t) = \sum_{\omega} p(\omega) C_{\mathbf{q}}(\omega) \exp\left[-i(\omega + E_i)t\right], \quad \mathbf{q} = 0, \mathbf{k}, \qquad (18.307)$$

where

$$p(\omega) = C_0^*(\omega)C_0(t = 0) + \sum_{\mathbf{k}} C_{\mathbf{k}}^*(\omega)C_{\mathbf{k}}(t = 0) = C_0^*(\omega). \qquad (18.308)$$

By substituting from (18.307) in (18.305) and (18.306) we obtain

$$\omega C_0(\omega) = e \sum_{\mathbf{k}} \frac{\beta(k)}{\sqrt{2k}} C_{\mathbf{k}}(\omega), \tag{18.309}$$

and

$$\omega C_{\mathbf{k}}(\omega) = (E_f + k - E_i) C_{\mathbf{k}}(\omega) + \frac{e\beta^*(k)}{\sqrt{2k}} C_0(\omega). \tag{18.310}$$

Now if we eliminate $C_{\mathbf{k}}(\omega)$ between (18.309) and (18.310) and replace the summation over $\mathbf{k}$ by integration

$$\sum_{\mathbf{k}} \rightarrow \frac{2}{(2\pi)^3} \int d^3 k, \tag{18.311}$$

we find the eigenvalue equation for ω

$$\omega + \frac{e^2}{2\pi^2} \int_0^\infty \frac{|\beta(k)|^2 k\, dk}{(E_f + k - E_i - \omega)} = 0. \tag{18.312}$$

This equation is similar to the eigenvalue equation that we obtained for Ullersma's model, Eq. (9.10), and the roots of (18.312) have the same properties that we mentioned in Chapter 9 for the roots of Eq. (9.10).

In order to calculate $C_0(\omega)$, we first normalize $C_0(\omega)$ and $C_{\mathbf{k}}(\omega)$ by requiring

$$C_0^*(\omega) C_0(\nu) + \sum_{\mathbf{k}} C_{\mathbf{k}}^*(\omega) C_{\mathbf{k}}(\nu) = \delta_{\omega,\nu}, \tag{18.313}$$

and then we solve (18.310) for $C_{\mathbf{k}}(\omega)$ and substitute the result in (18.313) and set $\omega = \nu$. This gives us

$$C_0(\omega) = \left(\frac{dG(\omega)}{d\omega} \right)^{-\frac{1}{2}}, \tag{18.314}$$

where the function $G(\omega)$ is given by

$$G(\omega) = \omega + \frac{e^2}{2\pi^2} \int_0^\infty \frac{|\beta(k)|^2 k\, dk}{(E_f + k - E_i - \omega)}. \tag{18.315}$$

Similarly for $C_{\mathbf{k}}(\omega)$ we find

$$C_{\mathbf{k}}(\omega) = \frac{e\beta^*(k)}{(\omega - E_f - k + E_i)\sqrt{2k}} \left(\frac{dG(\omega)}{d\omega} \right)^{-\frac{1}{2}}. \tag{18.316}$$

Now from Eqs. (18.307), (18.308), (18.314) and (18.315) we have

$$C_0(t) = \sum_{\omega} \exp\left[-i(\omega + E_i)t \right] \left(\frac{dG(\omega)}{d\omega} \right)^{-1}. \tag{18.317}$$

Changing the variable in the integrand, ω, to z, where

$$z = \omega + E_i - E_f, \tag{18.318}$$

we can write (18.317) as a contour integral

$$C_0(t) = \frac{1}{2\pi i} e^{-iE_f t} \oint_C \frac{e^{-izt}}{G(z)} dz, \tag{18.319}$$

where the contour C contains all the roots of $G(z)$. We expect the amplitude $C_0(t)$ to decay more or less exponentially, and thus we want to consider the roots of $G(z)$ with negative imaginary part. For this we must look onto the second Reimann sheet for the desired root. If in the first Reimann sheet $G^I(z)$ represents the analytic continuation of $G(z)$, then

$$G^I(z) = z - E_i + E_f + \frac{e^2}{2\pi^2} \int_0^\infty \frac{|\beta(k)|^2 k dk}{k - z}, \tag{18.320}$$

where $0 < \arg z < 2\pi$. By analytic continuation of $G(z)$ in the second Reimann sheet we find $G^{II}(z)$, and for $0 > \arg z > -2\pi$, we have

$$G^{II}(x - i\epsilon) = G^I(x + i\epsilon). \tag{18.321}$$

Thus

$$G^{II}(z) = G^I(z) + \frac{ie^2}{\pi} z |\beta(z)|^2. \tag{18.322}$$

If we assume that only one of the roots of $G^{II}(z)$ at $z = z_0$ makes the major contribution to the integral (18.319) [6], i.e.

$$G^{II}(z_0) = 0, \quad \text{for} \quad z_0 = E_R - E_f - \frac{i}{2}\Gamma, \tag{18.323}$$

then from Eqs. (18.315), (18.322) and (18.323) it follows that

$$E_R - E_i + \left(\frac{e^2}{2\pi^2}\right) \int_0^\infty \frac{(k + E_f - E_R)|\beta(k)|^2 k dk}{[(k + E_f - E_R)^2 + \frac{1}{4}\Gamma^2]}$$
$$+ \left(\frac{e^2 \Gamma}{2\pi}\right) \left|\beta\left(E_R - E_f - \frac{i}{2}\Gamma\right)\right|^2 = 0, \tag{18.324}$$

and

$$\frac{\Gamma}{2} \left\{ 1 + \frac{e^2}{2\pi^2} \int_0^\infty \frac{|\beta(k)|^2 k dk}{[(k + E_f - E_R)^2 + \frac{1}{4}\Gamma^2]} \right\}$$
$$- \left(\frac{e^2}{\pi}\right)(E_R - E_f)\left|\beta\left(E_R - E_f - \frac{i}{2}\Gamma\right)\right|^2 = 0. \tag{18.325}$$

These solutions are obtained by equating the real and imaginary parts of $G^{II}(z)$. The quantity $E_R - E_i$ presents the shift in the energy of the electron and $\Gamma/2$ is the line width of the emitted electromagnetic wave.

A very good approximation to the coupled nonlinear integral equations for E_R and Γ can be found by noting that

$$E_R \approx E_i + \mathcal{O}\left(e^2\right), \quad \text{and} \quad \Gamma \approx \mathcal{O}\left(e^2\right), \tag{18.326}$$

where $\mathcal{O}\left(e^2\right)$ means of the order e^2, which in proper units is $e^2/(\hbar c) \approx (1/137)$. Using these we can find the approximate solution to (18.324) and (18.325);

$$E_R - E_i \approx -\left(\frac{e^2}{2\pi^2}\right)\mathcal{P}\int_0^\infty \frac{|\beta(k)|^2 k dk}{k + E_f - E_i}, \tag{18.327}$$

and

$$\Gamma \approx \frac{e^2}{\pi}(E_i - E_f)|\beta(E_i - E_f)|^2, \tag{18.328}$$

where $\mathcal{P}$ stands for the principal value of the integral.

These results agree with the results obtained from the first-order perturbation theory [64]. Evaluating the contour integral (18.319) following the same technique used in solving Ullersma's model, Chapter 9, we find

$$C_0(t) = \frac{\exp\left(-iE_R t - \frac{1}{2}\Gamma t\right)}{\left(\frac{dG^{II}(z)}{dz}\right)_{z_0}} - \left(\frac{e^2}{2\pi^2}\right)e^{-iE_f t}\int_{-\infty}^0 \frac{|\beta(x)|^2 e^{-ixt}x dx}{G^I(x)G^{II}(x)}. \tag{18.329}$$

Again if we ignore terms of the order $\mathcal{O}\left(e^2\right)$ we have

$$\frac{dG^{II}(z)}{dz} = 1 + \left(\frac{e^2}{2\pi^2}\right)\int_0^\infty \frac{|\beta(k)|^2 k dk}{(k - z)^2} + \frac{ie^2}{\pi}\frac{d}{dz}[z|\beta(z)|^2] \approx 1 + \mathcal{O}\left(e^2\right). \tag{18.330}$$

Therefore

$$C_0(t) \approx \exp\left[-iE_R t - \frac{1}{2}\Gamma t\right] + \mathcal{O}\left(e^2\right), \tag{18.331}$$

and

$$C_{\mathbf{k}}(t) \approx -e\beta^*(k)\frac{\left\{\exp\left(-iE_R t - \frac{1}{2}\Gamma t\right) - \exp\left[-i(k + E_f)t\right]\right\}}{\sqrt{2k}\left(E_f + k - E_R + \frac{i}{2}\Gamma\right)} + \mathcal{O}\left(e^2\right). \tag{18.332}$$

Thus by keeping only the first order term in $C_0(t)$ we have the approximate result

$$|C_0(t)|^2 \approx e^{-\Gamma t}, \tag{18.333}$$

or the decay is purely exponential. However a more careful analysis of the contour integral (18.319) shows that both the initial form of the decay and the decay after a long time $t \gg \Gamma^{-1}$ are not exponential as we observed in Sec. 18.1 [6],[7].

A two-level system similar to the one that we have studied in this section has been discussed by Morse and Feshbach [68].

18.10   Spin-Boson System: Another Model of Dissipative Two-Level Quantum System

In condensed matter physics one encounters systems which effectively can be described by a single spin half particle $\boldsymbol{\sigma} = (\sigma_x, \sigma_y, \sigma_z)$ coupled to a set of independent harmonic oscillators. In addition to the problem of a spin in cavity Sec. 18.3, an important example of such a system is that of a particle which is initially trapped in one of the two wells of a confining symmetric double-well potential, and is coupled to a heat bath of harmonic oscillators. In the absence of coupling this particle can tunnel back and forth between the two wells with a given frequency [19]. If the potential is such that the ground and the first excited states are far from the second and higher excited states, then the period of the coherent oscillations in the absence of any external force is given by $\frac{2\pi}{\Delta}$, where Δ is the energy difference between the two states. In the case of two asymmetric wells, the two lowest energy states are close to each other and far from other excited states only when there is resonant tunneling. For a full account of this type of tunneling the reader is referred to ref. [19].

Now if such a system is coupled to a heat bath, composed of an infinite number of harmonic oscillators, then the period of oscillation may change, if the coupling is weak, just like an underdamped oscillator. For strong coupling the motion exhibits the characteristics of an overdamped motion. These systems can be viewed as two-level systems coupled to a bath of many bosons, or spin-boson systems. Choosing the units so that $\hbar = 1$, we write the Hamiltonian as [20],[69]

$$H = -\frac{1}{2}\Delta\sigma_x + \sum_k \omega_k a_k^\dagger a_k + \frac{1}{2}\sum_k g_k \left(a_k + a_k^\dagger \right)\sigma_z, \qquad (18.334)$$

In this relation $\mathbf{S} = \frac{1}{2}\boldsymbol{\sigma}$ ($\hbar = 1$), and $\boldsymbol{\sigma} = (\sigma_x, \sigma_y, \sigma_z)$ are Pauli spin matrices and in the subsequent discussion we will use σ_i s, and reserve S for the unitary transformation. The Hamiltonian (18.334) differs from our earlier formulation of the two level system, Eq. (18.136), since here we are dealing with a many body problem, whereas in (18.136) we had eliminated the many-boson part and had replaced it by a Cardiola-Kanai type Hamiltonian.

When at $t = 0$ the spin is not coupled to the oscillators, we have an isolated spin in the presence of a "magnetic field" proportional to Δ, or the level splitting caused by tunneling in a double-well potential. Initially the spin is an eigenstate of σ_x, the magnetic field induces transitions and the expectation value $\langle\sigma_x(t)\rangle$ oscillates with frequency Δ. Now for a general coupling constant g_k we will not get a simple linear damping (Ohmic dissipation). But we obtain a damping proportional to velocity if we impose the condition

$$\sum_k g_k^2 \delta(\omega - \omega_k) = 2\lambda\theta(\omega_c - \omega), \qquad (18.335)$$

where λ and ω_c are constants, both having the dimension of time^{-1}.

We can find a solution of (18.334) by perturbation in the following way: Let us transform the Hamiltonian H by a unitary transformation to H' [70],

$$H' = SHS^{-1}, \tag{18.336}$$

where S, not to be confused with the spin $\mathbf{S}$ denotes the transformation

$$S = \exp\left\{ \sigma_z \sum_k \frac{g_k}{2\omega_k} \xi_k \left(a_k^\dagger - a_k \right) \right\}. \tag{18.337}$$

In this expression for S we have introduced a function ξ_k which will be determined later. This transformation leads to the Hamiltonian

$$H' = H_0' + H_1' + H_2', \tag{18.338}$$

where

$$H_0' = -\frac{1}{2}(\eta\Delta)\sigma_x + \sum_k \omega_k a_k^\dagger a_k - \sum_k \frac{g_k^2}{4\omega_k}\xi_k(2 - \xi_k), \tag{18.339}$$

$$H_1' = \frac{1}{2}\sigma_z \sum_k g_k(1 - \xi_k)\left(a_k^\dagger + a_k \right) - \frac{i}{2}(\eta\Delta)\sigma_y \sum_k \frac{g_k}{\omega_k}\xi_k \left(a_k^\dagger - a_k \right), \tag{18.340}$$

and

$$H_2' = -\frac{1}{2}\sigma_x\Delta\left\{ \cosh\left[\sum_k \frac{g_k}{\omega_k}\xi_k \left(a_k^\dagger - a_k \right) \right] - \eta \right\}$$
$$- \frac{i}{2}\Delta\sigma_y\left\{ \sinh\left[\sum_k \frac{g_k}{\omega_k}\xi \left(a_k^\dagger - a_k \right) \right] - \eta\sum_k \frac{g_k}{\omega_k}\xi_k \left(a_k^\dagger - a_k \right) \right\}. \tag{18.341}$$

In Eqs. (18.339)-(18.341), the number η denotes the quantity

$$\eta = \exp\left(-\sum_k \frac{g_k^2}{2\omega_k^2}\xi_k^2 \right). \tag{18.342}$$

We note that in H_0' the spin operator σ_x and the boson operators are decoupled and thus as the result of this transformation the energy splitting Δ is reduced to $\eta\Delta$, i.e. a lowering in the characteristic frequency of oscillation between the two wells. This change is caused by the dragging of the polarization cloud of boson excitation [24]. While H_0' is separable, H_1' and H_2'. the operators $\boldsymbol{\sigma}$ and $a_k^\dagger$ and a_k are coupled.

The eigenstates of H_0' should describe the spin state of the system as well as the number of bosons in each mode k. Thus any state can be represented by the ket $|s; \{n_k\}\rangle$, where s is either s_1 or s_2

$$|s_1\rangle = \frac{1}{\sqrt{2}}\begin{bmatrix} 1 \\ 1 \end{bmatrix}, \quad \text{and} \quad |s_2\rangle = \frac{1}{\sqrt{2}}\begin{bmatrix} 1 \\ -1 \end{bmatrix}, \tag{18.343}$$

and by $\{n_k\}$, where n_k is the number of bosons in the state k. The initial state of the system is the one where the spin is up $s = s_1$ and all n_k s are zero;

$$|G\rangle = |s_1; \{0_k\}\rangle. \tag{18.344}$$

Now we fix the parameter ξ_k in Eq. (18.340) by requiring that

$$H'_1|G\rangle = H'_1|s_1; \{0_k\}\rangle = 0, \tag{18.345}$$

i.e. the perturbation due to H'_1 must be as small as possible [69]. By substituting H'_1 from (18.340) in (18.345) we find that the parameter ξ_k is related to ω_k by

$$\xi_k = \frac{\omega_k}{\omega_k + \eta\Delta}. \tag{18.346}$$

This result has also been found by Silbey and Harris using a variational method to obtain the minimum value for the ground state energy [70]. By substituting (18.346) in H'_1, Eq. (18.340), we have

$$H'_1 = \frac{1}{2}\eta\Delta \sum_k \frac{g_k \xi_k}{\omega_k} \left[b_k^\dagger (\sigma_z - i\sigma_y) + b_k (\sigma_z + i\sigma_y) \right]. \tag{18.347}$$

By writing H'_1 in this form it is easy to verify Eq. (18.345). In addition, using Eq. (18.342) we find the following relations:

$$\langle s_1; \{0_k\}|H'_2|s_1; \{0_k\}\rangle = 0, \tag{18.348}$$

$$\langle s_2; \{0_k\}|H'_2|s_1; \{0_k\}\rangle = 0, \tag{18.349}$$

$$\langle s_1; \{1_k\}|H'_2|s_1; \{0_k\}\rangle = 0, \tag{18.350}$$

and

$$\langle s_2; \{0_k\}|H'_2|s_1; \{1_k\}\rangle = 0. \tag{18.351}$$

The Diagonal Form of the Hamiltonian — As we notice, when only the lowest order excitation is taken into account, then all of the off-diagonal matrix elements of H' vanish. Therefore we can write the Hamiltonian H' in terms of its diagonal elements plus higher order terms, whch include the non-zero off-diagonal elements. Thus the Hamiltonian can be written as

$$H' = -\frac{1}{2}\Delta|G\rangle\langle G| + \sum_\omega \omega|\omega\rangle\langle\omega| + \cdots\cdots, \tag{18.352}$$

where dots denote diagonal and off-diagonal terms belonging to higher excited states. We have seen earlier that $|G\rangle = |s_1, \{0_k\}\rangle$ is the ground state of H'. The next two states $|s_1; 1_k\rangle$ and $|s_2; \{0_k\}\rangle$ will be parts of $|\omega\rangle$

$$|\omega\rangle = |G\rangle + C_0(t)|s_2; \{0_k\}\rangle + \sum_k C_k(t)|s_1; \{1_k\}\rangle + \text{higher order terms.} \tag{18.353}$$

From the eigenvalue equation

$$H'|\omega\rangle = \omega|\omega\rangle, \tag{18.354}$$

where H' is given by (18.352) and $|\omega\rangle$ by (18.353), we find the following set of coupled linear equations for $C_0(\omega)$ and $C_k(\omega)$:

$$\left(\omega - \frac{1}{2}\eta\Delta\right)C_0(\omega) = \sum_k V_k C_k(\omega), \tag{18.355}$$

$$\left(\omega + \frac{1}{2}\eta\Delta - \omega_k\right)C_0(\omega) = V_k C_0(\omega), \tag{18.356}$$

where

$$V_k = \frac{\eta\Delta g_k \xi_k}{\omega_k} \tag{18.357}$$

By eliminating C_k between (18.355) and (18.356) we get the eigenvalue equation for ω;

$$\omega - \frac{1}{2}\eta\Delta = \sum_k \frac{V_k^2}{\omega + \frac{1}{2}\eta\Delta - \omega_k}. \tag{18.358}$$

and this is similar to Eq. (18.312). Using the normalization condition for the eigenvectors, viz,

$$C_0^2(\omega) + \sum_k C_k^2(\omega) = 1, \tag{18.359}$$

and Eqs. (18.356)-(18.359) we find these eigenvectors to be

$$C_0(\omega) = \left[1 + \sum_k \frac{V_k^2}{\left(\omega + \frac{1}{2}\eta\Delta - \omega_k\right)^2}\right]^{-\frac{1}{2}}, \tag{18.360}$$

and

$$C_k(\omega) = \frac{V_k}{\left(\omega + \frac{1}{2}\eta\Delta - \omega_k\right)}C_0(\omega). \tag{18.361}$$

Time-Dependence of the Expectation Value of σ_z — Suppose that when $t < 0$ the spin system and the heat bath are uncoupled, and for these times the spin system is observed to have a value of $\sigma_z = +1$, and that the environment being undisturbed by the observation. Then if at $t = 0$ the coupling between the spin and the heat bath is switched on, we want to find the expectation value of σ_z at t. Denoting this expectation value by $P(t)$, we have [72]

$$P(t) = \left\langle (\{0_k\}; +1)\left|e^{iH't}\sigma_z e^{-iH't}\right|(+1; \{0_k\})\right\rangle. \tag{18.362}$$

From Eqs. (18.352)-(18.361) we obtain [69]

$$P(t) = \frac{1}{2}\sum_\omega C_0^2(\omega)\exp\left[-i\left(\omega + \frac{1}{2}\eta\Delta\right)t\right]$$
$$+ \frac{1}{2}\sum_\omega C_0^2(\omega)\exp\left[i\left(\omega + \frac{1}{2}\eta\Delta\right)t\right]. \tag{18.363}$$

The summation over the eigenvalues can be replaced by contour integration over $\omega' = \omega + \frac{1}{2}\eta\Delta$

$$P(t) = \frac{1}{4\pi i} \oint e^{-i\omega' t} \left(\omega' - \eta\Delta - \sum_k \frac{V_k^2}{\omega' - \omega_k + i\epsilon} \right)^{-1} d\omega'$$

$$+ \frac{1}{4\pi i} \oint e^{-i\omega' t} \left(\omega' - \eta\Delta - \sum_k \frac{V_k^2}{\omega' - \omega_k - i\epsilon} \right)^{-1} d\omega'. \qquad (18.364)$$

Let us denote the real and imaginary parts of the sum in (18.364) by

$$\sum_k \frac{V_k^2}{\omega' - \omega_k \pm i\epsilon} = R(\omega') \mp i\gamma(\omega') \qquad (18.365)$$

The two functions $R(\omega)$ and $\gamma(\omega)$ found from (18.365) are

$$R(\omega) = -\frac{2\lambda(\eta\Delta)^2}{\omega(\omega + \eta\Delta)} \left\{ \frac{\omega_c}{\omega_c + \eta\Delta} - \frac{\omega}{\omega + \eta\Delta} \ln\left[\frac{|\omega|(\omega_c + \eta\Delta)}{\eta\Delta(\omega_c - \omega)} \right] \right\}, \qquad (18.366)$$

and

$$\gamma(\omega) = \frac{2\lambda\pi(\eta\Delta)^2}{(\omega + \eta\Delta)^2}. \qquad (18.367)$$

The constants λ and ω_c were defined by Eq. (18.335).

Following the same argument advanced in Sec. 18.9, we observe that if ω_0 is the smallest real root of

$$\omega - \eta\Delta - R(\omega) = 0, \qquad (18.368)$$

that makes the major contribution to the integral in Eq. (18.365), then the decay is exponential and we have

$$P(t) = \cos(\omega_0 t)e^{-\gamma t}. \qquad (18.369)$$

In the case of weak damping, we can ignore $R(\omega)$ in (18.368), then $\omega_0 = \eta\Delta$ and to the first order in the friction coefficient λ, we get

$$\gamma = \gamma(\eta\Delta) = \frac{\lambda\pi\eta\Delta}{2\omega}. \qquad (18.370)$$

Now the condition for Eq. (18.368) to have a real root is that

$$\frac{2\lambda}{\omega} < 1 + \frac{\eta\Delta}{\omega_c}. \qquad (18.371)$$

If, however, the parameters $\frac{\lambda}{\omega}$ and ω_c are such that the inequality in (18.371) is reversed, then the root of (18.369) becomes imaginary, and $P(t)$ becomes a damped exponential. Thus the critical damping occurs when

$$\frac{\lambda}{\omega} = \frac{1}{2}\left(1 + \frac{\eta\Delta}{\omega_c} \right). \qquad (18.372)$$

The long-time deviation from exponential decay in a model very similar to the one discussed in this section has been derived by Knight and Millonni [58]. These authors find that the asymptotic form of the amplitude $C_0(t)$ for times $t \gg \gamma^{-1}$ is of the form

$$C_0(t) \to \exp\left(-\frac{1}{2}\gamma t\right) + \left(\frac{\gamma}{2\pi\omega_0}\right)\left(\frac{1}{\omega_0 t}\right)^2 e^{i\omega_0 t}. \tag{18.373}$$

18.11   Gisin's Nonlinear Wave Equation

A phenomenological nonlinear wave equation with complex interaction has been proposed by Gisin to account for decaying states in wave mechanics [73]-[77]. This model is completely quantum mechanical and does not correspond to a simple classical dissipative system of the types that we have seen in earlier chapters (except, possibly, the dynamical matrix formulations of Secs. 3.14 and 11.10). But within the framework of quantum mechanics one can show that under certain conditions, Gisin's equation can be derived from Pauli's master equation [74]. The fact that a complex potential with negative imaginary part can account for the damped motion of a particle suggests that a more general complex interaction may also be used as a model of dissipation (Chapter 20).

Gisin's model is described by the wave equation

$$i\hbar \frac{\partial \psi}{\partial t} = \left(1 - i\frac{\kappa}{2}\right) H\psi + i\frac{\kappa}{2} \langle \psi | H | \psi \rangle \psi, \tag{18.374}$$

where H is the Hamiltonian of the system and κ is a dimensionless positive real damping constant. This nonlinear wave equation has the following properties:

(1) The norm of the wave function ψ is independent of time, and therefore once it is normalized it will remain so at later times.

(2) When ψ is an eigenfunction of H then $(\langle \psi | H | \psi \rangle - H)\psi$ is zero and (18.374) reduces to the Schrödinger equation.

(3) The rate of change of $\langle \psi | H | \psi \rangle$ is negative definite, i.e.

$$\frac{d}{dt} \langle \psi | H | \psi \rangle = \left\langle \frac{\partial \psi}{\partial t} | H | \psi \right\rangle - \left\langle \psi | H | \frac{\partial \psi}{\partial t} \right\rangle = -\frac{\kappa}{\hbar} (\Delta H)^2 \le 0. \tag{18.375}$$

In this relation $(\Delta H)^2$ is defined as

$$(\Delta H)^2 = \left(\langle \psi | H | \psi \rangle^2 - \langle \psi | H^2 | \psi \rangle \right). \tag{18.376}$$

The equality in (18.375) holds only when ψ is an eigenfunction of H. This result is obtained by substituting for $\frac{\partial \psi}{\partial t}$ and $\frac{\partial \psi^*}{\partial t}$ from the nonlinear equation (18.374) in (18.375).

(4) For a Hamiltonian with discrete spectrum the solution of the nonlinear equation (18.374) can be expressed in terms of the eigenfunctions $\psi_{j,\alpha}(\mathbf{r})$ of H.

Wave Function in Gisin's Formulation — Consider the eigenvalue equation

$$H\psi_{j,\alpha}(\mathbf{r}) = E_j\psi_{j,\alpha}(\mathbf{r}), \tag{18.377}$$

where α labels various states with common E_j and $\psi_{j,\alpha}(\mathbf{r})$ s form an orthonormal set

$$\langle\psi_{j,\alpha}|\psi_{k,\beta}\rangle = \delta_{jk}\delta_{\alpha\beta}. \tag{18.378}$$

By expanding $\psi(\mathbf{r},t)$ in terms of the complete set of states $\psi_{j,\alpha}$, i.e.

$$\psi(\mathbf{r},t) = \sum_{j,\alpha} C_{j,\alpha}(t)\psi_{j,\alpha}(\mathbf{r}), \tag{18.379}$$

and substituting in (18.374) we obtain the time-dependent coefficients of expansion

$$C_{j,\alpha}(t) = \frac{C_{j,\alpha}(0)}{\sqrt{N(t)}}\exp\left[-\frac{(2i+\kappa)E_jt}{2\hbar}\right], \tag{18.380}$$

where

$$N(t) = \sum_{j,\alpha}|C_{j,\alpha}(0)|^2\exp\left[-\frac{\kappa E_jt}{\hbar}\right]. \tag{18.381}$$

The solution given by Eq. (18.380) is interesting since it shows that all $|C_{j,\lambda}|^2$ s tend to zero as $t \to \infty$, except for the one corresponding to the lowest energy state (when the ground state is non-degenerate), i.e.

$$\psi(\mathbf{r},t) \to \exp\left(-\frac{iE_0t}{\hbar}\right), \quad \text{as} \quad t \to \infty. \tag{18.382}$$

For the general case of H we have the formal solution of (18.374) with the initial $\psi(\mathbf{r},t) = \psi_0(\mathbf{r})$ which is given by

$$\psi(\mathbf{r},t) = \frac{\exp\left[-\frac{(2i+\kappa)}{2\hbar}Ht\right]\psi_0(\mathbf{r})}{\langle\psi_0|\exp\left(-\frac{\kappa}{\hbar}Ht\right)|\psi_0\rangle^{\frac{1}{2}}}. \tag{18.383}$$

Motion of a Wave Packet in a Harmonic Potential According to Gisin's Model — As an example let us consider the motion of a damped harmonic oscillator in Gisin's formulation. Here we assume that the initial wave function is of the form of a displaced normalized Gaussian [75]

$$\psi(x,0) = \frac{\sqrt{\alpha}}{\pi^{\frac{1}{4}}}\exp\left[-\frac{\alpha^2}{2}(x-\xi_0)^2\right], \tag{18.384}$$

where $\alpha = \sqrt{\frac{m\omega}{\hbar}}$. We expand (18.384) in terms of the complete set of eigenfunctions of the harmonic oscillator $\psi_n(x)$

$$\psi(x,0) = \sum_{n=0}^{\infty} C_n(0)\psi_n(x). \tag{18.385}$$

Using Eqs. (18.384) and (18.385) we can determine $C_n(0)$ analytically;

$$C_n(0) = \int_{-\infty}^{\infty} \psi(x,0)\psi_n(x)dx = \frac{1}{\sqrt{2^n n!}} \left\{ (\alpha\xi_0)^2 \exp\left[-\left(\frac{\alpha\xi_0}{2}\right)^2\right] \right\}. \quad (18.386)$$

The time dependent normalization can be found from Eqs. (18.381) and (18.386);

$$N(t) = \sum_{0}^{\infty} |C_n|^2 \exp\left[-\kappa\left(n+\frac{1}{2}\right)\omega t\right] \exp\left[-\frac{1}{2}\alpha^2\xi_0^2\right]$$

$$= \exp\left(\frac{-\lambda t}{2}\right) \exp\left[-\frac{1}{2}\alpha^2\xi_0^2\left(1 - e^{\frac{-\lambda t}{2}}\right)\right], \quad (18.387)$$

where we have introduced a new damping constant $\lambda = \kappa\omega$.

Having found $N(t)$, we can determine $\psi(x,t)$ by utilizing Eqs. (18.379) and (18.380);

$$\psi(x,t) = \sum_{n=0}^{\infty} C_n \exp\left[\left(i\omega + \frac{\lambda}{2}\right)t\right]\psi_n(x)$$

$$= \frac{\sqrt{\alpha}}{\pi^{\frac{1}{4}}} \sum_{n=0}^{\infty} \frac{C_n(0)}{\sqrt{2^n n!}} \exp\left[-\left(i\omega + \frac{\lambda}{2}\right)t\right] H_n(\alpha x) \exp\left[-\frac{\alpha^2 x^2}{2}\right].$$

$$(18.388)$$

By substituting for $C_n(0)$ from (18.386) in (18.388) we can calculate $\psi(x,t)$ in closed form:

$$\psi(x,t) = \frac{\sqrt{\alpha}}{\pi^{\frac{1}{4}}} \exp\left[-\frac{1}{2}\alpha^2\left(x - \xi_0 e^{-\frac{\lambda t}{2}}\cos(\omega t)\right)^2\right.$$

$$\left. - i\left(\frac{1}{2}\omega t + \alpha^2 x\xi_0 \sin(\omega t)e^{-\frac{\lambda t}{2}} - \frac{1}{4}\alpha^2\xi_0^2 e^{-\lambda t}\sin(2\omega t)\right)\right].$$

$$(18.389)$$

The absolute square of $\psi(x,t)$ which is the probability density for this problem is

$$|\psi(x,t)|^2 = \frac{\alpha}{\pi^{\frac{1}{2}}} \exp\left[-\alpha^2\left(x - \xi_0 e^{\frac{-\lambda t}{2}}\cos(\omega t)\right)^2\right], \quad (18.390)$$

and this relation shows that the center of the packet oscillates as a damped harmonic oscillator;

$$\langle x \rangle = \xi(t) = \xi_0 \exp\left[-\frac{\lambda t}{2}\right]\cos(\omega t). \quad (18.391)$$

In the same way the momentum of the center of the wave packet can be obtained from the wave function (18.389)

$$\langle p \rangle = \int_{-\infty}^{\infty} \psi^*(x,t)\left(-i\hbar\frac{\partial \psi(x,t)}{\partial x}\right)dx = -m\omega\xi_0 \exp\left[-\frac{\lambda t}{2}\right]\sin(\omega t). \quad (18.392)$$

Thus $\langle x \rangle$ and $\langle p \rangle$ satisfy the equations of motion [73]

$$\frac{d\langle x \rangle}{dt} = -\frac{\lambda}{2}\langle x \rangle + \frac{1}{m}\langle p \rangle, \tag{18.393}$$

and

$$\frac{d\langle p \rangle}{dt} = -\frac{\lambda}{2}\langle p \rangle - m\omega^2\langle x \rangle. \tag{18.394}$$

By eliminating $\langle p \rangle$ between (18.393) and (18.394) we find the equation of motion for $\langle x \rangle$,

$$m\frac{d^2\langle x \rangle}{dt^2} + m\lambda\frac{d\langle x \rangle}{dt} + m\left(\omega^2 + \frac{\lambda^2}{4}\right)\langle x \rangle = 0, \tag{18.395}$$

and a similar equation for $\langle p \rangle$.

If we apply Gisin's formulation to a two-level system having energy levels E_i and E_f ($E_i > E_f$), we find that the energy of the system decreases with time as

$$\langle \psi(x,t)|H|\psi(x,t)\rangle = \frac{1}{2}(E_f+E_i) + \frac{1}{2}(E_f-E_i)\tanh\left[\frac{\kappa}{\hbar}(E_f - E_i)t\right] + \ln\left|\frac{C_i(0)}{C_f(0)}\right|. \tag{18.396}$$

This equation is found from Eqs. (18.377)-(18.381).

18.12   Gisin's Formulation of Dissipation for Systems Periodic in Time

Suppose that the Hamiltonian of a system is time-dependent, i.e.

$$H(t) = H(t+T), \tag{18.397}$$

and let us define the unitary evolution operator $U(t,t_0)$ by

$$i\frac{d}{dt}U(t,t_0) = H(t)U(t,t_0), \tag{18.398}$$

where we have set $\hbar = 1$. We also note that the solution of (18.398) must be subject to the condition

$$U(t,t) = 1. \tag{18.399}$$

From (18.397) and (18.398) we deduce the periodicity of U;

$$U(t+T, t_0+T) = U(t, t_0). \tag{18.400}$$

According to Floquet theorem for the family of unitary operators $U(t,t_0)$ satisfying (18.398) and (18.400) one can find a self-adjoint operator Q and a family of unitary periodic operators $P(t)$ such that

$$P(t+T) = P(t), \qquad P(0) = I, \tag{18.401}$$

$$U(t, t_0) = P(t) \exp\left[-iQ(t - t_0)\right] P^\dagger(t_0), \tag{18.402}$$

and

$$\left(H(t) - i\frac{d}{dt}\right) P(t) = P(t)Q, \tag{18.403}$$

where I is the unit operator. In these relations Q and $P(t)$ are related to the operator U by [80]

$$\exp(-iQT) = U(T, 0), \tag{18.404}$$

and

$$P(t) = U(t, 0) \exp(iQt). \tag{18.405}$$

Now if we introduce $|\Phi(t)\rangle$ by

$$|\Phi(t)\rangle = P(t)|\psi(t)\rangle, \tag{18.406}$$

and substitute it in the Schrödinger equation

$$i\frac{\partial |\psi(t)\rangle}{\partial t} = H(t)|\psi(t)\rangle, \tag{18.407}$$

we obtain

$$i\frac{\partial |\Phi(t)\rangle}{\partial t} = Q\,|\Phi(t)\rangle. \tag{18.408}$$

This result shows that Q is a "reduced Hamiltonian", and that it is not unique, due to the fact that we can add any multiple of $\frac{2\pi}{T}$ to its eigenvalues. On the other hand the eigenvectors of Q, (or of $U(T,0)$) are unique. We define the steady state $|\psi^r(t)\rangle$ by

$$|\psi^r(t)\rangle = e^{-irt} P(t)|\Phi^r(t)\rangle. \tag{18.409}$$

The phase e^{-irt} does not affect the state, and these steady states are periodic in time.

The Effect of Friction — In Gisin's model the friction is introduced in the wave equation by the addition of the nonlinear terms as in Eq. (18.374). Thus for the periodic Hamiltonian $H(t)$ we obtain

$$\frac{\partial}{\partial t}|\Phi(t)\rangle = -iQ|\Phi(t)\rangle + \frac{\kappa}{2}\left[\langle\Phi(t)|P^\dagger(t)H(t)P(t)|\Phi(t)\rangle - P^\dagger(t)H(t)P(t)\right]|\Phi(t)\rangle. \tag{18.410}$$

In the following subsection we try to formulate this problem as a time-independent eigenvalue problem using a method advanced by Shirley [76] for the solution of the undamped time-dependent Schrödinger equation when the Hamiltonian is periodic in time [77]-[79]. To this end we transform (18.403) into an infinite-dimensional matrix eigenvalue equation. Let us denote the atomic states in the matrix elements by Greek letters and denote the Fourier components by the Roman letters, and write

$$P_{\alpha,\beta}(t) = \sum_n P_{\alpha,\beta}^n e^{in\omega t} \tag{18.411}$$

and

$$\langle\alpha|H_F|\beta\rangle = \sum_n (H_F^n)_{\alpha,\beta}\, e^{in\omega t}. \tag{18.412}$$

In this relation H_F is the time-independent Floquet Hamiltonian written in the Floquet-state basis $|\alpha n\rangle = |\alpha\rangle \times |n\rangle$. Thus the operator equation (18.403) can be written as a matrix equation

$$\sum_{\beta,j}\langle\alpha,n|H_F|\beta,j\rangle P^j_{\beta,k} = q_k P^n_{\alpha,k}. \tag{18.413}$$

In these relations ω is the frequency associated with the period T defined by the Hamiltonian, $\omega = \frac{2\pi}{T}$, and q is the quasienergy eigenvalue. The Fourier transform of (18.412) yields the following set of equations

$$\langle\alpha,n|H_F|\beta,j\rangle = H^{n-j}_{\alpha,\beta} + n\omega\delta_{\alpha,\beta}\delta_{n,j}. \tag{18.414}$$

where both n and j are integers running from $-\infty$ to ∞.

The operator $\langle\alpha,n|H_F|\beta,j\rangle$ is an infinite matrix with rows defined by the pair α and j and columns by β and n. For instance for the Hamiltonian of a two-level system given by $H(t) = H_0 - \boldsymbol{\mu}\cdot\mathbf{E}\cos(\omega t)$, (see Eq. (18.436) below), the Floquet Hamiltonian can be written as [76]

$$
\begin{bmatrix}
\ddots & \cdots & \cdots & \cdots & \cdots & \cdots & \cdots & \cdots & \cdots \\
\cdots & E_\beta - 2\omega & g & 0 & 0 & 0 & 0 & 0 & 0 & \cdots \\
\cdots & g & E_\alpha - \omega & 0 & 0 & 0 & 0 & 0 & 0 & \cdots \\
\cdots & 0 & 0 & E_\beta - \omega & g & 0 & 0 & 0 & 0 & \cdots \\
\cdots & 0 & 0 & g & E_\alpha & 0 & 0 & g & 0 & \cdots \\
\cdots & 0 & g & 0 & 0 & E_\beta & g & 0 & 0 & \cdots \\
\cdots & 0 & 0 & 0 & 0 & g & E_\alpha + \omega & 0 & 0 & \cdots \\
\cdots & 0 & 0 & 0 & g & 0 & 0 & E_\beta + \omega & g & \cdots \\
\cdots & 0 & 0 & 0 & 0 & 0 & 0 & g & E_\alpha + 2\omega & \cdots \\
\cdots & \cdots & \cdots & \cdots & \cdots & \cdots & \cdots & \cdots & \ddots
\end{bmatrix}
\tag{18.415}
$$

Next let us consider the eigenvectors associated with the quasienergy eigenvalue

$$q_{kn} = q_k + n\omega, \tag{18.416}$$

and let us denote it by $|\lambda_{k,n}\rangle$, then

$$P^j_{\beta,k} = \langle\beta,j|\lambda_{k,0}\rangle. \tag{18.417}$$

These eigenvectors satisfy the periodicity condition

$$\langle\beta,j+l|\lambda_{k,n+l}\rangle = \langle\beta,j|\lambda_{k,n}\rangle. \tag{18.418}$$

Now let us change $|\Phi(t)\rangle$ to $|\chi(t)\rangle$, where

$$|\chi(t)\rangle = e^{iQt}|\Phi(t)\rangle, \tag{18.419}$$

and substitute for $|\Phi(t)\rangle$ in (18.410) to obtain the following nonlinear equation for $|\chi(t)\rangle$,

$$\frac{\partial}{\partial t}\chi(t) = \kappa\left\{\langle\chi(t)|B(t)|\chi(t)\rangle - B(t)\right\}|\chi(t)\rangle, \qquad (18.420)$$

where $B(t)$ is defined by

$$B(t) = e^{iQt}P^+(t)H(t)P(t)e^{-iQt}. \qquad (18.421)$$

In terms of the state basis $\{|\lambda_{k,0}\rangle\}$ the matrix elements of the operator B are

$$B_{k,k'} = \sum_{\alpha,\beta}\sum_{l,n,j} H^l_{\alpha,\beta}\langle\lambda_{k,0}|\alpha,n\rangle\langle\beta,j|\lambda_{k',0}\rangle \exp[i(l+j-n)\omega t]\exp[i(q_k-q_{k'})t],$$
$$(18.422)$$

where we have replaced $P_{\alpha,k(t)}$ in terms of its Fourier components;

$$P_{\alpha,k} = \sum_n P^n_{\alpha,k}e^{in\omega t} = \sum_n \langle\alpha,n|\lambda_{k,0}\rangle e^{in\omega t} \qquad (18.423)$$

Equation (18.420) is exact, but cannot be solved analytically. An approximate solution in closed form can be found if we take a long-time average on $B(t)$;

$$\bar{B}_{k,k'} = \sum_{\alpha,n}|\langle\alpha,n|\lambda_{k,0}\rangle|^2(q_k-n\omega)\delta_{k,k'}. \qquad (18.424)$$

This approximation is valid only if the damping constant κ is small [79],[80]. In this case $[\bar{B},\,Q] = 0$, and $|\psi(t)\rangle$ can be expressed by

$$|\psi(t)\rangle = \frac{P(t)\left[e^{-iQt}e^{-i\kappa\bar{B}(t-t_0)}e^{iQt_0}P^\dagger(t_0)\right]}{\sqrt{\langle\psi(t_0)P(t_0)e^{-iQt_0}e^{-\kappa\bar{B}(t-t_0)}e^{iQt_0}P^\dagger(t_0)|\psi(t_0)\rangle}}|\psi(t_0)\rangle. \qquad (18.425)$$

From Eqs. (18.424) and (18.425) we can construct the density matrix from $\psi(t)$

$$\begin{aligned}\langle\alpha|\rho|\beta\rangle &= \langle\alpha|\psi(t)\rangle\langle\psi(t)|\beta\rangle \\ &= N\sum_{k,k'}\sum_{n,j}\langle\alpha,n|\lambda_{k,0}\rangle\langle\lambda_{k',0}|\beta,j\rangle e^{-i(q_k-q_{k'})t}e^{-i(n-j)\omega t} \\ &\quad \times e^{-\frac{\kappa}{2}(\bar{B}_{k,k}+\bar{B}_{k',k'})t}C_kC^*_{k'},\end{aligned} \qquad (18.426)$$

where we have set $t_0 = 0$ and N and C_k are given by

$$N(t) = \left[\sum_k |C_k|^2 e^{-\kappa\bar{B}_{k,k}t}\right]^{-1}, \qquad (18.427)$$

and

$$C_k = \sum_{\alpha,n}\langle\lambda_{k,0}|\alpha,n\rangle\langle\alpha|\psi(0)\rangle. \qquad (18.428)$$

The approximate wave function and also the density matrix gives us the complete solution of the problem.

Forced Harmonic Oscillator — Consider a harmonic oscillator with angular frequency ω which is subject to a force $Fq\cos\omega t$. The Hamiltonian for such system is given by [80]

$$H(t) = \frac{1}{2}\omega_0\left(p^2 + q^2\right) + Fq\cos(\omega t), \qquad (18.429)$$

where q and p are the canonical position and momentum operators. For this problem the coherent state of the oscillator is preserved when the dissipation of the form (18.374) is introduced, and then the coherent states evolves in time as the corresponding classical problem [80].

For the Hamiltonian (18.429) the operators Q and P in Eq. (18.403) are known to be [80],[81],[82]:

$$Q = \frac{1}{2}\omega_0\left(p^2 + q^2\right) + \frac{\omega_0^2}{\omega_0^2 - \omega^2}q, \qquad \omega_0 \neq \omega, \qquad (18.430)$$

and

$$P(t) = e^{i\varphi(t)}\exp\left[\frac{iF}{\omega_0^2 - \omega^2}[\omega\sin(\omega t)q] + \omega_0[\cos(\omega t) - 1]p\right], \qquad (18.431)$$

where $\varphi(t)$ is a global phase of angular frequency ω. In this case the dissipative operator B given by (18.421) can be calculated from Eqs. (18.429) -(18.431) with the result that $B = Q$. Also here we find that the asymptotic solutions of (18.374) for the Hamiltonian (18.429) are the eigenstates of Q translated in position and momentum by the periodic operator $P(t)$. For instance the ground steady state of this system is a Gaussian with minimum spread centered around the oscillatory point, i.e. $\langle q(t)\rangle$ and $\langle p(t)\rangle$ are given by

$$\langle q(t)\rangle = -\frac{\omega_0 F}{\omega_0^2 - \omega^2}\cos(\omega t) \qquad (18.432)$$

and

$$\langle p(t)\rangle = \frac{\omega F}{\omega_0^2 - \omega^2}\sin(\omega t). \qquad (18.433)$$

When $\omega \to \omega_0$, i.e. the case of resonance, Q and $P(t)$ have the following forms

$$Q = \frac{1}{2}Fq - \frac{1}{8\omega}F^2 \qquad (18.434)$$

and

$$P(t) = e^{i\varphi(t)}\exp\left[-\frac{i\omega}{2}\cdot\left(p^2 + q^2\right)t\right]\exp\left\{\frac{iF}{4\omega}[(\cos(2\omega t) - 1)p - \sin(2\omega t)q]\right\}. \qquad (18.435)$$

In this case no operator corresponding to B of Eq. (18.421) exists. The reason is that when the friction is assumed to be small, at the resonance, the amplitude of oscillations will tend to infinity.

Dissipative Two-Level System in Gisin's Formulation — As a second example we want to consider a damped two-level system where the Hamiltonian is periodic and is given by [78]

$$H(t) = H_0 - \boldsymbol{\mu} \cdot \mathbf{E} \cos(\omega t), \tag{18.436}$$

where H_0 is the unperturbed Hamiltonian for the two-level system having the eigenvalues $E_\alpha = -\frac{1}{2}\omega_0$ and $E_\beta = \frac{1}{2}\omega_0$ and the eigenvectors $|\alpha\rangle$ and $|\beta\rangle$ respectively.

In Eq. (18.436) $\boldsymbol{\mu}$ is the dipole moment and $\mathbf{E}$ is the electric field amplitude. In the matrix representation $H(t)$ takes the form of

$$H(t) = \begin{bmatrix} -\frac{1}{2}\omega_0 & 2g\cos(\omega t) \\ 2g\cos(\omega t) & \frac{1}{2}\omega_0 \end{bmatrix}, \tag{18.437}$$

where g is the coupling constant

$$g = -\frac{1}{2}\langle\alpha|\left(-\boldsymbol{\mu}\cdot\mathbf{E}\right)|\beta\rangle. \tag{18.438}$$

Resonant transition between $|\alpha\rangle$ and $|\beta\rangle$ occurs whenever

$$E_\beta - E_\alpha = (2n+1)\omega, \quad n = 0, 1, 2, \cdots \tag{18.439}$$

Under this resonance condition we can apply Van Vleck's degenerate perturbation theory to reduce the infinite dimensional matrix (18.414) into 2×2 matrix [83]. Using this perturbation theory for $\omega_0 \approx \omega$ we find that H_F to the first order in wave function, and will reduce to the third order in energy reduces to

$$H_V = \begin{bmatrix} -\frac{1}{2}\omega_0 - \frac{g^2}{\omega+\omega_0} & g - \frac{g^3}{(\omega+\omega_0)^2} \\ g - \frac{g^3}{(\omega+\omega_0)^2} & \frac{1}{2}\omega_0 - \omega + \frac{g^2}{\omega+\omega_0} \end{bmatrix}, \tag{18.440}$$

The eigenvalues of H_V can be found from (18.413) and (18.440). They are given by $q_\pm$ where

$$q_\pm = -\frac{1}{2}\omega \pm \Omega. \tag{18.441}$$

In this relation Ω is the Rabi frequency [84],[85],

$$\Omega = \left\{ \Delta^2 + \left[g - \frac{g^3}{(\omega+\omega_0)^2} \right]^2 \right\}^{\frac{1}{2}}, \tag{18.442}$$

and Δ which is the detuning parameter is given by

$$\Delta = \frac{1}{2}(\omega - \omega_0) - \frac{g^2}{\omega + \omega_0}. \tag{18.443}$$

The long-time average of $B_{k,k'}$, i.e. $\bar{B}_{k,k'}$, Eq. (18.424) in this approximation reduces to $\bar{B}_{++}$ and $\bar{B}_{--}$ where

$$\bar{B}_{++} = -\bar{B}_{--} = \left\{ \Omega - \frac{\omega\Delta}{2\Omega} \left[1 + 2\left(\frac{g}{\omega + \omega_0}\right)^2 \right] \right\} \left[1 + \left(\frac{g}{\omega + \omega_0}\right)^2 \right]^{-1}. \tag{18.444}$$

An interesting result of the present approach is that there exists a frequency $\omega = \omega_c$ at which both $\bar{B}_{k,k'}$ and $\bar{B}_{k,k'}$ are zero and thus the effect of quantum dissipation disappears. If we set $\kappa = 0$, i.e. in the absence of frictional force and calculate the transition probability $\bar{P}_{\alpha \to \beta}$ as a function of $\frac{\omega_0}{\omega}$, we find that $\bar{P}_{\alpha \to \beta}$ is always less than or equal to $\frac{1}{2}$.

The problem of coupling of a charged particle to the electromagnetic field when the damping is assumed to be of Gisin's form is discussed in ref. [86]

18.13 Nonlinear Generalization of the Wave Equation

So far we have seen two completely different nonlinear wave equations describing damped quantum mechanical systems. First in Chapter 12 we obtained the Schrödinger-Langevin equation with logarithmic nonlinearity, and second, in this chapter we discussed Gisin's model with cubic nonlinearity. Now in this section we will consider other types of nonlinear wave equations for quantum dissipative motion.

We start with the Doebner-Goldin wave equation which has the following form [87]-[89]:

$$i\hbar \frac{\partial \psi}{\partial t} = \hat{H}\psi + iD\hbar G(\psi), \tag{18.445}$$

where $\hat{H} = -\frac{\hbar^2}{2m}\nabla^2 + V$, D is the diffusion constant and $G(\psi)$ is given by

$$G(\psi) = \nabla^2 \psi + \frac{|\nabla \psi|^2}{|\psi|^2}\psi. \tag{18.446}$$

Defining the probability density ρ and the current $\mathbf{j}$ in the usual way by

$$\rho = |\psi|^2, \tag{18.447}$$

and

$$\mathbf{j} = \frac{i\hbar}{2m}\left(\psi \nabla \psi^* - \psi^* \nabla \psi\right), \tag{18.448}$$

from (18.445) and its complex conjugate we find that the continuity equation in the case of this nonlinear equation is

$$\frac{\partial \rho}{\partial t} + \nabla \cdot \mathbf{j} = D\nabla^2 \rho, \tag{18.449}$$

that is here the continuity equation is given by the Fokker-Planck equation, and this is why this particular form of $G(\psi)$ was chosen [87]-[93].

One of the remarkable properties of Doebner-Goldin equation is that it is time reversible provided that the signs of t and D are changed simultaneously, therefore the sign of D determines the direction of flow of time in this equation. Equation (18.449) shows that $\int \rho d^3 r$ remains constant in time and also from (18.446) it follows that

$$\int \psi^* G(\psi)\, \psi d^3 r = 0. \tag{18.450}$$

From the wave equation we can determine $\langle \mathbf{r} \rangle$ and $\langle \mathbf{p} \rangle$ with the help of Ehrenfest's theorem

$$\frac{d}{dt}\langle \mathbf{r} \rangle = \frac{1}{m}\langle \mathbf{p} \rangle, \tag{18.451}$$

$$\frac{d}{dt}\langle \mathbf{p} \rangle = -\langle \nabla V \rangle + 2D \operatorname{Re} \int \psi^* (-i\hbar\nabla)\, G(\psi) d^3 r. \tag{18.452}$$

and

$$\frac{d}{dt}\left\langle i\hbar\frac{\partial}{\partial t} \right\rangle = \left\langle \frac{\partial V}{\partial t} \right\rangle = 2D \operatorname{Re} \int \psi H G(\psi) d^3 r \tag{18.453}$$

The additional term in (18.453) is somewhat similar to the effect of an imaginary external potential (Chapter 20) which acts on the particle. An examination of the two extra terms both proportional to D in $d\langle \mathbf{p} \rangle/dt$, Eq. (18.453), i.e.

$$-2m\int \psi^* \left(\frac{\mathbf{j}}{\rho}\right)\left(\frac{D\nabla^2\rho}{\rho}\right)\psi d^3 r + 2m\int \psi^* \left(\frac{D\nabla\rho}{\rho}\right)\left(-\frac{\nabla\cdot\mathbf{j}}{\rho}\right)\psi d^3 r \tag{18.454}$$

shows that in this approach there are two different kinds of frictions. The first term in (18.454) is a friction due to the velocity density $\left(\frac{\mathbf{j}}{\rho}\right)$, while the second term is the friction due to the diffusive velocity density $-D\frac{\nabla\rho}{\rho}$. The last term in Eq. (18.452) is related in a complicated way to the expectation value of the momentum operator $(-i\hbar\nabla)$ showing the presence of a damping term proportional to the diffusion coefficient D [87].

Solution of One-Dimensional Doebner-Goldin Equation — Now let us consider the one-dimensional motion where $V(x)$ is

$$V(x) = \frac{1}{2}m\omega_0 x^2 - f(t)x, \tag{18.455}$$

and a Gaussian wave function ψ given by

$$\psi(x,t) = N(t) \exp\left[-a(t)x^2 + b(t)x\right]. \tag{18.456}$$

Substituting (18.456) in (18.445) we find that $a(t)$ satisfies the nonlinear differential equation

$$\frac{da(t)}{dt} = \frac{2\hbar}{im}a^2(t) + \frac{im\omega_0^2}{2\hbar} - 8D\,a(t)\mathrm{Re}\,a(t). \qquad (18.457)$$

Equation (18.457) shows that $a(t)$ is a complex function. By writing it as

$$a(t) = \alpha(t) + i\beta(t), \qquad (18.458)$$

we can separate the real and imaginary parts in (18.457) and obtain two first order coupled nonlinear equations for $\alpha(t)$ and $\beta(t)$;

$$\frac{d\alpha(t)}{dt} = \frac{4\hbar}{m}\alpha(t)\beta(t) - 8D\alpha^2(t), \qquad (18.459)$$

and

$$\frac{d\beta(t)}{dt} = \frac{m\omega_0^2}{2\hbar} - \frac{2\hbar}{m}\left(\alpha^2(t) - \beta^2(t)\right) - 8D\alpha(t)\beta(t). \qquad (18.460)$$

We can also write for $\alpha(t)$ and $\beta(t)$ in terms of variances of the coordinate and momentum corresponding to this Gaussian wave function. To this end we introduce three functions, $\sigma_{xx}(t), \sigma_{xp}(t)$ and $\sigma_{pp}(t)$, by

$$\sigma_{xx}(t) = \frac{1}{4\alpha(t)}, \quad \sigma_{pp}(t) = \hbar^2\frac{\alpha^2(t) + \beta^2(t)}{\alpha(t)} \quad \text{and} \quad \sigma_{xp}(t) = -\frac{\hbar\beta(t)}{2\alpha(t)}, \qquad (18.461)$$

and substitute these for $\alpha(t)$ and $\beta(t)$ to obtain

$$\frac{d\sigma_{xx}(t)}{dt} = \frac{2}{m}\sigma_{xp}(t) + 2D, \qquad (18.462)$$

$$\frac{d\sigma_{xp}(t)}{dt} = \frac{1}{m}\sigma_{pp}(t) - m\omega_0^2\sigma_{xx}(t), \qquad (18.463)$$

and

$$\frac{d\sigma_{pp}(t)}{dt} = -2m\omega_0^2\sigma_{xp}(t) - 2D\frac{\sigma_{pp}(t)}{\sigma_{xx}(t)}. \qquad (18.464)$$

Of course these three functions $\sigma_{xx}(t), \sigma_{xp}(t)$ and $\sigma_{pp}(t)$ are not independent and from their definitions it is easy to show that

$$\sigma_{xx}(t)\sigma_{pp}(t) - \sigma_{xp}^2(t) = \frac{\hbar^2}{2m}. \qquad (18.465)$$

By eliminating $\sigma_{xp}(t)$ and $\sigma_{pp}(t)$ between Eqs. (18.462)-(18.464) we find that $\sigma_{xx}(t)$ satisfies a third order nonlinear differential equation

$$\frac{d^3\sigma_{xx}}{dt^3} + \frac{2D}{\sigma_{xx}}\frac{d^2\sigma_{xx}}{dt^2} + 4\omega_0^2\frac{d\sigma_{xx}}{dt} = 0. \qquad (18.466)$$

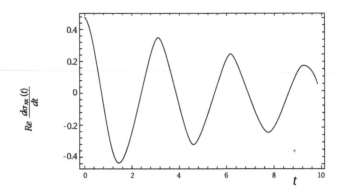

Figure 18.3: For the case of weak damping e.g. $D = 0.1$, the exact solution of the differential equation (18.466) for $\text{Re}[(d\sigma_{xx}/dt)]$ is plotted as a function of time.

We want to find a solution of (18.466) which deviates slightly from the stationary solution. For this we note that if we ignore the time dependence of $a(t)$, Eq. (18.457), as a first approximation we have

$$a(t) = \frac{m\omega_0}{2\hbar} \frac{1 - i\Gamma}{\sqrt{1 + \Gamma^2}}, \tag{18.467}$$

where $\Gamma = 2mD/\hbar$ is a dimensionless number. Thus to this order, from (18.457), (18.458) and (18.461) we find that σ_{xx} is given by

$$\sigma_{xx} = \frac{1}{\alpha} = \frac{\hbar \left(1 + \Gamma^2\right)^{\frac{1}{2}}}{2m\omega_0}. \tag{18.468}$$

Substituting for σ_{xx} in the coefficient $(2D/\sigma_{xx})$ in Eq. (18.466), and linearizing the resulting equation, for the approximate linearized solution we find

$$\frac{d\sigma_{xx}(t)}{dt} \approx \left(\frac{d\sigma_{xx}(t)}{dt}\right)_{t=0} \exp\left[\frac{-\omega_0 t}{\sqrt{1+\Gamma^2}}\left\{\Gamma \pm i\left(4 + 3\Gamma^2\right)^{\frac{1}{2}}\right\}\right], \tag{18.469}$$

i.e. the damping constant in this approximation is given by

$$\lambda = \frac{\omega_0 \Gamma}{\sqrt{1 + \Gamma^2}}. \tag{18.470}$$

In Fig. (18.3) the exact solution of the nonlinear differential equation (18.466) for $\text{Re}[\frac{d\sigma_{xx}}{dt}]$ is plotted as a function of time. This graph shows that for weak damping, $\text{Re}[\frac{d\sigma_{xx}}{dt}]$ follows the motion of a damped oscillator. A plot of the approximate solution given by (18.469) is indistinguishable from the exact solution for the range of t shown in the figure.

A Class of Nonlinear Wave Equation for the Description of Dissipative Systems — Earlier in this section we observed that by generalizing the continuity equation by replacing the standard continuity equation with the Fokker-Planck equation we arrive at a new set of nonlinear Schrödinger equation which can describe a dissipative motion. For an extension of this approach let us assume that any nonlinear generalization of the wave equation, on physical grounds, must be subject to the following restrictions, [90], [91]:

(a) The separability property, i.e. two noninteracting particle subsystems must remain uncorrelated. We have seen that this is also the property of the Schrödinger-Langevin equation (see Eq. (12.207)).

(b) The nonlinear equation must remain invariant under homogeneous transformation of the wave function, i.e. if ψ is a solution, $\alpha\psi$ (α is a constant) must also be a solution.

(c) When the potential V is zero, then the equation must obey Euclidean and time-reversal invariance.

(d) The locality condition, viz, the nonlinear part cannot contain space derivatives of orders higher than two.

Now let us consider a general type of the nonlinear Schrödinger equation of the form

$$i\hbar\frac{\partial\psi}{\partial t} = -\frac{\hbar^2}{2m}\nabla^2\psi + V\psi + F\left(\psi^*, \psi\right)\psi, \tag{18.471}$$

and let us determine the functional $F\left(\psi^*, \psi\right)$ in such a way that all of the above conditions (a)-(d) are satisfied [90],[91].

We first consider the general case where $F\left(\psi^*, \psi\right)$ is given by a complex function of ψ^* and ψ (or ρ and $\mathbf{j}$) subject to homogeneity and locality conditions (b) and (d);

$$\text{Re } F\left(\psi^*, \psi\right) = Dc_1\frac{\nabla\cdot\mathbf{j}}{\rho} + Dc_2\frac{\nabla^2\rho}{\rho} + Dc_3\frac{\mathbf{j}^2}{\rho^2}$$
$$+ Dc_4\frac{\mathbf{j}\cdot\nabla\rho}{\rho^2} + Dc_5\frac{(\nabla\rho)^2}{\rho^2}, \tag{18.472}$$

$$\text{Im } F\left(\psi^*, \psi\right) = \frac{1}{2}D\left(\frac{\nabla^2\rho}{\rho}\right). \tag{18.473}$$

Clearly if we set all c_i s in (18.472) equal to zero we find the Doebner-Goldin equation (18.445)-(18.446). On the other hand if we choose c_2, c_3. c_4 and c_5 all equal to zero, and also choose $c_1 = -\frac{\gamma\hbar}{D}$, and then let $D \to 0$, we obtain a real function for $F\left(\psi^*, \psi\right)$,

$$F\left(\psi^*, \psi\right) = -\gamma\hbar\left(\frac{\nabla\cdot\mathbf{j}}{\rho}\right). \tag{18.474}$$

In addition we want that the nonlinear wave equation to have the same stationary solutions as that of the linear equation, i.e. when $F\left(\psi^*, \psi\right) = 0$, and the condition for having stationary states is $\frac{\partial\rho}{\partial t} = 0$, therefore $\nabla\cdot\mathbf{j} = 0$. Thus the functional expression $F\left(\psi^*, \psi\right)$ has to be proportional to $\nabla\cdot\mathbf{j}$ for these states. At the same time $F\left(\psi^*, \psi\right) = 0$ should not contain higher powers or higher derivatives of $\nabla \cdot \mathbf{j}$, otherwise it would violate the condition of locality (d). Thus we choose $F\left(\psi^*, \psi\right)$ given by (18.474) to represent the nonlinearity in the wave equation.

By substituting for ρ and $\mathbf{j}$ from Eqs. (18.447) and (18.448) we find the nonlinear wave equation to be

$$i\hbar \frac{\partial \psi}{\partial t} = -\frac{\hbar^2}{2m} \nabla^2 \psi + V \psi - \frac{\hbar^2}{m} \gamma \, \text{Im} \left[\nabla^2 \ln \psi + (\nabla \ln \psi)^2 \right] \psi \qquad (18.475)$$

An Equivalent Form for the Nonlinear Wave Equation — By combining Eqs. (18.471) and (18.475) we find the following equation for ψ

$$i\hbar \left[\frac{\partial \psi}{\partial t} + i\gamma \left(\frac{1}{\rho} \frac{\partial \rho}{\partial t} \right) \psi \right] = -\frac{\hbar^2}{2m} \nabla^2 \psi + V \psi. \qquad (18.476)$$

The expression in the square bracket in (18.476) can be written as

$$\left[\frac{\partial \psi}{\partial t} + i\gamma \left(\frac{1}{\rho} \frac{\partial \rho}{\partial t} \right) \psi \right] = \frac{\partial \psi}{\partial t} + \rho^{-i\gamma} \frac{\partial \rho^{i\gamma}}{\partial t} \psi = \frac{1}{\rho^{i\gamma}} \frac{\partial}{\partial t} \left(\rho^{i\gamma} \psi \right). \qquad (18.477)$$

Thus if we multiply (18.476) by $\rho^{i\gamma}$ and change ψ to $\Psi = \rho^{i\gamma} \psi$, we obtain the following equation

$$i\hbar \frac{\partial \Psi}{\partial t} = -\frac{\hbar^2}{2m} \left(\nabla - i\gamma \mathbf{A}' \right)^2 \Psi + V \Psi, \qquad (18.478)$$

where $\mathbf{A}'$ is defined by

$$\mathbf{A}' = \nabla \ln \left(\Psi^* \Psi \right) = \nabla \ln \left(\psi^* \psi \right) = \nabla \ln \rho. \qquad (18.479)$$

Equation (18.478) is similar to the Schrödinger equation for a charged particle moving in an external electromagnetic field $\mathbf{A}'$. For stationary solutions of (18.478)

$$\frac{\partial \rho}{\partial t} = \frac{\partial}{\partial t} |\psi|^2 = \frac{\partial}{\partial t} |\Psi|^2 = 0, \qquad (18.480)$$

and in this case $\mathbf{A}'$ can be removed by a gauge transformation Eq. (12.122), and then Eq. (18.478) reduces to the standard Schrödinger equation.

When $\frac{\partial}{\partial t} \rho \neq 0$, the equation remains nonlinear, but it can be cast into a Schrödinger like equation when an "external electric field" $\mathbf{E} = -\nabla \left(\frac{\nabla \cdot \mathbf{j}}{\rho} \right)$ is present. But from the definition of $\mathbf{A}'$ it is evident that the "external magnetic field" is always zero. If we define the probability density and probability current by

$$R = \Psi^* \Psi = \psi^* \psi = \rho, \qquad (18.481)$$

and

$$\mathbf{J} = -\frac{i\hbar}{2m} \left(\Psi^* \nabla \Psi - \Psi \nabla \Psi^* \right), \qquad (18.482)$$

just as in Eq. (18.449) we obtain a new Fokker-Planck equation

$$\frac{\partial R}{\partial t} = -\nabla \cdot \mathbf{J} + \gamma \, \nabla^2 R. \tag{18.483}$$

This equation describes the diffusion of the probability density, thus γ can be regarded as the diffusion coefficient, and Eq. (18.475) can have dissipative solutions.

Ehrenfest's Theorem for the Nonlinear Wave Equation — Defining the mean values of coordinate, momentum and energy by $\langle \mathbf{r} \rangle$, $\langle \mathbf{p} \rangle = \langle -i\hbar \nabla \rangle$ and $\langle i\hbar \frac{\partial}{\partial t} \rangle$ respectively, then from Eqs. (18.471) and (18.475) we obtain

$$\frac{d}{dt} \langle \mathbf{r} \rangle = \frac{1}{m} \langle \mathbf{p} \rangle, \tag{18.484}$$

$$\frac{d}{dt} \langle \mathbf{p} \rangle = -\langle \nabla V \rangle - \gamma \left\langle \nabla \left(\frac{\nabla \cdot \mathbf{J}}{\rho} \right) \right\rangle, \tag{18.485}$$

and

$$\frac{d}{dt} \langle E_{quantum} \rangle = \frac{d}{t} \left\langle i\hbar \frac{\partial}{\partial t} \right\rangle = -\gamma \left\langle \left(\frac{\nabla \cdot \mathbf{J}}{\rho} \right)^2 \right\rangle. \tag{18.486}$$

This last relation shows that if $\nabla \cdot \mathbf{J} \neq 0$, then the mean energy decreases with time, and the motion is damped.

Wave Function for the Damped Nonlinear Oscillator — A number of solvable cases of Doebner-Goldin equation has been given in ref. [90]. Here as a simple example let us consider the one-dimensional problem where $V(x) = \frac{1}{2} K x^2$. To simplify the equations we choose the units where $\hbar = m = 1$. Then the wave equation for this problem becomes

$$i \frac{\partial \psi(x,t)}{\partial t} + \frac{1}{2} \frac{\partial^2 \psi(x,t)}{\partial x^2} - \frac{1}{2} \gamma \left\{ \frac{\partial}{\partial x} \left(\frac{\psi'(x,t)}{\psi(x,t)} \right) + \left(\frac{\psi'(x,t)}{\psi(x,t)} \right)^2 \right\} \psi(x,t)$$
$$= \frac{1}{2} K x^2 \psi(x,t), \tag{18.487}$$

where prime denotes derivative with respect to x. We use the trial Gaussian wave function

$$\psi(x,t) = \exp \left\{ -\frac{1}{2} \Phi(t) x^2 + \Gamma(t) x - \frac{1}{2} \Omega(t) \right\}, \tag{18.488}$$

and substitute this expression for $\prod \psi(x)$ in Eq. (18.487). By matching the factors on the two sides of (18.487) we obtain the following result:

$$\psi(x,t) = \left(\frac{1}{\sqrt{q(t)}} \right) \exp \left\{ -\frac{1}{2} i \left(\dot{s}(t) s(t) + I(t) + \frac{\dot{q}(t)}{q(t)} s^2(t) \right) \right\}$$
$$\times \exp \left\{ -\frac{1}{2} \left(\frac{1}{q^2(t)} - i \frac{\dot{q}(t)}{q(t)} \right) [x - s(t)]^2 + i\dot{s}(t) x \right\}, \tag{18.489}$$

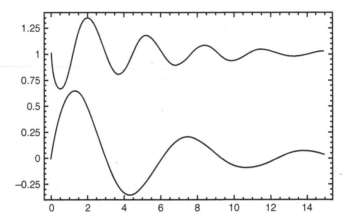

Figure 18.4: Plots of $q(t)$ (upper curve) and $s(t) = \langle x \rangle$ (lower curve), in units where $\hbar = m = 1$. In this figure the solutions of Eqs. (18.491) and (18.492) are displayed using the parameters $\gamma = 0.2$ and $K = 1$.

where

$$I(t) = \int_0^t \frac{1 + \gamma - \gamma \dot{s}(t') \, s(t')}{q^2(t')} dt'. \tag{18.490}$$

In Eq. (18.489), $q(t)$ and $s(t)$ are real functions and are solutions of the differential equations

$$\ddot{q}(t) + \frac{2\gamma}{q^2(t)} \dot{q}(t) - \frac{1}{q^3(t)} + Kq(t) = 0, \tag{18.491}$$

$$\ddot{s}(t) + \frac{\gamma}{q^2(t)} \dot{s}(t) + Ks(t) = 0, \tag{18.492}$$

where time derivatives are denoted by dots.

These equations for $q(t)$ and $s(t)$ are the classical equations of motion for particles with unit mass moving in the potentials

$$U_1(q) = \frac{1}{2q(t)} + \frac{1}{2} Kq^2, \quad \text{and} \quad U_2 = \frac{1}{2} Ks^2, \tag{18.493}$$

respectively and at the same time are subject to frictional forces. These particles tend to their equilibrium positions at the point $q = 1$ and $s = 0$, where the two potentials $U_1(q)$ and $U_2(s)$ have their minima (see Fig. 18.4).

Mean Position, Mean Momentum and Mean Energy — We can calculate the mean values of the coordinate, momentum and energy with the help of the exact wave function (18.489). Thus we find

$$\langle x \rangle = s(t), \tag{18.494}$$

$$\langle p \rangle = \dot{s}(t), \tag{18.495}$$

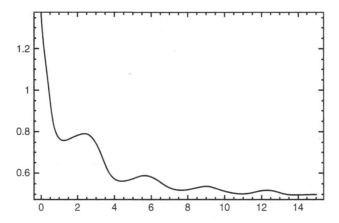

Figure 18.5: The mean energy, $\langle H \rangle$, calculated from Eq. (18.496), using the same values of γ and K of the previous figure.

and

$$\langle H \rangle = \left\langle i\hbar \frac{\partial}{\partial t} \right\rangle = \frac{1}{2} \left[\dot{s}^2(t) + K s^2(t) \right]$$

$$+ \frac{1}{4} \left[\dot{q}^2(t) + \frac{1}{q^2(t)} + K q^2(t) \right] - \frac{1}{2} \left(\gamma \frac{\dot{q}(t)}{q(t)} \right). \tag{18.496}$$

We note that in this model the mean energy oscillates with decreasing amplitude. Now asymptotically as $t \to \infty$, $s(t) \to 0$, $q(t) \to K^{-\frac{1}{4}}$, the latter being the position of the minimum of the potential $U_1(q)$. Also $\dot{s}(t) \to 0$, $\dot{q}(t) \to 0$ and $\langle H \rangle \to \frac{1}{4} K$, therefore the wave function (18.489), in this limit becomes

$$\psi(x,t) \to \exp\left(-\frac{1}{2}\sqrt{K}x^2\right) \exp\left(-\frac{i}{2}\sqrt{K}t\right), \tag{18.497}$$

i.e. the ground state wave function of the oscillator.

In addition to the potential $V(x) = \frac{1}{2}Kx^2$, for the motion of a particle in the absence of an external potential, the wave function can be found analytically [91].

18.14   Decaying States in a Many-Boson System

In solvable models designed to explain the decay of an unstable state that we have seen in this chapter, there is a central particle which is coupled to a group of noninteracting particles. When a charged particle is coupled to its own and/or to an external electromagnetic field, this assumption is valid. But what kind of decay is possible when the many-body system consists of a large number of particles interacting in pairs? As a solvable example we will study a many-boson system, and following Bassichis and Foldy we assume that N bosons are particles interacting in pairs? As a solvable example we will study a many-boson system, and following Bassichis and Foldy we assume that N bosons are constrained to move on a ring of radius L [94]-[97]. The many-body Hamiltonian for this system can be written as

$$H = \sum_{i=1}^{N} \left\{ \frac{1}{2mL^2} \frac{\partial^2}{\partial \theta_i^2} + \sum_{j \neq i} v(\theta_i - \theta_j) \right\}, \tag{18.498}$$

where θ_i and θ_j are the angular coordinates of the particles i and j. To simplify the problem we limit our discussion to the case where

$$v(\theta_i, \theta_j) = v(|\theta_i - \theta_j|). \tag{18.499}$$

The many-body wave function Ψ which is the solution of the Schrödinger equation

$$H\Psi = i\hbar \frac{\partial}{\partial t} \Psi, \tag{18.500}$$

has to be symmetric with respect to an interchange of the coordinates. We can write the Hamiltonian (18.498) in the second quantized form as

$$H = \sum_k \frac{\hbar^2 k^2}{2m} a_k^\dagger a_k + \frac{1}{2} \sum_{k_1, k_2, k_3, k_4} \langle k_2 k_4 | v(\theta_1, \theta_2) | k_1, k_2 \rangle a_{k_1}^\dagger a_{k_2}^\dagger a_{k_3} a_{k_4}, \tag{18.501}$$

where

$$\langle k_2 k_4 | v(\theta_1, \theta_2 | k_1 k_2) \rangle a_{k_1}^\dagger a_{k_2}^\dagger a_{k_3} a_{k_4}$$
$$= \frac{1}{(2\pi)^2} \int \int e^{-i[(k_1 - k_4) L \theta_1 + (k_2 - k_3) L \theta_2]} v(\theta_1 - \theta_2) d\theta_1 d\theta_2, \tag{18.502}$$

and where $\hbar k$ is the momentum of the particle. Now let us consider the special case of

$$v(\theta_1, \theta_2) = v(|\theta_1 - \theta_2|), \tag{18.503}$$

then the matrix element of the potential becomes

$$\langle k_3 k_4 | v(\theta_1, \theta_2 | k_1 k_2 \rangle = u(k_1 - k_2) \delta_{k_1 + k_2, \, k_3 + k_4}. \tag{18.504}$$

By substituting (18.504) in (18.502) we find a simpler Hamiltonian

$$H = \sum_k \frac{\hbar^2 k^2}{2m} a_k^\dagger a_k + \frac{1}{2} \sum_{k_1, k_2, k_3, k_4} u(k_1 - k_4) \, \delta_{k_1 + k_2, \, k_3 + k_4} a_{k_1}^\dagger a_{k_2}^\dagger a_{k_3} a_{k_4}, \tag{18.505}$$

If in addition we consider a situation where the contribution of all matrix elements of the potential except those related to the lowest single particle energies are ignorable, then we have to investigate a system with only zero momentum state and a degenerate excited state with equal and opposite momenta. Setting the kinetic energy of a single particle in the excited state equal to unity and denoting the excited momentum states by 1 and 3 corresponding to positive and negative momenta, we can reduce the Hamiltonian (18.505) to the following form:

$$
\begin{aligned}
H = {}& a_1^\dagger a_1 + a_3^\dagger a_3 + \frac{1}{2}\left(u(1) + u(3)\right) \\
& \times \left[a_0^\dagger a_0 \left(a_1^\dagger a_1 + a_3^\dagger a_3 \right) + a_0^{\dagger\,2} a_1 a_3 + a_0^2 a_1^\dagger a_3^\dagger \right] \\
& + \frac{1}{2} u(0) \left[a_0^\dagger a_0 \left(2 a_1 a_1 + a_0 a_0 + 2 a_3^\dagger a_3 \right) \right] \\
& + \frac{1}{2} u(0) \left(2 a_1^\dagger a_3^\dagger a_1 a_3 + a_1^\dagger a_1^\dagger a_1 a_1 + a_3^\dagger a_3^\dagger a_3 a_3 \right).
\end{aligned}
\tag{18.506}
$$

We will consider two special cases of this Hamiltonian. In the first case we set $u(0) = 0$, and choose $\frac{1}{2}(u(1)+u(3)) = g$, where g is a dimensionless coupling constant. From this Hamiltonian we derive an equation for the time-development of the wave function Ψ, Eq. (18.500). Then we will consider the effect of scattering of the particles in the 1 state by those in 2 state and also the particles in 2 by 1, by retaining only the term $Fg\left(a_1^\dagger a_3^\dagger a_1 a_3\right)$ in the Hamiltonian (18.506). Thus the reduced Hamiltonian can be written as

$$
H = a_1^\dagger a_1 + a_3^\dagger a_3 + g\left[a_2^\dagger a_2\left(a_1^\dagger a_1 + a_3^\dagger a_3\right) + a_2^2 a_1^\dagger a_3^\dagger + a_2^{\dagger\,2} a_1 a_3\right] - Fg a_1^\dagger a_1 a_3^\dagger a_3,
\tag{18.507}
$$

This Hamiltonian admits two constant of motion:

(a) The number operator N

$$
N = a_1^\dagger a_1 + a_2^\dagger a_2 + a_3^\dagger a_3,
\tag{18.508}
$$

and

(b) The difference between the number of particles in the states 1 and 3, i.e.

$$
\Delta = a_1^\dagger a_1 - a_3^\dagger a_3.
\tag{18.509}
$$

The time evolution of this many-body system is given by the time-dependent Schrödinger equation Eq. (18.500). This wave function depends on the number of particles in three different states. Let n be the number of particles in the state 3, i.e.

$$
a_3^\dagger a_3 \Psi = n\Psi,
\tag{18.510}
$$

then the wave function which depends on the number of particles in each of the three states can be written as

$$
|\Psi\rangle = \sum_{n=0}^{\frac{1}{2}(N-\Delta)} C_n |N, \Delta, n\rangle.
\tag{18.511}
$$

Substituting for H from Eq. (18.507) in (18.500) we find that the Schrödinger equation (18.500) can be transformed into a linear differential-difference equation,

$$g\left[(N-\Delta-2n+2)(N-\Delta-2n+1)(\Delta+n)n\right]^{\frac{1}{2}}C_{n-1}$$
$$+\left[\Delta+2n+g(N-\Delta-2n)(\Delta+2n)-Fgn(n+\Delta)-i\frac{d}{dt}\right]C_n$$
$$+g\left[(N-\Delta-2n-1)(N-\Delta-2n)(\Delta+n+1)(n+1)\right]^{\frac{1}{2}}C_{n+1}=0.$$
$$(18.512)$$

This equation for fixed Δ and N can be solved numerically. Here let us study the following simple case:

(a) We set $\Delta=0$, i.e. when the number of particles in the states 1 and 3 are equal.

(b) We take the limit of $N\to\infty$ and $g\to0$ in such a way that gN remains finite.

Then Eq. (18.512) will reduce to

$$i\frac{dC_n}{dt}=2n(1+gN)C_n+gN\left[nC_{n-1}+(n+1)C_{n+1}\right].\qquad(18.513)$$

We solve this equation with the initial conditions

$$C_0(t=0)=1,\quad\text{and}\quad C_n(t=0)=0,\quad\text{for}\ \ n\neq0.\qquad(18.514)$$

These initial conditions imply that at $t=0$ all of the N particles are in the state 2. Because of the interaction, n particles move to the state 3 with equal number moving to the state 1. The probability of having $N-2n$ particles left in 2 at t is given by $|C_n(t)|^2$.

In order to solve (18.513) we first obtain the solution of the first order partial differential equation

$$\frac{\partial G}{\partial t}+\left[\frac{1}{L}+\frac{2L}{\pi}gN\sin\left(\frac{\pi x}{L}\right)\right]\frac{\partial G}{\partial x}+gN\exp\left(\frac{i\pi x}{L}\right)G=0,\qquad(18.515)$$

subject to the boundary conditions

$$G(L,t)=G(-L,t),\quad G(x,t=0)=(2L)^{-\frac{1}{2}}.\qquad(18.516)$$

The partial differential equation (18.515) is found by the method of generating function which we used in Chapter 7 in connection with the Schr"odinger chain.

If ω^2, which is defined by

$$\omega^2=\frac{\pi^2}{4L^2}-N^2g^2,\qquad(18.517)$$

is positive, then $G(x,t)$ will be a periodic function of time and is given by

$$G(x,t)=\frac{(2L)^{-\frac{1}{2}}\exp\left(\frac{i\pi t}{2L^2}\right)}{\left\{\cos\omega t+\frac{gN}{\omega}\left[\exp\left(\frac{i\pi x}{L}\right)+\frac{i\pi}{2L^2gN}\right]\sin\omega t\right\}}.\qquad(18.518)$$

By expanding $G(x,t)$ in terms of the orthogonal set $(2L)^{-\frac{1}{2}} \exp\left(\frac{in\pi x}{L}\right)$, i.e.

$$G(x,t) = (2L)^{-\frac{1}{2}} \sum_{n=0}^{\infty} D_n(t) \exp\left(\frac{in\pi x}{L}\right), \qquad (18.519)$$

and substituting in (18.515) we find that D_n must satisfy the differential-difference equation;

$$\frac{dD_n}{dt} = -\frac{in\pi}{L^2} D_n + gN\left[(n+1)D_{n+1} - nD_{n-1}\right], \qquad (18.520)$$

thus showing that $G(x,t)$ is the generating function for the differential-difference equation (18.520). At the same time we observe that if we expand (18.518) in powers of $\exp\left(\frac{i\pi t}{2L^2}\right)$ and compare the result with (18.520) we obtain D_n for the case when $\frac{\pi}{2L^2} > gN$,

$$D_n(t) = \left(-\frac{gN}{\omega}\right)^n \frac{\exp\left(\frac{i\pi t}{2L^2}\right)\sin^n \omega t}{\left\{\cos\omega t + \frac{i\pi}{2\omega L^2}\sin\omega t\right\}^{n+1}}. \qquad (18.521)$$

Now we return to our original set of equations for $C_n(t)$ and compare (18.513) and (18.520) to find $C_n(t)$

$$C_n(t) = i^n D_n(t) = \left(-\frac{igN}{\omega}\right)^n \frac{\exp\left(i(1+gN)t\right)\sin^n \omega t}{\left\{\cos\omega t + \frac{i(1+gN)}{\omega}\sin\omega t\right\}^{n+1}}, \qquad (18.522)$$

where

$$\omega^2 = (1+gN)^2 - g^2 N^2 = 1 + 2gN. \qquad (18.523)$$

For g negative and less than $\left[-(2N)^{-1}\right]$, ω^2 in (18.523) becomes negative

$$-\omega^2 \to \nu^2 = 2|g|N - 1, \qquad (18.524)$$

where ν^2 is a positive number. For this case $C_n(t)$ satisfies the equation

$$i\frac{dC_n}{dt} = -2\nu^2 n C_n + |g|N\left[(n+1)C_{n+1} + nC_{n-1}\right]. \qquad (18.525)$$

We can solve (18.525) as before with the result that

$$C_n(t) = \left(-\frac{|g|N}{\nu}\right)^n \frac{\exp\left(i(|g|N-1)t\right)\sinh^n \nu t}{\left\{\cosh\nu t + \frac{i(|g|N-1)}{\nu}\sinh\nu t\right\}^{n+1}}. \qquad (18.526)$$

A simpler version of this model where the Hamiltonian is given by

$$H = ig\left(a_2^{\dagger 2} a_3 a_1 - a_3^{\dagger} a_1^{\dagger} a_2^2\right) \qquad (18.527)$$

is also exactly solvable. Here we have the same constants of motion N and Δ as in (18.512). Again for $\Delta = 0$ and in the limit of large N, and gN finite the time-dependent Schrödinger equation reduces to

$$\frac{dC_n}{dt} = gN\left[(n+1)C_{n+1} - nC_{n-1}\right]. \tag{18.528}$$

This equation is a special case of (18.520) and for this model the initial state $C_0(t)$ decays with the probability

$$|C_0(t)|^2 = \frac{1}{\cosh^2(gNt)}. \tag{18.529}$$

An interesting feature of the model described by the Hamiltonian (18.507) is that for finite but large N it has a degenerate ground state, and this corresponds to the value of F for which the classical limit of the system possesses a special symmetry [97]. The quantal Hamiltonian has an approximate symmetry for the same value of F. Thus if we choose $\Delta = 0$, set F equal to $\frac{4}{gN}$, and use the approximations

$$[(N - 2n + 2)(N - 2n + 1)]^{\frac{1}{2}} \approx (N - 2n + 2), \tag{18.530}$$

and

$$[(N - 2n - 1)(N - 2n)]^{\frac{1}{2}} \approx (N - 2n), \tag{18.531}$$

in Eq. (18.512), then the resulting equation which is

$$gn(N - 2n + 2)C_{n-1} + \left[2n(N - 2n)\left(g + \frac{1}{N}\right) - i\frac{d}{dt}\right]C_n$$
$$+ g(n+1)(N - 2n)C_{n+1} = 0, \tag{18.532}$$

will remain invariant under the transformation

$$n \to \frac{1}{2}N - j, \quad C_{n\pm1} \to C_{\frac{1}{2}N-(j\mp1)} = C_{j\mp1}. \tag{18.533}$$

Therefore if we find the solution of (18.512) with $\Delta = 0$ and $F = \frac{4}{gN}$ and with the initial conditions (18.514) the result would be the same as that of solving (18.512) for the same Δ and F but with the initial conditions

$$C_{\frac{N}{2}}(t = 0) = 1, \quad C_n(t = 0) = 0, \quad \text{for} \quad n \neq 0. \tag{18.534}$$

For the general case with N finite and large, the system will return to its original state but after a very long time. The decay except for very short time is exponential. Let us calculate the probability of finding the many-boson system in its original state where all of the particles are in the state 2, or zero momentum state, this is given by $|C_0(t)|^2$. The curves shown in Fig. (18.6) are calculated for $N = 200$, $g = -0.05$ and $F = 0$. For these values the numerical solution of (18.512) is very close to the approximate but analytic solution given

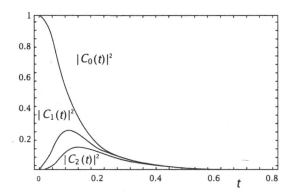

Figure 18.6: The probability of decay of the initial state $|C_0(t)|^2$ is plotted as a function of time. For comparison $|C_1(t)|^2$ and $|C_2(t)|^2$ are also shown.

by (18.526). In addition to $|C_0(t)|^2$, the time-dependence of $|C_1(t)|^2$ and $|C_2(t)|^2$ are also displayed.

The equations of motion derived from the Bassichis-Foldy model are very similar to the classical equations describing the modulational stability of a Langmuir condensate [98].

References

[1] For a detailed discussion of these aspects of decay and examples from quantum tunneling see M. Razavy, *Quantum Theory of Tunneling*, Second Edition, (World Scientific, Singapore, 2014) Chapter 2.

[2] L. Fonda, G.C. Ghirardi, A. Ramini and T. Weber, Nuovo Cimento, 15 A, 689 (1973).

[3] L. Fonda, G.C. Ghirardi, and A. Ramini, Rep. Prog. Phys. 41, 587 (1978).

[4] G.N. Fleming, Nuovo Cimento, 16 A, 232 (1973).

[5] R.E.A.C. Paley and N. Wiener, *Notes on the Theory and Application of Fourier Transform*, Trans. Am. Math. Soc. 35 (1933).

[6] M.W. Goldberger and K.M. Watson, *Collision Theory*, (John Wiley & Sons, New York, N.Y. 1964).

[7] L.A. Khalfin, Zh. eksper. teor Fiz. 33, 1371 (1957).

[8] K.J.F. Gaemers and T.D. Visser, Physica, A 153, 234 (1988).

[9] K. Bhattacharya, J. Phys. A 16, 2993 (1983).

[10] J. Uffink, Am. J. Phys. 61, 935 (1993).

[11] A comprehensive discussion of different forms of time-energy uncertainty relations can be found in P. Busch's article "The time-energy uncertainty relation" in *Time in Quantum Mechanics*, edited by J.G. Muga, R. Sala Mayato and I.L. Egusquiza (Springer, Berlin, 2002).

[12] V.V. Dodonov and A.V. Dodonov, preprint, arXiv:1504.00862 (1015).

[13] L. Mandelstam and I. Tamm, J. Phys. (USSR) 9, 249 (1945).

[14] M. Bauer and P.A. Mello, Ann. Phys. (NY), 111, 38 (1978).

[15] A. Peres, Ann. Phys. (NY), 129, 33 (1980).

[16] M.D. Arthur, H. Brysk, S.L. Paveri-Fontana and P.F. Zweifel, Nuovo Cimento, 63 B, 565 (1981).

[17] W. Band, Am. J. Phys. 30, 646 (1962).

[18] S. Bloom, J. Appl. Phys. 28, 800 (1957).

[19] M. Razavy, *Quantum Theory of Tunneling*, Second Edition, (World Scientific, Singapore, 2014) Chapter 8.

[20] For a detailed discussion of the two-level system coupled to a heat bath see U. Weiss, *Quantum Dissipative Systems*, Fourth Edition, (World Scientific, Singapore, 2012) Part IV.

[21] E.C. Behrman and P.G. Wolynes, J. Chem. Phys. 83, 5863 (1985).

[22] A. Amann, J. Chem. Phys. 96, 1317 (1992).

[23] A. Caldeira and A. Leggett, Phys. Rev. Lett. 46, 211 (1981).

[24] W. Zwerger, Z. Physik B 53, 53 (1983).

[25] K.W.H. Stevens, Proc. Roy. Soc. 72, 1027 (1958).

[26] K.W.H. Stevens, and B. Josephson Jr. Proc. Phys. Soc. 74, 561 (1959).

[27] K.W.H. Stevens, Proc. Phys. Soc. 77, 515 (1961).

[28] J.E. Carroll, Int. J. Electronics, 54, 139 (1983).

[29] J.D. Jackson, *Classical Electrodynamics*, Third Edition (John-Wiley & Sons, New York, N.Y. 1999) Sec. 6.7.

[30] See for instance G. Baym *Lectures on Quantum Mechanics*, (W.A. Benjamin, Reading, 1976) p. 315.

[31] A.M. Kowalski, A. Plastino and A.N. Proto, Phys. Rev. E 52, 165 (1995).

[32] V. Weisskopf and E. Wigner Z. Phys. 63, 54 (1930).

[33] The English translation of the paper of Weisskopf and Wigner is reprinted in W.R. Hindmarch, *Atomic Spectra*, (Pergamon Press, Oxford, 1967).

[34] J.L. Martin, Proc. Camb. Phil. Soc. 60, 587 (1964).

[35] R.J. Newton, *Scattering Theory of Waves and Particles*, Second Edition, (Springer-Verlag, New York, N.Y. 1982) Sec. 17.3.3.

[36] M. Razavy, Can. J. Phys. 45, 1469 (1967).

[37] D.J. Diestler and H.D. Ladouceur, Opt. Commun. 20, 6 (1977).

[38] G.W. Ford, J.T. Lewis and R.F. O'Connell, Phys. Rev. Lett. 55, 2273 (1985).

[39] G.W. Ford, and R.F. O'Connell, J. Stat. Phys. 57, 803 (1989).

[40] G.W. Ford and R.F. O'Connell, Appl. Phys. B 60, 301 (1995).

[41] G.W. Ford and R.F. O'Connell, Phys. Lett. 158, 31 (1991).

[42] G.W. Ford and R.F. O'Connell, Phys. Lett. 157, 217 (1991).

[43] G.W. Ford and R.F. O'Connell, Phys. Lett. 174, 182 (1993).

[44] J. McConnell, *Quantum Particle Dynamics*, Second Edition (North-Holland, Amsterdam, 1960) Chapters VII and VIII.

[45] G. Baym, *Lectures on Quantum Mechanics*, (W.A. Benjamin, Reading, 1976) Chapter 23.

[46] M.Razavy, *Heisenberg's Quantum Mechanics*, (World Scientific, Singapore, 2011) Chapter 18.

[47] N. Rosen, Phys. Rev. 87, 940 (1952).

[48] N. Rosen, Phys. Rev. 88, 1434 (1952).

[49] R.P. Feynman, Phys. Rev. 97, 660 (1955).

[50] R.P. Feynman and A. Hibbs, *Quantum Mechanics and Path Integrals*, (McGraw-Hill, New York, N.Y. 1965).

[51] D. P. L. Castrigiano and N. Kokiantonis, Phys. Lett. 96 A, 55 (1983).

[52] D. P. L. Castrigiano and N. Kokiantonis, Phys. Rev. A 35, 4122 (1987).

[53] D. P. L. Castrigiano and N. Kokiantonis, Phys. Rev. A 38, 527 (1988).

[54] D. P. L. Castrigiano, N. Kokiantonis, Phys. Lett. 104 A. 123 (1984).

[55] D. P. L. Castrigiano, N. Kokiantonis, and H. Stierstorfer, Nuovo Cimento 108 B, 765 (1993).

[56] See for instance, M. Razavy, *Heisenberg's Quantum Mechanics*, (World Scientific, Singapore, 2011) p. 590.

[57] A.O. Caldeira and A. Leggett, Ann. Phys. (NY) 149, 374 (1983).

[58] P.L. Knight and P.W. Milonni, Phys. Lett. 56 A, 275 (1976).

[59] G. Morpurgo, Nuovo Cimento, 9, 808 (1952).

[60] A. Loinger, Nuovo Cimento, 2, 511 (1955).

[61] M.D. Crisp, Phys. Rev. A 42, 3703 (1990).

[62] M. Ozaki and S. Sasabe, Phys. Rev. A 80, 024102 (2009).

[63] J.D. Jackson, *Classical Electrodynamics*, Third Edition (John Wiley & Sons, New York, N.Y. 1999) Sec. 6.5.

[64] W. Heitler, *The Quantum Theory of Radiation*, Third Edition (Oxford University Press, London 1954) p. 181.

[65] M. Razavy and E.A. Henley Jr., Can. J. Phys. 48, 2439 (1970).

[66] H.H. Chan and M. Razavy, Chem. Phys. Lett. 2, 202 (1968).

[67] F. Haake and W. Weidlich, Z. Phys. 213, 445 (1968).

[68] P.M. Morse and H. Feshbach, *Methods of Theoretical Physics*, Vol. II (McGraw-Hill, New York, N.Y. 1953) p. 1754.

[69] H. Zhang, Eur. Phys. J. B 38, 559 (2004).

[70] R. Silbey and R.A. Harris, J. Chem. Phys. 80, 2615 (1984).

[71] R.A. Harris and R. Silbey, J. Chem Phys. 78, 7330 (1983).

[72] A.J. Leggett, S.A. Chakravarty. A.T. Dorsey, M.P.A. Fisher, A.Garg and W. Zwerger, Rev. Mod. Phys. 59, 1 (1987).

[73] N.Gisin, J. Phys. A 14, 2259 (1981).

[74] N. Gisin, Physica A 111, 364 (1982).

[75] See for example L.I. Schiff, *Quantum Mechanics*, Third Edition (McGraw-Hill, New York, N.Y. 1968).

[76] J.H. Shirley, Phys. Rev. 138 B, 979 (1965).

[77] Y. Huang, S-I. Chu and J.O. Hirschfelder, Phys. Rev. A 40, 4171 (1989).

[78] J.O. Hirschfelder, Adv. Chem. Phys. 73, 1 (1989).

[79] K. Aravind and J.O. Hirschfelder, I. Phys. Chem. 88, 4788 (1984).

[80] N. Gisin, Found. Phys. 13, 643 (1983).

[81] G.S. Agarwal and E. Wolf, Phys. Rev. 2 D, 2206 (1970).

[82] R.P. Feynman, Phys. Rev. 84, 108 (1951).

[83] H. Van Vleck, Phys. Rev. 33, 467 (1929).

[84] I.I. Rabi, Phys. Rev. 51, 652 (1937).

[85] M. Fox, *Quantum Optics: An Introduction*, (Oxford University Press, Oxford, 2006) p. 178.

[86] A. Pimpale and M. Razavy, Phys. Rev. A 36, 2739 (1987).

[87] H.-D. Doebner and G.A. Goldin, Phys. Lett. A 162, 397 (1992).

[88] H.-D. Doebner and G.A. Goldin, Phys. Rev. A 54, 3764 (1996).

[89] P. Guerrero, J.L. López, J. Montejo-Gámez and J. Nieto, J. Nonlinear Sci. 22, 631 (2012).

[90] P. Nattermann, W. Scherer and A.G. Ushveridze, Phys. Lett. A 184, 234 (1994).

[91] A.G. Ushveridze, Phys. Lett. A 185, 123 (1994).

[92] V.V. Dodonov, J. Korean Phys. Soc. 26, 111 (1993).

[93] V.V. Dodonov and S.S. Mizrahi, Physica, A 214, 619 (1995).

[94] W.H. Bassichis and L.L. Foldy, Phys. Rev. A 133, 935 (1964).

[95] W.E. Bassichis, Ph.D. Thesis, Case Institute of Technology, (1963).

[96] M. Razavy Int. J. Theor. Phys. 13, 237 (1975).

[97] M. Razavy and R.B. Ludwig, Phys. Rev. A 33, 1519 (1985).

[98] R.O. Dendy and D. ter Haar, J. Plasma Phys. 31, 67 (1984).

Chapter 19

Dissipation Arising from the Motion of the Boundaries

In quantum mechanics a change in the boundaries of a system has an effect similar to the action of an external force. There are many examples where the motion of the boundaries play an important role in the dynamical behavior of the system. These include problems from relativity [1],[2], quantum physics [3] statistical mechanics [4], and condensed matter physics [5].

Mathematically, these wave mechanical problems are similar to the diffusion problems with moving boundary conditions, but with imaginary diffusion coefficient. Therefore the methods used for solving diffusion problems can also be applied to these problems. For a list of exactly solvable cases the reader is referred to [6] and [7].

The simplest model of this type of decay is the one where there are N noninteracting bosons occupying different energy levels of a cavity of length L_0 at $t = 0$. Now if the length of the cavity increases (or decreases) as a function of time according to the relation

$$L(t) = \zeta(t)L_0, \tag{19.1}$$

where $\zeta(t)$ is a continuous and twice differentiable function of time, then the particles will make transition from the original levels to lower or higher levels by exchanging energy with the moving walls of the cavity. Thus in this problem the coupling is between each individual particle and the walls of the cavity. The Hamiltonian for these noninteracting particles which we assume to satisfy the Schrödinger equation is given by

$$H_0 = -\frac{i\hbar}{2m} \int_0^{L(t)} \frac{\partial \pi(x,t)}{\partial x} \frac{\partial \psi(x,t)}{\partial x} dx, \tag{19.2}$$

503

where $\psi(x,t)$ is the field amplitude and $\pi(x,t)$ is its conjugate momentum density. These operators must satisfy the boundary conditions at the walls of the cavity

$$\psi(0,t) = \psi(L(t),t) = 0, \tag{19.3}$$

and

$$\pi(0,t) = \pi(L(t),t) = 0, \tag{19.4}$$

as well as the canonical commutation relation [8]

$$[\psi(x',t), \pi(x'',t)] = i\hbar\delta(x' - x''). \tag{19.5}$$

As it was mentioned earlier imposing these types of boundary conditions within the framework of quantum theory can be problematic, Sec. 1.11, but here we will use them in order to get a manageable system of equations.

The Hamiltonian (19.2) is defined over a length $L(t)$ which depends on time. To remove the time-dependence from the boundary at $x = L(t)$ and show that the motion of the boundary is equivalent to the action of a force, we use two sets of unitary transformations and we denote these by $U_1(t)$ and $U_2(2)$ respectively. The first operator acting on the Schrödinger equation

$$H_0\psi = i\hbar\frac{\partial\psi}{\partial t}, \tag{19.6}$$

changes the Hamiltonian H_0 to H_1 where

$$H_1 = \exp(iU_1(t))\left(H_0 - i\hbar\frac{\partial}{\partial t}\right)\exp(-iU_1(t)). \tag{19.7}$$

We choose $U_1(t)$ to be given by the following integral:

$$U_1(t) = \frac{1}{2\hbar}\ln\zeta(t)\int_0^{L(t)} 2x'\frac{\partial\pi(x',t)}{\partial x'}\psi(x')\,dx'. \tag{19.8}$$

Using the Baker-Campbell-Hausdorff theorem, Eq. (11.116), we find

$$[U_1(t), \pi(x)] = i\ln(\zeta(t))x\frac{\partial\pi}{\partial x}, \tag{19.9}$$

and therefore from (19.8) it follows that

$$\exp(iU_1(t))\pi(x)\exp(-iU_1(t)) = \exp\left[-(\ln\zeta(t))x\frac{\partial}{\partial x}\right]\pi(x) = \pi\left(\frac{x}{\zeta(t)}\right). \tag{19.10}$$

Similarly for the $\psi(x)$ we have

$$\exp(iU_1(t))\psi(x)\exp(-iU_1(t)) = \frac{1}{\zeta(t)}\psi\left(\frac{x}{\zeta(t)}\right). \tag{19.11}$$

The factor $\frac{1}{\zeta(t)}$ in front of $\frac{1}{\zeta(t)}\psi\left(\frac{x}{\zeta(t)}\right)$ in (19.11) introduces an asymmetry between $\psi(x,t)$ and $\pi(x,t)$ fields. We can make the transformed ψ and π symmetrical by means of a second unitary transformation $\exp(iU_2(t))$, where

$$U_2(t) = \frac{1}{2\hbar}\ln\zeta(t)\int_0^{L(t)}\pi\left(\frac{x'}{\zeta(t)}\right)\psi\left(\frac{x'}{\zeta(t)}\right)\frac{dx'}{\zeta(t)}. \qquad (19.12)$$

Evaluating the transformed field operators using (11.116) we get

$$\exp(iU_2(t))\frac{1}{\zeta(t)}\psi\left(\frac{x}{\zeta(t)}\right)\exp(-iU_2(t)) = \frac{1}{\sqrt{\zeta(t)}}\psi\left(\frac{x}{\zeta(t)}\right), \qquad (19.13)$$

and

$$\exp(iU_2(t))\pi\left(\frac{x}{\zeta(t)}\right)\exp(-iU_2(t)) = \frac{1}{\sqrt{\zeta(t)}}\pi\left(\frac{x}{\zeta(t)}\right). \qquad (19.14)$$

Thus the transformation $\exp(iU_2(t))\exp(iU_1(t))$ is a unitary transformation which leaves the commutation relation (19.5) invariant:

$$\left[\frac{1}{\sqrt{\zeta(t)}}\psi\left(\frac{x'}{\zeta(t)},t\right),\frac{1}{\sqrt{\zeta(t)}}\pi\left(\frac{x''}{\zeta(t)},t\right)\right] = i\hbar\delta\left(x'-x''\right). \qquad (19.15)$$

Next let us consider the effect of the transformation $\exp(iU_1(t))$ on the operator $\left[H_0 - i\hbar\frac{\partial}{\partial t}\right]$;

$$H_1 = e^{iU_1(t)}\left[H_0 - i\hbar\left(\frac{\partial}{\partial t}\right)\right]e^{-iU_1(t)}$$

$$= \int_0^{L(t)}\left[\frac{-i\hbar}{2m\zeta(t)}\left\{\frac{\partial}{\partial x}\pi\left(\frac{x}{\zeta(t)}\right)\right\}\left\{\frac{\partial}{\partial x}\psi\left(\frac{x}{\zeta(t)}\right)\right\}\right.$$

$$\left. - \frac{\dot{\zeta}(t)}{\zeta^2(t)}x\left\{\frac{\partial}{\partial x}\pi\left(\frac{x}{\zeta(t)}\right)\psi\left(\frac{x}{\zeta(t)}\right)\right\}\right]dx. \qquad (19.16)$$

By changing x to $\xi = \frac{x}{\zeta(t)}$, we note that the integrand in (19.16) can be written in terms of ξ alone, and then the upper limit becomes L_0 independent of time. Now let us consider the effect of the second transformation:

$$H_2 = e^{iU_2(t)}\left[H_1 - i\hbar\left(\frac{\partial}{\partial t}\right)\right]e^{-iU_2(t)}$$

$$= \int_0^{L_0}\left[\frac{-i\hbar}{2m\zeta^2(t)}\left(\frac{\partial\pi}{\partial\xi}\frac{\partial\psi}{\partial\xi}\right) - \frac{\dot{\zeta}(t)}{\zeta(t)}\left\{\xi\frac{\partial\pi(\xi)}{\partial\xi}\psi(\xi) + \frac{1}{2}\pi(\xi)\psi(\xi)\right\}\right]d\xi. \qquad (19.17)$$

Finally to simplify the result we perform a third unitary transformation of the form $\exp(iS(t))$;

$$S(t) = -\frac{i}{\hbar}\int_0^{L_0}\beta(t)\eta^2\pi(\eta)\psi(\eta)d\eta, \qquad (19.18)$$

where $\beta(t)$ is a function of time which will be determined later. The final operator, a Hamiltonian which we denote by H_{eff}, is obtained by applying this last transformation:

$$
H_{eff} = e^{iS(t)} \left[H_2 - i\hbar \left(\frac{\partial}{\partial t} \right) \right] e^{-iS(t)} =
$$

$$
= \int_0^{L_0} \left[\frac{-i\hbar}{2m\zeta^2(t)} \left(\frac{\partial \pi(\xi)}{\partial \xi} \frac{\partial \psi(\xi)}{\partial \xi} \right) - \left(\frac{\hbar\beta(t)}{m\zeta^2(t)} + \frac{\dot\zeta(t)}{2\zeta(t)} \right) \right.
$$

$$
\times \left\{ 2\xi \frac{\partial \pi(\xi)}{\partial \xi} \psi(\xi) + \pi(\xi)\psi(\xi) \right\}
$$

$$
\left. - i \left(\frac{2\hbar\beta^2(t)}{m\zeta^2(t)} + \frac{2\beta(t)\dot\zeta(t)}{\zeta(t)} - \frac{d\beta(t)}{dt} \right) \xi^2 \pi(\xi)\psi(\xi) \right] d\xi. \tag{19.19}
$$

Since $\beta(t)$ is an arbitrary function of time we choose it is such a way that H_{eff} takes a simple form. Thus we set the coefficient of $\left(\frac{\partial \pi(\xi)}{\partial \xi} \right) \psi$ equal to zero to get

$$
\beta(t) = -\frac{m}{2\hbar}\zeta(t)\dot\zeta(t), \tag{19.20}
$$

and with this choice of $\beta(t)$, H_{eff} becomes [9]

$$
H_{eff} = \int_0^{L_0} \left[\frac{-i\hbar}{2m\zeta^2(t)} \left(\frac{\partial \pi(\xi)}{\partial \xi} \frac{\partial \psi(\xi)}{\partial \xi} \right) - \frac{im}{2\hbar}\zeta(t)\ddot\zeta(t)\xi^2\pi(\xi)\psi(\xi) \right] d\xi. \tag{19.21}
$$

Now we write this Hamiltonian in terms of the creation and annihilation operators. To this end we first define a complete set of eigenfunctions by solving the differential equation

$$
-\frac{\hbar^2}{2m}\frac{d^2\phi_j}{d\xi^2} + \frac{1}{2}\xi^2\Omega^2\phi_j = \varepsilon_j\phi_j, \tag{19.22}
$$

with the boundary conditions

$$
\phi_j(0) = \phi_j(L_0) = 0. \tag{19.23}
$$

In the differential equation (19.22), Ω^2 is a nonnegative constant. Since $\{\phi_j\}$ s form a complete set, we expand $\psi(\xi, t)$ and $\pi(\xi, t)$ in terms of $\phi_j(\xi)$;

$$
\psi(\xi, t) = \sum_j a_j(t)\phi_j(\xi), \tag{19.24}
$$

and

$$
\pi(\xi, t) = \sum_j i\hbar a_j^\dagger(t)\phi_j^*(\xi). \tag{19.25}
$$

By substituting (19.24) and (19.25) in (19.21), we find H_{eff} in terms of $a_j^\dagger(t)$ and $a_j(t)$;

$$H_{eff} = \frac{1}{\zeta^2(t)} \sum_j \varepsilon_j a_j^\dagger(t) a_j(t) + \frac{m}{2\zeta^2(t)} \left[\zeta^3(t)\ddot{\zeta}(t) - \Omega^2 \right] \sum_{j,k} I_{kj} a_k^\dagger(t) a_j(t),$$

(19.26)

where

$$I_{kj} = I_{jk}^* = \int_0^{L_0} \xi^2 \phi_k^*(\xi) \phi_j(\xi) d\xi.$$

(19.27)

The set of eigenfunctions $\{\phi_j(\xi)\}$ are given in terms of confluent hypergeometric functions [10]

$$\phi_j(\xi) = N_j \xi \exp\left[-\frac{m\Omega}{2\hbar^2}\xi^2 \right] {}_1F_1\left(\frac{3}{4} - \frac{\varepsilon_j}{2\hbar\Omega}, \frac{3}{2}, \frac{m\Omega}{\hbar}\xi^2 \right),$$

(19.28)

with the corresponding eigenvalues which are the roots of

$$ {}_1F_1\left(\frac{3}{4} - \frac{\varepsilon}{2\hbar\Omega}, \frac{3}{2}, \frac{m\Omega}{\hbar}L_0^2 \right) = 0.$$

(19.29)

In Eq. (19.28) N_j is the normalization constant. In this model for arbitrary $\zeta(t)$, the number of particles in each state will change, but the total number of particles in the cavity remains constant, i.e.

$$ i\hbar \frac{dN}{dt} = \sum_j \left[a_j^\dagger(t) a_j(t), H_{eff} \right] = 0.$$

(19.30)

But for the special case of

$$ \zeta^3(t)\ddot{\zeta}(t) = \Omega^2,$$

(19.31)

or

$$ \zeta(t) = \left[\left(\frac{c^2}{L_0^2} + \Omega^2 \right) t^2 + \frac{2c}{L_0}t + 1 \right]^{\frac{1}{2}},$$

(19.32)

where c is an arbitrary constant, the particles in each level stay in that level, and only the energy of the system decreases (or increases) in time. Thus the total energy of the particles when $\zeta(t)$ is given by (19.32) is

$$ E(t) = \sum_j \frac{\varepsilon_j n_j}{\zeta(t)}.$$

(19.33)

Decay of a Single Particle — As a special case of this model let us consider the decay of a single particle originally located at the ith level of a cavity to the lower states. We write the time-dependent wave function for this particle as

$$ |\Psi(t)\rangle = C_i(t)|1, 0, 0\cdots\rangle + C_{i-1}(t)|0, 1, 0\cdots\rangle + C_{i-2}|0, 0, 1, \cdots\rangle + \cdots, \quad (19.34)$$

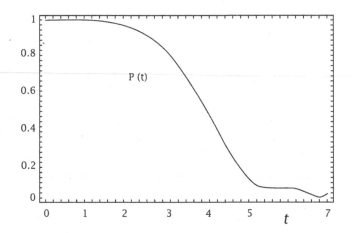

Figure 19.1: The probability of finding a particle in its initial state when the particle is confined to be inside a cavity with a moving wall.

where $|C_{i-k}(t)|^2$ is the probability of the particle to be at the $(i-k)$th level at the time t. This wave function together with the Hamiltonian (19.26) gives us the set of coupled equations

$$i\hbar\frac{dC_j(t)}{dt} = \frac{1}{\zeta^2(t)}C_j(t) + \frac{m}{2\zeta^2(t)}\left[\zeta^3(t)\ddot{\zeta}(t) - \Omega^2\right]\sum_{k=1}^{i}I_{jk}C_k(t), \quad j = 1,2\cdots i.$$

(19.35)

These equations are subject to the boundary conditions

$$C_i(0) = 1, \quad C_{i-k}(0) = 0, \quad k = 1,\cdots i - 1.$$

(19.36)

For instance if we choose $\Omega = 0$ and $L_0 = 1$, then

$$I_{jk} = 2\int_0^1 \xi^2 \sin(j\pi\xi)\sin(k\pi\xi)d\xi,$$

(19.37)

and if we choose $\zeta(t)$ to be given by

$$\zeta(t) = \left[1 + \frac{a^4t^4}{1 + a^2t^2}\right],$$

(19.38)

then we can integrate Eq. (19.35) and calculate $|C_j(t)|^2$. For $a = 0.5$ and $i = 4$, there are four $C_j(t)$'s and all of them satisfying the initial conditions (19.36). In Fig. (19.1) the probability $P(t) = |C_4(t)|^2$ is plotted as a function of time.
Massive Particles Satisfying the Klein-Gordon Equation — We can formulate this problem for massive particles satisfying the Klein-Gordon equation [11], or for photons [12],[13], i.e. Maxwell's equations.

For the particles satisfying the Klein-Gordon equation we start with the Hamiltonian

$$H = \frac{1}{2} \int_0^{L(t)} \left[\pi^2(x,t) + \left(\frac{\partial \psi(x,t)}{\partial x} \right)^2 + m^2 \psi^2(x,t) \right] dx, \qquad (19.39)$$

where we have set $\hbar = c = 1$. The fields $\psi(x,t)$ and $\pi(x,t)$ satisfy the boundary conditions (19.3) and (19.4) and the canonical commutation relation (19.5). Using the unitary transformations $\exp(iU_1(t))$ and $\exp(iU_2(t))$ on the operator $H - i\frac{\partial}{\partial t}$ we obtain the effective Hamiltonian for this field which is given by

$$H_{eff} = \int_0^{L_0} \left[\frac{1}{2} \left\{ \pi^2 + \frac{1}{\zeta^2(t)} \left(\frac{\partial \psi}{\partial \xi} \right)^2 + m^2 \psi^2 \right\} \right.$$

$$\left. - \frac{\dot{\zeta}(t)}{\zeta(t)} \left\{ \xi \frac{\partial \pi}{\partial \xi} \psi + \frac{1}{2} \pi \psi \right\} \right] d\xi. \qquad (19.40)$$

From the effective Hamiltonian density, i.e. the integrand in (19.40) we obtain the equations for $\psi(\xi,t)$ and $\pi(\xi,t)$. The above unitary transformation preserves the reciprocal symmetry of the field, viz, the equations for $\psi(\xi,t)$ and $\pi(\xi,t)$ are identical in form. Thus for $\psi(\xi,t)$ we find

$$\frac{1}{\zeta^2(t)} \left(1 - \dot{\zeta}^2(t) \xi^2 \right) \frac{\partial^2 \psi}{\partial \xi^2} - \frac{\partial^2 \psi}{\partial t^2} + \frac{2\dot{\zeta}(t)}{\zeta(t)} \xi \frac{\partial^2 \psi}{\partial t \partial \xi} + \frac{\dot{\zeta}(t)}{\zeta(t)} \frac{\partial \psi}{\partial t}$$

$$+ \frac{1}{\zeta^2(t)} \left(\ddot{\zeta}(t) \zeta(t) - 3\dot{\zeta}^2(t) \right) \xi \frac{\partial \psi}{\partial \xi} + \left[\frac{\ddot{\zeta}(t)}{2\zeta(t)} - \frac{3}{4} \left(\frac{\dot{\zeta}(t)}{\zeta(t)} \right)^2 - m^2 \right] \psi = 0.$$

$$(19.41)$$

We can also write H_{eff} in terms of creation and annihilation operators. For the present case it is convenient to take $L_0 = \pi$ and expand both $\psi(\xi,t)$ and $\pi(\xi,t)$ as

$$\psi(\xi,t) = \left(\frac{\zeta(t)}{\pi} \right)^{\frac{1}{2}} \sum_{k=1}^{\infty} \frac{1}{\sqrt{\omega_k(t)}} \left(a_k^\dagger + a_k \right) \sin(k\xi), \qquad (19.42)$$

and

$$\pi(\xi,t) = \frac{i}{(\pi \zeta(t))^{\frac{1}{2}}} \sum_{k=1}^{\infty} \sqrt{\omega_k(t)} \left(a_k^\dagger - a_k \right) \sin(k\xi), \qquad (19.43)$$

where in these relations

$$\omega_k(t) = \left[k^2 + m^2 \zeta^2(t) \right]^{\frac{1}{2}}. \qquad (19.44)$$

Substituting (19.42) and (19.43) in (19.40) and carrying out the integration over ξ, we obtain

$$H_{eff} = \frac{1}{\zeta(t)} \sum_k \omega_k(t) \left(a_k^\dagger a_k + \frac{1}{2} \right) + \frac{i\dot{\zeta}(t)}{\zeta(t)} \sum_k \sum_{j \neq k} (-1)^{k+j}$$

$$\times \frac{jk}{j^2 - k^2} \left(\frac{\omega_k(t)}{\omega_j(t)} \right)^{\frac{1}{2}} \left(a_k^\dagger a_j^\dagger - a_k^\dagger a_j + a_k a_j^\dagger - a_k a_j \right). \tag{19.45}$$

As in the non-relativistic problem we can find the equation of motion for a_k and $a_k^\dagger$ from (19.45). Thus for da_k/dt we obtain

$$i \frac{da_k}{dt} = \frac{\omega_k(t)}{\zeta(t)} a_k + i R_k(t), \tag{19.46}$$

where

$$R_k(t) = \frac{\dot{\zeta}(t)}{\zeta(t)} \sum_{j \neq k} \frac{(-1)^{k+j} jk}{j^2 - k^2} \left[\left(\frac{\omega_j(t)}{\omega_k(t)} \right)^{\frac{1}{2}} \left(a_j - a_j^\dagger \right) + \left(\frac{\omega_k(t)}{\omega_j(t)} \right)^{\frac{1}{2}} \left(a_j + a_j^\dagger \right) \right]. \tag{19.47}$$

If at $t = 0$ there are no particles in the state k, i.e. if

$$a_k^\dagger(0) a_k |0\rangle = 0, \tag{19.48}$$

then the number of particles created in this state between $t = 0$ and $t = \infty$ is given by

$$\langle 0|N_k|0\rangle = \int_0^\infty \left\langle 0 \left| a_k^\dagger R_k + R_k^\dagger a_k \right| 0 \right\rangle dt. \tag{19.49}$$

The asymptotic form of the Hamiltonian as $t \to \infty$ can be expressed in terms of $\langle 0|N_k|0\rangle$;

$$H_{eff}(\infty) \to \sum_{k=1}^\infty \left(m^2 + \frac{k^2}{\zeta^2(\infty)} \right)^{\frac{1}{2}} \left(\langle 0|N_k|0\rangle + \frac{1}{2} \right), \tag{19.50}$$

and, therefore, the change in the energy of the system is

$$E_f - E_i = \sum_1^\infty \left[\left(m^2 + \frac{k^2}{\zeta^2(\infty)} \right)^{\frac{1}{2}} \left(\langle 0|N_k|0\rangle + \frac{1}{2} \right) - \frac{1}{2} \left(m^2 + k^2 \right)^{\frac{1}{2}} \right]. \tag{19.51}$$

This change in the energy can be finite or infinite depending on how $\zeta(t)$ evolves in time [11].

19.1   Dissipative Forces Arising from Vacuum Fluctuations

A purely quantum mechanical dissipative force which can be calculated exactly is the drag force on a sphere which is moving in vacuum, but first let us consider the simpler problem of motion of a conducting plate in vacuum. The motion causes a change in the zero-point energy of the electromagnetic field which in turn reacts back by exerting a damping force on the plate.

By calculating the change in the zero-point energy, Casimir showed that there is an attractive force between two parallel plates given by

$$F = -\frac{\hbar \pi^2 c A}{240 \; L^4},$$ (19.52)

where A is is the area of each of the plate and L is the distance between them [14],[15]-[18].

In the first approach, due to Casimir, one can assume the existence of a field (e.g. electromagnetic field) whose properties are modified by mechanical constraint, and this gives rise to the forces between material bodies even in the vacuum [19]. A second interpretation can be given in the following way:

The quantum fluctuations of the distributions of spontaneously polarized material is responsible for the forces between bodies. This explanation is similar to the explanation of the origin of the van der Waals forces in atomic physics [20].

The third way of interpreting this force is to consider a coupled system of matter and field, where there exists a vacuum state of the matter and field which determines the ground state properties of the electromagnetic field. In this way the Casimir force is fundamentally a vacuum state of the coupled system in which the interaction between the plates is mediated by zero-point energy [19].

In the dynamical theory, the Casimir pressure on a plate (or mirror) can be obtained as the difference between the mean values of the Maxwell stress tensor

$$T_{ij} = \frac{1}{4\pi} \left[E_i E_j + B_i B_j - \frac{1}{2} \delta_{ij} \left(\mathbf{E}^2 + \mathbf{B}^2 \right) \right],$$ (19.53)

calculated on each side of the mirror. When there is a single mirror these stresses cancel each other, but only on the average.

In the following approach we will use the fluctuation-dissipation theorem to find the force of friction of mirrors moving in vacuum. This theorem provides us with an interface between microscopic and macroscopic properties of a system i.e. it gives us a prescription of how to proceed when we want to relate the microscopic and the macroscopic properties. Starting with the microscopic description of dynamical models we can calculate equilibrium correlation functions for the properties of interest. Then the fluctuation-dissipation theorem enables us to relate this correlation function with the results found from the measurement of the corresponding response function.

Let us write the fluctuations of the force on the mirror in terms of the correlation function $C_{FF}(t)$ [21],[22]

$$C_{FF}(t) = \langle F(t) F(0) \rangle - \langle F \rangle^2.$$ (19.54)

Since the Maxwell stress tensor is quadratic in electromagnetic field variables, we expect that in its quantized form, this tensor should depend quadratically

on creation and annihilation operators. Therefore for the field in vacuum state
we have

$$C_{FF}(t) = \frac{1}{2} \sum_{n,n'} \langle 0|F(t)|n,n'\rangle \langle n,n'|F(0)|0\rangle , \qquad (19.55)$$

where the sum in (19.55) runs over all normal modes n and n'. Thus $|n,n'\rangle$
is a state containing one photon in the mode n and the other in the mode n',
and $|0\rangle$ is the vacuum state. Since the part of $F(t)$ which contributes, when it
operates on the vacuum state, depends on the annihilation operator, therefore
we can write the time-dependence of the matrix element $\langle 0|F(t)|n,n'\rangle$ as

$$\langle 0|F(t)|n,n'\rangle = \exp\left[-i\left(\omega_n + \omega_{n'}\right)t\right]\langle 0|F(0)|n,n'\rangle , \qquad (19.56)$$

where ω_n and $\omega_{n'}$ are positive frequencies. Using this result we can also write
the Fourier transform of $C_{FF}(t)$ as

$$\tilde{C}_{FF}(\omega) = \pi \sum_{n,n'} \delta\left(\omega - \omega_n - \omega_{n'}\right)|\langle n,n'|F(0)|0\rangle|^2. \qquad (19.57)$$

To determine the connection between fluctuations and dissipation we first consider the spectral density $\xi(t)$ given by [22];

$$\xi(t) = \frac{1}{2\hbar}\langle 0\left[F(t),\,F(0)\right]0\rangle , \qquad (19.58)$$

i.e. the vacuum expectation value of the commutator of $F(t)$ and $F(0)$. Thus
from (19.54) and (19.58) we find

$$\tilde{\xi}(\omega) = \frac{1}{2\hbar}\left[\tilde{C}_{FF}(\omega) - \tilde{C}_{FF}(-\omega)\right], \qquad (19.59)$$

where $\tilde{\xi}(\omega)$ is the Fourier transform of $\xi(t)$. The response to an applied perturbation, according to the linear response theory is given by the susceptibility
function $\xi(t)$ defined by (19.58). Suppose that the mirror (or plate) is displaced
by a small distance $\delta q(t)$, then the reaction of the electromagnetic field to this
displacement is given by a force $\delta F(t)$, where

$$\delta F(t) = \int \chi\left(t - t'\right)\delta q\left(t'\right)dt', \qquad (19.60)$$

or expressed in spectral domain, this expression becomes

$$\delta\tilde{F}(\omega) = \tilde{\chi}(\omega)\delta\tilde{q}(\omega). \qquad (19.61)$$

The linear response theory gives us a relation between $\tilde{\xi}(\omega)$ and $\tilde{\chi}(\omega)$ [23];

$$\tilde{\chi}(\omega) = -\frac{1}{\pi}\int \frac{\tilde{\xi}\left(\omega'\right)}{\omega - \omega' + i\epsilon}d\omega'. \qquad (19.62)$$

Then from Eqs. (19.57), (19.59) and (19.62) we can relate $\tilde{\chi}(\omega)$ to the matrix elements of $F(0)$;

$$\tilde{\chi}(\omega) = -\frac{1}{\hbar} \sum_{n,n'} \frac{\omega_n + \omega_{n'}}{(\omega_n + \omega_{n'})^2 - (\omega + i\epsilon)^2} |\langle n, n'|F(0)|0\rangle|^2 . \tag{19.63}$$

Also from the well-known result [24]

$$\frac{1}{\pi}\left(\frac{1}{\omega - \omega' + i\epsilon}\right) = \frac{\mathcal{P}}{\pi}\left(\frac{1}{\omega - \omega'}\right) - i\delta\left(\omega - \omega'\right) \tag{19.64}$$

and Eq. (19.62) we obtain

$$\mathrm{Re}\,\tilde{\chi}(\omega) = \frac{\mathcal{P}}{\pi} \int \frac{\tilde{\xi}(\omega')}{\omega' - \omega} d\omega', \tag{19.65}$$

and

$$\mathrm{Im}\,\tilde{\chi}(\omega) = \tilde{\xi}(\omega) \tag{19.66}$$

The two relations (19.65) and (19.66) enable us to calculate $\tilde{\chi}(\omega)$ from $\tilde{\xi}(\omega)$.

The Dynamical Casimir Effect — The original result found by Casimir, Eq. (19.53), shows both the magnitude and the direction of the force arising from the change in zero point energy when the two plates are motionless. When one of the plates moves relative to the other, then the vacuum fluctuations of the electromagnetic field are modified. In this case the coupling between vacuum fluctuations and a moving boundary may give rise to a dissipative force which acts on the boundary. In addition because of the motion of the plate photons can be created. The presence of such an effect can also be seen in the motion of a single perfectly reflecting plate (mirror).

For the sake of simplicity let us calculate the force exerted by quantum fluctuations of a massless scalar field, $\phi(x,t)$, on a moving boundary (e.g. a moving mirror) where it is assumed that at the moving boundary, the field $\phi(x,t)$ satisfies a Robin boundary condition [29]. To solve this problem we will use the perturbation method advanced by Ford and Vilenkin [30]. This approach has been applied to the calculation of of more realistic problems, such as determining the force generated by the motion of a perfectly conducting plate [31]-[40] and the force on an oscillating cavity formed by two parallel perfectly conducting plates [35]. In both cases the fluctuating field is assumed to be electromagnetic.

Let us denote the displacement of the plate at the time t by $\delta q(t)$. We want to consider the nonrelativistic problem, therefore we assume that $\left|\frac{d}{dt}\delta q(t)\right| \ll c$, and in addition $|\delta q(t)|$ is small compared with some characteristic field wavelength [36] (c is the velocity of light which we set equal to one). Thus we have the scalar wave equation

$$\frac{\partial^2 \phi(x,t)}{\partial x^2} - \frac{\partial^2 \phi(x,t)}{\partial t^2} = 0, \tag{19.67}$$

with the boundary condition

$$\left\{ \left[\frac{\partial}{\partial x} + \delta \dot{q}(t) \frac{\partial}{\partial t} \right] \phi(x,t) \right\}_{x=\delta q(t)} = \frac{1}{\beta} \phi(x,t) \bigg|_{x=\delta q(t)} . \tag{19.68}$$

Now we write $\phi(x,t)$ as a sum of two terms

$$\phi(x,t) = \phi^0(x,t) + \delta\phi(x,t), \tag{19.69}$$

where $\delta\phi(x,t)$ is a small motion-induced perturbation of the field

$$\delta\phi(x,t) = -\delta q(t) \left[\frac{\partial}{\partial x} \phi^0(x,t) \right]_{x=0} + \mathcal{O}\left(\delta^2 q(t)\right). \tag{19.70}$$

The xx component of the energy-momentum tensor for this scalar field is given by

$$T_{xx} = -\frac{1}{2} \left[\left(\frac{\partial \phi}{\partial x} \right)^2 - \left(\frac{\partial \phi}{\partial t} \right)^2 \right], \tag{19.71}$$

and the perturbation to this stress caused by the motion of the plate, δT_{xx}, to the first order is

$$\delta T_{xx} = -\frac{1}{2} \left\{ \left(\frac{\partial \phi^0}{\partial x} \right) \left(\frac{\partial \delta\phi}{\partial x} \right) + \left(\frac{\partial \delta\phi}{\partial x} \right) \left(\frac{\partial \phi^0}{\partial x} \right) \right\}, \tag{19.72}$$

where we have symmetrized δT_{xx} since both $\frac{\partial \phi^0}{\partial x}$ and $\frac{\partial \delta\phi}{\partial x}$ are quantized. The total force acting on the moving boundary can be obtained from the energy-momentum tensor T_{xx}:

$$\delta F(t) = \langle 0|\delta T_{xx}(t, \delta q^+(t)) - \delta T_{xx}(t, \delta q^-(t))|0 \rangle, \tag{19.73}$$

where the superscripts + and - refer to the operator δT_{xx} tensor on the two sides of the moving boundary. The Fourier transform of Eq. (19.73) gives us the susceptibility $\tilde{\chi}(\omega)$ defined by Eqs. (19.60) and (19.61)

$$\tilde{\chi}(\omega) = \frac{\delta \tilde{F}}{\delta \tilde{q}(\omega)}. \tag{19.74}$$

If we impose either Dirichlet or Neumann boundary conditions on $\phi(x,t)$ (19.68), i.e. either choose $\beta = 0$ or $\beta = \infty$, then the susceptibility found from (19.74) will be purely imaginary;

$$\tilde{\chi}_D(\omega) = \tilde{\chi}_N(\omega) = \frac{i\hbar\omega^3}{6\pi c^2}, \tag{19.75}$$

but for Robin boundary condition $\tilde{\chi}(\omega)$ is complex [29]. Since Im $\tilde{\chi}(\omega) > 0$, the vacuum fluctuations take energy from the moving boundary and thus the force is dissipative. The work done by this force is

$$\int_{-\infty}^{\infty} \delta F(t) \delta \dot{q}(t) dt = -\frac{1}{\pi} \int_0^{\infty} \omega \, \text{Im} \, \tilde{\chi}(\omega) |\delta \tilde{q}(\omega)|^2, \tag{19.76}$$

i.e. this work is done by dissipative effects and is related to the total energy converted into particles. This subject will be discussed in the next section.

However if we impose Robin boundary condition in (19.68) then $\tilde{\chi}(\omega)$ can be complex and in this case Re $\tilde{\chi}(\omega)$ contributes only to dispersive effects. From the Fourier transform of (19.74) we obtain the relation between $\delta F(t)$ and $\delta q(t)$:

$$\delta F(t) = \frac{\hbar}{6\pi c^2} \frac{d^3}{dt^3} \delta q(t). \tag{19.77}$$

Thus there are no frictional forces proportional to velocity nor proportional to acceleration appearing in the expression of the radiative reaction of the vacuum fields on the moving mirror. This happens because of the spatial symmetry of the vacuum, i.e. vacuum fields are invariant under the action of Lorentz boost while they appear as thermal fields to a uniformly accelerating observer [32]-[34]. The coupling of the mirror to the field is such that it brings about a vanishing field amplitude at the "surface" of the mirror.

19.2 Frictional Force on an Atom Moving with Constant Velocity above a Planar Surface

Among the problems which can be solved by applying fluctuation-dissipation theorem to nonequilibrium processes is the calculation of the drag force acting on an atom which is moving parallel to the planar surface of bulk matter.

Consider the case of a polarizable particle, e.g. an atom which is interacting with the vacuum electromagnetic field. In such a system the electromagnetic field fluctuates not only by itself, but the fluctuations are modified by the presence of the dipole moment of the atom. On the other hand the dipole moment of the atom fluctuates on its own, and these fluctuations are also influenced by the fluctuations of the electromagnetic field. These fluctuations between the macroscopic object with planar surface and the atom generates a force, the Casimir-Polder force, similar to the van der Waals force, and this force depends on the distance between the atom and the surface [42]-[48].

When a polarizable particle is placed at a position $\mathbf{r}_a$ in an electric field $\mathbf{E}$, then the energy of the system will change by $\delta\hat{\mathcal{E}}$ (a hat on a symbol indicates that we are dealing with a quantum mechanical operator)

$$\delta\hat{\mathcal{E}} = -\frac{1}{2}\langle\hat{\mathbf{d}}(t)\cdot\hat{\mathbf{E}}(\mathbf{r}_a, t)\rangle, \tag{19.78}$$

where $\hat{\mathbf{d}}(t)$ and $\hat{\mathbf{E}}(\mathbf{r}_a, t)$ are the dipole moment of the atom and the electric field respectively, and both are expressed as Heisenberg operators. The symbol $\langle\cdots\rangle$ denotes that the expectation value is to be taken over the initial state of the uncoupled (free field+free dipole) system. One can use the normalized form of the product of operators on the right-hand side of (19.78) when it is necessary (e.g. when second or higher order terms in perturbation are used in the calculation) [49]. The Hamiltonian for this system is composed of two parts $\hat{H} = \hat{H}_0 + \hat{H}'$, where $\hat{H}_0$ is the sum of the Hamiltonians of two isolated subsystems, the atom and the field, and $\hat{H}'$ is the interaction between them.

The formal solution of the equation of motion for any operator $\hat{O}$ in this case is given by the integral equation

$$\hat{O}(t) = \hat{O}_0(t) + \frac{i}{\hbar} \int_0^t \exp\left(\frac{i}{\hbar}\hat{H}_0(t-\tau)\right) \left[\hat{H}', \hat{O}(\tau)\right] \exp\left(-\frac{i}{\hbar}\hat{H}_0(t-\tau)\right) d\tau.$$
(19.79)

Here it is assumed that at $t = 0$, the two parts are uncoupled and the subscript 0 indicates that the operator $\hat{O}$ evolves with respect to $\hat{H}_0$, i.e.

$$\hat{O}_0(t) = \exp\left(\frac{i}{\hbar}\hat{H}_0 t\right) \hat{O} \left(\frac{i}{\hbar}\hat{H}_0 t\right).$$
(19.80)

The interaction Hamiltonian $\hat{H}'$ describes the coupling between the atom and the field

$$\hat{H}' = -\hat{\mathbf{d}} \cdot \hat{\mathbf{E}}(\mathbf{r}_a, t)$$
(19.81)

and therefore in the first order perturbation, from (19.79) we obtain the following expressions for $\hat{\mathbf{d}}$ and $\hat{\mathbf{E}}(\mathbf{r}_a, t)$:

$$\hat{\mathbf{d}}(t) \approx \hat{\mathbf{d}}_0(t) + \hat{\mathbf{d}}_{in}(t),$$
(19.82)

$$\hat{\mathbf{d}}_{in}(t) = \int_{-\infty}^{\infty} \left\{\frac{i}{\hbar}\left[\hat{\mathbf{d}}_0(t), \hat{\mathbf{d}}_0(t-\tau)\right]\theta(\tau)\right\} \cdot \hat{\mathbf{E}}_0(\mathbf{r}_a, t) \, d\tau,$$
(19.83)

and

$$\hat{\mathbf{E}}(\mathbf{r}_a, t) \approx \hat{\mathbf{E}}_0(\mathbf{r}_a, t) + \hat{\mathbf{E}}_{in}(\mathbf{r}_a, t),$$
(19.84)

$$\hat{\mathbf{E}}_{in}(\mathbf{r}, t) = \int_{-\infty}^{\infty} \left\{\frac{i}{\hbar}\left[\hat{\mathbf{E}}_0(\mathbf{r}, t), \hat{\mathbf{E}}_0(\mathbf{r}_a, t-\tau)\right]\theta(\tau)\right\} \cdot \hat{\mathbf{d}}_0(t-\tau) \, d\tau.$$
(19.85)

where the subscript (in) stands for induced field and induced electric dipole moment. Here we have assumed that when τ becomes large the system becomes stationary. Thus to the first order of perturbation the dipole is the sum of two parts: (a) - the unperturbed time-dependent part $\hat{\mathbf{d}}_0(t)$, and (b) - the linear response to the external perturbation $\hat{\mathbf{E}}(\mathbf{r}_a, t)$ which contributes the term $\hat{\mathbf{d}}_{in}(t)$ to $\hat{\mathbf{d}}(t)$. Similarly to the first order, $\hat{\mathbf{E}}(\mathbf{r}, t)$, has two terms and these are given by (19.84). We can evaluate the term inside the curly bracket in (19.83) over a particular state $|b\rangle$ and the result gives us the susceptibility tensor. The Fourier transform of this susceptibility is the polarizability tensor which we denote by $\alpha(\omega)$

$$\alpha(\omega) = \int_{-\infty}^{\infty} \frac{i}{\hbar} \left\langle b \left|\left[\hat{\mathbf{d}}_0(t), \hat{\mathbf{d}}_0(0)\right]\right| b \right\rangle \theta(t) e^{i\omega t} \, dt.$$
(19.86)

Regarding the perturbation of $\hat{\mathbf{E}}_0$ caused by the presence of dipole, i.e. $\hat{\mathbf{E}}_{in}$, Eq. (19.85), we note that

$$\int_{-\infty}^{\infty} \frac{i}{\hbar} \left[\hat{\mathbf{E}}_0(\mathbf{r}, t), \hat{\mathbf{d}}_0(\mathbf{r}_a, t)\right] \theta(t) e^{i\omega t} \, dt = G(\mathbf{r}, \mathbf{r}_a, \omega),$$
(19.87)

where $G(\mathbf{r}, \mathbf{r}_a, \omega)$ is the electric field Green tensor. This tensor can be obtained from Maxwell's equation for the electric field $\hat{\mathbf{E}}$

$$\nabla_{\mathbf{r}} \wedge \nabla_{\mathbf{r}} \wedge \hat{\mathbf{E}}(\mathbf{r}, \omega) - \frac{\omega^2}{c^2} \mathcal{E}(\mathbf{r}, \omega) \hat{\mathbf{E}}(\mathbf{r}, \omega)(\mathbf{r}, \omega) = -\frac{\omega^2}{\mu_0} \hat{\mathbf{d}} \, \delta(\mathbf{r} - \mathbf{r_a}), \quad (19.88)$$

and is a solution of the inhomogeneous partial differential equation

$$-\nabla_{\mathbf{r}} \wedge \nabla_{\mathbf{r}} \wedge G(\mathbf{r}, \mathbf{r}_a, \omega) + \frac{\omega^2}{c^2} \mathcal{E}(\mathbf{r}, \omega) G(\mathbf{r}, \mathbf{r}_a, \omega) = -\frac{\omega^2}{\epsilon_0 c^2} \delta(\mathbf{r} - \mathbf{r}_a) I, \quad (19.89)$$

where I is the unit tensor and $\mathcal{E}(\mathbf{r}, \omega)$ is the local dielectric function, assumed to be scalar for simplicity. With the help of the two tensors $\alpha(\omega)$ and $G(\mathbf{r}, \mathbf{r}_a, \omega)$, we can write $\hat{\mathbf{d}}_{in}(\omega)$ and $\hat{\mathbf{E}}(\mathbf{r}, \omega)$ in terms of linear response function

$$\hat{\mathbf{d}}_{in}(\omega) = \alpha(\omega) \cdot \hat{\mathbf{E}}_0(\mathbf{r}_a, \omega), \quad (19.90)$$

and

$$\hat{\mathbf{E}}_{in}(\mathbf{r}, \omega) = G(\mathbf{r}, \mathbf{r}_a, \omega) \cdot \hat{\mathbf{d}}_0(\omega). \quad (19.91)$$

Thus each of $\hat{\mathbf{d}}(t)$ and $\hat{\mathbf{E}}(\mathbf{r}, \omega)$ can be written as the sum of a free fluctuating part which describes the free intrinsic fluctuations, and the induced part which, in the perturbation theory, arises from the coupling of the field to the particle. By substituting (19.91) in (19.84) we find $\hat{\mathbf{E}}$ in frequency domain

$$\hat{\mathbf{E}}(\mathbf{r}, \omega) = \hat{\mathbf{E}}_0(\mathbf{r}, \omega) + G(\mathbf{r}, \mathbf{r}_a, \omega) \cdot \hat{\mathbf{d}}(\omega). \quad (19.92)$$

from the inverse Fourier transform of (19.92) we obtain

$$\hat{\mathbf{E}}(\mathbf{r}, t) = \hat{\mathbf{E}}_0^{(+)}(\mathbf{r}, t) + \frac{i}{\pi} \int_0^\infty d\omega \int_0^t e^{-i\omega\tau}$$
$$\times \operatorname{Im} G(\mathbf{r}, \mathbf{r}_a, \omega) \cdot \hat{\mathbf{d}}(t - \tau) \, d\tau$$
$$+ \text{Hermitian conjugate}, \quad (19.93)$$

where G is the electric Green tensor of the surface and $\hat{\mathbf{E}}^{(+)}$ denotes the positive frequency solution for the electric field in vacuum in the absence of the atom. The vector $\hat{\mathbf{E}}(\mathbf{r}, t)$ contains the contributions from the atom represented by a dipole as well as the field of surface and the bulk.

Let us consider an atom which can be modelled by a two-state system, the ground state and an excited state [42]. That is, in this formulation the atom is approximated by its electric dipole operator which is given by

$$\hat{\mathbf{d}} = \mathbf{d}\sigma_1. \quad (19.94)$$

Such an atom has a ground state $|g\rangle$ and a single excited state $|e\rangle$. Therefore σ_1 as an operator can be written as

$$\sigma_1 = |e\rangle\langle g| + |g\rangle\langle e|. \quad (19.95)$$

The total force on the atom normal to the surface at any time t is given by

$$F_z(t) = \left\langle \hat{\mathbf{d}}(t) \cdot \frac{\partial}{\partial z_a} \hat{\mathbf{E}}(\mathbf{r}, t) \right\rangle, \tag{19.96}$$

where we have denoted the position of the electron by $\mathbf{r}_a$, and the symbol $\langle \cdots \rangle$ in (19.96) denotes the expectation value over the initial state. In averaging over the initial states in (19.96) we assume that the initial atom plus field-matter states are factorizable, i.e.

$$\hat{\rho}(0) = \hat{\rho}_a(0) \bigotimes \hat{\rho}_{f-m}(0), \tag{19.97}$$

with the joint field-matter subsystem in its vacuum state. Also we assume that the position of the atom $\mathbf{r}_a(t)$ can be given by its classical equation of motion along x;

$$m_a \ddot{x}_a(t) = F_{ext} + F_f. \tag{19.98}$$

In this relation F_{ext} denotes the external classical force which is responsible for moving the atom which is at rest in its initial position and giving it a constant velocity v_x above the surface. Thus $\mathbf{r}_a$ can be written as

$$\mathbf{r}_a = (x_a + v_x t)\mathbf{i} + y_a \mathbf{j} + z_a \mathbf{k}. \tag{19.99}$$

Now, after a long time the stationary frictional force becomes

$$F_f = -\,\mathrm{Re}\left\{ \frac{2}{\pi} \int \frac{d^2\mathbf{k}}{(2\pi)^2} k_x \int_0^\infty d\omega \int_0^\infty d\tau \right.$$
$$\left. \times\, e^{-i(\omega - k_x v_x)\tau} \mathrm{tr}\left[\bar{C}(\tau, v_x) \cdot \mathrm{Im}\, G(\mathbf{k}, z_a, z_a, \omega) \right] \right\}, \tag{19.100}$$

where

$$\bar{C}_{i,j}(\tau, v_x) = \mathrm{tr}\left\{ \hat{\mathbf{d}}_i(\tau)\hat{\mathbf{d}}_j(0)\hat{\rho}(\infty) \right\}, \tag{19.101}$$

is the two-time correlation function in the equilibrium nonstationary state $\hat{\rho}(\infty)$ of the coupled moving atom plus field and matter. Starting from Eq. (19.100) Intravaia et al. using the results of nonequilibrium statistical mechanics obtain the following expression for the near field approximation [42]

$$F_f \approx -\frac{90}{\pi^3} \frac{\hbar\rho^2 \alpha_0^2}{(2z_a)^{10}} v_x^3, \tag{19.102}$$

where ρ is the electrical resistivity of the surface and α_0 is the static atomic polarizability [42].

In the vast literature on the Casimir-Polder force between an atom which is moving relative to a surface, one can find results with varying dependence on v_x and z_a. Barton compares a few of these forms of F_f with each other and with his own result where for a slowly moving atom he finds the force to be proportional to (v_x/z_a^8) [44]-[46].

19.3 Motion of the Boundaries and Production of Particles in Two-Dimensional Space-Time

Let us consider two neutral plates separated by a distance $L(t)$ and assume that at $t = 0$ one of the plates starts moving relative to the other and after a time $t = T$ it comes back to rest in its original position without $L(t)$ going to zero during this time. The force per unit area, to the first order of approximation, has the form like the force in the static case, i.e.

$$F(L(t)) = -\frac{\pi \hbar c}{240 L^4(t)}. \tag{19.103}$$

This result is valid in the nonrelativistic limit of the relative motion of the plates, that is when $\frac{dL(t)}{dt} \ll c$. The motion of the boundary acts in the same way as a nonstationary external field. Again, for the sake of simplicity we consider a massless scalar field in the space between the two plates and we also set $\hbar = 1$. The boundary conditions on the wave equation (19.67) now will be

$$\phi(0, t) = \phi(L(t), t) = 0. \tag{19.104}$$

For $t \leq 0$, the solution of the operator wave equation (19.67) is given by

$$\phi(x, t) = \sum_n \frac{1}{\sqrt{\pi n}} \left\{ f_n^{(-)}(x, t) a_n + f_n^{(+)}(x, t) a_n^\dagger \right\}, \tag{19.105}$$

where a_n and $a_n^\dagger$ are the creation and annihilation operators for the scalar field $\phi(x, t)$ and $f_n^{(+)}(x, t)$ and $f_n^{(-)}(x, t)$ are the solutions of the free field equation:

$$f_n^{(\pm)}(x, t \leq 0) = \frac{1}{\sqrt{\pi n}} e^{\pm i \omega_n t} \sin \frac{n \pi x}{L_0}, \qquad L_0 = L(0). \tag{19.106}$$

The motion of one of the plates starts at $t = 0$, and in order to satisfy the boundary conditions in (19.104) we express $f_n^{(\pm)}(x, t)$ as an infinite series

$$f_n^{(-)}(x, t) = \frac{1}{\sqrt{\pi n}} \sum_k Q_{kn}(t) \sqrt{\frac{L_0}{L(t)}} \sin \left(\frac{n \pi x}{L(t)} \right), \tag{19.107}$$

and $f_n^{(+)}(x, t)$ is the complex conjugate of $f_n^{(-)}(x, t)$. In (19.107) $Q_{kn}(t)$, $n, k = 1, 2, 3 \cdots$ are time-dependent coefficients to be determined. From (19.107) and the boundary conditions (19.104) we obtain the initial conditions for the coefficients $Q_{kn}(t)$:

$$Q_{kn}(0) = \delta_{kn}, \qquad \dot{Q}_{kn}(0) = -i \omega_n \delta_{kn}. \tag{19.108}$$

If we substitute (19.107) in (19.67) and use the orthogonality of the set of functions $\left\{ \sin \frac{\pi k x}{L(t)} \right\}$ in the space bounded by $0 \leq x \leq L(t)$, we obtain an infinite set of differential equations for $Q_{kn}(t)$;

$$\frac{d^2 Q_{kn}(t)}{dt^2} + \omega_k^2(t)Q_{kn}(t) - \sum_j \left[2\nu(t)h_{kj}\frac{dQ_{jn}(t)}{dt} \right.$$

$$\left. + \frac{d\nu(t)}{dt}h_{kj}Q_{jn}(t) + \nu^2(t)\sum_l h_{jk}h_{jl}Q_{ln}(t) \right] = 0, \qquad (19.109)$$

where

$$\omega_k(t) = \frac{ck\pi}{L(t)}, \qquad \nu(t) = \frac{1}{L(t)}\frac{dL(t)}{dt} \qquad (19.110)$$

and

$$h_{kj} = -h_{jk} = (-1)^{k-j}\left(\frac{2jk}{j^2 - k^2} \right), \qquad j \neq k. \qquad (19.111)$$

After a time T, the moving plate returns to its original position. We assume that for $t > T$, $\dot{L}(t)$ and $\dot{\nu}(t)$ are zero. Then the right-hand side of (19.109) vanishes and the solution of this equation with the initial conditions (19.108) will be a linear combination of complex exponentials:

$$Q_{kn}(t) = \alpha_{kn}e^{-i\omega_k t} + \beta_{kn}e^{i\omega_k t}. \qquad (19.112)$$

Denoting the coefficient of $e^{-i\omega_k t}$ by b_k and the term multiplying $e^{i\omega_k t}$ by $b_k^\dagger$, we find new annihilation and creation operators in terms of the old ones, i.e. a_k and a_k;

$$b_k = \sum_n \sqrt{\frac{k}{n}}\left(\alpha_{kn}a_n + \beta_{kn}^* a_n^\dagger \right), \qquad (19.113)$$

and $b_n^\dagger$ which is the Hermitian adjoint of b_k. Thus, because of the motion of the boundary, we have different vacuum states for $t < 0$ and for $t > T$

$$a_k|0_{in}\rangle = 0, \quad \text{for} \quad t < 0, \qquad (19.114)$$

$$b_k|0_{out}\rangle = 0, \quad \text{for} \quad t > T. \qquad (19.115)$$

Since a_k, $a_k^\dagger$, b_k and $b_k^\dagger$ satisfy the commutation relations

$$\left[a_k,\, a_k^\dagger \right] = \left[b_k,\, b_k^\dagger \right] = 1, \qquad (19.116)$$

and b_k, $b_k^\dagger$ transform to a_k, $a_k^\dagger$ by Bogolyubov transformation [53],[54] from (19.116) and (19.113) and its Hermitian conjugate we get

$$\sum_k k\left(|\alpha_{kn}|^2 - |\beta_{kn}|^2 \right) = n. \qquad (19.117)$$

The number of particles created in the kth mode is obtained from the vacuum-vacuum matrix element in the in-vacuum of the out-operator for the number of particles. Using Eqs. (19.113), (19.114) and (19.115) we obtain n_k, the number of particles in the kth mode

$$n_k = \left\langle 0_{in} \left| b_k^\dagger b_k \right| 0_{in} \right\rangle = k \sum_{n=1}^{\infty} \frac{1}{n} |\beta_{nk}|^2 . \tag{19.118}$$

The total number of the particles created in the time T is the sum of n_k s;

$$N = \sum_{k=1}^{\infty} k \sum_{n=1}^{\infty} |\beta_{kn}|^2 \frac{1}{n} . \tag{19.119}$$

Number of Particles Created in One Period — Rather than finding the exact solution of (19.109) we will discuss the approximate solution of the simple case where the motion is sinusoidal with small amplitude [37],[38].

Let us consider an example where $L(t)$ is given by

$$L(t) = L_0 \left[1 + \varepsilon \sin(2\omega_1 t) \right] , \tag{19.120}$$

where $\omega_1 = \frac{c\pi}{L_0}$ is the lowest angular frequency which can be associated with the initial position of the two plates Eq. (19.110). The amplitude of oscillations ε which is dimensionless is assumed to be small $\varepsilon \ll 1$. By calculating $\nu(t)$, $\dot{\nu}(t)$ and $\nu^2(t)$ from (19.110) and (19.120) and substituting in (19.109) we find a complicated set of differential equations that we want to solve approximately. We observe that in the time interval $0 < t < T$, $\alpha_{kn}(t)$ and $\beta_{kn}(t)$ vary slowly, when compared to $e^{i\omega_k t}$ and $e^{-i\omega_k t}$, $k > 1$. Therefore by substituting (19.110) in (19.109), neglecting all the terms of order ε^2, and averaging over the fast oscillations with frequencies ω_k, we find the following set of differential equations

$$\frac{d\alpha_{1n}}{d\tau} = -\beta_{1n} + 3\alpha_{3n}, \tag{19.121}$$

$$\frac{d\beta_{1n}}{d\tau} = -\alpha_{1n} + 3\beta_{3n}, \tag{19.122}$$

$$\frac{d\alpha_{kn}}{d\tau} = (k+2)\alpha_{k+2n} - (k-2)\alpha_{k-2n}, \quad k \geq 2, \tag{19.123}$$

and

$$\frac{d\beta_{kn}}{d\tau} = (k+2)\beta_{k+2,n} - (k-2)\beta_{k-2,n}, \quad k \geq 2, \tag{19.124}$$

where τ is the dimensionless time variable

$$\tau = \frac{1}{2}\varepsilon\omega_1 t. \tag{19.125}$$

The initial conditions for these equations which can be found form (19.108) are given by

$$\alpha_{kn}(0) = \delta_{kn}, \qquad \beta_{kn}(0) = 0. \tag{19.126}$$

We note that in this set of differential equations even modes are not coupled to the odd modes and with the initial conditions (19.126) we have

$$\beta_{2k,n}(t) = \beta_{k,2l}(t) = 0. \tag{19.127}$$

We also note that by multiplying (19.123) and (19.124) by $k\alpha_{kn}$ and by $k\beta_{kn}$ respectively and summing the results over k we have

$$\sum_k k\alpha_{kn}\frac{d\alpha_{kn}}{d\tau} = \sum_k k\beta_{kn}\frac{d\beta_{kn}}{d\tau} = -\alpha_{1n}\beta_{1n}. \tag{19.128}$$

Equations (19.121)–(19.124) show that α_{kn} s are not only coupled to $\alpha_{k+j,n}$ s but also to β_{kn} s. By introducing new time-dependent functions $\eta_{kn}(t)$ and $\zeta_{kn}(t)$ by the relations

$$\eta_{kn} = \alpha_{kn} + \beta_{kn}, \tag{19.129}$$

$$\zeta_{kn} = \alpha_{kn} - \beta_{kn}, \tag{19.130}$$

we get two sets of equations, one for η_{kn} s and the other for ζ_{kn} s

$$\begin{cases} \frac{d}{d\tau}\eta_{1n} = -\eta_{1n} + 3\eta_{3n} \\ \\ \frac{d}{d\tau}\eta_{kn} = (k+2)\eta_{k+2,n} - (k-2)\eta_{k-2,n}, \quad k \geq 3 \end{cases}, \tag{19.131}$$

$$\begin{cases} \frac{d}{d\tau}\zeta_{1n} = \zeta_{1n} + 3\zeta_{3n} \\ \\ \frac{d}{d\tau}\zeta_{kn} = (k+2)\zeta_{k+2,n} - (k-2)\zeta_{k-2,n}, \quad k \geq 3 \end{cases}, \tag{19.132}$$

These equations are subject to the boundary conditions

$$\eta_{kn}(0) = \zeta_{kn}(0) = \delta_{kn}, \tag{19.133}$$

obtained from (19.126). We will solve these sets by introducing two generating functions following the method that we used to solve the problem of the Schrödinger chain in Chapter 7.

Consider the first set Eqs. (19.131), and let $M_n(\tau, z)$ be defined by [41]

$$M_n(\tau, z) = \sum_{k=1}^{\infty} \eta_{kn}(\tau)z^k \tag{19.134}$$

where summation is over all odd values of k (see Eq. (19.131)). By multiplying (19.131) by z^k and simplifying we get

$$\frac{\partial M_n}{\partial z} = \left(\frac{1}{z} - z^3\right)\frac{\partial M_n}{\partial z} - \eta_{1n}(\tau)\left(z + \frac{1}{z}\right). \tag{19.135}$$

The solution of this linear partial differential equation subject to the initial condition $M_n(0, z) = z^n$ is given by

$$M_n(\tau, z) = \left[\frac{1 + z^2 - \exp(-4\tau)\left(1 - z^2\right)}{1 + z^2 + \exp(-4\tau)\left(1 - z^2\right)}\right]^{\frac{n}{2}}$$

$$- 2 \int_0^\tau \left[1 - \left(\frac{1 - z^2}{1 + z^2}\right)^2 \exp(-8x)\right]^{-\frac{1}{2}} \eta_{1n}(\tau - x)\, dx. \qquad (19.136)$$

Setting $z = 0$ in (19.136) and noting that from (19.134), $M_n(\tau, 0) = 0$, we find the following integral equation for η_{1n} [41],[50];

$$\int_0^\tau \frac{\eta_{1n}(\tau - x)dx}{\sqrt{1 - \exp(-8x)}} = \frac{1}{2}\left(\frac{1 - \exp(-4\tau)}{1 + \exp(-4\tau)}\right)^{\frac{n}{2}}. \qquad (19.137)$$

In similar way we obtain an integral equation for ζ_{1n};

$$\int_0^\tau \frac{\exp[4(\tau - x)]\zeta_{1n}(\tau - x)dx}{\sqrt{1 - \exp(-8x)}} = \frac{1}{2}e^{4\tau}\left(\frac{1 - \exp(-4\tau)}{1 + \exp(-4\tau)}\right)^{\frac{n}{2}}. \qquad (19.138)$$

The left side of these two integral equations are of convolution type, therefore we can use the method of Laplace transform applied to these equations and find the solutions. Let $\bar{\eta}_{1n}(s)$ and $\bar{\zeta}_{1n}(s)$ denote the Laplace transforms of $\eta_{1n}(\tau)$ and $\zeta_{1n}(s)$ respectively, then taking the transforms of the two sides of Eqs. (19.137) and (19.138) we find [41]

$$\bar{\eta}_{1n}(s) = \frac{\Gamma\left(\frac{s}{8} + \frac{1}{2}\right)}{\Gamma\left(\frac{s}{8}\right)\Gamma\left(\frac{1}{2}\right)}I(s, n), \qquad (19.139)$$

and

$$\bar{\zeta}_{1n}(s) = \frac{\Gamma\left(\frac{s}{8} + 1\right)}{\Gamma\left(\frac{s}{8} + \frac{1}{2}\right)\Gamma\left(\frac{1}{2}\right)}I(s, n), \qquad (19.140)$$

where

$$I(s, n) = \int_0^1 x^{\frac{s}{4} - 1}\left(\frac{1 - x}{1 + x}\right)^{\frac{n}{2}} dx$$

$$= \frac{\Gamma\left(\frac{s}{4}\right)\Gamma\left(1 + \frac{n}{2}\right)}{\Gamma\left(\frac{s}{4} + 1 + \frac{n}{2}\right)}\, {}_2F_1\left[\frac{s}{4}, \frac{n}{2}; \frac{s}{4} + 1 + \frac{n}{2}; -1\right]. \qquad (19.141)$$

Here ${}_2F_1[a, b; c; z]$ is the Gauss hypergeometric function. For the special case of $n = 1$ we can express ${}_2F_1$ in terms of Γ functions and thus simplify $I(s, 1)$ [51]. By substituting the resulting expression in Eqs. (19.139)–(19.141) we obtain the two functions $\bar{\eta}_{11}(s)$ and $\bar{\zeta}_{11}(s)$ in closed forms:

$$\bar{\eta}_{11}(s) = \frac{1}{2} - \frac{s}{16}\left[\frac{\Gamma\left(\frac{s}{8} + \frac{1}{2}\right)}{\Gamma\left(\frac{s}{8} + 1\right)}\right]^2, \qquad (19.142)$$

$$\bar{\zeta}_{11}(s) = -\frac{1}{2} + \frac{s}{16} \left[\frac{\Gamma\left(\frac{s}{8}\right)}{\Gamma\left(\frac{s}{8}+1\right)} \right]^2. \tag{19.143}$$

We can also find the inverse Laplace transforms of $\bar{\eta}_{11}(s)$ and $\bar{\zeta}_{11}(s)$ analytically [41]. These are given by

$$\eta_{11}(\tau) = -\frac{1}{\pi} \frac{\partial}{\partial \tau} \left[e^{-4\tau} K \left(\sqrt{1 - \exp(-8\tau)} \right) \right], \tag{19.144}$$

$$\zeta_{11}(\tau) = \frac{1}{\pi} \frac{\partial}{\partial \tau} \left[K \left(\sqrt{1 - \exp(-8\tau)} \right) \right], \tag{19.145}$$

where K is the elliptic integral of the first kind [41],[51]

$$K(\kappa) = \int_0^{\frac{\pi}{2}} \frac{d\theta}{\sqrt{1 - \kappa^2 \sin^2 \theta}} \tag{19.146}$$

Now we change the variable τ to κ, where κ and $\bar{\kappa}$ are defined by the following relations

$$\kappa = \sqrt{1 - \exp(-8\tau)}, \qquad \bar{\kappa} = \exp(-4\tau), \tag{19.147}$$

and then we calculate the derivative of $K(\kappa)$ with respect to κ;

$$\frac{dK(\kappa)}{d\kappa} = \frac{E(\kappa)}{\kappa \bar{\kappa}^2} - \frac{K(\kappa)}{\kappa}. \tag{19.148}$$

In this relation $E(\kappa)$ is the complete elliptic integral of the second kind [51],[52]

$$E(\kappa) = \int_0^{\frac{\pi}{2}} \sqrt{1 - \kappa^2 \sin^2(\theta)} \, d\theta. \tag{19.149}$$

The derivative of $E(\kappa)$ is also related to $E(\kappa)$ and $K(\kappa)$ by

$$\frac{dE(\kappa)}{d\kappa} = \frac{E(\kappa) - K(\kappa)}{\kappa}. \tag{19.150}$$

By substituting for K and its derivative as functions of κ in (19.144) and (19.145), we find η_{11} and ζ_{11}, and from these and Eqs. (19.129) and (19.130) we obtain α_{11} and β_{11};

$$\alpha_{11}(\kappa) = \frac{2}{\pi} \left(\frac{E(\kappa) + \bar{\kappa} K(\kappa)}{1 + \bar{\kappa}} \right), \tag{19.151}$$

$$\beta_{11}(\kappa) = -\frac{2}{\pi} \left(\frac{E(\kappa) - \bar{\kappa} K(\kappa)}{1 - \bar{\kappa}} \right), \tag{19.152}$$

Having found $\alpha_{11}(\tau)$ and $\beta_1(\tau)$, we want to show that there are recurrence relations which will enable us to determine $\eta_{1n}(\tau)$ and $\zeta_{1n}(\tau)$ and consequently $\alpha_{1n}(\tau)$ and $\beta_{1n}(\tau)$. To this end we differentiate both sides of Eqs. (19.137) and

(19.138) with respect to τ and use the initial conditions $\eta_{1n}(0) = \zeta_{1n}(0) = 1$ to obtain

$$\eta_{13} = -\eta_{11} - \dot{\eta}_{11}, \tag{19.153}$$

$$\zeta_{13} = \zeta_{11} + \dot{\eta}_{11}e^{-4\tau}, \tag{19.154}$$

and

$$\dot{\eta}_{11} + e^{4\tau}\dot{\zeta}_{11} = 0. \tag{19.155}$$

In these and the following equations a dot over a symbol denotes differentiation with respect to τ. Replacing η_{1n} and ζ_{1n} by α_{1n} and β_{1n} using Eqs. (19.129) and (19.130) we have

$$\beta_{13} = -\alpha_{11} - \dot{\beta}_{11}, \tag{19.156}$$

$$\alpha_{13} = -\beta_{11} - \dot{\alpha}_{11}. \tag{19.157}$$

For $n \geq 3$, using the same procedure and the initial conditions $\eta_{1n}(0) = \zeta_{1n}(0) = 0$, after some algebra we get the recurrence relations

$$2n\eta_{1\ n+2} = -2n\eta_{1n} - 4e^{4\tau}\zeta_{1n} - \dot{\eta}_{1n} + \frac{d}{dt}\left(e^{4\tau}\zeta_{1n}\right), \tag{19.158}$$

$$2ne^{4\tau}\zeta_{1\ n+2} = (2n+4)e^{4\tau}\zeta_{1n} + \dot{\eta}_{1n} - \frac{d}{dt}\left(e^{4\tau}\zeta_{1n}\right). \tag{19.159}$$

These lead us to the relation

$$\eta_{1\ n+2} + e^{4\tau}\zeta_{1\ n+2} = e^{4\tau}\zeta_{1n} - \eta_{1n}, \tag{19.160}$$

and this relation is valid for $n \geq 1$. If we replace n by $n-2$ in (19.158) and then subtract the result from (19.158), and also replace the difference $e^{4\tau}(\zeta_{1n} - \zeta_{1,n-2})$ by the sum $\eta_{1n} + \eta_{1n-2}$, we obtain the following recurrence relation

$$n\eta_{1,\ n+2} = n\eta_{1\ n-2} - \dot{\eta}_{1\ n}, \tag{19.161}$$

and also find a similar relation for $\zeta_{1\ n+2}$

$$n\zeta_{1\ n+2} = n\zeta_{1\ n-2} - \dot{\zeta}_{1\ n}. \tag{19.162}$$

In terms of the original coefficients α_{1n} and β_{1n} Eqs. (19.161) and (19.162) can be written as

$$n\alpha_{1\ n+2} = n\alpha_{1\ n-2} - \dot{\alpha}_{1\ n}, \qquad n \geq 3, \tag{19.163}$$

$$n\beta_{1\ n+2} = n\beta_{1\ n-2} - \dot{\beta}_{1\ n}, \qquad n \geq 3. \tag{19.164}$$

From Eqs. (19.156) and (19.164) we can find a closed expression for the rate of generation of particles (photons) in the principle cavity mode, i.e. with $k = 1$ in (19.118). Denoting this rate by $\frac{d\mathcal{P}_1}{d\tau}$, we have

$$\frac{d}{d\tau}\mathcal{P}_1 = \sum_{\text{odd } n} \frac{2}{n}\beta_{1n}\dot{\beta}_{1n} = -2\beta_{11}(\tau)\alpha_{11}(\tau), \tag{19.165}$$

Also the total rate of creation of particles in all of the modes can be found from

$$\frac{d}{d\tau}\mathcal{P} = \sum_{k \ odd} \sum_{n} \frac{2k}{n}\beta_{kn}\dot{\beta}_{kn} = -\sum_{n} \frac{2}{n}\beta_{1n}(\tau)\alpha_{1n}(\tau). \tag{19.166}$$

Rate of Creation of Particles In the First Cavity Mode — We have already calculated $\alpha_{1n}(\tau)$ and $\beta_{1n}(\tau)$, Eqs. (19.151) and (19.152). Now we substitute these in Eq. (19.165) to find $\left(\frac{d\mathcal{P}_1}{dt}\right)$;

$$\frac{d\mathcal{P}_1}{dt} = \frac{4\varepsilon\omega_1}{\pi^2}\left(\frac{E^2(\kappa) - \bar{\kappa}^2 K^2(\kappa)}{\kappa^2}\right). \tag{19.167}$$

For the two asymptotic cases. viz, small τ (or t) and large τ (or t) we can find simple expressions for the massless particle (photon) generation rate in the first cavity mode [41],[50].

(1) When $\tau \ll 1$, then $\kappa = \sqrt{8\tau} \ll 1$, and we can use the power expansions of the elliptic integrals to find [52]

$$K(\kappa) = \frac{\pi}{2}\left[1 + \frac{1}{4}\kappa^2 + \frac{9}{64}\kappa^4 + \cdots\right], \tag{19.168}$$

$$E(\kappa) = \frac{\pi}{2}\left[1 - \frac{1}{4}\kappa^2 - \frac{3}{64}\kappa^4 - \cdots\right], \tag{19.169}$$

and therefore

$$\frac{d\mathcal{P}_1}{dt} \approx \frac{1}{2}\varepsilon^2\omega_1^2 t, \qquad \varepsilon\omega_1 t \ll 1. \tag{19.170}$$

(2) In the long-time limit $\tau \gg 1$ and we have $\bar{\kappa} \to 0$, $K(\kappa)$ and $E(\kappa)$ have the following asymptotic expansions [52]:

$$K(\kappa) \approx \ln\frac{4}{\bar{\kappa}} + \frac{1}{4}\left(\ln\frac{4}{\bar{\kappa}} - 1\right)\bar{\kappa}^2 + \cdots \tag{19.171}$$

$$E(\kappa) \approx 1 + \frac{1}{2}\left(\ln\frac{4}{\bar{\kappa}} - \frac{1}{2}\right)\bar{\kappa}^2 + \cdots. \tag{19.172}$$

Substituting these two asymptotic forms in (19.167) we get

$$\frac{d\mathcal{P}_1}{dt} = \frac{4\varepsilon\omega_1}{\pi^2}, \qquad \varepsilon\omega_1 t \gg 1. \tag{19.173}$$

Now by integrating (19.167) over τ, noting that $d\tau = \frac{\kappa d\kappa}{4\bar{\kappa}^2}$, we obtain the total number of massless particles (photons) emitted to be

$$\mathcal{P}_1(\kappa) = \frac{2}{\pi^2}E(\kappa)K(\kappa) - \frac{1}{2}. \tag{19.174}$$

This follows from the fact that the right-hand side of (19.167) is a total differential, as can be seen from Eqs. (19.148) and (19.150). Let us write $\mathcal{P}_1$ as a function of time, using Eq. (19.125);

$$\mathcal{P}_1(t) = \frac{2}{\pi^2} E[\sqrt{1 - \exp(-4\varepsilon\omega_1 t)}] \times K[\sqrt{1 - \exp(-4\varepsilon\omega_1 t)}] - \frac{1}{2}. \qquad (19.175)$$

For very short time, $\varepsilon\omega_1 t \ll 1$, from the expansion of (19.175) we get

$$\mathcal{P}_1(t) \approx \frac{1}{4}(\varepsilon\omega_1 t)^2, \qquad (19.176)$$

and after a long time, $\varepsilon\omega_1 t \gg 1$, again by expanding (19.175) we obtain

$$\mathcal{P}_1(t) \approx \frac{4}{\pi^2}\varepsilon\omega_1 t + \frac{2}{\pi^2}\ln 4 - \frac{1}{2} + \mathcal{O}\left(t\exp(-4\varepsilon\omega_1 t)\right). \qquad (19.177)$$

The Total Rate of Creation of Particles (Photons) — We can calculate the total number of particles created in all the modes, $\mathcal{P}$, first by summing $\frac{d\mathcal{P}_k}{d\tau}$ over the index k as was done in (19.166). Now by differentiating (19.166) with respect to τ we find

$$\frac{d^2\mathcal{P}(\tau)}{d\tau^2} = -\sum_{odd\,n}\left(\frac{2}{n}\right)\left[\dot{\beta}_{1n}(\tau)\alpha_{1n}(\tau) + \beta_{1n}(\tau)\dot{\alpha}_{1n}(\tau)\right] = 2\left\{\alpha_{11}^2(\tau) + \beta_{11}^2(\tau)\right\},$$
$$(19.178)$$

where we have used Eqs. (19.156), (19.157), (19.163) and (19.164) to replace the derivatives $\dot{\alpha}_{11}(\tau)$ and $\dot{\alpha}_{11}(\tau)$ and sum over n to get the right side of (19.178). Since $\alpha_{11}(\tau)$ and $\alpha_{22}(\tau)$ are known analytically, Eqs. (19.151) and (19.152), we obtain a closed form for the second derivative of $\mathcal{P}(\tau)$;

$$\frac{d^2\mathcal{P}(\tau)}{d\tau^2} = \frac{16}{\pi^2\bar{\kappa}^4}\left\{\left[E(\kappa) - \bar{\kappa}^2 K(\kappa)\right]^2 + \bar{\kappa}^2\left[E(\kappa) - K(\kappa)\right]^2\right\}. \qquad (19.179)$$

The next step is to integrate (19.179) twice, and both integrals can be done analytically, and thus we find an exact expression for $\mathcal{P}(\tau)$

$$\mathcal{P}(\tau) = \frac{1}{\pi^2}\left[\left(1 - \frac{1}{2}\kappa^2\right)K^2(\kappa) - E(\kappa)K(\kappa)\right]. \qquad (19.180)$$

From this expression we can derive the asymptotic forms of $\mathcal{P}(\tau)$ for short and long times. Thus for $\tau \ll 1$ we get $\mathcal{P}(\tau) \approx \tau^2$, and for $\tau \gg 1$ we have [41]

$$\mathcal{P}(\tau) \approx \frac{8}{\pi^2}\tau^2 + \frac{4\tau}{\pi^2}(\ln 4 - 1) - \frac{\ln 4}{\pi^2}(1 - \ln 2) + \mathcal{O}\left(\tau\, e^{-8\tau}\right). \qquad (19.181)$$

We observe that in the limit of small τ, $\mathcal{P}(\tau) \approx \mathcal{P}_1(\tau)$, whereas when τ becomes large $\mathcal{P}(\tau) \gg \mathcal{P}_1(\tau)$, as is expected.

The Energy Associated with the Generation of Particles — Having found the total number of the particles produced due to the motion of the boundaries, we want to calculate the energy which is spent for creating these massless particles (photons). Remembering that we have set $\hbar = 1$, the energy of the lowest mode is $\mathcal{E}_1 = \omega_1 \mathcal{P}_1(\tau)$ and for the total energy in all modes we have

$$\mathcal{E} = \omega_1 \sum_{k=1}^{\infty} \sum_{n=1}^{\infty} \frac{k^2}{n} |\beta_{kn}|^2. \tag{19.182}$$

We can calculate this energy without a knowledge of the explicit form of $\beta_{kn}(\tau)$. For this let us introduce a function $S_n(\tau)$ which we define in the following way:

$$S_n(\tau) = \sum_{k=1}^{\infty} k^2 |\beta_{kn}(\tau)|^2. \tag{19.183}$$

Now if we differentiate $S_n(\tau)$ with respect to τ and substitute for $\dot{\beta}_{kn}(\tau)$ from Eqs. (19.122) and (19.124) we obtain

$$\dot{S}_n = -2\alpha_{1n}\beta_{1n} - 4 \sum_{k=1}^{\infty} k(k+2)\beta_{kn}\beta_{k+2\,n}. \tag{19.184}$$

By differentiating $\dot{S}_n$ once more we find

$$\ddot{S}_n = 2\alpha_{1n}\dot{\beta}_{1n} - 2\dot{\alpha}_{1n}\beta_{1n} + 4\alpha_{1n}^2 - 4\beta_{1n}^2 + 16 S_n. \tag{19.185}$$

The total energy according to (19.182) is

$$\mathcal{E} = \omega_1 \sum_{n}^{\infty} \sum_{n=1} \frac{1}{n} S_n, \tag{19.186}$$

and from Eqs. (19.156), (19.157), (19.163) and (19.164) we have the identity

$$\sum_{n=1}^{\infty} \frac{1}{n} \left(\dot{\alpha}_{1n}\beta_{1n} - \alpha_{1n}\dot{\beta}_{1n} \right) = 1. \tag{19.187}$$

By combining (19.187) with (19.117) we obtain a simple differential equation for $\mathcal{E}$

$$\ddot{\mathcal{E}} - 16\mathcal{E} = 2\omega_1, \tag{19.188}$$

which is subject to the initial conditions

$$\mathcal{E}(0) = \dot{\mathcal{E}}(0) = 0. \tag{19.189}$$

The solution of (19.188) with these initial conditions is given by

$$\mathcal{E}(t) = \frac{1}{4}\omega_1 \sinh^2(2\tau), \tag{19.190}$$

and this total energy grows faster than the number of photons. Therefore there is a rapid pumping in the high frequency modes at the expense of the lower frequencies [41].

19.4 Dissipative Force Acting on a Spherical Mirror Moving in Vacuum

For the case of a perfectly reflecting spherical mirror moving in vacuum, we can use Eqs. (19.57) and (19.59) and the fluctuation-dissipation theorem to find the force on the sphere. We start our inquiry by finding the solution of the wave equation in free space written for the vector potential and in Coulomb gauge and then we impose the boundary conditions on the surface of the sphere. Let $\mathbf{A}(\mathbf{r}, t)$ denote the vector potential which satisfies the wave equation

$$\nabla^2 \mathbf{A}(\mathbf{r}, t) - \frac{\partial^2}{\partial t^2} \mathbf{A}(\mathbf{r}, t) = 0, \tag{19.191}$$

where we have set $c = 1$. The solution of this wave equation in free space can be written as

$$\mathbf{A}(\mathbf{r}, t) = (\mathbf{r}) e^{-i\omega t}, \tag{19.192}$$

where ω is the frequency, and $(\mathbf{r})$ is a solution of the vector Helmholtz equation

$$\nabla^2 \mathbf{A}(\mathbf{r}) + \omega^2 \mathbf{A}(\mathbf{r}) = 0. \tag{19.193}$$

Because the boundary conditions are spherically symmetric, we can write the solution of (19.193) in terms of the scalar Helmholtz equation [25]-[28]

$$\Phi_{\mathbf{k}}(r, \theta, \phi) = \sum_{\ell=0}^{\infty} e^{-i\delta_\ell} i^\ell \left[\cos \delta_\ell \, j_\ell(\omega r) + \sin \delta_\ell \, n_\ell(\omega r) \right]$$

$$\times \sum_{m=-\ell}^{\ell} Y_{\ell m}^*(\hat{\mathbf{k}}) Y_{\ell m}(\theta, \phi) \tag{19.194}$$

In this relation $Y_{\ell m}(\theta, \phi)$ s are spherical harmonics, and $j_\ell(\omega r)$ and $n_\ell(\omega r)$ are spherical Bessel functions of the first and second kind respectively. Since we have set $c = 1$, ω is related to $\mathbf{k}$ by $\omega = |\mathbf{k}|$. This equation describes the scattering of a plane wave which is incident in the direction $\mathbf{k}$. As is well-known in scattering theory, these phase functions, $\delta_\ell(\omega)$, contain all the information about the scatterer, and these phases are determined by imposing the boundary condition(s) on the surface of the scattering object [26].

We can find the solution to the vector equation in the Coulomb gauge by applying the vector angular momentum operator $\mathbf{L}$ and also $\nabla \times \mathbf{L}$ on $\Phi_{\mathbf{k}}(r, \theta, \phi)$ [22]-[28]. Since $\mathbf{E}$ and $\mathbf{B}$ are related to $\mathbf{A}$ by

$$\mathbf{E}(\mathbf{r}, t) = -\frac{\partial}{\partial t} \mathbf{A}(\mathbf{r}, t), \tag{19.195}$$

and

$$\mathbf{B}(\mathbf{r}, t) = \nabla \times \mathbf{A}(\mathbf{r}, t), \tag{19.196}$$

from $\mathbf{L}\Phi_{\mathbf{k}}(r, \theta, \phi)$ and $(\nabla \times \mathbf{L})\Phi_{\mathbf{k}}(r, \theta, \phi)$ we get transverse electric and magnetic modes (perpendicular to $\hat{\mathbf{r}}$) [25],[28].

Using normal modes for the scattering problem given by (19.194), we can find the quantum operators $\mathbf{A}^{TE}(\mathbf{r},t)$ and $\mathbf{A}^{TM}(\mathbf{r},t)$, which we can express as a linear combination of creation and annihilation operators. Thus we write

$$\mathbf{A}^{TE}(\mathbf{r},t) = \int \left[a_{\mathbf{k}}^{TE}(t)\mathbf{A}^{TE}(\mathbf{r}) + \text{Hermitian conjugate} \right] d^3k, \qquad (19.197)$$

and a similar relation for $\mathbf{A}^{TM}(\mathbf{r},t)$. From Eq. (19.192) we find that the operators $a_{\mathbf{k}}^{TE}(t)$ and $a_{\mathbf{k}}^{TM}(t)$ have simple time dependence:

$$a_{\mathbf{k}}^{TE(TM)}(t) = a_{\mathbf{k}}^{TE(TM)}(0)e^{-i\omega t}. \qquad (19.198)$$

Now let us find the spectral density of the force on a perfectly reflecting spherical mirror which is at rest. The two-photon matrix element $\langle n, n'|F(0)|0\rangle$ for this can be written in terms of the wave vectors for TE or TM mode in the following way:

$$\langle \mathbf{k}(TE), \mathbf{k}'(TM)|F(0)|0\rangle = -\frac{\hbar}{\pi\sqrt{\omega\omega'}} \sum_{\ell=1}^{\infty} \left(\frac{df_{\ell}^{(TE)}(z)}{dz} \right)_{z=\omega R} f_{\ell}^{(TM)}(\omega'R)$$

$$\times \sum_{m=-\ell}^{\ell} c_{\ell m} Y_{\ell,m}\left(\hat{\mathbf{k}}\right) Y_{\ell,-m}\left(\hat{\mathbf{k}}'\right), \qquad (19.199)$$

where

$$c_{\ell m} = \frac{im(-1)^{\ell+m+1}}{\ell(\ell+1)}, \qquad (19.200)$$

and

$$f_{\ell}^{\epsilon}(\omega R) = \exp\left(i\delta_{\ell}^{\epsilon}\right) \omega R \left[\cos\delta_{\ell}^{\epsilon} \, j_{\ell}(\omega R) + \sin\delta_{\ell}^{\epsilon} \, n_{\ell}(\omega R)\right], \qquad (19.201)$$

and ϵ stands for TE or TM.

When the photons have the same polarization, then the matrix elements take the form [22],[25]

$$\langle \mathbf{k}(\epsilon), \mathbf{k}'(\epsilon)|F(0)|0\rangle = -\frac{\hbar}{\pi\sqrt{\omega\omega'}} \sum_{\ell=1}^{\infty} \sum_{m=-\ell}^{\ell} d_{\ell m} \left[F_{lm}^{\epsilon}(\mathbf{k}, \mathbf{k}') + F_{lm}^{\epsilon}(\mathbf{k}', \mathbf{k}) \right]$$

$$(19.202)$$

where

$$d_{mn} = (-i)^{2\ell+1}(-1)^{m+1}\frac{\sqrt{\ell^2-1}}{\ell} \left[\frac{\ell^2 - m^2}{(2\ell+1)(2\ell-1)} \right]^{\frac{1}{2}}. \qquad (19.203)$$

For TE and TM modes $F_{lm}^{TE}(\mathbf{k}', \mathbf{k})$ and $F_{lm}^{TE}(\mathbf{k}', \mathbf{k})$ are given by

$$F_{lm}^{TE}(\mathbf{k}, \mathbf{k}') = \left(\frac{df_{\ell}^{TE}(z)}{dz} \right)_{z=\omega R} \left(\frac{df_{\ell-1}^{TE}(z)}{dz} \right)_{z=\omega'R} Y_{\ell,m}\left(\hat{\mathbf{k}}\right) Y_{\ell-1,-m}\left(\hat{\mathbf{k}}'\right),$$

$$(19.204)$$

and

$$F_{\ell m}^{TM}\left(\mathbf{k},\mathbf{k}'\right) = \left[1 + \frac{\ell^2}{\omega\omega' R^2}\right] f_{\ell}^{TM}\left(\omega R\right) f_{\ell-1}^{TM}\left(\omega' R\right) Y_{\ell,m}\left(\hat{\mathbf{k}}\right) Y_{\ell-1,-m}\left(\hat{\mathbf{k}}'\right).$$
(19.205)

The phase shifts δ_{ℓ}^{TE} and δ_{ℓ}^{TM} are determined by the boundary condition. Once the calculation is done we substitute for these phases using the following relations [21],[22]

$$\tan\delta_{\ell}^{TE} = -\frac{j_{\ell}(\omega R)}{n_{\ell}(\omega R)},$$
(19.206)

$$\tan\delta_{\ell}^{TM} = -\left\{\frac{\frac{d}{dz}[z\,j_{\ell}(z)]}{\frac{d}{dz}[z\,n_{\ell}(z)]}\right\}_{z=\omega R}.$$
(19.207)

Having determined $\langle\mathbf{k}(\epsilon)\mathbf{k}'(\epsilon)|F(0)|0\rangle$ and $\langle\mathbf{k}(\epsilon)\mathbf{k}(\epsilon)|F(0)|0\rangle$ from Eqs. (19.204) and (19.205) we can calculate $\tilde{C}_{FF}(\omega)$ from (19.57). We write the function $\tilde{C}_{FF}(\omega)$ as the sum of three terms

$$\tilde{C}_{FF}(\omega) = I_{AA} + I_{BB} + I_{AB},$$
(19.208)

where

$$I_{AA} = \frac{\hbar\omega^3}{3\pi}\sum_{\ell=1}^{\infty}\frac{\ell(\ell+2)}{\ell+1}\int_0^1 x\,(1-x)\,A_{\ell+1}^{-1}\left(R\omega x\right) A_{\ell}^{-1}\left[(R\omega(1-x)\right]dx,$$
(19.209)

$$I_{BB} = \frac{\hbar\omega^3}{3\pi}\sum_{\ell=1}^{\infty}\frac{\ell(\ell+2)}{\ell+1}\int_0^1\frac{\left[x(1-x)+\left(\frac{\ell+1}{\omega R}\right)^2\right]^2}{x(1-x)}$$
$$\times\, B_{\ell+1}^{-1}\left(R\omega x\right) B_{\ell}^{-1}\left[(R\omega(1-x)\right]dx,$$
(19.210)

and

$$I_{AB} = \frac{\hbar\omega^3}{3\pi}\sum_{\ell=1}^{\infty}\frac{2\ell+1(\ell+2)}{\ell+1}\int_0^1 x\,(1-x)\,A_{\ell}^{-1}\left(R\omega x\right) B_{\ell}^{-1}\left[(R\omega(1-x)\right]dx$$
(19.211)

In these relations the functions A_{ℓ}^{-1} and B_{ℓ}^{-1} are defined by

$$A_{\ell}^{-1}(z) = \frac{1}{[z\,j_{\ell}(z)]^2 + [z\,n_{\ell}(z)]^2},$$
(19.212)

and

$$B_{\ell}^{-1}(z) = \frac{1}{[\frac{d}{dz}(z\,j_{\ell}(z))]^2 + [\frac{d}{dz}(z\,n_{\ell}(z))]^2}.$$
(19.213)

Once $\tilde{C}_{FF}(\omega)$ is determined, from Eqs. (19.208)–(19.213) we can find $\tilde{\xi}(\omega)$ from (19.59) and $\tilde{\chi}(\omega)$ from (19.62). The inverse Fourier transform of $\tilde{\chi}(\omega)$ gives us

$\chi(t)$ and finally from Eq. (19.60) we obtain the relation between $\delta F(t)$ and $\delta q(t)$.

While for any value of $R\omega$, which is proportional to the ratio of radius of the sphere to the wavelength, $\tilde{C}_{FF}(\omega)$ can be calculated numerically, we can find asymptotic forms of $\tilde{\xi}(\omega)$ for $\omega R \ll 1$ and for $R\omega \gg 1$.

Quantum Friction for the Motion of a Small Sphere in Vacuum — When $R\omega \ll 1$, only $\ell = 1$ partial wave will be important in the sums I_{AA}, I_{BB} and I_{AB}, Eqs. (19.209)–(19.211). This becomes evident if we consider the expansions of $A_\ell^{-1}(z)$ and of $B_\ell^{-1}(z)$ for small z;

$$A_\ell^{-1}(z) \approx \frac{z^{2\ell}}{[(2\ell-1)!!]^2} \{1 + \mathcal{O}(z)\}, \quad z \ll 1, \tag{19.214}$$

and

$$B_\ell^{-1}(z) \approx \frac{z^{2\ell+2}}{\ell^2[(2\ell-1)!!]^2} \{1 + \mathcal{O}(z)\}, \quad z \ll 1. \tag{19.215}$$

Now if we only keep the term $\ell = 1$ in the sums used in Eqs. (19.209)-(19.211), and also substitute for A_1^{-1} and B_1^{-1} from (19.214) and (19.215) in Eqs. (19.209)–(19.211) we find $\tilde{\xi}(\omega)$ to be

$$\tilde{\xi}(\omega) = \frac{\hbar\omega^9 R^6}{648\pi}, \quad \omega R \ll 1. \tag{19.216}$$

We can obtain the force $\delta F(t)$ acting on a small sphere from the Fourier transform of Eq. (19.61);

$$\delta F(t) = -\frac{\hbar R^6}{648\pi c^8} \delta^{(9)} q(t), \tag{19.217}$$

where $\delta^{(9)} q(t)$ denotes the ninth derivative of $\delta q(t)$.

The Effective Force on a Large Sphere in Vacuum — For a perfectly reflecting sphere with large radius, $\omega R \gg 1$, for the calculation of $\chi(\omega)$ or $\xi(\omega)$ we have to include partial waves up to $\ell \approx \omega R$. When we are summing a large number of partial waves, we replace the summation by integration as it is well-known in scattering theory [55]. This is done by introducing a sequence of $\delta\left(\lambda - \ell - \frac{1}{2}\right)$ functions;

$$\sum_{\ell=0}^{\infty} f_{\ell+\frac{1}{2}} = \sum_{\ell=0}^{\infty} \int_0^\infty f(\lambda)\delta\left(\lambda - \ell - \frac{1}{2}\right) d\lambda$$

$$= \int f(\lambda) \left(\sum_{M=-\infty}^{\infty} (-1)^M e^{2iM\pi\lambda}\right) d\lambda, \tag{19.218}$$

where

$$f\left(\lambda = \ell + \frac{1}{2}\right) = f_{\ell+\frac{1}{2}}. \tag{19.219}$$

We also replace $A_\lambda^{-1}(z)$ and $B_\lambda^{-1}(z)$ by their Debye's asymptotic forms [28],[52]

$$A_\lambda^{-1}(z) \approx B_\lambda^{-1}(z) \approx \sqrt{1 - \frac{\lambda^2}{z^2}}, \qquad (19.220)$$

and these are valid for $1 \ll \ell \le z$. When the stress tensor is integrated over the surface of the sphere to give the force, most of the contribution comes from the $M = 0$ term. Thus neglecting terms with $M \ne 0$ and using the indicated approximation to calculate $\tilde{C}_{FF}(\omega)$, Eq. (19.208), we obtain $\tilde{\xi}(\omega)$ to be given by

$$
\tilde{\xi}(\omega) = \frac{\hbar R^2 \omega^5}{3\pi} \int_0^{\frac{1}{2}} \sigma d\sigma \int_\sigma^{1-\sigma} \left\{ x(1-x) \left[1 - \frac{\sigma^2}{x^2}\right]^{\frac{1}{2}} \left[1 - \frac{\sigma^2}{(1-x)^2}\right]^{\frac{1}{2}} \right.
$$
$$
\left. + \frac{[x(1-x) + \sigma^2]^2}{x(1-x)} \left[1 - \frac{\sigma^2}{x^2}\right]^{-\frac{1}{2}} \left[1 - \frac{\sigma^2}{(1-x)^2}\right]^{-\frac{1}{2}} \right\}. \qquad (19.221)
$$

By evaluating the integral we find $\tilde{\xi}(\omega)$ to be

$$\tilde{\xi}(\omega) = \frac{\hbar \omega^5 R^2}{45\pi}. \qquad (19.222)$$

We calculate the force on a large sphere in the same way as we found the force on a small sphere, but now with $\tilde{\xi}(\omega)$ given by (19.222). This force in the case of a large sphere is of the form

$$\delta F(t) = -\frac{\hbar R^2}{45\pi c^4} \delta^{(5)} q(t), \qquad (19.223)$$

where $\delta^{(5)} q(t)$ is the fifth derivative of $\delta q(t)$.

References

[1] S.A. Fulling and P.C.W. Davies, Proc. R. Soc. London A 348, 393 (1976).

[2] B.S. De Witt, Phys. Rep. C 19, 295 (1975).

[3] M. Razavy, Phys. Rev. A 48, 3486 (1993).

[4] Y. Takahashi and H. Umezawa, Nuovo Cimento 6, 1324 (1957).

[5] E.P. Gross, in *Mathematical Methods in Solid States and Superfluid Theory*, edited by R.C. Clark and G.H. Derrick, (Edinburgh 1969).

[6] J.R. Ockendon and W.R. Hodgkins, *Moving Boundary Problems in Heat Flow and Diffusion*, (Clarendon Press, Oxford, 1974).

[7] J.H. Hill, *Solution of Differential Equations by Means of One-Parameter Group*, (Pitman Advanced Publishing Program, London, 1982) Chapter 6.

[8] L.I. Schiff, *Quantum Mechanics*, Third Edition (McGraw-Hill, New York, N.Y. 1968).

[9] M. Razavy, Lett. Nuovo Cimento, 37, 449 (1983).

[10] M. Abramowitz and I.A. Stegun, *Handbook of Mathematical Functions*, (Dover Publications, New York, 1965) p. 503.

[11] M. Razavy and J. Terning, Lett. Nuovo Cimento, 41, 561 (1984).

[12] G.T. Moore, J. Math. Phys. 11, 2679 (1970).

[13] M. Razavy and J. Terning, Phys. Rev. D 31, 307 (1985).

[14] H.B.G. Casimir, Proc. K. Ned. Akad. Wet. 51, 793 (1948).

[15] A simple derivation of this force from quantum electrodynamics can be found in E.G. Harris's book, *A Pedestrian Approach to Quantum Field Theory*, (John Wiley & Sons, New York, N.Y. 1972).

[16] D. Dalvit, P. Milonni, D. Roberts and F. de Rosa, *Casimir Physics*, in Lecture Notes in Physics, Vol. 834 (Springer-Verlag, Berlin 2011).

[17] M. Kimball, *The Casimir Effect: Physical Manifestations of Zero-Point Energy*, (World Sientific, Singapore 2001).

[18] V.M. Mostepaneko and N.N. Trunov, *The Casimir Effect and Its Applications* (Clarendon Press, Oxford, 1997).

[19] W.M.R. Simpson, *Surprises in Theoretical Casimir Physics*, (Springer International, Switzerland, 2015) Chapter 3.

[20] R.H.S. Winterton, Contemporary Phys. 11, 559 (1970).

[21] P.A. Maia Neto, S. Reynaud and M.T. Jaekel, in *Perspectives in Neutrinos, Atomic Physics and Gravitation*, Edited by J. Trân Thanh Vân, T. Damour, E. Hinds and J. Wilkerson (Edition Frontières, Gif-sur-Yvette, 1993).

[22] P.A. Maia Neto and S. Reynaud, Phys. Rev. A 47, 1639 (1993).

[23] See, for instance J. Rammer, *Quantum Field Theory of Non-Equilibrium States*, (Cambridge University Press, Cambridge, 2007), Chapter 6.

[24] H.M. Nussenzveig, *Causality and Dispersion Relations*, (Academic Press, New York, N.Y. 1972).

[25] C. Eberlein, J. Phys. A 25, 3015 (1992).

[26] See for instance, R.G. Newton's *Scattering Theory of Waves and Particles*, Second Edition (Dover Publications, New York, N.Y. 2013).

[27] L.P. Bayvel and A.R. Jones, *Electromagnetic Scattering and Its Applications*, (Applied Science Publisher, London, 2012), Sec. 1.5.1.

[28] L.C. Biedenharn and J.D. Louck, *Angular Momentum in Quantum Physics*, (Addison-Wesley, Reading, 1981).

[29] For a detailed discussion of this as well as other boundary conditions for partial differential equations see K. Eriksson, D. Estep and C. Johnson, *Applied Mathematics: Body and Soul*, Vol. 2. (Springer-Verlag, Berlin, 2008).

[30] L.H. Ford and A. Vilenkin, Phys. Rev. D 25, 2569 (1980).

[31] P.A. Maia Neto, J. Phys. A 27, 2167 (1994).

[32] M.T. Jaekel and S. Reynaud, Phys. Lett. A 172, 319 (1993).

[33] T.H. Boyer, Sci. Am. 253, 56 (1985).

[34] T.H. Boyer, Am. J. Phys. 71, 990 (2003).

[35] D.F. Mundarian and P.A. Maia Neto, Phys. Rev. A 57, 1379 (1998).

[36] L.A.S. Machado, P.A. Maia Neto and C. Farina, Phys. Rev. D 66, 105016 (2002).

[37] M. Bordag, U. Mohideen and V.M. Mostepanenko, Phys. Rep. 353, 1 (2001).

[38] M. Bordag, G.L. Klinchitskaya, U. Mohideen and V.M. Mostepanenko *International Series of Monographs on Physics, 145: Advances in Casimir Effect*, (Oxford University Press, Oxford, 2009) Chapter 7

[39] M. Kardar and R. Golestanian, Rev. Mod. Phys. 71, 1233 (1991).

[40] C. Farina, Braz. J. Phys. 36, 1137 (2006).

[41] V.V. Dodonov, in *Advances in Chemical Physics, Modern Nonlinear Optics*, Part I, Edited by M.W. Evans, (John Wiley & Sons, New York, N.Y. 1999).

[42] F. Intravaia, R.O.Behunin and D.A.R. Dalvit, Phys. Rev. A 89. 050101 (2014).

[43] F. Intervaia, C. Henkel and M. Antezza, Induced forces between atoms and surfaces: the Casimir-Polder interaction in *Casimir Physics*, Lecture Notes in Physics, Vol. 834 (Springer-Verlag, Berlin, 2011).

[44] G. Barton, New J. Phys. 12, 113045 (2010).

[45] G. Barton, New J. Phys. 14, 079502 (2014).

[46] G.V. Dedkov and A.A. Kyasov, Surface Sci. 605, 1077 (2011).

[47] H.B.G. Casimir and D. Polder, Phys. Rev. 73, 360 (1948).

[48] B-S. Skagerstam, P.K. Redkal and A.H. Vaskinn, Phys. Rev. A 80, 022902 (2009).

[49] See, for instance, M. Razavy *Heisenberg's Quantum Mechanics*, (World Scientific, Singapore, 2011).

[50] V.V. Dodonov in *Advances in Chemical Physics Series*, Vol. 119, Part 1, Edited by M.E. Evans (John Wiley & Sons, New York, N.Y. 2001) p. 309.

[51] I.S. Gradshetyn and I.M. Ryzhik, *Tables of Integrals, Series and Products*, (Academic Press, New York, N.Y. 1994).

[52] See for instance: M. Abramowitz and I.A. Stegun, *Handbook of Mathematical Functions*, (Dover Publications, New York, N.Y. 1965) p. 366.

[53] N.N. Bogoliubov, J. Phys. (USSR) 11, 77 (1947).

[54] N.N. Bogoliubov, *Selected Works of N.N. Bogoliubov*, (World Scientific, Singapore, 2015) Chapter 1.

[55] R.K.B. Helbing, J. Chem. Phys. 50, 493 (1969).

Chapter 20

The Optical Potential

In Chapter 3 we observed that one of the ways of describing the damped harmonic motion in classical dynamics was by introducing complex potentials. The phenomenological optical (or complex) potential can also be utilized in quantum theory of dissipative system as was done in Sec. 15.2. The complex potential is called "optical potential" or "optical model" by analogy with the treatment of the scattering of light by a refracting or absorbing medium, and this is the term that is used to describe the forces used in nucleon-nucleus and nucleus-nucleus scattering. In this section we want to show that such a potential can also be derived from a conservative many-body system S by isolating the motion of a part of the system S_1 from the rest of the system S_2 and study its dependence on the coupling between different particles in the system. However in this case, unlike the coupling of a particle to a heat bath the inter-particle forces are assumed to go to zero as the distance between the two particles tends to infinity. These forces can have a long range, such as the Coulomb force or can have a short range, like nuclear potentials or be a mixture of the two. But in all cases a convenient way to describe the collision between the projectile and the target, when the collision is not purely elastic, is by means of optical potentials. For instance, in atomic physics, we can consider electrons scattering from the hydrogen atom in its ground state. If the electron has a kinetic energy larger than the excitation energy of the $n = 2$ hydrogen state ($10.2\ eV$), then the electron impact, in addition of producing elastic scattering effects, causes the inelastic excitation of the $n = 2$ state [1]-[3]. That is in this and similar inelastic scatterings some particles have been removed from the incident (elastic) channel, or absorption of particles have taken place. This absorption can be represented phenomenologically by a complex or optical potential.

In this chapter we want to study the theory of the optical potential for systems where the inter-particle forces are of short range. Since this type of complex potential has been extensively used in nuclear reaction theory, the simplest model which we want to study is the description of the scattering of a nucleon which is the subsystem S_1 from the target nucleus, S_2, which contains a number of nucleons. Thus we are concerned with a system composed of $A+1$ nucleons [4]–[7]. This approach has also been used to explain the scattering of an electron from an atom, but in this case one has to generalize the formulation so that one can deal with the long range Coulomb force [2],[3].

In the following discussion we follow Feshbach's theory [4], however there is an alternative approach to the problem of scattering by optical potential advanced by Bell and Squires [8].

Let us denote all of the degrees of freedom of the jth nucleon including its spin and isospin by $\mathbf{r}_j$, and the jth state of the target by $\psi_j(\mathbf{r}_1, ...\mathbf{r}_A)$. We write the Hamiltonian of the total system as

$$H\Psi = E\Psi, \tag{20.1}$$

where

$$H = H_A(\mathbf{r}_1, ...\mathbf{r}_A) - \frac{\hbar^2}{2m}\nabla_0^2 + V(\mathbf{r}_0, \mathbf{r}_1, ...\mathbf{r}_A). \tag{20.2}$$

In Eqs. (20.1) and (20.2) E is the total energy and H_A is the Hamiltonian operator of the target nucleus. The operator $\left(-\frac{\hbar^2}{2m}\nabla_0^2\right)$ is the kinetic energy operator of the incoming nucleon and $V(\mathbf{r}_0, \mathbf{r}_1, ...\mathbf{r}_A)$ is the interaction potential between the incoming particle and the nucleons in the target.

Now we expand the total wave function Ψ in terms of the complete set of states, j, of the target

$$\Psi = \sum_j u_j(\mathbf{r}_0)\,\psi_j(\mathbf{r}_1, ...\mathbf{r}_A). \tag{20.3}$$

By substituting (20.3) in (20.2) and using the orthogonality of ψ_j s we find

$$\left(-\frac{\hbar^2}{2m}\nabla_0^2 + V_{jj} + \epsilon_j - E\right)u_j(\mathbf{r}_0) = -\sum_{k\neq j} V_{jk}(\mathbf{r}_0)\,u_k(\mathbf{r}_0), \tag{20.4}$$

where

$$V_{jk}(\mathbf{r}_0) = \int \psi_j^* V \psi_k\, d^3 r_1 \cdots d^3 r_A, \tag{20.5}$$

and

$$H_A\psi_j = \epsilon_j\psi_j. \tag{20.6}$$

Let $j = 0$ term in (20.3) denote the state where the target is in its ground state and a nucleon of energy E is in the incident channel and has a wave function $u_0(\mathbf{r}_0)$. We want to drive an uncoupled Schrödinger equation for $u_0(\mathbf{r}_0)$ by eliminating all of the other channels. To this end we introduce the column matrix $\boldsymbol{\Phi}$ by

$$\boldsymbol{\Phi} = \begin{bmatrix} u_1 \\ u_2 \\ \vdots \end{bmatrix}, \tag{20.7}$$

and a matrix operator $\hat{H}$ with the matrix elements

$$H_{jk} = -\frac{\hbar^2}{2m}\nabla_0^2\delta_{jk} + V_{jk} + \epsilon_j\delta_{jk}, \quad i,j \neq 0. \tag{20.8}$$

For the coupling between $\boldsymbol{\Phi}$ and u_0 we define the matrix potentials

$$\mathbf{V}_0 = (V_{01}, V_{02}, \cdots), \tag{20.9}$$

and

$$\mathbf{V}_0^\dagger = \begin{bmatrix} V_{01}^* \\ V_{02}^* \\ \vdots \end{bmatrix}. \tag{20.10}$$

With the introduction of these potentials we can write (20.4) in the simpler form of

$$\left(-\frac{\hbar^2}{2m}\nabla_0^2 + V_{00} - E\right)u_0 = -\mathbf{V}_0\boldsymbol{\Phi}, \tag{20.11}$$

and

$$\left(\hat{H} - E\right)\boldsymbol{\Phi} = -\mathbf{V}_0^\dagger u_0. \tag{20.12}$$

We can solve (20.12) for $\boldsymbol{\Phi}$ to get

$$\boldsymbol{\Phi} = \left(\frac{1}{E^+ - \hat{H}}\right)\mathbf{V}_0^\dagger u_0, \tag{20.13}$$

where

$$E^+ = E + i\varepsilon, \quad \varepsilon \to 0^+. \tag{20.14}$$

The positive small number ε in (20.14) guarantees that we have only outgoing waves in the exit channels u_i $(i \geq 1)$. Now by inserting (20.13) in (20.11) we obtain

$$\left[-\frac{\hbar^2}{2m}\nabla_0^2 + V_{00} + \mathbf{V}_0\left(\frac{1}{E^+ - \hat{H}}\right)\mathbf{V}_0^\dagger - E\right]u_0(\mathbf{r}_0) = 0, \tag{20.15}$$

and as this equation shows there is an effective potential $\mathcal{V}$ which acts on the particle:

$$\mathcal{V} = V_{00} + \mathbf{V}_0\left(\frac{1}{E^+ - \hat{H}}\right)\mathbf{V}_0^\dagger. \tag{20.16}$$

Note that because of the coupling potential, $\mathbf{V}_0^\dagger$, the incoming particle leaves the channel u_0 and is then emitted in the exit channel u_i which is a part of $\boldsymbol{\Phi}$ in accordance with the Eq. (20.11), and that is why $\mathcal{V}$ is complex. This does not happen unless the energy E is larger than the lowest eigenvalue of the unperturbed Schrödinger equation i.e. $E > \epsilon_1$ (see below).

In general the spectrum of $\hat{H}$ consists of a discrete part and a continuum. Denoting the wave function for the discrete spectrum by Φ_n and for the continuum states by $\Phi(\mathcal{E}, \alpha)$, where α labels various states having a common $\mathcal{E}$, we have

$$\hat{H}\Phi_n = \mathcal{E}_n \Phi_n, \tag{20.17}$$

and

$$\hat{H}\Phi(\mathcal{E}, \alpha) = \mathcal{E}\Phi(\mathcal{E}, \alpha). \tag{20.18}$$

If we want to express V in terms of Φ_n and $\Phi(\mathcal{E}, \alpha)$, we introduce a complete set of states and write

$$V = V_{00} + \sum_n \frac{|\mathbf{V}_0 \Phi_n\rangle \langle \Phi_n \mathbf{V}_0^\dagger|}{(E - \mathcal{E}_n)}$$

$$+ \int d\alpha \int_{\epsilon_1}^\infty \frac{|\mathbf{V}_0 \Phi(\mathcal{E}', \alpha)\rangle \langle \Phi(\mathcal{E}', \alpha) \mathbf{V}_0^\dagger|}{(E^+ - \mathcal{E}')} d\mathcal{E}'. \tag{20.19}$$

In order to show that V is nonlocal, we find the result of action of V on u_0. Consider the operator sum in (20.19) when it operates on u_0;

$$\sum_n \frac{|\mathbf{V}_0 \Phi_n\rangle \langle \Phi_n |\mathbf{V}_0^\dagger| u_0\rangle}{(E - \mathcal{E}_n)} = \int K(\mathbf{r}_0, \mathbf{r}') u_0(\mathbf{r}') d^3 r', \tag{20.20}$$

then the kernel $K(\mathbf{r}_0, \mathbf{r}')$ which is defined by

$$K(\mathbf{r}_0, \mathbf{r}') = \sum_n \sum_{j,k \neq 0} \frac{V_{0j}(\mathbf{r}_0) u_j^{(n)}(\mathbf{r}_0) \left[u_k^{(n)}(\mathbf{r}')\right]^* V_{k0}(\mathbf{r}')}{E - \mathcal{E}_n}, \tag{20.21}$$

shows that V is nonlocal.

In addition to nonlocality the optical potential V has the following general properties:

(1) The numerators of the two terms in the expansion in (20.19) are positive definite, i.e. for any arbitrary function w we have

$$\langle w \mathbf{V}_0 \Phi_n\rangle \langle \Phi_n \mathbf{V}_0^\dagger w\rangle = |\langle w \mathbf{V}_0 \Phi_n\rangle|^2 \geq 0 \tag{20.22}$$

and

$$|\langle w \mathbf{V}_0 \Phi(\mathcal{E}', \alpha)\rangle|^2 \geq 0. \tag{20.23}$$

From (20.19) it readily follows that

$$\operatorname{Im} V = -\pi \int |\mathbf{V}_0 \Phi(E, \alpha)\rangle \langle \Phi(E, \alpha) \mathbf{V}_0^\dagger| \, \theta(E - \epsilon_1) \, d\alpha, \tag{20.24}$$

where $\theta(x)$ is a step function

$$\theta(x) = \begin{cases} 1 & \text{for } x > 0 \\ 0 & \text{for } x < 0 \end{cases}, \tag{20.25}$$

and where ϵ_1 is the eigenvalue of the uncoupled Schrödinger equation

$$\left(-\frac{\hbar^2}{2m}\nabla_0^2 + V_{11}\right) u_1 = -\left(\epsilon_1 - \mathcal{E}\right) u_1. \tag{20.26}$$

We observe that Im $\mathcal{V} \leq 0$ as it should be since $\mathcal{V}$ is an absorptive potential.

(2) Equation (20.19) shows that Re $\mathcal{V}$ has simple poles at $E = \mathcal{E}_n$ and a branch line for $E > \epsilon_1$. By taking the derivative of Re $\mathcal{V}$ with respect to E we find that

$$\frac{\partial}{\partial E}\left(\text{Re } \mathcal{V}\right) \leq 0. \tag{20.27}$$

(3) The real and the imaginary parts of $\mathcal{V}$ are related to each other. Thus if we substitute (20.24) in (20.19) we find

$$\text{Re } \mathcal{V} = V_{00} + \sum_n \frac{|\mathbf{V}_0 \mathbf{\Phi}_n\rangle \langle \mathbf{\Phi}_n \mathbf{V}_0^\dagger|}{E - \mathcal{E}_n} - \frac{\mathcal{P}}{\pi}\int_{\epsilon_1}^{\infty}\frac{\text{Im } \mathcal{V}\left(\mathcal{E}'\right)}{E - \mathcal{E}'}d\mathcal{E}', \tag{20.28}$$

where $\mathcal{P}$ denotes Cauchy principal value. For nuclear matter there is no summation over discrete states, therefore (20.28) will reduce to [4]

$$\text{Re } \mathcal{V} = V_{00} - \frac{\mathcal{P}}{\pi}\int_0^{\infty}\frac{\text{Im } \mathcal{V}\left(\mathcal{E}'\right)}{E - \mathcal{E}'}d\mathcal{E}'. \tag{20.29}$$

The Feshbach theory of optical potential has been very successful in explaining nucleon-nucleus and nucleus-nucleus scattering [4],[9]–[13]. Since the wave function is in general complex, the presence of a complex (optical potential) in the Schrödinger equation arises naturally from the interaction of a particle (or particles) from a system of particles. But the idea of a complex force law is not compatible with the real position, real velocity, and the real trajectory of the particle in classical mechanics. The nonlocality of the interaction also adds to the problem of finding the classical limit of an optical potential.

The Case of a Single Resonance — As we have seen from Eq. (20.19) the effective potential $\mathcal{V}$ can have a number of poles at the energies $E = \mathcal{E}_n$, and for these values of E we have resonances. Let us assume that E is very close to one of the discrete eigenvalues, say $\mathcal{E}_n$, so that only the singular term in $\mathcal{V}$ varies with the energy E. Thus the main contribution to $\mathcal{V}$ comes from

$$\mathcal{V} = U_n + \frac{|\mathbf{V}_0 \mathbf{\Phi}_n\rangle \langle \mathbf{\Phi}_n \mathbf{V}_0^\dagger|}{E - \mathcal{E}_n}, \tag{20.30}$$

where

$$U_n \approx V_{00} + \int \frac{d\mathcal{E}}{\mathcal{E}_n - \mathcal{E}'} \int |V_0 \Phi\left(\mathcal{E}', \alpha\right)\rangle\langle \Phi\left(\mathcal{E}', \alpha\right) V_0^{\dagger}| d\alpha$$
$$+ \sum_{j \neq n} \frac{|V_0 \Phi_j\rangle\langle \Phi_j V_0^{\dagger}|}{\mathcal{E}_n - \mathcal{E}_j}. \tag{20.31}$$

We have omitted $i\varepsilon$ in (20.31) since only elastic scattering is possible. The Schrödinger equation for u_0 when E is close to $\mathcal{E}_n$ becomes

$$(H_0 - E)u_0 = -\Lambda_n V_0 \Phi_n, \tag{20.32}$$

where the constant Λ_n is defined by

$$\Lambda_n = \frac{\langle \Phi_n | V_0 | u_0 \rangle}{E - \mathcal{E}_n}, \tag{20.33}$$

and

$$H_0 = -\frac{\hbar^2}{2m} \nabla^2 + U_0. \tag{20.34}$$

The formal solution of the wave equation (20.32) is

$$u_0 = f_0^{(+)} + \Lambda_n \frac{1}{E^+ - H_0} V_0 \Phi_n, \tag{20.35}$$

where $f_0^{(+)}$ is the outgoing wave solution of

$$(H_0 - E)f_0^{(+)} = 0. \tag{20.36}$$

We can obtain a linear equation for Λ_n if we substitute (20.35) in (20.33);

$$\Lambda_n = \frac{\langle \Phi_n V_0^{\dagger} f_0^{(+)} \rangle}{E - \mathcal{E}_n - \left\langle \Phi_n V_0^{\dagger} \frac{1}{E^+ - H_0} V_0 \Phi_n \right\rangle}, \tag{20.37}$$

and once Λ_n is determined then the transition matrix T can be found from the asymptotic behaviour of the wave function u_0, Eq. (20.35)

$$T = T_p + \Lambda_n \left\langle f_n^{(-)} V_0 \Phi_n \right\rangle, \tag{20.38}$$

where T_p is the transition matrix describing the asymptotic behavior of $f_0^{(+)}$, i.e. the scattering caused only by the potential U_n. Now if we substitute (20.37) in (20.38) we get an expression for the full transition matrix

$$T = T_p + \frac{\langle f_0^{(-)} V_0 \Phi_n \rangle \langle \Phi_n V_0^{\dagger} f_0^{(+)} \rangle}{E - \mathcal{E}_n - \left\langle \Phi_n V_0^{\dagger} \frac{1}{E^+ - H_0} V_0 \Phi_0 \right\rangle}. \tag{20.39}$$

Next let us consider the specific case of the S-wave scattering for which the solution of (20.36) is simple

$$f_0 = e^{i\delta} g(r_0), \qquad (20.40)$$

where δ is the phase shift for the angular momentum $\ell = 0$ and $g(r_0)$ is given by

$$g(r_0) = \left(\frac{2}{\pi} k^2 \frac{dk}{dE}\right)^{\frac{1}{2}} \frac{\sin(kr_0 + \delta)}{kr_0}. \qquad (20.41)$$

For this wave function T_p which is proportional to the scattering amplitude is a function of δ alone [14]

$$T_p = -\frac{1}{\pi} e^{i\delta} \sin \delta. \qquad (20.42)$$

Now substituting for $f_0^{\pm}$, we can write the last term in the denominator of T-matrix as

$$\left\langle \Phi_n V_0^\dagger \left(\frac{1}{E^+ - H_0}\right) V_0 \Phi_0 \right\rangle = \left\langle \Phi_n V_0^\dagger \mathcal{P} \left(\frac{1}{E - H_0}\right) V_0 \Phi_0 \right\rangle - i\pi \left| \langle \Phi_n V_0^\dagger g \rangle \right|^2, \qquad (20.43)$$

where $\mathcal{P}$ denotes the principle value integral. From (20.39) and (20.43) we find T;

$$T = -\frac{1}{\pi} \left[\sin \delta e^{i\delta} - \pi \frac{e^{2i\delta} \langle g V_0 \Phi_n \rangle \langle \Phi_n V_0^\dagger g \rangle}{E - E_n - \Delta_n + i\pi \langle g V_0 \Phi_n \rangle \langle \Phi_n V_0^\dagger g \rangle} \right]. \qquad (20.44)$$

The denominator in (20.44) shows that the resonant energy is shifted by Δ_n where

$$\Delta_n = \left\langle \Phi_n V_0^\dagger \mathcal{P} \left(\frac{1}{E - H_0}\right) V_0 \Phi_0 \right\rangle, \qquad (20.45)$$

and also that the width of the resonance is given by

$$\Gamma = 2\pi \left| \langle \Phi_n V_0^\dagger g \rangle \right|^2. \qquad (20.46)$$

Once the T-matrix is known, we can calculate the cross section for scattering from the optical potential [14]. For instance the cross section of elastic scattering of the S-wave neutrons by a target nucleus of spin I is

$$\sigma_{elastic} = \frac{\lambda^2}{\pi} \frac{2J+1}{(2J+1)(2s+1)} \left| \sin \delta \, e^{i\delta} - \frac{\frac{1}{2}\Gamma_n^{(J)} e^{2i\delta}}{E - E_0 + \frac{1}{2}i\Gamma_c^{(J)}} \right|^2, \qquad (20.47)$$

where λ is the wavelength of the incident neutron and s is its spin. In (20.47) E_c, J and $\Gamma_c^{(J)}$ are the resonance energy, the spin and the width of the compound nucleus respectively, and $\Gamma_n^{(J)}$ is the neutron width. Thus the energy dependence of the cross section in the neighborhood of a resonance is given by Breit-Wigner formula which gives an excellent fit to the empirical data [15].

In the following section we study the classical limit of the wave equation with the optical potential, assuming that this potential is nonlocal and independent of the energy of the particle. But first we will consider the question of the classical form of a nonlocal real potential.

20.1   The Classical Analogue of a Nonlocal Interaction

As we have seen in Feshbach's theory, the optical potential is not only a complex function of $\mathbf{r}$ but it is also nonlocal. Here we will investigate the classical analogue of a symmetric and real nonlocal potential, and then consider how the complex nature of the potential manifests itself in the classical limit [16]-[18]. For a detailed discussion of the semi-classical treatment of nonlocal potentials and their equivalent local potential see [17].

We write the Schrödinger equation in the simple form of

$$\left(\nabla^2 + k^2\right)\psi(\mathbf{r}) = \int K\left(\mathbf{r},\mathbf{r}'\right)\psi\left(\mathbf{r}'\right)d^3r', \tag{20.48}$$

where $K\left(\mathbf{r},\mathbf{r}'\right)$ is a symmetric kernel

$$K\left(\mathbf{r},\mathbf{r}'\right) = K\left(\mathbf{r}',\mathbf{r}\right). \tag{20.49}$$

Now we expand $K\left(\mathbf{r},\mathbf{r}'\right)$ in the following way [16],[18]:

$$\int K\left(\mathbf{r},\mathbf{r}'\right)d^3r' = U_0(r), \tag{20.50}$$

$$\int K\left(\mathbf{r},\mathbf{r}'\right)\left(x - x'\right)d^3r' = -xU_1(r), \tag{20.51}$$

$$\cdots \cdots \cdots \cdots$$

$$\int K\left(\mathbf{r},\mathbf{r}'\right)\left(x - x'\right)\left(y - y'\right)d^3r' = xyU_2(r), \tag{20.52}$$

$$\cdots \cdots \cdots \cdots$$

Thus the right hand side of (20.48) can be written as

$$\int K\left(\mathbf{r},\mathbf{r}'\right)\psi\left(\mathbf{r}'\right)d^3r = U_0\psi\left(\mathbf{r}\right) + U_1\left(\sum x\frac{\partial}{\partial x}\right)\psi\left(\mathbf{r}\right)$$

$$+ \frac{1}{2}U_2\sum\sum xy\frac{\partial^2}{\partial x\partial y}\psi\left(\mathbf{r}'\right) + \cdots, \tag{20.53}$$

where the sum is over the components x, y and z. Rearranging the terms on the right hand side of (20.53) we get

$$V(\mathbf{r}, \nabla)\psi(\mathbf{r}) = \left[U_0 + \frac{1}{2}U_1 (\mathbf{r} \cdot \nabla + \nabla \cdot \mathbf{r}) + \frac{1}{4}U_2 (\mathbf{r} \cdot \nabla + \nabla \cdot \mathbf{r})^2 + \cdots \right] \psi(\mathbf{r}),$$
(20.54)

where the potential $V(\mathbf{r}, \nabla)$ is an infinite differential operator.

We can write $V(\mathbf{r}, \nabla)$ as a momentum (or velocity) dependent potential by noting that $\mathbf{p} = -i\hbar\nabla$, and therefore

$$V(\mathbf{r}, \mathbf{p})$$
$$= \left[U_0 + \frac{i}{2\hbar}U_1 (\mathbf{r} \cdot \mathbf{p} + \mathbf{p} \cdot \mathbf{r}) - \frac{1}{4\hbar^2}U_2 \left\{ \mathbf{r}^2\mathbf{p}^2 + 2(\mathbf{r} \cdot \mathbf{p})(\mathbf{p} \cdot \mathbf{r}) + \mathbf{p}^2\mathbf{r}^2 \right\} \right] + \cdots .$$
(20.55)

Thus the classical momentum-dependent potential is

$$V(\mathbf{r}, \mathbf{p}) = \left[U_0 + \frac{i}{\hbar}U_1 (\mathbf{r} \cdot \mathbf{p}) - \frac{1}{\hbar^2}U_2 (\mathbf{r} \cdot \mathbf{p})^2 + \cdots \right].$$
(20.56)

If we impose the requirement of invariance under rotation, then $V(\mathbf{r}, \mathbf{p})$ must be a function of r^2, p^2, $(\mathbf{r} \times \mathbf{p})^2$, and $(\mathbf{r} \cdot \mathbf{p} + \mathbf{p} \cdot \mathbf{r})$ [19]. The time-reversal invariance implies that only even powers of $(\mathbf{r} \cdot \mathbf{p} + \mathbf{p} \cdot \mathbf{r})$ should appear in $V(\mathbf{r}, \mathbf{p})$. Thus the most general velocity-dependent local potentials satisfying these requirements will have the general form of $U\left(r^2, p^2, L^2\right)$, where $\mathbf{L} = \mathbf{r} \times \mathbf{p}$ is the angular momentum.

A potential of the form

$$V(\mathbf{r}, \mathbf{p}) = U_0(r) + \frac{1}{2m}\mathbf{p} \cdot W(r)\mathbf{p},$$
(20.57)

which is a special case of (20.55), has been used to describe the interaction between two nucleons [20]. Note the similarity between (20.57) and the Hamiltonian for the dissipative motion with quadratic dependence on velocity [21]. The potential $V(\mathbf{r}, \mathbf{p})$ given in (20.56) is, in general, an infinite series in powers of $\mathbf{r} \cdot \mathbf{p}$ however one can construct models of nonlocal potentials where for a specific partial wave scattering, e.g. S-wave the Schrödinger equation which is an integro-differential equation can be transformed into an ordinary differential equation of fourth, sixth, $\cdots$ order in $\frac{d}{dr}$ [22],[23].

Ehrenfest's Theorem in the Presence of Complex Potential — Next we will study the classical limit of a local optical potential. We assume that the potential is independent of the energy of the particle and is of the form

$$V_0(r) = v(r) + i\mathcal{V}(r).$$
(20.58)

Now according to Feshbach's theory $\mathcal{V}$ must be negative definite, i.e. $\mathcal{V}(r) \leq 0$ for all r. One of the possible ways of studying the classical limit of the motion in a complex potential field is by examining the time evolution of the expectation values of the radial coordinate and momentum of the particle. To this end we start with the Schrödinger equation with the complex potential

$$i\hbar\frac{\partial\psi}{\partial t} = \left[-\frac{\hbar^2}{2m}\nabla^2 + v(r) + iV(r)\right]\psi. \tag{20.59}$$

Noting that for any operator $\hat{O}(\mathbf{r}, -i\hbar\nabla)$, the rate of change of the expectation value is given by

$$\frac{d}{dt}\left\langle\hat{O}\right\rangle = \frac{i}{\hbar}\left(\int H^*\psi^*\hat{O}\psi d^3r - \int \psi^*\hat{O}H\psi d^3r\right), \tag{20.60}$$

where H is the operator given in the square bracket in (20.59), we want to determine the expectation values of r and p_r.

The last equation can be written in terms of a Hermitian and an anti-Hermitian operator [24],[25]. Let us write

$$H = H_H + H_A = \frac{1}{2}(H + H^*) + \frac{1}{2}(H_H - H_A^*), \tag{20.61}$$

where the first term $(1/2)(H + H^*)$ is the Hermitian part. Then (20.60) becomes

$$\frac{d}{dt}\left\langle\hat{O}\right\rangle = \int \psi^*\left(\frac{i}{\hbar}\left[H_H, \hat{O}\right] - \frac{i}{\hbar}\left[H_A, \hat{O}\right]_+\right)\psi\, d^3r, \tag{20.62}$$

where $[\ ,]$ and $[\ ,]_+$ denote the commutator and the anti-commutator respectively. Substituting for $\hat{O}$ first r and then p_r in Eq. (20.62), we find

$$\frac{d}{dt}\langle r\rangle = \frac{\langle p_r\rangle}{m} + \frac{\langle 2rV(r)\rangle}{\hbar}, \tag{20.63}$$

and

$$\frac{d}{dt}\langle p_r\rangle = -\left\langle\frac{\partial v(r)}{\partial r} + i\frac{\partial V}{\partial r}\right\rangle + 2\frac{\langle V(r)p_r\rangle}{\hbar}. \tag{20.64}$$

These relations reduce to the usual form of the Ehrenfest's theorem when $V(r)$ is zero [26],[27]. From these relations we draw the following conclusions [28]:

(1) For the existence of a well-defined classical limit $V(r)$ has to be proportional to $\hbar^\alpha, (\alpha \geq 1)$. We can also reach the same result by considering the hydrodynamical formulation of the Schrödinger equation [29],[30]. In the latter formulation we start with the wave function $\psi(\mathbf{r}, t)$ and write it as

$$\psi(\mathbf{r}, t) = \sqrt{\rho}\exp\left(\frac{iS}{\hbar}\right). \tag{20.65}$$

By substituting (20.65) in (20.59) and separating the real and imaginary parts we find two equations for ρ and S;

$$\frac{\partial\rho}{\partial t} + \nabla\cdot\left(\frac{\rho\nabla S}{m}\right) - \frac{2\rho V}{\hbar} = 0, \tag{20.66}$$

and

$$\frac{\partial S}{\partial t} + \frac{1}{2m} \left(\nabla S \right)^2 - \hbar^2 \left[\frac{\nabla^2 \rho}{\rho} - \left(\frac{\nabla \rho}{\rho} \right)^2 \right] + v(r) = 0. \tag{20.67}$$

If we integrate (20.66) over all space and observe that $\rho \nabla S \to 0$ as $r \to \infty$, we obtain

$$\frac{d}{dt} \int \rho d^3 r = \frac{2}{\hbar} \int \rho \mathcal{V}(r) d^3 r \leq 0. \tag{20.68}$$

The negative-definiteness in (20.68) follows from the fact that $\rho(r)$ is positive or zero and $\mathcal{V}(r)$ is negative or zero for all r. This relation shows that $\mathcal{V}(r)$ has to go to zero as $\hbar^a (a \geq 1)$ in order to have a well-defined classical limit.

(2) Equation (20.63) shows that the canonical momentum is not the same as the mechanical momentum. As we have seen before this creates a problem with using the minimal coupling rule when the particle is charged.

(3) The time evolution equations for $\langle x \rangle$ and $\langle p \rangle$ show that both of these are complex quantities. Thus the three-dimensional motion of a particle corresponds to six degrees of freedom. This confirms the result of Sec. 2.6 where we noticed that the motion of a spring with a complex spring constant can be decomposed into the motion of two real damped harmonic oscillators.

20.2 Minimal and/or Maximal Coupling when the Potential is Real and Nonlocal

In the case of the optical potential found in the previous section there is no general rule for the coupling of a charged particle to the electromagnetic field which is compatible with the gauge transformation. In what follows we will study the electromagnetic field coupling when the potentials are real, energy-independent and are either separable or are non-separable.

Minimal Coupling for Separable Potentials — First let us write Eq. (20.48) as

$$\left(\frac{\hbar^2}{2m} \nabla^2 + E \right) \psi(\mathbf{r}) = \int \mathcal{V} \left(\mathbf{r}, \mathbf{r}' \right) \psi \left(\mathbf{r}' \right) d^3 r', \tag{20.69}$$

with $\mathcal{V} \left(\mathbf{r}, \mathbf{r}' \right)$ being a real energy independent potential acting between, say, a proton and a neutron. If in addition we assume that this potential is separable then we can write it as

$$\mathcal{V} \left(\mathbf{r}, \mathbf{r}' \right) \to v \left(\mathbf{r}_p - \mathbf{r}_n \right) v \left(\mathbf{r}'_p - \mathbf{r}'_n \right). \tag{20.70}$$

In this relation $\mathbf{r}_p$ and $\mathbf{r}_n$ are the coordinates of the proton and neutron respectively, and $\mathbf{r} = \mathbf{r}_p - \mathbf{r}_n$ and $\mathbf{r}' = \mathbf{r}'_p - \mathbf{r}'_n$. Now if $\mathbf{R}$ denotes the center of mass coordinate of this two-body problem

$$\mathbf{R} = \frac{1}{2} \left(\mathbf{r}_p + \mathbf{r}_n \right) = \frac{1}{2} \left(\mathbf{r}'_p + \mathbf{r}'_n \right), \tag{20.71}$$

then in the presence of the electromagnetic field $\mathbf{A}$, the potential $\mathcal{V}(\mathbf{r}, \mathbf{r}')$ is replaced by

$$\mathcal{V}(\mathbf{r}, \mathbf{r}', \mathbf{A}) = v(\mathbf{r}_p - \mathbf{r}_n) v(\mathbf{r}'_p - \mathbf{r}'_n)$$

$$\times \exp\left\{ ie \int_{\mathbf{R}}^{\mathbf{r}_p} \mathbf{A} \cdot d\mathbf{s} + ie \int_{\mathbf{r}_{p'}}^{\mathbf{R}} \mathbf{A} \cdot d\mathbf{s} \right\}. \tag{20.72}$$

The line integral is taken along a straight line connecting the lower to the upper limit [31].

It should be pointed out that (20.72) is not the most general form of the coupling which is possible. Thus if F and G are two gauge invariant quantities such that as the electromagnetic field tends to zero, $F \to 1$ and $G \to 0$, then (20.72) can be replaced by [32],[33]

$$\mathcal{V}(\mathbf{r}, \mathbf{r}', \mathbf{A}) F + G. \tag{20.73}$$

Nonseparable and Nonlocal Interactions — In Now let us consider the Schrödinger equation for two particles where only one of the particles (e.g. particle 1) is charged. For this case the charge and current densities are given by

$$\langle \Psi_f \,|\rho_1(\mathbf{r})|\, \Psi_i \rangle = e \int \Psi_f^*(\mathbf{r}, \mathbf{r}_2) \Psi_i(\mathbf{r}, \mathbf{r}_2) d^3 r_2, \tag{20.74}$$

and

$$\langle \Psi_f \,|\mathbf{j}_1(\mathbf{r})|\, \Psi_i \rangle = \frac{-ie}{2m} \int \left[\Psi_f^*(\mathbf{r}, \mathbf{r}_2) \nabla \Psi_i(\mathbf{r}, \mathbf{r}_2) - \Psi_i(\mathbf{r}, \mathbf{r}_2) \nabla \Psi_f^*(\mathbf{r}, \mathbf{r}_2) \right] d^3 r_2, \tag{20.75}$$

respectively. To find the conservation law for the charge and current density we start with the Schrödinger equation with the nonlocal potential

$$-\frac{\hbar^2}{2m} \left(\nabla_1^2 + \nabla_2^2 \right) \Psi(\mathbf{r}_1, \mathbf{r}_2, t) + \int \mathcal{V}(\mathbf{r}_1, \mathbf{r}_2; \mathbf{r}'_1, \mathbf{r}'_2) \Psi(\mathbf{r}'_1, \mathbf{r}'_2, t) d^3 r'_1 d^2 r'_2$$

$$= i\hbar \frac{\partial}{\partial t} \Psi(\mathbf{r}_1, \mathbf{r}_2, t), \tag{20.76}$$

and a similar equation for $\Psi^*(\mathbf{r}_1, \mathbf{r}_2, t)$. Here we have assumed that the masses of the two particles are equal. The general case of unequal masses have been discussed by Heller [34]. The potential $\mathcal{V}(\mathbf{r}_1, \mathbf{r}_2; \mathbf{r}'_1, \mathbf{r}'_2)$ depends on the relative coordinates of the two particles, therefore

$$\mathcal{V}(\mathbf{r}_1, \mathbf{r}_2; \mathbf{r}'_1, \mathbf{r}'_2) = \delta(\mathbf{R} - \mathbf{R}') \langle \mathbf{r} \,|\mathcal{V}|\, \mathbf{r}' \rangle, \tag{20.77}$$

where $\mathbf{R}$ is the center of mass coordinate and $\mathbf{r}$ is the relative coordinate of the two particles.

From (20.76) and its complex conjugate we find

$$\frac{\partial}{\partial t}(e\Psi^*\Psi) = -\left(\frac{ie\hbar}{2m}\right) \nabla_1 \cdot (\Psi \nabla_1 \Psi^* - \Psi^* \nabla_1 \Psi) - Y(\mathbf{r}_1, \mathbf{r}_2, t), \tag{20.78}$$

where

$$Y\left(\mathbf{r}_1, \mathbf{r}_2, t\right) = \left(\frac{ie}{\hbar}\right) \int \mathcal{V}\left(\mathbf{r}_1, \mathbf{r}_2; \mathbf{r}_3, \mathbf{r}_4\right)$$
$$\times \left\{\Psi\left(\mathbf{r}_3, \mathbf{r}_4, t\right) \Psi^*\left(\mathbf{r}_1, \mathbf{r}_2, t\right) - \Psi^*\left(\mathbf{r}_3, \mathbf{r}_4, t\right) \Psi\left(\mathbf{r}_1, \mathbf{r}_2, t\right)\right\} d^3 r_3 d^3 r_4.$$

(20.79)

The charge density in this case is expressible as an integral

$$\rho(\mathbf{r}_1, t) = e \int \Psi^*\left(\mathbf{r}_1, \mathbf{r}_2, t\right) \Psi\left(\mathbf{r}_1, \mathbf{r}_2, t\right) d^3 r_2,$$

(20.80)

and since $\rho(\mathbf{r}_1, t)$ is nonlocal we expect a nonlocal expression for the current density also. Now the divergent of a part of current, $\mathbf{j}_1\left(\mathbf{r}_1\right)$, is given by

$$\nabla_1 \cdot \mathbf{j}_1\left(\mathbf{r}_1\right) = -\left(\frac{ie\hbar}{2m}\right)$$
$$\times \nabla_1 \cdot \int \left\{\nabla_1 \Psi\left(\mathbf{r}_1, \mathbf{r}_2, t\right) \Psi^*\left(\mathbf{r}_1, \mathbf{r}_2, t\right) - \Psi\left(\mathbf{r}_1, \mathbf{r}_2, t\right) \nabla_1 \Psi^*\left(\mathbf{r}_1, \mathbf{r}_2, t\right)\right\} d^2 r_2.$$

(20.81)

In order to have conservation law for the charge, in addition to $\mathbf{j}_1\left(\mathbf{r}_1\right)$ we must have a current $\mathbf{j}_2\left(\mathbf{r}_1\right)$ where

$$\nabla_1 \cdot \mathbf{j}_2\left(\mathbf{r}_1\right) = \left(\frac{ie}{2\hbar}\right) \int \mathcal{V}\left(\mathbf{r}_1, \mathbf{r}_2; \mathbf{r}_3, \mathbf{r}_4\right)$$
$$\times \left[\Psi\left(\mathbf{r}_3, \mathbf{r}_4, t\right) \Psi^*\left(\mathbf{r}_1, \mathbf{r}_2, t\right) - \Psi^*\left(\mathbf{r}_3, \mathbf{r}_4, t\right) \Psi\left(\mathbf{r}_1, \mathbf{r}_2, t\right)\right] d^3 r_2 d^3 r_3 d^3 r_4.$$

(20.82)

From $\nabla_1 \cdot \mathbf{j}_2(\mathbf{r}_1)$ we find $\mathbf{j}_2(\mathbf{r}_1)$ to be [34]

$$\mathbf{j}_2\left(\mathbf{r}_1\right) = \left(\frac{ie}{2\hbar}\right) \int \mathbf{K}\left(\mathbf{r}_1 - \mathbf{r}_3; \mathbf{r}_1 - \mathbf{r}_5\right) \mathcal{V}\left(\mathbf{r}_2, \mathbf{r}_3; \mathbf{r}_4, \mathbf{r}_5\right)$$
$$\times \Psi^*\left(\mathbf{r}_3, \mathbf{r}_2, t\right) \Psi\left(\mathbf{r}_4, \mathbf{r}_5, t\right) d^3 r_2 d^3 r_3 d^3 r_4 d^3 r_5,$$

(20.83)

where

$$\nabla_1 \cdot \mathbf{K}\left(\mathbf{r}_1 - \mathbf{r}_3; \mathbf{r}_1 - \mathbf{r}_5\right) = \delta\left(\mathbf{r}_1 - \mathbf{r}_3\right) - \delta\left(\mathbf{r}_1 - \mathbf{r}_5\right).$$

(20.84)

Note that the Hermiticity requires that

$$\mathbf{K}^*\left(\mathbf{r}_1 - \mathbf{r}_3; \mathbf{r}_1 - \mathbf{r}_5\right) = -\mathbf{K}\left(\mathbf{r}_1 - \mathbf{r}_5; \mathbf{r}_1 - \mathbf{r}_3\right).$$

(20.85)

For a local potential from (20.77) we have

$$\mathcal{V}\left(\mathbf{r}_1, \mathbf{r}_2; \mathbf{r}_1', \mathbf{r}_2'\right) = \delta\left(\mathbf{r}_1 - \mathbf{r}_1'\right) \delta\left(\mathbf{r}_2 - \mathbf{r}_2'\right) V(|\mathbf{r}_1 - \mathbf{r}_2|),$$

(20.86)

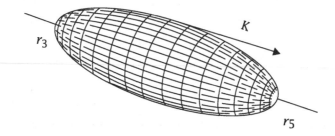

Figure 20.1: An elementary solution of Eq. (20.84). For any point on the surface, $\mathbf{K}$ is a unit vector in the direction of current multiplied by a two-dimensional delta function in the orthogonal plane.

and therefore $\mathbf{j}_2(\mathbf{r}_1)$ is zero. Having obtained the total current $\mathbf{j}_1(\mathbf{r}_1) + \mathbf{j}_2(\mathbf{r}_1)$ we can express the conservation law as

$$\frac{\partial \rho}{\partial t} + \nabla_1 \cdot \mathbf{j}(\mathbf{r}_1) = 0, \qquad (20.87)$$

where

$$\mathbf{j}(\mathbf{r}_1) = \mathbf{j}_1(\mathbf{r}_1) + \mathbf{j}_2(\mathbf{r}_1). \qquad (20.88)$$

The interaction between the charged particle and the electromagnetic field can be written as H_{int} where

$$H_{int} = \int \left[\rho(\mathbf{r}_1)\phi(\mathbf{r}_1, t) - \frac{1}{c}\mathbf{j}(\mathbf{r}_1) \cdot \mathbf{A}_1(\mathbf{r}_1, t) \right] d^3 r_1. \qquad (20.89)$$

In this equation $\phi(\mathbf{r}_1, t)$ and $\mathbf{A}_1(\mathbf{r}_1, t)$ are the electric and magnetic potentials respectively.

If $\phi(\mathbf{r}_1, t)$ is zero, then H_{int} becomes

$$H_{int} = -\frac{1}{c} \int [\Psi^* \nabla_1 \Psi - \Psi \nabla_1 \Psi^*] \cdot \mathbf{A}_1 d^3 r_1 - H_I, \qquad (20.90)$$

where H_I is defined by

$$H_I = \frac{1}{c} \int \mathbf{K}(\mathbf{r}_1 - \mathbf{r}_3;\ \mathbf{r}_1 - \mathbf{r}_5) \cdot \mathbf{A}_1(\mathbf{r}_1, t)\mathcal{V}(\mathbf{r}_2, \mathbf{r}_3;\ \mathbf{r}_4, \mathbf{r}_5)$$
$$\times \Psi^*(\mathbf{r}_2, \mathbf{r}_3, t)\, \Psi(\mathbf{r}_4, \mathbf{r}_5, t)\, d^3 r_1 d^3 r_2 d^3 r_3 d^3 r_4 d^3 r_5. \qquad (20.91)$$

Let us consider a surface of revolution which is obtained by rotating an arbitrary smooth curve which joins $\mathbf{r}_3$ and $\mathbf{r}_5$ about $\mathbf{r}_3 - \mathbf{r}_5$ axis (Fig. 20.1). Any current which is distributed uniformly on this surface will be a solution of (20.84). In particular we can join $\mathbf{r}_3$ and $\mathbf{r}_5$ by a straight line. This is the case of "minimal coupling", which corresponds to the shortest path connecting these two points. Thus if the effect of an arbitrary nonlocal potential (20.77) is expressed as a superposition of displacement operators, and if each $-i\hbar\nabla_r$ is replaced by $-i\hbar\nabla_r - \frac{ie}{c}\mathbf{A}(r)$, then the resulting electromagnetic current is the one where $\mathbf{K}$ follows a straight line from $\mathbf{r}$ to $\mathbf{r}'$ [32].

A simple solution of (20.84) corresponding to minimal coupling can be found if we choose $\mathbf{K}$ to be the unit vector in the direction of $\mathbf{r}_3 - \mathbf{r}_5$ multiplied by a two-dimensional δ-function in the plane orthogonal to $\mathbf{r}_3 - \mathbf{r}_5$, i.e.

$$\mathbf{K} = \frac{\mathbf{r}_3 - \mathbf{r}_5}{|\mathbf{r}_3 - \mathbf{r}_5|}\delta^2(\mathbf{X}), \quad \mathbf{X} \cdot (\mathbf{r}_3 - \mathbf{r}_5) = 0. \tag{20.92}$$

Next let us study the "maximal coupling" solution. We can find the solution of (20.84) by analogy with the equation for the electric field due to a dipole with charges located at $\mathbf{r}_3$ and $\mathbf{r}_5$. Thus we write $\mathbf{K}$ as the sum of two vectors

$$\mathbf{K} = \nabla_r \Lambda + \nabla_r \wedge \mathbf{\Gamma}, \tag{20.93}$$

and substitute it in (20.84) to obtain

$$\nabla_1 \cdot \mathbf{K}(\mathbf{r}_1 - \mathbf{r}_3; \mathbf{r}_1 - \mathbf{r}_5) = \nabla_r^2 \Lambda = \delta(\mathbf{r}_1 - \mathbf{r}_3) - \delta(\mathbf{r}_1 - \mathbf{r}_5). \tag{20.94}$$

Equation (20.94) can be solved for Λ subject to the boundary condition that Λ should go to zero as $r \rightarrow \infty$ with the result that

$$\Lambda = -\frac{1}{4\pi}\left(\frac{1}{|\mathbf{r} - \mathbf{r}_3|} - \frac{1}{|\mathbf{r} - \mathbf{r}_5|}\right). \tag{20.95}$$

From Eqs. (20.93) and (20.95) we find $\mathbf{K}$:

$$\mathbf{K}(\mathbf{r} - \mathbf{r}_3; \mathbf{r} - \mathbf{r}_5) = \frac{1}{4\pi}\left(\frac{\mathbf{r}_1 - \mathbf{r}_3}{|\mathbf{r}_1 - \mathbf{r}_3|^3} - \frac{\mathbf{r} - \mathbf{r}_5}{|\mathbf{r} - \mathbf{r}_5|^3}\right) + \nabla_r \wedge \mathbf{\Gamma}. \tag{20.96}$$

The last term in (20.96) is arbitrary but it must satisfy the same boundary condition as Λ does.

20.3 Optical Potential and the Classical Velocity-Dependent Frictional Force

In Chapter 2 we suggested that in the classical regime, for two heavy ion scattering, we can use a frictional force which is proportional to the relative velocity of the two, to explain the trapping and fusion of these systems. This frictional force determines the amount of the energy dissipated from the relative motion into the internal degrees of freedom.

Microscopic Theory of Nuclear Damping Force — We also discussed the quantum theory of scattering of a nucleon from a target nucleus and we found that there the effective interaction is an optical potential. Therefore we expect that the classical frictional force must in some way be related to the classical limit of the optical potential [35]-[38]. In other words we are looking for a microscopic theory of nuclear friction.

We start our quantum formulation with the time-dependent density matrix $\rho(t)$ obeying Liouville-von Neumann equation [39],[40]

$$i\hbar\frac{\partial\rho(t)}{\partial t} - [H, \rho(t)] = 0. \tag{20.97}$$

In this relation, H the total Hamiltonian, consists of two parts $H = H_0 + U$, the part H_0 which describes the motion of the two ions plus an interaction U which leads to inelastic excitations. The Hamiltonian H_0 can also be split into two parts, a part H_0^{intr}, the intrinsic Hamiltonian and H_0^{rel} which is

$$H_0^{rel} = -\frac{\hbar^2}{2M}\nabla^2 + V(r). \tag{20.98}$$

In this relation $V(r)$ is the first order (Hartree-Fock) potential in the elastic channel which acts between the ground state of the centers of two ions, and r which is the distance between the centers of the two parts (For definition of the Hartree-Fock interaction see [41],[42]). The eigenvalues of H_0^{rel} and H_0^{intr} the latter for the ground state of the two ions will be denoted by E_0^{rel} and E_0^{intr} respectively.

Now let us examine the operator $\rho(t)$. The density operator $\rho(t)$ given by (20.97) is subject to the initial condition

$$\lim_{t\to-\infty} \| \rho(t) - \rho_0(t) \| \to 0, \tag{20.99}$$

where $\rho_0(t)$ is the density matrix for the motion of the two ions without the interaction U, far before the two interact and $\| O \|$ is the norm of the operator O.

Now let us express $\rho_0(t)$ in terms of the wave packet. The free particle density matrix is given by

$$\rho_0(t) = |\psi_0(t)\rangle\langle\psi_0(t)| = \exp\left(-\frac{i}{\hbar}H_0 t\right)\rho_0\exp\left(\frac{i}{\hbar}H_0\right). \tag{20.100}$$

where $|\psi_0(t)\rangle$ is the unperturbed initial wave packet (see Eq. (20.113) below).

Let us first consider the formal solution of Liouville-von Neumann equation (20.97) which we write as

$$\rho^\epsilon(t) = \epsilon\int_{-\infty}^t e^{\epsilon(\tau-t)}\exp\left[\frac{i}{\hbar}H(\tau-t)\right]\rho_0(\tau)\exp\left[-\frac{i}{\hbar}H(\tau-t)\right]d\tau. \tag{20.101}$$

This ϵ-dependent operator $\rho^\epsilon(t)$ is a solution of Liouville-von Neumann equation up to source term proportional to ϵ:

$$i\hbar\frac{\partial}{\partial t}\rho^\epsilon(t) = i\hbar\epsilon\rho_0^\epsilon(t) + [H, \rho^\epsilon(t)]. \tag{20.102}$$

By partial integration of (20.101) using (20.100) we find

$$\rho^\epsilon(t) = \rho_0(t) + \frac{1}{i\hbar} \int_{-\infty}^t e^{\epsilon(\tau-t)} \exp\left[\frac{i}{\hbar}(\tau-t)\right]$$
$$\times [U, \rho_0(\tau)] \exp\left[-\frac{i}{\hbar}(\tau-t)\right] d\tau. \qquad (20.103)$$

Therefore we have

$$\parallel \rho^\epsilon(t) - \rho_0(t) \parallel \le \frac{1}{\hbar} \int_{-\infty}^t \parallel [U, \rho_0(\tau)] \parallel d\tau. \qquad (20.104)$$

Thus if

$$\lim_{t \to -\infty} \int_{-\infty}^t |\langle \psi_0(\tau)|U^2|\psi_0(\tau)\rangle|^{\frac{1}{2}} \, d\tau = 0, \qquad (20.105)$$

then the left-hand side of (20.104) goes to zero as $t \to -\infty$.

Once $\rho(t)$ is constructed we can get a semiclassical description of this quantum problem. For this we find it convenient to use the Ehrenfest theorem and write the following equations for the mean position, the mean velocity, and the mean acceleration of the components of the relative distance

$$\langle x_j \rangle_t = \mathrm{tr}\{x_j \rho(t)\}, \qquad (20.106)$$

$$\frac{d}{dt}\langle x_j \rangle_t = \frac{1}{i\hbar}\mathrm{tr}\{x_j[H, \rho(t)]\} = \frac{1}{M}\langle p_j \rangle_t, \qquad (20.107)$$

and

$$M\frac{d^2}{dt^2}\langle x_j \rangle_t + \left\langle \frac{\partial V(r)}{\partial x_j} \right\rangle_t - M\langle F_j \rangle_t = 0. \qquad (20.108)$$

In Eq. (20.107), p_j represents the momentum conjugate to the coordinate x_j and

$$\langle F_j \rangle_t = -\left\langle \frac{\partial}{\partial x_j} U(r) \right\rangle_t, \qquad (20.109)$$

is the force responsible for changing the relative motion of the two ions due to the inelastic excitation of their internal degrees of freedom. The density matrix $\rho(t)$ in this case should describe the motion of a sufficiently narrow wave packet for relative motion of the two ions. By imposing this condition we can test the local property of the system and at the same time the chosen wave packet must be wide enough so that its spreading can be ignored. Having found $\rho^\epsilon(t)$ to the first order of perturbation, Eq. (20.103), we can calculate $\langle F_j \rangle_t$:

$$\langle F_j \rangle_t = \mathrm{tr}\{F_j \rho^\epsilon(t)\} = \mathrm{tr}\{F_j \rho_0(t)\}$$
$$+ \frac{1}{i\hbar} \int_{-\infty}^t e^{\epsilon(\tau-t)} \, \mathrm{tr}\left\{F_j \exp\left[\frac{i}{\hbar}H_0(\tau-t)\right]\right.$$
$$\left. \times [U, \rho_0(\tau)] \exp\left[-\frac{i}{\hbar}H_0(\tau-t)\right]\right\} d\tau, \qquad (20.110)$$

The first term on the right-hand side of Eq. (20.110) vanishes since it is the diagonal element of U between two intrinsic ground states. This mean value, $\langle F_j \rangle_t$, can be simplified

$$\langle F_j \rangle_t = \frac{1}{i\hbar} \int_{-\infty}^{0} e^{\epsilon \tau} \text{tr} \left\{ \left[F_j, \hat{U}(\tau) \right] \rho_0(t) \right\} d\tau, \tag{20.111}$$

with

$$\hat{U}(\tau) = \exp\left(\frac{i}{\hbar} H_0 \tau\right) U \exp\left(-\frac{i}{\hbar} H_0 \tau\right). \tag{20.112}$$

If $|\mathbf{p}_i(t)\rangle$ denotes the initial wave packet of the relative motion, which is an eigenstate of H_0, and $|0\rangle$ is the product of the intrinsic ground states, then we can write

$$\rho_0(t) = |\mathbf{p}_i(t), 0\rangle \langle 0, \mathbf{p}_i(t)|. \tag{20.113}$$

Using this density matrix Eq. (20.111) reduces to

$$\begin{aligned}
\langle F_j \rangle_t &= \frac{1}{i\hbar} \int_{-\infty}^{0} e^{\epsilon \tau} \langle \mathbf{p}_i(t), 0| \left[F_j, \ \hat{U}(\tau) \right] |\mathbf{p}_i(t), 0\rangle d\tau \\
&= \frac{1}{i\hbar} \int_{-\infty}^{0} e^{\epsilon \tau} \left\{ \langle \mathbf{p}_i(t), 0| F_j \exp\left[\frac{i}{\hbar}(H_0^{intr} - E_0^{intr})\tau\right] \hat{U}^{rel}(\tau) |\mathbf{p}_i(t), 0\rangle \right\} d\tau \\
&\quad - \frac{1}{i\hbar} \int_{-\infty}^{0} e^{\epsilon \tau} \left\{ \langle \mathbf{p}_i(t), 0| \hat{U}(\tau)^{rel} \exp\left[-\frac{i}{\hbar}(H_0^{intr} - E_0^{intr})\tau\right] F_j |\mathbf{p}_i(t), 0\rangle \right\} d\tau,
\end{aligned} \tag{20.114}$$

where $\hat{U}^{rel}(\tau)$ is defined by (20.112). By partial integration of the two terms in (20.114) and making use of

$$\frac{d}{d\tau} \hat{U}^{rel} = \frac{i}{\hbar} \left[H^{rel}, \ \hat{U}^{rel}(\tau) \right], \tag{20.115}$$

we find

$$\begin{aligned}
\langle F_j \rangle_t &= -\left\langle \mathbf{p}_i(t), 0| F_j \ \frac{1}{H_0^{intr} - E_0^{intr} - i\hbar\epsilon} U |\mathbf{p}_i(t), 0 \right\rangle \\
&\quad - \left\langle \mathbf{p}_i(t), 0| U \frac{1}{H_0^{intr} - E_0^{intr} + i\hbar\epsilon} F_j |\mathbf{p}_i(t), 0 \right\rangle \\
&\quad - \frac{1}{i\hbar} \int_{-\infty}^{0} e^{\epsilon} \left\langle \mathbf{p}_i(t), 0| F_j \frac{\exp\left[\frac{i}{\hbar}\left(H_0^{intr} - E_0^{intr}\right)\tau\right]}{H_0^{intr} - E_0^{intr} - i\hbar\epsilon} \left[H^{rel}, \hat{U}^{rel}(\tau) \right] |\mathbf{p}_i(t), 0 \right\rangle d\tau \\
&\quad - \frac{1}{i\hbar} \int_{-\infty}^{0} e^{\epsilon} \left\langle \mathbf{p}_i(t), 0| \left[H^{rel}, \hat{U}^{rel}(\tau) \right] \frac{\exp\left[-\frac{i}{\hbar}\left(H_0^{intr} - E_0^{intr}\right)\tau\right]}{H_0^{intr} - E_0^{intr} + i\hbar\epsilon} F_j |\mathbf{p}_i(t), 0 \right\rangle d\tau.
\end{aligned} \tag{20.116}$$

By adding the first two terms in Eq. (20.116) we obtain the following relation

$$\langle F_j \rangle_t^I = - \left\langle \mathbf{p}_i(t), 0 \left| \nabla_j \left\{ U \frac{\mathcal{P}}{H_0^{intr} - E_0^{intr}} U \right\} \right| \mathbf{p}_i(t), 0 \right\rangle, \qquad (20.117)$$

and this is a second order correction which can be added to the real potential $V(r)$. Here $\mathcal{P}$ denotes the principal value of the integral. We can also combine the last two terms of (20.116) if we use

$$[H_0^{rel}, U] = \frac{\hbar}{i} \sum_j \left(F_j \frac{p_j}{2M} + \frac{p_j}{2M} F_j \right) \qquad (20.118)$$

This expression can be simplified by neglecting the commutator $[p_j, F_j]$ which is of the order of the momentum transfer Δp and is small compared to the other terms that we are keeping. We can choose the wave packet $|\mathbf{p}_i(t)\rangle$ narrow enough so that

$$\frac{p_j}{M} |\mathbf{p}_i(t)\rangle = \langle v_{jt} \rangle |\mathbf{p}_i(t)\rangle = v_j(t) |\mathbf{p}_i(t)\rangle \qquad (20.119)$$

and in addition to get

$$\exp\left[\frac{i}{\hbar} H^{rel} \tau \right] |\mathbf{p}_i(t)\rangle = |\mathbf{p}_i(t+\tau)\rangle \approx |\mathbf{p}_i(t)\rangle \exp\left(-\frac{i}{\hbar} E^{rel} t \right). \qquad (20.120)$$

This last relation is a good approximation if for the duration of time τ where the dominant contribution to the integrals in (20.116) comes from, the external force $\nabla_r V(r)$ does not change the velocity $v(t)$ significantly. Thus the second part of the contribution to $\langle F_j \rangle_t$ is the sum of the following two terms

$$[\langle F_j \rangle_t^{II}]_1 = \sum_k v_k \int_{-\infty}^0 e^{\epsilon t} \left\{ \left\langle \mathbf{p}_i(t), 0 \left| F_j \frac{\exp\left[\frac{i}{\hbar}(H_0 - E_0)\tau\right]}{H^{intr} - E_0^{intr} - i\hbar\epsilon} F_k \right| \mathbf{p}_i(t), 0 \right\rangle \right\} d\tau \qquad (20.121)$$

$$[\langle F_j, t \rangle^{II}]_2 = \sum_k v_k \int_{-\infty}^0 e^{\epsilon t} \left\{ \left\langle \mathbf{p}_i(t), 0 \left| F_k \frac{\exp\left[-\frac{i}{\hbar}(H_0 - E_0)\tau\right]}{H^{intr} - E_0^{intr} + i\hbar\epsilon} F_j \right| \mathbf{p}_i(t), 0 \right\rangle \right\} d\tau, \qquad (20.122)$$

where E_0 is the sum of two energies;

$$E_0 = E_0^{intr} + E_0^{rel}. \qquad (20.123)$$

Let us note that the intrinsic spectrum near the ground state, E_0^{intr}, and the ground state $|0\rangle$ do occur in the sum over intermediate states. We can write $\langle F_j \rangle_t^{II}$ as the sum of two tensors, one symmetric and the other antisymmetric;

$$\langle F_j O F_k \rangle^s = \frac{1}{2} \{ \langle F_j O F_k \rangle + \langle F_k O F_j \rangle \}. \qquad (20.124)$$

and

$$\langle F_j O F_k \rangle^a = \frac{1}{2} \{ \langle F_j O F_k \rangle - \langle F_k O F_j \rangle \}. \qquad (20.125)$$

From the symmetric part we obtain

$$\langle F_j \rangle^{II\ s} = \sum_k k_{jk}(t) v_k(t), \tag{20.126}$$

where the friction tensor k_{jk} which is normally anisotropic is given by

$$k_{jk}(t) = 2\pi\hbar \left\langle \mathbf{p}_i(t), 0 \left| F_j \delta(H_0 - E_0) \frac{P}{H^{intr} - E_0^{intr}} F_k \right| \mathbf{p}_i(t), 0 \right\rangle^s, \tag{20.127}$$

(see also Eq. (2.35)). Thus we have derived a classical damping force which is proportional to velocity from a large quantum mechanical system using the perturbation theory. We note that it is the coupling between the internal degrees of freedom and the relative motion of the ions which leads to the frictional force.

20.4 A Solvable Model Showing the Energy Transfer from the Collective to Intrinsic Degrees of Freedom

To get a clearer picture of the problem of energy transfer we want to study a solvable model which describes the coupling between the collective and the intrinsic degrees of freedom, and to investigate the way in which the energy is transferred from the former to the latter. The collective degrees of freedom can be the relative coordinates as we have seen in the previous section. One of the aims of studying such a model is that we can test the approximate methods that have been used to calculate the rate of energy dissipation of more realistic systems [43]-[44].

The model consists of two oscillators in the $x - y$ plane, forming the intrinsic coordinates, and the translational collective degree of freedom is z. Thus the total Hamiltonian of the system is

$$H = H^{coll}(z) + H^{intr}(x, y, z), \tag{20.128}$$

where

$$H^{coll}(z) = -\frac{\hbar^2}{2M} \frac{\partial^2}{\partial z^2}, \tag{20.129}$$

and

$$H^{intr}(z) = -\frac{\hbar^2}{2m} \left(\frac{\partial^2}{\partial x^2} + \frac{\partial^2}{\partial y^2} \right) + \frac{1}{2} K(x - z)^2 + \frac{1}{2} K(y - z)^2. \tag{20.130}$$

In these relations M is the mass of the free particle and m is the mass of each of the oscillators. Since the Hamiltonian is quadratic in coordinates and momenta, we can find the normal modes of vibration by diagonalizing H. We observe that H, Eq. (20.128), can be written in terms of the matrices $\hat{T}$ and $\hat{V}$, where

$$\hat{T} = \begin{bmatrix} m & 0 & 0 \\ 0 & m & 0 \\ 0 & 0 & M \end{bmatrix}, \quad \hat{V} = \begin{bmatrix} K & 0 & -K \\ 0 & K & -K \\ -K & -K & 2K \end{bmatrix}. \tag{20.131}$$

From these matrices we find the normal frequencies of vibration, ω, as the roots of the secular equation

$$\det\left[\hat{V} - \omega^2\hat{T}\right] = 0. \tag{20.132}$$

The roots of (20.132) are given by

$$\begin{cases} \omega_1^2 = \frac{K}{mM}(2m + M) \\ \omega_2^2 = \frac{K}{m} \\ \omega_3^2 = 0. \end{cases} \tag{20.133}$$

Let us define the normal coordinate ξ, η and ζ in the following way: ξ which is the relative coordinate between z and the center of the intrinsic coordinate at $\frac{1}{2}(x + y)$ is given by

$$\xi = \frac{1}{2}(-x - y + 2z). \tag{20.134}$$

We denote the relative coordinate between the intrinsic coordinates by η

$$\eta = (x - y), \tag{20.135}$$

and we use ζ for the center-of-mass coordinate

$$\zeta = \frac{m(x + y) + Mz}{M + 2m}. \tag{20.136}$$

For the corresponding masses we find

$$\begin{cases} \mu_\xi = \frac{2mM}{2m+M} \\ \mu_\eta = \frac{1}{2}m \\ \mu_\zeta = M + 2m. \end{cases} \tag{20.137}$$

The total Hamiltonian written in terms of these normal coordinates becomes separable

$$H = -\frac{\hbar^2}{2}\left[\frac{1}{\mu_\xi}\frac{\partial^2}{\partial\xi^2} + \frac{1}{\mu_\eta}\frac{\partial^2}{\partial\eta^2} + \frac{1}{\mu_\zeta}\frac{\partial^2}{\partial\zeta^2}\right]$$
$$+ \frac{1}{2}\left(\mu_\xi\omega_\xi^2\xi^2 + \mu_\eta\omega_\eta^2\eta^2\right), \tag{20.138}$$

where $\omega_\xi^2 = \omega_1^2$ and $\omega_\eta^2 = \omega_2^2$.

Solution of the Wave Equation for the Collective and Intrinsic Motions — The Schrödinger equation for the Hamiltonian H is separable and its solution can be written as a product of three wave functions;

$$\Psi_{kss'}(\xi, \eta, \zeta) = \psi_s(\xi)\psi_{s'}(\eta)u_k(\zeta), \tag{20.139}$$

where

$$u_k(\zeta) = \frac{1}{\sqrt{2\pi}}e^{ik\zeta}, \tag{20.140}$$

$$\psi_s(\xi) = N_s H_s(\alpha\xi) \exp\left(-\frac{1}{2}\alpha^2\xi^2\right), \tag{20.141}$$

and

$$\psi_{s'}(\eta) = N_{s'} H_{s'}(\beta\eta) \exp\left(-\frac{1}{2}\beta^2\eta^2\right). \tag{20.142}$$

Here H_s is the Hermite polynomial of order s and α and β are defined by

$$\alpha = \sqrt{\frac{\mu_\xi \omega_\xi}{\hbar}}, \quad \text{and} \quad \beta = \sqrt{\frac{\mu_\eta \omega_\eta}{\hbar}}. \tag{20.143}$$

The two quantities N_s and $N_{s'}$ are the normalization constants for the wave functions of the harmonic oscillators. From Eqs. (20.140)-(20.142) it follows that the total wave function $\Psi_{kss'}$ satisfies the normalization condition

$$\int \Psi^*_{k_1 s_1 s'_1} \Psi_{k_2 s_2 s'_2} d\xi d\eta d\zeta = \delta(k_1 - k_2)\delta_{s_1 s_2}\, \delta_{s'_1\, s'_2}. \tag{20.144}$$

The energy eigenvalues for the whole (uncoupled) system is given by

$$H\Psi_{kss'}(\xi,\eta,\zeta) = E_{kss'}\Psi_{kss'}(\xi,\eta,\zeta), \tag{20.145}$$

where

$$E_{kss'} = \frac{\hbar^2 k^2}{2\mu_\zeta} + \hbar\omega_\xi\left(s + \frac{1}{2}\right) + \hbar\omega_\eta\left(s' + \frac{1}{2}\right). \tag{20.146}$$

That is the initial energy of the system is $E_{kss'}$.

The Eigenvalues of the Intrinsic Hamiltonian — By changing the variables y and z to x' and y' defined by

$$x' = x - z \quad \text{and} \quad y' = y - z, \tag{20.147}$$

we find the eigenvalue equation

$$\left[-\frac{\hbar^2}{2m}\left(\frac{\partial}{\partial x'^2} + \frac{\partial}{\partial y'^2}\right) + \frac{1}{2}K\left(x'^2 + x'^2\right)\right]\Phi_{nn'} = \mathcal{E}_{nn'}\Phi_{nn'}, \tag{20.148}$$

where

$$\mathcal{E}_{nn'} = \hbar\omega_\eta\left(N + 1\right), \quad N = n + n' \tag{20.149}$$

and

$$\Phi_{nn'}(x',y') = \psi_n(x')\psi_{n'}(y'), \quad n,\, n' = 0,\, 1\, 2\cdots. \tag{20.150}$$

Here $\psi_n(x')$ is given by

$$\psi_n(x') = \sqrt{\frac{\nu}{\sqrt{\pi}2^n n!}} \exp\left(-\frac{1}{2}\nu^2 x'^2\right) H_n(\nu x'), \quad \nu = \sqrt{\frac{m\omega_\eta}{\hbar}} \tag{20.151}$$

and a similar relation for $\psi_{n'}(y')$.

The Collective Wave Function — The eigenfunctions of the intrinsic Hamiltonian $\Phi_{nn'}(x', y')$ form a complete set and we can expand the total wave function in terms of members of this set

$$\Psi_{kss'}(x, y, z) = \sum_{n,n'} f_{nn'}^{kss'}(z)\Phi_{nn'}(x - z, y - z). \tag{20.152}$$

Here s and s' are the quantum numbers of the oscillators when they are coupled and n and n' are their quantum numbers when they remain uncoupled. From this relation we can find the coefficients of the expansion:

$$f_{nn'}^{kss'}(z) = \int_{-\infty}^{\infty}\int_{-\infty}^{\infty} \Phi_{nn'}(x - z, y - z)\Psi_{kss'}(x, y, z)\,dxdy. \tag{20.153}$$

By substituting $\Psi_{kss'}$ from (20.139)-(20.142) and $\Phi_{nn'}$ from (20.150),(20.151) in Eq. (20.153) we can express $f_{nn'}^{kss'}(z)$ as a double integral

$$f_{nn'}^{kss'}(z) = \frac{1}{\sqrt{2\pi}}N_n N_{n'} N_s N_{s'} \int_{-\infty}^{\infty}\int_{-\infty}^{\infty} dxdy$$

$$\times \exp\left\{-\frac{\nu^2}{4}\left(3 + \frac{1}{\delta}\right)(x^2 + y^2) - \nu^2\left(1 + \frac{1}{\delta}\right)z^2\right.$$

$$+ \left[z\nu^2\left(1 + \frac{1}{\delta}\right) + b\right](x + y) + \frac{1}{2}\nu^2\left(1 - \frac{1}{\delta}\right)xy + \frac{bMz}{m}\right\}$$

$$\times H_n\left[\nu(x - z)\right] H_{n'}\left[\nu(y - z)\right]H_{s'}\left[\frac{\nu}{\sqrt{2}}(x - y)\right]$$

$$\times H_s\left[\frac{\nu}{\sqrt{2\delta}}(-x - y + 2z)\right], \tag{20.154}$$

where the constants δ and b are defined by

$$\delta = \frac{\omega_\xi}{\omega_\eta} = \sqrt{\frac{2m}{M} + 1} \quad \text{and} \quad b = \frac{imk}{2m + M}. \tag{20.155}$$

Using the addition theorem for Hermite polynomials [45]

$$H_n(x + y) = 2^{-\frac{n}{2}}\sum_{j=0}^{n} \binom{n}{j} H_j(x\sqrt{2})H_{n-j}(y\sqrt{2}), \tag{20.156}$$

we can carry out the integration over x and y to find an analytic expression for $f_{n,n'}^{kss'}(z)$. Thus we find the final form of this collective wave function to be

$$f_{nn'}^{kss'}(z) = \mathcal{F}(ss'; nn')(\delta)\exp\left\{ikz - \frac{1}{4}\left(1 + \frac{1}{\delta}\right)\left(\frac{k}{\nu}\right)^2\left(1 - \frac{1}{\delta}\right)^2\right\}. \tag{20.157}$$

The factor $\mathcal{F}(ss'; nn')$ is a constant which can be evaluated in terms of the factorials and Hermite polynomials [46]. We also note that the amplitude $f_{nn'}^{kss'}(z)$ satisfies the normalization condition

$$\sum_{ss'} \left| f_{nn'}^{kss'}(z) \right|^2 = \sum_{nn'} \left| f_{nn'}^{kss'}(z) \right|^2 = \frac{1}{2\pi}. \tag{20.158}$$

Coupled Differential-Difference Equations for the Collective Wave Function — While, as we have seen in this section, the collective wave function can be found analytically, to show the similarity with the optical potential that we have used earlier, Sec. 20.3, we will look at an alternative way of finding $f_{nn'}^{kss'}(z)$. Thus, multiplying the Schrödinger equation for the whole system, $H\Psi = E\Psi$, by $\Phi_{j'j'}(x-z, y-z)$ and integrating over x and y, using Eqs. (20.148)-(20.150) we find

$$\{T(n, n'; z) + \mathcal{E}_{nn'} - E\} f_{nn'}^{kss'}(z) = - \sum_{j,j' \neq n,n'} T(n, n'; j, j'; z) f_{jj'}^{kss'}(z). \tag{20.159}$$

In this equation $T(n, n'; j, j')$ is given by

$$T(n, n'; j, j'; z) = \int \int \Phi_{nn'}^*(x-z, y-z) \left(-\frac{\hbar^2}{2M} \frac{\partial^2}{\partial z^2} \right) \Phi_{jj'}(x-z, y-z). \tag{20.160}$$

Using the properties of the eigenfunctions of the harmonic oscillator, we can evaluate the integrals over x and y in (20.160) to find

$$\left(-\frac{2M}{\hbar^2} \right) T(n, n'; j, j'; z) = \delta_{nn',jj'} \left[\frac{d^2}{dz^2} - \nu^2(n+j+1) \right]$$

$$+ \frac{\nu^2}{2} \delta_{jj'} \left[\sqrt{n'(n'-1)} \delta_{n,n'-2} + \sqrt{(n'+1)(n'+2)} \delta_{n,n'+2} \right]$$

$$+ \frac{\nu^2}{2} \delta_{nn'} \left[\sqrt{j'(j'-1)} \delta_{j,j'-2} + \sqrt{(j'+1)(j'+2)} \delta_{j,j'+2} \right]$$

$$- \sqrt{2}\nu \delta_{jj'} \left(\sqrt{n'} \delta_{n,n'-1} - \sqrt{n'+1} \delta_{n,n'+1} \right) \frac{d}{dz}$$

$$- \sqrt{2}\nu \delta_{nn'} \left(\sqrt{j'} \delta_{j,j'-1} - \sqrt{j'+1} \delta_{j,j'+1} \right) \frac{d}{dz}$$

$$+ \nu^2 \left[\left(\sqrt{n'} \delta_{n,n'-1} - \sqrt{n'+1} \delta_{n,n'+1} \right) \left(\sqrt{j'} \delta_{j,j'-1} - \sqrt{j'+1} \delta_{j,j'+1} \right) \right]. \tag{20.161}$$

If we substitute (20.161) in (20.159) we obtain the following set of coupled differential-difference equations for $f_{nj}(z)$:

$$\left[-\frac{d^2}{dz^2} + \frac{2M\omega_\eta}{\hbar}\left(1 + \frac{m}{2M}\right)(n+j+1) - \frac{2ME}{\hbar^2}\right]f_{nj}(z)$$

$$= \frac{\nu^2}{2}\sqrt{(n+1)(n+2)}f_{n+2,j}(z) + \sqrt{(n-1)n}f_{n-2,j}(z)$$

$$+ \frac{\nu^2}{2}\sqrt{(j+1)(j+2)}f_{n,j+2}(z) + \sqrt{(j-1)j}f_{n,j-2}(z)$$

$$+ \nu^2\left[\sqrt{(n+1)(j+1)}f_{n+1,j+1}(z) - \sqrt{(n+1)j}f_{n+1,j-1}(z)\right]$$

$$+ \nu^2\left[\sqrt{nj}f_{n-1,j-1}(z) - \sqrt{n(j+1)}f_{n-1,j+1}(z)\right]$$

$$- \sqrt{2}\nu\frac{d}{dz}\left[\sqrt{n+1}f_{n+1,j}(z) - \sqrt{n}f_{n-1,j}(z) + \sqrt{j+1}f_{n,j+1}(z) - \sqrt{j}f_{n,j-1}(z)\right].$$

$$(20.162)$$

We have already encountered a similar, but simpler, differential-difference equations for the solution of the Schrödinger equation when the spring constant was assumed to be complex (leaky spring constant).

From the analytical expression for $f_{nn'}^{kss'}(z)$, Eq. (20.157) we can calculate the probability for the intrinsic excitation $P_{n,N}^{ss'}(k)$ which is given by

$$P_{nN}^{ss'}(k) = 2\pi\left|f_{nn'}^{kss'}\right|^2. \tag{20.163}$$

Noting that for a given N, we have $n' = N - n$, Eq. (20.149), therefore we can denote the probability by P_{nN}. This probability will also be dependent on k and thus to the translational degree of freedom. Then according to Eq. (20.146) we have

$$E_{kss'} = \frac{\hbar^2 k^2}{2\mu_s} = \frac{\hbar^2 k^2}{2(M+2m)}. \tag{20.164}$$

We can measure this energy in units of $\hbar\omega_\eta$, and plot $P_{nN}^{ss'}$ as a function of $\tilde{E}_k$, where

$$\tilde{E}_k = \frac{\hbar^2 k^2}{2(M+2m)\hbar\omega_\eta}. \tag{20.165}$$

In Fig. 20.2, some of these probabilities are plotted as functions of $\tilde{E}_k$. For this calculation we have assumed $m = M$. Also the eigenvalues of the total Hamiltonian H in the uncoupled state were assumed to be E^{k00}, i.e. the oscillators were in the ground state of H_{intr}. The coupling between the collective and the intrinsic degrees of freedom allows for the oscillators to make transitions into the excited states, and therefore have energy eigenvalues $\mathcal{E}_n$ given by (20.149). For this calculation N was varied between 0 and 10 [46]. In Fig. 20.3 plots of the same probabilities are given when the intrinsic wave function is not the ground state wave function but is in the state $s = s' = 1$. In this case we observe that the state $n = n' = 0$ cannot be excited. This result follows from the integral

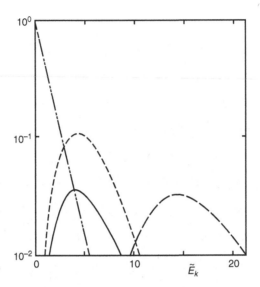

Figure 20.2: The probability $P^{00}_{nn'}$ as a function of $\tilde{E}_k = \frac{\hbar^2 k^2}{2\mu_\varsigma \hbar \omega_\eta}$ for different integers n and n'. Here P^{00}_{00} is shown by dashed-dotted line, P^{00}_{12} by short dashed line, P^{00}_{03} by solid line and P^{00}_{46} by long dashed line [46].

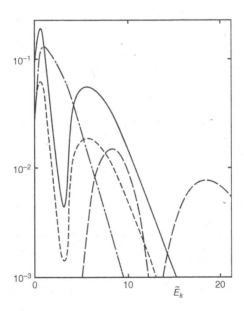

Figure 20.3: The probability $P^{11}_{n,n'}$ as a function of $\tilde{E}_k$ for different integers n and n'. In this case $P^{11}_{00} = 0$, P^{11}_{01} is shown by dashed-dotted line, P^{11}_{12} by short dashed line, P^{11}_{03} by solid line and P^{11}_{46} by long dashed line [46].

representation of $f_{00}^{k11}(z)$, Eq. (20.154). In addition we note that for the quantum numbers $s = s' = 1$ the probabilities oscillate more rapidly [46].

The graphs shown in these two figures are the results of calculation done for strong coupling between the collective and the intrinsic parts of the system. Thus when the coupling is strong $m \geq M$, as $\tilde{E}_k$ increases, $P_{nn'}^{ss'}$ increases rapidly and reaches a maximum, Fig. 20.2. This rapid increase is followed by a sharp decrease to very small values of the probability. On the other hand if $m \ll M$, i.e. for the case of weak coupling $P_{nn'}^{ss'}$ grows gradually and reaches a maximum at higher energies, and then decreases slowly as $\tilde{E}_k$ is increased.

An extension of this work to the cases where more oscillators form the intrinsic degrees of freedom, and also when the masses are different has been carried out in ref. [44].

20.5 Damped Harmonic Oscillator and Optical Potential

In Section 2.6 we found the Lagrangian (2.192) for the motion of two harmonic oscillators. Let us now consider the quantum mechanical version of that problem where the optical potential is given by $\frac{m}{2}\omega^2 (1 - i\alpha)^2 z^2$. The classical Lagrangian for this system can be written either as (2.192) or as a real Lagrangian

$$L = \frac{m}{4} \left[\left(\frac{dz}{dt}\right)^2 + \left(\frac{dz^*}{dt}\right)^2 \right] - \frac{m\omega^2}{4} \left[(1 - i\alpha)^2 z^2 + (1 + i\alpha)^2 z^{*2} \right]. \quad (20.166)$$

The Hamiltonian for this system found by the standard method is

$$H = \frac{1}{4m} \left(p_z^2 + p_{z^*}^2 \right) + \frac{m\omega^2}{4} \left[(1 - i\alpha)^2 z^2 + (1 + i\alpha)^2 z^{*2} \right], \quad (20.167)$$

Expressing this H in terms of the original variables x and y we have

$$H = \frac{1}{2m} \left(p_x^2 - p_y^2 \right) + \frac{m\omega^2}{2} \left[(1 - \alpha^2) \left(x^2 - y^2 \right) + 4\alpha xy \right]. \quad (20.168)$$

Now the Hamiltonian (20.168) can be quantized with the resulting wave equation

$$\left(\frac{\partial^2}{\partial x^2} - \frac{\partial^2}{\partial y^2} \right) \psi + \left[\mathcal{E} - \beta^2 \left\{ (1 - \alpha^2) \left(x^2 - y^2 \right) + 4\alpha xy \right\} \right] \psi = 0, \quad (20.169)$$

where

$$\mathcal{E} = \frac{2mE}{\hbar^2}, \quad \text{and} \quad \beta = \frac{m\omega}{\hbar}, \quad (20.170)$$

and E is the energy eigenvalue.

Equation (20.169) becomes separable if $\alpha = 0$, and then $\psi(x, y)$ will be a product of the wave functions for two harmonic oscillators. For the damped oscillator we expand $\psi(x, y)$ in terms of the normalized harmonic oscillator wave function $\psi_n(y)$,

$$\psi(x, y) = \sum_n X_n(x)\psi_n(y). \tag{20.171}$$

By substituting (20.171) in (20.169) and using the relations

$$\int_\infty^\infty \psi_j^*(y)y\psi_n(y)dy = \sqrt{\left(\frac{j+1}{2\beta}\right)}\delta_{n,j+1} + \sqrt{\left(\frac{j}{2\beta}\right)}\delta_{n,j-1}, \tag{20.172}$$

and

$$\int_\infty^\infty \psi_j^*(y)y^2\psi_n(y)dy = \left(\frac{j+1}{2\beta}\right)\delta_{n,j}, \tag{20.173}$$

we find that $X_n(x)$ satisfies the differential-difference equation

$$\frac{d^2 X_n(x)}{dx^2} + \left[\mathcal{E} + \beta\left(n + \frac{1}{2}\right)(2 - \alpha^2) - \beta^2\left(1 - \alpha^2\right)x^2\right]X_n(x)$$
$$- (2\beta)^{\frac{3}{2}}\alpha x\left(\sqrt{n+1}X_{n+1}(x) + \sqrt{n}X_{n-1}\right) = 0. \tag{20.174}$$

This equation with the boundary conditions

$$X_n(x) \to 0, \quad \text{as} \quad x \to \pm\infty, \tag{20.175}$$

is an eigenvalue equation with the solution $X_{nj}(x)$ for the characteristic values $\mathcal{E}_{nj}$.

Earlier in Sec. 20.1 we showed that the motion of a single particle coupled to a many-body system can be reduced to the motion of that particle in an optical potential. The present case shows that the motion in an optical potential field can be viewed as the interaction between a particle and a many-channel problem.

In the optical potential model that we have considered so far, a state or a particle decays into other states or particles, therefore the norm of the central particle (or state) is not preserved. We can use a model similar to Gisin's which preserves the norm of the wave function for the central particle (or state).

In the present problem the potential is complex, but we use it in a non-linear evolution equation [47],[48]

$$i\frac{\partial\psi}{\partial t} = H\psi + i\left[V(x) - \langle\psi|V(x)|\psi\rangle\right]\psi, \tag{20.176}$$

where we have set $\hbar = 1$. In Eq. (20.176), H is a Hermitian Hamiltonian. From (20.176) and its complex conjugate we find that

$$\frac{\partial}{\partial t}\langle\psi|\psi\rangle = 0, \tag{20.177}$$

and therefore ψ can be normalized.

We now change the wave function ψ to ϕ where

$$\psi(x,t) = \phi(x,t) \exp\left[-\int_0^t \langle\psi|\mathcal{V}(x)|\psi\rangle dt\right]. \tag{20.178}$$

By substituting (20.178) in (20.176) we find the partial differential equation for ϕ:

$$i\frac{\partial\phi}{\partial t} = H\phi + i\mathcal{V}(x)\phi. \tag{20.179}$$

Next we define the set of complex eigenfunctions $\phi_n(x)$ by the relation

$$[H + i\mathcal{V}]\,\phi_n(x) = \left(E_n - \frac{i}{2}\Gamma_n\right)\phi_n(x). \tag{20.180}$$

These ϕ_n s satisfy the orthogonality condition

$$\int_{-\infty}^{\infty} \phi_n(x)\phi_j(x)dx = 0, \quad n \neq j, \tag{20.181}$$

but since ϕ_j s are complex functions we do not have an orthonormal set. We write $\phi(x,t)$ in terms of $\phi_n(x)$;

$$\phi(x,t) = \sum_n C_n \exp\left[-i\left(E_n - \frac{i}{2}\Gamma_n\right)t\right]\phi_n(x), \tag{20.182}$$

where C_n s are the coefficients of the expansion and are related to $\psi(x,0)$ by

$$C_n = \frac{\int_{-\infty}^{\infty}\psi(x,0)\phi_n(x)dx}{\int_{-\infty}^{\infty}(\phi_n(x))^2\,dx}. \tag{20.183}$$

Using the time-dependent wave packet we can determine the position of the center of the wave packet as a function of time, i.e.

$$\langle x(t)\rangle = \frac{\int_{-\infty}^{\infty} x|\psi(x,t)|^2 dx}{\int_{-\infty}^{\infty} |\psi(x,t)|^2 dx}. \tag{20.184}$$

Substituting for $\psi(x,t)$ we get

$$\langle x(t)\rangle = \frac{\sum_{n,j}\langle\phi_j|x|\phi_n\rangle C_k C_j^* \exp(-i\omega_{nj}t)\exp[-\frac{1}{2}(\Gamma_n + \Gamma_j)t]}{\sum_{n,j}\langle\phi_j|\phi_n\rangle C_k C_j^* \exp(-i\omega_{nj}t)\exp[-\frac{1}{2}(\Gamma_n + \Gamma_j)t]}, \tag{20.185}$$

where

$$\omega_{nj} = E_n - E_j. \tag{20.186}$$

If Γ_i is the smallest member of the set of Γ_n's then as $t \to \infty$, $\langle x(t)\rangle$ tends to the limit

$$\langle x(t)\rangle \to \langle\phi_i|x|\phi_i\rangle, \tag{20.187}$$

and thus the final state of the system is given by the wave function $\phi_j(x)$.

Motion of the Center of a Displaced Gaussian Wave Packet in a Harmonic Oscillator with Complex Spring Constant — As an example of the model that we discussed in this section let us consider an oscillator with the harmonic optical potential $v(x) + iV(x) = \frac{m}{2}\omega^2 (1 - i\alpha)^2 x^2$ where $\alpha = \lambda/2\omega$ (see Eq. (4.25). The Schrödinger equation for this potential can be written as

$$\frac{d^2\psi}{dx^2} + \left[\varepsilon - \beta^2 x^2\right]\psi = 0, \tag{20.188}$$

with

$$\beta = m(1 - i\alpha)\omega, \tag{20.189}$$

and the energy eigenvalues

$$E = \left(\frac{\omega}{2\beta}\right)\varepsilon. \tag{20.190}$$

The solution of (20.188) is given by the parabolic cylinder function

$$\psi_\nu(x) = \frac{1}{\sqrt{n!}}\left(\frac{\beta}{\pi}\right)^{\frac{1}{4}} D_\nu\left(x\sqrt{2\beta}\right). \tag{20.191}$$

This function has the asymptotic property [49]

$$D_\nu\left(x\sqrt{2\beta}\right) \to \begin{cases} \left(x\sqrt{2\beta}\right)^\nu e^{-\frac{1}{2}\beta x^2} & \text{as } x \to \infty \\ \frac{\sqrt{2\pi}}{\Gamma(-\nu)}\left[\frac{e^{\frac{1}{2}\beta x^2}}{\left(-x\sqrt{2\beta}\right)^{\nu+1}}\right] & \text{as } x \to -\infty \end{cases}, \tag{20.192}$$

and ν is related to ε by

$$\nu = \frac{\varepsilon}{2m(1 - i\alpha)\omega} - \frac{1}{2}, \tag{20.193}$$

and $\Gamma(z)$ denotes the Gamma function. Now for $\psi(x)$ to be square integrable we must have

$$\Gamma(-\nu) = \infty, \quad \text{or} \quad \nu = n, \tag{20.194}$$

where n is an integer. Thus for negative values of x we have

$$D_n\left(-x\sqrt{2\beta}\right) = (-1)^n D_n\left(x\sqrt{2\beta}\right). \tag{20.195}$$

From Eqs. (20.190), (20.193) and (20.194) we find the energy eigenvalues to be

$$E_n - \frac{i}{2}\Gamma_n = \left(n + \frac{1}{2}\right)(1 - i\alpha)\omega = \left(n + \frac{1}{2}\right)\left(\omega - \frac{i\lambda}{2}\right). \tag{20.196}$$

This result is similar to the energy eigenvalues of H_n^-, Eq. (11.143).

For the initial wave packet of the form of a displaced Gaussian,

$$\psi(x,0) = \left(\frac{\beta^2}{\pi}\right)^{\frac{1}{4}}\exp\left[-\frac{\beta}{2}(x - x_0)^2\right], \tag{20.197}$$

we find that the center of the wave packet oscillates with decreasing amplitude as is shown in Fig. 20.4. The time-dependence of $\langle x(t)\rangle$ can be found from Eq. (20.185).

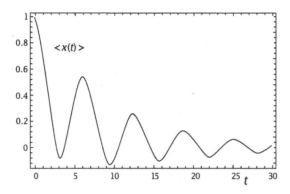

Figure 20.4: Damped motion of the center of a displaced Gaussian wave packet calculated from Eq. (20.185).

20.6 Quantum Mechanical Analogue of the Raleigh Oscillator

We have seen that the motion of a tuning fork in a viscous fluid can be described by a velocity-dependent frictional force and by the fluid accreting to the oscillator and thus changing its effective mass (Sec. 1.3). The quantum analogue of this problem can be formulated in the following way [50]:
We write the total Hamiltonian as the sum of two terms $\hat{H} = \hat{H}_0 + \hat{H}'$, where

$$\hat{H}_0 = \frac{\hat{p}^2}{2m} + \frac{1}{2}\omega_0^2 x^2 \tag{20.198}$$

and

$$\hat{H}' = \frac{\hat{p}^2}{2}\left(\frac{1}{m'} - \frac{1}{m}\right) - \frac{i\lambda\hbar}{4}. \tag{20.199}$$

In these relations λ is the damping constant and $\hat{p} = -i\hbar\frac{\partial}{\partial x}$ is the linear momentum operator and m' is defined by

$$\frac{m}{m'} = 1 - \frac{\lambda^2}{4\omega_0^2}. \tag{20.200}$$

The total Hamiltonian is that of an optical potential Hamiltonian with a constant $\mathcal{V}$.

We note that the classical motion generated by the Hamiltonian H does not depend on λ. But the imaginary part of $\hat{H}$ changes the eigenvalues of (Re H) by the amount $\left(\frac{-i\hbar\lambda}{4}\right)$. If we denote the initial eigenstates of Re $\hat{H}$ by $\phi_n(x)$, then the time evolution of the wave function is expressed by

$$\psi_n(x,t) = \exp\left(-\frac{iHt}{\hbar}\right)\phi_n(x) = \exp\left(-\frac{\lambda t}{4}\right)\exp\left(-\frac{i\mathcal{E}_n t}{\hbar}\right)\phi_n(x). \tag{20.201}$$

Here $\mathcal{E}_n$ is the nth eigenvalue of Re $\hat{H}$, and is given by

$$\mathcal{E}_n = \hbar\omega\left(n + \frac{1}{2}\right), \quad \omega = \left(\omega_0^2 - \frac{\lambda^2}{4}\right)^{\frac{1}{2}}. \tag{20.202}$$

The expectation value of $x(t)$ for any arbitrary time-dependent wave function $\Psi(x,t)$ can be found from (20.201) and is given by

$$\langle\Psi|x|\Psi\rangle = e^{\frac{-\lambda t}{2}} \int_{-\infty}^{\infty} \phi^*(x,t)x\phi(x,t)dx, \tag{20.203}$$

where $\phi(x,t)$ is the solution of

$$-i\hbar\frac{\partial\phi(x,t)}{\partial t} = \left[\text{Re } \hat{H}\right]\phi(x,t). \tag{20.204}$$

From this expression we can verify that the expectation value satisfies the equation of motion

$$m'\langle\ddot{x}(t)\rangle + \lambda m'\langle\dot{x}(t)\rangle + m'\omega_0^2\langle x(t)\rangle = 0. \tag{20.205}$$

We can also find Eq. (20.205) from the Ehrenfest theorem for complex potentials, Eqs. (20.63) and (20.64).

As we noted in the general formulation of the optical potential problem, this model can be viewed as the coupling of the oscillator to an infinite number of channels or to a sink. A sink here represents any process by which the quantum system can leak out from the original state. The probability of the system remaining in its initial state being proportional to $e^{-\lambda t}$.

References

[1] P.G. Burke and C.J. Joachain, *Theory of Electron-Atom Collisions: Part 1, Potential Scattering*, (Plenum Press. New York, N.Y. 1995).

[2] I.E. McCarthy, in *Microscopic Optical Potentials*, Lecture Notes in Physics, Vol. 89, (Springer-Verlag, Berlin, 1979) p. 447.

[3] I.E. McCarthy, C.J. Noble, B.A. Phillips and A.D. Turnbull, Phys. Rev. A 15, 2173 (1977).

[4] H. Feshbach, Ann. Phys. (NY), 5, 357 (1958).

[5] L.L. Foldy and J.D. Walecka, Ann. Phys. (NY), 54, 447 (1969).

[6] P.E. Hodgson, *The Nucleon Optical Model*, (World Scientific, Singapore, 1994)

[7] N.F. Mott and H.S.W. Massey, *The Theory of Atomic Collisions*, Third Edition (Oxford University Press, London, 1971) p. 404.

[8] J.S. Bell, and E.J. Squires, Phys. Rev. Lett. 3, 96 (1959).

[9] W.K. Weigel, On the general theory of optical potentials in *Microscopic Optical Potential*, Edited by H.V. von Geramb, Lecture Notes in Physics, Vol. 89 (Springer-Verlag, Berlin, 1979).

[10] M. Bertero and G. Passatore, Nuovo Cimento, 2 A, 579 (1971).

[11] G. Passadore, A sketch of the various formulations of the theoretical optical potential for scattering processes in *Nuclear Optical Model Potential*, Edited by S. Boffi and G. Passatore, Lecture Notes in Physics (Springer-Verlag, Berlin, 1976).

[12] P.E. Hodgson, *Nuclear Reactions and Nuclear Structure*, (Oxford University Press, London, 1971) Chapter 6.

[13] G.R. Satchler, *Introduction to Nuclear Reactions*, (MacMillan, London, 1980) Chapter 4.

[14] M.L. Goldberger and K.M. Watson, *Collision Theory*, (John Wiley & Sons, New York, N.Y. 1964) Sec. 6.3.

[15] J.M. Blatt and V.F. Weisskopf, *Theoretical Nuclear Physics*, (John Wiley & Sons, New York, N.Y. 1952).

[16] N.F. Mott and H.S.W. Massey, *Theory of Atomic Collisions*, Third Edition, (Oxford University Press, London, 1965) p. 181.

[17] H. Horiuchi, Prog. Theor. Phys. 64, 184 (1980).

[18] N.K. Glendenning, *Direct Nuclear Reactions*, (World Scientific, Singapore, 2004) p. 42.

[19] The most general form of velocity-dependent nucleon-nucleon interaction in nonrelativistic approximation is derived in a paper by S. Okubo and R.E. Marshak, Ann. Phys. (NY), 4, 166 (1958).

[20] M. Razavy, Phys. Rev. 125, 269 (1962).

[21] M. Razavy, Phys. Rev. 171, 1201 (1968).

[22] M. Razavy, Nucl. Phys. Nuc. Phys. 78, 256 (1966).

[23] V. de la Cruz, B. Orman and M. Razavy, Can. J. Phys. 44, 629 (1966).

[24] J.S. Eck and W.J. Thompson, Am. J. Phys. 45, 161 (1977).

[25] E.V. Graefe, M. Höning and H.J. Korsch, J. Phys. A 43, 075306 (2010).

[26] See for instance A. Messiah, *Quantum Mechanics*, Vol. I (John Wiley & Sons, New York, N.Y. 1958) Chapter VI.

[27] Y. Takahashi and T. Toyoda, Physica 138 A, 501 (1986).

[28] M. Razavy, Hadronic J. 11, 75 (1988).

[29] E. Madelung, Z. Physik 40, 322 (1926).

[30] D. Schuch and K.M. Chung, Int. J. Quantum Chem. xxix, 1561 (1986).

[31] Y. Yamaguchi, Phys. Rev. 95, 1628 (1954).

[32] R.G. Sachs, Phys. Rev. 74, 433 (1948).

[33] K.K. Osborne and L.L. Foldy, Phys. Rev. 79, 795 (1948).

[34] L. Heller in *Symposium on the Two-Body Forces in Nuclei*, edited by S.M. Austin and C.M. Cawley, (Plenum Press, New York, N.Y. 1972) p. 79.

[35] R. Beck and D.H.E. Gross, Phys. Lett. 47 B, 143 (1973).

[36] B. Sinha, Phys. Lett. 71, 243 (1977).

[37] D.H.E. Gross, Nucl. Phys. A 240, 472(1975).

[38] R.W. Hasse, Nucl. Phys. A 318, 480 (1979).

[39] R. McWeeny, Rev. Mod. Phys. 32, 335 (1960).

[40] D. ter Haar, Rep. Prog. Phys. 24, 304 (1961).

[41] P. Ring and P. Schuk, *The Nuclear Many-Body Problem*, (Springer-Verlag, Berlin, 1980) Chapter 7.

[42] F. Schwabl *Quantum Mechanics*, Third Edition, (Springer-Verlag, Berlin 2013) Sec. 13.3.

[43] E.D. Mshelia, W. Scheid, and W. Greiner Nuovo Cimento, 30 A, 589 (1975).

[44] E.D. Mshelia and Y.H. Ngadda, J. Phys. G, 15, 1281 (1989).

[45] I.S. Gradshetyn and I.M. Ryzhik, *Tables of Integrals, Series and Products*, (Academic Press, New York, N.Y. 1965) p. 1035.

[46] E.D. Mshelia, D. Hahn and W. Scheid, Nuovo Cimento, 61 A, 28 (1981).

[47] M. Razavy and A. Pimpale, Phys. Rep. 168, 305 (1988).

[48] M. Razavy, Can. J. Phys. 73, 131 (1994).

[49] P.M. Morse and H. Feshbach, *Methods of Theoretical Physics*, Part II (McGraw-Hill, New York, N.Y. 1953) p. 1641.

[50] R.L. Anderson, Am. J. Phys. 61, 343 (1993).

Index

Printed in the United States
By Bookmasters